Seed Pathology

VOLUME I

Seed Pathology

VOLUME I

PAUL NEERGAARD
Director, Danish Government Institute of Seed Pathology for Developing Countries, Copenhagen, Denmark

M

First published 1977 by
THE MACMILLAN PRESS LTD
London and Basingstoke
Associated companies in New York Dublin
Melbourne Johannesburg and Madras

ISBN 0 333 19273 7 (Volumes I and II)

Printed in Great Britain by
Unwin Brothers Limited
The Gresham Press, Old Woking, Surrey, England

CONTENTS OF VOLUME I

Contents of Volume II

Preface

This book has a dual purpose. On the one hand it is intended as a textbook covering the principles of seed pathology, for use in university courses at different levels. On the other, the intention is to provide a comprehensive reference book on seed-borne diseases, their detection, evaluation and control, for established workers in the following fields: university teaching and research; research at experimental stations; practical seed testing; formulation of quarantine regulations; techniques of quarantine inspection; development and testing of seed-treatment materials and procedures in laboratory and field; seed certification schemes; seed production; and the seed trade.

For some years I have taught seed pathology to students at the graduate, post-graduate and post-doctoral levels. For these courses I have used draft texts of this book. At the Danish Government Institute of Seed Pathology for Developing Countries in Copenhagen, I have had the very great privilege of co-operating with colleagues in developing countries, and I feel that this continuous contact with the problems and the potentials of the 'third world' has been most beneficial in my endeavours to outline the principles, as well as the leading philosophies, of this new branch of plant pathology and seed technology.

Nevertheless, my thoughts also go to my colleagues in the 'old world', in particular to those with whom I have co-operated in the development of seed health testing techniques. Without the numerous contacts and fruitful co-operation (especially within the framework of the International Seed Testing Association and its Plant Disease Committee) in comparative testing schemes, international workshops, etc., I would not have been able to write this book.

Through these contacts I have received much valuable information, some of which has been acknowledged as 'personal communications'. Through the years I have been encouraged by many of my colleagues who have urged me to bring my book to completion. In this connection I cannot forget the late W. C. Moore, formerly director of the Phytopathological Laboratory of the Ministry of Agriculture, Fisheries and Food, at Harpenden, England, for his persistent stimulation.

In extending my thanks to the many colleagues who have helped me by checking the texts that fell within the realms of their specialities, I must first pay tribute to my closest colleague, Dr. Mary Noble, Department of Agriculture and Fisheries for Scotland, Edinburgh. Since the 1950s we have had the closest co-operation, and our joint efforts, together with other colleagues, in developing seed health testing techniques, and in formulating seed certification schemes and quarantine regulations, all advocate the incontestable principle that prevention is better than cure. Dr. Noble has read the whole book critically, and made numerous comments and suggestions. I can find no adequate expression for my gratitude to her for her careful perusal of the manuscript and for her inspiration and encouragement throughout the years in whatever I may have contributed to seed pathology.

The wide scope of the theme of this book has made consultation with specialists essential. I am grateful indeed for the prompt readiness of many outstanding colleagues for their valuable co-operation in reading chapters or sections of the book and for communicating their criticism and suggestions. I express my deep gratitude to the following for their co-operation: Dr. Joan Moore, Ministry of Agriculture, Fisheries and Food, London, who perused and gave most useful critical comments on Chapter 1 and the whole of Part III. Dr. H. C. Phatak, at present in Bangkok, Thailand, who checked Chapter 3. Professor Ernst Hellmers, Royal Veterinary and Agricultural University, Copenhagen, and Dr. D. C. Graham, Department of Agriculture and Fisheries for Scotland, Edinburgh, who checked Chapter 4. Dr. M. J. Thirumalachar, India, over a period associated with the Institute of Seed Pathology for Developing Countries, Copenhagen, who checked Chapter 5. Professor Clyde M. Christensen, University of Minnesota, U.S.A., who checked Chapter 7. Dr. Palle Krogh, Royal Veterinary and Agricultural University, Copenhagen, Denmark, and Dr. Chester Mirocha, University of Minnesota, U.S.A., who checked Chapter 8. Dr: Dalbir Singh, Reader, University of Rajasthan, India, who checked Chapter 10, Professor R. K. S. Wood, Imperial College, London, England, who checked Chapter 18. Dr. J. T. Mills, Canada Department of Agriculture, Winnipeg, Manitoba, who checked Chapters 21, 22 and 23. Dr. J. de Tempe, Government Seed Testing Station, Wageningen, the Netherlands, and Mr. R. B. Maude, National Vegetable Research Station, Warwickshire, England, who checked Table 22.3. Dr. Johs. Jørgensen, Danish Government Seed Testing Station, Copenhagen, who checked Chapter 24. Dr. S. B. Mathur, Institute of Seed Pathology for Developing Countries, Copenhagen, who checked Chapters 28 and 29. And finally, Professor Gaston Waringhien, Paris, for checking the Greek and Latin etymology of the terms included in the Glossary.

I have worked on this book for many years, but particularly intensively during the past eight years when I greatly needed texts for teaching seed pathology at the Institute of Seed Pathology in Copenhagen. I would like to thank my colleagues at the Institute for their interest and co-operation.

The teaching and research activities at this Institute have given me a unique opportunity to study many details pertaining to routine testing for seed-borne pathogens. In these activities Dr. S. B. Mathur, my closest colleague through this period, has taken a most active part; his initiative and enthusiasm should be recorded here.

I have derived considerable benefit from the excellent technical assistance at the Institute in the preparation of the illustrations for the book, and I would like to thank Mrs. Olga Kongsdal for producing or redrawing many of the line drawings, and Mr. Magd El-din Ragab for his devoted and skilful co-operation in producing most of the original photographs.

My final thanks go to the excellent secretarial help I have received, and I am indebted to Mrs. Rigmor Hjælmhof for her patient copying of the thousands of items for the index. In the practical work on the manuscript I have received the greatest support and most efficient help from Mrs. Ellen Hersdorf. Her assistance in the typing of thousands of pages at all stages of the manuscript, and especially her efforts at reducing the 'stealing-in' of errors that is the ever-threatening spectre in the life of an author, have been invaluable. I wish to express my truly deep gratitude for her steady zeal in the elaboration and final accomplishment of this book.

The wide scope of the book, covering as it does a considerable range of aspects of plant pathology including the many categories of pathogens that may be transmitted through seeds, makes this work an ambitious one in the hand of one author. There are advantages and disadvantages of one-man authorship of a book of this type, which is always vulnerable to specialist criticism, and thus finally I wish to make a sincere request for critical comments and corrections.

Copenhagen, August, 1976 Paul Neergaard

Some suggestions on the use of the book

There is a range of possibilities for the use of the book in university teaching. The texts to be included in a course depend, of course, on the scope of the particular course, and the time available.

Part I (Pathogens, Diseases, Hosts) which is the 'special' part is primarily a compendium of seed-borne pathogens and seed disorders, and also includes a review of their harmful effects on crop production and, in one chapter, on human and animal health. While the first eight chapters may be used mainly for reference, the last two chapters may be suitable for some courses, in particular Chapter 10 which deals with the fundamental subject of morphology and anatomy of seeds in relation to transmission of pathogens. This chapter at least should be included for graduate and post-graduate students. The introductory sections of some of the chapters may also be considered for inclusion.

Part II (Mechanisms of Seed Transmission) is considered as basic for any academic study of seed pathology. It outlines the 'natural laws' governing the establishment of pathogens in seeds, their transmission, and their further development in crops. I would suggest that in any comprehensive course at the graduate and post-graduate levels this part should be included as a whole, with some omissions. For courses at the undergraduate level I would certainly concentrate on Chapter 14, largely dealing with the major types of cycles of seed-borne diseases. It might be useful in these elementary courses also to include Chapter 11 describing the influence of environment on establishment of pathogens in the seed crop.

Part III (Control) can be used as a whole for graduate and post-graduate courses, with omissions according to the time available and the scope of the course. For undergraduate courses I would select Chapter 19 on seed crop management; also Section 20.1, the Conclusions of Section 20.2, and Chapter 21 dealing in a general way with seed treatment procedures and equipment.

Part IV (Testing Methods) should support courses in practical seed health testing at any level, and would provide texts for thorough studies at the graduate and post-graduate levels.

Part V (Assessment of Seed-borne Inoculum) is, I think, truly basic for any interpretation of results recorded in seed health testing, and funda-

Table 0.1 Summary of suggested sections of the book to be included in university and other courses

Part	*Undergraduate*	*Graduate*	*Post-graduate*	*Technicians*
I *Pathogens, Diseases, Hosts* (1–10)	None	Sections 1.3, 3.1–3.4. Introduction to Chapter 4. Sections 4.1, 5.1, 5.2. Introduction to Chapter 6. Chapter 10	As for Graduate, plus Chapter 7	Subsections of Chapter 2
II *Mechanisms of Seed Transmission* (11–18)	Chapters 11 and 14	All chapters	All chapters	Chapters 13 and 14, with some omissions
III *Control* (19–24)	Introduction. Chapters 19 and 21, Sections 20.1 22.3, 22.6, 22.10 and 24.1	Introduction. Chapters 19, 21–24, and Section 20.1 and Subsection 20.2.6	Introduction and all chapters	Chapter 23, with some omissions
IV *Testing Methods* (25–31)	Chapters 25–28 for reference; Sections 31.2 and 31.3	All chapters	All chapters	Chapters 25–30
V *Assessment of Seed-borne Inoculum* (32)	None	Chapter 32	Chapter 32	None

mental for the establishment of rational disease tolerances in seed certification schemes. This part, I believe, should be included in any course at the graduate and post-graduate levels.

The book may also be used in courses for training technicians in seed health testing. For such purposes, I suggest the selection of Chapter 2 on diseases and injuries of seed (with omissions), and most of Chapter 13 dealing with modes of infection and contamination of seed; the essential sections of Chapter 14, discussing disease cycles, should perhaps also be taught to convey a general understanding of seed-borne diseases. Important in courses for technicians would be the whole of Part IV, dealing with methods of testing (with omission of some special discussions), and possibly also Chapter 23, on testing of seed germicides and treated seed, with some omissions.

I have summarised these suggestions in Table 0.1

List of Colour Plates

(appearing between pages 232/233 and 680/681

PART I

Pathogens — Diseases — Hosts

1 Economic Significance of Seed-borne Diseases

Allegedly, about 90 per cent of all food crops grown on earth are propagated by seed. Nine crops are of major importance, wheat, rice, maize, barley, sorghum, sugarbeet, common (Phaseolus) bean, soybean and groundnut, representing by far the greater part of food production in the world. All these crops are attacked by devastating seed-borne diseases.

The information available on crop losses varies considerably in precision. Assessments are often based on loose 'guestimates', and only in recent years have more precise standard methods of disease loss appraisal been developed (Chester, 1950*b*; Padwick, 1956; Miller, 1960; Large, 1966; Moore, 1967; Chiarappa, 1970; Tarr, 1972; James, 1974).

In the following sections different conceptions of disease losses are discussed and defined, reports of crop losses caused by seed-borne diseases in some major crops are summarised and the economic impact of seed transmission is compared with the role of other means of dissemination of seed-borne pathogens.

1.1 DEFINITIONS

1.1.1 Calculation of loss

Terms used in reporting yield losses from diseases are far from being uniform, a fact that makes comparison between reports difficult. Usually crop losses are reported in *units of yield,* such as bushels, kilogrammes or metrotons, or as per cent *reduction of the 'potential' yield,* i.e. of the yield assumed to be that of a healthy crop. Other indications of losses refer to *monetary units,* often without reference to the value of the 'potential' yield, thus leaving complications of calculating values from one currency to others and from year to year. Losses in monetary units depend, naturally, on the fluctuating market value which is sometimes even influenced by the disease under consideration, especially by reduction in supply or quality. Finally, losses may be calculated in terms of *area of crop* which would be released had there been no disease (Ordish, 1952; Padwick, 1956).

Percentage of loss related to the 'normal' yield is probably the most workable term in most cases, although difficulties may arise in defining precisely 'normal' yield, especially if the 'normal' yield does not represent the full potential. It is an advantage that loss percentages can be related to other percentages, those of the effects of disease, such as percentage of leaf area lost at any stage of the crop, percentage of lost plants, percentage of infection in the seed used for the crop, and other effects based on percentage of a normally developed crop.

1.1.2 Types of disease loss reports

There are three categories of assessment: (1) case histories, (2) general 'guestimates' and (3) assessments based on analytical and experimental methods.

(1) *Case histories*. These are reports which usually give details of losses in individual crops or in crops of a defined region or locality, and the losses are often related to an accepted level of normal yields. They often deal with exceptionally severe attacks in contrast to 'normal cropping' or 'normal years'. They may refer to losses in individual fields or in individual nurseries.

(2) *General 'guestimates'*. These are attempts to assess levels of losses on an annual basis in crops of a region or a country. Reports are often based on information from collected general estimates made regionwise by observers who unavoidably have different experience and personal views. Most of the information available about crop losses due to diseases is accumulated in this way but if the collator of such regional or cropwise reports or observations has practical experience in the surveyed area and inclines towards conservative estimates, valuable and fairly reliable information may well be provided by such co-operative efforts.

(3) *Assessments based on analytical and experimental methods*. There are three lines of approach in the more objective assessment of plant diseases, as excellently discussed by Chester (1950*a*); Large (1966); Moore (1967) and Tarr (1972): (1) assessing damage in crops by standard procedures designed to provide reproducible results; (2) using experimental data (infection experiments, pesticide control experiments, comparison between susceptible and resistant cultivars, etc.) for analysing disease intensity, linking disease records with loss in yield, thus providing data for assessing crop losses from field observations; (3) surveying statistically acceptable, randomly selected, representative samples of commercial crops, stores, etc.

For reproducible assessments of disease damage in crops the primary requirement is precision in defining the stage of development of the crop, as stressed by Large (1954*b,* 1966) for wheat (cf. Section 11.1, Fig. 11.1), and later followed by other workers for other crops. An attempt towards standardisation of keys for growth stages of crop plants has been made by FAO (Chiarappa, 1970). Another essential aspect is standardised descriptions of disease severity, usually based on diagrams or photographs ('standard diagrams'), which can be linked to the growth stage of the host, and which specify the parts of the plant involved, such as roots and tubers (severity rating, per cent area affected), leaves (percentage area affected and, when appropriate, including specifications such as the important flag leaf in cereals, etc.), stems, flowers, fruits, ears of Gramineae, etc. For seed-borne diseases, naturally, assessment must begin with the seed-borne inoculum (cf. Section 32.2).

Crop losses, as related to disease intensity, present a considerable range of aspects, of which quantitative losses of the harvested product in terms of weight, number of units or volume are usually the most simple to record but not always economically the most important. Qualitative losses may, for instance, involve degradation of baking quality of wheat or malting quality of barley, or the effect of foliage diseases on the nutritional value of grasses (e.g. *Colletotrichum graminicola* in Sudan grass), spoiling of cut flowers by leaf spot diseases (different species of *Alternaria* in carnation, *Senecio cruentus, Zinnia* and others), celery becoming unfit for eating (*Septoria apiicola*), tobacco leaves unsuitable for curing (blue mould, *Peronospora tabacina*).

In surveying disease losses the present tendency is to analyse minor sampled areas, or individual plants, comparing healthy plants with those diseased to different degrees, and to record accurately details of damage and effects thus to supersede the traditional subjective assessment by objective procedures providing reproducible results.

1.1.3 Loss related to surplus in crop production

Losses are usually estimated in relation to the total value of the crop but it may be more realistic to relate them to the net output. Profits in crop production vary considerably, depending on many factors which determine the cost of production and the market value of the crop. Calculations are complicated and economists reluctant to give general figures. Let us assume that the net income from crop production in small farms fluctuates between perhaps 10 and 20 per cent, the surplus after deduction of expenses for seed, fertilisers, labour, equipment, apart from money invested in land. According to such calculations, loss of 5 per cent of the yield would be equal to 25–50 per cent of the potential profit. This illustrates why in subsistence agriculture, as in developing countries depending almost exclusively on rice, yield losses of not more than 10–20 per cent may lead to famine. In a normal crop only a small amount of seed is left over for sowing and there is nothing to replace the food lost.

1.1.4 Costs of control measures

The costs of control must be added to the economic losses caused by diseases. They are often regarded as a regular part of the production costs though they may be avoidable. It is always cheaper to use healthy seed than later to carry out an expensive spraying programme.

The importance of cost of control measures becomes obvious when an epidemic disease is introduced to a new area. When, in 1959–1962, tobacco blue mould spread like a steppe fire throughout Europe and the Mediterranean countries, expensive spraying programmes became necessary everywhere.

1.2 CROP LOSSES CAUSED BY SEED-BORNE DISEASES

Old records of former heavy losses in countries where a disease has been controlled can be used for comparison with the situation in less developed agricultural countries where control has not yet begun.

1.2.1 Cereal diseases

Wheat, rice and maize are the major cereal crops of the world, the annual production being at present approximately 350, 300 and 300 million tons. If, following general estimates of the direct feeding value of these crops to man, about a quarter of a ton is needed annually per person, which equals about 90,000 calories (Borgström, 1970), then the total production of these cereals constitutes food in terms of calories for 3,000–4,000 million people, potentially enough for almost the total world population; seed-borne diseases take a heavy toll.

In **wheat** the six major groups of seed-transmitted pathogens are bunt and smut fungi, *Fusarium, Drechslera, Septoria* and nematodes.

Bunt and loose smut of wheat, both of world-wide distribution, are the most important smuts. Bunt (*Tilletia caries, T. foetida*) may cause field losses of grain in varying percentages, to which may be added lowering in quality of 'smutty' grain. Old records are dramatic. In the U.S.A. Heald (1918) recorded, in individual fields, infection up to 87 per cent by actual head counts; in 1927 local losses were more than 80 per cent, the national loss being 4 per cent (Stevens, 1940). In Denmark Gram and Rostrup (1922) reported up to 79 per cent, in Roumania Savulescu *et al.* (1937) reported in 1935–1936 50–60 per cent, in Germany Klemm (1940) 20–68 per cent, in Yugoslavia Vukčevic (1954) in 1949–1953 70–90 per cent, Mitra (1935) in 1932 up to 33 per cent and in 1933 up to 30–40 per cent losses of grain in India. In the U.S.A. the losses in the national wheat crop progressively decreased from 4 per cent in 1926 to 1 per cent in 1950

(Chester, 1950*a*), and were estimated at 0.4 per cent for the period 1951–1960 (USDA, 1965), the decrease owing to seed treatments and certification. But in 1955 losses in the Pacific Northwest were from 3 to 34 per cent, here both common bunt and dwarf bunt (*Tilletia contraversa*) were combined (Fischer and Holton, 1957). In many developing countries losses are still high. In Nepal and Afghanistan recently, fields with 40–60 per cent infection have been observed (Saari and Wilcoxson, 1974).

Dwarf bunt, which became important comparatively recently, is both seed- and soil-borne, and in some countries more severe than common bunt, in the U.S.A. having caused up to 95 per cent losses in some fields (Young, 1935; Bamberg, 1941), in British Columbia, Canada, up to 5 per cent (Foster and MacSwan, 1949), and in Germany up to 30 per cent (Wagner, 1950), in 1958 in some fields up to 75–80 per cent (Stauber, 1958; Bachthaler, 1958*a*, 1958*b*); in many localities in Bulgaria losses of 30–60 per cent have been recorded (Todorova, 1958). The disease reached a climax in the 1950s, and has since then decreased considerably in severity, probably because of improved control measures, including strict crop rotation where the pathogen is endemic.

Fischer and Holton (1957) have shown how losses due to smuts can be calculated. Application of their formula, which is based on total production of wheat in relation to bushels being graded as 'smutty' and on dockage assessment, is cited in full (Table 1.1).

In the U.S.A. estimates of losses somewhat less spectacular than those from bunt have been attributed to loose smut (*Ustilago tritici*). Chester (1950*a*) refers to average losses of 1–4 per cent in the main wheat belt with national losses between 0.5–1 per cent; however, losses of 5 per cent were not uncommon in individual fields, and fields with as much as 40 per cent have been observed. In the 1940s in most countries of Eastern Europe, eastern States of the U.S.A. and Canada, and in major wheat-growing countries of South America, an annual average loss of 2–4 per cent or more had to be anticipated (Niemann, 1962). For the period 1951–1960 the continental average annual loss in the U.S.A. went down to 0.8 per cent (USDA, 1965) and the achievements in controlling wheat smuts in other technically advanced agricultural countries are comparable to those obtained in the U.S.A.

However, in large areas of the world, in many developing countries, the situation is different for both types of smut, in some places being indeed similar to the situation in the first decades of this century in North America and Europe. In an extensive survey of North West India in 1967–1971 losses up to 40 per cent were encountered; an average loss of 7 per cent for 1967–1968, 2 per cent for 1968–1969, while for the two last years average losses dropped to less than 0.5 per cent, probably because of the extensive use of a resistant cultivar, 'Kalyansona' (Joshi, Gera and Saari, 1973). In some developing countries losses of 5 per cent due to one

Table 1.1 Wheat smut (bunt) loss calculation for the 1955 crop year in the Pacific Northwest of the U.S.A. — an example (from Fischer and Holton, 1957)

Loss in yield

1. Total production in survey area=86,439,000 bushels
2. Survey shows that 34 per cent of crop graded smutty
3. 34 per cent of the crop=29,000,000 bushels of smutty wheat
4. Average smut grade was 0.75 per cent
5. A smut grade of 1 per cent is believed to reflect a field incidence of 10 per cent smut on a head-count basis
6. 10 per cent×0.75 per cent=7.5 per cent average smut incidence in the field
7. 7.5 per cent smutted heads in the field approximates an equivalent loss in yield
8. 7.5 per cent×28.0 bushel average acre yield=2.1 bushels per acre loss
9. Total acreage=3,800,000 of which 34 per cent produced smutty wheat
10. 34 per cent of 3,800,000=1,292,000 acres that lost 2.1 bushels of wheat each
11. 1,292,000×2.1=2,713,200 bushels destroyed by smut
12. 2,713,200×$1.85 (1955 support price)=$5,019,420, loss in income

Market discounts

13. 29,000,000 bushels of smutty wheat with average smut grade of 0.75 per cent
14. Dockage for weight on freight=0.75 per cent
15. 0.75×29,000,000=217,000 bushels of smut
16. 217,000 bushels of smut×25 cents a bushel-freight=$54,250
17. Washing charge for smutty wheat=2 cents a bushel
18. 29,000,000×$0.02=$580,000

Summary of losses for 1955

Yield	$5,019,420.00
Dockage	54,250.00
Washing charge	580,000.00
Total	$5,653,670.00

or the other of these smuts are regarded as rather unimportant, according to statements by officials to the author during recent visits in South Asia and Latin America, where too often only losses approaching 10 per cent or more are taken seriously.

In wheat, four species of *Fusarium* are important pathogens: *Fusarium avenaceum, F. culmorum, F. graminearum* and *F. nivale*. They cause pre-emergence and post-emergence death of seedlings, foot rot and ear infection (headblight, scab). The two last pathogens are the most severe, while *F. avenaceum* is probably the weakest. *Fusarium graminearum* prevails in warm temperate, and *F. nivale* in cool temperate climates. Heavy losses due to scab, *F. graminearum,* have been encountered in wheat and maize in the U.S.A., e.g. in Iowa in 1919 causing 25 per cent loss (Chester, 1950*a*), and in the U.S.A. even in recent years the prevalence of infection may vary from a trace up to virtually 100 per cent of the heads in a field and losses may vary correspondingly (Schroeder and

Christensen, 1963). In the Pacific Northwest of the U.S.A. *Fusarium culmorum* predominates as a cause of foot and root rot, resulting in yield reduction of 10 per cent or more (Cook, 1968).

In a survey of overwintering crops in Finland Blomqvist (1970*b*) established that *F. nivale,* snow mould, in 1968–1969 caused about 5 per cent damage to the stands of wheat in southern Finland and about 15 per cent in central Finland. In Scotland this pathogen often causes severe leaf blight and lodging and heavy losses (Richardson and Zillinsky, 1972; Mary Noble, personal communication, 1973). In Finland Blomqvist and Jamalainen (1968) demonstrated experimentally that in winter wheat up to 85 per cent could be killed.

Drechslera sorokiniana has assumed increasing importance in Europe as a cause of foot rot in wheat. In the U.S.A. and Canada it has been known for decades as the most dangerous 'dry-land foot rot' pathogen on this host. In Kansas losses caused by foot rots of wheat, in which *D. sorokiniana* plays the major role, were 4.5 per cent in both 1931 and 1932, 6 per cent in 1935 and 1937, and average losses ranged from 0.5 to 1.5 per cent of the national crop (Chester, 1950*a*). Systematic surveying in Manitoba, Canada, showed that in 1939–1942 the average yield reduction was 12 per cent, and in some fields 30–70 per cent of the plants were killed before harvest (Chester, 1950*a*); and a later survey in Canada showed that the estimated average annual loss due to common root rot, *D. sorokiniana* and *Fusarium* spp., for three years, 1969–1971, was 5.7 per cent (Ledingham *et al.,* 1973). In laboratory seed tests *D. sorokiniana* is known to be very active on seedlings, attacking the roots, and Skou (1966) showed that in the field the symptoms may easily be confused with or hidden by those of the take-all disease, *Gæumannomyces graminis* (syn. *Ophiobolus graminis*). *Drechslera sorokiniana* is a common cause of kernel blight in North America; Dickson (1956) reported an average of over 5 per cent and in some seasons as high as 75 per cent infection may occur in some seed lots. However, there is little concrete information indicating percentages of loss due to kernel infection. In North Dakota, U.S.A., heavy losses were due to a complex leaf spot disease caused by *D. tritici-repentis* (perf. state *Pyrenophora tritici-repentis*), and *Septoria avenae* f.sp. *triticea* (perf. state *Leptosphaeria avenaria* f.sp. *triticea*). The disease develops under extremely moist conditions and the losses in grain yield averaged 13 per cent in 1970 and 1971 (Hosford and Busch, 1974).

In recent years *Alternaria triticina,* leaf blight, has caused serious damage on wheat crops in India, particularly in eastern and central India, where the disease has limited the use of rust-resistant cultivars (Prasada and Prabhu, 1963; Prabhu and Prasada, 1966). Mexican cultivars and their derivatives are susceptible. In field experiments, Chenulu, Singh and Joshi (1970) established losses in yield from 36 to 93 per cent depending on the time of inoculation — maximum losses occur when the plants are infected at the boot stage. In individual fields losses up to 60 per cent may occur (Sokhi, 1974). The disease has not yet been reported outside India.

Glume blotch of wheat, *Septoria nodorum*, has attracted increasing attention during the last decades as one of the worst pathogens on wheat, previously underestimated because the symptoms often were believed to be of physiogenic nature. Experiments have shown that under suitable conditions infection just before heading can cause yield losses up to about 65 per cent (Brönnimann, 1968, 1969; Brönnimann, Sally and Sharp, 1972) or 36 per cent, as established by an intensive spraying programme in England (Jenkins and Morgan, 1969). In Georgia, U.S.A., Nelson, Morey and Brown (1974) estimated that wheat yields for 1972–1973 were reduced by 20 per cent or more in some areas, and the disease was regarded as second to leaf rust in reduction of yield in this State. In some crops, losses of 50 per cent or more have been encountered (Weber, 1922; Schmiedeknecht, 1967). In India the pathogen may be severe in individual fields at higher elevations (Joshi *et al.*, 1970). In 1973 catastrophic losses of about 50 per cent on average in the national wheat production of Paraguay were reported to be due to the combined effect of two seed-borne pathogens: *Drechslera sorokiniana* at the beginning of the growth season, and *Septoria nodorum* when the crops were in the heading stage (Torres, personal communication, 1973).

Ear cockle, *Anguina tritici*, often combined with *Corynebacterium tritici* in tundu disease, has previously been one of the worst pests of wheat in Europe and North America; it is still devastating where modern equipment for cleaning seed is not available. Losses of 30–70 per cent are common where the nematode is uncontrolled (Thorne, 1961). In Europe ear cockle has been practically eliminated but in developing countries it is still rather common. In India infections up to 50 per cent have been recorded, and in Delhi State the damage average was 1–2 per cent (Vasudeva and Hingorani, 1952*b*). Ear cockle and tundu disease account for annual losses from 2 to 5 per cent in Northern India and took a toll of 50–60 per cent in grain during 1965–1966 in Uttar Pradesh (Raychaudhuri, 1968). In 1966–1967 losses from 25 to 40 per cent were recorded in badly affected fields in Pakistan where the disease is increasing (Kamal and Moghal, 1968); in Iran it often causes losses of 50 per cent (Eng-Daflasi, personal communication, 1967), and in Ethiopia in 1968 50 per cent losses were also encountered (Hingorani and Bekele, 1969).

In **rice** the three most important pathogens are *Pyricularia oryzae*, blast, *Drechslera oryzae*, brown spot, and *Xanthomonas oryzae*, bacterial blight. In the U.S.A. the first two account for 30 per cent of losses due to all rice diseases (USDA, 1965), which means about 2 per cent loss of the total annual rice production under the advanced technical conditions of that country. However, North America contributes less than 2 per cent to world production. In the major rice-producing countries, such as India, the situation is different. In India (about 20 per cent of world production) and other Asian countries, blast and brown spot are common and devastating. *Pyricularia oryzae* often kills seedlings or plants at the tillering stage and may largely destroy the panicles. In Japan, blast has been regarded as by far the most serious biological menace to food production and was responsible for famine in several districts in the 1930s and 1940s (Walde and Roberts in Padwick, 1950). Annual losses in Japan during 1953–1960 were from 1.4 to 7.3 per cent with an average of 3 per cent of total in spite of extensive chemical control (Goto, 1965). In India, where losses up to 66 per cent have been recorded in Bombay State (Kulkarni, 1959), 0.8 per cent of the total yield was lost in 1960–1961 according to Padmanabhan (1965) who estimated average yield losses in different States from 0.01 per cent (Bengal) to 9.8 per cent (Jammu and Kashmir) for this year; Singh, Sardana and Khosla (1972) estimated, in 1959–1967, average field losses of about 2 per cent. In many areas in the Philippines more than 50 per cent losses have been recorded (Ou, 1972). Padwick (1956) reported the following estimates of losses in Africa: approximately 1 per cent in Gold Coast and Togo, and 5 per cent in Kenya, Malawi and Sierra Leone, and Awoderu and Onuorah (1974) reported losses ranging from 12 to 40 per cent in Nigeria. In Italy losses due to blast range between 5 and 100 per cent (Baldacci and Corbetta, 1960), and Sibilla (personal communication, 1960) and Gambogi (personal communication, 1974) have suggested as a conservative estimate about 10 per cent average reduction of yield. Mathur *et al.* (1964), who correlated percentage levels of blast infection to percentage crop loss, found that each 10 per cent infection leads to approximately 4.5 per cent loss.

In many areas of the world losses comparable to those caused by blast are due to *Drechslera oryzae* which is prevalent under poor farming conditions. Up to 58 per cent dead plants have been reported from the Philippines (Ocfemia, 1924), and in Punjab State, India, damage to the leaves resulted in 5–29 per cent losses of grain weight (Bedi and Gill, 1960). Moreover, infected grains lead to seedling blight in the next crop resulting again in weakening of younger and older plants. The most notorious case of brown spot was the outbreak in Bengal, India, in 1942–1943 when between 50 and 90 per cent of the rice crops were destroyed thus contributing as a major factor to the Bengal famine in which two million people died of starvation (Ghose, Ghatge and Subrahmanyan, 1956; Padmanabhan, 1973). Singh, Sardana and Khosla (1972) recorded in India, in 1959–1967, field losses from 3 to 13 per cent, and in a detailed survey carried out in Tamil Nadu, South India, based on disease severity related to yield losses, Vidhyasekaran and Ramadoss (1974) showed that severe infections in the cultivar IR5 resulted in quantitative yield losses from 20 to 40 per cent and, moreover, in qualitative losses in terms of a 25–55 per cent decrease in starch content and a 71–89 per cent decrease in sugar content of the grain. In the U.S.A. the average loss caused by *Drechslera oryzae* is only approximately 0.5 per cent (Cramer, 1967); in Africa, in Zanzibar according to the survey made by Padwick (1956), it is about 5 per cent, whereas in Nigeria losses range from 12 to 43 per cent (Awoderu and Onuarah, 1974).

There are some other important fungi parasitising rice, the effects of which appear not to be fully recognised. *Trichoconis padwickii* is widely distributed in most rice-growing areas — in a survey of 388 tested rice seed samples from 11 countries in Asia and Africa about 75 per cent were infected, some individual samples up to 80 per cent (Mathur, Mallya and Neergaard, 1972). Seed infection leads to loss of seedlings; there is little doubt that this pathogen is responsible for considerably more damage in rice fields than has been recorded. Another fungus pathogen, *Fusarium moniliforme,* bakanae disease or foot rot, may be considerably more harmful than realised. It is distributed in all rice-growing areas. From Japan losses of 20 per cent have been reported (Ito and Kimura, 1931); from India 15 per cent losses in Uttar Pradesh (Pavgi and Singh, 1964) and 5 per cent of the total Indian rice crops (Singh *et al.,* 1958, according to Cramer, 1967); in Thailand, losses may vary from 4 to 15 per cent (Kanjanasoon, 1965). In Northern Australia in 1963, foot rot, caused by *F. moniliforme,* resulted in up to 70 per cent loss of early maturing cultivars (Heaton and Morschel, 1965). Losses are difficult to evaluate because infected plants may exhibit a range of disease intensities, making distinction between healthy and diseased plants difficult or virtually impossible; infected plants may be elongated (bakanae effect), or they may be stunted, they may appear normal, and sometimes only slightly stunted. Careful investigations taking into account these different responses to infection are needed to evaluate the true economic effects. Losses due to other fungous diseases should not be disregarded. In the U.S.A. they may account for about 45 per cent of the total of disease-induced losses (cf. Cramer, 1967); relatively much less, however, in countries where the major diseases, particularly blast, are prevalent, as in Asia.

The most serious bacterial disease in rice is leaf blight, *Xanthomonas oryzae*. In Japan, losses in severely infected fields range from 20 to 30 per cent, sometimes up to 50 per cent, and in the Philippines and Indonesia losses are equally high or even higher (Ou, 1972). In India, too, the losses range from 6 to 60 per cent (Srivastava, 1967). At the International Rice Research Institute in the Philippines (IRRI, 1967) artificial inoculation of a susceptible and a moderately resistant cultivar showed about 75 per cent and 47 per cent losses, respectively.

Some seed-borne nematodes are extremely harmful in rice. *Aphelenchoides oryzae,* white tip, has caused losses of 10–35 per cent in Japan (Yoshii and Yamamoto, 1950) but is now less important because of seed treatment. Losses from 29 to 46 per cent have been encountered in Taiwan (Hung, 1959). Stem nematode, 'ufra' or 'dak pora', *Ditylenchus angustus,* is important in some regions in South Asia, having caused losses of 50 per cent in Uttar Pradesh in India (Singh, 1953) and 20–90 per cent in Thailand (Hashioka, 1963).

The major seed-borne pathogens of **maize** are the following: *Diplodia* spp. (*D. macrospora, D. maydis, D. frumenti*), dry or white ear rot, stalk rot; *Drechslera maydis,* southern leaf spot or blight; *Drechslera turcica,* northern leaf spot, charred ear mould; *Fusarium graminearum,* seedling blight, cob rot; *F. moniliforme,* Gibberella ear rot, kernel rot, seedling blight; and *Erwinia stewartii,* bacterial wilt, Stewart's disease. Almost half of the world's maize is grown in North and Central America. Two-thirds of losses due to diseases are caused by seed-borne diseases (cf. Cramer, 1967), and of these *Diplodia* and *F. graminearum* account for about 25 per cent and *Drechslera* spp. for almost 20 per cent of the total loss due to diseases. The total loss in the U.S.A. in 1954 was about 4.7 per cent, in 1965 about 9.4 per cent of the potential maize production (Cramer, 1967) and during an earlier nine-year period *Diplodia* spp. caused losses of 9 per cent of the total yield (McCallan, 1946).

Drechslera maydis, southern leaf blight, and *D. turcica,* northern leaf blight, cause considerable yield loss in the U.S.A. The latter has caused yield losses of 30 per cent or more over fairly large areas (Hooker, 1972). Southern leaf blight, widespread in tropical and subtropical areas of the world, until 1969 was considered a relatively minor disease causing an average annual loss of less than 1 per cent. Beginning in 1969, and reaching devastating proportions in 1970 in the U.S.A., a new race attacked cultivars carrying the Texas type of male sterile cytoplasm (TMS cytoplasm developed to avoid manual detasseling), cultivars grown at that time in more than 80 per cent of the maize-growing areas of the States. The estimated losses amounted to approximately one billion dollars, or more than 12–15 per cent loss of yield (Tatum, 1971). In Illinois losses went up to 25 per cent, in some States to 100 per cent, as many crops were not harvested (Hooker, 1971). In the South the average loss in yield was 20–30 per cent and in parts of southern Illinois and Indiana there were losses from 50 to 100 per cent (Ullstrup, 1972). As a consequence, seed production was almost completely shifted to maize with other than TMS cytoplasm and to manual detasseling.

Stalk rot of maize is a disease complex, in which *Diplodia* spp. and also *Fusarium moniliforme* and *F. graminearum* are involved. Estimates of losses often apply to the complex. In most of the maize-growing States of the U.S.A. losses are frequently much higher than 7.5 per cent (Christensen and Wilcoxson, 1966), often reported as high as 18–20 per cent or more, as in Virginia (Roane, 1950), in Ohio (Williams and Schmitthenner, 1963), in

Minnesota (Devay, Covey and Linden, 1957) and in Iowa (Worf and Foley, 1963); in Illinois, Diplodia and Gibberella stalk rots were responsible for annual State-wide losses of 71 million dollars or 8.6 per cent of the total production (Hooker and Britton, 1962), and the average annual loss for 1951–1960 for the country was 3 per cent (USDA, 1965). In Africa *Diplodia* and *Fusarium* cause losses of more than 10 per cent in Lesotho and Rhodesia, less than 10 per cent in Tanzania (Padwick, 1956), and in Rhodesia a conservative estimate of losses was about 10 per cent of the total crop (Kerr, personal communication, 1960). In Zambia stalk rot due to these fungi is often confined to a few patches in a field but can be severe, losses up to 75 per cent having been recorded; the effect of cob rot fungi is in some years slight, in others very important — in several farms losses up to 50 per cent have been encountered (Logan, 1974). *Fusarium moniliforme* is extremely widespread and in some countries, and in seed, particularly uncertified seed, infection may be 100 per cent as revealed by laboratory testing. In addition to stalk rot this pathogen causes seed rot, wilt and stunt, and it may be difficult or impossible in the field to identify infected, including slightly stunted, plants. There is little doubt that such infections also make depreciation of yield difficult to estimate.

Stewart's bacterial wilt, *Erwinia stewartii,* primarily a disease of sweet maize, is widespread in the maize-growing regions, particularly in North and Central America, and in more restricted regions in other parts of the world. In the U.S.A. losses were greatest in 1932 when 13 per cent of sweet maize crops were lost (Pepper, 1967). The disease decreased gradually in subsequent years but was, however, destructive in Kentucky in 1958 (Johnson and Chapman, 1958) and since then serious losses have not been reported in the U.S.A. (Pepper, 1967). In Italy losses have amounted from 40 to 90 per cent of the crops (Pasinetti, 1937).

The total of the other five cereals, barley, rye, oats, millet and sorghum, accounts for a world production only comparable to each of the three major cereal crops, wheat, rice and maize, the production of all five cereals amounting to about 320 tons. Their feeding value is less than that of wheat and rice; only rye is used almost exclusively for bread. The other four cereals are used partly for feeding animals, although in some areas, e.g. Sudan, sorghum and millet constitute the main source of human food.

Barley is used for malting and for stock feeding; the proportion between these two uses is not well established in available international statistics. Two-thirds of the world production takes place in Europe and the U.S.S.R. In the U.S.A. estimated annual yield losses due to diseases between 1951 and 1960 were 14 per cent of the potential production (USDA, 1965), seed-borne diseases accounting for about two-thirds of these losses, making about 9 per cent of total production. This estimate is probably fairly applicable also to the main production areas. There is a considerable range of serious seed-borne diseases in barley: loose smut, *Ustilago nuda*; covered smut, *U. hordei*; leaf stripe disease, *Drechslera graminea*; net blotch, *D. teres*; foot rot, *D. sorokiniana*; seedling and head blight, *Fusarium* spp., including *F. culmorum, F. graminearum* and *F. nivale*; scald, *Rhynchosporium secalis;* and barley stripe mosaic. Some of these diseases, which previously have caused heavy losses in many countries, are now largely checked by seed certification and seed treatments, although not eradicated. In some countries of the main growing areas outbreaks of some of these diseases have reached considerable proportions even in recent years. In Eastern Germany in 1965 8 per cent of the winter barley acreage was severely infected by loose smut, 27 per cent moderately, and of the spring barley acreage 0.5 per cent was severely infected, 4 per cent moderately infected (Masurat, Peschel and Stephan, 1966). In Western Germany loose smut has largely been checked by certification schemes and more than 10 per cent infection is infrequent (Niemann, 1962). Serious losses have been commonly encountered in Kansas, U.S.A. (Schafer and Hansing, 1950); in some fields up to 50 per cent, however, of semi-loose smut, *Ustilago nigra*, may have taken an unknown part and in this State smut is or has been a limiting factor for barley production. In Cyprus, losses of 10 per cent due to *U. hordei* have been recorded (Padwick, 1956).

In Denmark in recent years there has been a marked increase in the prevalence of net blotch, *Drechslera teres*, perhaps because barley is grown continuously in many areas. Smedegård-Petersen (1974) assessed losses in plots from naturally infected seeds of two cultivars and found yield reduction of 9–11 per cent. He concluded from field observations that wherever untreated seed is used a yield reduction of 10 per cent is to be expected due to this pathogen. He found *D. teres* in 70 per cent of 148 barley fields surveyed in Denmark (Smedegård-Petersen, 1971), and in Finland the pathogen was observed in more than 50 per cent of 180 fields examined in 1970 and 1971 (Mäkelä, 1972*b*). In Canada McDonald and Buchannon (1964) recorded yield losses of about 11 per cent comparing a susceptible with a resistant cultivar, and Shipton (1966*a*) also established in spraying experiments a yield reduction attributed to net blotch of 17 per cent. In Germany Rintelen (1969) estimated yield reduction of about 13 per cent in crops in which 30 per cent of the plants were infected. He found that if shoots were infected in the 3–5 leaf stage, grain weight was reduced by about 35 per cent. In addition to quantitative yield reduction, the quality of the grain may be spoiled. Kamal and Naguib (1957) found that *D. teres* reduced the brewing quality by diminishing the carbohydrate content and thereby the malt yield. Previously *Drechslera graminea* caused heavy losses with up to 70 per cent recorded in Germany (Braun and Riehm, 1945), and 12 per cent in India (Mathur, Mathur and Bajpai, 1964). For many years seed treatment with organomercury compounds has reduced leaf stripe disease as well as covered smut, but the more restricted use of less poisonous materials in recent years has resulted in an increase, particular of *D. graminea,* for instance in Scandinavia. In Denmark losses in a few heavily infected crops from untreated seed have been estimated at 20 per cent (Statens Plantepatologiske Forsøg, 1974). As in wheat, *D. sorokiniana,* common root rot, has assumed increasing importance on barley crops in several European countries, causing both foot rot and leaf blight. In Canada Piening (1973) recorded from 5 to 42 per cent yield loss in ten cultivars of barley, due to common root rot.

Scald, *Rhynchosporium secalis,* now recognised to be seed-borne although predominantly carried over on plant residues, is a major disease of barley in cool semi-humid regions (Shipton, Boyd and Ali, 1974). Yield losses may be as high as 35–40 per cent but are more commonly between 1 and 10 per cent, mainly through reduction of grain weight. Losses of 20–35 per cent have been reported in California, up to 45 per cent in Japan, and in Victoria, Australia, a yield loss of about 18 per cent has been demonstrated in a field experiment.

Barley stripe mosaic may reduce yields substantially. In an infection experiment Catherall (1972) found that the virus reduced the plant height of barley cv. Akka by 21 per cent and reduced grain yield by 62 per cent, Hagborg (1954) found a reduction of 64 per cent, and in infection experiments in Denmark the virus caused yield reduction of 25–60 per cent in nine cultivars (Rønde Kristensen, personal communication, 1974). In Denmark infections in fields are no longer common and now of little economic importance. In North Dakota, U.S.A., yields in some cultivars have been reduced by 17–24 per cent due to seed-transmitted infection (Timian and Sisler, 1955).

In **rye,** used for baking bread, three pathogens (*Fusarium nivale,* snow mould; *Urocystis occulta,* stem smut; and *Claviceps purpurea,* ergot), are particularly important.

There is little doubt that in cool temperate climates snow mould is the most serious, being responsible for winter killing and lodging as in Germany where in some years from 6 to 12 per cent of the acreage of rye had to be ploughed down due to winter killing (Braun and Riehm, 1945).

Previously, flag smut, *Urocystis occulta,* caused very heavy losses in Europe, 20–40 per cent in some fields in Denmark in 1911 (Ravn, 1912), 10–30 per cent in Germany (Pape, 1928; Anonymous, 1948*a*), or even up to 60 per cent (Pape, 1927), and in the U.S.A. an annual loss of some 5 per cent was usual in Minnesota (Stakman and Levine, 1916). Now the pathogen is perhaps of some importance only in Eastern Europe (Schuhmann, 1962). Ergot, *Claviceps purpurea,* has been known as a plague in rye for many centuries, indeed since the time of the Caesars. The direct yield losses have been as high as 20 per cent in Russia, and similar heavy losses have been recorded in the U.S.A. (Chester, 1950*a*).

However, the greatest damage caused by ergot is human poisoning. In the Middle Ages, the 'holy fire' or 'St. Anthony's fire' assumed catastrophic proportions, claiming hundreds of thousands of victims. Today ergotism is rare although still occurring in some regions (cf. Chapter 8). Also livestock, including cattle, horses and sheep, have been and are still sometimes victims.

In technically advanced countries seed-borne diseases of rye have been reduced to relatively minor importance. In the U.S.A. average annual losses from 1951 to 1960 due to such diseases, including ergot, stalk smut, leaf spot diseases and root rots, were about 1–2 per cent (USDA, 1965). In cold temperate countries *Fusarium nivale* and other species of *Fusarium,* as well as *Drechslera* spp., may be responsible for considerable losses, generally underestimated because the diseases are inconspicuous and difficult to assess precisely. *Colletotrichum graminicola,* cereal 'anthracnose', a high temperature fungus is often neglected; it may be destructive under humid and warm conditions. In Minnesota, U.S.A., losses in some fields amounted in 1918 to 33 per cent, in Wisconsin to 50 per cent, in Oklahoma to 50–60 per cent; in 1923 the anthracnose was the worst rye disease in several States, including Mississippi and Ohio, in 1924 10 per cent losses were encountered in the southern rye areas, and generally in the 1920s the disease was regarded as one of the most destructive diseases in Ohio (Chester, 1950*a*). However, for the period 1951–1960 the average annual loss was 0.4 per cent of the national production of rye in the U.S.A. (USDA, 1965) but anthracnose reached epidemic proportions in Florida in 1961, 1962 and 1963 resulting in sharp decreases in grain yield of rye, sometimes by two-thirds (Luke and Sechler, 1963).

Oats, predominantly used for fodder, are attacked by a considerable range of serious pathogens, including *Drechslera avenae,* seedling blight and leaf spot; *D. sorokiniana,* seedling blight, root rot; *D. victoriae,* Victoria leaf blight; *Fusarium graminearum,* scab; *F. nivale,* snow mould, brown foot rot; *Septoria avenae,* leaf blotch, black stem; and two smuts, *Ustilago avenae,* loose smut, and *U. hordei,* covered smut. In the U.S.A. the estimated annual loss for the period 1959–1960 due to all oat diseases was 21 per cent (USDA, 1965), of which seed-borne diseases may account for at least one-third. Septoria leaf blotch is in many regions the most important of these; in Iowa it caused losses of 15 per cent in 1952. Victoria blight caused losses of only 0.4 per cent on an average in the above 10-year period, but was a true disaster from 1946 to 1948 causing losses of 25–32 per cent in the crown rust resistant oat cultivars with Victoria parentage then widely grown in Iowa (USDA, 1965). Earlier in this century the two oat smuts were the most serious pathogens, in the years before and after 1900 causing an annual average loss of about 8 per cent of the national crop of the U.S.A., according to conservative estimates

(Fischer and Holton, 1957); Kansas suffered the heaviest losses — in 1936 the average was 20 per cent, and some fields averaged 40–65 per cent (Heyne *et al.*, 1947). In the period 1930–1939 the national losses from smut ranged from 2.5 to 4.7 per cent, mean 3.7 per cent (McCallan, 1946). In Switzerland losses due to the two smuts were in the 1940s in many fields about 25 per cent (Rapin, 1947). Covered smut of oats is less prevalent than loose smut in most parts of the world, exceptions being Ghana and some parts of India (Thirumalachar, personal communication, 1975).

Sorghum is extensively used for human consumption in some regions, such as most of Africa. The following may be considered as major seed-borne pathogens: *Claviceps microcephala,* ergot, sugary disease; *Fusarium moniliforme,* seed rot, stalk rot; *Gloeocercospora sorghi,* zonate leaf spot; *Periconia circinata,* milo disease; *Sphacelotheca cruenta,* loose smut; *S. sorghi,* kernel smut or covered smut; *Tolyposporium ehrenbergii,* long smut; and *Xanthomonas holcicola,* bacterial leaf streak. Of these the smuts are the most destructive. Heavy losses due to kernel smut have been reported from different countries: 20–30 per cent in Manchuria (Takasugi and Akaishi, 1933), 25 per cent in Tanzania (Wallace, 1934), 60 per cent in Burma (Anonymous, 1943), 30–60 per cent in Italy (Goidanich and Scardovi, 1946), more than 30 per cent in Venezuela (Ciccarone and Malaguti, 1950), more than 10 per cent in Lesotho (Padwick, 1956) and an average of 6.6 per cent for kernel smut alone in West and East Africa, and losses up to 70 per cent in Kenya (Padwick, 1956), in India in some fields up to 65 per cent (Mehta *et al.*, 1953), and for the whole of India an average of 10 per cent (Kumar *et al.*, 1963). Milo disease is in some regions so devastating that it becomes a limiting factor, sometimes causing complete failure of the crop (Elliott, Wagner and Melchers, 1932); damage reaching 50–60 per cent losses in yield has also been reported in susceptible cultivars of sorghum (Ramakrishnan, 1963).

The contribution of regularly seed-transmitted diseases to the total of diseases in the U.S.A. (9 per cent being grain sorghum, 15 per cent sweet sorghum — USDA, 1965) is about a third. For other parts of the world losses are likely to be considerably higher. Vallega and Chiarappa (1964) suggest that the loss of sorghum due to all diseases in Africa is equivalent to crops covering 2 million hectares, and at the prevailing yield rate this was reckoned to equal food for 5.5 million people. Assuming the above share of the seed-borne diseases which nevertheless is probably far too low for Africa where the smuts are dominant, this means that they cause annual losses equivalent to food for 2–3 million people. Losses from *Fusarium moniliforme* are reported in a general way to be 'serious'; precise estimates are difficult and practically never given. As in rice and maize this pathogen reduces stands, causes stalk rot, top rot, mouldy ears and, what is almost entirely neglected, stunt which may depreciate yields to an unknown extent, much damage being unnoticed. Other diseases, regarded as minor because they are inconspicuous and often unrecorded, take a steadily accumulating toll: *Ascochyta sorghi,* rough leaf spot, is particularly severe on fodder crops and may affect 10–15 per cent of the plants during the rainy season in India (Singh, Mehta and Singh, 1951). Downy mildew, *Sclerospora sorghi,* causes heavy losses in India (Safeeulla, 1971). *Drechslera turcica,* leaf blight, is severe in the U.S.A., Africa and Eastern Europe, but precise estimates are not available; other species of *Drechslera,* very common in seed lots of sorghum tested in the laboratory in Denmark (Chidambaram, Mathur and Neergaard, 1973), may be assumed to cause some of these unappreciated losses. The bacterial diseases of sorghum, in particular leaf streak, *Xanthomonas holcicola,* is prevalent in humid, somewhat cooler regions in the U.S.A., Australia, South Africa and the U.S.S.R., and causes defoliation and hence loss, the extent of which, however, also is not precisely recorded.

For **millets,** comprising different species of *Panicum, Setaria, Echinochloa, Leersia, Eleusine,* and other genera of Gramineae, many pathogens are responsible for considerable losses but again precise estimates are often lacking. For *Ustilago crameri* in *Setaria italica* losses in China have varied from 8 to 50 per cent, in Manchuria up to 50 per cent (Ramakrishnan, 1963), and in India approximately 75 per cent infection has been observed (Sundararaman, 1921). In *Panicum miliaceum,* proso millet, *Sphacelotheca destruens,* head smut, is most important having caused losses up to 100 per cent in Eastern and Southern Europe (Ramakrishnan, 1963), and in *Coix lachryma-jobi,* Job's tears, 15–35 per cent losses in yield due to *Ustilago coicis* have been reported from India (Chowdhury, 1946*a*). Downy mildew or green ear, *Sclerospora graminicola,* is sometimes devastating in India in *Pennisetum typhoides,* pearl millet or bajra, and *Panicum miliaceum,* proso millet, but actual figures are lacking; in *Setaria italica,* foxtail millet, however, losses up to 50 per cent have been reported from India (Ramakrishnan, 1963). Blast, *Pyricularia setariae,* causes heavy losses of grain in *Eleusine coracana,* finger millet, which is cultivated especially for local consumption in India — losses of 80–90 per cent have been reported (Venkatarayan, 1947); and another blast, *Pyricularia grisea,* has caused from 20 to 60 per cent reduction in grain yield of *Panicum miliaceum* in Queensland, Australia (Simmonds, 1948). Ergot or sugary disease, *Claviceps paspali* and *C. microcephala,* is sometimes 'serious' in *Paspalum scrobiculatum,* kodo millet, grown both for the grain and fodder, as well as in *Pennisetum typhoides,* pearl millet (Ramakrishnan, 1963). In India the pathogen caused damage in pearl millet in recent years to such an extent that many crops had to be burned in order to prevent further damage and spread (Raychaudhuri, 1968; Lambat, personal communication, 1968).

1.2.2 Diseases in leguminous crops

The leguminous crops, in particular soybean, groundnut, bean, pea and chickpea, constitute the most important source of plant proteins in the human diet. While 250 kg of rice or wheat is sufficient to feed a man for one year in terms of calories, only a little more than half of his need for proteins would be met by this amount (assuming the relatively low but common protein percentage of 8–9 for these two cereals). On the other hand, in terms of calories, 300 kg soybeans would feed a man for one year while in terms of proteins this quantity would be enough for three persons, the protein content being about 40 per cent. In some regions of the world, such as parts of India and Africa, people cannot afford animal proteins and are largely dependent on plant sources. These facts emphasise the economic importance of the disease problems in leguminous food crops.

Soybean is attacked by a wide range of diseases, most of them seed-borne. In the U.S.A., in the period 1951–1960, 12 named diseases, all seed-borne, were responsible for annual average crop losses of 13.6 per cent (additional 0.4 per cent were classified as 'others'), according to USDA (1965). The list includes fungi, bacteria and one virus, tobacco ringspot virus, the cause of bud blight; other important soybean viruses, such as soybean mosaic virus and soybean stunt mottle virus, were not included. In the U.S.A. Ross (1968) and Schmitthenner and Gordon (1969) demonstrated yield reduction in different cultivars from 20 to 40 per cent due to soybean mosaic virus and much more severe losses when this virus was in combination with another. Tobacco ringspot virus occurs sporadically but may cause losses from 25 to 100 per cent (Nyvall *et al.*, 1972). In Canada, in southwestern Ontario, stem canker, *Diaporthe phaseolorum* var. *caulivora*, caused losses in yield of at least 8 per cent when infected seed was used (Wallen and Seaman, 1963), and in the U.S.A. brown stem rot, *Cephalosporium gregatum,* may cut yield by as much as 25 per cent or more when all plants are infected (Nyvall *et al.*, 1972); the disease causes losses each year in Illinois (Gray and Sinclair, 1973). Target spot, *Corynespora cassiicola,* causes measurable yield reduction in 6 out of 10 years, and some losses may reach 50 per cent (Sinclair, personal communication, 1974). There is no reason to believe that losses in technically less advanced countries are less than 14 per cent of the average loss in the U.S.A., although precise information is scarce. In India downy mildew, *Peronospora manshurica*, is destructive in the major soybean areas (Safeeulla, 1971). According to Dunleavy (1973) selections of soybean that are resistant to *Xanthomonas phaseoli* var. *sojense,* bacterial pustule, which prevails in warmer climatic conditions, yield 11 per cent more than susceptible selections, and Sinclair (personal communication, 1974) showed experimentally that bacterial blight, *Pseudomonas glycinea,* can cause yield losses of more than 50 per cent.

Bean, *Phaseolus vulgaris,* is like soybean rich in proteins, with 15–31 per cent, average 24 per cent. Most of the diseases causing loss are seed-borne. In the U.S.A. the average annual loss for 1951–1960 was 20 per cent (USDA, 1965) of which 16–17 per cent may be assigned to seed-borne diseases, one-third of these to Fusarium root rot, *Fusarium solani* f.sp. *phaseoli;* in several counties in New York State as many as 90 per cent of the fields used to be infected, resulting in heavy losses (Burkholder, 1919). In 1939 Fusarium root rot caused 95 per cent losses in New South Wales, Australia (Anonymous, 1941), and as recently as 1971 84 per cent yield reduction was recorded in Southwestern Colorado, where in 1971 and 1972 no field sampled was free of the disease (Keenan *et al.,* 1974).

Colletotrichum lindemuthianum, anthracnose, being almost world-wide in distribution, causes greater losses in the temperate zones than in the tropics. In the nineteenth century it was exceedingly devastating, e.g. in parts of Italy bean crops were completely destroyed (Voglino, 1892), and in the first decades of this century it was the most serious bean disease in the U.S.A. east of the Mississippi River. In 1915 in New York losses were from 30 to 100 per cent. In 1965 5 per cent of acreage of bean was 'severely infected' in Eastern Germany (Masurat, Peschel and Stephan, 1966). In the U.S.S.R. Chang (1960) reported losses of 54 per cent green bean yield and 50 per cent of ripe seed. In experiments on control in Sweden Kolk (1958) demonstrated losses of 50–60 per cent. In North Spain annual average losses amount to 15 per cent, in all Spain 2–3 per cent (M.-Sagasta, personal communication, 1974). In Kenya 18 per cent of 150 fields were infected in 1973 and the disease was very severe in Central Kenya (Mukunya, 1974). In Malawi the disease is the biggest limiting factor to production of good quality beans in the main rain crop where almost complete crop loss can occur; by intensive fungicide spraying total bean yield was doubled by the best treatment in one year, and even more in another (Peregrine, 1971). In the U.S.A. losses due to anthracnose are now less because of production of certified seed in semi-arid areas (Zaumeyer and Thomas, 1957), and these authors found during the survey that losses amount to only 0.5 per cent of the national production. On the other hand, *Sclerotinia sclerotiorum,* white mould, causes heavy losses both in the field and during transportation. Losses up to 50 per cent have been reported in the U.S.A. (Zaumeyer and Thomas, 1957) and in recent years average losses have increased alarmingly; in Western Nebraska the average in dry beans was about 19 per cent in 1973 while the 1970–1972 losses were 13, 6, and 14 per cent, respectively (Kerr and Steadman, 1974), and in New York fresh bean losses were 10–20 per cent (Lumsden *et al.,* 1974); in Colorado Oshima and Dickens (1974) obtained yield increases between 25 and 100 per cent by sprays of benomyl.

Bacterial diseases of bean are responsible for a considerable share of the losses. In the U.S.A. the two bacterial blights, *Pseudomonas phaseolicola,* halo blight, and *Xanthomonas phaseoli,* common bacterial blight, caused average losses from 1930 to 1939 ranging from 1.2 to 7.9 per cent, mean 4.1 per cent (McCallan, 1946), and during the period 1951–1960 about 2 per cent of the national crop was lost (USDA, 1965). In the U.S.S.R. reduction in bean yield caused by *X. phaseoli* was up to 37 per cent in 1958 and up to 65 per cent in 1959 (Korsakov, 1960). In the U.K., Taylor (1969) found 43 per cent yield reduction by halo blight when in early summer 2.6 per cent of the plants showed primary infection, 12.5 per cent reduction when 0.4 per cent showed primary infection; in North Spain an annual average loss of 20 per cent is usual, and for the whole of Spain 1–2 per cent (M.-Sagasta, personal communication, 1974). In many places the losses from bean wilt, *Corynebacterium flaccumfaciens,* must be added, in Wyoming, U.S.A. designated as 'serious reductions', and in 1951 there was almost total loss in many fields in Colorado (Zaumeyer and Thomas, 1957). Bean common mosaic, accounting for an annual average of 3 per cent loss of total bean production in the U.S.A., from 1951 to 1960, according to USDA (1965), is still widespread but was previously more severe because more susceptible cultivars were grown. In New York State, U.S.A., Horsfall (1930) in a survey of canning beans found from a trace to 100 per cent infection and estimated an average loss of about 10 per cent, and in Colorado Zaumeyer (1946) found 75 per cent infection resulting in losses of 33 per cent, i.e. roughly 1 per cent loss from each 2–2.5 per cent infection. Burke and Silbernagel (1974) found in infection experiments yield reductions from 35 to 65 per cent. In Morocco seed-borne infection of this virus has been identified as one of the principal causes of yield loss (Lockhart and Fischer, 1974), growers' estimates of loss being 50 per cent in cases of highest incidence. In trials of imported seed lots in Algeria in 1974, the author found bean samples almost totally infected and virtually destroyed by the virus indicating heavy losses in crops from such seeds, similar to those reported from Morocco. In Iran this virus disease is the most prevalent disease in most bean-growing regions, seed transmission ranging up to 50 per cent in some local cultivars, and yields reduced by 80 per cent or more (Kaiser, personal communication, 1974).

The major seed-borne diseases of **pea** are those of the so-called 'Ascochyta complex', the pathogens being *Ascochyta pisi, Phoma medicaginis* var. *pinodella* (syn. *A. pinodella*), and *A. pinodes* (perf. state *Mycosphaerella pinodes*); further serious diseases are wilt and near-wilt, caused by *Fusarium oxysporum* f.sp.*pisi,* race 1 and race 2, respectively, and bacterial blight, *Pseudomonas pisi.*

In the U.S.A. average annual losses of the national production of green peas for the period 1951–1960 (USDA, 1965) were estimated at 2 per cent for 'Ascochyta blight' which, however, may not include Mycosphaerella foot rot, a part of the Ascochyta complex, a further 2 per cent for

Fusarium wilt and 1 per cent for bacterial blight, totalling 5 per cent. Root rots caused further 10 per cent losses, of which *A. pinodes* may be responsible for a considerable part being the most severe of the three pathogens of the complex. Total for seed-borne diseases may thus be 7–8 per cent. In Canada Wallen (1965) found experimentally that *A. pisi*, under certain conditions, caused yield reduction of 11 per cent, *Phoma medicaginis* var. *pinodella* 25 per cent, and *A. pinodes* 45 per cent. Bacterial blight which usually is not destructive, may sometimes be so, having caused losses up to 25 per cent in the U.S.A. (Chupp and Sherf, (1960). Fusarium wilt has been destructive in the north central States of the U.S.A. (Chupp and Sherf, 1960) but has generally been controlled through development of resistant cultivars, although still local losses are encountered in susceptible cultivars.

Chickpea, or gram, *Cicer arietinum,* is another important high protein pulse crop grown world-wide, though 90 per cent of the production is concentrated in the Middle East and South East Asia. Major pathogens are *Ascochyta rabiei (Phyllosticta rabiei)*, Ascochyta blight, or leaf blight, and sometimes *Sclerotinia sclerotiorum,* stem and crown blight.

Kaiser (1972*a*) has published a survey of the damage caused by these pathogens in Iran, where high rainfall provides favourable conditions for epidemics. In 1969 *A. rabiei* resulted in severe dieback of foliage, and yields were so 'greatly reduced' that the disease was a limiting factor in cultivation of this food crop under conditions favourable for infection. A serious epidemic occurred one year in yield trials in Khuzeston Province; most chickpea lines were completely destroyed (Kaiser, personal communication, 1974). *Sclerotinia sclerotiorum* killed a considerable number of plants thus contributing to yield reductions but no figures are given. In Israel Halfon-Meiri (1970) found that the weight of seeds from pods infected by *Ascochyta* was only 40 per cent of that from healthy pods, cf. Table 19.1; emergence from diseased seed was faulty, and the dry weight of the one-month-old diseased plants was only one-third of that of healthy plants! In Azerbajan, U.S.S.R., *A. rabiei* attacks all cultivars causing losses in yield from 15 to 83 per cent (Askerov, 1968). In Morocco the disease causes up to 100 per cent loss; under humid conditions a crop may be destroyed totally in three days (Schlüter, personal communication, 1975). In Canada the pathogen, apparently introduced with seed, was destructive in yield trials in 1973, where in eight cultivars all plants were killed in 18 per cent of the plots and severely attacked in 42 per cent of the plots (Morall and McKenzie, 1974). In India the leaf blight causes considerable damage necessitating breeding for resistance; and in Spain this pathogen is the most important on chickpea, causing up to 30–50 per cent losses in wet years, 5–8 per cent being an annual average (M.-Sagasta, personal communication, 1974). The pathogen is quite common in seed lots from different countries tested for health, which is indicative of its wide distribution and economic importance.

Groundnut, or **peanut,** suffers from a range of diseases. In the U.S.A. over the period 1951–1960 a total of about 28 per cent has been recorded as an annual average (USDA, 1965), and half of the diseases found in this survey are seed-borne. On a rough estimate a total of 15 per cent loss is due to them, among which seed rots are particularly destructive, such as those caused by species of *Aspergillus, Fusarium,* and *Rhizoctonia solani,* also pathogens producing dry rot and collar rot, including *Physalospora rhodina (Diplodia gossypina); Cercospora arachidicola (Mycosphaerella arachidis)* and *Cercospora personata (Mycosphaerella berkeleyi),* both cause tikka disease or Cercospora ear spot which very often results in defoliation and accounts for 1.5 per cent of the total loss in the above survey. The *Cercospora* spp. persist on pods and may thus be spread to new areas but seed transmission is of minor importance where the disease is endemic. In India tikka disease causes losses in yield from 15 to 50 per cent (Lewin, Natarajan and Govindarajan, 1972), and in Zaire up to 50–60 per cent (Gerlagh and Bokdam, 1973). In Nigeria losses in yield of pods may reach 40 per cent in most years, and application of fungicides to well-grown crops has increased pod yields by up to 150 per cent but the very low basic yields obtained on small farms do not permit economic control of the disease by fungicides (McDonald and Misari, 1974). Seed-infecting fungi may have a considerable impact on the quality of groundnuts: in India *Aspergillus flavus, Botryodiplodia* sp. and *Cladosporium herbarum* were found to cause reduction in oil content altering its colour and making it smell rancid (Lalithakumari, Govindaswamy and Vidhyasekaran, 1971). In Virginia, U.S.A., groundnut stunt virus reduced yield and quality in three commercial fields, loss of sound mature nuts being more than 80 per cent (Culp and Troutman, 1967).

1.2.3 Diseases in oil crops

Soybean and groundnut are major sources of oil production but they have been discussed above because they are important in providing food for consumption in the daily diet. Soybean also contains 20 per cent oil which is composed of 80 per cent unsaturated fatty acids and 7–8 per cent linolenic, and groundnut seed contains 46–52 per cent oil and 25–30 per cent protein. Other major oil crops, to be discussed below, are sunflower, rape, linseed (oil flax), and sesame.

Sunflower is assuming increasing importance for oil production, the oil being a rich source of polyunsaturated fatty acids for human consumption; the seed contains 25–32 per cent oil. There are three very important seed-borne pathogens; *Plasmopara halstedii,* causing downy mildew, *Sclerotinia sclerotiorum,* white rot or stem rot, and *Verticillium albo-atrum,* wilt; all three are destructive under favourable conditions. Downy mildew can kill 70 to 80 per cent of the plants in the field and germination of seeds from infected plants may be considerably reduced, and the disease is destructive for commercial production of sunflower in Canada (Sackston, in Leppik, 1962*b*), Roumania and the U.S.S.R. (Savulescu and Vanky, 1955–1956). In Canada 95 per cent of young plants have been found systemically infected so that the whole crop had to be ploughed down.

Rape is the major oil crop in cool temperate zones, being important in Northern Europe and in cooler parts of India and Pakistan. The seed yields about 40 per cent oil and 25–35 per cent protein. In seed crops heavy losses may be caused by *Alternaria brassicicola* but generally rape is less seriously affected by pathogens than cruciferous vegetables such as cabbage and cauliflower. No precise data are available on yield losses.

Linum usitatissimum is grown for oil **(linseed)** as well as a fibre crop (flax); the seed contains 35–45 per cent oil and 24 per cent protein. A considerable range of seed-borne pathogens causes heavy losses of seed and fibre. Major pathogens are *Colletotrichum lini,* stem canker, *Fusarium oxysporum* f.sp.*lini,* wilt, *Aureobasidium lini,* stem break, browning, *Septoria linicola* (perf. state *Mycosphaerella linorum),* pasmo, *Phoma exigua* var. *linicola,* foot rot, and *Melampsora lini,* rust. In the U.S.A., 1951–1960, pasmo caused an average annual loss of 3 per cent, wilt 1 per cent, seedling blight 1.5 per cent and rust 2 per cent, making a total of about 7 per cent. Losses in Africa seem to be on a similar level. According to Padwick (1956) the diseases stem break, stem canker, wilt and powdery mildew, cause losses of 5.3 per cent, and as mildew is not seed-borne a 4 per cent loss due to these seed-borne diseases would be a fair estimate. In Northern Ireland treatment of seed infected up to 50 per cent by *Aureobasidium lini*, increased yield of fibre by 40 per cent in one year, 370 per cent in another (Muskett and Colhoun, 1947). In India losses due to Fusarium wilt reached up to 30–55 per cent (Sharma, Mathur and Gupta, 1971).

Sesame, *Sesamum indicum,* is used for confectionery and oil for human consumption; the seed contains about 50 per cent oil and 20–25 per cent protein. It is grown in tropical and subtropical areas. In the U.S.A. more than 5 per cent annual average losses were caused by bacterial leaf spot, *Pseudomonas sesami* and *Xanthomonas sesami,* during the period 1951–1960 (USDA, 1965). Other destructive seed-borne pathogens are *Alternaria sesami*, Alternaria blight, and (cf. Leppik and Sowell, 1964), also *Corynespora cassiicola*, Corynespora blight, and *Drechslera sesami*, leaf blotch. For total loss in the U.S.A. due to seed-borne diseases 8 per cent appears to be a conservative estimate. In India Alternaria blight causes defoliation and severe damage (Mohanty and Behera, 1958), and in south eastern U.S.A. these blights are production hazards; in 1961 and 1963 average yield losses due to Alternaria and Corynespora blights at Stoneville, Mississippi, were more than 50 per cent (Culp and Thomas, 1964). In Venezuela *Cercospora sesami* and *Alternaria sesami* reduced yields in ten cultivars by from 18 to 55 per cent (Barboza, Mazzani and Malaguti, 1966).

1.2.4 Diseases in vegetables

Three vegetables, bean, pea and chickpea, have been discussed in Subsection 1.2.2; now another five vegetables of major importance will be considered: cabbage, carrot, celery, lettuce and tomato. Cramer (1967) points out that statistical information on world-wide yields are not available for the majority of vegetables, but as the bulk of vegetables are grown in many smallholdings for direct consumption, if the figures from these were to be added to available data, the total losses would be 'most enormous'.

In **cabbage** (world production 15 million tons on 775,000 ha), two seed-borne pathogens are of dominant importance, *Phoma lingam (Plenodomus lingam*, perf. state *Leptosphaeria maculans)*, dry rot, black leg, and *Xanthomonas campestris*, black rot, to which may be added a third pathogen often destructive in this crop plant, *Sclerotinia sclerotiorum*, watery soft rot, white blight. They may all be devastating; in the U.S.A. the first two pathogens combined caused losses in 1973, estimated roughly at a level of 20 million dollars, and an average yield loss for the nation of 10 per cent. In New York State losses went up to 100 per cent in some fields, and in Minnesota about 70 per cent of the crops were lost on an average (Gabrielson, personal communication, 1974). Previously such losses were not uncommon in the U.S.A., as they were in Europe; the situation improved, however, considerably after seed production in the 1920s and 1930s had been shifted to the semi-arid western States such as Washington State. For the period 1951–1960 the total average annual loss caused by the three pathogens combined in commercial crops were estimated at only 2.5 per cent (USDA, 1965). The writer has seen cauliflower seed crops in India in 1967 and in Israel in 1972 heavily infected by *Sclerotinia sclerotiorum*. In a large field in the Negev Desert the losses were assessed to be at least 40 per cent (author's record). In seed production *Alternaria brassicicola* may cause heavy losses, quantitatively as well as qualitatively. In Denmark control programmes in seed crops indicate quantitative seed yield losses from 17 to 42 per cent (Nielsen, 1933; Wraae-Jensen, 1938), and qualitative losses are equally important: during the period 1935–1941 more than 90 per cent of 746 seed lots of cabbage and other cruciferous vegetables were infected, often reducing appreciably the quality of the seed (Neergaard, 1945).

The following may be considered as major seed-borne pathogens of **carrot:** *Alternaria porri* f.sp. *dauci,* Alternaria blight, *Alternaria radicina,* black root rot, *Cercospora carotae,* Cercospora blight, and *Xanthomonas carotae,* bacterial blight. In the estimates of annual average losses in the U.S.A., collated by USDA (1965) for 1951–1960, bacterial blight accounts for 2 per cent, Alternaria and Cercospora blights for another 2 per cent, the first pathogen being predominant in California, one or both of the last two blight pathogens occurring wherever carrot is grown. Black root rot is not mentioned in the survey of USDA from which a total of 5 per cent may be assessed as an annual average. Bacterial blight reduces seed yields quantitatively and qualitatively. In 1933 Humphrey and Wood (1935) recorded losses up to 50 per cent in Massachusetts due to Alternaria leaf blight. *Alternaria radicina* causes a black rot on roots, reducing their market value in quality as well as quantity (Lauritzen, 1926). A similar root rot is caused by *Alternaria porri* f.sp. *dauci* (Schneider, 1961). Seed yields are considerably reduced by *A. radicina* in Denmark. Neergaard (1945) recorded in the period 1934–1943 a commercial depreciation of 30 carrot seed lots grown for a seed company in Copenhagen. On account of damage to germination, the depreciation in seed grower's price for these seed lots was from 5 to 55 per cent. Sometimes infection renders seed lots quite worthless, resulting in a shortage of seed for sowing.

Of the seed-borne pathogens of **celery** and **celeriac,** particularly *Septoria apiicola,* late blight, is still important being responsible for frequent heavy losses which could be prevented by efficient seed certification schemes and seed treatment. In experiments late blight losses amounting from 50 to 60 per cent have been recorded in Denmark (Jørgensen, 1934), from 10 to 90 per cent in Germany (Elssmann, 1936; Wenk, 1939; Baumann, 1957), and a loss of 80 per cent has been encountered in the U.S.A., Florida (Darby, 1959). In Chile the average annual losses have been from 10 to 15 per cent of the national production, or — at that time — about 200,000 U.S. dollars (Mujica, 1943), and in the U.S.A. the annual losses averaged 4 per cent for the period 1951–1960 (USDA, 1965). Occasionally *Phoma apiicola,* Phoma root and crown rot of celery, or scab of celeriac, may cause losses in the cooler temperate zones. Previously the disease was a serious problem in Germany (Klebahn, 1910) but is probably now of little importance in commercial production. The average annual loss ascribed to bacterial leaf spot, *Pseudomonas apii,* was 1 per cent of the national production in the U.S.A. during 1951–1960 (USDA, 1965) making a total of 5 per cent for *Septoria* and bacterial blights. Storage rot of celery and celeriac is greatly dependent on infection introduced from the field. In Israel 26 per cent of winter celery was affected by petal and end rot due to *Sclerotinia sclerotiorum* and *Botrytis cinerea* (Barkai-Golan and Aharoni, 1972).

Field and glasshouse grown **lettuce** is particularly damaged by lettuce mosaic, a disease of world-wide distribution, largely being dependent on seed transmission. There are several reports of complete failure due to this pathogen because head formation in cabbage lettuce is prevented. In the U.S.A. estimated average annual loss caused by lettuce mosaic is 4 per cent for the period 1951–1960 (UDSA, 1965).

Tomato (about 1.5 million ha with a world production of 30 million tons) is subject to a considerable range of serious seed-borne diseases. In fresh market tomatoes in the U.S.A. diseases accounted for average annual losses of 21 per cent during the period 1951–1960, two-thirds being seed-borne (USDA, 1975). Damage included leaf spots and defoliation, fruit rot, stem rot, wilt and mosaic. Major pathogens were (in parenthesis percentage loss in the U.S.A. according to the above survey) tobacco mosaic virus (5 per cent in field crops and 7 per cent in greenhouse tomatoes), *Alternaria porri* f.sp. *solani*, early blight (1 per cent), *Fusarium oxysporum* f.sp. *lycopersici*, Fusarium wilt (1 per cent), *Phytophthora infestans*, late blight (0.5 per cent), *Stemphylium solani*, grey leaf spot (3 per cent), *Verticillium dahliae*, Verticillium wilt (3 per cent), *Corynebacterium michiganense*, bacterial canker (0.5 per cent), and *Xanthomonas vesicatoria*, bacterial spot (1 per cent), totalling approximately 15 per cent for seed-borne diseases. Tobacco mosaic virus, estimated as involving more losses in the U.S.A. than any other pathogen, may be considered among the most serious pathogens of tomato in all parts of the world. The tomato strain of the virus is widely and frequently seed-borne and extensively transmitted when seedlings are transplanted because inoculation from contaminations of the virus, carried on the seed coat, takes place during this manipulation. Indicative of losses was the increase in yield obtained in the U.S.S.R. by field cultivation *without* transplanting, this procedure resulting in 19 per cent quantitative increase as well as improved fruit quality, cf. Subsection 29.5.4 (Nikitina, 1966). Early infection may kill young plants or cause severe dwarfing with few or no saleable fruits. In Germany Klinkowski (1958) estimated the average loss of tomatoes from tobacco mosaic to be 10 per cent, and in the U.S.A. Crill *et al.* (1973) recorded yield losses by weight of 15 and 25 per cent in two cultivars.

Early blight or Alternaria leaf spot, which is more important in field crops than under glass, is often destructive having occasionally caused losses on tomato up to 50 per cent in Louisiana and Florida, U.S.A. (Edgerton and Moreland, 1914; Burger, 1924), more than 50 per cent in Germany (Richter, 1943) as well as in Denmark (Neergaard, 1945). More recently McNew (1963) in the U.S.A. found losses of 25–30 per cent in spray experiments. Fusarium wilt has been destructive in the U.S.A. and in many other countries, since the turn of the century. In the U.S.A. 50 per cent of the crop in Louisiana was attacked in 1907 (Orton and Ames, 1908), and in 1916 80 per cent in a field in Virginia (Fromme and Thomas, 1916); in many fields there is a total loss (Chester, 1950*a*). Fusarium wilt is widespread and destructive in the Middle East being a limiting factor in tomato production in many places, e.g. in Jordan, Iraq and Syria (information during author's visit to these countries in 1971). Walker (1971) reviewed the history and international distribution of the disease. In cool temperate zones Verticillium wilt is considerably more widespread than Fusarium wilt.

Late blight is destructive on tomato, both foliage and fruits being destroyed but inoculum is not usually seed-borne. In 1959 the disease caused a loss of 40 per cent on 90 hectares of tomato crops in Baden, Germany (Pflanzenschutzamt Karlsruhe, 1959). In Eastern Germany the disease was severe in 1965 in 10 per cent of the acreage of tomato, moderate in 24 per cent (Masurat, Peschel and Stephan, 1966); if it is assumed, according to Cramer (1967), that severe infection leads to a loss of 20 per cent, moderate to 10 per cent loss, the average loss for 1965 for Eastern Germany would be 4.4 per cent. Grey leaf spot, *Stemphylium solani,* may be serious particularly in southern States of the U.S.A., known first in Florida, and has become a serious problem in Hawaii (Chupp and Sherf, 1960).

Bacterial spot, *Pseudomonas punctulans,* in Tennessee and States to the north, attacks the fruits of tomato and pepper rendering them unsaleable. In Florida less than 1 per cent seed-borne infection can cause an epidemic (Cox, 1966). Bacterial canker, *Corynebacterium michiganense,* has caused serious losses in North America, Europe, Australia and New Zealand. Since fermentation of pulp in commercial extractions of tomato seed in 1926 was replaced by cyclonic machines as used for extraction of tomato juice in canning, seed is obtained without fermentation. The omission of fermentation allows perpetuation of the pathogen and, probably for this reason, it has become widely disseminated and increased in importance, being particularly destructive in the U.S.A. (Walker, 1969). In western North Carolina losses have been as high as 70–80 per cent and by 1962 threatened to destroy the industry (Strider, 1969). *Phoma destructiva,* fruit and stem rot, is most destructive on ripened fruits and has been quite serious in India. During the season 1966–1967 about 7–10 per cent of the harvested tomatoes of the winter crop and 2–3 per cent of the summer crop were infected, and thus actually lost, in Punjab and Uttar Pradesh (Srivastava and Tandon, 1966). In the southern States of the U.S.A. this pathogen is also destructive as a leaf and stem blight (Obrero, Trujillo and Aragaki, 1968). Stem canker, *Didymella lycopersici,* is a major disease in some regions, e.g. in England (Moore, 1959) and Germany, and in 1969 Fusarium wilt, *Fusarium oxysporum* f.sp. *lycopersici,* caused in commercial tomato crops in Ireland yield losses from 5 to 40 per cent (Staunton, 1971).

1.2.5 Diseases in fibre crops

Cotton is by far the world's most important fibre crop, world production amounting to 37 million tons in 1972.

The best information on crop losses is from U.S.A. where annual reports on yield reduction due to diseases are reported by the Cotton Disease Council. For the period 1951–1960 the estimated average losses caused by all diseases were 12 per cent, of which about half were seed-borne (USDA, 1965). These include *Ascochyta gossypii,* seedling blight, *Fusarium oxysporum* f.sp. *vasinfectum,* Fusarium wilt, *Colletotrichum gossypii* (perf. state *Glomerella gossypii*), seedling blight, boll rot, *Physalospora rhodina (Diplodia gossypina),* Diplodia rot, black boll rot, *Verticillium alboatrum* and *V. dahliae,* Verticillium wilt, and *Xanthomonas malvacearum,* black arm, angular leaf spot. Until about 1950 an average of 5 per cent was the annual loss due to Fusarium wilt in the U.S.A., in some States 10 per cent or more, and in individual farms up to 60 per cent (Chester, 1950*a*). The Cotton Disease Council (1954–1972) reported for the period 1953–1972 that the annual national loss of cotton in the U.S.A. averaged 1 per cent for Fusarium wilt while it was 2 per cent for Verticillium wilt for the period 1953–1962, and 3 per cent for 1963–1972 thus increasing in importance. In some areas where irrigation and heavy applications of nitrogen fertilisers were used, losses due to Verticillium wilt varying from 10 to 50 per cent are not uncommon in some places (Presley, 1953).

In some areas of Central Asia Verticillium wilt is widespread, causing as much as 25 per cent loss; in the U.S.S.R. the disease is considered a national problem in cotton production (Kamal and Naim, 1969), experimental control increased yield by 12 per cent (Mirpulatova, 1961) and recently by 46 per cent (Tkachenko and Yunusov, 1973), while Fernandes in 1961 recorded 13 per cent yield loss and Bȳchenko (1961) up to 50 per cent loss. In Uzbekistan, the main cotton area in the U.S.S.R., the diseased plants lose about 73 per cent of the cotton bolls and the average weight of the bolls is reduced by 50 per cent, or, in the case of late infection, 32 per cent (Kamal and Naim, 1969). In Iran Verticillium wilt causes losses from 10 to 14 per cent (Cauquil and Rahmani, 1964). The boll rot complex, in which a range of pathogens participate, includes *Colletotrichum gossypii, Fusarium moniliforme, Rhizoctonia solani, Physalospora rhodina* and other fungi, including some usually regarded as saprophytes; also *Xanthomonas malvacearum*, may be devastating causing serious losses. According to Chester (1950*a*), estimated annual losses in 1937 and 1938 in the U.S.A. were 15 per cent in North Carolina, 13 per cent in Texas, and from 10 to 20 per cent in several other States. For the period 1930–1939 the national loss ranged from 0.2 to 11.6 per cent, average 5.1 per cent in the U.S.A. (McCallan, 1946), but losses have since then been somewhat reduced, partly through development of cultivars with the desired boll shape, but without normal bracts; thus for the period 1951–1960 average annual losses were 2.2 per cent.

In the principal cotton-growing countries of Central America boll rot is serious, according to Müller (in Cramer, 1967) causing about 25 per cent loss of production. In Iran damping-off, mainly due to *Rhizoctonia solani*, causes severe damage to yield, losses being from 15 to 25 per cent in Masandaran and up to 50 per cent in Gorgan area (Moshir-Abadi, 1974), and in Syria an average yield loss of 3.6 per cent is due to *Verticillium dahliae* (Khoury and Billar, 1970). In Central Asia grey rot of lint, *Nigrospora gossypii*, causes 30 per cent weight reduction of the boll (Kamal and Naim, 1969). Bacterial blight or black arm disease, *Xanthomonas malvacearum*, has world-wide distribution, and is in many places devastating. In the south west of the U.S.A. up to 10 per cent of the crops have been destroyed (Chester, 1950*a*), and now the average annual losses are of the order of 2 per cent (Cotton Disease Council, 1954–1973). In Africa losses may go up to 25 per cent, and are often more than 10 per cent (Padwick, 1956; Clinton, 1959). In Sudan Wickens (1961) recorded 17 per cent losses, Tarr (1972) 8 per cent in an area of several thousands hectares; Padwick (1956) estimated yield reduction of up to 60 per cent in some years and an average of 25 per cent in Sudan, whereas Logan (1961) recorded about 20 per cent when seed is not treated. In Nigeria losses are more than 10 per cent, according to Padwick (1956), or 26 per cent according to Wickens (1961) who further estimated losses from bacterial blight or black arm in Tanzania to be from 25 to 29 per cent.

According to Tarr (1972), under conditions favourable to black arm 0.017 per cent infected seeds, equivalent to 43 infected seedlings per ha, caused an epidemic that reduced the yield of seed cotton by about 17 per cent. (This is a seed infection/yield reduction rate of no less than 1:1000!). In India Mathur, Daftar and Ihamaria (1973) obtained from 32 to 67 per cent increase in yield by reducing intensity of the disease testing five different seed treatments, and up to 122 per cent increase applying seed treatment plus spray. Average losses from 20 to 30 per cent are encountered in Sudan, India, China, the U.S.S.R. and Pakistan (Tarr, 1959; Singh and Verma, 1973).

Flax, an important fibre crop in some countries of Europe, including particularly the U.S.S.R., has been discussed as an oil crop in Section 1.2.3. During World War II, when the U.K. was dependent on fibre production from flax, seed-borne diseases were effectively controlled by a well-organised seed health testing scheme in Northern Ireland and by seed treatment. Muskett (1948) reports results of his and co-workers' investigations on 3,000–4,000 flax seed lots from all regions of the U.K. He points out that all of these seed lots had a germination of over 90 per cent and were of fine quality also with respect to purity; in ordinary routine seed testing they would be classified as first class seed for sowing. However, had this seed been used uncritically without disinfection, to control seed-borne fungi especially *Colletotrichum lini,* stem canker, and *Aureobasidium lini,* the production of flax fibres during World War II would have been reduced to an extent representing — at that time! — about two million pounds sterling.

1.3 SIGNIFICANCE OF SEED TRANSMISSION COMPARED WITH OTHER MEANS OF TRANSMISSION

1.3.1 Predominant transmission by seed

Many plant pathogens are disseminated practically wholly or predominantly by seed, and the effective control measures are exclusion through certification and seed treatment. Typical of this category are smut fungi in cereals, the economic significance of which has been greatly reduced by these measures. Barley leaf stripe disease, *Drechslera graminea,* was previously of great economic importance in many countries but due to the routine seed treatment consistently carried through its importance has decreased. *Septoria apiicola,* late blight of celery and celeriac, is predominantly seed-borne, plant residue transmission being subordinate. In the U.K. substantial progress has been made in recent years in controlling this disease through a well-organised seed health certification scheme involving laboratory testing, including a test of treated seed to confirm efficacy of treatment, a possibility which still is largely neglected in most other countries where the disease, therefore, is still serious. Seed health certification schemes in the U.S.A., Canada and Australia where bacterial blight of bean, *Xanthomonas phaseoli,* and bean anthracnose, *Colletotrichum lindemuthianum,* were widespread and destructive, have reduced inoculum in seed for sowing to such an extent that epidemics of these diseases are no longer common. Lettuce mosaic previously caused more serious losses but infections have now been substantially reduced through well-organised certification schemes (Zink, Grogan and Welch, 1956; Greathead, 1964).

More complicated from a control point of view are those seed-borne pathogens for which sources of inoculum other than seed are important, or other agents of dissemination prevail. This applies chiefly to soil, wind, water, insects and nematodes.

1.3.2 Seed and soil transmission

If seed, infected or contaminated by a pathogen that is also soil-borne, is sown in non-infested soil, the pathogen may be established for many years. Dwarf bunt of wheat, *Tilletia contraversa,* being both seed- and soil-borne, may be established in new areas through use of contaminated seed. Seed transmission of *Sclerotinia sclerotiorum* and *Rhizoctonia solani* has for many years been overlooked or underestimated because of the predominance of soil transmission. In testing 1,045 seed lots of *Brassica oleracea* and other *Brassica* spp. in Denmark mycelial infection of *Sclerotinia sclerotiorum* was found in 23 seed lots, and *Rhizoctonia solani* recorded in 68 seed lots, and further in 10 per cent of 179 seed lots of *Zinnia elegans* and 6.5 per cent of 203 seed lots of spinach (Neergaard, 1958*a*, 1958*b*). Both fungi are world-wide in distribution and almost ubiquitous. However, where soil becomes infested through infected or contaminated seed a chain of disease cycles may develop, in *Sclerotinia sclerotiorum* due to effective air-borne dissemination of ascospores as discussed below, Subsection 1.3.3; and seed transmission of soil invaders or soil inhabitants may introduce new pathotypes into a field, such as those of *Rhizoctonia solani,* dealt with in Section 1.4.

Tomato stem rot, *Didymella lycopersici,* sometimes serious on tomato crops particularly in greenhouses, is seed-borne as well as soil-borne. Ogilvie (1945) reports incidences in which seed transmission resulted in serious attacks on such crops in England; in 1942, 7 of 10 seed lots from Evesham district were infected. Seed disinfection with mercury compounds prevented the disease in the seedlings, but later 80 per cent of the plants showed stem rot. Infected seed may give rise to infected seedlings, but some of them die before emergence and contaminate the soil from which the inoculum may produce stem rot near the soil surface. Under the conditions often prevailing in greenhouses the disease may occur on older plants producing stem and fruit rot, although the disease started from the seed through soil transmission.

Plasmodiophora brassicae, club root of crucifers, is a soil invader, known to exist without the host in soil for more than 10 years. Yet Warne (1943) established three cases of seed transmission in rape. Presumably the seed may be contaminated by contact with infested soil and thus be introduced to new areas.

Fusarium oxysporum is another persistent soil invader, crop rotation of 10–12 years between crops of the specific host being required to starve out the parasite, or at least to reduce inoculum substantially. Many of its *formae speciales* are known also to be seed-borne, causing wilt, e.g. *Fusarium oxysporum* f.sp. *callistephi,* aster wilt (Baker, 1953), and in *Tagetes erecta* its race 2 causing marigold wilt (Olsen, 1960), *F. oxysporum* f.sp. *cumini,* cumin wilt (Singh, Choudhary and Patel, 1972). *F. oxysporum* f.sp. *lini*, Fusarium wilt of flax (Bolley and Manns, 1932),

F. oxysporum f.sp. *lycopersici*, Fusarium wilt of tomato (Kendrick, 1944; Katan, personal communication, 1973), *F. oxysporum* f.sp. *pisi,* pea wilt and pea near wilt (Snyder, 1932; Kerling, 1952), *F. oxysporum* f.sp. *vasinfectum,* Fusarium wilt of cotton (Perry, 1961; Gubanov and Sabirov, 1972), and others. Though predominantly soil-borne, even a low incidence of seed transmission of these soil invaders may be of great importance if the seed is sown in uninfested soil.

Corynebacterium fascians, cause of fasciation, leafy gall or cauliflower disease in many plants, occurs in certain soils, perhaps as a soil inhabitant. Seed may be contaminated as in nasturtium, *Tropaeolum majus,* the long peduncles of which droop and allow the developing seeds to rest on soil (Baker, 1950).

1.3.3 Seed transmission and dissemination by air

In beet there are three seed-borne pathogens for which *air transmission* is the chief means of spread. Beet rust, *Uromyces betae,* is commonly encountered on beet seeds, both uredo- and teliospores occurring in sori or as spores loosely adhering to the surface of the seeds. Downy mildew of beet, *Peronospora farinosa,* has been reported as seed-borne in low percentages, and finally Cercospora leaf spot of beet, *Cercospora beticola,* frequently occurs as mycelial infection and spore contamination in seed lots produced in warmer temperate regions, in Europe and in the central and southern parts rather than in Northern Europe, in the United States chiefly east of the Continental Divide. All three fungi overwinter in infected plant debris and in overwintering two-year crops, and from these sources spores are disseminated, chiefly by air currents, to the young plants (see e.g. Hochapfel, 1950 on *Peronospora farinosa* and air dissemination). In regions where these fungi are endemic, seed transmission undoubtedly is relatively unimportant. In Northern Europe beet rust often starts in spring in overwintering crops on which the aecial stage is developed; later the uredial spores spread and infect the leaves of the first year plants. Seed transmission is of no importance where the disease is endemic but of great importance where spreading occurs from one continent to another, across natural barriers. Beet rust has been spread as teliospores in seed from the Mediterranean area into South Africa, Australia, New Zealand (Branquinho de Oliveira, 1963) and Canada (Newton and Peturson, 1943). Gassner (1952*a*) demonstrated that *Cercospora beticola* in Turkey occurs early in the growing season on plants from contaminated or infected seeds, but that the inoculum from refuse and overwintering plants soon leads to such severe infections on plants from uncontaminated seeds that the attacks are comparable with those originating from infected seeds. Gassner concludes that although the disease is commonly seed-transmitted, this means of transmission is less important under the climatic conditions of Turkey, where the fungus is endemic. In Austria where the disease has been thoroughly investigated by Wenzl (1959*a*, 1959*b*) the situation is quite different, seed transmission being essential for development of epidemics.

Plasmopara halstedii, downy mildew of sunflower, is another example of downy mildews in which the unquestionable prominence of air transmission suggests underestimation of seed transmission. *Plasmopara halstedii* causes a destructive disease of sunflower. The mycelium of the fungus was observed in the seed by Nicolic (1952) and complete proof of transmission was established by Leppik (1962*b*) in the U.S.A. and by Novotel'nova (1963) in the U.S.S.R., later confirmed by Cohen and Sackston (1973*a*, 1973*b*) in Canada. Leppik, at the Regional Introduction Station in Ames, Iowa, reported the presence of downy mildew on sunflower seeds imported from ten countries in Europe, Africa and Asia, cf. Section 24.2.

Seed-borne inoculum of *Sclerotinia sclerotiorum*, introduced to a field, may infect subsequent crops of the same or other plant species. In Scotland sclerotia of the pathogen were found in seed of early turnips, the crop having been infected from apothecia on harvested artichokes (Noble, 1951). Gray and Noble (1965) reported that mycelial inoculum from pea seed caused infection not only in peas but potatoes and carrots. In continuous cool moist weather, such as prevails in Scotland, the ascospores may be discharged over a considerable period and be carried several miles by wind (Brown and Butler, 1936).

Another example is loose smut of barley and wheat, *Ustilago nuda* and *U. tritici*, predominantly seed-transmitted, but with an epidemic phase dependent on only a few days during the flowering of the host when dissemination of the teliospores occurs by air contents and wind over several hundred metres (cf. Fig. 1.1).

Under wet field conditions *Phoma lingam*, black leg, rapidly produces pycnidia, the conidia are disseminated by spattered rain, seedlings are killed resulting in increase of inoculum, and even a low percentage of seed infection is effective for establishment of the disease; in row-sown crops such as swede, 'the fungus runs in the rows' it is said in Denmark. However, where the disease is endemic as in New Zealand (Allen and Smith, 1961), and under suitable weather conditions, the perf. state, *Leptosphaeria maculans*, releases ascospores which are disseminated by air to considerable distances and provide a potent inoculum. Brassica crop residues in the soil such as stubbles of colza may harbour the pathogen for 3–4 years thus acting as source of ascospore inoculum, as in France (Dept. Chez-Indre) where colza is sown in October when ascospores are produced and infect the seedlings (Lacoste *et al.*, 1969), or as in the southern part of Sweden where rape is sown in late August when ascospores are produced and infect seedlings (Ndimande, personal communication, 1974). Cabbage, in contrast, is in France sown in March and is

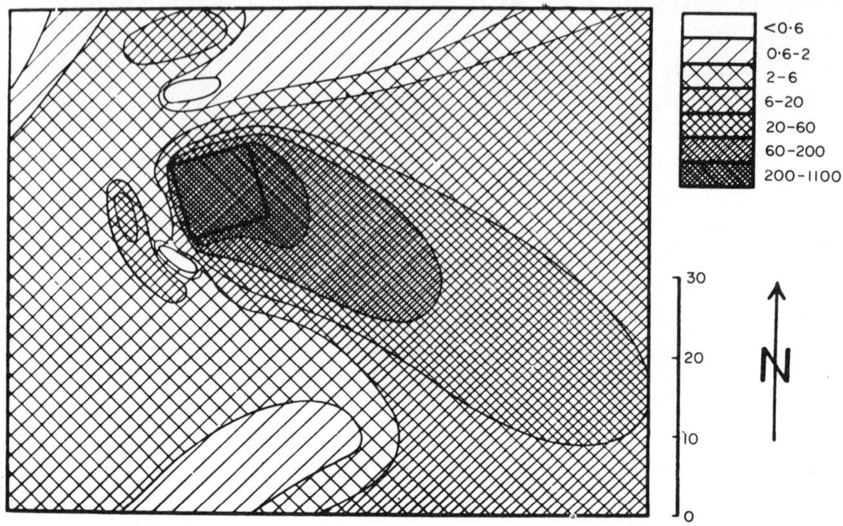

Fig. 1.1. Wind-borne dissemination of *Ustilago tritici* in wheat. The map is based on seed samples drawn from plots at different distances from a heavily infected plot placed in the mother crop which otherwise was free from loose smut (Oort 1940b).

totally dependent on seed transmission. Where seed-borne inoculum has been reduced to a very low level, below 0.5 per cent as in New Zealand, and humidity is moderate in the field, the recurrence of the disease in swede and cabbage may occur more by ascospores than from seed-borne inoculum (Allen and Smith, 1961).

Fusarium nivale, snow mould, brown root rot, in cereals and grasses, is seed- and soil-borne or plant residue-borne. Seed produced in cool temperate climates such as in Northern Europe is commonly infected, severe seed infection results in poor emergence and surviving plants suffer a root and stem rot and become unthrifty, retarded in development and susceptible to lodging. The perf. state *Griphosphaeria nivalis (Calonectria nivalis, Micronectriella nivalis)* occurs in many parts of Europe (Schaffnit, 1912; Lind, 1913; Dobrozakova, 1928; Müller, 1952; Noble and Montgomerie, 1956). In Scotland the perithecia appear on oat plants in late May when the plants are about 15 cm high and have five–six leaves; as the stem elongates, the perithecia continue to be produced on the sheathing leaf-bases and the culm; they quickly shed their ascospores, which are disseminated widely by air currents and immediately germinate producing infection on leaves, stems and flowers. Ascospores are held responsible for a part of seed infection. In North America, where *Fusarium nivale* appears to be more soil-borne than seed-borne, in contrast to Europe, the *Griphosphaeria*-state occurs in the field in the Pacific Northwest but late in the development of the crop, as first established by Cook and Bruehl (1966).

The three members of the 'Ascochyta complex' of pea, *Ascochyta pisi*, *Phoma medicaginis* var. *pinodella (Ascochyta pinodella)*, and *Ascochyta pinodes* (perf. state *Mycosphaerella pinodes*), are considered to be chiefly seed-transmitted, although they may survive for 1–5 years in soil not cropped with peas, as shown for *Phoma medicaginis* var. *pinodella* in Canada (Wallen, Wong and Jeun, 1967). For *A. pinodes*, however, the perfect state constitutes a powerful epidemic mechanism, the ascospores being responsible for effective dissemination over distances of kilometres. In contrast, seed-transmitted inoculum of the imperfect state of the three fungi which consists of slimy conidia, depends on rain and wind, and distances of spread are very limited; Dekker (1957) found under rainy conditions in the Netherlands that the maximum spread from a source of 50 diseased plants during about six weeks was only 3.5 m in the direction of prevailing winds.

1.3.4 Seed and insect transmission

Bean common mosaic virus, lettuce mosaic virus and many other viruses are transmitted by insects, aphids acting as vectors, but seed transmission which in these two viruses may occur by up to 90 and 17 per cent, respectively, is of crucial importance even at very low rates. Seed infection of 0.5 per cent or less may lead to total loss of crop if a vector population is active early in the growth season.

In bacterial wilt of maize, *Xanthomonas stewartii*, seed transmission occurs but the transfer of the bacteria from infected seeds to seedlings is by gnawing insects (Section 16.1) inoculating the pathogen. The occurrence of the pathogen in or on the seed ensures the presence of the pathogen in the seedlings, being the first prerequisite of establishing the disease; the second being the presence of the vectors. Primary infection independent of seed transmission may, however, also occur. Elliott and Poos (1934) established that the bacteria commonly overwinter in the intestinal tracts of the brassy flea beetle, *Chaetocnema pulicaria* Melsh., and that overwintered adults feeding on healthy maize plants may produce the wilt.

1.3.5 Seed and nematode transmission

Tobacco rattle virus which has more than 100 different hosts is soil-borne, or rather transmitted in the soil by nematodes (*Trichodorus pachydermus*, *T. primitivus* and others) which act as vectors. But the virus is also seed-transmitted in *Stellaria media* (very frequently so in Denmark), *Capsella bursa-pastoris* (Lister and Murant, 1961) and *Myosotis arvensis* (Lister and Murant, 1961). Tobacco ringspot virus is seed-transmitted in soybean (Desjardins *et al.*, 1954), tobacco (Valleau, 1932*a*, 1932*b*, 1941), petunia

(Henderson, 1931), and dandelion, *Taraxacum officinale* (Tuite, 1960). In the last host the virus is symptomless, detectable only by the indexing method. Fulton (1962) established that the virus may be transmitted by the soil-inhabiting nematode *Xiphinema americanum* in soybean, in which this relationship may multiply the effect of seed transmission.

Lister and Murant (1967) demonstrated seed transmission of three more soil-borne viruses, tomato black ring virus in 19 species (13 botanical families), including cowpea, soybean, strawberry, *Capsella bursa-pastoris, Senecio vulgaris* and *Nicotinia rustica*; raspberry ringspot virus (syn. raspberry Scottish leaf curl virus) in six species (five families), including soybean and strawberry; and arabis mosaic virus in 13 species (11 families), including soybean and petunia. Often seedlings appeared healthy but infections could be revealed using *Chenopodium amaranticolor* as an indicator plant.

The dispersal of virus-infected weed seeds, often by wind (or indeed in commerce!), is one of the important ways certain seed-borne and nematode-borne viruses spread over considerable distances.

These cycles are comparable to the disease cycles of seed-borne viruses transmitted by insects. Both categories exemplify the fact that a disease may have two or more different concurrent cycles, cf. Chapter 14, Introduction.

1.4 SIGNIFICANCE OF SEED TRANSMISSION IN SELECTION OF VIRULENT PATHOGENS

The old conception of *Rhizoctonia solani* (perf. state *Thanatephorus cucumeris,* syn. *Corticium solani*) as a very polyphagous pathogen with an almost unlimited host range, and as a soil inhabitant, may have to be modified. There is growing evidence of some host specialisation in that the pathogen comprises a huge number of unnamed biotypes each varying in its host range (Storey, 1941; Person, 1944; Papavizas and Davey, 1962; El Zarka, 1965), and it has been established that this range of biotypes may be classified into a number of anastomosis groups, characterised by heterocaryon compatibility [cf. Subsection 17.2 (4.3)]; these groups are distinct in host range and some other characters (Richter and Schneider, 1953; Anderson *et al.,* 1972; Ogoshi, 1973). Seed transmission of *Rhizoctonia solani* ensures that the fungus is present in a susceptible host (Baker, 1947) and probably the more often the pathogen is transmitted in this way the more susceptible is the host, and the greater is the risk for subsequent crops. There is a possibility, not yet established, that some biotypes of *R. solani* may be dependent on the presence of their host for subsistence in the soil, and for maintenance of virulence; if this holds true such biotypes are soil invaders rather than soil inhabitants. Seed transmission may have a similar selective function in the collective species *Botrytis cinerea*. In the seed field there is probably a strong preferential selection towards strains of a pathogen which are particularly virulent for a given host species or variety (Baker, 1972). Eggplant seedlings inoculated with a virulent strain of *Verticillium dahliae* produced 2.5 per cent seeds infected by the fungus, whereas none of the seeds from plants inoculated with a mild strain was infected (Ribeiro, 1972).

Seed transmission of pathogenic races is an important consequence of this preferential selection and is extremely important from a quarantine point of view — such races are easily spread by seed over long distances and across natural barriers (for further discussion, cf. Section 17.2).

2 Diseases and Injuries of Seed

2.1 DEFINITIONS

The fact that seeds are both vehicles and victims of disease calls for some clarification of concepts. So do the facts that seeds, without being diseased, may be vehicles of pathogens, as well as they may be victims of detriments other than disease, such as insect damage.

First, the following tentative definitions of disease and injury may be considered.

Disease is a factor that interferes with the normal functions or structure of a plant. It may be physiogenic, caused by the direct effect of unfavourable environmental factors, or biogenic, caused by the effect of pathogens, including animal parasites and viruses.

Injury is damage to the structure of a plant, brought about by physical, chemical or animal force: wounding, mechanical injury, insect injury, frost injury.

The concepts of disease and injury may overlap: 'cold', encountered in thermophilic plants by low temperatures, for instance in orchids, would be regarded as a physiogenic disease, frost damage usually as an injury. Mercury injury to seeds may lead to seedlings that are abnormal due to induced chromosome disorders, and it may be justified to consider such seedlings as diseased.

There has been some controversy about the use of terms such as 'seed-borne disease', 'seed-transmitted disease', and 'seed-borne' and 'seed-transmitted' pathogen. The writer holds views that are in agreement with those of Leach (1940), expressed in his definition of insect transmission of plant diseases. Leach found ample justification, both etymological and traditional, for the use of the expression 'disease transmission' and accepts this term. On the other hand, he reserved 'dissemination' for the use only in connection with the pathogen. He holds the concept that the pathogen may be disseminated widely but the disease is not transmitted until inoculation and infection have occurred. Insects — and seeds — may disseminate a pathogen without transmitting the disease. Moreover, a disease is 'borne' in the seed in the sense that potentially it is brought forth or given support by the seed. Here the sense of 'bear' is by tradition broad enough to make 'borne' acceptable as a general indication of seed as an agency potentially conveying disease. Distinction is thus implied between seed-borne disease, the seed being an agency, and seed disease, the seed being itself a victim of disease, not necessarily conveying this disease to the subsequent crop. For distinction between 'seed-borne pathogen' and 'seed-transmitted pathogen', see Part II, Introduction.

2.2 PHYSIOGENIC SEED DISEASES

2.2.1 Deficiency diseases

Nitrogen deficiency. The so-called 'yellow berry' of wheat is an important degrading factor in this cereal. Grains of wheat are normally translucent but kernels suffering this anomaly have light yellowish spots affecting part or all of the grain (Fig. 2.1). The affected parts are high in starch and low in protein content, and the grains are low in specific gravity. The disease is prevalent in soils high in potash and phosphorus and is prevented by application of nitrates and by crop management favouring the nitrogen level of the soil, such as including legumes in rotations, introducing green manure crops and by using nitrogenous fertilisers. The disease is economically important, lowering the market value of grain. It is common in North America, Argentina and North Africa.

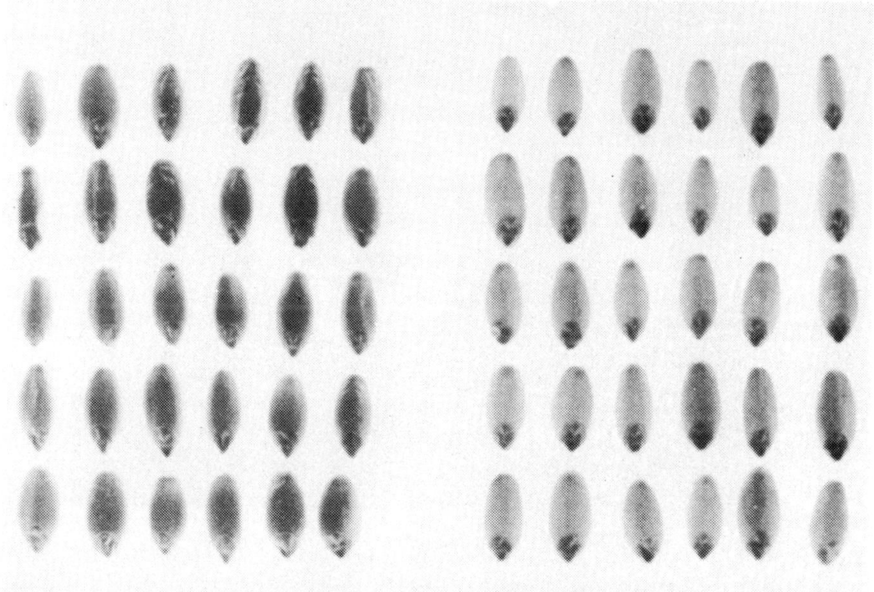

Fig. 2.1. Yellow berry of wheat, nitrogen deficiency. Left, affected kernels; right, unaffected kernels. Photographed by transmitted light. Note semi-transparency of healthy kernels

Potassium deficiency. This may cause substantial quantitative yield reduction of seed (Eckstein, Bruno and Turrentine, 1937; Stapel and Bovien, 1943) and may also affect the quality. Seed often fails to mature and becomes small. De Bruijn (1933) regards potassium deficiency as a factor contributing to development of marsh spot in pea, discussed below. Hoffman (1933) has shown that shortage of potassium may cause seeds of cucumber to assume an abnormally tapering shape.

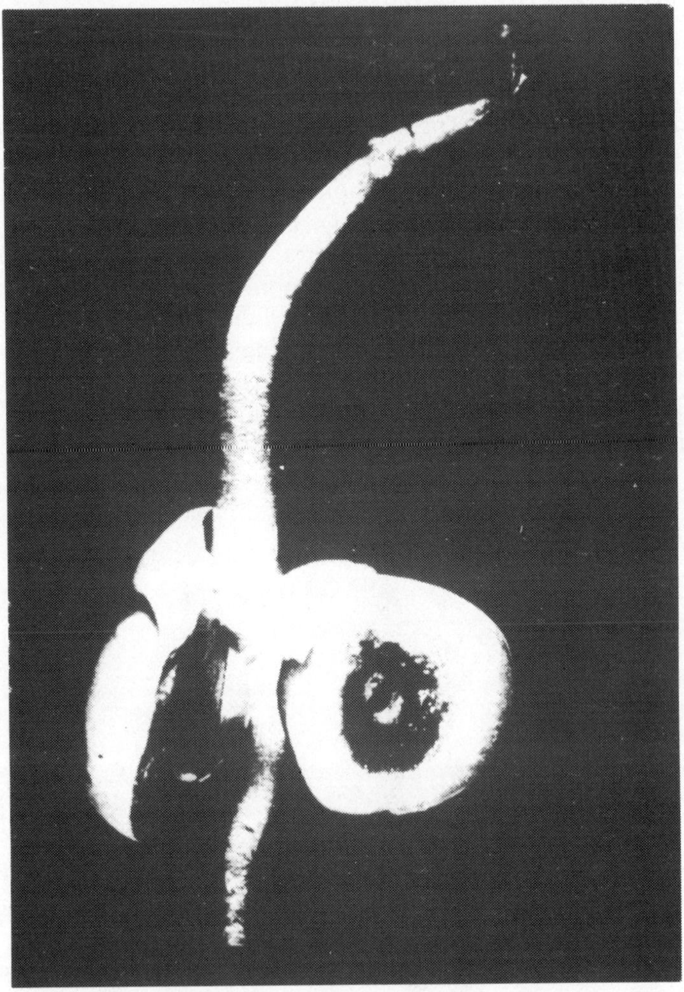

Fig. 2.2. Marsh spot, manganese deficiency, in pea

Manganese deficiency. This is a rather common disorder in pea seed, to a less extent in bean seed. Symptoms appear as an extended internal brown necrosis in the seed, the central part of the cotyledons being affected (Fig. 2.2). When these are severely affected cracks may appear. Some seeds may show brown, sunken spots in the seed coat and these necroses continue into the two cotyledons beneath. The internal symptoms are detectable two to three days after the seeds have been sown. The seedling is weak and will later develop poorly; once the plumule has emerged, the top dies and a double replacement plumule develops. The parent plants, on the other hand, often show only a slight chlorosis, even when they yield severely affected seeds.

The disease has been known since the past century and described by Mansholt (1894); later the disease was studied in the Netherlands by De Bruijn (1933), in Britain by Lacey (1934), and in Finland by Jamalainen (1936). Manganese deficiency was first detected as cause of the disease by Löhnis (1936) and Pethybridge (1936), and manganese sulphate as a measure of control was established by these and other authors (Koopman, 1937; Ovinge, 1937; Von Berenberg-Gossler, 1943). Von Berenberg-Gossler recorded 41 mg manganese per kg dry matter in healthy seed but only 13 mg in affected seeds. She found that all or only some of the seeds in one pod may be affected, and that the last developed pods yielded the highest percentage of diseased seeds.

Marsh spot is common in pea grown in the polder districts of the Netherlands (Ovinge, 1935; Doyer, 1938). Marrow-fat peas are very susceptible, particularly in alkaline soil which is rich in humus as in Romney Marsh, England (Furneaux and Glasscock, 1936), hence the name 'marsh' spot, and in newly reclaimed sea silt soils in the Netherlands (Ovinge, 1935, 1937). In England late maturing pea cultivars with large seeds have been found to be more severely affected than early maturing cultivars with small seeds (Glasscock, 1941). De Bruijn (1933) observed that a protracted ripening period encouraged the development of marsh spot, while the trouble was less in the seeds of plants in which maturation had been hastened. The disease is only rarely encountered in North America (Cuddy, 1959).

Marsh spot in Phaseolus bean and broad bean is milder and less frequent (Hewitt, 1945). This condition has been confused with hollow heart, see below.

Boron deficiency. This is known in Florida, U.S.A., as cause of a type of 'hollow heart' in groundnut, leading to discoloration and rotting of the seed. Soil amendments with boron eliminates the defect (Harris and Gilman, 1957).

Pea seeds appearing normal may suffer boron deficiency; seedlings develop pale stunted shoots without plumule bud. Addition of boron completely overcomes the abnormality (Leggatt, 1948).

Molybdenum deficiency. In New South Wales, Australia, interveinal mottle, 'scald' or necrosis of interveinal and marginal regions of the leaves developed in plants of Phaseolus bean, grown from seed, produced in plants grown in acid soil and sown in similar soil. The plants were cured by application of sodium molybdate (Wilson, 1949).

Calcium deficiency. Wallace (1951) showed, in his colour atlas of deficiency diseases, the effect of calcium deficiency in broad bean pods and seeds. Pods are deformed, wilted and blackened, and seeds fail to develop. In general, calcium deficiency occurs most commonly in strongly acid soils. Often deficiency of manganese and magnesium may be involved.

2.2.2 Temperature effects on seed

Low temperature effects. Dry seed withstands extremely low temperatures (−185 to −192°C) but wet seed will usually be killed at a few degrees below zero. This effect of temperature applies also to fungi and bacteria. Frost damage on seed crops during maturation may reduce or destroy the viability of unripe seed. The damage induced may appear as symptoms on the seedlings such as necrotic spots in the cotyledons.

In North Europe (Germany) late frost in spring may cause severe damage to maturing seeds of winter crops. The siliquas of rape, *Brassica napus,* may, after frost, still look normal but the seeds inside are brown, shrunken and dried-up. Yield losses may be up to 100 per cent (Pape and Härle, 1943). Frost may cause reduced germination in seeds of lupin and beets, and early autumn frost injures or destroys maturing of seeds of crucifers, particularly in cauliflower (Stapel and Bovien, 1943). In Norway frost damage is rather common in seeds of barley, oats and rye. Seedlings from damaged seeds often have a stunted, swollen root and a curled sprout (Overaa, personal communication, 1973).

High temperature effects. Air-dried seed tolerates rather high temperature of dry heating, e.g. barley 64°C, rye and wheat 67°C, pea 71°C, but in water or after soaking, seed may be killed at temperatures of about 50°C (Ferdinandsen and Buchwald, 1936*a*). Self-heating of seed during storage, particularly when the seed is moist or insufficiently mature, may lead to serious damage. The seed may become brownish, wrinkled, the embryo weakened, brown, and roots of the seedling may not develop root hairs, or the embryo may be killed.

Hot water treatment (Subsection 21.3.1) may reduce the germination capacity of seed; only a few degrees separate *dosis tolerata* and *dosis curativa*.

2.2.3 Humidity effects on seed

The 'normal' water content of seeds varies from one plant species to another and depends greatly on the chemical composition of the seed. While 14 per cent water content is low for oats it is high for cabbage and radish. The maximal limit for seeds of grass, beet and spinach is 15 per cent, but for cabbage and radish only 9 per cent (Franck, 1940). Onion seed must be dried down to 8 per cent water content to keep well in storage.

The air humidity during storage of seed is an essential factor for keeping the viability of seed, see Section 20.1.

High humidity effects. At high relative atmospheric humidity the lifetime of seed is substantially reduced. The germination capacity

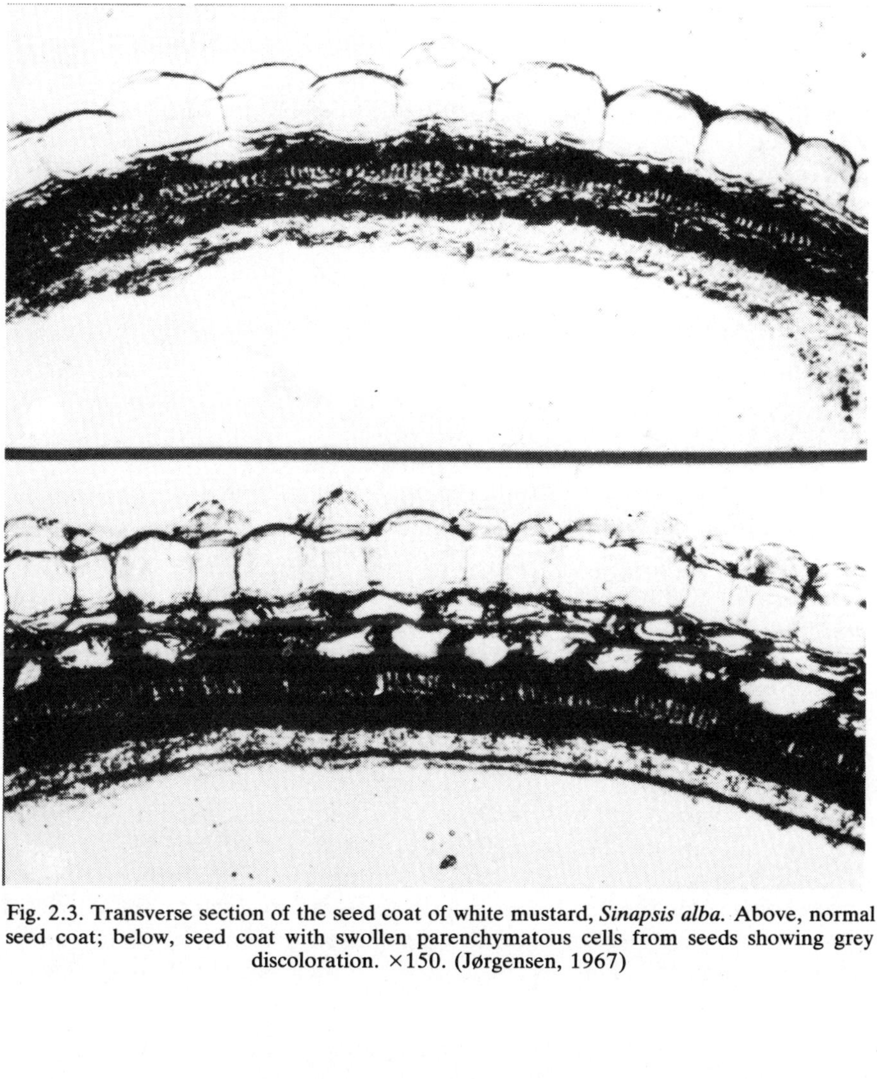

Fig. 2.3. Transverse section of the seed coat of white mustard, *Sinapsis alba.* Above, normal seed coat; below, seed coat with swollen parenchymatous cells from seeds showing grey discoloration. ×150. (Jørgensen, 1967)

decreases while the number of abnormal seedlings increases. In cucumber, lettuce, spinach and other crop seeds such poorly developed seedlings, induced by high humidity, include 'stubbed roots' involving development of a secondary root (Toole, personal communication, 1958). Voisenat (*fide* Doyer, 1938) produced abnormal seedlings with stubbed roots by keeping lettuce seed for some time in humid air, then drying it. Such seedlings cannot survive in the field.

In Denmark grey discoloration of seeds of radish, *Raphanus sativus* (Neergaard, 1945), and white mustard, *Sinapis alba* (Jørgensen, 1967), is very common. Jørgensen demonstrated that the direct cause of the discoloration is swelling of the subepidermal parenchyma of the seed coat which leads to a distortion or breaking of the epidermis (Fig. 2.3). The swelling occurs under humid conditions. In the field the disorder occurs only when the crop is close to maturity and the weather humid. The breaking of epidermis may facilitate invasion of *Alternaria* spp. which may penetrate deeply into the seed. *A. brassicae* is particularly frequent in affected white mustard seeds (Jørgensen, 1967), *A. raphani* in affected radish seeds (Neergaard, 1945). The quality of grey seeds of white mustard in terms of oil quality (per cent fatty acids), germination capacity and seed weight, is lower than that of normal, yellow seeds. The effect on oil quality may be due to easier access of oxygen to the outer cell layers of affected seeds.

Low humidity effects. A common phenomenon in spinach is 'broken seedlings'. However, close examination reveals no true rupture; the necrosis is caused by a drying effect, not by mechanical damage (Toole, personal communication, 1958).

McCollum (1953), Pirson (1966), Pollock, Roos and Manalo (1969) and Kietreiber (1969) have described seedlings of bean and pea, abnormal due to the effect of drought. The cotyledons may have transverse cracks and the primary leaves may be small and deformed. The defect resembles mechanical damage. The tetrazolium test reveals a thin untinged layer at the outer and inner sides of the cotyledons and untinged spots of the primary leaves of the embryo. Affected seeds show poor emergence, and the seedlings are very weak. Dickson (1973) believes that this disorder is due to a combination of factors: the genotype, the rate of seed drying at maturation and imbibition rate at germination. In the U.S.A. transverse cracking is common in most Phaseolus bean cultivars and can occur as much as 100 per cent in some cultivars. Dickson classed a number of cultivars in consistently resistant, intermediate and susceptible cultivars.

Another drought reaction is 'physiological drought' resulting in necrotic spots in cotyledons of lettuce due to moisture stress at temperatures of 32°C and above, over an extended period. In extreme cases the entire embryo may be killed. In Idaho, U.S.A., lettuce seed fields may be kept very dry and not irrigated because of the danger of losses due to *Sclerotinia sclerotiorum*. The etiology of the malady was experimentally confirmed (Finley, 1959).

In soybean, cracking and wrinkling of the seed coat may be due to desiccation during the maturation of the seed; field emergence of such seeds is lower than in seeds with sound coat (Moore, 1960). Agraval and Menon (1973) observed that rapid desiccation during the last one to two weeks of crops maturity results in cracking, moderate desiccation in wrinkling. Some cultivars were highly resistant, others highly susceptible to cracking, whereas none of the cultivars were resistant to seed coat wrinkling.

Hollow heart or *cavitation* in pea closely resembling marsh spot was first described by Myers (1948) in New South Wales, Australia, who knew that it was not marsh spot, and later work has confirmed that the two disorders are distinct.

Seeds with hollow heart are usually normal in outward appearance but have sunken or cracked areas in the centre of the adaxial cotyledon face; there is no discoloration as in marsh spot (Fig. 2.4). However, the cells lining the hollow are abnormal (Moore, 1964; Perry and Howell, 1965), and these are readily colonised by pathogenic and saprophytic fungi. Gane and Biddle (1973) found that affected seeds often were triangular in cross-section, rather than square, and the seedlings were less vigorous and more readily killed in adverse conditions and weather. Affected seeds usually germinate well on blotter but produce weak plants in pot tests. In

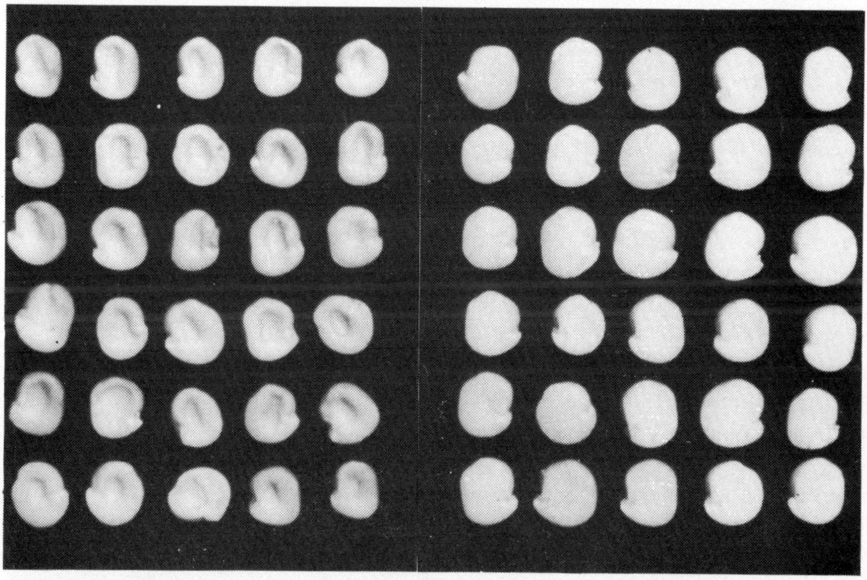

Fig. 2.4 Hollow heart or cavitation. Adaxial cotyledon face of affected seeds (left) and healthy seed (right)

the field the affected cotyledons quickly rot. Myers established that hollow-hearted seeds yield plants of significantly lower weight than normal seeds, Perry (1967) that seed lots with a high incidence emerged less well than normal seeds, and Harrison and Perry (1973) found that the disease delayed germination and reduced the size of seedlings in a sterile medium; in glasshouse experiments it was associated with smaller plants which flowered later and yielded 32 per cent less than plants from normal seed. Heydecker and Kohistani (1969) did not obtain such effects. Perry and Howell (1965) found the disorder in seeds from many countries and that it was prevalent in seed stocks received for sowing in Scotland.

It has been suggested that hollow heart in pea may be due to too quick drying of immature seed, such as may be caused by hot dry winds, confirmed by Perry and Harrison (1973) who showed experimentally that predisposition to hollow heart may be caused by high ambient temperatures during maturation of the seeds on the plant, and by drying them when immature. The incidence of hollow heart increased with increasing drying temperature from 35 to 45°C and decreased as seed maturity advanced. They found also, in accordance with R. P. Moore (1964) and Heydecker and Feast (1969), that the incidence of hollow heart ultimately depends on the rate and quantity of water imbibed by dry seeds, rapid softening increasing the condition; high soil moisture increases the damage when affected seed is sown.

Damage in seeds may be due to improper drying. Wheat grains damaged during the process of drying may produce abnormal seedlings (Wellington, 1968), see Table 2.1 later in the chapter.

2.2.4 Effects of poisons on seed

Chemical damage to seed is common because of the wide use of chemical seed treatment. Phytotoxicity of some seed fungicides is discussed in Section 22.4; and the importance of condition of seed to be treated, in Section 22.7. Some aspects of injuries due to seed fungicides will briefly be discussed here, pertaining to two categories of classical seed germicides: copper sulphate (blue vitriol) and formaldehyde.

Injury caused by copper sulphate solution has been studied already by Nobbe (1872) and Kühn (1873). Hurd (1921) found that in mechanically damaged wheat kernels the radicle was injured by seed treatment much more frequently than the embryo, resulting in 'stunted roots', because threshing usually caused wounding of the pericarp just above the radicle. The detrimental effect of copper sulphate may not be completed until after a period of storage. Jørgensen (1934) improved germination of celery seed immediately after 24 hours of treatment, as compared with untreated, but after storage the germination capacity of the treated seed went down, and after three years the germination was half of that of untreated seed.

Formaldehyde has a similar or even worse effect. When treated seed is dried, formaldehyde is converted to paraformaldehyde which persists on and in the seed coat or pericarp. If the seed is sown immediately under humid conditions, there is no injury at adequate dosage (Hurd, 1921; Braun, 1922). Jørgensen (1934) recorded almost total loss of germination capacity in celery seed treated in 0.25 per cent formaldehyde for six hours and then stored for three years.

The effect of mercury compounds in producing abnormal seedlings is dealt with in Subsection 22.1.4.

Growth substances. Detrimental effects of plant hormones applied to seed crops have sometimes been encountered in germination of seed. For instance, treatment of a seed crop of cotton with 2,4-D resulted in poor germination and abnormal seedlings, exhibiting swollen root tip and hypocotyl. The young seedlings were not much damaged but the first and second true leaves were substantially malformed, partly filiform, and the growth of the seedling was much slower than normal (Dunlap, 1948). Kjær (1950) reported damage to germination of pea seed from a lot which had been stored close to a leaky jar with P.46 (4 chloro-2 methyl-fenoxyacetic acid). One sack of the pea lot had been moistened at the bottom. Laboratory testing showed that the presence of only a few of the hormone-treated seeds was enough to prevent the other seeds from germinating.

In grass seed crops, in England, application of herbicides often causes damage to seeds, resulting in abnormal seedlings (Mary Noble, personal communication, 1968).

2.3 SEED DISEASES CAUSED BY VIRUSES

Tomato aspermy virus is the best known example of a virus producing seed abortion, and barley stripe mosaic virus may cause seed abortion or flower sterility in wheat (Inouye, 1962) and barley (Inouye, 1962; Sandfaer, 1973). A mite-transmitted virus, pigeon pea sterility mosaic virus, which is not seed-transmitted, causes sterility in pigeon pea, *Cajanus cajan* (Capoor, 1952; Seth, 1962). Some amount of abortion is produced by several viruses, e.g. bean common mosaic virus in bean (cf. Schippers, 1963) and Prunus necrotic ringspot virus in cherry (Attafuah, 1965). Seed abortion also occurs in several types of citrus infected with citrus stubborn disease virus which is not seed-transmitted (Calavan and Christiansen, 1966).

Embryo-borne viruses provide many examples of effects reducing the viability of seeds, particularly in leguminous seed. Seeds of pea, infected by pea early browning virus, become wrinkled, the seed coat showing a greenish-grey discoloration (Bos and Van Der Want, 1962), cowpea seeds infected by cowpea aphid-borne mosaic virus are small and shrivelled

(Phatak and Summanwar, 1967), pea seeds, infected by pea seed-borne mosaic virus, become small and cracked, the seed coat subsequently healing, and generally only such abnormal seeds transmit the virus (Stevenson and Hagedorn, 1969, 1973). Soybean seeds, infected by soybean mosaic virus, have a brown or blackish mottled seed coat and are substantially smaller than healthy seeds (Koshimizu and Iizuka, 1963; Ross, 1963; Van Niekerk and Lombard, 1967; Kennedy and Cooper, 1967; Phatak, 1974), see Colour Plate IV; frequency and intensity of seed mottling vary according to strain and temperature, and effects on seed size and mottling are increased by added infection of bean pod mottle virus which is not seed-borne (Ross, 1969*a*, 1969*b*). Soybean seeds infected by soybean stunt virus also have a discoloured seed coat mottled with concentric rings (Koshimizu and Iizuka, 1963).

Seed of mungbean may become wrinkled and shrivelled due to mungbean mosaic virus (Phatak, 1974). Kuhn (1965) established that groundnut mottle virus occurs more frequently in small and in discoloured seeds than in normal. Barley seeds infected by barley stripe mosaic virus are also abnormally small and shrivelled (Inouye, 1962; Błaszczak, 1963; Phatak and Summanwar, 1967). In barley this virus causes a considerable number of triploid and aneuploid seeds; the frequency of triploids seems to depend on the rate of infection. The triploid seeds are very light, shrivelled, and all or nearly all of them will be discarded in a normal seed-cleaning procedure. The triploid plants are usually a few days later in germination, they are weak in the initial stages and are unlikely to survive to maturity (Sandfaer, 1973). In sorghum the sugarcane mosaic virus (strain maize dwarf mosaic virus) causes panicle necrosis and shrunken, undersized seeds in plants that otherwise are symptomless. Affected seeds are light, have chalky endosperms, break easily and are more heavily attacked by *Alternaria* and *Fusarium* than normal seeds. Reductions in yield may be as high as 40 per cent (Edmunds and Niblett, 1973). Also seeds of squash, *Cucurbita pepo*, infected by squash mosaic virus, are lighter than healthy seed, poorly filled and deformed (Middleton, 1944). Other examples of reduced seed size are onion seed infected by onion dwarf virus (Härdtl, 1960), and beet seed infected by beet yellows virus which is not seed-transmitted (Sŭtić, Jončić and Djordjević, 1959; Blaszyk, 1963).

In England broad bean stain virus produces a stain (called 'Evesham Stain') or dark discoloration in the form of peripheral, brown necrosis of the seed coat, see Fig. 2.5 (Lloyd, Smith and Jones, 1965). The presence of the disease renders affected crops unsuitable for processing, and the disease is therefore of considerable economic importance.

Stones of apricot, infected by plum pox virus, which is not seed-transmitted, show distinct brown ringspots (Baumann, 1958; Kegler, 1962; Bojnansky *et al.*, 1963).

An attempt to separate out lettuce seeds according to weight in order to remove seeds infected by lettuce mosaic virus failed (Grogan and Bardin,

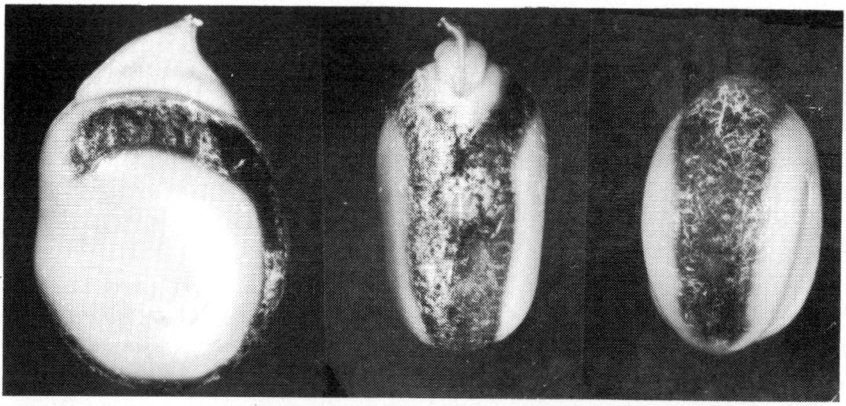

Fig. 2.5. Broad bean seeds showing brown necrotic, peripheral bands typical of 'Evesham Stain', caused by broad bean stain virus. (Lloyd, Smith and Jones, 1965)

1950). Rohloff (1962) also found that large, well-developed seeds in an infected sample did not yield fewer infected seedlings than did small seeds. In many cultivars, however, the virus produces necroses in the seeds which then have much reduced germination capacity (Szirmay, 1957).

In *Lupinus luteus,* infection of cucumber mosaic virus as well as bean yellow mosaic virus results in a different effect on seed size. The viruses cause shredding of flowers, few pods are formed, and as a consequence these contain few big seeds which largely carry the virus (Troll, 1957; Błaszczak, 1963).

Tobacco mosaic virus (tomato strain) may cause necroses in tomato seeds which become black (Broadbent, 1965).

The longevity of many viruses in seed is remarkable, see Subsection 20.2.4, but there seems to be no information on the effect of viruses on the longevity of seeds and there is little information on the longevity of virus in seeds which have lost their germination capacity. However, Pierce and Hungerford (1929) established that bean common mosaic virus was infective after 30 years of storage of bean seed, and Laviolette and Athow (1971) found that soybean after five years of storage at 16–32°C lost most of its germination capacity, whereas a high infection percentage of tobacco ringspot virus remained constant, outliving the seed.

Some viruses may directly influence the viability of seed, as in lettuce mosaic virus. A mild strain of seed-transmitted hop virus (which may have been Prunus necrotic ringspot virus) caused a reduction of 20 per cent in germination, while a severe strain reduced germination by no less than 90 per cent (Blattny and Osvald, 1954). Seeds of spurrey, *Spergula arvensis,* infected by tomato black ring virus, germinate more slowly than do healthy seeds but there is little or no effect on the seedlings (Lister and Murant, *fide* Cadman, 1963).

2.4 SEED DISEASES CAUSED BY BACTERIA

The direct pathogenic effect of bacteria on seeds may be referred to the following four types of disease condition: (1) seed abortion, (2) seed rot, (3) seed discoloration, and (4) 'slime disease'. According to pathogen and weather conditions during seed maturation one or another of these disease conditions may be prevalent in the seed or may occur in combinations.

Seed abortion. Black chaff of wheat, *Xanthomonas translucens* f.sp. *undulosa,* may reduce seed yield heavily, being very serious in some localities. Seed formation may be totally stopped or seed formed but much reduced in size, and the seeds shrivelled (Bamberg, 1936; Wallin, 1946*a*, 1946*b*; Boosalis, 1952). A disease caused by *X. panici* prevents seed formation of millets in Hungary (Klement, 1954).

Seed rot. Angular leaf spot or black arm of cotton, *Xanthomonas malvacearum,* produces water-soaked lesions on the boll, and the bacteria may penetrate from this position into the seed primordia, often effectively assisted by rot fungi, including *Colletotrichum gossypii*. The bacteria may cause a general rotting of the young seed which decomposes in bacterial slime but it is usually only present in the seed coat of mature seeds. Young seedlings are rapidly decayed, the cotyledons showing first water-soaked, then dead lesions. Fibres of infected seeds are often discoloured by bacterial ooze (Tarr, 1961; Brinkerhoff and Hunter, 1963; Schnathorst, 1964).

Seeds of cabbage and cauliflower may be so affected by black rot, *X. campestris,* that they are completely rotted or the embryo so infected that on germination typical black rot lesions become apparent in the cotyledons (Srinivasan, Neergaard and Mathur, 1973).

In Japan *X. itoana* has been found to produce a black rot of rice grain (Tochinai, 1932).

Seed discoloration. Discoloration of the seed coat may be due to lesions produced by pathogenic bacteria. This is common in bean infected by bacterial blight, *Xanthomonas phaseoli,* and halo blight, *Pseudomonas phaseolicola*. On the pods reddish-brown, water-soaked slightly sunken lesions are formed, and from these there is subsequent invasion of the bacteria into the developing seeds. If bean seeds are infected, for instance by fuscous blight (*Xanthomonas phaseoli* var. *fuscans*) at a very early stage, they completely rot while later infections produce discoloured seeds in the form of yellow patches, readily seen in white-seeded cultivars. In this disease also a small yellowish resin-like mass of bacteria is sometimes present at the hilum.

Slime disease. In many developing countries, including India, Egypt and China, yellow slime disease, or in India 'tundu disease' involving also *Anguina tritici* or 'tannau disease', *Corynebacterium tritici*, is widespread and important in wheat (Sabet, 1954*a*, 1954*b*). A similar disease, yellow slime disease or Rathay's disease, *C. rathayi*, is prevalent in Denmark in seed crops of cocksfoot, *Dactylis glomerata* (Skou, 1965) but has been encountered occasionally also in New Zealand (Johnston, 1956) and the U.S.A. Both pathogens which perhaps are identical or varieties of one species (Sabet, 1954*c*; Vasudeva, 1957) cause rotting of developing seeds and produce excessive amounts of bacterial slime which on drying becomes a varnish-like cover of the seed. A third yellow slime disease has been encountered on wheat in Iran, *C. iranicum*, where it causes abortion of grain and bacterial slime in embryos (Scharif, 1961). On *Agropyron smithii*, yellow gum disease, *C. agropyri*, is similar to the above diseases, and the pathogen may be identical with *C. rathayi* (Elliott, 1951).

2.5 SEED DISEASES CAUSED BY FUNGI

The direct impact of fungi on seed is considerable. Many fungi are serious parasites of seed primordia and maturing seeds and reduce yields of seed both quantitatively and qualitatively. Other fungi, including saprophytes and very weak parasites, may lower the quality of seeds by causing discoloration which may seriously depreciate the commercial value of seeds, particularly of grain when graded for consumption. The following types of disease and disorder are encountered, often in combination: (1) seed abortion; (2) shrunken seeds, reduced seed size; (3) seed rot; (4) sclerotisation or stromatisation of seed; (5) seed necrosis; (6) seed discoloration; (7) reduction or elimination of germination capacity; (8) physiological alterations in seed.

Seed abortion. The most prominent examples of fungi producing abortion are the smut fungi that infect cereals and grasses systemically, and the ergot fungi. The flower organs of the hosts are replaced by the fructifications of the parasites.

Certain parasitic fungi, belonging to Fungi Imperfecti, are particularly pathogenic to flowers and young seed structures, but lose their infectivity as seeds approach maturity; other parasitic fungi of this group attack maturing seeds and may be of little or no importance as pathogens in the first stages of development (e.g. *Drechslera sorokiniana*, cf. Section 11.3). Still other of these fungi may have little or no preference. Parasites of the first category are those that cause deterioration of ovules and young seed primordia, thus reducing seed yield quantitatively. Important examples are *Gloeotinia temulenta* causing 'blind' (ungerminating) seed in ryegrass and other grasses (Section 11.5), different species of *Fusarium*, particularly in cereals, including *F. culmorum* (Section 11.5), *F. graminearum* in wheat, maize and rice (Section 11.4) and *F. moniliforme* in maize. In flax, seed formation may be prevented by *Aureobasidium lini*, stem-break or browning (Section 11.5), and *Septoria linicola*, pasmo (Section 11.5); in California poppy, *Eschscholtzia californica*, *Acroconidiella eschscholtziae* (Section 11.5), and in chickpea, *Ascochyta rabiei* (Section 11.9) may kill the young seeds. In brome grass and in wheat, *Pyrenophora semeniperda* (perf. state; imp. state: *Podosporiella verticil-*

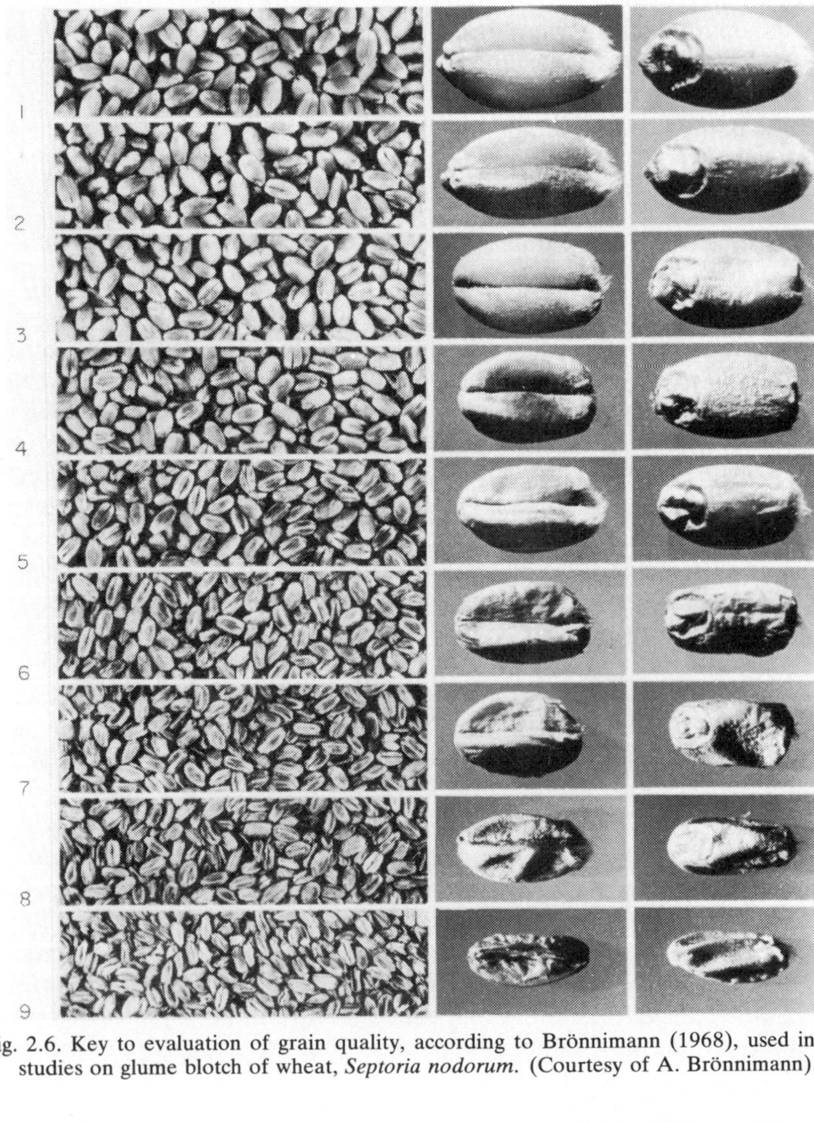

Fig. 2.6. Key to evaluation of grain quality, according to Brönnimann (1968), used in his studies on glume blotch of wheat, *Septoria nodorum*. (Courtesy of A. Brönnimann)

lata) causes death of seed primordia, hence the name of the perfect state of the fungus, and the vernacular name of the disease, 'sterility disease' (Cooke and Shaw, 1952). *Sclerospora sorghi,* downy mildew of sorghum, infects maize systemically, and usually causes sterility (Jones, Leeper and Frederiksen, 1972).

Shrunken seeds, reduced seed size. Examples of more or less heavy reduction of seed size have been given in the discussion on seed processing, including removing of poorly developed shrivelled seeds from seed lots (Section 19.9): *Alternaria brassicicola* and *Phoma lingam* in crucifers, *Ascochyta rabiei* in chickpea (Halfon-Meiri, 1970), *Phoma medicaginis* in lucerne, *Septoria linicola* in flax, *Drechslera teres* in barley (Smedegård-Petersen, 1974), and *Septoria nodorum* in wheat (Brönnimann, 1968), cf. Fig. 2.6.

Reduction of yields as a result of poor development of seeds is caused by many fungi, which are not necessarily seed-transmitted. Pathogens attacking foliage may severely impair photosynthesis with adverse consequences to seed development. Important examples are black stem rust, *Puccinia graminis,* and other rust fungi in cereals, grasses and other crops; downy mildew, e.g. *Peronospora destructor* in onion, *Sclerophthora macrospora* and *Sclerospora philippinensis* in maize, *S. sorghi* in sorghum, *Albugo candida* in crucifers; powdery mildews, e.g. *Erysiphe graminis*, in cereals and grasses; *Puccinia helianthi, Plasmopara halstedii* and *Verticillium alboatrum*, reducing also the oil content in sunflower (Zimmer and Zimmerman, 1972).

Seed rot. Many seed-borne fungi produce seed rot either in the crop or during germination. Examples are *Fusarium* in cereals, including *Fusarium avenaceum, F. culmorum, F. graminearum, F. moniliforme, F. nivale* and *F. semitectum* [Fig. 2.7(1)]. The last fungus may cause dry rot in rice grains. *F. avenaceum* is a common seed rot organism in a wide range of hosts of different families. Many species of *Drechslera* cause seed rot, e.g. *D. oryzae* in rice, *D. maydis* in maize, *D. sorokiniana* in cereals and in many other kinds of crop seed, *D. papaveris* in poppy. *Diplodia maydis* and *Fusarium graminearum* produce seed rot in maize (Ullstrup, 1961), *Trichoconis padwickii* in rice, and *Curvularia penniseti* in pearl millet. *Alternaria porri* f.sp. *dauci* and *Alternaria radicina* are causes of seed rot in carrot, *Colletotrichum graminicola* in soybean [Fig. 2.7(2)], *Phoma lingam* in cabbage, *Phoma medicaginis* in lucerne and clover, *Mycosphaerella pinodes* in pea, *Diaporthe phaseolorum* in soybean (Wallen and Cuddy, 1960), *Botrytis cinerea* in seeds of many hosts, including forest trees. Seed rot is produced in oak by *Ciboria batschiana, Sclerotinia sclerotiorum, Phomopsis quercella, Valsa intermedia* and *Gloeosporium quercinum* (Urošević, 1961*a*). In cacao, seed deterioration is caused by *Monilia roreri* and *Botryodiplodia theobromae* (Wellman, 1954).

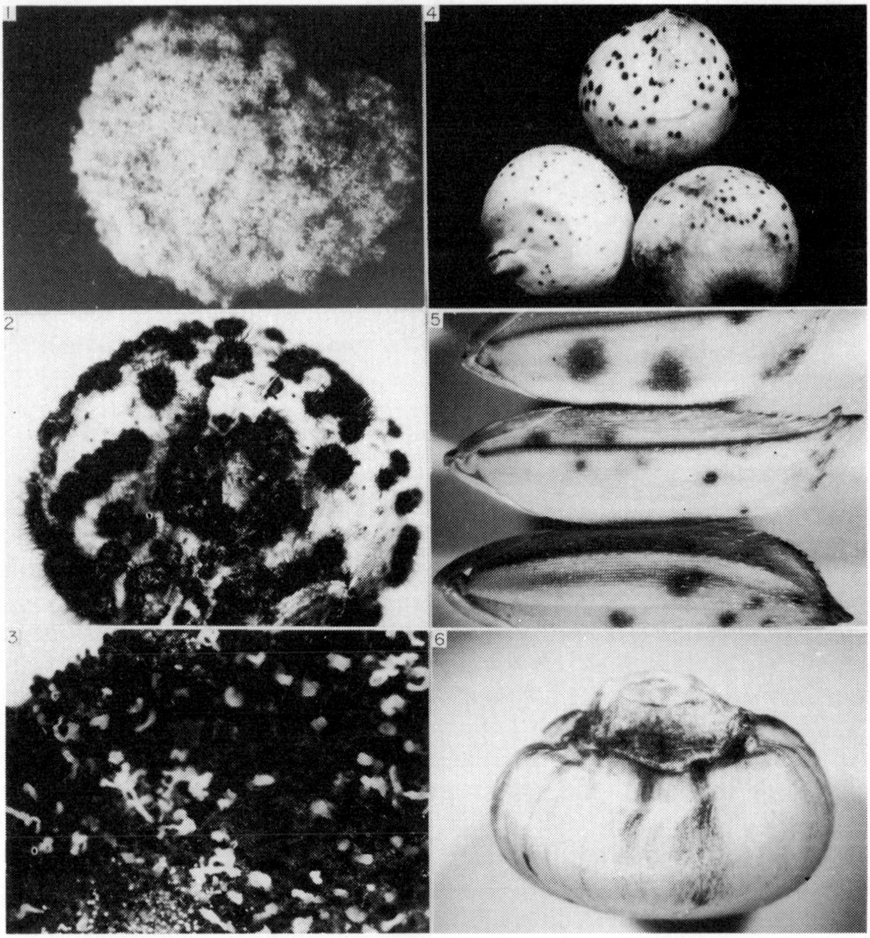

Fig. 2.7. Examples of seed rots and seed discolorations. Seed rot in (1) eggplant (*Fusarium semitectum*); (2) soybean (*Colletotrichum graminicola*); (3) *Phaseolus mungo* (*Macrophomina phaseolina*). Seed discoloration in (4) sorghum (*Colletotrichum graminicola*); (5) rice (*Drechslera oryzae*); and (6) maize (*Drechslera maydis*)

Trichothecium roseum causes rot in Scots pine, *Pinus sylvestris,* Norway spruce, *Picea abies*, black locust, *Robinia pseudacacia*, ash, *Fraxinus excelsior,* birch, *Betula*, spindle tree, *Euonymus*, European beech, *Fagus sylvatica*, and oak, *Quercus robur* (Urošević, 1961*b*). *Schizophyllum commune (Agaricaceae)*, a common wood-destroying fungus, produces rapid deterioration of cones and seeds of Douglas fir, *Pseudotsuga taxifolia* (Shea and Rediske, 1964). Also, acorns may be infected (Urošević, 1957). Under unfavourable conditions of collecting seeds and in storage *Mucor mucedo* and *M. racemosus* contribute to deterioration of seeds of oak and beech. The fungi may originate from the ground and penetrate through cracks in the fruit coat. Seed rot produced by storage fungi is discussed in Chapter 7.

Sclerotisation, stromatisation. Transformation of floral organs or seed into sclerotia or stromata is an important disease condition in certain categories of fungi and hosts. Ergots produced by *Claviceps purpurea* and other species of *Claviceps* in cereals and grasses exemplify sclerotia of this type. Different species of *Ciboria* stromatise forest seeds and seeds of grasses, e.g. *C. alni* in alders, *Alnus incana* and *A. glutinosa*; *C. batschiana* in oak, *Quercus robur*, and chestnut, *Castanea vesca*; *C. betulae* in birch, *Betula pubescens* and *B. verrucosa*; *C. carunculoides*, known as popcorn disease in mulberry, *Morus* spp.; and *C. granigena* in brome grasses, *Bromus* spp. (Buchwald, 1949). Closely related to these fungi is *Gloeotinia temulenta*, cause of blind seed disease in different grasses, particularly in *Lolium perenne*. Another example of sclerotisation or stromatisation of seed is *Phomopsis viterbensis* in chestnut, *Castanea vesca*.

Seed necroses. Many seed-rotting fungi produce superficial necroses in the seed; other fungi never penetrate deeply into the tissues, most seed-borne fungi usually not beyond the protective layers, the seed coat or pericarp. In leguminose seeds anthracnose fungi, *Colletotrichum* spp., as well as *Ascochyta* spp., often penetrate into the fleshy cotyledons, producing conspicuous necrotic lesions in seeds of bean, soybean, pea, chickpea, cowpea and other hosts.

Seed discoloration. Discoloration of seeds is a very important degrading factor. In seeds for sowing, such disorders may indicate that seed-transmitted parasites are present, although it must be emphasised that this is not necessarily so; in seeds for consumption (e.g. grain) or for industrial purposes (e.g. oil seed) it may be a general indication of poor quality. Although in many cases blemishes in seed have no real effect on the quality of seed for sowing, they usually do depreciate, often seriously, its market value. There are at least three categories of seed discoloration when causes and effects are considered: (1) superficial necrotic lesions, (2) fungous coatings and (3) pigmentation.

Many seed-borne parasitic fungi infect the seed coat causing conspicuous necrotic black, brown to grey discolorations. Well-known examples are the effects of *Ascochyta pisi* in pea; *Colletotrichum lindemuthianum* in bean; *Drechslera oryzae* in rice, 'pecky rice' [Fig. 2.7(5) and Colour Plate IV] in which the grain not only becomes poor in appearance but also low in milling quality. Other examples are *Fusarium equiseti, F. semitectum and Macrophomina phaseolina* in mungbean, *Phaseolus aureus* (Ram Nath, Neergaard and Mathur, 1970); *Phaesisariopsis griseola* in bean (Orozco-

Sarria and Cardona-Alvarez, 1959); in sorghum dark brown to black spots on the seeds are associated with *Colletotrichum graminicola* [Fig. 2.7(4)], *C. gloeosporioides* and *Phoma* spp.; and in maize white streaks, often radiating from the embryo part of the seed, have been found due to infection of *Fusarium moniliforme* and *Cephalosporium acremonium,* while greyish, dark brown to black streaks are associated with *Drechslera maydis* [Fig. 2.7(6)], *Macrophomina phaseolina* and *Botryodiplodia* sp. Black streaks or specks beneath the pericarp of maize may be caused by *Physalospora zeae,* grey ear rot (Ullstrup, 1961).

'Black point' or 'kernel smudge' in cereals is a combination of profuse contamination of fungi, as discussed below, and discoloration of the tissues of the pericarp and testa due to infection. In certain parts of the world, e.g. Canada, *Drechslera sorokiniana* is the dominant infection. Dark brownish spots prevail at the pointed end and/or the crease near the round end, as it has been shown for wheat. Under the microscope the attacked tissues appear yellowish-brown, while the aleurone cells underneath are not damaged (Kietreiber, 1972). Affected kernels are, surprisingly, heavier than normal (Brentzel, 1944; Crosier, 1953; Karpejuk, 1969; Kietreiber, 1972). Kietreiber found for 96 samples of durum wheat an average of 49 g for normal, 53 g for discoloured kernels. Affected kernels are also larger, but there is practically no difference in specific gravity. The number of discoloured kernels may increase during storage. Kietreiber (1972) studied in Austria the development of discoloured kernels in the field and found the condition caused by *Alternaria tenuis*, infection beginning at the stage of milk-ripening, about one month before harvest. The discoloration started later, at the stage of yellow ripening, about 10–14 days before the harvest. She showed that a short advance of the harvest date reduces the percentage of discoloured kernels appreciably. Cultivars exhibited marked differences in susceptibility to the condition.

The second category of seed discoloration, coatings by mycelium and sporulating structures of fungi, often saprophytes, is another degrading factor of utmost importance. In cereals this kind of contamination is frequent and known as part of the syndrome in 'black point' or 'kernel smudge'. The disorder is caused sometimes by profuse growth of pathogens such as *Drechslera sorokiniana* in different cereals, *D. teres* in barley, *D. oryzae* in rice (Kondō and Okamura, 1927), as well as by common saprophytes such as *Alternaria tenuis* and *Cladosporium cladosporioides*, to a less extent by *Curvularia* spp., particularly in rice, and *Nigrospora* spp. Contaminations of *Alternaria tenuis* and *Cladosporium cladosporioides* are common also in many kinds of seeds other than cereals and may depreciate their market value heavily, because of their sooty appearance.

Pathogenic fungi may be so profusely present in seeds that these appear discoloured: heavy sporulation of *Fusarium graminearum*, scab or head blight in wheat, makes the kernels pink or orange; other examples are the blackening effect of heavy smut or bunt contaminations in grains, and oospore crusts of *Sclerospora sorghi* on the seed surface and glumes of sorghum. Machacek (1953) reported loss in Canada from a lowering of grade in grain due to heavy smut contaminations; at that time about three carloads of wheat out of each thousand inspected at Winnipeg were so degraded.

Finally, pigments produced by various fungi may indicate the presence of pathogens or saprophytes. *Cercospora kikuchii* produces a purple stain in seeds of soybean, see Colour Plate IV, and clusterbean, *Cyamopsis tetragonoloba* (Kilpatrick, 1957; Johnson and Jones, 1962; Laviolette and Athow, 1972), *Fusarium moniliforme* a pink discoloration in seeds of maize, *Stemphylium botryosum* a condition of red discoloration, 'red nose', in bean seeds in the region around the micropyle (Brinkman, 1931), and *Trichoconis padwickii* a pink stain in rice (Ou, 1963). However, not all seeds that are infected or contaminated with these fungi show discoloration. The capacity of producing pigments varies markedly from strain to strain, as shown for *Fusarium moniliforme*, and may probably also vary according to environmental conditions. *Alternaria* (probably *A. tenuis*) produced a blue strain on the lint in cotton.

Reduction or elimination of germination capacity, lowered viability. Obviously, necroses or more deeply penetrating rots in seeds reduce the viability of seeds, their longevity in storage and their emergence in field. But also infections which do not cause rot are known to lower the viability of seeds. Seeds of barley or wheat in which *Ustilago nuda* and *U. tritici* are present as dormant mycelium in the embryo, show less resistance to severe adverse field conditions such as excessively deep sowing or excessive high seeding rate, germination and seedling development being prevented (Mathur and Hansing, 1962); see Subsection 19.6.2. Barley seeds infected by *U. nuda* are slightly smaller than healthy ones, another indication that infection interferes with normal development of the seed. Wheat kernels, infected by *Podosporiella verticillata* fail to germinate, or they produce stunted seedlings, possibly due to a toxin produced by the fungus (Wallace, 1959).

Physiological alterations or effects in seed. Metabolic products of seed-borne microorganisms may affect the seed itself or may have other sometimes serious consequences such as toxicity to animals and human beings. *Aspergillus flavus*, which produces aflatoxin both toxic and carcinogenic in groundnut (Brook and White, 1966) and other seeds, may cause albinism in seedlings of citrus (Durbin, 1959) and virescence in seedlings of maize (Koehler and Woodworth, 1938); *Penicillium oxalicum* causes yellowing of the leaves in maize (Johann, Holbert and Dickson,

1931). The implications of mycotoxins, etc., produced by seed-borne pathogens are discussed in Chapter 8.

2.6 SEED DISEASES CAUSED BY NEMATODES

The genus *Anguina* includes some species which are specific endoparasites of flower primordia and produce 'seed-galls'. Most important is *A. tritici* in wheat, the seed galls of which are known under a variety of vernacular names: earcockles, purples, peppercorns, cockle wheat, hard smut, eelworm galls and nematode galls (Leukel, 1924; Thorne, 1961). Also rye, *Secale cereale*, emmer, *Triticum dicoccum*, and spelt, *T. spelta*, are hosts. The galls are produced mainly from undifferentiated flower buds (Marcinowski, 1910). Leukel (1924) showed that up to five galls may replace one kernel. The galls are different from kernels by shape and colour. They are usually 3.5–4.5×2–3 mm — some, however, are smaller; others may be larger than wheat kernels. They are almost black, often with beaklike projections, furrowed on one side. Galls formed in rye are longer in proportion to their width and more irregular in form. The life-cycle of *Anguina tritici* is described in Section 14.8.

Seed galls are also produced by *A. agrostis* (Fig. 6.3) in different species of *Agrostis*, *Arctagrostis*, *Dactylis*, *Dupontia*, *Festuca*, *Holcus*, *Koeleria*, *Phalaris*, *Phleum* and *Poa* (detailed list in Chapter 6, on Nematodes) and by *Anguina agropyronifloris* in western wheatgrass, *Agropyron smithii* (Lakon, 1953; Norton and Sass, 1966) and in golden oatgrass, *Trisetum flavescens*, up to 28 per cent by weight (Lakon, 1953). The seed galls in grasses are very conspicuous and much bigger than the seeds.

Undetermined nematode seed galls have been encountered in western wheatgrass, *Agropyron smithii* (Collins, 1966), *Arrhenatherum elatius* (Mühle, 1953), *Bromus inermis* (Mühle, 1953), and golden oat grass, *Trisetum flavescens* (Goodey, 1959).

Lesions and discolorations on seeds or fruits, caused by nematodes, characterise another possible category. *Pratylenchus brachyurus* has been associated with injury to peanuts in several southern States of the U.S.A. Minton, Hammons and Parham (1970) reported lesions and discolorations in peanut of six cultivars grown in soil infested with the nematode.

2.7 EFFECTS OF AGEING IN SEEDS

In Subsection 20.2.1 longevity of seeds is discussed. Old seeds gradually decline in germination capacity and develop plants of reduced vigour. Categories of abnormal seedlings encountered in testing aged seeds for germination, according to Wellington (1968), are listed in Table 2.1.

2.8 SEED DISORDERS DUE TO UNKNOWN FACTORS

Disorders similar to hollow heart (Subsection 2.2.3) have been recorded in the U.S.A. in groundnut (Harris and Gilman, 1957) and in Venezuela in cacao (Cifferi, 1951).

In cacao the cavity is recognisable only in sections of the fresh bean. It is found only in Cirollo subspecies or in high-quality hybrids. All pods from a single diseased tree contain abnormal seeds. Dry seasons seem to be favourable to development of the disease. Affected cacao beans fail to germinate.

2.9 MECHANICAL INJURIES IN SEEDS

The *primary sources* of mechanical damage are the operations of (1) harvesting, (2) threshing, (3) processing, including cleaning and transportation within the warehouse, and (4) handling during planting (drilling).

Three levels of mechanical damage are recognised (Pollock and Roos, 1972): (1) slight injuries in the form of microscopic breaks, particularly in the seed coat, which make the seed susceptible to attack by microorganisms; (2) gross damage to seed and seed coat, easily visible; (3) internal damage, often injury to the embryo, being visible usually only after the seed has germinated.

Factors in the seed affecting susceptibility to mechanical damage include: (1) genetic constitution, (2) growth conditions of the seed crops, (3) moisture content of the seed.

Consequences of mechanical damage are primarily: (1) loss or damage of tissues and hence of viability, (2) loss of capacity in regulating the water content of the seed, (3) increased susceptibility to invasion by microorganisms, and (4) increased susceptibility to phytotoxic effects.

Classification of abnormal seedlings. In germination tests seriously damaged seedlings are classed as 'abnormal'. Wellington (1968) described categories of abnormal seedlings with reference to mechanical damage of the seed, listed in Table 2.1.

Table 2.1. Types of abnormal seedlings according to Wellington (1968)
Causal factors: desiccation = D, ageing = A, mechanical = M

Host	*Root (r.)*	*Coleoptile*	*Hypocotyl (h.)*	*Cotyledons (c.)*	*Plumule epicotyl, primary leaves*
Bean			Open splits, lesions and breaks in h. (M)		No primary leaves (M)
Beet	No primary r. (AM) Stunted primary r. (AM) Stunted r. (M)		Short h. (M)	Decayed c. (A)	Completely decayed or blackened seedlings (M)
Cabbage	No primary r. (A) Radicle with no root hair (A) Weak and spindly primary r. (A)		Short h. (A) Split h. (A) Decayed h. (A)	Decayed c. (A)	Watery seedlings (A)
Carrot	Stunted primary r. (A) Watery r. (A)		Short h. (A) Brown h. (A) Watery h. (A)	Decayed c. (A)	
Flax	No primary r. (A) Stunted primary r. with and without secondary r.s. (A) Stunted r. (A)		Short h. (A)	Decayed c. (A)	
Italian ryegrass	No seminal r.s (A) Stunted r. (A) One weak spindly seminal r. (A)	Coleoptile or leaf split (A)			Decayed plumule (A) No green leaf within coleoptile (A) Short leaf less than half way up coleoptile (A)
Lettuce	No primary r. (AM) Stunted primary r. (AM) Stunted r. (A)		Short h. (A)	No c. (AM) Discoloured c. (AM)	Decayed seedlings (AM)
Lupine	No primary r. (M) No well-developed secondary r.s (M)		Open splits, lesions and breaks in h. (M)		Open splits, lesions and breaks in epicotyl (M) Decayed epicotyl (M)
Onion	No primary r. (M) Stunted primary r. (AM)			Decayed c. (M) Short c. with poorly developed crook (A)	Completely decayed seedlings (M)

Table 2.1 *continued*

Host	*Root (r.)*	*Coleoptile*	*Hypocotyl (h.)*	*Cotyledons (c.)*	*Plumule epicotyl, primary leaves*
Radish	No primary r. (M) Stunted primary r. (M) Weak and spindly primary r. (M)		Short h. (M) Split h. (M) Decay at point of attachment between h. (M)	Spots on c. (M) Decay at point of attachment between c. (M) Decayed c. (M)	
Red clover	No primary r. (AM) Stunted primary r. (AM) Weak and spindly primary r. (M) Primary r. with longitudinal split (M)		Short h. (M) Decay between c. and h. (A)	Decayed c. (AM) No c. (A) Enlarged c. (M) One c. and damage to shoot apex (AM)	Watery seedlings (AM) Completely shaftered seedlings (M)
Wheat	No seminal r.s (DM) Stunted seminal r.s (D) One spindly seminal r. (DM)	Coleoptile or leaves split (DM)			Decayed plumule (DM) Leaves less than half way up the coleoptile (D)

2.9.1 Injuries in seed due to threshing

Leguminous seeds are particularly sensitive to thresher injury which in beans may produce up to 30 per cent abnormal seedlings (Harter, 1930). 'Baldhead' or 'snakehead' is a type of abnormality caused by such injury, the epicotyl being fractured just below the plumule. Similar damage occurs in pea and clover. Cultivars with hard thick coats are more resistant to thresher injury (Zaumeyer and Thomas, 1957). Baldhead rarely occurs in seed threshed by hand (Harter, 1930; Borthwick, 1932). Bainer and Borthwick (1934) showed for lima beans that by increasing the speed of the revolving cylinder threshing was improved but mechanical damage increased, and they showed that the higher the moisture content is, the less is the damage. Troelsen (1947), however, found in Denmark, where harvest conditions often are humid, that seeds of pea, bean, lupin and similar kinds of seed are easily damaged by hard threshing when the moisture content of the seed is high. Injured seeds have poor keeping quality. The more injured are Phaseolus bean seeds, the more rapidly they lose viability in storage (Toole and Toole, 1960). During the threshing operation the machinery should be adjusted according to the moisture content of the seed.

In dry regions in the U.S.A. Phaseolus bean may become too dry for combining without damage to the seed coat. To prevent this, bean windrows may be sprayed with water to prolong or induce seed coat toughening, and after combining, any desired amount of moisture may be added by mixing damped sawdust with the beans, followed by screening after some intervals. This precaution protects bean during the processes of unloading, conveying, screening and binning (Dexter, 1966).

Flax seeds are also sensitive to thresher injury. The seed coat is brittle and may easily rupture when the seeds are knocked against a hard surface. The seed coat may be cracked or fractured, injuries are visible as scratches or splits, or the seeds may be more or less broken (Kommedahl *et al.*, 1955). Injury is caused by high rotation speed of the drum treater. Forsyth and Vogel (1945) found that the amount of injury increased with cylinder speed, and Van Es (1959) that the speed should not be above 500 rotations per minute. He established other factors that contributed to damage: wrong shape of the cylinder, the fact that seeds are thrown with high velocity against metal parts of the machine, and fall during the pneumatic transport of the seeds within the machine. Flax seeds are particularly easily injured after dry weather during the maturing and harvesting period; the fractures are generally minute and invisible to the unaided eye and, consequently, their presence does not affect the market grade of the seed (Machacek and Brown, 1945). They also found that large seeds were more injured than smaller, and that injury was more serious in large-seeded flax cultivars. Frederiksen (1948) emphasised that evaluation of flax seed should be based on normal seedlings — abnormal and weakened seedlings are unable to penetrate a soil layer 2 cm thick and are therefore worthless. In the purity test all severely injured and broken seeds should be referred to inert matter (Härtel, 1936).

Burema (1959) investigated the influence of the peripheral speed of the threshing drum on seed damage and germination of various crop seeds. He established that the adequate peripheral speed was 18–22 m/sec for cabbage seed and 20–22 m/sec for poppy seed (Fig. 2.8); higher speed results in damage. The speed of the drum had hardly any influence on the seed quality of meadow grass, *Poa trivialis*.

In most seeds thresher injury increases with decrease of the moisture content but in maize injury to the pericarp increases with seed moisture content from just above 12 per cent. Generally, to reduce risk of damage threshing is often carried out in the early morning while the seeds are still moderately moist, as is especially so for California flower seed crops.

In cereals broken and bruised kernels may considerably reduce the market value of seed grain. This was already pointed out many years ago and modern threshing machinery is still the cause of seed damage, especially when improperly used. Damage occurs particularly when the grain is threshed after a dry period of maturation and harvesting (Zundel, 1921).

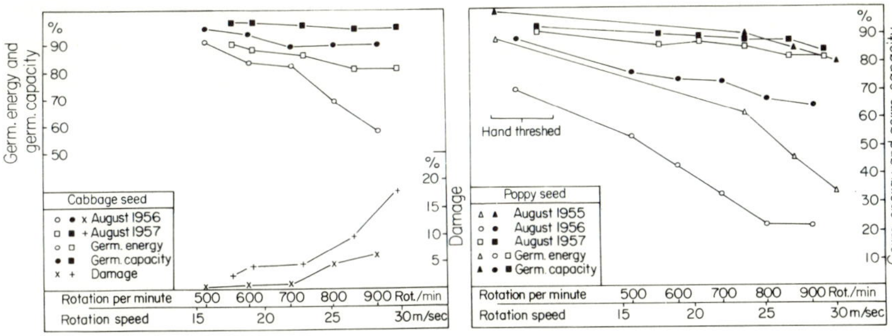

Fig. 2.8. Influence of the thresher cylinder speed on the germination energy and germination capacity of cabbage and poppy. (Burema, 1959)

Walldén (1916), Machacek and Greaney (1933) and Koehler (1957) demonstrated that mechanically injured grains are much more susceptible to moulds, saprophytes as well as pathogens, than are uninjured seed, both during storage and under field conditions. Already Nobbe (1872) and later Hurd (1921) have shown that machine-threshing injury in the seed leads to chemical injury after treatment with copper sulphate, and this still holds true for modern chemical treatments (cf. Section 22.7). The localisation of the injury in the seed is important. Hurd showed that if the injury of the pericarp in wheat is above the endosperm there is 100 per cent fatal infection when *Penicillium* and *Rhizopus* spores are present, while if the rupture is above the embryo the kernel is practically unaffected. Mechanical damage, moreover, shortens the storage life of seeds as Alberts (1927) has shown for maize. Seeds with injured pericarps are subject to excessive exchange of water vapour and they are more susceptible to storage fungi.

2.9.2 Injuries in seed during cleaning and other seed processing

Preparing seed for marketing comprises a series of separate processes: separating by air blast, specific gravity, sorting according to size and colour, polishing, and other processes (Harmond, Klein and Brandenburg, 1961). Seeds may be handled more than a dozen times in plants fully equipped for processing and treating. This involves transportation, often from the top of a multistoreyed building, the seeds flowing by gravity through several machines to the lower floors. When processing is on one level the seeds may be elevated through a tank, then passed through a machine by gravity and then re-elevated for the next process. As seeds may be mechanically damaged by minor falls and increasingly so by repeated falls (Asgrow, 1949) the effects accumulate during such transportation. Much has been done to reduce handling injuries by improving cleaning and conveying equipment in seed-processing plants. A rather recent development is dense-phase or fluidised conveying of seeds (Brandenburg and Harmond, 1964). In this pneumatic operation seed movement consists of aerated seed sludge that form, move, dissipate and reform in a cyclical manner throughout the pipe run. In this way easily damaged seed can be transported at low velocities, without damage, through flexible small-pipe systems.

2.9.3 Injuries in seed inflicted during planting (drilling)

While damage to seeds inflicted during harvesting and processing is assessed in seed testing, assessment of damage which may occur during sowing cannot of course be anticipated. Hulbert and Whitney (1934) found up to 10 per cent injury in large and smooth seeded pea during drilling and they recommend mixing seeds with graphite. Bean seeds with low moisture content may be severely damaged during planting. Dexter (1966) obtained, by increasing the seed moisture content from 11 to 16 per cent, increased emergence from 39 to 78 per cent.

2.10 CONGENITAL DISORDERS

In seeds of maize, particularly those of certain inbred lines which have a weak pericarp, two types of disorder may occur. 'Popped kernels' appear as irregular breaks over the crown of the kernel, a condition resembling partially expanded popcorn kernels (Fig. 2.9). 'Silk cut' is characterised by a horizontal cut or split in the pericarp over the sides of the kernel (Fig. 2.10). Both break the seed at the weakest point, exposing the starch of the endosperm which then may be subject to infection by fungi, e.g. *Fusarium moniliforme* (Ullstrup, 1961).

In flax the seed coats split because of a genetic disorder (Fig. 2.11); the two halves separate at the small end of the seed, exposing the cotyledons. The percentage of affected seed may be considerable — seed lots with 10–60 per cent 'split seeds' have been common in Minnesota, U.S.A. The anomaly is worst in yellow-seeded cultivars (Kommedahl *et al.*, 1955). The same type of disorder is encountered in some cultivars of bean (Zaumeyer and Thomas, 1957).

2.11 INSECT INJURIES

2.11.1 Embryoless seeds

Various umbelliferous crops, including carrot, celery, coriander, dill, fennel, parsley and parsnip, often produce normal-looking but embryoless seeds, i.e. seeds in which the embryo is replaced by a cavity, while the

Fig. 2.9. 'Popped kernel' in maize. (Courtesy of A. J. Ullstrup)

endosperm and seed coat is undamaged (Flemion and Waterbury, 1941; Flemion and Uhlmann, 1946; Flemion and Olson, 1950). Handford (1949) and Flemion and Olson (1950) have shown that this injury is produced by various species of *Lygus*, including *L. campestris*, *L. elisus*, *L. hesperus*, *L. oblineatus* and *L. sallei*. In carrot similar damage has been observed in Denmark, due to *L. calmei*, *L. campestris*, *L. pratensis*, *L. pubescens* and *Echoris norvegicus* (Wagn, 1954), and in Holland due to *L. campestris* (Kho and Braak, 1956).

Lygus bugs feed preferentially on meristematic tissues. When the insects, nymphs or adults, are allowed to feed on the developing seeds they destroy the embryo at almost any stage, without affecting other parts of the seed. When they feed on flowers or ovaries, the entire fruit is destroyed and the seed yield thereby reduced. Flemion, Weed and Miller (1951) and Flemion, Ledbetter and Kelley (1954) demonstrated that secretions are injected by the insects into the tissues but they did not determine whether these were phytotoxic. Sorenson (1939) showed that *L. hesperus*

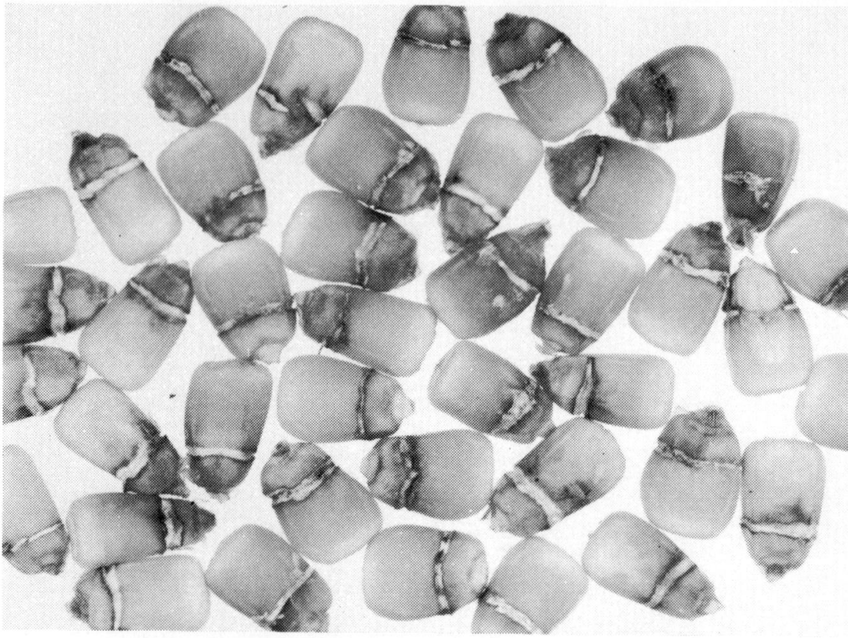

Fig. 2.10. 'Silk cut' in maize. (Courtesy of A. J. Ullstrup)

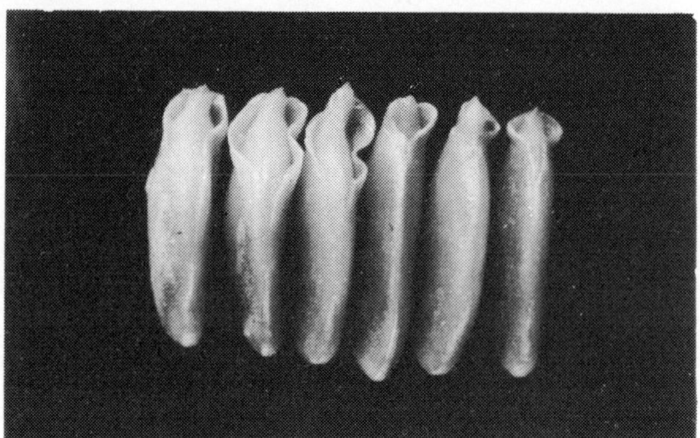

Fig. 2.11. Split seed coats ('split seed') in flax. (Kommedahl, 1975)

and *L. elisus* may attack immature seeds within the pods of lucerne causing them to become misshapen or shrivelled. Also in beet different species of *Lygus* may cause severe damage of the seed, involving loss of viability (Hills and Taylor, 1950; Carlson, 1961).

2.11.2 Seed pitting (stigmatosis)

In the U.S.A. and other regions *Lygus hesperus* and *L. elisus* have been shown to cause destructive pitting of several kinds of bean, including lima bean, also in cowpea (Davis, 1897; Hawley, 1922; Shull, 1933; Baker, Snyder and Holland, 1946). Young pods may be shed as a result of their feeding; insect punctures through older pods produce lesions varying from tiny sunken pinpoints to large, irregular crater-like yellow or brown lesions in the cotyledons over which the seed coat is destroyed. A brown granular mass of dead cells fills the cavity.

'Stigmatomycosis', sunken necrotic lesions produced on seeds of cotton and legumes, has been ascribed to certain yeast fungi of the genus *Nematospora*, claimed to be inoculated by different plant bugs, in particular southern green stink bug, *Nezara viridula* (Wingard, 1925). The lesions may, however, be produced independently by the insects (cf. Section 16.2). *Nezara viridula*, southern green stink bug, produces 'black pit' in pecan, *Carya illinoensis*, causing the nuts to drop prematurely if injured before the shells have become hard, and later 'kernel spot', in which the spots are decidedly bitter (Demaree, 1922; Weber, 1933; Phillips and Cole, 1945; Phillips, Large and Cole, 1960).

2.11.3 Damage caused by insects feeding in seeds

The larvae of a considerable range of insects destroy seeds by feeding in immature or mature seeds. Many are important storage insects. Well-known examples of world-wide importance are the corn earworm, *Heliothis zeae*, in maize; the pink bollworm, *Pectinophora gossypiella*, in cotton; various species of *Dioryctrya*, coneworms, in pines and firs; boll weevil, *Anthomonus grandis*, in cotton; alfalfa seed chalcid, *Brucophagus roddi*, in lucerne; pea weevil, *Bruchus pisorum*, in peas; bean weevil, *Acanthoscelides obtectus*, in stored seeds of bean, cowpea, vetch, clover, lentil and lupine; bean weevil, *Bruchus granarius*, in beans and lentil; lentis bruchid, *Bruchus lentis*; the embryo-eating beetle, *Cryptolestes ferrugineus*. In peanut the principal pest is *Carydon serratus*. *Sitophilus* spp., including grain borer, *S. granarius*, and rice weevil, *S. oryzae*, attack many kinds of seed in warehouses; the lesser grain borer, *Rhizopertha dominica*, is an important pest of paddy rice; the Khapra beetle, *Trogoderma granarium*, probably native to India, is excluded by quarantine from most of the world — it has spread in South Asia and most of Africa, but has been kept out of South America, North America and Australia.

Secondary grain pests can be serious in stored grain. Most granivorous insects are tropical or subtropical species and breed over a fairly wide range of temperatures, with optima from 30 to 38°C. They can cause heating in stored grain bulks. Important are saw-toothed grain beetle, *Oryzaephilus surinamensis*; grain weevil, *Sitophilus (Calandra) granarius*; rust-red flour beetle, *Tribolium castaneum*; confused flour beetle, *Tribolium confusum*; rust-red grain beetle or embryo-eating beetle,

Cryptolestes ferrugineus; flat grain beetle, *Cryptolestes pusillus*; and Khabra beetle, gram borer, rice weevil and lesser grain borer, *Rhizopertha dominica*.

Various moths are serious pests; *Sitotroga cerealella*, which attacks barley, wheat, maize and sorghum, is prevalent in warmer parts of the world, and different species of *Ephestia*, which eat the embryo of stored wheat, are common in most tropical produce, including peanut and maize.

Reviews on seed-damaging insects have recently been given by Howe (1972) and Bohart and Koerber (1972).

3 Seed-borne Viruses

Although most viruses are not seed-borne the number of viruses for which seed transmission has been recognised is steadily growing, and in some genera within the phanerogams seed transmission is rather frequent. This holds true for certain genera of leguminous plants (Leguminosae), stone fruits (Amygdalaceae) and composites (Compositae).

3.1 FACTORS PREVENTING PASSAGE OF VIRUS THROUGH SEEDS

Much discussion has been devoted to the question of why seed transmission is relatively infrequent in spite of the systemic nature of most virus infections. The following explanations have been suggested.

(1) The virus may not invade the ovule or embryo. The ovule is isolated by lack of plasmodesmal connections with the surounding tissues, as are usually also the guard cells of stomata.

(2) The virus may not be capable of establishing a compatible relationship with the gametes or the embryo sac of the host. Viruses are commonly found in flower parts and immature seeds but become inactivated in the mature seed.

(3) The virus is lethal to the gametes thus causing sterility and preventing production of infected seed.

(4) The virus is not capable of infecting male and female gametes or the young embryo, either due to lack of virulence to these stages of the host or due to the resistance of these.

Bennett (1969), who reviewed literature on factors related to seed transmission of viruses, discussed in detail the restricting factors. He found evidence for an effective but not necessarily very complicated mechanism for exclusion, that prevents embryos from being infected throughout their development. He points to the fact that when an embryo is initiated by union of nuclei in an infected embryo sac, the embryo begins development in a virus-free medium. There are no protoplasmic connections between cells of the embryo and cells of adjacent tissue. The embryo can absorb food materials from virus-infected areas without becoming infected. Only a few viruses are able to pass through cell walls. Kassanis, Tinsley and Quak (1958) found that tobacco mosaic virus could move through cultivated host tissues without plasmodesmata at a rate of 1 mm per week (approximately the same rate as it moves through cells of leaf parenchyma) but that usually injuries were required for invasion by the virus. Bennett (1969) concludes that 'much evidence suggests that the cell wall in the absence of penetrating cytoplasmic strands usually is a formidable barrier to virus passage'.

However, in some hosts, sometimes restricted to certain cultivars, some viruses may invade the young embryo. Barley stripe mosaic virus may be seed-transmitted when the host is inoculated as late as in the dough stage of seed development (Eslick and Afanasiev, 1955; cf. Crowley, 1959); and some viruses infecting ovules through pollen may escape and invade the mother plant, e.g. prune dwarf virus in cherry (Gilmer and Way, 1963), Prunus necrotic ringspot virus in cherry (George and Davidson, 1963), and apple chlorotic leaf spot virus in raspberry cv. Lloyd George (Cadman, 1965). Nevertheless, all present evidence indicates that the usual path of movement is through the plasmodesmata connecting protoplasts of neighbouring cells (Bennett, 1969). Esau, Cronshaw and Hoefert (1967) established the presence of beet yellows virus particles in sieve elements, pores of sieve plates, plasmodesmata connecting sieve with parenchyma cells, and in plasmodesmata connecting parenchyma cells. The virus can move from cell to cell in the form of complete particles.

Inactivation of virus in seed has been found by Kausche (1940) for tobacco mosaic virus in seeds of tobacco, by Stelzner (1942) for potato virus X in seeds of *Datura* and potato, and by Zaumeyer and Harter (1943*b*) for bean southern mosaic virus in bean. The virus may be recovered from unripe or freshly ripened seeds but not from ripened seeds or seeds that have been stored for only a short time as in the case of the last-mentioned virus in bean.

Tomato aspermy virus in tomato provides an example of a direct lethal effect on both pollen and ovules. The virus interferes with the normal meiotic processes of both megaspore and microspore, and seed formation is prevented (Caldwell, 1952). This is apparently a hypersensitive reaction. Hypersensitivity is also assumed in the lettuce cultivar Cheshunt Early Giant in which floral shoots are killed by lettuce mosaic virus (Couch, 1955), and in the bean cultivar Top Crop by bean common mosaic virus.

Several viruses cause pollen sterility, i.e. tobacco ringspot virus in tobacco (Valleau, 1932*b*), lettuce mosaic virus in lettuce (Ryder, 1964), both exerting no comparable effect on the ovule. Tobacco streak virus produces almost complete pollen sterility in *Datura stramonium* (Blakeslee, 1921), and barley stripe mosaic virus causes high percentages of pollen sterility in barley (Inouye, 1962).

Resistance of gametes to infection has been studied for some viruses. Evidence of such resistance was found by Medina and Grogan (1961) in the bean cultivars Red Mexican and Idaho Refugee, both of which are resistant to bean common mosaic virus. When pollinated from infected susceptible cultivars, no seed transmission was established, while infected pollen from susceptible cultivars transferred to susceptible plants resulted in seed transmission. The virus was introduced into the ovules, and failed in resistant embryos but succeeded in susceptible ones.

The capacity of seed transmission is specific to both virus and host. Cucumber mosaic virus is commonly transmitted through the fleshy seeds of wild cucumber, *Echinocystis lobata*, but not through seeds of squash and only rarely through seeds of cucumber (Doolittle and Walker, 1925). Tomato aspermy virus is seed-transmitted in *Stellaria media* but not in tomato. Broad bean true mosaic virus and broad bean stain virus are readily seed-transmitted in broad bean (Brunt, 1970; Gibbs and Paul, 1970), whereas broad bean mottle virus and broad bean vascular wilt virus are not. Bean common mosaic virus is commonly seed-transmitted in bean but bean yellow mosaic virus and bean southern mosaic virus are not. The same contrast exists between lettuce mosaic virus, which is commonly seed-transmitted in lettuce, and lettuce big vein virus and lettuce necrotic yellows virus neither of which is seed-transmitted. Coffee ringspot virus is transmitted through seeds of *Coffea excelsa* but not through seeds of *C. arabica* (Reyes, 1961).

There is considerable variation in the rate of seed transmission between strains of individual viruses in a certain host as well as between different hosts, including different cultivars of the same species, in relation to a certain virus.

Among strains of barley stripe mosaic virus, McKinney and Greeley (1965) found great variation in amount of seed transmission in barley and wheat, ranging from more than 50 per cent down to zero, and in oats from 10 per cent to zero. One strain of tobacco ringspot virus was transmitted in 3 per cent of the seeds of cv. Paris Island Cos lettuce, while another strain was not seed-transmitted in this cultivar (Grogan and Schnathorst, 1955). Different strains of bean common mosaic virus are transmitted at different rates through seeds of each of different bean cultivars but in any one cultivar each strain of the virus is transmitted at a different rate (Hubbeling, 1972).

Quantitative differences in seed transmission of a definite virus is quite common between hosts. Seed transmission of squash mosaic virus was found to range in different types of cucurbits from 0 to 20 per cent (Grogan, Hall and Kimble, 1959). Raspberry ringspot virus is seed-transmitted by 2–3 per cent in *Capsella bursa-pastoris* but by more than 50 per cent in *Beta vulgaris*, and tomato black ring virus by about 5 per cent in *Rubus idaeus* but by up to 100 per cent in *Cerastium vulgatum*, *Fumaria officinalis*, *Myosotis arvensis* and *Polygonum persicaria* (Lister and Murant, 1967). Two strains of cowpea mosaic virus and one strain of cucumber mosaic virus were all distinctly more readily seed-transmitted in cowpea cv. White Acre pea than in three other cultivars, and in cowpea cv. Yardlong Bean only cucumber mosaic virus was transmitted (Anderson, 1957). Soybean mosaic virus in soybean has been found transmitted in 1 per cent of seeds of cv. Hill but in 11 per cent of cv. Lee (Ross, 1961), and in 20 per cent of cv. Harosoy but not at all in seeds of cv. Merit (Kennedy and Cooper, 1967). Also, between cultivars of lettuce there are considerable differences in amount of seed transmission of lettuce mosaic virus, ranging from 16 per cent (Kietreiber, 1973AR) through 11 per cent (Ogilvie, Mulligan and Brian, 1935; Herold, 1956) and 3 per cent (Newhall, 1923) down to zero in the cultivar Cheshunt Early Giant (Kassanis, 1947; Couch, 1955).

3.2 OTHER MEANS OF TRANSMISSION

Viruses differ widely in the means by which they are transmitted. The method of transmission has bearing not only on the epidemiology of virus diseases but is intimately related to the specific nature of the different groups of viruses, and consequently will have to be taken into account in any system of classification of viruses attempting an approach towards a natural system.

Disregarding seed transmission discussed elsewhere, methods of transmission may be referred to the following categories: (1) simple sap transmission, (2) insect transmission, (3) mite transmission, (4) nematode transmission, (5) transmission through fungi, (6) transmission through bacteria, (7) transmission through parasitic phanerogams, (8) transmission through grafting—vectors or mode of transmission unknown. Carter (1966) and Fritzsche *et al.* (1972) published comprehensive lists of virus vectors, arranged according to virus as well as according to vector.

(1) *Simple sap transmission.* This mode of transmission is purely mechanical, transfer being accomplished by contact, rubbing or under special circumstances injecting. Mechanical transmission almost always involves wounding of the inoculated plant, such as breaking of hairs or epidermal cells, which is sufficient to bring about infection by tobacco mosaic virus. This highly infectious virus may be spread by hands, workers, clothes, implements, and the virus may remain infectious after two years on farm equipment (cf. Subsection 19.6.4). Potato virus X is transmitted from plant to plant by contact (Loughnane and Murphy, 1938). Potato spindle tuber virus may be spread by farm implements, such as

contaminated tractor wheels, or by knife-cutting seed potatoes (Bonde and Merriam, 1951). Cucumber mosaic virus may be spread by pickers.

Sap-transmitted viruses usually infect shoots, they occur in relatively high concentrations in pressed plant sap and are usually rather stable *in vitro*. There is a considerable range of types of particle within this group.

(2) *Insect transmission.* Most viruses are transmitted by insects. The great majority of virus vectors belongs to two families, *Aphididae* (aphids) and *Cicadellidae* (leaf hoppers). For seed-borne viruses aphids are by far the most important. Distinction is made between two principal categories of insect-transmitted viruses: *stylet-borne*, usually termed non-persistent and semi-persistent viruses, and *circulative*, usually termed persistent viruses (Watson and Roberts, 1939, 1940). The stylet-borne viruses are retained for only a few seconds, minutes or perhaps hours, in semi-persistent up to weeks, still being infective, while circulative viruses may be retained for hours, days or months, still being infective. Viruses of the first category are aphid-borne, the last category comprises a few aphid-borne viruses and many viruses transmitted by leaf hoppers, mites, white flies and thrips. Some circulative viruses may multiply in the vector; they are referred to as *propagative* viruses.

Circulative (persistent) viruses may remain infective for a very long period of time, often throughout the lifetime of the vector; moreover, usually the vector may transmit the virus only after elapse of a respite, termed the cessation or circulation time, required to re-establish the infectivity of the virus. This period may range from less than an hour to more than two weeks. A circulative vector acts as an alternate host. In a vector one virus may be persistent while another virus may be non-persistent.

Most vectors of mosaic viruses are aphids. Some chewing insects are vectors, although acting as agents of mechanical transmission; squash mosaic virus and a cowpea mosaic virus are transmitted by beetles.

Some viruses are transmitted by a considerable number of aphid species, e.g. cucumber mosaic virus by about 75 species of aphids (Fritzsche *et al.*, 1972), and conversely, some species of aphids may act as vector for many viruses, e.g. *Myzus persicae* for more than 100 different viruses, *Aphis fabae* for about 40 different viruses, *Macrosiphum euphorbiae* for some 50 viruses (Carter, 1966; Fritzsche *et al.*, 1972).

(3) *Mite transmission.* A few viruses are transmitted by mites including wheat streak mosaic virus (Slykhuis, 1953, 1955) and wheat spot mosaic virus (Slykhuis, 1956). Recently, wheat streak mosaic virus has been shown to be seed-transmitted in maize (Hill, Martinson and Russell, 1974).

(4) *Nematode transmission.* Some very widespread viruses, which have considerable host ranges, are transmitted by different soil-inhabiting

nematodes, including the genera *Xiphinema*, *Longidorus* and *Trichodorus*. The relationships between nematodes and seed-borne viruses are dealt with in Section 16.8.

(5) *Transmission through fungi*. A few viruses but none seed-borne are transmitted by soil-inhabiting fungi, lettuce big vein virus by *Olpidium brassicae*, carried within the zoospores into fresh roots (Fry, 1958; Campbell and Grogan, 1963; Tomlinson and Garrett, 1964; Teakle, 1969). Tobacco stunt virus is transmitted by the same fungus.

(6) *Transmission through bacteria*. In the synergistic effect of *Xanthomonas phaseoli* on bean common mosaic virus in bean, the bacterium acts as a true vector, see Section 16.13. It seems to be the only known case of bacterial virus transmission.

(7) *Transmission through parasitic phanerogams*. Various species of *Cuscuta*, dodder, which parasitise different higher plants, are susceptible to viruses of these plants, and being in intimate connection through their haustoria with their hosts they may transmit the virus to them (Bennett, 1940). However, dodder transmission in nature has so far not been reported. Several seed-borne viruses are transmissible by dodder, *Cuscuta campestris*, namely those of clover yellow mosaic (Smith, 1972), tobacco streak (Fulton, 1948), white clover mosaic (Bos, Delević and Van Der Want, 1959), tomato spotted wilt (Bennett, 1944) and cucumber mosaic (Bennett, 1940); alfalfa mosaic virus is transmitted by *C. subinclusa* (Schmelzer, 1957*a*).

(8) *Transmission through grafting — vectors or mode of transmission unknown*. All systemically infecting viruses may be transmitted by means of grafting. However, in experimental practice this procedure is used only for those viruses for which a vector is either unknown or unavailable, and which are not sap-transmissible.

Seed-borne viruses, for which vectors are unknown and for which transmission has been effected by grafting or budding, are avocado pear sunblotch virus (Wallace and Drake, 1953), Citrus psorosis virus (Wallace, 1957) and Citrus xyloporosis virus (Childs, 1956). Another seed-borne virus for which vectors are unknown and sap transmission not feasible is groundnut marginal chlorosis virus (Van Velsen, 1961).

3.3 CLASSIFICATION AND NOMENCLATURE OF PLANT VIRUSES

It is the general view of virologists that the time has not yet come for establishing a 'natural' classification of viruses and for building up a nomenclature system of plant viruses. A latinised binomial nomenclature, it is said, reminds one of a phylogenetic system as used for classification of higher organisms. At present most virologists prefer vernacular names for virus groups and for single viruses.

Different systems of classification and nomenclature have been proposed, the most well known of which have been reviewed by Carter (1966). They include the following which here are briefly characterised by indication of the name for tobacco mosaic virus, according to the system concerned:

J. Johnson's system: Tobacco virus 1 Johnson
K. M. Smith's system: *Nicotiana* virus 1 Smith
H. S. Fawcett's system: *Nicotianaevir commune*
F. O. Holmes's system: *Marmor tabaci* Holmes
H. H. McKinney's system: *Marmor tabaci* (Holmes ex Valleau) McKinney
H. H. Thornberry's system: *Phytovirus nicomosaicum* Thornberry
W. D. Valleau's system: *Musivum tabaci* Valleau

None of these systems is classification involving concepts of phylogenetic relationship; all of them are systems of simple indexing and cataloguing viruses.

The main criteria used for identification of viruses are the following, mentioned not necessarily in sequence of importance: (1) natural method of transmission, including vector if known, and preferential affinity to host tissue region of inoculation-infection; (2) host range, symptomatology and pathogenesis; (3) immunological properties; (4) physical and chemical properties; (5) type of virus particle; (6) susceptibility to physical and chemical agents (applicable for those viruses that are stable enough *in vitro* and easily mechanically transmitted).

Hansen (1956, 1957*a*, 1957*b*, 1966, 1968, 1969, 1970, 1972) has made an attempt at a rational classification and nomenclature of plant viruses, based on the main criteria used for identification, i.e. the principal physical and biological properties of viruses. His system of code-name designation is summarised in Table 3.1. Generic names of viruses are composed by a set of synthesised prefixes and suffixes, latinised and constructed in such a way that they indicate some basic characters of the virus concerned: normal site of infection, type of vector, if known, shape and size of particles, if any. Quoting Carter (1966): 'the classification itself is independent of any theory of the origin of viruses and is, to date, the best integrated system to be proposed and the one most consistent with the biological approach to the virus problem'.

In the following survey of seed-borne viruses, these will be grouped according to Hansen's classification, and generic names of his system will

Table 3.1 Summary of the system of code-name designation of plant viruses, according to Hansen (1970)

The initial indicates the normal regional site of infection

Transmission, whether direct or by vectors:

M(e)	(mechanical): to leaf surface region.
D(a)	(deep): to central region.
G(e)	(soil): to root surface region.

*The middle symbol indicates specific vector, if any**

aca	= mites	(Acarina)
ale	= white flies	(Aleurodidae)
aphi	= aphids	(Aphididae)
cica	= leafhoppers	(Cicadina)
coc	= scale-insects	(Coccidae)
het	= plant bugs	(Heteroptera)
in	= no specific vector known	(incognitum)
nema	= nematodes	(Nematoda)
plano	= fungal zoospores	(planospores)
psylli	= psyllids	(Psyllidae)
thy	= thrips	(Thysanoptera)
x(e)	= multiple vector types	(xenos = strange)

* Biting insects (e.g. beetles), mining flies, etc. are not considered specific vectors, notwithstanding their (mechanical) spread of certain viruses.

The ending symbol indicates shape and size of particles, if any

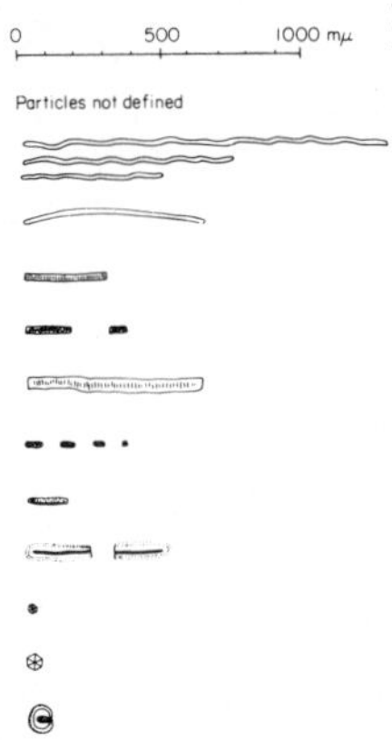

Ending	Some examples	
virus	Daphiflexus	betae
flexus	Maphiflexus	solani
	Minflexus	solani
lancea	Maphilancea	dianthi
chorda	Minchorda	nicotianae
pachus	Genemapachus	nicotianae
codex	Daphicodex	piceae
brevia	Maphibrevia	medicaginis
media	Dacocmedia	theobromae
physa	Maphiphysa	lactucae
globus	Minglobus	lycopersici
sphaera	Maphisphaera	brassicae
	Dacicasphaera	trifolii
sagum	Methysagum	lycopersici
myca: viruslike agents (mycoplasma)	Dacicamyca	callistephi

be used as code-names for subgroups. For the viruses dealt with below and elsewhere in this book the generally accepted standard names published by the Commonwealth Mycological Institute (CMI), have been used whenever available. These names are those adopted in *Plant Virus Names, an Annotated List of Names and Synonyms of Plant Viruses and Diseases*, edited by Martyn (1968, 1971). However, later alterations in names, adopted by the CMI (cf. CMI/AAB *Descriptions of Plant Viruses*)

Table 3.2. Seed-borne viruses grouped according to Hansen's (1970) classification, cf. Table 3.1. Number of seed-transmitting hosts, more than one, is indicated in brackets. Scale of particles outlined 0 500 1000

PREFIX Transmission (vector) / type of particles SUFFIX	MECHANICAL (*to leaf surface region*) Min	Maphi	Male	Methi	Mexe	DEEP (*to central region*) Din	Daphi	GEO(SOIL) (*to root surface region*) Genema
	vectors unknown (*incognitum*)	vectors *Aphididae*	vectors *Aleurodidae*	vectors *Thysanoptera*	vectors different (*xenos*)	vectors unknown (*incognitum*)	vectors *Aphididae*	vectors *Nematoda*
3.4.1 globus	**3.4.1.1** Apple mosaic Bean southern mosaic Black raspberry latent (2) Broad bean mottle (2) Broad bean stain Chicory yellow mottle Cowpea mosaic beetle transmit. (2) Cowpea ringspot Dulcamara mottle Echtes Ackerbohnenmosaik Eggplant mosaic (2) Elm mottle Prune dwarf (2) Prunus necrotic ringspot (8) Sowbane mosaic (5) Squash mosaic (7) Tobacco streak (4) Tomato bushy stunt	**3.4.1.2** Barley mottle mosaic Cowpea isometric Cucumber mosaic (17) Soybean mild mosaic Tomato aspermy					**3.4.1.3** Groundnut stunt	**3.4.1.4** Arabis mosaic (17) Cherry leaf roll (6) Grapevine fanleaf (3) Raspberry ringspot (8) Strawberry latent ringspot (8) Tobacco ringspot (12) Tomato black ring (22) Tomato ringspot (>12)
3.4.2 flexus	**3.4.2.1** Apple chlorotic leafspot Broad bean mosaic Guar symptomless virus Hippeastrum mosaic Potato virus X Raspberry bushy dwarf White clover mosaic White clover yellow mosaic	**3.4.2.2** Bean common mosaic (3) Bean yellow mosaic (5) Cowpea aphid-borne mosaic (3) Groundnut mottle Lettuce mosaic (4) Mungbean mosaic Onion yellow dwarf Pea leaf rolling mosaic Pea seed-borne mosaic Red clover mosaic (2) Soybean mosaic Sugarcane mosaic Watermelon mosaic						
3.4.3 chorda	**3.4.3.1** Cucumber green mottle mosaic (2) Tobacco mosaic (7)							
3.4.4 pachus	**3.4.4.1** Barley stripe mosaic (31) Lychnis ringspot (6)							**3.4.4.2** Pea early browning Tobacco rattle (4)
3.4.5 lancea	**3.4.5.1** Cowpea mild mottle (3)	**3.4.5.2** Red clover vein mosaic (2)						
3.4.6 brevia		**3.4.6.1** Alfalfa mosaic (2)						
3.4.7 sagum				**3.4.7.1** Tomato spotted wilt				
3.4.8 nudum					**3.4.8.1** Potato spindle tuber (4)			
3.4.9 virus (particles not defined)	**3.4.9.1** Asparagus latent Dodder latent mosaic (2) Hop chlorotic disease Runner bean mosaic Sincamas mosaic Tomato bunchy top (3)	**3.4.9.2** Broad bean mild mosaic Cowpea mosaic (2) Soybean stunt (2)	**3.4.9.3** Abutilon mosaic			**3.4.9.4** Avocado pear sunblotch Cherry necrotic rusty mottle Citrus psorosis (2) Citrus xyloporosis Falsches Blattroll-virus der Erbse Groundnut marginal chlorosis Loganberry degeneration	**3.4.9.5** Beet 41 yellows Carrot red leaf Coffee ringspot	

have been accepted. Reference is made to the above list for synonyms. For viruses, named in other languages than English, the isonym (translated name) in English is indicated.

3.4 SURVEY OF SEED-BORNE VIRUSES

Seed-borne viruses, grouped according to Hansen's classification, have been listed in Table 3.2. In the survey below, the principal grouping is based on shape and size of virus particles, including (1) *spherical* or *polyhedric* particles (Virus Group Globus — Virus Group Sphaera disregarded, as seed transmission is not known to occur); (2) *rod-shaped* particles (Virus Groups Flexus, Chorda, Pachus, Lancea and Brevia — Virus Groups Media and Physa disregarded for the reason stated above); (3) *pleomorph, mostly roundish* particles (Virus Group Sagum); (4) *minute particles of the 'viroid' type* (Virus Group Nudum); and (5) *Particles unknown* (see Fig. 3.1).

For this system of classification it is relevant that serological reactions greatly correspond with type of particle. Positive reaction is easily obtained within strains of the same virus, but with high antibody titres positive serological reaction may be obtained also between different viruses possessing the same type of particle. For instance, the viruses of bean common mosaic, bean yellow mosaic, cowpea aphid-borne mosaic, soybean mosaic and watermelon mosaic, all falling within the potato Y group, are serologically interrelated (Matthews, 1970), watermelon mosaic virus being more distantly related (Martyn, 1968). They all belong to the Virus Group Maphiflexus, described below. Bawden (1955) suggested grouping viruses into 'collective species' for those groups that share common antigens, e.g. tobacco mosaic virus, including the strains common and masked tobacco mosaic, tomato aucuba mosaic, enation mosaic, streak and rib-grass.

In the survey below indications on percentage of seed transmission usually refer to the rate of transmission from plants infected by the virus concerned. Some indications, however, may refer to percentages found by testing seeds from crops with unknown rate of infected plants. An attempt has been made to specify such cases.

Surveys of seed-borne viruses, listed according to hosts, have been provided by Carter (1966) and Noble and Richardson (1968).

3.4.1 Virus Group Globus

Viruses possessing isometric, polyhedral or spherical particles, usually ranging from 20 to 40 mμ, see Fig. 3.1. Includes a subgroup of viruses

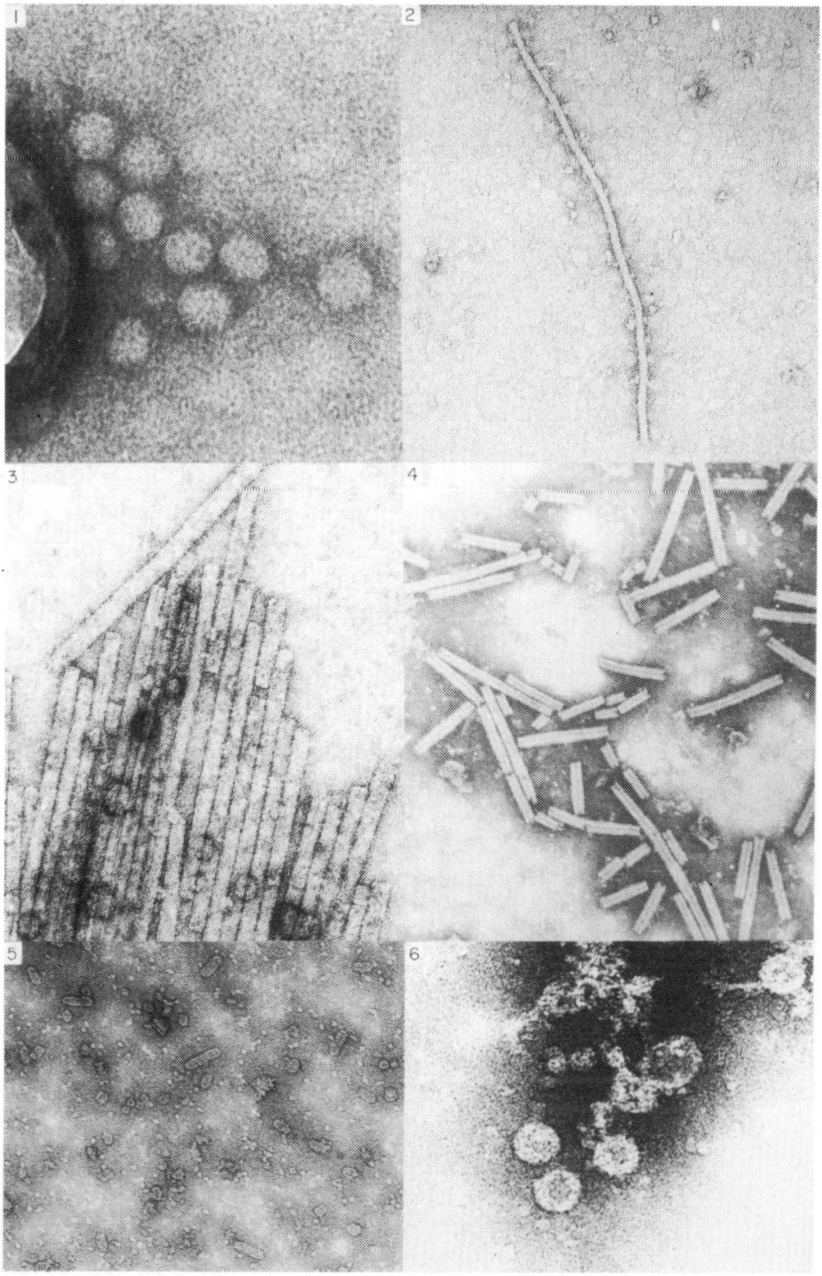

Fig. 3.1. Virus particles. (1) Tobacco ringspot virus *(globus)*, ×300,000; (2) bean yellow mosaic virus (*flexus*), ×15,000; (3) tobacco mosaic virus (*chorda*), ×30,000; (4) barley stripe mosaic virus (*pachus*), ×60,000; (5) alfalfa mosaic virus (*brevia*), ×80,000; (6) tomato spotted wilt virus (*sagum*), ×70,000. (Courtesy of Jens Begtrup)

which are principally only mechanically transmitted, and subgroup members transmitted by aphids or by nematodes.

3.4.1.1 VIRUS SUBGROUP MINGLOBUS

Vectors unknown for most viruses; beetles are vectors for broad bean true mosaic, squash mosaic and eggplant mosaic viruses. Simple sap transmission.

Apple mosaic virus

Gotlieb and Berbee (1973) recorded 2 per cent transmission through seeds of cowpea, *Vigna unguiculata*. They established the virus in pistils and pollen of white birch, *Betula pendula*, suggesting the possibility that the virus is pollen- and seed-transmitted in this host.

The virus produces conspicuous creamy or yellow spots in apple leaves but different strains produce different categories of symptoms in apple, including vein-banding and mild mosaic. Rose mosaic virus is identical with apple mosaic virus. CMI/AAB Description 83.

The virus is known in North and South America (scattered), Europe, South and East Africa, Asia (India), Australia, New Zealand.

Apricot gummosis virus

See Prunus necrotic ringspot virus.

Bean southern mosaic virus

This virus is transmitted through 1–3 per cent of seeds from infected plants of cv. Early Wilt Resistant Ramshorn of cowpea, *Vigna unguiculata* (Shepherd and Fulton, 1962; Lamtsey and Hamilton, 1974). In seeds of bean the virus is unstable, losing infectivity after a few months of storage.

A strain of the virus, bean southern mosaic virus 2, is seed-transmitted in bean at a rate of 5 per cent, *fide* Smith (1972).

Crowley (1959) recorded 21 per cent pollen transmission in bean.

In Phaseolus bean symptoms are very similar to those of bean common mosaic, vein-banding is common, and leaves may be puckered and blistered. Symptoms are particularly marked on the pods (Zaumeyer and Harter, 1943*b*). CMI/AAB Description 57.

Distribution: North America and Africa.

Black raspberry latent virus

The virus is seed-transmitted in *Rubus* spp., 10 per cent (Converse and Lister, 1969), and in *Stellaria media*, 8–26 per cent (Murant, Taylor and Chambers, 1968). The virus is also pollen-borne in raspberry.

Occurs frequently but symptomless in many black and in some red raspberry cultivars in the U.S.A. CMI/AAB Description 106.

The virus is poorly immunogenic.

Broad bean mottle virus

Seed-transmitted in *Phaseolus mungo*, 6–7 per cent of seed from crops with unknown infection rate (Phatak, 1974).

Severe infection may lead to reduction in size of the host. French bean, crimson clover and pea are also susceptible. CMI/AAB Description 101.

Distribution: England and Germany. Present in India in blackgram, *P. mungo*.

Broad bean stain virus

Seed-transmitted in broad bean, *Vicia faba*, 1 per cent (Lloyd, Smith and Jones, 1965), 7 per cent (Cockbain, Cook and Vorra-Urai, 1973), and up to 10 per cent (Gibbs and Smith, 1970). Infected seeds are stained.

Severity of symptoms is very variable, being severe at lower, mild at higher temperatures. In addition to broad bean, some cultivars of bean and pea may be attacked. The virus is serologically related to cowpea mosaic virus and squash mosaic virus. CMI/AAB Description 29.

Distribution: Europe and Northwest Africa.

Chicory yellow mottle virus

Transmitted through seed of chicory, *Cichorium intybus*, at a rate of 3 per cent (Vovlas, 1973).

Yellow mottle, ringspot and line pattern in leaves of chicory. CMI/AAB Description 132.

Distribution: Italy.

Cowpea mosaic beetle-transmitted virus

Embryo-borne and transmitted at various rates in seed of Vigna (asparagus) bean, *Vigna sesquipedalis*, 8 per cent (Dale, 1949) and in cowpea, *V. sinensis*, 10 per cent (Shepherd, 1964).

Symptoms are usually mild but seed setting is reduced appreciably (Rothamsted, personal communication, 1973). In soybean, infected plants are stunted, may produce only a few pods and may fail to set seed. CMI/AAB Description 47.

Distribution: North and South America, Cuba, the Netherlands, Trinidad, Nigeria.

Cowpea ringspot virus

Phatak (1974) demonstrated transmission in 10–30 per cent seeds of cowpea, *Vigna unguiculata* (Fig. 3.2).

Symptoms may be local or systemic chlorotic or necrotic ringspots and

Fig. 3.2. Cowpea ringspot. Mixed infection by two seed-borne viruses, cowpea aphid-borne mosaic virus and cowpea ringspot virus in cowpea. (Phatak, 1974)

mottle; vein-clearing in young trifoliate leaves; reduced leaf size; reduction in number of flowers and pods; these may be twisted and contain fewer seeds. The virus has a fairly wide host range, even in *Leguminosae*.

It is serologically unrelated to over 35 other viruses with isometric particles.

Dulcamara mottle virus

In *Solanum dulcamara* this virus was found to be seed-transmitted in trace amount (Gibbs *et al.*, 1966).

Infected plants show mild mottling and leaf puckering.

Virus is serologically closely related to potato (Andean) latent virus.

Echtes Ackerbohnenmosaikvirus

(Isonym: *Broad bean true mosaic virus*)

Up to 15 per cent seeds of broad bean, *Vicia faba*, may transmit the virus (Gibbs and Smith, 1970), 3–15 per cent according to Quantz (1953) and Cockbain, Cook and Vorra-Urai (1973).

Symptoms are similar to those produced by broad bean stain virus. The two viruses are similar as well but distinguishable serologically. CMI/AAB Description 20.

Distribution: Europe and Northwest Africa.

Eggplant mosaic virus

(Strain: *Potato (Andean) latent virus*)

Gibbs *et al.* (1966) recorded seed transmission in *Nicotiana clevelandii* and *Petunia hybrida*. The infected seedlings were symptomless.

In *N. glutinosa* symptoms include vein-clearing, flecking and mosaic. CMI/AAB Description 124.

The virus is closely serologically related to Dulcamara mottle virus.

Distribution: Central and South America.

Elm mottle virus

Transmitted through seed of *Ulmus glabra* (Jones and Mayo, 1973). Pollen-borne in *Syringa vulgaris* (Schmelzer, 1969).

Mottle and ringspot symptoms in elm. CMI/AAB Description 139.

Distribution: Europe (Germany, United Kingdom, Bulgaria, Czechoslovakia, U.S.S.R.).

Prune dwarf virus

In sour cherry, *Prunus cerasus*, the virus is transmitted in 9–15 per cent of the seeds (Cation, 1952; Gilmer and Way, 1960), and has been shown to be embryo-borne (Gilmer and Way, 1961) in *P. mahaleb*, 9 per cent (Cation, 1949). Transmitted in pollen of *P. cerasus,* 28 per cent (Way and Gilmer, 1958; Gilmer and Way, 1960). The virus has also been recorded in pollen of *P. avium* and *P. mahaleb* (Schimanski and Schade, 1974).

In cherry the leaves are reduced in size, distinctly narrowed, become thickened, rugose and exhibit mottling. In spite of profuse flowering there is heavy reduction in fruit setting (Thomas and Hildebrand, 1936). An important disease is produced in Italian prune. Sweet cherry, peach and apricot are also infected. CMI/AAB Description 19.

Distribution: North America, and probably wherever *Prunus* spp. are cultivated.

Prunus necrotic ringspot virus

This virus is established as embryo-borne in cherry (Gilmer and Way, 1961) and has been shown to be widely seed-transmitted in a range of hosts at rates of 6 per cent in a commercial seed lot of *Prunus avium*

(Cochran, 1946) or 37 per cent (Megahed and Moore, 1967), in *P. cerasus* 20–56 per cent (Cation, 1949, 1952) or up to 91 per cent (Megahed and Moore, 1967), 10 per cent in *P. mahaleb* (Cation, 1949) or 70 per cent (Megahed and Moore, 1967), 3–9 per cent in *P. persica* (Cochran, 1950), 16 per cent (Millikan, 1959) or 1–12 per cent (Wagnon *et al.*, 1960), in *P. pennsylvanica* 37 per cent (Megahed and Moore, 1967), and in *P. americana* (Hobart, 1956). The virus is also seed-transmitted in squash, *Cucurbita maxima,* to an extent of 2–3 per cent (Das, Milbrath and Swenson, 1961), and in *Rubus idaeus,* 40–50 per cent (Cadman, 1965). In the U.S.A. Gilmer (1955), using an indexing procedure, established presence of the virus in seedlings of *P. mahaleb* grown from seed imported from Europe. Seedlings from infected seeds of *P. mahaleb* may be symptomless, and may show symptoms only from the sixth year (Baumann, 1959). In addition to seed transmission of the type virus, Fridlund (1966) demonstrated that a closely related virus which may be a strain, apricot gummosis virus, is seed-transmitted in *P. serrulata* and *P. armeniaca*; Hobart (1956) recorded that this virus was transmitted in 15 per cent seeds in *P. avium*, and Traylor *et al.* (1963) that it was transmitted in *P. domestica*.

Pollen transmission is common, 28 per cent in *P. cerasus* (Way and Gilmer, 1958); also established in *P. avium* and *P. mahaleb* (Schimanski and Schade, 1974).

Symptoms and severity vary according to cultivar. Affected trees are retarded, buds and terminal growth may be killed and necrotic cankers at the nodes and base of lateral buds are common in peach. The virus is widely spread in sour and sweet cherries in the U.S.A. in combination with another seed-borne virus, Arabis mosaic virus; Prunus necrotic ringspot virus and Arabis mosaic virus produce the hop nettlehead disease in *Humulus lupulus*. Infected plants have shortened internodes, leaves stand out stiffly and fall off (Bock, 1966). CMI/AAB Description 5.

The virus shows serological response to strains from different hosts. The agar-gel diffusion test is suitable using crude sap from buds (Schade, 1967).

Distribution: North America and United Kingdom, and is probably world-wide.

Sowbane mosaic virus

The virus is transmitted in seeds of the following hosts at different rates: *Atriplex pasifica* in 21 per cent (Bennett and Costa, 1961), *Chenopodium album* in 30 per cent (Bennett and Costa, 1961), *C. amaranticolor* in 14–62 per cent (Dias and Waterworth, 1967; Kado, 1967), *C. murale* in 20–70 per cent (Bennett and Costa, 1961) and *C. quinoa* in 2–46 per cent (Bancroft and Tolin, 1967; Dias and Waterworth, 1967).

In addition to species of *Atriplex* and *Chenopodium*, also spinach, *Spinacia oleracea*, and *Beta vulgaris* are susceptible. CMI/AAB Description 64.

The virus shows serological response in gel double-diffusion tests.

Distribution: North and South America, Europe, South Africa, Japan, and Australia.

Fig. 3.3 Squash mosaic on *Curcurbita maxima*. (Courtesy of E. E. Leppik)

Squash mosaic virus

The virus is seed-transmitted in melon and different species of cucurbits as follows: in *Citrullus vulgaris*, 2 per cent (Nelson and Knuhtsen, 1969, 1973), *Cucumis melo*, 12–94 per cent from diseased plants (Rader, Fitzpatrick and Hildebrand, 1947), 7–21 per cent (Grogan, Hall and Kimble, 1959), 3 per cent (Nelson and Knuhtsen, 1973*a*), 9 per cent in a commercial seed lot (Kemp, Wiebe and Patrick, 1972), 3 per cent in seed of Rock Melon Honeydew, imported to New Zealand from the U.S.A. (Thomas, 1973), 4 per cent in seeds from crops of unknown rate of infection (Phatak, 1974), *Cucurbita flexuosus* (Rader, Fitzpatrick and Hildebrand, 1947), *C. maxima*, 0.2–1.5 per cent (Grogan, Hall and Kimble, 1959), see Fig. 3.3, *C. mixta*, 0.3 per cent (Grogan, Hall and Kimble, 1959), *C. moschata* (Rader, Fitzpatrick and Hildebrand, 1947; Campbell, 1971), and *C. pepo*, up to 2 per cent (Middleton, 1944), low percentage (Rader, Fitzpatrick and Hildebrand, 1947), up to 5 per cent (Grogan, Hall and Kimble, 1959; Nelson and Knuhtsen, 1973*a*). Nelson and Knuhtsen (1973*b*) established two serological groups of strains with different seed transmissibility.

In Chile a scheme of regular post-entry quarantine screening of imported melon seed lots is carried out by serological tests, about 15 per cent of the lots having been rejected, as was reported to the author while visiting Chile in 1972. In Canada Kemp, Wiebe and Patrick (1972) warned that the virus in muskmelon seed is distributed commercially in Ontario. CMI/AAB Description 43.

The virus is known to occur in the U.S.A., mostly California, in Venezuela, Chile and Israel. In Israel the virus is very common on squirting cucumber, *Ecbalium elaterium*, a weed present throughout the year in all agricultural regions (Cohen and Nitzany, 1963). Muskmelon mosaic virus (Rader, Fitzpatrick and Hildebrand, 1947) is generally considered as a strain.

Tobacco streak virus
(Datura 'quercina' virus)

Seed-borne in *Chenopodium quinoa*, *Datura stramonium* and *Phaseolus vulgaris* (Thomas and Graham, 1951; Brunt, 1969), in the last host seed transmission varying from 1 to 26 per cent, in *D. stramonium* up to 94 per cent (Blakeslee, 1921; Edwardson and Purcifull, 1974); moreover seed transmission in *Glycine max*, depending on cultivar, from 3 to 31 per cent (Ghanekar and Schwenk, 1974); not seed-transmitted in tobacco. Blakeslee recorded almost non-viability of infected pollen of *Datura*.

Causes diseases in tobacco, tomato and sweet clover, producing in tobacco necrotic lesions spreading along the veins. Seeds from infected tomato plants show poor germination. CMI/AAB Description 44.

Distribution: North and South America, Europe, New Zealand.

Tomato bushy stunt virus

Allen (1969) demonstrated seed transmission, up to 17 per cent in apple, *Malus pumila*.

Tomato may be severely infected, circular necrotic lesions are formed, young plants are killed; in older plants the growing points may be killed, followed by growth of secondary shoots producing the bushy or rosetted plants characteristic of the disease. In tobacco infections are less serious. CMI/AAB Description 69.

The virus is serologically very reactive (Bawden and Pirie, 1938).

Distribution: North and South America, Europe.

CMI Distribution Map No. 178.

3.4.1.2 VIRUS SUBGROUP MAPHIGLOBUS

Aphids act as vectors. Sap-transmissible.

Barley mottle mosaic virus

Transmitted in 2–45 per cent seeds of *Hordeum vulgare* (Dhanraj and Raychaudhuri, 1969).

Affected plants show mosaic, chlorosis and stunt.

Cowpea isometric virus

In cowpea, *Vigna sinensis*, the virus is transmitted in 5–16 per cent seeds (Chenulu, Sachchidananda and Mehta, 1968) and Phatak (personal communication, 1973) recorded transmission in 1–5 per cent seeds.

Symptoms are typical mosaic mottling, yellowing and distortion of leaf lamina in very susceptible cultivars; downward leaf curling and cupping of leaf margins. The virus has a restricted host range infecting only a few members of Leguminosae systemically.

Present in India.

Cucumber mosaic virus

This virus has a wide host range and is seed-transmitted in a range of hosts, many of which are important and widespread weeds. Seed transmission has been demonstrated in ash gourd, *Benincasa hispida*, 1 per cent (Sharma and Chohan, 1974), *Cerastium holosteoides*, 2 per cent (Tomlinson and Carter, 1970*b*), *Cucumis melo*, although with conflicting evidence (Molnar, 1963), *Cucumis sativus*, in trace amounts (Doolittle, 1920), pumpkin, *Cucurbita moschata*, 0.7 per cent (Sharma and Chohan, 1974), vegetable marrow, *Cucurbita pepo* (Reddy and Nariani, 1963; Rai and Chohan, *fide* Phatak, 1974; Sharma and Chohan, 1971, 1974), wild cucumber, *Echinocystis lobata* (Horvath and Szimai, 1973), 9 per cent (Doolittle and Gilbert, 1919), 55 per cent (Doolittle and Walker, 1925), *Lamium purpureum*, 4 per cent (Tomlinson and Carter, 1970*b*), *Lupinus luteus* (Troll, 1957), 14 per cent (Porembskaya, 1964), *Lycopersicon esculentum*, 0.2 per cent (Van Koot, 1949), *Phaseolus aureus*, 5 per cent of seeds from a crop with unknown rate of infection (Phatak, 1974), *Phaseolus vulgaris*, 7 per cent (Bos and Maat, 1974), *Spergula arvensis*, 2 per cent (Tomlinson and Carter, 1970*b*), *Stellaria media*, 21–40 per cent (Tomlinson and Carter, 1970*a*, 1970*b*), or 1–30 per cent (Häni, Pelet and Kern, 1970; Häni, 1971), *Vigna cylindrica* cv. Monarch, but no transmission in another cultivar (Brantley, Kuhn and Sowell, 1965), *V. sesquipedalis* (Anderson, 1957), and *V. unguiculata*, 4–28 per cent (Anderson, 1957). Häni (1971) established that the virus is borne in the embryo of chickweed, *Stellaria media*, and Crowley (1957) that it is borne in the endosperm of wild cucumber, *Echinocystis lobata*.

The virus, comprising a number of allied strains, has a very large host range including, according to Smith (1972), diseases in 31 families of phanerogams. Important diseases are spinach blight in spinach, *Spinacia oleracea*, a dwarf disease in pigweed, *Amaranthus retroflexus*, mosaic in cucumber, in which flowers are reduced in number, and fruits often may become distorted; a disease in muskmelon, in which fruits remain nearly normal (Doolittle, 1920); stunt in watermelon, with severe malformation of leaves (Walker, 1933); in tomato the well-known 'fern leaf' exhibiting reduced lamina of the leaf (tobacco mosaic virus may have the same effect). In Switzerland the virus is serious in tobacco, causing severe damage. In these and many other hosts the virus may cause considerable yield reduction. CMI/AAB Description 1.

The virus is a poor immunogen (Scott, 1968; Häni, 1971) which, however, can be improved by purification according to Scott (1963).

Distribution: World-wide.

Soybean mild mosaic virus

(Syn. *Soybean chlorotic spot virus*)

Seed-transmitted in different cultivars of soybean from 22 to 76 per cent, in two cultivars no seed transmission (Takahashi, Tanaka and Tsuda, 1974).

Symptoms in soybean are mild mottling, slight crinkling of leaves and reduction in plant size.

Distribution: Japan.

Tomato aspermy virus

The virus is seed-transmitted in *Stellaria media* (Noordam *et al.*, 1965) but not in tomato.

In tomato the virus affects the growing points of the main stem and, as a consequence, the plant assumes a characteristic bushy appearance. Fruit setting is reduced and the fruits may become small and distorted, and seed formation is suppressed (Blencowe and Caldwell, 1949). *Nicotiana tabacum* shows mosaic mottle but no distortion, whereas *N. glutinosa* develops great distortion, including enations. *Chrysanthemum indicum* is commonly and seriously infected, and symptoms may vary and include flower distortion, leading to loss of commercial value of the crop which is extensively grown at high cost under glass. CMI/AAB Description 79.

The virus is reported to be serologically related to cucumber mosaic virus by some authors (Govier, 1957; Hollings, 1964; Mink, 1969) but not by others (Grogan, Uyemoto and Kimble, 1963; Dunez and Monsion, 1967). Broad-spectrum antisera to these viruses react with the heterologous viruses while narrow-spectrum antisera do not (Hollings, Stone and Brunt, 1968). The virus is also serologically related to groundnut stunt virus (Mink, 1969).

Distribution: North America, Europe, Asia, Australia, New Zealand.

3.4.1.3 VIRUS SUBGROUP DAPHIGLOBUS

The viruses are transmitted by aphids but normally not transmissible by mechanical inoculation. The viruses have preferential affinity to assimilation tissue and phloem, poor affinity to epidermis. In the vector the viruses are either semi-persistent or persistent.

Groundnut (peanut) stunt virus

Troutman, Bailey and Thomas (1967) have recorded transmission in 0.2 per cent seeds in groundnut, *Arachis hypogaea*, harvested in a field having a very high incidence of stunt infection.

The virus has a wide host range. It has caused severe losses to groundnut in Virginia, U.S.A. Pods in severely stunted plants are malformed and small, and the size of the seeds and their germination capacity heavily reduced. The virus may infect white clover and overwinter in this host. CMI/AAB Description 92.

The virus has been found to be serologically related to cucumber mosaic virus and tomato aspermy virus (Mink, 1969).

Distribution: North America and Japan.

3.4.1.4 VIRUS SUBGROUP GENEMAGLOBUS

The virus is transmitted by soil-inhabiting nematodes of the genera *Longidorus* and *Xiphinema* (cf. Section 16.8). The viruses dealt with below are known under the designation 'NEPO viruses', *sensu* Cadman (1963), indicating *ne*matode transmission and *po*lyhedric particle shape.

Arabis mosaic virus

Lister and Murant (1967) demonstrated seed transmission in a considerable range of hosts: *Beta vulgaris*, 13 per cent; *Capsella bursa-pastoris*, 33 per cent; *Chenopodium album*, 80 per cent; *Fragaria* × *ananassa*, 7 per cent; *Lamium amplexicaule*, 1–25 per cent; *Lycopersicon esculentum*, 2 per cent; *Myosotis arvensis*, 19–95 per cent; *Plantago major*, 5–28 per cent; *Polygonum persicaria*, 21–100 per cent; *Senecio vulgaris*, 2 per cent; and *Stellaria media*, 57 per cent. Other records are *Glycine max*, 6 per cent (Lister, 1960); *Humulus* spp., 10 per cent (Kriz, 1959); *Lactuca sativa*, 60–100 per cent (Walkey, 1967); *Petunia hybrida*, 20 per cent (Lister, 1960) and *P. violacea*, up to 34 per cent (Phatak, 1974); and *Poa annua*, 4 per cent (Taylor and Thomas, 1968).

Diseases are caused by this virus in *Arabis hirsuta*, and in tobacco, cucumber and bean. There are many slightly different strains of the virus attacking a large range of hosts. In lettuce a strain produces a severe stunt in England leading to complete failure of the crop (Walkey, 1967), in raspberry a strain produces a severe disease, raspberry yellow dwarf (Harrison, 1958), in cucumber another produces a disease known as cucumber stunt mottle (Hollings, 1963), and in strawberry, *Fragaria chiloensis*, a severe stunt has been recorded (Posnette, 1956). The virus can be identified serologically. CMI/AAB Description 16.

Distribution: Europe and New Zealand.

Cherry leaf roll virus

Seed transmission has been demonstrated for *Glycine max* at the rate of 100 per cent (Lister and Murant, 1967); *Nicotiana tabacum*, 1 per cent (Schmelzer, 1965, 1966); *Phaseolus vulgaris*, 12–40 per cent; *Sambucus racemosa*, 13–44 per cent (Schimanski and Schmelzer, 1972); and *Viola tricolor*, 1.6 per cent. Elm mosaic virus, which may be regarded as a strain, is transmitted in 1–4 per cent of the seeds of *Ulmus americana* (Bretz, 1950) and also through pollen. Schmelzer (1966) detected the virus in pollen of *Sambucus racemosa*.

In sweet cherry the virus causes delay in development of leaves and flowers, upward rolling of leaves, decline in vigour of trees, and gummosis eventually leading to death of branches or entire trees (Cropley, 1961). In cucumber the virus produces a stunt. In diseased American elm, leaves may be abnormally large or small, stiff, distorted, mottled; the trees generally decline in vigour, with thinning of leaves and death of branches. Serological testing of trees for the virus has been carried out. The type virus is serologically related to elm mosaic virus (Jones and Murant, 1971). CMI/AAB Description 80.

Distribution: North America, Britain, Holland, Germany.

Grapevine fanleaf virus

Seed transmission has been recorded in *Chenopodium amaranticolor* in about 1 per cent (Dias, 1963), *C. quinoa* (Brückbauer and Rüdel, 1961; Cory and Hewitt, 1968), and grapevine, *Vitis vinifera* (Cory and Hewitt, 1968); however, Hevin *et al.* (1973) obtained no seed transmission in the last host from sowing open-pollinated seeds from infected plants.

The main host is *V. vinifera* which is severely stunted, the shoot internodes becoming short, and there is an abnormal development of secondary shoots. The leaves may be strongly distorted, the incisions deeper than normal, and due to the reduction of the area between the main veins the leaves assume the appearance of a half-closed fan; vein-clearing is common. The disease causes poor setting of fruits. Seed transmission in this host has not been established but is likely to occur. CMI/AAB Description 28.

The virus is serologically distantly related to arabis mosaic virus (Dias and Harrison, 1963; Taylor and Hewitt, 1964).

Distribution: Probably wherever grapevine is grown.

Raspberry ringspot virus

This widespread virus is seed-transmitted in *Beta vulgaris*, 50–55 per cent; *Capsella bursa-pastoris*, 2–3 per cent; *Rubus idaeus*, 18 per cent; *Stellaria media*, 29 per cent (all *fide* Lister and Murant, 1967); moreover in *Petunia violacea*, 8–20 per cent (Phatak, 1974); *Fragaria* × *ananassa*, 35 per cent; and in *Glycine max*, 7 per cent (Lister, 1960). Established as embryo-borne in *F. chiloensis* (Lister and Murant, 1967).

Transmitted in 5–6 per cent of pollen of *F. virginiana* and *Rubus* sp.

The virus produces a ringspot disease in red raspberry, *Rubus idaeus*, and a necrotic blotch disease in tomato, which becomes retarded in growth but later may recover reaching normal size. CMI/AAB Description 6.

The virus is unrelated to other NEPO viruses (Murant, 1970).

Distribution: Europe.

Strawberry latent ringspot virus

Seed-transmitted at a rate of 98–100 per cent in celery, *Apium graveolens* (Walkey and Whittingham-Jones, 1970); in *Capsella bursa-pastoris,* 4 per cent; in 65–100 per cent seeds of *Chenopodium quinoa* (Schmelzer, 1969; Allen, Davidson and Briscoe, 1970); *Lamium amplexicaule* (Murant and Goold, 1969); in 6 per cent seeds of *Mentha arvensis* (Taylor and Thomas, 1968); and in *Rubus idaeus* 75 per cent, *Senecio vulgaris*, 20 per cent and *Stellaria media*, 97 per cent (Murant and Goold, 1969).

Natural infections have been encountered in cultivated and wild rosaceous species, including strawberry, raspberry, cherry, plum and black-currant, and on elder (Lister, 1964). CMI/AAB Description 126.

Serological test for detection of the virus in peach tree (Richter, 1969).

Tobacco ringspot virus

Seed-transmitted in the following hosts: *Cucumis melo*, 3–7 per cent (McLean, 1962); *Glycine max*, 70 per cent (Desjardins, Latterell and Mitchell, 1954), 40 per cent (Kahn, Scott and Monroe, 1962), up to 100 per cent (Owusu, Crowley and Francki, 1968) and 100 per cent (Iizuka, 1973); *Gomphrena globosa*, 25–50 per cent (Kahn, Scott and Monroe, 1962), 46 per cent (Iizuka, 1973); *Lactuca sativa*, 3 per cent (Grogan and Schnathorst, 1955), 21 per cent (Iizuka, 1973); *Nicotiana glutinosa* (Stace-Smith, *fide* Phatak, 1974); *N. tabacum*, 5 per cent (Valleau, 1932*a*, 1932*b*); *Pelargonium hortorum*, 4–6 per cent (Scarborough and Smith, 1975); *Petunia violacea*, 20 per cent (Henderson, 1931); *Senecio vulgaris*, 12 per cent (Tomlinson and Carter, 1971); *Taraxacum officinale*, 9–36 per cent (Tuite, 1960); *Vigna sinensis*, 82 per cent (Kahn, 1956); and in *Zinnia elegans*, 5 per cent (Iizuka, 1973).

The virus has been shown to be embryo-borne in soybean and tobacco (Athow and Bancroft, 1959; Crowley, 1959).

This virus has a very large host range. Important diseases are produced in spinach as chlorotic, later necrotic blotches in leaves; in cucumber, pumpkin and cantaloupe, which are all very susceptible, as ringspot symptoms appearing on leaves and fruits, and in the fruits as a sunken spot eventually surrounded by definite rings; also in French bean severe ringspot symptoms are produced, and the plant may be killed; in tobacco necrotic concentric rings in normal-looking interveinal leaf tissue are characteristic, but symptoms may later disappear; *Petunia* may be severely affected, dwarfed, stunted, flowering may be profuse but few seeds develop; snapdragon, *Antirrhinum majus*, may also be badly attacked, necrotic rings and spots are produced which may coalesce, leaves or the whole plant may be killed. CMI/AAB Description 17.

Serological gel diffusion test on microscope slides at high concentrations of sera and antigen preparations have been used for detection of the virus (Hollings, 1965*a*).

Distribution: North America, Europe, Australia.

CMI Distribution Map No. 144.

Tomato black ring virus

This very widespread virus is seed-transmitted in many hosts, including, according to Gibbs and Harrison (1964) in *Beta vulgaris*, 3–27 per cent, Lister and Murant (1967), 56 per cent; *Cerastium vulgatum*, 33–100 per cent; *Chenopodium album*, 33–100 per cent; *Fumaria officinalis*, 100 per cent; *Lamium amplexicaule*, 10–48 per cent; *Ligustrum vulgare*, 6–8 per cent; *Lycopersicon esculentum*, 19 per cent; *Myosotis arvensis*, 100 per cent; *Nicotiana rustica*, 5–9 per cent; *Poa annua*, 3 per cent; *Polygonum persicaria*, 21–100 per cent; *Rubus idaeus*, 1–6 per cent; *Spergula arvensis*, 63 per cent; *Stellaria media*, 65 per cent. Moreover, Lister (1960) found 90 per cent seed transmission in *Capsella bursa-pastoris*; 40 per cent in *Fragaria* × *ananassa*; 83 per cent in *Glycine max*; 14 per cent in *Senecio vulgaris*; and 23 per cent in *Vigna sinensis*; and Phatak (1974) 9–30 per cent in *Petunia violacea*. Embryo-borne in *Rubus* sp. and soybean, *Glycine max*, as established by Lister and Murant (1967). A sero-type of the virus, cocoa necrosis virus, was transmitted in up to 24 per cent seeds of *Phaseolus vulgaris* (Kenten, 1972). The virus is also pollen-transmitted in *Beta vulgaris* (Gibbs and Harrison, 1964), and in *Rubus* sp., 13 per cent (Lister and Murant, 1967).

The virus is the cause of important diseases in tomato, tobacco, spinach, sugarbeet and celery. In tomato it produces a streak disease, black necrotic rings are formed in leaves, sometimes also in the stem, and young plants may be killed. In tobacco severe infections produce ring-shaped lesions that coalesce and may kill the leaf. High temperatures suppress symptoms. Also in spinach necrotic rings are characteristic, as well as chlorotic rings. Affected sugarbeets show a chlorotic blotching mottle (Gibbs and Harrison, 1964), and in celery a chrome yellow vein-banding is characteristic, and leaves may be asymmetrically malformed (Hollings, 1965*b*). CMI/AAB Description 38.

Serological precipitin test used by Hollings (1965*b*).

Distribution: Europe.

Tomato ringspot virus

Seed transmission has been demonstrated in strawberry, *Fragaria vesca*, 68 per cent (Mellor and Stace-Smith, 1963); *Glycine max*, 76 per cent (Kahn, 1956); *Gomphrena globosa*, 76 per cent (Keplinger and Braun, 1973); *Lycopersicon esculentum*, 3 per cent (Hollings, Stone and Dale, 1972); *Nicotiana tabacum*, *Pelargonium hortorum* (Hollings, Stone and Dale, 1972), 11 per cent, also being pollen-borne, producing 1 per cent transmission (Scarborough and Smith, 1975); *Rubus idaeus* (Stace-Smith, *fide* Phatak, 1974; Braun and Keplinger, 1973); *Sambucus* spp., 11 per cent (Uyemoto, Gilmer and Williams, 1971); and *Trifolium pratense*, 3–7 per cent (Hampton, 1967). A strain, grape vine yellow vein virus, in *Chenopodium amaranticolor*, *Glycine max* and *Vitis vinifera* (Cory and Hewitt, 1968).

Ringspot symptoms are produced in tomato, Turkish tobacco and Jimson weed, *Datura stramonium*, but in advanced stages of the disease symptoms fade. In red raspberry the virus may cause considerable reduction in fruit yield, third-year plants decline rapidly and most of the fruiting canes may be killed. Freeman and Stace-Smith (1968) suggest that the virus may be the cause of 'winter injury'. Serological test procedures have been used (Wright and Stace-Smith, 1966). CMI/AAB Description 18.

Distribution: North America and United Kingdom.

3.4.2 Virus Group Flexus

Viruses having particles of the flexus-type, i.e. thread-like or flexuous, measuring generally around 750 mμ, while the diameter may range from 10 to 25 mμ, according to virus; see Fig. 3.1(2). The group includes subgroups of viruses, some confined to mechanical transmission, others being transmitted by aphids.

3.4.2.1 VIRUS SUBGROUP MINFLEXUS

Vectors are unknown. Simple sap transmission.

Apple chlorotic leaf spot virus

Converse (1967) demonstrated seed transmission in *Rubus* spp.

The virus does not cause symptoms in most apple cultivars, but in some there are conspicuous symptoms, including stunt, chlorotic spotting and leaf distortion. CMI/AAB Description 30.

The virus is widespread in Europe and North America, and possibly wherever apple is grown.

Broad bean mosaic virus

The virus is seed-transmitted in *Vicia faba* (Brunt, 1969).

The disease expression is a mild, diffuse mottling and slight elongation of the leaves.

Guar symptomless virus

Transmitted in about 20 per cent seed of guar, *Cyamopsis tetragonoloba* (Hansen and Lesemann, 1975).

Distribution: U.S.A.

Hippeastrum mosaic virus

Seed transmission in *Hippeastrum hybridum* (Brants and Van Den Heuvel, 1965).

Symptoms in *H. equestre* are red, necrotic spots at the margins of the leaves; these often become chlorotic and die. CMI/AAB Description 117.

Distribution: probably world-wide.

Potato virus X

Seed transmission ranging from 0.6 to 2.3 per cent in potato has been recorded by Darozhkin and Grabenshchykava (1974).

Common in commercial potato stocks, known as 'healthy potato virus', often without producing symptoms. It produces diseases, some severe, in a number of other solanaceous plants. CMI/AAB Description 4.

Distribution: Widespread, wherever potato is grown.

Raspberry bushy dwarf virus

Cadman (1965) recorded transmission in 40–50 per cent seeds of *Rubus idaeus*; the virus enters the ovule through infected pollen.

On red raspberry the virus produces downward curling of leaves and stunting of canes, and the plant becomes dwarfed and bushy. The virus has a wide host range comprising 12 families of phanerogams (Barnett and Murant, 1970).

White clover mosaic virus

Transmission in 6 per cent of seed in *Trifolium pratense*, red clover (Hampton, 1963).

Various leguminous plants, including clovers, pea and broad bean, are natural hosts. In clover mosaic and mottling is prevalent, in pea and broad bean local necrotic lesions and vein-clearing. CMI/AAB Description 41.

Serologically unrelated to clover yellow mosaic virus.

Distribution: North America, Europe, New Zealand.

White clover yellow mosaic virus

Hampton (1963) recorded transmission through 8 per cent seeds of *Trifolium pratense*.

In clovers more or less stunting, growth of crimson clover and subterranean clover may be severely retarded. Pronounced mottling in pea and broad bean.

The virus is serologically distantly related to white clover mosaic virus and potato virus X. Pratt (1961) found no serological relationship between clover yellow mosaic virus and white clover mosaic virus, whereas Bercks and Brandes (1963) found distant relationship. CMI/AAB Description 111.

Distribution: North America.

3.4.2.2 VIRUS SUBGROUP MAPHIFLEXUS

Aphids act as vectors, and the viruses are sap-transmissible although some with difficulty [Fig. 3.1(2), showing particles].

Bean common mosaic virus

The virus is embryo-borne (Crowley, 1957; Quantz, 1962) and seed-transmitted in bean, *Phaseolus vulgaris*, at varying rates, greatly dependent on cultivar, 50 per cent transmission recorded by Reddick and Stewart (1919); 23 and 43 per cent according to cultivar established by Archibald (1921); 10 and 25 per cent by Kendrick and Gardner (1924); 50 per cent by Burkholder and Müller (1926); 21–51 per cent by Merkel (1929); 33 and 48 per cent by Pierce and Hungerford (1929); 36 and 57 per cent by Fajardo (1930); 18, 29 and 50 per cent by Nelson (1932); 20–60 and 55 per cent in commercial seed samples of which none was completely free of infection, by Harrison (1935); 2–66 per cent by Smith and Hewitt (1938); 25–86 per cent by Medina and Grogan (1961); up to 15 per cent in cultivar Beka according to Schippers (1963), cf. Section 13.1; and, by testing seed samples of different origin, up to 40 per cent (Rohloff, 1964) or up to 70 per cent (Quantz, 1962); more than 50 per cent in many bean cultivars (Kaiser *et al.*, 1968); 5–33 per cent according to cultivar (Ordosgoitty, 1972); 7–20 per cent (Phatak, 1974), and in seed from inoculated plants, up to 89 per cent (Hagita *et al.*, 1975). In *Phaseolus vulgaris* var. *nanus* Van Der Meulen (1928) found 32 per cent transmission; in *Vigna sesquipedalis* Snyder (1942) recorded 37 per cent transmission; and in cowpea, *V. sinensis,* Sachchidananda *et al.* (1973) recorded 25–40 per cent. Pollen transmission in bean has been demonstrated by Reddick (1931), Nelson and Down (1933) who established the rate of 25 per cent transmission, Medina and Grogan (1961) who found 18–76 per cent transmission according to strain and cultivar combination, and who recorded no pollen transmission in resistant cultivars, and finally by Schippers (1963).

Fig. 3.4. Bean common mosaic in bean, *Phaseolus vulgaris*

Bean (western) mosaic virus is a strain for which Skotland and Burke (1961) recorded seed transmission of 2–3 per cent in bean.

In bean the virus causes leaf mottle (Fig. 3.4), stunting and necrosis. The virus occurs ubiquitously in bean; the seed is often infected in high percentages in commercial cultivars. The symptoms are very variable and depend on the cultivar, the age of the plant and environmental conditions. In susceptible cultivars vein-clearing and mottle are produced; the leaves may be crinkled, chlorotic, with mosaic formation of light and dark green areas, and with downward rolling of the leaflets. (It should be noticed that downward leaf rolling is generally not produced by bean yellow mosaic virus, and this virus is not seed-borne in bean.) The following symptoms may occur: green portions of the leaves decidedly puckered and in some cultivars strongly malformed, the petioles shortened, and the plant becomes generally stunted, excessive branching of the stems, flowers and pods deformed, flowers drop, and pods may become shorter than normal. In certain cultivars, including Corbett Refugee, an extreme necrosis of leaves, stems and roots, a disease known as black root, has been described (Klinkowski and Behr, 1953), the affected plants having brown vascular discoloration, and, in the field, a sudden wilt with subsequent breakdown of the plant (Quantz, 1961). Black root usually occurs only at higher temperatures and only in cultivars which are hypersensitive and therefore normally resistant to the virus. It is a hypersensitive reaction taking place when the host becomes systemically infected through the phloem. This has been established by grafting hypersensitive plants with scions from virus-infected plants (Grogan and Walker, 1948; Klinowski and Behr, 1953), and by studying the influence of higher temperatures after sap inoculation (Thomas and Fischer, 1954; Quantz, 1957). Black root is rare but important when it does occur, and it must be taken into account in breeding for resistance. In general, the virus causes considerable yield reduction, in some cases total loss of crop.

Bean yellow mosaic virus which is not seed-borne in bean (*Phaseolus*), reduces or prevents seed transmission of bean common mosaic virus in bean when present in combined infection (Grogan and Walker, 1948).

Different strains of the virus may produce different symptoms and forms of disease in the same cultivar, e.g. one strain may produce ordinary mosaic and related symptoms of systemic infection, and another only local necroses and local infections (Silbernagel, 1969; Hubbeling, 1972).

The host range seems to be narrow, but certain other species of *Phaseolus* are susceptible, including *P. lunatus*, and *P. aureus* in which infection may cause heavy losses; also *Vicia faba* may be susceptible (Smith, 1972). The virus is serologically related to bean yellow mosaic virus and other viruses of the potato virus Y group (Brandes and Wetter, 1959). CMI/AAB Description 73.

CMI Distribution Map No. 213, world-wide.

Bean yellow mosaic virus

This virus is not seed-borne in French bean, *Phaseolus vulgaris*, but in *Lupinus albus* (Zschau, 1962; Błaszczak, 1963, 1965), *L. luteus*, 6 per cent (Corbett, 1958; Zschau, 1962), or 7–21 per cent (Porembskaya, 1964); in *Melilotus alba* (Frandsen, 1952), at rates of 3–5 per cent (Phatak, 1974); *Pisum sativum* (Inouye, 1967), 10–30 per cent; and in broad bean, *Vicia faba*, from 0.1 to 2.4 per cent (Kaiser *et al.*, 1968; Kaiser, 1972*b*, 1973; Evans, 1973).

The virus produces a typical mosaic pattern in Phaseolus bean and pea, bean plants may become stunted and bushy, and pod production reduced. Pea leaves become smaller than normal, upward curling of the margins; the plants are slightly distorted. In sweet clover, *Melilotus alba*, severe infections may cause dwarfing and ruffling of the leaves (Smith, 1972), and broad bean plants may become mildly chlorotic and later stunted. Seventy-seven leguminous species have been shown to be natural hosts (Kovachevsky, 1968). The virus is serologically related to bean common mosaic virus and other viruses of the potato virus Y group (Brandes and Wetter, 1959; Bercks, 1960). CMI/AAB Description 40.

Distribution: Asia, Europe, North America, Brazil, China, Tanzania, probably world-wide.

Cowpea aphid-borne mosaic virus

Seed-transmitted in *Phaseolus angularis* (Tsuchizaki, Yora and Asuyama, 1970), in about 1 per cent of the seed in *Vigna sinensis* (Lovisolo and Conti, 1966), but in Thailand in over 20 per cent in commercial seed lots and in 35 per cent of seed from infected mother plants (Phatak, 1975, personal communication); and in cowpea, *V. unguiculata*, 27 per cent (Snyder, 1942), 0–30 per cent (Kaiser *et al.*, 1968; Khatri and Chohan, 1972; Phatak, 1974). The virus has been recovered from both pollen, anthers and ovaries of cowpea (Tsuchizaki *et al.*, 1970).

Severe mosaic of cowpea, often chloroses, leaf distortion, stunt. See Fig. 3.2. CMI/AAB Description 134.

The virus is widespread in northern Italy, Egypt, India, Nigeria and East Africa (Phatak, 1974), also in Japan and probably China. In Iran yield reduction ranges from 13 to 83 per cent.

Groundnut (peanut) mottle virus

The virus is seed-transmitted at the rate of 2 per cent in *Arachis hypogaea* (Kuhn, 1965; Anonymous, 1972), up to 20 per cent (Bock, 1973), in commercial seed 0.3 per cent (Paguio and Kuhn, 1974), or 5 per cent, and in seed from infected mother plants 9 per cent (Phatak, 1975, personal communication).

Infected plants show leaf mottling and upward curling of leaflets. Seeds may be discoloured, 25 per cent yield reduction. Pea and soybean may be infected. CMI/AAB Description 141.

Distribution: North America, East Africa, Australia.

Lettuce mosaic virus

Established as embryo-borne in lettuce (Ryder, 1964). Seed transmission in *Chenopodium quinoa*, from zero to about 1 per cent (Phatak, 1974); lettuce, *Lactuca sativa*, different percentages according to cultivar, 3 per cent (Newhall, 1923), 5 per cent (Ryder, 1964), 10 per cent (Ogilvie, Mulligan and Brian, 1935), 2–8 per cent (Ainsworth and Ogilvie,

1939), 6–15 per cent (Kramer, Orlando and Silberschmidt, 1945), and in tested samples of commercial seed lots in Germany up to 11 per cent (Herold, 1956), up to 13 per cent (Rohloff, 1962), or even up to 16 per cent (Kietreiber, 1973AR); in the U.S.A. 1–8 per cent (Grogan and Bardin, 1950), rather frequently encountered and intercepted in Denmark in seed for export (Andersen, 1974 — index). Also in wild or prickly lettuce, *Lactuca scariola*, 1–50 per cent, all of 28 seed batches being infected (Van Hoof, 1959); and in groundsel, *Senecio vulgaris* (Ainsworth and Ogilvie, 1939), 2–3 per cent (Phatak, 1974). Ryder (1964) demonstrated transmission in 0.5 per cent of pollen of lettuce.

The virus is extremely harmful in lettuce, particularly cabbage lettuce, *Lactuca sativa* var. *capitata*, in which the effects include defective hearting and loss of commercial value. 25 per cent yield reduction has been common in the U.S.A. Although screening of seed lots and field inspection for the disease in recent years have been carried out in many countries and by many seed companies, the virus is still widespread, but usually in low percentages in seed produced by reputable companies. The symptoms comprise vein-clearing, mottling and yellowing, and later in the season necrosis and scorching, distortion of leaves, dwarfing and lack of head formation. Necrosis may be in the form of numerous minute spots between veins or as definite vein necrosis, and sometimes as scorching of leaf margins. The plants are stunted, flowers may fail to open, production of seed is reduced. A badly infected crop may be a total loss. Sweet pea, *Lathyrus odoratus*, exhibits streak symptoms in the stems, and shoots may die. Certain weeds are commonly infected and may serve as reservoirs, including wild or prickly lettuce, *Lactuca scariola*, prickly sowwhistle, *Sonchus asper*, and groundsel, *Senecio vulgaris*. Also *Chenopodium quinoa*, used as a crop plant at higher altitude in Bolivia and Peru and as an indicator test plant for many viruses, is infected. CMI/AAB Description 9.

Distribution: World-wide.

CMI Distribution Map No. 376.

Mungbean mosaic virus

Seed transmission ranges from 8 to 32 per cent in mungbean, *Phaseolus aureus* (Kaiser *et al.*, 1968; Kaiser and Mossahebi, 1974); a seed sample from a crop of unknown infection rate showed 3 per cent infection (Phatak, 1974).

It is one of the economically most important and widespread diseases of mungbean in Iran. Yields from virus-infected plants may be reduced by losses from 30 to 75 per cent. Symptoms are similar to those of bean common mosaic virus with which it has a serological relationship. However, unlike bean common mosaic virus it induces local lessions in *Chenopodium amaranticolor*, *C. quinoa*, and local as well as systemic reactions in cowpea, *Vigna unguiculata*, hence identified as a distinct virus (Phatak, 1974).

Distribution: Iran.

Onion yellow dwarf virus

Seed transmission in *Allium cepa* has been established by Härdtl (1964, 1972*a*). The virus is transmitted by pollen (Louie and Lorbeer, 1966).

In onion, *Allium cepa*, short yellow streaks appear in the leaves, later the leaves become yellow throughout, crinkled, flat and fall over, flower stalks become twisted, curled and the entire plant dwarfed, the bulbs are underdeveloped and of little or no commercial value. Infected flower clusters are abnormally small, have few flowers, and consequently seed setting is much reduced.

The disease is widespread in the U.S.A. and Europe.

Pea leaf rolling mosaic virus

(Valid isonym: *Blattroll-Mosaikvirus der Erbse*)

Musil (1970) found in *Pisum sativum* from 2 to 15 per cent transmission in seeds from plants of four cultivars of unknown infection rate, and 55 per cent transmission through seeds from experimentally infected plants of a fifth cultivar, Raman. The virus may be a strain of pea seed-borne mosaic virus, encountered in Japan (Inouye, 1967).

Leaves of infected pea plants roll downward, become somewhat narrowed and sometimes show faint mottling; the plants may be stunted. Broad bean may be severely affected, with leaf rolling and yellowing, defoliation and general stunt. Severe infections in pea, bean, broad bean, cowpea and lentil may lead to complete failure of crops (Kaiser, 1972*b*).

Distribution: Europe, Middle East (widely distributed and destructive in Iran) and New Zealand.

Pea seed-borne mosaic virus

(Syn. *Pea fizzletop virus*)

Seed transmitted in *Pisum sativum*, 8–30 per cent (Inouye, 1967), 0–88 per cent, seed transmission having been recorded in 33 of 38 tested cultivars, the rate of transmission differing very much between cultivars (Stevenson and Hagedorn, 1969, 1973), 20–80 per cent (Hampton, 1969), 65–90 per cent (Mink *et al.*, 1969), 4–32 per cent (Hampton, 1972). The virus is embryo-borne in the plumule bud and pollen-transmitted in pea (Stevenson and Hagedorn, 1973).

In the U.S.A. field symptoms in pea include stunting, rosetting, curling, leaf cupping and severe reduction in yield. Flowering is retarded but profuse, most of the blossoms are aborted, and few seeds are produced. CMI/AAB Description 146.

Distribution: North America, Europe, possibly world-wide.

Red clover mosaic virus

Hampton (1967) established transmission through 12–18 per cent seeds of *Trifolium pratense*; also, seed-transmitted in *Vicia faba* (Sander, 1959).

Mosaic, vein mosaic, streak and stunt in legumes. CMI/AAB Description 22.

The virus was found to be serologically unique.

Encountered in North America, Europe, South Africa.

Soybean mosaic virus

Transmitted from infected plants of *Glycine max* in various percentages depending on cultivar and time of infection of the mother plants, 13 per cent (Kendrick and Gardner, 1924), 0–68 per cent (Kendrick and Gardner, 1924), about 40 per cent (Heinze and Köhler, 1941), 1–18 per cent (Ross, 1963), 1–24 per cent (Kennedy and Cooper, 1967), 34 per cent (Iizuka, 1973), or 4–30 per cent (Phatak, 1974). Infected seed may become distinctly discoloured showing mottling (Kennedy and Cooper, 1967; Van Niekerk and Lombard, 1967; Ross, 1970; Porto and Hagedorn, 1975), brown or black according to strain (Phatak, 1974), cf. Section 19.9 and Colour Plate IV. Infected plants develop the highest percentage of seed coat mottling at 20°C, whereas mottling is reduced or suppressed at 30°C. However, seeds of normal appearance as well as those which are mottled transmit the virus equally (Ross, 1970).

In soybean the virus produces yellowish vein-clearing, the leaves become rugose, dark green, puffy areas appear through upward proliferation of the blade, leaf margins curve downward at the sides, upward at the tip, and infected plants may become stunted and produce fewer and abnormal pods, and seed setting is heavily reduced. Symptoms are severe at 18.5°C but masked at 29.5°C (Conover, 1948). CMI/AAB Description 93.

CMI Distribution Map No. 390, common where soybean is grown.

Sugarcane mosaic virus

The maize dwarf mosaic strain of the virus is transmitted through seeds of maize, *Zea mays*, from trace amounts up to about 1 per cent (Shepherd and Holdeman, 1965; Williams *et al.*, 1968; Baudin, 1969). The virus seems, however, to reduce the germination capacity of seed from infected maize plants (Baudin, 1969).

In sugarcane the virus produces light blotches or chlorotic streaking on the leaves and different types of mottling, and stunting of plants. The disease is destructive; yield reduction of sugar in affected plants may range from 40 to 60 per cent. Much effort has been made to produce resistant cultivars, so that the disease is not now so serious. CMI/AAB Description 88.

Serologically distantly related to the potato virus Y group.
CMI Distribution Map No. 299.

Watermelon mosaic virus

Lindberg, Hall and Walker (1956) have recorded transmission through 2 per cent seeds of wild cucumber, *Echinocystis lobata*.

In watermelon, *Citrullus vulgaris*, the virus causes mottle, chlorosis, stunting and distortion.

Serologically distantly related to bean yellow mosaic virus and potato virus Y (Van Regenmortel, Brandes and Bercks, 1962). CMI/AAB Description 63.

Distribution: North America, Europe, South Africa and Japan.

3.4.3 Virus Group Chorda

The particles of viruses belonging to this group are rigid and rod-shaped, measuring about 250–300 mμ with diameter of about 15 mμ [Fig. 3.1(3)].

3.4.3.1 VIRUS SUBGROUP MINCHORDA

Readily sap-transmitted. The virus, kept *in vitro*, may keep infectivity for years, tobacco mosaic virus in extracted plant juice for 50 years. Efficient vectors are not known.

Cucumber green mottle mosaic virus

Transmission by 44 per cent seeds of *Cucumis sativus* has been established by Yakovleva (1965); moreover the virus is seed-transmitted in bottle gourd, *Lagenaria siceraria*, which is used as a root stock for watermelon (Komuro *et al.*, 1971; Sharma and Chohan, 1974).

In cucumber the symptoms are vein-clearing, light or dark green mottle, blistering and distortion of leaves, stunt. Symptoms are severe, particularly in slowly growing plants. Similar symptoms in melon and watermelon. CMI/AAB Description 154.

The virus is serologically related to tobacco mosaic virus [Knight, 1955; Hollings, 1967, in litt. (Martyn, 1968)]. Kozlow and Porembs'ka (1972) described a serodiagnostic method for detection of the virus in seeds, using precipitin in agar.

Distribution: Europe, Asia.

Tobacco mosaic virus

Seed transmission has been established in *Capsicum frutescens*, 22 per cent (Ikeno, *fide* Phatak, 1974; McKinney, 1952); *Lycopersicon esculentum*, e.g. in 4 per cent (Doolittle and Beecher, 1937); other records are established by Broadbent (1965), John and Sova (1955), Taylor, Grogan and Kimble (1961), Procter and Fry (1965) and Van Winckel, 1965. Gilmer and Wilks (1967) recorded transmission through 38 per cent in *Malus platycarpa*, 3–37 per cent in *M. pumila (Pyrus sylvestris)*, 35 per cent in *P.*

communis, and Allen (1969) through up to 21 per cent of seeds from apple cv. McIntosh. Gilmer and Kelts (1968) recorded 20 per cent in *Vitis vinifera*, and in *Vigna unguiculata* 1–4 per cent of seeds from crops of unknown rate of infection was found by Phatak (1974). Seed transmission in tomato has been questioned and much discussed but it is now well established (Broadbent, 1965) that the virus is carried as contamination or infection on or within the seed coat, and that the inoculum from these sites may be inoculated into young seedlings during handling, especially transplanting seedlings (Crowley, 1957; Taylor, 1962; Gilmer and Wilks, 1967), cf. Subsection 19.6.3, see Fig. 3.5. Also in pungent pepper, tobacco and in pear and apple the site of infection is the seed coat (cf. Section 13.2, see Fig. 13.1). Gilmer and Wilks (1967) found, however, that in the apple cultivars Antonowka and Wealthy the virus was carried in the embryo.

This highly infectious virus has an immense host range covering widely different plant families. It is a major factor in tobacco production, losses in severely attacked tobacco crops in the U.S.A. being about 30–35 per cent. Losses in tomato crops are often less, but may go up to about 40 per cent. The symptoms are mottling, leaf distortion, stunting of the plants, dwarfing, local or widespread necroses, the symptoms varying with strain and host cultivar. The virus may produce stunts in *Tagetes patula*, in spinach and in larkspur, *Delphinium consolida*. CMI/AAB Description 151.

Distributed wherever natural hosts are grown.

3.4.4 Virus Group Pachus

Virus particles rod-shaped, of varying length, with a diameter of about 20–25 mμ. In barley stripe mosaic virus particles range from 20 to 280 mμ, the most common length being 128 mμ, and particles less than 111 mμ not being infective (Harrison, Nixon and Wood, 1965); see Fig. 3.1(4). Viruses of Group Genemapachus may have five different lengths of particles, in pea early browning virus 102 mμ and 205 mμ (Smith, 1972). For the first of these subgroups effective vectors are not known, while viruses of the second subgroup are transmitted by soil-inhabiting nematodes.

3.4.4.1 VIRUS SUBGROUP MINPACHUS

Vectors are unknown, the viruses are mechanically transmitted.

Barley stripe mosaic virus

(Syn. *False stripe virus*)

The virus is transmitted in *Agropyron elongatum* (Inouye, 1962), *Avena fatua*, 22 per cent of seed from naturally infected plants (Chiko, 1975), *A. sativa* by 0–10 per cent of the seed (McKinney and Greeley, 1965); *Bromus inermis*, 8 per cent (Inouye, 1962); *Commelina communis*, 4 per cent (Inouye, 1962); *Hordeum depressum*, 3 per cent, *H. glaucum*, 2 per

Fig. 3.5. Tomato streak. Example of a disease which may be caused by simultaneous infection of different viruses, e.g. tobacco mosaic virus and potato virus X

cent, *H. stebbinsii* (Inouye, 1962); *H. vulgare*, 58 per cent (McKinney, 1951), 4–64 per cent (Eslick and Afanasiev, 1955), 38–86 per cent (Inouye, 1962), 3–53 per cent according to strain (McKinney and Greeley, 1965; Carroll, 1972), varying percentages in two cultivars (Carroll and Chapman, 1970), 38 per cent (Catherall, 1972), 5–60 per cent (Phatak, 1974); in three species of *Lolium*, 3–8 per cent (Inouye, 1962); and in *Triticum aestivum*, 7–81 per cent (McNeal and Afanasiev, 1955). Moreover, Nitzany and Gerechter (1962) demonstrated seed transmission in the following grasses: *Aegilops kotschyi*, *A. peregrina*, *Agrotriticum*, *Avena barbata. A. strigosa, Bromus alopecurus, B. rigens, B. scoparius, B. fasciculatus*, *Cutandia memphitica*, *Cynosurus coloratus*, *Hordeum murinum, H. spontaneum, Koeleria phleoides, Phalaris canariensis, Pholiurus filiformis*, *P. incurvus*, and *Poa exilis*. In barley and wheat the virus is carried in the embryo as well as in the endosperm (Gold *et al.*, 1954). This virus is pollen-transmitted in barley by 10 per cent, according to Gold *et al.* (1954), 35 per cent according to Inouye (1962). Carroll (1974) found the virus particles in both the sperm and vegetative cells of barley pollen. Carroll (1972) established that some strains are not seed-transmitted; he found that a non-transmitted strain was absent in pollen, embryos and endosperm of seed in contrast to seed-transmitted strains.

Characteristic of the disease in barley are the necrotic brown stripes on the leaves; in wheat and maize mainly yellowish continuous or broken stripes. Inouye (1962) established that out of more than 2,000 cultivars of barley only one cultivar, Wien, was not systemically infected, and that 140 cultivars of wheat were susceptible. He found that early infection leads to retardation of heading time, shortening of plant height, incomplete emergence from the sheaths, reduction of number of heads, of kernel weight and proportion of large seeds. The virus is transmissible to small-grain cereals and a number of grasses. Italian types of millet have been found very susceptible (Slykhuis, 1952). CMI/AAB Description 68.

The virus is highly immunogenic. Scott (1961) and Hamilton (1965) showed that the virus is detectable in single seeds by means of the gel precipitin tests (Preer, 1956).

The virus is widely internationally spread.

CMI Distribution Map No. 391.

Lychnis ringspot virus

Bennett (1959) established transmission in 10 per cent seeds in *Beta vulgaris*; 9 per cent in *Capsella bursa-pastoris*; 27 per cent in *Cerastium viscosum*; 58 per cent in *Lychnis divaricata*; 28 per cent in *Silene gallica*; and 41 per cent in *S. noctiflora*. He found pollen transmission to the extent of 19 per cent in *Lychnis divaricata* and *S. noctiflora*, but the virus was predominantly transmitted through the ovule.

In *S. noctiflora* and sugarbeet the virus produces leaf distortion and plants are stunted; in tobacco necrotic lesions are developed on leaves and stems of cv. Samsun. Serologically the virus is distantly related to barley stripe mosaic virus (Gibbs *et al.*, 1963).

Geographical distribution seems to be confined to California, U.S.A.

3.4.4.2 VIRUS SUBGROUP GENEMAPACHUS

Nematodes of the genus *Trichodorus* act as vectors. The viruses are mechanically transmissible.

Pea early browning virus

Transmission in about 50 per cent of seeds of *Pisum sativum* has been recorded by Bos and Van Der Want (1962), 1–2 per cent in several cultivars (Harrison, 1973, in CMI/AAB Description 120).

In pea the symptoms are purplish-brown necrotic discoloration starting as vascular necrosis. Infected plants are often stunted, somewhat distorted and slightly chlorotic (Bos and Van Der Want, 1962). CMI/AAB Description 120.

The virus is serologically distantly related to tobacco rattle virus (Maat, 1963).

The virus has been found in Britain and the Netherlands.

Tobacco rattle virus

Lister and Murant (1967) found 2 per cent seed transmission in *Capsella bursa-pastoris*; 2 per cent in *Lamium amplexicaule*; 6 per cent in *Myosotis arvensis*; and 1 per cent in *Papaver rhoeas*.

The host range comprises several hundred species, more than 350 according to Schmelzer (1957*b*). The virus causes harmful diseases such as tulip white streak by which the plant is stunted and the flower misshapen, or internal tuber necrosis in potato, known as corky ringspot, a disease forming corky tissue deep in the flesh of the tubers (Cadman, 1959; Harrison, 1968) — shoots become stunted, with necrotic spots. Rattle disease is quite destructive in tobacco in which the virus causes severe systemic necrosis. CMI/AAB Description 12.

Distribution: North and South America, Europe, Japan.

3.4.5 Virus Group Lancea

Particles of the viruses are long, filiform and sometimes slightly curved but rather stiff, 620–700 mμ long, 12–15 mμ wide. Vectors are unknown (Subgroup Minlancea) or aphids (Subgroup Maphilancea).

3.4.5.1 VIRUS SUBGROUP MINLANCEA

Vectors unknown, the viruses are mechanically transmissible.

Cowpea mild mottle virus

Brunt and Kenten (1973) recorded transmission of the virus through over 90 per cent seeds of *Glycine max*; about 6 per cent seeds of *Phaseolus vulgaris*; and about 90 per cent seeds of *Vigna unguiculata*.

In cowpea the virus produces a mild systemic chlorotic mottle or, occasionally, a conspicuous chlorosis but often the host is symptomless.

The virus shows a distant serological relationship to carnation latent virus, but not to ten other morphologically similar viruses, including red clover vein mosaic virus. CMI/AAB Description 140.

The virus has been encountered in the eastern region of Ghana where cowpea is often infected.

3.4.5.2 VIRUS SUBGROUP MAPHILANCEA

Aphids act as vectors; the viruses are also mechanically transmissible.

Red clover vein mosaic virus

Seed transmission in red clover, *Trifolium pratense* (Matsulevich, 1957) and in broad bean, *Vicia faba*, up to 100 per cent (Sander, 1959).

In red clover vein-clearing, sometimes interveinal spotting, but normal growth. In contrast, crimson clover, *T. incarnatum*, reacts by severe stunting, and the plant usually dies. In pea as well as in broad bean the virus causes severe stunt. CMI/AAB Description 22.

Distribution: North and South America, Europe, South Africa.

3.4.6 Virus Group Brevia

Virus particles are short, bacilliform, 35–60 mμ long, about 18 mμ wide; see Fig. 3.1(5). The particles of the individual virus, discussed below, show a range of length such as 36, 48 and 58 mμ (Gibbs, Nixon and Woods, 1963; Hull, 1969).

3.4.6.1 VIRUS SUBGROUP MAPHIBREVIA

Vectors are aphids. Viruses mechanically transmissible, although with some difficulty.

Alfalfa mosaic virus

Seed-transmitted in *Capsicum annuum*, 1–5 per cent (Sŭtič, 1959) and lucerne, *Medicago sativa*, 55 per cent (Zschau and Janke, 1962), up to 6 per cent (Belli, 1962); Frosheiser (1970) recorded from 0.2 to 6 per cent infection in 56 seed lots of lucerne from five States of the U.S.A., and the rate of seed transmission varies considerably according to strain and cultivar (Frosheiser, 1974). Pollen transmission up to 27 per cent in lucerne (Frosheiser, 1974).

The virus has a wide host range, including species outside the Leguminosae, a total of more than 300 species known as hosts. In lucerne it causes dwarf, the leaves being mottled and wrinkled. In broad bean, *Vicia faba*, there may develop a general necrosis leading to the death of the plant. In pea the disease may develop as general wilting and stunting and the plant may be killed, and the same fatal development may take place in sweet pea, *Lathyrus odoratus*. Up to 75 per cent yield reduction has been recorded in cowpea (Kaiser *et al.*, 1968). The virus comprises many specific strains, including the potato calico strain, the pepper strain and alfalfa mosaic virus (cf. Smith, 1972). CMI/AAB Description 46.

Serologically detectable by the gel double-diffusion test (Van Vloten-Doting, Kruseman and Jaspars, 1968).

CMI Distribution Map No. 388, world-wide.

3.4.7 Virus Group Sagum

The virus particles are roundish, essentially spherical, and, in tomato spotted wilt virus, 68–100 mμ in diameter; see Fig. 3.1(6).

3.4.7.1 VIRUS SUBGROUP METHYSAGUM

Various species of thrips in which the virus is circulative act as vectors. The virus is also mechanically transmissible if careful techniques are used.

Tomato spotted wilt virus

Jones (1944) established seed transmission in cineraria, *Senecio cruentus*, up to 70 per cent in commercial seed lots.

In tomato the characteristic symptom is bronzing of the leaves, which curl slightly downward and inward. On mature fruits pale, yellow, rarely white areas occur in the normal red skin. Young plants are often killed. In cineraria the leaves become rugose, curled, with red discoloration of the veins and necrotic spots on the leaf blade; the leaves may become chlorotic and wilt (Cineraria streak disease). In greenhouse cultures heavy infection may lead to 50 per cent loss of the crop. In *Dahlia* the virus produces a ringspot disease, in *Chrysanthemum* bronzing of leaves, and in pineapple, *Ananas comosus*, yellow later watersoaked leaf spots, followed by rotting. The virus causes a serious decline in *Ageratum houstonianum* (Gumpf and Weathers, 1972). The host range is extensive, according to Best (1968) comprising 165 species in 34 families, including 60 solanaceous species. Feldman and Boninsegna (1968) established a serological procedure. CMI/AAB Description 39.

Distribution: World-wide in temperate and subtropical regions.

CMI Distribution Map No. 8.

3.4.8 Virus Group Nudum

The virus in this category forms no recognisable particles but has been described as present in the host as free ribonucleic acid. Raymer and

Diener (1969) described the extremely small 'particles' of potato spindle tuber virus, but these are really molecules of ribonucleic acid. These entities are very stable and have a molecular weight of 50,000, thus being extremely small as compared with other known virus particles, 80 times smaller than the smallest known virus, and they multiply in living cells. Diener (1971) designates this type as 'viroid'. A 'viroid' is different from ordinary viruses in size and in lacking a protective protein coat which in ordinary viruses surrounds a central core of ribonucleic acid, deoxy-ribonucleic acid or an allied nucleic acid. Thus in the 'viroid' type of viruses the 'particles' are 'naked', hence the designation of the virus group suggested by the writer, Nudum.

3.4.8.1 VIRUS SUBGROUP MEXENUDUM

This subgroup comprises, provisionally, the virus indicated below. The virus possesses very different vectors, aphids and beetles. In practice only mechanical transmission of the virus is important.

Potato spindle tuber virus

Seed transmission in tomato, *Lycopersicon esculentum*, 2–11 per cent (Singh, 1970); *Physalis peruviana*, 29 per cent (McClean, 1948, *fide* Phatak, 1974); *Solanum incanum*, 53 per cent (McClean, *fide* Phatak, 1974), and potato, *S. tuberosum*, 6–12 per cent (Singh, *fide* Phatak, 1974), or even up to 100 per cent (Hunter, Darling and Beale, 1969; Fernow, Peterson and Plaisted, 1970). The virus is also pollen-borne in potato.

The most characteristic symptoms produced by this virus are those in potato tubers. These are abnormally long, cylindrical, spindle-shaped, with tapering ends; affected tubers often have more or less irregular outline, the 'eyebrows' of the buds are flattened, sometimes even somewhat protruding; the skin is more smooth than normal, often shining. The sprouts from infected tubers are more erect and spindling than in normal plants, the leaves are smaller, more erect, often darker green than normal, slightly rugose; the plant is late in development and remains weak throughout the growth. CMI/AAB Description 66.

The disease is important in North America; outside this continent it seems to have been observed occasionally only in the Soviet Union and South Africa.

See remark under tomato bunchy top virus below.

3.4.9 Group of viruses in which particles have not been defined

Subgroups defined according to vectors and mode of transmission, including viruses for which vectors are unknown; and viruses with the following vectors: aphids carying non-persistent viruses; aphids carrying persistent viruses; white flies; and nematodes.

3.4.9.1 VIRUS SUBGROUP MINVIRUS

Vectors, if any, have not been identified. Mechanical transmission.

Asparagus latent virus

Paludan (1964) recorded up to 65 per cent seed transmission in *Asparagus officinalis*.

No symptoms are produced in *A. officinalis*. The virus produces systemic infection in *Nicotiana tabacum* cv. Samsun.

Encountered in Denmark and Germany.

Dodder latent mosaic virus

Bennett (1944) established transmission in 2 per cent seeds of *Cuscuta californica*, 5 per cent seeds of *C. campestris*.

The virus causes chlorotic spots and dwarfed, wrinkled leaves in sugar-beet; chlorotic spots in melon; dark necrotic lesions in potato and tomato.

Distribution: North America.

Hop chlorotic disease virus

In *Humulus lupulus* Salmon and Ware (1935) recorded 27 per cent seed transmission.

In hop plants leaves may become partly or almost totally chlorotic, and in severely affected plants distorted.

Runner bean mosaic virus

Vashisth and Nagaich (1965) found 42 per cent seed transmission of this virus in *Phaseolus coccineus*.

Diseased plants are stunted, but symptoms masked in winter season.

Encountered in India.

Sincamas mosaic virus

In sincamas, *Pachyrrhizus angulatus (P. erosus)*, the virus is transmitted in 5–26 per cent of the seeds in different lots (Fajardo and Marañon, 1932).

Chlorosis, mottling, blistering of leaves which may be reduced in size, early infected plants are usually stunted or spindly, the fleshy root of affected plants contains less sugar, and reduced dry matter.

Encountered on cultivated and wild sincamas in the Philippines.

Tomato bunchy top virus

The virus is seed-transmitted in *Physalis peruviana*, *Solanum duplosinuatum* and *S. incanum* (McClean, *fide* Phatak, 1974).

In tomato the virus causes severe retarding in growth of shoots resulting in close crowding of leaves which become small and downwardly curled at the margins, and the fruits become small, seedless or with few seeds. A number of other solanaceous plants may be infected.

The distribution of the virus is limited to South Africa.

Benson *et al.* (1965) have suggested a relationship between this virus and potato spindle tuber virus.

3.4.9.2 VIRUS SUBGROUP MAPHIVIRUS

Aphids act as vectors. Mechanical transmission.

Bean western mosaic virus

See bean common mosaic virus.

Broad bean mild mosaic virus

Devergne and Cousin (1966) recorded seed transmission in *Vicia faba*. The virus produces symptoms similar to 'Evesham Stain', cf. Fig. 2.5.

Cowpea mosaic virus

The virus is transmitted through 17 per cent seeds of *Vigna catjang* and 23 per cent seeds of cowpea, *V. sinensis* (Capoor and Varma, 1956).

In cowpea the symptoms are chlorotic lesions, vein-clearing, mosaic, with dark green areas on pale green background, subsequent mottling. Similar symptoms in mungbean, *Phaseolus aureus*. Soybean, *Glycine max*, becomes stunted, and infected plants produce few pods and may fail to set seed. CMI/AAB Description 47.

Distribution: North and South America, the Netherlands, Trinidad, Nigeria.

Soybean stunt virus

The virus is transmitted in 50 per cent seeds of *Glycine max* (Koshimizu and Iizuka, 1963), 95 per cent (Iizuka, 1973); and in 5 per cent seeds of cowpea, *Vigna unguiculata* (Iizuka, 1973).

3.4.9.3 VIRUS SUBGROUP MALEVIRUS

Vectors are white flies (for Abutilon mosaic virus, *Bemisia tabaci*). Otherwise only transmissible by grafting; mechanical transmission only with difficulty.

Abutilon mosaic virus

Transmission has been recorded in 0.1 per cent seeds of *Abutilon thompsoni* × *A. mulleri* (Keur, 1934).

The disease is the classical 'infectious chlorosis', the attractive yellow-green variegation in *Abutilon*, much exploited in commercial ornamental plants. The plants grown in Europe originate from the first infected plants imported by a British nursery from West India in 1868, and then vegetatively propagated. By 1869 transmission of 'variegation' through grafting was detected (Morren, 1869). Symptoms may disappear, particularly under subdued light conditions or in darkness. In cotton the virus causes a mosaic. Many malvaceous species are susceptible, as well as a number of species of other families.

Distribution: World-wide in the 'variegated' *Abutilon*.

3.4.9.4 VIRUS SUBGROUP DINVIRUS

Vectors unknown, transmission by grafting only.

Avocado pear sunblotch virus

Wallace and Drake (1953) recorded transmission in 75 per cent of seeds of avocado pear, *Persea americana*. The infected seedlings are symptomless carriers and when used as rootstocks, scions become infected.

Symptoms on young trees: discoloured longitudinal grooves or streaks on stems or on shoots, frequently fading out to obscure mottling; shoots may become decumbent, somewhat twisted, older stems uneven, rough, with prominent lenticles. Considerable yield reduction.

Important in California where a health certification scheme is implemented, based on testing mother trees, that may be symptomless carriers, by applying indexing techniques. Possibly introduced from the U.S.A. into Israel.

Cherry necrotic rusty mottle virus

The virus is seed-borne in *Prunus cerasus* (Nyland, 1962).

In affected *P. avium* development of leaves and flowers are delayed in the spring, brown rusty, necrotic and chlorotic spots appear followed by shot-holes, and in some cultivars leaves become prematurely senescent and fall. There may occur cankers or blister-like lesions on the cortex of young branches.

Distribution: United Kingdom and U.S.A. (Utah).

Citrus psorosis virus

Seed-transmitted in citrange, *Citrus sinensis* × *Poncirus trifoliata* (Childs and Johnson, 1966) at the rate of 6 per cent, and in trace amount in *Citrus* sp. (Wallace, 1957).

The virus infects predominantly lemon, *Citrus limon*, grapefruit, *C. paradisi*, and orange, *C. sinensis*; only members of the *Rutaceae* seem to be susceptible. Leaves become small and chlorotic. In orange and grapefruit bark symptoms develop as formation of scales and small pustules in the outer layer and as brown spots beneath these malformations; gum deposits are formed, particularly between the layers of the wood; gum may plug the vessels; later appears discoloration of the wood, from which starch disappears, and eventually the trees are killed.

CMI Distribution Map No. 65.

Citrus xyloporosis virus

Childs (1956) established 66 per cent of seed transmission in *Citrus aurantifolia*, sweet lime.

The disease may be severe in sweet lime, grapefruit, mandarin and orange, all when grafted on the very susceptible Shamouti orange. Infected trees become stunted, trunk and roots decay, followed by disintegration of the cambium, xylem and phloem, the lesions being impregnated by gum.

CMI Distribution Map No. 374.

Falsches Blattrollvirus der Erbse

(Valid isonym: *Pea false leafroll virus*)

The virus is transmitted in 40 per cent of seeds of *Pisum sativum* (Thottappilly and Schmutterer, 1968).

Causes leafroll, chlorotic, later necrotic leaf spots, and vein necrosis. In addition to aphid transmission (*Myzus persicae*), the virus seems to be transmissible also through the zoospores of *Pythium* sp.

Groundnut (peanut) marginal chlorosis virus

Seed transmission at the rate of 30 to 100 per cent in groundnut, *Arachis hypogaea* (Van Velsen, 1961).

The disease, known since 1958 in New Guinea, is characterised by yellowing of the leaf margins, leaf crinkle and severe stunting.

Loganberry degeneration virus

Ormerod (1970) established pollen and seed transmission in loganberry, *Rubus loganobaccus*.

The virus occurs in weak, 'degenerate' plants of loganberry, transmissible to *R. henryi*, in which it has produced chlorosis, weak growth and stunt.

3.4.9.5 VIRUS SUBGROUP DAPHIVIRUS

Persistent viruses transmitted by aphids. The individual viruses may not be mechanically transmitted.

Beet 41 yellows virus

In an inbred line of sugarbeet, No. 41, Clinch, Loughnane and McKay (1948) recorded in Eire 48 per cent seed transmission of the virus investigated by them. However, there has been no confirmation of seed transmission (cf. Ernould, 1950) of beet yellows virus which on account of the Irish incidence has entered the quarantine regulations of several countries.

The Irish strain is undoubtedly not a strain of beet yellows virus. Particles of the seed-borne virus seem not to have been detected (particles of true beet yellows virus are filiform, flexuous and more than 1,000 mμ long, hence this virus should be referred to Virus Group Daphiflexus).

The disease produced by the true beet yellows virus in beets, particularly sugarbeet, is widespread and serious. Mosaic appears predominantly in older leaves; furthermore, a chlorosis develops, the affected areas appearing waxy, subsequently dry, and finally the leaves become brittle. The symptoms produced by the seed-borne virus are very similar.

The seed-borne virus is not serologically related to the true beet yellows virus (Watson, Hull and Hartsuijkeri, 1949).

Carrot red leaf virus

Watson and Serjeant (1962) have recorded transmission in 25 per cent seeds of carrot, *Daucus carota*.

The virus produces reddening or yellowing of the leaves, but little or no stunting.

Coffee ringspot virus

The virus has been found transmitted in 10–11 per cent seeds of *Coffea excelsa* (Reyes, 1961).

The disease, affecting coffee trees in Brazil, spreads from tree to tree very slowly. Symptoms are ringspot and line patterns on leaves.

Distribution: Brazil, Costa Rica, Philippines.

3.5 SEED TRANSMISSION OF BACTERIOPHAGES

Bacteriophages infecting seed-borne bacteria may themselves be seed-borne. Katznelson and Sutton (1951) demonstrated in seeds of Phaseolus bean the presence of phages, active on *Xanthomonas phaseoli* and *Pseudomonas phaseolicola*, respectively; Sutton and Katznelson (1953) isolated a phage for *P. atrofaciens* from 55 of 56 wheat seed samples, and Sutton (1966) obtained 68 isolates of phages for this bacterial pathogen from a total of 156 seed samples of barley, rye, buckwheat, timothy, oats and wheat, all of Canadian origin. The phages were lytic to one or more of four indicator strains of the bacterium, and on the basis of their cross-reaction the phage isolates were divided into three groups. Hayward (1964) reported that seed of cotton infected with *Xanthomonas malvacearum* frequently contains phages active against the bacterial pathogen.

Occurrence of bacteriophages in nature is likely to be of great consequence to development of epidemics such as those caused by certain seed-borne bacteria, e.g. *Xanthomonas oryzae*. Rao and Srivastava (1971) reported that in occasional cases they failed to isolate this bacterium from infected leaves because of the dominant presence of bacteriophages. In one gram leaf of a sample 6.4×10^{10} phages were recorded. Association of bacteriophages with their host bacteria on seed may be expected to have influence on the rate of seed transmission and further spread of the pathogen.

Actual seed transmission of bacteriophages seems, however, not yet to have been established, nor is there any information on longevity of bacteriophages on seed during storage. It may be speculated that the phage particles within infected bacterial cells would be preserved provided the latter are subjected to rapid drying. Such bacteria may in turn be borne on or within the seed. The International Rice Research Institute (1969) has reported that phages survive much longer than the bacteria and that they are more tolerant to high temperatures. After sowing of the seed, the bacteriophages may resume activity simultaneously with uninfected bacteria.

3.6 SEED TRANSMISSION OF MYCOPLASMAS

In 1967 Doi, Teranaka, Yora and Asuyama discovered a new category of plant pathogens by studying ultrathin sections of *Petunia*, affected by yellows, and of other plants infected with similar diseases, notably mulberry dwarf and potato witches' broom or Paulownia witches' broom. These organisms as well as similar organisms later encountered were all found in phloem; they showed substantial similarities with the so-called mycoplasmas, a group of organisms some of which are pathogenic, most of them apathogenic, encountered in animals and referred to the genus *Mycoplasma* of the class *Mollicutes*, a class parallel to *Schizomycetes*.Mycoplasmas in animals are intracellular or extracellular, the cell is pleomorph or filamentous and has no wall but a three-layered membrane, and they can be grown in a cell-free substrate. The 'mycoplasmas' encountered in plants are also provided with a three-layered membrane; they have been found only in phloem, where they are always located intracellularly; they have also been encountered in those leaf hoppers that transmit pathogens of yellows and similar diseases. They can be found in the leaf hoppers about two weeks after they have fed on plants infected by yellows. Reference is made to the review by Davis and Whitcomb (1971) on Mycoplasmas, Rhickettsiae and Chlamydiae, as well as to the review by Maramorosch, Granados and Hirumi (1970).

No yellows pathogens have so far been found to be seed-transmitted. It seems unlikely that these phloem-infecting pathogens are embryo-transmissible and even less probable that these intracellular organisms can be transmitted as contaminants with seed.

4 Seed-borne Bacteria

INTRODUCTION

Bacteria are commonly carried on seed; potentially, any bacterial plant pathogen may probably be seed-transmitted.

Plant pathogenic bacteria do not form endospores. With the exceptions of *Agrobacterium*, *Corynebacterium* and *Streptomyces*, which are soil inhabitants, most members of the other genera are not well fitted for survival in soil being largely dependent on their hosts for survival through critical seasons; generally they are rather soil invaders than soil inhabitants. In perennial hosts as well as in vegetative propagation material the bacteria survive in infected tissues, in galls (*Agrobacterium tumefaciens*), in tubers, e.g. of potatoes (*Corynebacterium sepedonicum*), in roots, e.g. of lucerne (*C. insidiosum*), and in lesions in the bark, e.g. of trees such as plum (*Xanthomonas pruni*) and pear (*Erwinia amylovora*). In annual host plants, on the other hand, the bacteria are more dependent on seed transmission regarding normal crop rotation, and, in practice, seed transmission is of particular importance in cultivated annuals such as Phaseolus bean, soybean, pea, cucumber, cabbage, cotton, tomato and cereals. Crop rotation with only one or two years of non-host cropping is effective against many bacterial pathogens while for many soil-invading pathogenic fungi a considerably longer span of years is needed, such as *Fusarium oxysporum* requiring five years or more (Subsection 19.2.1).

The bacteria are often carried on the surface of the seed (cf. Fig. 13.13) but bacteria causing vascular or systemic infections are frequently found in the seed coat or other tissues of the seed. For instance, *Pseudomonas phaseolicola* and *Xanthomonas phaseoli* in Phaseolus bean are found in the hilum region of the seed, into which they penetrate from the vascular system through the funiculus (Section 12.1).

Bacteria borne on the surface of the seed may keep alive for a limited period of time only, perhaps one or two years, whereas bacteria harboured within seed tissues may show a surprising extended longevity, such as *Corynebacterium flaccumfaciens* which has survived 24 years in Phaseolus bean seed kept under laboratory conditions (Subsection 20.2.3).

4.1 CLASSIFICATION

The classification of bacteria is based on morphological and physiological, and especially on biochemical, characters. These include size and shape of the cell, number and position of flagella; gram stain (a basic criterion); growth characters as determined on certain media, production of pigment, growth rate of colonies in response to temperature; biochemical reactions which include proteolysis using casein and gelatin as substrates, hydrolysis of starch, lipolysis, oxidation or fermentation of hexoses, hexose alcohols, mono- and disaccharides and glycosides; nitrate reduction; growth in milk; and many others which can be found in standard texts on bacterial identification. Serological procedures are increasingly used for identification of plant pathogenic bacteria, as well as bacteriophages, following the more extensive use of these procedures in medical bacteriology.

Motility of bacteria can usually be determined under the microscope, if they are mounted in a drop of water, or better, in peptone water. Flagella in motile bacteria can be seen under the highest power when the bacteria have been killed, dried in a thin film and suitably stained (Fig. 4.1).

Some main characters of the five or six genera of plant pathogenic bacteria now generally recognised, are set out in Table 4.1.

4.2 SURVEY OF SEED-BORNE BACTERIA

Common characteristics of plant pathogenic bacteria: rod-shaped, usually actively motile, aerobic or facultatively anaerobic, not forming endospore. According to Bergey's *Manual of Determinative Bacteriology*, 8 ed. (Buchanan and Gibbons, 1974), the five genera concerned are classified as follows, *Pseudomonas* and *Xanthomonas* in the family Pseudomonadaceae, *Agrobacterium* in Rhizobiaceae, *Erwinia* in Enterobacteriaceae, while *Corynebacterium* is kept in a provisional 'Coryneform Group of Bacteria'. However, *Erwinia* sensu lato has been reclassified, distinction being made between two genera, *Erwinia* sensu stricto and *Pectobacterium* (see Subsection 4.2.3). In this survey the genera and species are listed alphabetically.

For detailed bacteriological descriptions of species, reference may be made to Bergey's manual, and to the manuals of Elliott (1951), Dowson (1957) and Stapp (1958).

4.2.1 Agrobacterium

Gram-negative, small short rods, motile (*A. tumefaciens*) or non-motile; when motile, with one polar or lateral flagellum or with up to a few

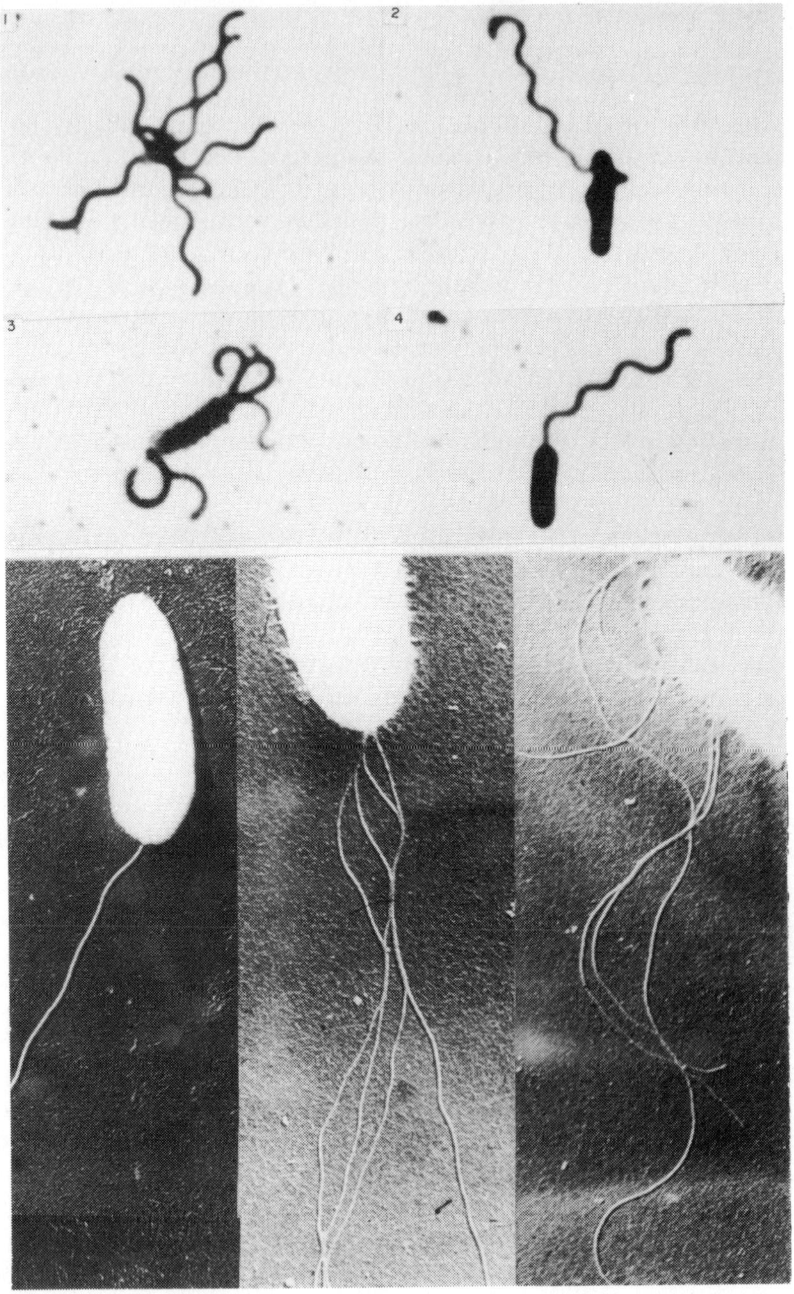

Fig. 4.1. Position of flagella in four principal genera of plant pathogenic bacteria. Above, (1) *Erwinia herbicola;* (2) *Corynebacterium flaccumfaciens*; (3) *Pseudomonas angulata*; and (4) *Xanthomonas campestris*. All ×3,500 (Leifson, 1960). Below, from the left, *Xanthomonas, Pseudomonas, Erwinia*. Electron micrography ×11,000. (Courtesy of E. Hellmers)

Table 4.1 Classification of plant pathogenic bacteria

Genus	*Gram staining*	*General morphology*	*Flagellation*	*Pigmentation on appropriate media*	*Pathogenicity*
Pseudomonas	Negative	Long straight to curved rods, non-sporulating	One to several polar flagella (lophotrichous) or without flagella	Often white or greyish, or yellow-green; often fluorescent	Most species causing parenchymatous infections, some vascular
Xanthomonas	Negative	Cylindrical or ellipsoid rods, non-sporulating	Usually one polar flagellum	Typically yellow	Parenchymatous, vascular infection
Agrobacterium	Negative	Short rods, non-sporulating	Usually one to four flagella peritrichous	Generally non-pigmented but may appear slimy opalescent, white to yellowish	Formation of galls
Erwinia and Pectobacterium	Negative	Short rods with blunt ends, non-sporulating	Several flagella peritrichous	White, grey or yellow	Vascular or parenchymatous infections or soft rots
Corynebacterium	Positive	Slender, straight to curved rods, sometimes club-shaped, non-sporulating	Usually without flagella	Pale yellow to buff or orange	Vascular infections; one species causes galls

peritrichous flagella. Commonly present in soil as inhabitants and in plant roots and stems. Cause hypertrophy and hyperplasia in plants, being similar in some respects to *Rhizobium*.

Agrobacterium tumefaciens

Crown gall

Seed-transmitted, stated to occur in hops in the U.S.S.R. (Komarova, 1959) and to be seed-borne by contamination in *Prunus* sp. in the U.S.A. (Cochran, Cooper and Blodgett, 1961). These scattered records may indicate that seed transmission is of minor importance. The organism is a persistent soil inhabitant.

Infection results in galls developing on stems, leaves or root, their size and structure depending partly on the host plant involved; older galls may decay from outside. Host range extremely wide, including hundreds of species belonging to about fifty families, CMI Description 42.

Distribution practically universal: North and South America, Europe, Africa, Asia, Australia and New Zealand.

CMI Distribution Map No. 137.

4.2.2 Corynebacterium

Gram-positive, straight to slightly curved rods, often club-formed, and often flagellate but technically lacking flagella. Two species, *C. fascians* and *C. rathayi*, are non-motile (Dowson, 1957). Cause often vascular infections, or galls (*C. fascians*).

Corynebacterium agropyri

Yellow gum disease

Possibly identical with *C. rathayi*.

According to Elliott (1951) massive contamination on the outer surface of the glumes of *Agropyron smithii*; normal seeds are rarely produced from infected inflorescence.

Infection vascular, leads to dwarfing and profuse yellow ooze in layers between the stem and the upper sheath.

Distribution: U.S.A. and possibly China.

Corynebacterium betae

Silvering of red beet

One per cent of seeds from infected plants have been infected and producing plants with systemic symptoms (Keyworth and Howell, 1961). Seed treatment with streptomycin gives almost complete control.

Vascular infection causing veinal or uniform silvering of the leaves followed by wilting and death.

Distribution: Eire and United Kingdom.

Corynebacterium fascians

Fasciation disease, witches' broom, leafy gall

Seed-borne in *Schizanthus retusus* but seed transmission not established (Lacey, 1939), and seed transmitted in nasturtium, *Tropaeolum majus*, infected seeds producing typical 'cauliflower growth' in sterile soil (Baker, 1950). Eight out of 13 cultivars had infected seeds. Hot water treatment effective. The pathogen is also soil-borne.

Infection causing fasciation of shoots and leafy galls. Wide host range, more than 35 host plant species. CMI Description 121.

Distribution: North America, Europe.

CMI Distribution Map No. 426.

Corynebacterium flaccumfaciens

Bacterial wilt of beans

Seed-transmitted externally or internally in soybean, *Glycine max* (Leonard, 1924; Burkholder, 1926; Dunleavy, 1962) and in Phaseolus bean (Leonard, 1924; Burkholder, 1926); *C. flaccumfaciens* var. *aurantiacum* also seed-borne in Phaseolus bean (Rickard, 1964). Wilson and Dunleavy (1964) described a destructive disease in Phaseolus bean caused by another as yet unnamed seed-borne species of *Corynebacterium*.

Infection vascular, causing wilt and stunt, and often severely reducing yield. CMI Description 43.

Distribution: North and Central America, Europe (Bulgaria, Hungary, Yugoslavia), Australia.

CMI Distribution Map No. 370.

Corynebacterium insidiosum

Bacterial wilt of lucerne

Seed-transmitted in lucerne, *Medicago sativa*, both by seed and by debris mixed with seed (Cormack and Moffart, 1955; Cormack, 1961; Golenia, 1965). Important quarantine object.

Vascular infection causing stunt and wilt, pale foliage, leaves roll upward, diseased plants may be killed the second year after infection. Very destructive but resistant cultivars are widely grown. CMI Description 13.

Distribution: North America, Europe, Asia.

CMI Distribution Map No. 67.

Corynebacterium iranicum

In wheat

Yellow bacterial slime developed from the ovaries, abortion of the grain (Scharif, 1961).

Profuse honey-yellow ooze in spikelets which are stuck together by the slimy secretion.

Distribution: Iran.

Corynebacterium michiganense

Bacterial canker of tomato

Seed-transmitted in tomato (Bryan, 1928, 1930; Blood, 1933, 1937; Ark, 1944*b*; W. C. Moore, 1947; Grogan and Kendrick, 1953; Israilski and Karpovskaya, 1957; Robbs, 1960; Volcani and Henis, 1964; Strider and Konsler, 1965; Layne and Rainforth, 1966; Basu, 1966; Patiño-Méndez, 1967; Ercolani, 1968; Thyr, 1969; Strider, 1969; Thyr *et al.*, 1973). Control by fermentation of the seed and fruit pulp at high acidity (Bryan, 1930; Blood, 1937). Important quarantine object, cf. Strider (1969).

Vascular infection. Symptoms on young seedlings, whitish pimple-like spots, cankers on the hypocotyl. Yellow-brown stripes on stems and petioles, mottling and malformation of fruits, blighting of leaves; the plant may be killed. CMI Description 19.

Distribution: North and South America, Europe, South Africa and Morocco, Asia, Australia, New Zealand.

CMI Distribution Map No. 26.

Corynebacterium nebraskense

Nebraska leaf freckle and wilt of maize

The pathogen invades the kernel of maize via the vascular system and contaminates the seed surface from the inner husk. Infected kernels have been found to give about 1 per cent transmission (Schuster *et al.*, 1973). Seed treatment in mercuric chloride recommended.

Vascular infection producing leaf freckle and wilt, symptomatology similar to Stewart's disease, *Erwinia stewartii*, but no relationship with flea beetles as in this latter disease established. Primary source of inoculum is maize stubble.

Distribution: U.S.A. (Nebraska).

Corynebacterium rathayi

Yellow slime disease, Rathay's disease

Seed-transmitted in cocksfoot, *Dactylis glomerata* (Smith, 1914; Hardison, 1945; Johnston, 1956; Williams and Taylor, 1957; Skou, 1965; Schlösser, 1969) and possibly in *Festuca rubra* var. *fallax* (Noble, 1957). 10–50 per cent of seed lots produced in Denmark infected.

Probably vascular infection. The main damage is caused by profuse development of yellow bacterial slime that may cover the uppermost leaves, upper parts of the stem and parts of the inflorescence (Fig. 4.2). Under dry conditions the slime becomes a hardened gum, and seeds from infected plants, 'bacterial grains', may be recognised in seed testing because of their characteristic appearance, involving a varnish-like cover of bacteria. Heavy infections may lead to sterility.

Distribution: Europe, North America, New Zealand.

CMI Distribution Map No. 156.

Corynebacterium tritici

Yellow slime disease, yellow ear rot

Constitutes a component in the disease complex 'tundu' or 'tannau disease' in India. Seed transmitted in wheat (Hutchinson, 1917; Fahmy and Mikhail, 1925; Carne, 1926; Cheo, 1946; Vasudeva and Higorani, 1952*b*; Sabet, 1954*a*, 1954*b*; Swarup and Singh, 1963; Mathur and Ahmad, 1964; Amani, 1969). May keep alive in the galls of the nematode, *Anguina tritici* for at least five years (Mathur and Ahmad, 1964).

Infection probably vascular, similar to that produced by *C. rathayi*, but the main damage is caused by the exudation of a yellow slime or gum over affected parts, some of which may become distorted because of interference with the growth of the plant. Infection closely associated with earcockle, *Anguina tritici* (Gupta and Swarup, 1972).

Distribution: Europe (Cyprus), Africa (Egypt), Asia, Australia.

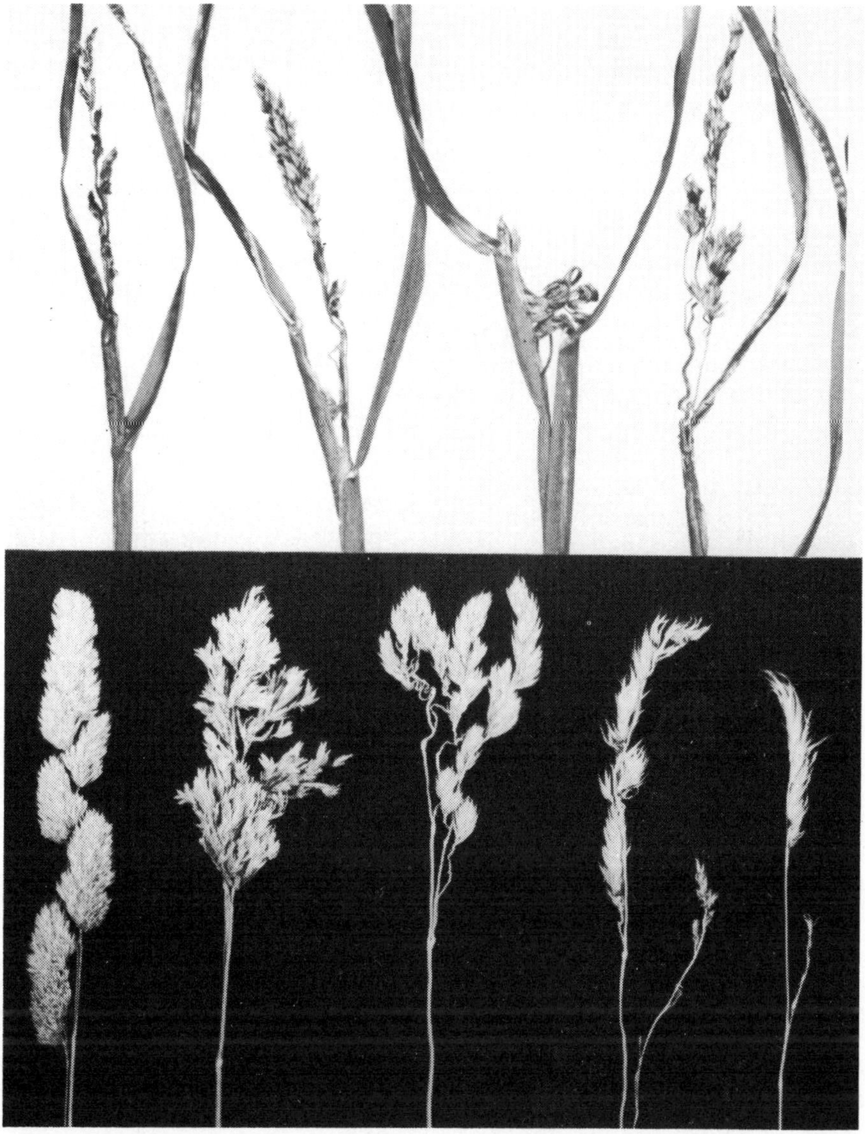

Fig. 4.2 Cocksfoot infected by yellow slime disease, *Corynebacterium rathayi*. Spike at left in the lower figure, healthy. (Figure above, courtesy of E. Hellmers; figure below, courtesy of Margot E. H. Johnston)

4.2.3 Erwinia

Gram-negative short rods with blunt ends, motile by a few peritrichous flagella (one exception). Waldee (1945) proposed to refer species that are capable of producing pectolytic enzymes to a separate genus, *Pectobac-*

terium, and this has been accepted by some bacteriologists while others maintain the original conception of *Erwinia* in its broader sense at present. The distinction between the two genera has recently been discussed by Fonnesbech (1974), and has been accepted in this book.

The species of the genus cause parenchymatous necroses, vascular diseases and soft rots.

Erwinia herbicola

(Syn. *E. lathyri*)

A very common seed-borne yellow-pigmented saprophyte which may complicate identification of pathogenic bacteria (see Graham and Hodgkiss, 1967; cf. Noble and Richardson, 1968). Among many organisms reported as seed-borne but confused with this species are *E. ananas*, *E. milletiae* and *Xanthomonas rubrisorghi*.

'Probably the most frequently isolated of the yellow secondary organisms that occur in and around diseased plant tissue', quoted from CMI Description 232.

Distribution: probably world-wide.

Erwinia stewartii

(Syn. *Xanthomonas stewartii*)

Bacterial wilt, Stewart's disease

Known from the past century to be seed-transmitted in maize but flea beetles are important in dissemination and in accomplishing transmission from seed, see Section 16.1 (Stewart, 1897; Smith, 1914; Ivanoff, 1933; Elliott, 1935; Robert, Jenkins and Findley, 1953). Difficult to control by chemical seed treatment (Williams, 1957).

Vascular infection, leading to dwarfing, water-soaked pale stripes on the leaves, the plant may be killed in the seedling stage; later plants may exhibit wilting or stunting, symptoms that may be confused with drought damage or nutritional injury. Cross-section of the stem may reveal yellow bacterial ooze from the vascular bundles. Premature development of the male inflorescence which withers. CMI Description 123.

Distribution: North and South America, Europe, Asia.

CMI Distribution Map No. 41.

4.2.4 Pectobacterium

The genus has the main characters described for *Erwinia* but is different particularly by its capacity of producing pectolytic enzymes.

Pectobacterium carotovorum

(Syn. *Erwinia carotovora*)

Soft rot, crater rot

Seed transmission reported in celery by Elliott (1951) who recommends hot water treatment unless seed has been stored for two years, Raciu, Mihalea and Păun (1954) recommend hot water treatment for seed of coriander, and Sabet (1954*d*) described a stalk and ear rot of maize which reduced grain production of maize in Egypt.

Parenchymatous infection of fleshy tissues in a wide range of host plants, in which the organism causes a rapidly progressive soft wet rot.

Distribution: very probably in all parts of the world, reports from North and South America, Europe, Africa, Asia, Australia.

Pectobacterium carotovorum var. **atrosepticum**
(Syn. *Erwinia atroseptica*)
Bacterial soft rot, black leg

The cause of black leg disease of potato, but Pape (1964) reports that the pathogen is serious in seed crops of larkspur, *Delphinium ajacis*, in North America; hot water seed treatment advised.

Parenchymatous infection in larkspur, causing black discoloration and softening of the stem, the plant may be stunted and killed back.

Distribution: the pathogen in all parts of the world wherever the potato is grown; the disease in larkspur prevalent in North America.

4.2.5 Pseudomonas

Gram-negative, long straight or curved rods with one or several polar flagella. Often a fluorescent yellowish-green or bluish, diffusible pigment is produced. Most diseases caused by members of this genus are parenchymatous, but some are systemic, and one of the species, *P. savastanoi*, causes hypertrophy. Some saprophytic species are common soil inhabitants and may contaminate seeds.

Pseudomonas angulata
Angular leaf spot of tobacco, black fire

Seed transmission reported from the U.S.A. (Fromme and Murray, 1919; Fromme, 1920; Fromme and Wingard, 1922). Hopkins (1956) reported that seed transmission in tobacco is important in Rhodesia. Silver nitrate used for seed treatment. Also soil-borne.

Infection parenchymatous. Small blackish-brown, angular leaf spots, the central, lighter coloured portions often fall out, and the leaves may assume a ragged appearance.

Distribution: very common in almost all tobacco growing countries. North America, Africa, Asia.

CMI Distribution Map No. 321.

Pseudomonas apii
Bacterial blight of celery

Chupp and Sherf (1960) indicate that the bacterium lives on and in the seed for one year after seed harvest.

Infection parenchymatous, causing yellow-brown leaf spots, rarely blight.

Distribution: North America.

Pseudomonas aptata

Bacterial blight, black streak, black spot of beet

Seed transmission in sugarbeet (Hirata, 1928; Carsner, 1944; Ark and Leach, 1946). Seed treatment by mercury compounds in California.

Infection parenchymatous, water-soaked light yellow leaf spots on seedlings, later black leaf spots, streaks on stem and seed stalks.

Distribution: North America, Europe, Asia (Korea), Australia.

Pseudomonas atrofaciens

Basal glume rot, spikelet rot

Affecting grains of barley and wheat (McCulloch, 1920; Elliott, 1951). Seed treatment advised (Noble, 1933; Staněk and Ujević, 1959).

Parenchymatous infection causing yellow-brown leaf spots, diseased grains brown to charcoal black.

Distribution: North America, Europe, South Africa, Australia, New Zealand.

Pseudomonas cichorii

Bacterial leaf blight

Seed transmission in lettuce (Krŭstev, 1965); seed treatment advised in Brazil (Freire, 1954).

Parenchymatous infection causing dark-brown leaf spots and leaf blight in lettuce.

Distribution: North and South America, Europe, Asia (Formosa).

Pseudomonas coronafaciens

Halo blight

Evidence of seed transmission in oats (Elliott, 1920, 1927; Sampson and Davies, 1923). Seed treatment by mercurials unsuccessful but streptomycin effective (Griffiths and Peregrine, 1960). *P. coronafaciens* var. *atropurpurea* suspected to be seed-transmitted in *Agropyron inerme* (Reddy and Godkin, 1923).

Parenchymatous infection. Leaf spots on the first leaves of seedlings. Light green, later brown leaf spots with pale green to yellowish halo-like margins, coalescence forming general leaf blight. Glumes may be infected. CMI Description 235.

Distribution: North America, South America (Argentina), Europe, possibly New Zealand, probably in most oat-growing areas.

CMI Distribution Map No. 356.

Pseudomonas cumini

Very common on seeds of *Cuminum cyminum* in Bulgaria, seeds usually 100 per cent infected (Kovačevski, personal communication, 1955).

Parenchymatous infection, all parts above ground infected, water-soaked, later greasy brown spots, bacterial ooze drying to a white crust; fruits turn dark and shrivel.

Distribution: Europe (Bulgaria).

Pseudomonas glycinea

Bacterial blight

Commonly seed-transmitted in soybean (Coerper, 1919; Kendrick and Gardner, 1921; Kent, 1945; Graham, 1953*a*; Krasnova, 1963; Muras, 1964; Nicholson and Sinclair, 1971; Parashar and Leben, 1972; White, Nicholson and Sinclair, 1972; Nicholson, Sinclair and Joshi, 1973; Nicholson, Sinclair and White, 1973; Leben, 1975). The seed may be completely covered by slimy bacterial growth (Elliott, 1951).

The disease prevails in cooler climatic conditions than bacterial pustule (*Xanthomonas phaseoli* var. *sojense*).

Parenchymatous infection, the seedlings may show the first symptoms: brown spots in cotyledons, particularly at the edge, seedlings may be killed. Angular, first water-soaked, later yellow to brown or black spots on leaves, stems and pods which may be severely affected.

Distribution: North and South America, Europe, Africa, Asia, Australia.

Pseudomonas lachrymans

Angular leaf spot of cucumber (Figs 4.3–4.5)

Seed transmitted in cucumber (Carsner, 1918; Gilbert and Gardner, 1918; Meier and Link, 1922; Gorlenko and Voronkevich, 1946; Wiles, 1951; Wiles and Walker, 1951, 1952; Van Gundy and Walker, 1957; Naumann, 1963; Walker, Chand and Wade, 1963). Antibiotic seed treatment (streptomycin) recommended (Bushkova, 1962). Ten of 13 seed lots grown in California and Oregon were found to carry the pathogen (Van Gundy and Walker, 1957). Seed treatment by mercuric chloride.

Parenchymatous infection. Often brown necroses on cotyledons. Water-soaked spots on leaves, stems and fruits, often with shot-hole effect, that may give the leaves a ragged appearance; in fruits the water-soaked areas may dry leaving chalky-white centres. Profuse bacterial exudates in the lesions. Premature fruits, when attacked, may fall off, or a severe fruit rot may develop and bacteria penetrate the seeds through the funiculus or contaminate them during seed extraction (Wiles, 1951). CMI Description 124.

Distribution: North and Central America, Europe, Asia, Australia.

CMI Distribution Map No. 355.

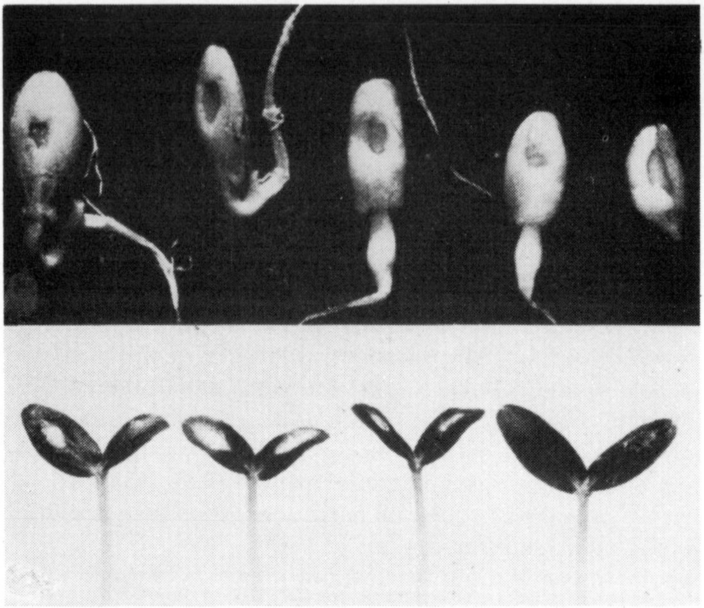

Fig. 4.3. Above, lesions, caused by *Pseudomonas lachrymans,* on cotyledons of infected seed of cucumber, in an early stage of germination. Below, lesions on cotyledons at a later stage. Note the tissue of some of the lesions has dropped out. Healthy seedling on right. (Wiles and Walker, 1951)

Pseudomonas maculicola

Bacterial leaf spot

Seed transmission in cabbage and cauliflower reported by Bakker (1951) and Kvíčala (1964). Seed treatment by organic mercury or by hot water shown to be effective (Bakker, 1951).

Infection parenchymatous, producing many small dark leaf spots, particularly in cauliflower; the leaf spots may coalesce forming larger areas.

Distribution: North America, Europe, Africa, Australia.

CMI Distribution Map No. 312.

Pseudomonas martyniae

Infections cause rot in developing capsules of *Proboscidea jussieui* (syn. *Martynia louisiana*) (Elliott, 1924); it is advised to use seed only from healthy pods (Elliott, 1924, 1951). However, actual seed transmission seems not to have been proved.

Systemic infection; round angular translucent leaf spots coalescing to form irregular light-brown patches; infected fruits assume a water-soaked appearance, later turn brown, shrivel and die. Plants may be killed.

Distribution: North America (Kansas).

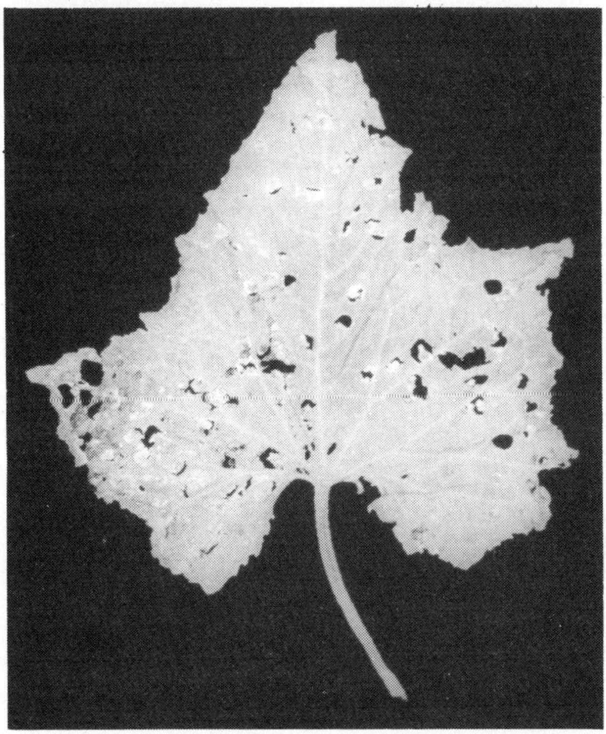

Fig. 4.4. Leaf spots and shot-hole effect, caused by *Pseudomonas lachrymans* in cucumber leaf. (Courtesy of E. Hellmers)

Pseudomonas mellea

Wisconsin leaf spot of tobacco

Seed-borne in tobacco (Gigante, 1950; Elliott, 1951). Seed treatment recommended (Hopkins, 1956).

Infection parenchymatous. The first symptoms appear on the seedlings as brown, angular spots surrounded by an ill-defined halo, later larger spots. Under favourable conditions the plants may be severely injured or killed.

Distribution: North America, Africa (Rhodesia), Europe.

Pseudomonas oryzicola

Brusone, glume blotch of rice, bacterial sheath brown rot

Podharadszky (1955) reported seed transmission on rice in Europe. In Hungary and Australia the pathogen has caused sterility (Klement, 1955; Cother, 1974).

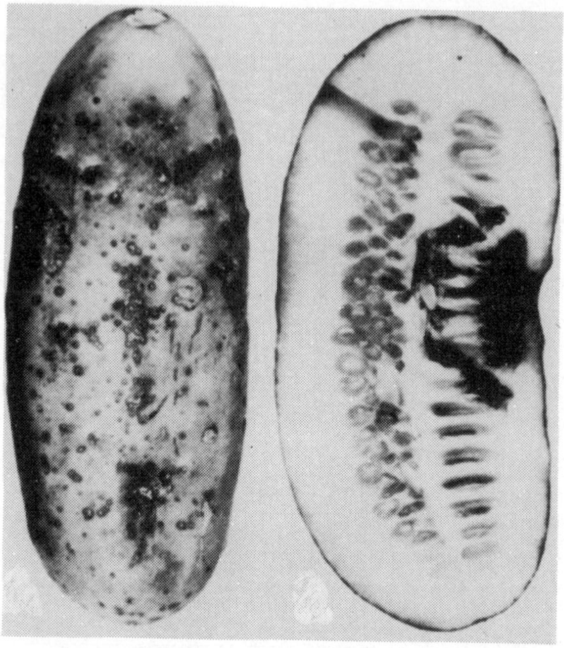

Fig. 4.5. Left, small water-soaked lesions following stomatal invasion of a cucumber fruit by *Pseudomonas lachrymans*. Right, longitudinal section of infected fruit showing restricted invasion of the hypodermal area and extensive breakdown of the middle and inner mesocarp. (Wiles and Walker, 1951)

Parenchymatous infection causing greyish-green to dark brown spots on sheath of the panicle, exceptionally on leaves. Panicles become infected, as they emerge. CMI Description 236.

Distribution: Europe (Hungary), Asia, Australia.

CMI Distribution Map No. 508.

Pseudomonas phaseolicola

Halo blight, grease spot of bean

Seed transmission very important in Phaseolus bean, taking place both from surface-borne contamination and from infected seeds (Burkholder, 1926, 1930; Higgins, 1930; Katznelson, Sutton and Bayley, 1954; Walker and Patel, 1964*a*; Guthrie, Fenwick and Huber, 1966; Oshima, Dickens and Counter, 1966; Grogan and Kimble, 1967; Guthrie, 1970; Taylor, 1970). In England commercial seed stocks have been found to contain 0.1–1 per cent infection compared with 1–10 per cent infection in seed from heavily infected crops (Taylor, 1970). Seed treatment with antibiotics (streptomycin) effective. Seed health certification schemes in operation in some regions, e.g. Iowa, U.S.A. (Guthrie *et al.*, 1975), Canada and Australia.

Infection systemic. Seedlings may show brown lesions on the cotyledons. Milder infections cause water-soaked or greasy leaf spots and mosaic-like symptoms (Figs 4.6 and 4.7), dark green spots on the pods; in mature leaves the spots have a characteristic light green halo, severe infections lead to wilt or stunt, plants may be killed. Round dark green, later reddish-brown and greasy spots on pods. The disease is very destructive. CMI Description 45.

Distribution: North and South America, Europe, Africa, Asia, Australia, New Zealand.

CMI Distribution Map No. 85.

Fig. 4.6. Leaf of Phaseolus bean infected by halo blight, *Pseudomonas phaseolicola*. (Courtesy of E. Hellmers)

Fig. 4.7. Leaf of Phaseolus bean showing mottling due to vascular infection of *Pseudomonas phaseolicola*. (Burkholder, 1926)

Pseudomonas pisi

Bacterial blight of pea

Frequently seed-transmitted, bacterial slime on the surface of pea seed or infection internal through the micropyle (Jennison, 1921; Skoric, 1927; Sutton and Katznelson, 1953; Hagedorn and Wade, 1964; Close, 1966; Watson and Dye, 1971). Seed health certification schemes in operation in New Zealand (Close, 1966). Important quarantine object.

Infection parenchymatous, green to brown water-soaked spots on all parts of the plant, coalescing under humid conditions. Badly infected pods may be covered by bacterial slime, young infected pods may dry. CMI Description 126.

Distribution: North and South America, Europe, Africa, Asia, Australia.

CMI Distribution Map No. 253.

Pseudomonas polycolor

Philippine leaf spot of tobacco

Reported as seed-borne from the Philippines (Clara, 1930).

Infection parenchymatous, causes a wet rot of seedlings with opaque or bleached white spots; white to brown leaf spots. Plants may be killed.

Seed treatment with silver nitrate recommended.

Distribution: The Philippines.

Pseudomonas punctulans
(Syn. *P. tomato*)
Tomato speck, bacterial leaf spot

In New Zealand the disease first appeared in 1944 in crop grown from imported seed, and young seedlings grown in boxes displayed extensive leaf infections (Reid, 1948).

Infection parenchymatous, yellowish-brown leaf spots or specks, with papery centres; severe stunting. On fruits minute, black slightly raised lesions.

Distribution: North America, Europe, New Zealand.

Pseudomonas solanacearum
(Syn. *Xanthomonas solanacearum*)
Brown rot, southern bacterial blight

Reported from the U.S.S.R. to be occasionally seed-borne in soybean (Krasnova, 1963; Muras, 1964). Possibly also seed-borne in *Capsicum* (Radulescu, according to Noble and Richardson, 1968), groundnut, *Arachis hypogaea* (Palm, 1922), and nasturtium, *Tropaeolum majus* (Wager, according to Noble and Richardson, 1968).

Vascular infection causing dwarfing and an often serious wilt in a wide host range, particularly within *Solanaceae*. The vascular bundles are stained brown. CMI Description 15.

Distribution: practically universal, in warmer regions, North and South America, Europe, Africa, Asia, Australia, New Zealand.

CMI Distribution Map No. 138.

Pseudomonas striafaciens
Bacterial stripe blight of oats

The glumes become infected, young seedlings are attacked but there seems to be no explicit demonstration of seed transmission which, however, seems likely on circumstantial evidence (Elliott, 1927, 1951).

Parenchymatous infection, sunken water-soaked, later rusty brown leaf spots, coalescing into long water-soaked stripes without any halo-like border. CMI Description 238.

Distribution: North America, South America (Argentina, Colombia), Europe (Germany, U.S.S.R.), Africa (Rhodesia), Asia (Japan, Korea), Australia.

CMI Distribution Map No. 363.

Pseudomonas syringae

Bacterial brown spot, bacterial canker of stone fruits, lilac blight

Seed transmission in *Hibiscus* sp. (Nakada and Takimoto, 1923), lima bean, *Phaseolus lunatus* (Thaung and Walker, 1957), French bean, *Phaseolus vulgaris* (Harrison and Freeman, 1965; Hoitink and Hagedorn, 1966; Hoitink, Hagedorn and McCoy, 1968), *Prunus* sp. by contamination (Cochran, Cooper and Blodgett, 1961), grain sorghum, *Sorghum vulgare* (Kendrick, 1926; Leukel and Martin, 1953; Tarr, 1962), catjang pea, *Vigna cylindrica* (Gardner and Kendrick, 1922, 1925; Gardner, 1925; Hoffmaster, 1944), and cowpea, *V. sinensis* (Gardner, 1925). Seed transmission in *Hibiscus* prevented by seed treatment with mercuric chloride or hot water.

Parenchymatous infection. In *Hibiscus* infection of cotyledons, severe blight may kill the plants. In clover, cowpea and sorghum the pathogen causes leaf spots, in cowpea shrivelled seeds, a reddish-brown leaf blotch in Phaseolus bean, and on the seed brown discoloration at hilum. Gum exudation on trunks, limbs, twigs and fruits of stone fruits. CMI Description 46.

Distribution: North and South America, Europe, Africa, Asia (Japan), Australia, New Zealand.

CMI Distribution Map No. 336.

Pseudomonas tabaci

Wildfire of tobacco

Graham (1953*a*) established that the pathogen may infect seeds of soybean and remain viable up to 18 months. Tobacco seeds may be infected, although seldom, and the pathogen may or may not survive for two years (Fromme, 1920; Clinton and McCormick, 1921; Wolf, 1922; Anderson, 1924; Johnson and Murvin, 1925; Valleau, Johnson and Diachun, 1943; Dowson, 1957). Silver nitrate seed treatment effective in eliminating infection in tobacco seed, in contrast to mercurial (Hopkins, 1956). Valleau *et al.* (1943) suggested hot water treatment evidently efficient; Dowson (1957) recommended seed treatment with mercuric chloride or silver nitrate. Also seed-transmitted in soybean (Graham, 1953*a*).

Infection parenchymatous. In crowded tobacco seedlings a wet rot stage may occur. Brown leaf spots with yellow-green halo; reddish-brown leaf spots coalescing into irregular blotches. Small spots on flowers and capsules. The most serious of bacterioses occurring in tobacco. In soybean *P. tabaci* causes secondary infection, producing brown leaf spots, after *Xanthomonas phaseoli* var. *sojense*. CMI Description 129.

Distribution: North, Central and South America, Europe, Africa, Asia, Australia, New Zealand.

CMI Distribution Map No. 293.

Pseudomonas tagetis

Bacterial leaf spot of African marigold

Seed transmission in African marigold, *Tagetes erecta* (Hellmers, 1955).

Parenchymatous infection. Seedlings show symptoms, necroses of cotyledons, and the seedlings may be killed. Angular, often oblong brownish-black leaf spots. Necroses develop from leaf margins thus indicating infection through hydathodes (Fig. 4.8).

Distribution: Europe (Denmark, Holland), New Zealand.

Fig. 4.8. Leaves of African marigold, *Tagetes erecta,* infected by *Pseudomonas tagetis.* (Courtesy of E. Hellmers)

4.2.6 Xanthomonas

Gram-negative cylindrical-elliptical rods with rounded ends, motile, usually with one polar flagellum; slow-growing yellow colonies. Most species produce vascular or systemic infection, some are parenchymatous. They occur rarely in soil.

Xanthomonas alfalfae

Bacterial leaf and stem spot, damping-off

Seed transmission in lucerne suggested by circumstantial evidence (Noble and Richardson, 1968).

Parenchymatous infection, water-soaked, later dark brown leaf spots coalescing forming blotches.

Distribution: North America, Asia (India).

Xanthomonas argemoneae

Bacterial blight, wilt

Seed transmission in Mexican poppy, *Argemone mexicana* (Srinivasan, Patel and Thirumalachar, 1961).

Vascular infection, V-shaped lesions extending into the leaves through the hydathodes. Petals, stamens and the stigmatic lobes may also have bacterial lesions.

Distribution: India.

Xanthomonas axonopodis

Gummosis of fodder grasses

Transmitted by seed in imperial grass, *Axonopus scoparius* (Castano and Thurston, 1964).

Vascular infection causing thin stems and few chlorotic leaves, contorted side shoots; the vascular bundles may be stained yellow or red, plants are killed, up to 75 per cent reduction in yield.

Distribution: Colombia.

Xanthomonas campestris

Black rot of crucifers

Seed transmission of great importance in cabbage (Harding, Stewart and Prucha, 1904; Walker and Tisdale, 1920; Monteith, 1921; Walker, 1934, 1941; Cook, 1952; Cook, Larson and Walker, 1952; Eddins, 1953; Klisiewicz and Pound, 1960; Smith, 1962; Shackleton, 1962; Pinto-Ganhão, 1962; Hunter, Abawi and Becker, 1975). Also transmitted in wallflowers, *Cheiranthus cheiri* (Pirone, 1970) and in garden stocks, *Matthiola incana* (Wilson, 1942). Hot water treatment commonly recommended. Seed health certification scheme recently organised in the U.S.A. because of country-wide epidemics in 1973 causing heavy national losses. Considered as quarantine object.

Vascular infection. Seedlings from infected seed may show typical necroses in cotyledons. Yellowing of leaves and black-brown discoloration of the leaf veins, V-shaped marginal lesions characteristic, indicating infection through hydathodes. Cross-section of stem shows a blackened ring of vascular tissue. CMI Description 47.

Distribution: North and South America, Europe, Africa, Asia, Australia, Oceania.

CMI Distribution Map No. 136.

Xanthomonas carotae

Bacterial blight, root scab

Seed transmission in carrot (Ark and Gardner, 1944; Watson, 1948; Chumaevskaya and Gorlenko, 1960; Burykhina and Buyanova, 1962). Spread in the U.S.S.R. by seeds. Hot water treatment effective.

Infection parenchymatous? Water-soaked, yellow to dark brown spots on leaves and stems, older leaves may dry up, floral parts may be heavily infected and the entire umbel killed before emergence, heavy gum exudates may develop. Serious losses in seed crops.

Distribution: North America, Europe (Hungary, U.S.S.R.), South Africa, Asia, Australia.

Xanthomonas citri

Citrus canker

Fruits and seeds may be infected and may carry the disease to new areas. Seed-borne infection controlled by seed treatment in hydrogen peroxide (Limber and Frink, 1953). Important quarantine object.

Parenchymatous infection, small oily yellowish-brown leaf spots, later rupturing giving a rough corky appearance, defoliation may follow, also twigs, limbs and fruits may be severely affected, fruits become scabby and are dropped. Heavy qualitative and quantitative losses in yield. CMI Description 11.

Distribution: Africa, Asia, Australia, Oceania. Complete eradication by cutting down trees and burning has been obtained in the U.S.A., South Africa, Mozambique and Australia (Northern Territory).

CMI Distribution Map No. 11.

Xanthomonas holcicola

Bacterial leaf streak

Seed-transmitted in sorghum (Elliott, 1930; Pastushenko, 1962; Ramakrishnan, 1963).

Infection causing narrow water-soaked, dark brown leaf streaks, often with oval tan centres, narrow dark margins, coalescing to irregular blotches (Fig. 4.9); abundant light yellow exudates drying to thin white scales.

Distribution: North America, South America (Argentina), Europe (Roumania, U.S.S.R.), Africa, Asia, Australia, New Zealand.

CMI Distribution Map No. 395.

Xanthomonas incanae

Bacterial blight of stocks, black rot, seedling wilt

The pathogen is carried on and within the seed (Kendrick and Baker, 1942; Anonymous, 1948*b*; Baker, 1956; Anonymous, 1963). Hot water seed treatment recommended. Organised production of disease-free seed in California (Baker, 1956).

Vascular infection, causing water-soaked wilting of young seedlings, stunting of older plants, dark sunken lesions on the main stem. Serious losses in seed crops. Very similar to *X. campestris* but differs from this in pathogenicity being confined to *Matthiola*.

Distribution: North America, Australia.

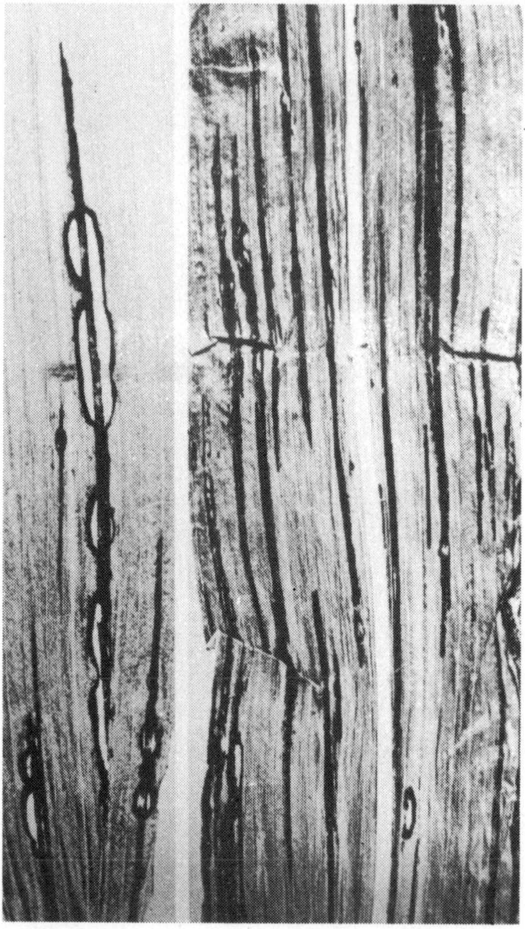

Fig. 4.9. Leaves of *Sorghum vulgare,* infected by bacterial streak disease, *Xanthomonas holcicola.* (Elliott, 1930)

Xanthomonas itoana

Black rot of rice grains

Seed-borne in rice, but transmission not established (Tochianai, 1932; Elliott, 1951; Ou, 1972). In the seed the pathogen penetrates the aleurone layer and upper part of the endosperm, the infected tissues turn black. The grains are particularly attacked at the milky stage, through wounds (Iwadara, 1936).

Distribution: Asia.

Xanthomonas juglandis

Bacterial blight of walnuts

Seed infection deep-seeded and seed transmission established.

(Alexandri *et al.*, 1965). Pollen transmission a possibility (Ercolani, 1962).

Vascular infection, irregularly shaped, brown to black leaf spots, young fruits fall prematurely, full-sized fruits partly or completely destroyed, blackened. Heavy losses. CMI Description 130.

Distribution: North and South America, Europe, South Africa, Asia, Australia, New Zealand.

CMI Distribution Map No. 133.

Xanthomonas lespedezae

Bacterial wilt of Lespedeza

Demonstrated to be carried both on and in seed of *Lespedeza stipulacea* and may thus be introduced into new areas (Ayers, Lefebvre and Johnson, 1939).

Vascular infection, water-soaked dark leaf spots, desiccation and curling of foliage, stems may crack and exude drops of bacterial slime. General wilting and death of plants.

Distribution: North America.

Xanthomonas malvacearum

Black arm, bacterial blight, gummosis, angular leaf spot of cotton

Seed transmission very important in cotton (Archibald, 1927; Tennyson, 1936; Bain, 1939; Brown, 1942; Taar, 1961; Brinkerhoff and Hunter, 1963; Hunter and Brinkerhoff, 1964; Dransfield, 1964; Hayword, 1964; Schnathorst, 1964, 1966, 1968, 1969; Nayudu, 1970). The organism very persistent in seed. Eradication programme successful in California (Schnathorst, 1966).

Parenchymatous infection, infected seeds may fail to germinate, in seedlings the cotyledons may show marginal invasion and primary leaves angular spots. Rarely infection penetrates veins with infection of leaves and water-soaked, then dark brown, 'angular' spots, coalescing to form large blotches (Fig. 4.10). Stem infections are sunken and turn black ('black arm'). Defoliation, shedding of flowers, premature drop of bolls, larger bolls and bracts may be attacked and bear large dark, water-soaked spots, the lint is often destroyed. Bacterial ooze on water-soaked lesions drying to form pale yellow scales. Devastating in East Africa. CMI Description 12.

Distribution: probably in all cotton-growing countries.

CMI Distribution Map No. 57.

Xanthomonas oryzae

(Syn. *X. kresek*)

Bacterial blight of rice, kresek disease of rice

There are conflicting views on the importance of seed transmission, although survival of the organism in the seed from season to season is well

Fig. 4.10. Angular leaf spot of cotton, *Xanthomonas malvacearum*. Top, typical symptoms; bottom, necroses caused by the bacteria moving through the vascular system from a stem lesion formed near to the meristem. (Courtesy of F. T. Last)

established. IRRI (1964) and Eamchit, Ou and Exconde (1969) found that inoculum in seed decreases during grain storage, and generally dies in a few days during soaking of seed for sowing, probably due to the rapid increase of bacteriophages (Kauffman and Reddy, 1975). Others share the view that seed transmission does occur, taking place particularly from infections of the glumes (Wakimoto, 1955; Fang, Lin and Chiu, 1956; Tagami *et al.*, 1963), and that seed transmission, as in India, may be important in the epidemiology of the disease under humid conditions (Srivastava and Rao, 1963, 1964). Seed lots may be infected up to 100 per cent. Selective media may be used for detection (Supriaman and Tantera, 1972). Should be considered as important in quarantine.

Parenchymatous and vascular infections. Seedlings show water-soaked leaf spots or streaks at the margin, leaves turn yellow and wither; later in the crop yellowish or whitish blotches or stripes, bacterial exudates are common, hardening to resinous granules. 'Kresek' is caused by systemic infection (Goto, 1964) and is common in many tropical countries; the leaves become greyish-green, roll up along the midrib and wither, the bacteria spread through the xylem vessels to the growing point of young plants, infect the base of developing leaves and, as a result, the plant dies. CMI Description 239.

Distribution: Asia, Australia.

CMI Distribution Map No. 304.

Xanthomonas panici

Bacterial streak of millet

The pathogen prevents seed formation (Klement, 1954), panicles are infected but transmission through seed seems not to have been established (Potlaĭchuk, 1962).

Narrow light green water-soaked, subsequently brown streaks on leaves, sheaths and culms, bacterial ooze under humid conditions.

Distribution: North America, Europe.

Xanthomonas papavericola

Bacterial blight of poppy

Internal infection of seeds which become discoloured and misshapen; seeds the chief source of infection (Christoff, 1931, 1933; Mráz, 1959).

Vascular infection. Dark, circular, often zonate leaf spots coalescing forming large black blotches, elongated black stem lesions, all floral parts are attacked and may be partly or entirely blackened, sepals and subsequently petals are attacked and further development of flowers is then stopped. Black conspicuous water-soaked spots on capsules. Many plants may be killed because of girdling of stems and by defoliation. In *Papaver rhoeas*, *P. orientale*.

Distribution: North America, Europe.

Xanthomonas phaseoli
Common bacterial blight of bean (Fig. 4.11)
and
Xanthomonas phaseoli var. **fuscans**
Fuscous blight of bean

Seed transmission very important. In white-seeded cultivars of Phaseolus bean infection visible as yellow spots on the dry seeds; in cultivars with coloured seeds infected seeds are indistinguishable from healthy by direct inspection, although often gum at the hilum can be seen. Transmission from infected seed in which the pathogen is harboured beneath the seed coat where it may survive for many years (Whetzel, 1906; Zaumeyer, 1929; Burkholder, 1930; Burke, 1957; Wallen *et al.*,

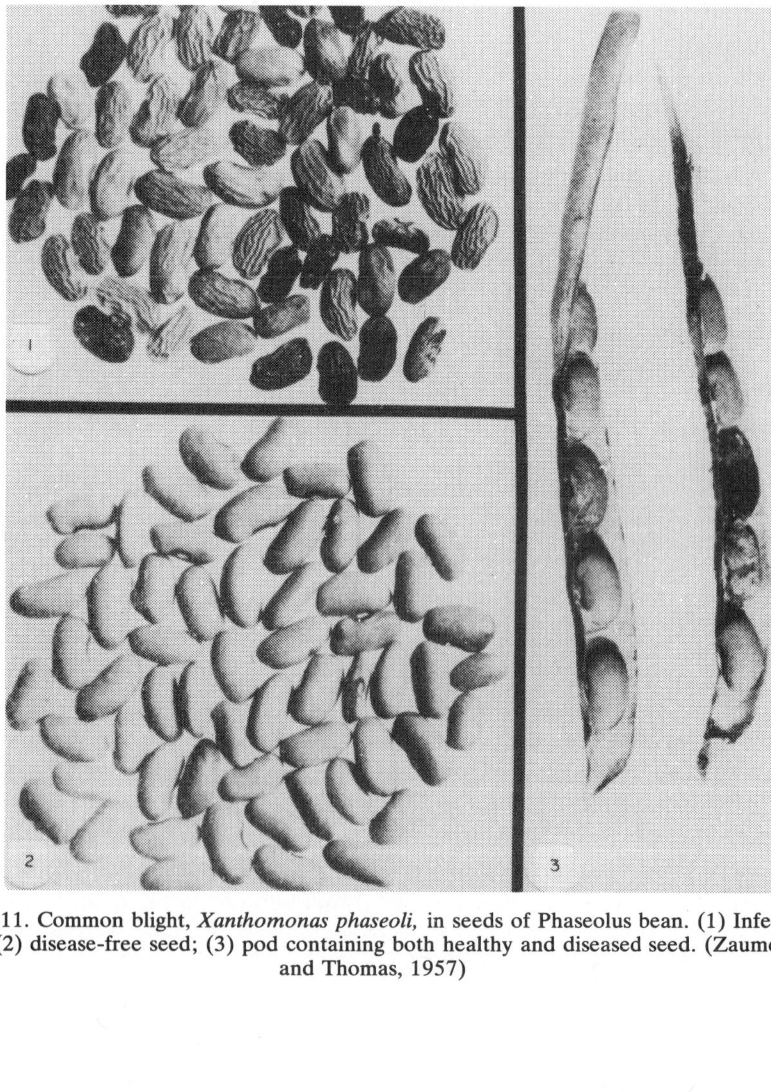

Fig. 4.11. Common blight, *Xanthomonas phaseoli,* in seeds of Phaseolus bean. (1) Infected seed; (2) disease-free seed; (3) pod containing both healthy and diseased seed. (Zaumeyer and Thomas, 1957)

1963, Oshima, Dickens and Counter, 1966; Harrison, Freeman and Smith, 1964; Wallen and Sutton, 1965; Basu and Wallen, 1966; Saettler and Perry, 1972). Also seed-transmitted in *Lablab niger* (Tarr, 1955*b*), and the variety *X. phaseoli* var. *sojense* in soybean (Graham, 1953*a*). Disease cycle is discussed in Section 14.1, mode of seed infection in Section 12.1, location in seed in Section 13.4, see Fig. 13.6.

Vascular infection. Symptoms similar to those produced by halo blight, *Pseudomonas phaseolicola*, leaves, stems, pods and seeds being infected. CMI Description 48.

The variety 'fuscans' is stated to be more virulent. CMI Description 49.

Distribution: probably in most areas where Phaseolus bean is grown, North America, Europe, Africa, Asia, Australia.

CMI Distribution Map No. 401 and (fuscous blight) 402.

Xanthomonas ricini

(Syn. *X. ricinicola*)

Bacterial leaf spot of castor bean

Field observations indicate seed transmission in castor bean, *Ricinus communis* (Calvert, Thomas and Van Horn, 1953).

Parenchymatous infection, causing water-soaked brown leaf spots, premature defoliation.

Distribution: In scattered areas in North and South America, East Africa, Asia.

CMI Distribution Map No. 394.

Xanthomonas rubefaciens

Purple spot of bean and pea

A saprophyte causing purple spotting of seeds in Phaseolus bean and in pea (Wieringa, 1954). The pigment is water-soluble.

Distribution: Europe.

Xanthomonas sesami

Bacterial leaf spot, Marad ed Dum disease in Sudan

Seed-borne infection in India (Rao and Durgapal, 1967; Habish and Hamad, 1971). Found to persist longer in seed than in soil.

Parenchymatous infection. Red-brown to black, slightly raised lesions on leaves, stems and capsules (Sabet and Dowson, 1960). CMI Description 16.

Distribution: Africa (Sudan), Asia (India).

CMI Distribution Map No. 398.

Xanthomonas translucens

Black chaff, leaf blight of cereals and grasses, bacterial streak of rice

Different formae speciales infect seeds of different genera of Gramineae which may transmit the disease at very varying rates. The formae speciales described by Hagborg (1942) and discussed by Wallin (1946*b*) include *X. translucens* f.sp. *hordei*, f.sp. *undulosa*, f.sp. *secalis*, f.sp. *hordei-avenae* and f.sp. *cereales*. Later *X. oryzicola* has been referred to this species as *X. translucens* f.sp. *oryzicola* (Bradbury, 1971). Seed infection and transmission reported for barley (Jones, Johnson and Reddy, 1917; Jones and Johnson, 1917; Wallin, 1946*b*; Leukel and Tapke, 1956), for rice (Faan and Wu, 1965; Shekhawat, Srivastava and Rao, 1969; Shekhawat and Srivastava, 1972); for wheat (Smith, 1917; Braun, 1920; Reddy, Godkin and Johnson, 1924; Elliott, 1930; Hagborg, 1942; Klykov, 1945; Wallin, 1946*b*; Boosalis, 1952; Tsilosani *et al.*, 1969). Seed treatment of rice with organomercury powder reduced but did not completely prevent transmission.

Parenchymatous infection producing translucent water-soaked stripes on leaves and culms, the lesions later turn yellow to brown or black; dark stripes or spots on the glumes. Infected kernels may become shrivelled, with bacterial pockets resulting in heavy yield losses. On rice similar symptoms, seed infection in the panicle shown to arise from bacterial exudates on the flag leaf (Shekhawat and Srivastava, 1972). Certain high-yielding rice cultivars are very susceptible. CMI Description 240.

Distribution: North America, Europe, Africa, Asia, Australia.

CMI Distribution Map No. 264.

X. translucens f.sp. *oryzicola* confined to tropical Asia.

CMI Distribution Map No. 463.

Xanthomonas vesicatoria

Bacterial scab or spot, black spot of tomato and pepper

Seed-transmitted in tomato (Doidge, 1920, 1921; Gardner and Kendrick, 1921*a*, 1923; Libman and Webb, 1963; Leben, 1963; Morton, 1965; Cox, 1966; Basu, 1966; Ercolani, 1968). In Florida less than 1 per cent seed-borne infection can cause an epidemic (Cox, 1966). In pepper, *Capsicum* (Patel *et al.*, 1950; Dempsey and Chandler, 1963; Crossan and Morehart, 1964). A variety *X. vesicatoria* var. *raphani* on radish and turnip is seed-transmitted in radish (White, 1930). Seed treatment in mercuric chloride recommended for tomato, pepper and radish.

Parenchymatous infection that is particularly injurious in producing small, black, scabby, roundish spots on the fruits of tomato and pepper (*Capsicum*), also black spotting of leaves and stems, and stem canker. CMI Description 20.

Distribution: North and South America, Europe, Africa, Asia, Australia.

CMI Distribution Map No. 269.

5 Seed-borne Fungi

Most of the seed-transmitted pathogens are fungi. Some classes or genera are frequent in seed, others occur only occasionally or perhaps not at all. Some are easily detected, others do occur but cannot be revealed by conventional testing procedures. The extent to which fungi occur in seeds depends on their capability to survive under the extreme dry conditions of seed as a carrier.

5.1 FACTORS LIMITING SEED TRANSMISSION OF FUNGI

To survive in seed, most fungi must be able to survive dehydration, yet there are two ecologically distinct groups of fungi forming a contrast regarding survival and longevity in seed: hydrophilic fungi and xerophilic fungi.

Hydrophilic fungi. These fungi are represented by downy mildews, in particular those unable to produce resting spores, oospores, or those doing so sparsely thus being dependent on a constant humid environment. The survival of these depends on the longevity of the mycelium as it may be harboured within tissues of seed. But regular seed transmission of downy mildews may occur when xerotolerant oospores are harboured on or in seed, as in *Peronospora manshurica* on soybean. Other downy mildews may survive in plant debris in which oospores are more likely to be established than in developing seeds; such debris may be a more or less effective vehicle which accompanies the seed, as *Sclerophthora macrospora* on grasses and oats (Section 13.8). Another possibility is establishment of dormant or semi-xerotolerant mycelium within the seed, as *Phytophthora parasitica* var. *sesami* in the embryo of *Sesamum orientale*; here the fungus survives at least for four months, scarcely long enough time to secure transmission which is in fact not yet proved for this pathogen (Section 13.2). Seed-borne dormant mycelium of downy mildews is a critical means of survival, partly accounting for the low rates of seed transmission in these fungi.

Xerophilic or xerotolerant fungi. These fungi are characteristically capable of producing xerotolerant propagules, often abundantly, such as chlamydospores, drought-resistant conidia, or other dormant structures, including dormant mycelium, sclerotia and microsclerotia. Examples are covered smuts of cereals and grasses, and certain genera of demataceous Hyphomycetes including *Alternaria*, *Cercospora*, *Curvularia*, *Drechslera* and *Stemphylium*.

Most xerotolerant conidia become quiescent if the relative humidity remains too low for germination, usually below 80 per cent. Chlamydospores are portions of hyphae that mostly become thick-walled under adverse conditions and rich in food contents. Both conidia and chlamydospores, as well as dormant mycelium, germinate immediately when enough humidity is provided; a period of post-maturation is not necessary for their germination.

Difference in longevity of fungi in stored seed depends essentially on their capacity to endure long desiccation. In Fungi Imperfecti phaeospores can stand long storage better than hyalospores, and dormant mycelium harboured in seed tissues survives longer than surface-borne conidia (Subsection 20.2.2; cf. Table 20.3).

5.2 IMPACT OF CLIMATE ON DISTRIBUTION OF SEED-BORNE FUNGI

Plants, as well as their infectious diseases, are generally distributed geographically according to climatic conditions. Leaf blights and leaf spot diseases, downy mildews, stem rot and root rot diseases are usually prevalent under humid conditions. Climatic conditions leading to severe field attacks of seed-borne pathogens usually result in increased seed-borne inoculum.

Tarr (1955*a*) surveyed major plant diseases in the Sudan and classified them according to rainfall areas (Table 5.1). According to this survey, few seed-borne pathogens are present in arid or semi-arid regions (Red Sea Coast, Desert), while they are abundantly present in rainy regions (Acasia Tall Grass Forest, Broad Leaf Woodland and Forest). Powdery mildews predominate in low-rainfall areas but are not present where rainfall is higher. *Alternaria* and *Cercospora* diseases prevail in the rainy parts of the country but are absent elsewhere. Smuts are less dependent on humidity because the infections are systemic, thus well protected within the host. This illustrates the fact that systemic infections, including Fusarium wilts, Verticillium wilts and smuts, are common under dry climatic conditions, which in turn are unfavourable to parenchymatous infections that require high level of humidity, such as leaf blights, leaf spot diseases and downy mildews.

Table 5.1. Influence of rainfall on distribution of different types of seed-borne fungi, as exemplified by fungi commonly encountered in the Sudan. Key = + presence of a disease, ± rare occurrence, − absence. (Tarr, 1955*a* — rearranged)

Fungus	*Host*	*Red Sea coast*	*Desert*	*Acasia short grass*	*Acasia tall grass forest*	*Broad leaf woodland and forest*
		<6	0–12	12–24	24–36	>36
PHYCOMYCETES:						
Sclerospora graminicola	Pennisetum typhoides	+	−	+	+	−
ASCOMYCETES, ERYSIPHACEAE:						
Erysiphe cichoracearum	Various	+	+	+	−	−
E. polygoni	Various	−	+	+	−	−
OTHER ASCOMYCETES:						
Glomerella cingulata	Various	−	−	+	+	+
BASIDIOMYCETES, USTILAGINALES:						
Sphacelotheca cruenta	Sorghum	−	+	+	+	+
S.reiliana	Sorghum	−	−	+	+	−
S. sorghi	Sorghum	+	+	+	+	+
Tolyposporium ehrenbergii	Sorghum	−	+	+	+	−
T.penicillariae	Pennisetum typhoides	+	−	+	+	+
Ustilago cynodontis	Cynodon	−	−	+	+	+
FUNGI IMPERFECTI, SPHAEROPSIDALES:						
Septoria lycopersici	Lycopersicon	−	−	−	−	+
MELANCONIALES:						
Colletotrichum graminicola	Sorghum	−	−	+	+	+
MONILIALES:						
Alternaria brassicicola	Brassica, Raphanus	−	−	−	+	+
A.cucumerina	Cucurbita	−	−	−	+	+
A. *porri* f.sp. solani	Solanaceae, Ipomoea	+	−	−	+	+
Cercospora arachidicola	Arachis	−	−	−	+	+
C.capsici	Capsicum	−	−	−	−	+
C.hibisci	Hibiscus	−	−	±	+	+
C.nicotianae	Nicotiana, Lycopersicon	−	−	+	−	+
C.personata	Arachis	−	−	±	+	+
C.ricinella	Ricinus	−	−	−	+	+
C.sesami	Sesamum	−	−	−	+	+
Cladosporium fulvum	Lycopersicon	−	−	±	+	+
C.herbarum	Various	+	+	+	+	+
Drechslera turcica	Sorghum, Zea	−	−	+	+	+
Nigrospora oryzae	Various	−	−	+	+	+

Location of seed production in areas with ideal climatic conditions is discussed in Subsection 19.1.1. Under semi-arid conditions infection or contamination of seed is largely prevented, resulting in (1) lower percentages of infected or contaminated seed lots, and (2) lower average infection percentages in seed lots. Semi-arid or less humid regions are therefore largely selected for seed production.

Frederiksen (1974) surveyed sorghum diseases according to their relative importance in relation to climatic zones (Table 5.2; cf. Fig. 5.1). Most of the seed-borne diseases tabulated occur in about equal frequency and severity in Subtropical Regions and Tropical Lowland, Summer. This indicates, perhaps, that humidity is more important than temperature in determining distribution and severity of these diseases.

Occurrence of pathogens on seed produced in a definite region may fluctuate considerably from year to year, reflecting yearly variation of weather conditions. Figures 5.2 and 5.3 illustrate such fluctuations in occurrence of seed-borne seedling pathogens of cereals, mostly *Fusarium* spp. and *Drechslera* spp. in Sweden over a period of 23 years. The results were based on the Hiltner test, and a total of several thousand seed samples were tested every year (Kolk, 1970). Seeds harvested in 'wet years' are more frequently and severely infected than those produced in 'dry years'.

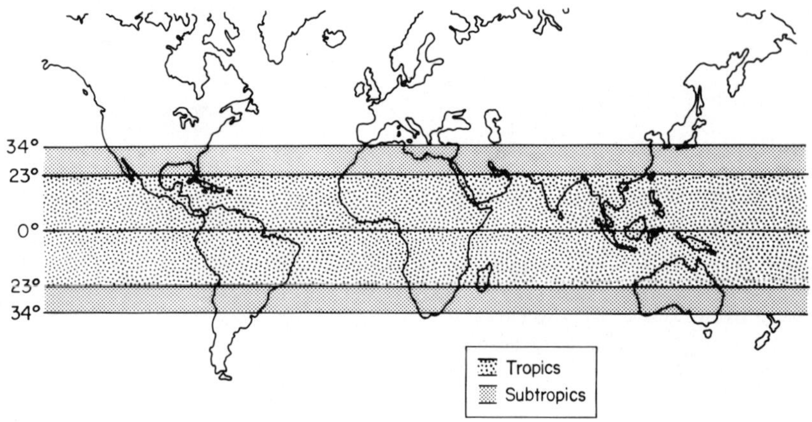

Fig. 5.1. The tropical and subtropical zones of the world

5.3 SURVEY OF SEED-BORNE FUNGI

In the survey below the fungi have been arranged in conventional taxonomic groups, mainly according to Ainsworth (1971). Fungi frequently found on seed in their imperfect state and only rarely in their perfect states are for practical reasons discussed under Fungi Imperfecti, but mentioned under their perfect state, while giving their classification.

5.3.1. Chytridiomycetes

These are generally aquatic fungi parasitic on fresh water algae or saprophytes in organic debris; they occur rarely in higher plants. The sporangium produces uniflegallate zoospores. *Synchytrium* belongs to this class.

Physoderma maydis, brown spot, may occur on maize seed but seed transmission is unimportant (Tisdale, 1920).

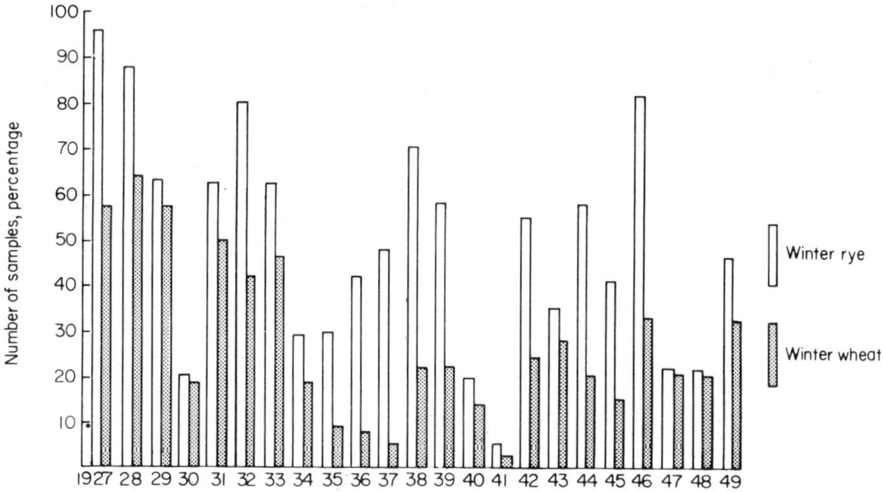

Fig. 5.2. Annual percentage of naturally infected seed samples of winter cereals in Sweden, 1927–1949. Infection of seedling-injuring fungi, mostly *Fusarium* spp. (particularly *F. nivale*), and *Drechslera* spp., exceeding 15 per cent. (Kolk, 1970)

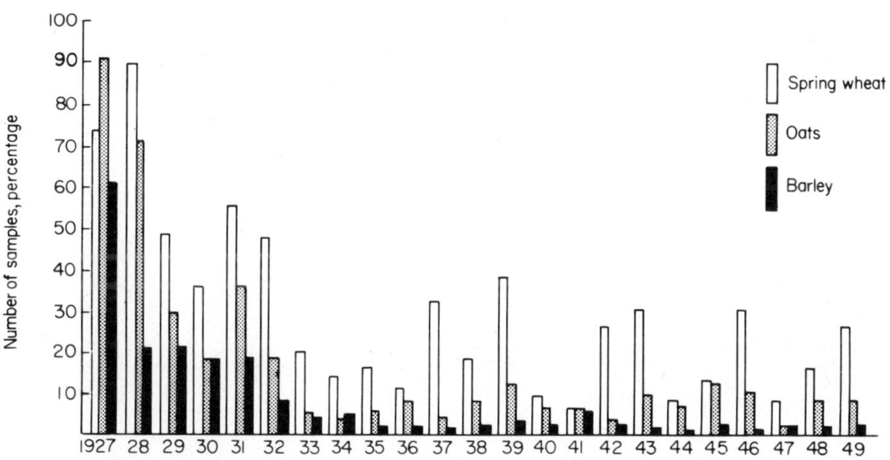

Fig. 5.3. Annual percentage of naturally infected seed samples of spring cereals in Sweden, 1927–1949. Infection of seedling-injuring fungi, mostly *Fusarium* spp. (particularly *F. avenaceum*), and *Drechslera* spp. (particularly *D. sorokiniana*; in barley *D. teres* and in oats *D. avenae*), exceeding 15 per cent. (Kolk, 1970)

Table 5.2. Relative prevalence and importance of seed-borne sorghum diseases of the world according to climatic zones (Frederiksen, 1974 — rearranged)

Frequency: * reported present; † occasional, †† commonly present; ††† generally found on most plants in most fields

Importance: 0 no loss; 1 minor importance; 2 moderate importance; 3 great importance, often causing heavy losses

Disease	*Temperate* (outside 34° lat.)	*Subtropical* (within 34° lat.)	*Tropical highland*	*Tropical lowland*	
			(within 23.15° lat.)	*winter*	*summer*
Sclerospora sorghi downy mildew	†† 2	††† 3	† 1	†† 2	††† 3
Colletotrichum graminicola red rot	†† 2	††† 3	† 1	† 1	††† 3
Sphacelotheca reiliana head smut	†† 2	††† 3	†† 2	† 1	†† 2
Fusarium sp. Fusarium stalk rot	†† 2	††† 3			†† 2
Macrophomina phaseolina charcoal rot	†† 2	††† 3			†† 2
Cercospora sorghi grey leaf spot	† 1	††† 2	†† 2	††† 2	††† 2
Gloeocercospora sorghi zonate leaf spot	* 0	††† 2		† 1	††† 2
Ascochyta sorghina rough spot	* 0	††† 2		* 0	††† 2
Ramulispora sorghi sooty stripe	†† 2	†† 2		* 0	†† 2
Colletotrichum graminicola anthracnose	† 1	†† 2			†† 2
Drechslera turcica leaf blight	†† 2	†† 1	†† 2	††† 3	†† 1
Sphacelotheca cruenta loose smut	† 1	† 2	† 1	† 0	† 1
Sclerophthora macrospora crazy top	†† 2	†† 2	† 1		
Sphacelotheca sorghi covered smut	† 1	† 1	†† 3	†† 3	†† 3
Phoma insidiosa leaf spot	† 1	† 1		* 0	†† 1
Ramulispora sorghicola leaf spot	* 0	† 1		* 0	†† 1
Drechslera sorghicola target leaf spot	* 0	† 1		† 1	† 1

5.3.2 Plasmodiophoromycetes

These are obligate endoparasites of phanerogams producing hypertrophy of the host tissue; in the infected cells the fungus develops as a plasmodium which gives rise to biflagellate zoospores.

Plasmodiophora brassicae, club-root, finger-and-toe disease of brassicas, is typically soil-borne but Warne (1943) demonstrated seed transmission in two cultivars of swede; the seeds were probably contaminated by contact with infested soil.

Distribution: North and South America, Europe, South Africa, East Asia, Australia, New Zealand.

CMI Distribution Map No. 101.

5.3.3 Oomycetes

Most members of this class are aquatic, but some, in particular Peronosporales, grow in soil, and some are parasites of higher plants. Sexual reproduction is by oospores but these are absent in some species. Usually biflagellate zoospores are formed but in some plant pathogenic Peronosporales the sporangia function as conidia only and germinate by means of a germ tube. Hyphae are aseptate.

5.3.3.1 PERONOSPORALES

Seed transmission occurs in some species within all three families, Albuginaceae (*Albugo*), Peronosporaceae (*Peronospora, Plasmopara*, *Sclerospora*), and Pythiaceae (*Pythium*).

Albugo

Conidia formed in basipetal chains in hyperdermal layers; they germinate by zoospores. Obligate plant parasites.

A. tragopogonis, white rust on *Tragopogon porrifolius*, salsify, vegetable oyster. Alcock (1931) recorded the pathogen on seed but actual transmission seems not to have been established.

Bremia

The sporangiophore is expanded at the tip as a club-shaped structure with four sterigmata bearing the sporangia. Germination is usually by means of a germ tube, occasionally by zoospores; oospores are rare.

B. lactucae, lettuce downy mildew. In China the fungus is claimed to be seed-transmitted (Tao, 1965). Wild (1947) recorded low infections developed by sowing seed from infected mother plants but concluded that there was ‘thus negligible evidence’ of seed transmission, a questionable conclusion. Systemic infection may occur.

Distribution: North and South America, Europe, Africa, Asia, Australia, New Zealand.

CMI Distribution Map No. 86.

Peronospora

Downy mildews, important pathogens, often producing severe leaf blights. Sporangia formed on dichotomously branched sporangiophores emerging singly or in groups from stomata.

P. arborescens, downy mildew of poppies. Seed-borne in *Meconopsis* spp., Chinese, Himalayan and satin poppies (Alcock, 1931), and in *Papaver somniferum* (Behr, 1956; Pietkiewicz, 1958), in which it is harboured in the endosperm.

Distribution: World-wide.

P. arthuri, downy mildew of *Clarkia elegans*, was probably introduced into Denmark with seed (Neergaard, 1943).

P. destructor, downy mildew of onion. Seed transmission has been claimed on experimental basis (Stuart and Newhall, 1935) but has not been recorded from seeds of severely infected crops (Yarwood, 1943).

Distribution: North and South America, Europe, Africa, Asia, Australia, New Zealand.

CMI Distribution Map No. 76.

P. farinosa (P. schachtii), downy mildew of beets, has been proved to be seed-transmitted in beet by Leach (1931), and this has later been confirmed by Darpoux, Lebrun and De La Tullaye (1963) who showed that oospores were an important source of inoculum. In spinach seed transmission has not been established but suggested (Wright and Yerkes, 1950), cf. Section 27.2.

Distribution: North and South America, Europe, Africa, Asia, Australia, New Zealand.

CMI Distribution Map No. 28.

P. manshurica, downy mildew of soybean. Heavy seed contamination and seed transmission is common, the surface covered by milk-white crusts of oospores (Pederson, 1958; Kingsley, 1960; Pietkiewicz, 1959; Nováková-Pfeiferová, 1964; Safeeulla and Shaw, 1964; Koretskiĭ, 1973). Serious pathogen in soybean.

Distribution: North America, Europe, Africa (Rhodesia), Asia.

CMI Distribution Map No. 268.

P. parasitica, downy mildew of cabbage. Seed transmission has been established in the U.S.S.R. by Polyakov and Vladimirskaya (1964), and the pathogen was observed on cabbage seed in Scotland (Alcock, 1931). The disease is particularly serious in seed beds.

Distribution: North America, Europe, Australia.

P. tabacina (P. hyoscyami), blue mould of tobacco. Evidence of seed transmission has been found in Australia (Pont and Hughes, 1961), and confirmed in the U.S.S.R. by Borovskaya (1965) who found oospores on the seed coat. A destructive pathogen in tobacco; there are several pathogenic races. Intercepted on seed in Portugal (see Section 24.2). 'Explosive' epidemic spread of a particular race in Europe and Middle East, see Fig. 24.3.

Distribution: North and South America, Europe, North Africa, Middle East, Australia.

CMI Distribution Map No. 23.

P. valerianellae, downy mildew. Transmitted by the oospores which may be found abundantly in seed of *Valerianella locusta* by the washing test. In France serious attacks in cultures when seed with 400 oospores per grain has been used (Champion, 1975, personal communication).

Distribution: Europe.

P. viciae (P. pisi), downy mildew. Found abundantly in inflorescences of pea (Safeeulla and Shaw, 1964). There is evidence that the pathogen was introduced into Italy with imported pea seed (Ciccarone, 1952). Also found in seed of vetch, *Vicia* sp. Usually mild but sometimes severe in pea.

Distribution: North America, Europe, Asia (e.g. India).

Phytophthora

Important plant pathogens, the best known being *P. infestans*, late blight of potato and tomato being seed-borne in the latter. Sympodial sporangia may germinate by zoospores or, under drier conditions, by germ tubes. Zoospores are laterally biflagellate; oospores may survive in soil for a long period. There is a considerable number of species and their pathogenic races.

P. capsici, Phytophthora blight, stem and fruit blight, may invade the seed coat of chilli pepper, *Capsicum annuum*, and infected seed is regarded as an important agent in dissemination (Leonian, 1922). Isolated from seed 30 days after fruit inoculation (Nakayama, 1960). In tomato the pathogen causes a fruit rot but it is uncertain whether this leads to seed transmission (Kreutzer and Bryant, 1946). The pathogen sometimes causes heavy losses in peppers and tomato.

Distribution: North, Central and South America, Europe, Asia.

CMI Distribution Map No. 277.

P. cinnamomi, avocado seedling blight, root rot. The embryo of avocado, *Persea americana*, may be invaded and the disease may be established in the nursery from infected seed (Zentmyer, personal communication, 1968). Hot water treatment is generally used in California, U.S.A. (Cook, 1959). Severe root rot predominates in trees grown on poorly drained and heavy soils.

Distribution: North and South America, Europe, Africa, East Asia, Australia, New Zealand.

CMI Distribution Map No. 302.

P. citrophthora, sweet orange gummosis, foot rot, brown rot, is transmitted from seed infected in the seed coat, and hot water treatment is recommended (Klotz, De Wolfe and Wong, 1957; Klotz *et al.*, 1960). The fungus produces foot rot, particularly when grapefruit, lemon, tangerine and sweet orange are grown on their own roots; sour orange is resistant and used as rootstock (Wellman, 1972).

Distribution: North and South America.

CMI Distribution Map No. 35.

P. infestans, late blight, fruit rot of tomato. The pathogen is harboured in the seed coat from which it may be transmitted (Boyd, 1935; Miller and Crosier, 1936). Cause of heavy losses in tomato. Pathogenic races.

Distribution: World-wide.

CMI Distribution Map No. 109.

P. megasperma var. *soyae*, root and stem rot of soybean. Transmission occurs both from infected seed and soil mixed with the seed (Klein, 1959). The disease is favoured by cold wet soil; young seedlings and older plants are killed, but usually only a few plants in a row.

Distribution: North America.

P. nicotianae var. *parasitica*, brown rot of *Citrus*. Seed transmission and pathogenesis as for *P. citrophthora*, see above. In sesame, mycelium of *P. nicotianae* var. *sesami* has been observed in the seed and could be isolated after four months' storage, but transmission has not been definitely proved (Sehgal and Prasad, 1966). The disease, sesame blight, is sometimes serious in India.

Distribution: America, Europe, Africa, Asia, Australia, Oceania.

P. phaseoli, downy mildew, collar rot of lima bean. In Roumania the fungus has been detected in seed incubated in very wet sand (Rădulescu, *fide* Noble and Richardson, 1968), and in the U.S.A. it was transmitted only when the seed contained oospores (Wester, Goth and Drechsler, 1966). The disease may seriously affect all parts of lima bean plants, but the pods are most severely attacked.

Distribution: North America, Europe.

Plasmopara

Many important pathogens, among which is *P. viticola*, downy mildew of grape vine. Sporangiophores are monopodially branched, sporangia are hyaline, germinating by zoospores or by germ tube.

P. halstedii, downy mildew of sunflower, Fig. 5.4. Commonly seed-transmitted on sunflower, harboured as mycelium in the seed (Novotel'nova, 1963). Leppik (1962*b*) made a world survey of the disease, based on seed samples introduced into the U.S.A. through the Regional Plant Introduction Station, Ames, Iowa (cf. Section 24.2). Mycelium present in the pericarp and embryo and oospores occur on or in the pericarp, rarely in the embryo (Cohen and Sackston, 1973*a*). Seed health testing procedure described by Tîrcomnicu and Iliescu (1973). Infected seed may produce systemically infected plants and symptomless root-infected plants; seed transmission probably important when infected seed from symptomless infected sunflowers are used in new fields (Cohen and Sackston, 1973*b*, 1974). A major disease of sunflower; the fungus has several pathogenic races, see Subsection 17.2.1(2).

Distribution: North and South America, Europe, Africa, South Asia.

CMI Distribution Map No. 286.

Fig. 5.4. Downy mildew of sunflower, *Plasmopara halstedii*. (Courtesy of E. E. Leppik)

Pythium

Many species occur in water, wet soil, and grow as saprophytes on plant debris. Some are serious seedling pathogens (damping-off). Sporangiophores are filamentous and difficult to distinguish from the vegetative hyphae. Sporangia terminate by formation of a thin-walled vesicle in which the cytoplasm splits up into laterally flagellate zoospores. In some species (e.g. *P. ultimum*) zoospores do not occur; the sporangia germinate directly forming a germ tube.

P. aphanidermatum causes rotting of seed of watermelon, *Citrullus vulgaris*, cucumber (both: Sherf, 1952) and tomato (Gattani and Kaul, 1951), and often severe pre- and post-emergence damping-off. Mainly soil-borne.

Distribution: North and South America, Europe, Africa, Asia, Australia.

CMI Distribution Map No. 309.

P. debaryanum, damping-off. Seed-borne in beet (Jovićević, *fide* Noble and Richardson, 1968; Von Plotho, 1953).

Distribution: World-wide.

Sclerophthora

Sporangiophores are like ordinary hyphae; in other respects the genus is similar to *Sclerospora*.

S. macrospora, downy mildew of cereals, crazy top of maize. Seed of oats is contaminated by plant debris containing oospores (Summers, 1952). In wheat the fungus may also occur in the seed, as in seed imported from Victoria to West Australia (Carne, 1924). In maize seed transmission has been reported from the U.S.A. (Ullstrup, 1952; Ullstrup and Sun, 1969). Serious in some tropical regions, including India.

Distribution: North America, Europe, Africa, Asia, Australia, New Zealand.

CMI Distribution Map No. 287.

Sclerospora

Downy mildew of grasses

Sporangiophores are often short, stout, with many upright branches near the end. Important pathogens on tropical cereals and pasture grasses. Original diagnoses and descriptions compiled by Waterhouse (1964).

S. graminicola, green ear of bajra, *Pennisetum typhoides*. Transmission by seed infected by mycelium in the embryo (Arya and Sharma, 1962; Sundaram, Ramasastry and Nayar, 1973). Singh and Pushpavathy (1965) found mycelium only in the ovary wall. Transmission may also occur from oospores in plant debris mixed with the seed (Tasugi, 1933; Suryanarayana, 1963). Systemically infected plants are stunted and chlorotic. Hypertrophy and proliferation of the flower organs result in a green ear, florets sometimes being completely replaced by a cluster of plantlets (Singh and Pushpavathy, 1965).

Distribution: North America, Europe, Africa, Asia.

CMI Distribution Map No. 431.

S. maydis, downy mildew of maize. Seed transmission in maize has been demonstrated, chlorotic seedlings developed from seed sown in sterile soil in quarantine house; primary infections result from mycelium in the embryo of kernels (Purakusumah, 1965). Destructive on maize in Indonesia.

Distribution: Indonesia only.

S. philippinensis, Philippine downy mildew, crazy top. Mycelium may be harboured in the seed coat and endosperm of maize (Weston, 1920), transmission up to 17 per cent from freshly harvested seed has been demonstrated by Advincula (1975). In the Philippines the disease can destroy up to 80 per cent of the crop (Exconde, Adversario and Advincula, 1968).

Distribution: Asia (Philippines, Indonesia, India, Nepal).

S. sorghi, sorghum downy mildew. Under extremely humid conditions the seed may become infected but when it has dried or has been stored for a few days the mycelium is incapable of infecting maize seedlings (Jones, Leeper and Frederiksen, 1972). Seed from systemically infected mother plant may transmit the disease (Bain and Alford, 1969). Occasional seed transmission may occur when seeds with glumes attached are harvested from systemically infected plants; the glumes carry the inoculum (Frederiksen *et al.*, 1973). Oospores may be air-borne, and exposed sorghum seed can be coated with air-borne oospores but this kind of contamination is less likely in maize with husk protection (Frederiksen *et al.*, 1973). The disease may be systemic or localised, the first type leads to chlorotic seedlings; later parallel white, ultimately brown, stripes or streaks are developed on leaves, then leaf shredding, stunt and, finally, premature death.

Distribution: U.S.A., South America, Europe (Italy), Africa, South Asia.

CMI Distribution Map No. 179.

5.3.4 Zygomycetes

Sexual reproduction takes place by conjugation of two gametangia forming a resting spore, the zygospore. Asexual reproduction takes place by means of non-motile spores formed in sporangia as well as mycelial chlamydospores. Most forms are saprophytes, some weak parasites and some parasites of animals. One of the two orders generally recognised is Mucorales, 'black moulds'.

Mucor

The members of this genus possess spherical sporangia on swollen sporangiophores. Some species occur in healthy-looking kernels of cereals from the field (Warnock and Preece, 1971), and may develop as storage fungi under suitable conditions.

Rhizopus

The mycelium develops stolons, and the sporangiophores arise above rhizoids. Occurs in soil, on fruits and all kinds of decaying materials, including moist stored grain. Common laboratory contaminants.

R. arrhizus, *R. nodosus* and *R. stolonifer (R. nigricans)*, occur in cotton as part of the boll rot organisms leading to internal infection of the seed (Venkatram, 1950; Halisky and Satour, 1964). *R. stolonifer* is often a troublesome laboratory contaminant.

5.3.5 Ascomycetes (Ascomycetina)

Sexual spores, ascospores, are borne within a single cell, ascus, and in the majority of ascomycetes, in groups of eight. In most forms asci develop in fruiting bodies, ascocarps, including cleistothecia, perithecia, pseudothecia or apothecia, but in some, regarded as more primitive, the asci are scattered; no fruiting structures are formed.

(1. Hemiascomycetes)

5.3.5.1 ENDOMYCETALES

This class includes the ascospore-forming yeasts in which a single cell is transformed directly into an ascus; other groups form a well-developed mycelium. Most forms are saprophytes.

Nematospora

Sexual reproduction by budding. The ascus contains eight spores, two bundles of four spores each.

N. coryli (N. phaseoli) occurs in different plants and is associated with stigmatomycosis as a secondary invader, following insect punctures (see Section 16.2); it has been found in seeds of coffee, *Coffea* spp. (Orton, 1931), Phaseolus bean (Moore, 1930), soybean (Johnson, Chamberlain and Lehman, 1954) and cotton (Moore, 1930; Pearson, 1947). Another species, *N. gossypii*, occurs also in cotton seed (Moore, 1930; Pearson, 1947).

5.3.5.2 PROTOMYCETALES

This is a class about which our knowledge is very fragmentary, and is the subject of speculation. The sac, named synascus, is regarded as a group of fused asci. Parasitic on higher plants.

Protomyces

Parasitic in Umbelliferae and other phanerogams. Segments of the hyphae enlarge, become multinucleate and surrounded by a thick wall, forming 'sporangia' or 'chlamydospores', Fig. 5.5, often remaining in the tissue of the host (Bessey, 1968).

P. macrosporus is common in coriander, *Coriandrum sativum*, in India. The fruits, cremocarps, become hypertrophied, see Fig.19.9, lose their germination power and condiment value. Gupta (1962) has described the pathological anatomy of fruits. The hypertrophied fruits may be removed by proper cleaning, but normal seeds may be heavily contaminated by sporangia ('chlamydospores'), and these can be detected in seed washings (Gupta and Neergaard, 1970), see Fig. 5.5. The quality of the plants is considerably lowered (Gupta, 1954), and viability of seed is substantially reduced (Pavgi and Mukhopadhyay, 1969).

Seed treatment by organomercurial preparations and other fungicides is probably effective if hypertrophied fruits are first removed (cf. Gupta, 1958). Nene (1971) indicated for seed treatment thiram, 2.5 g/kg seed.

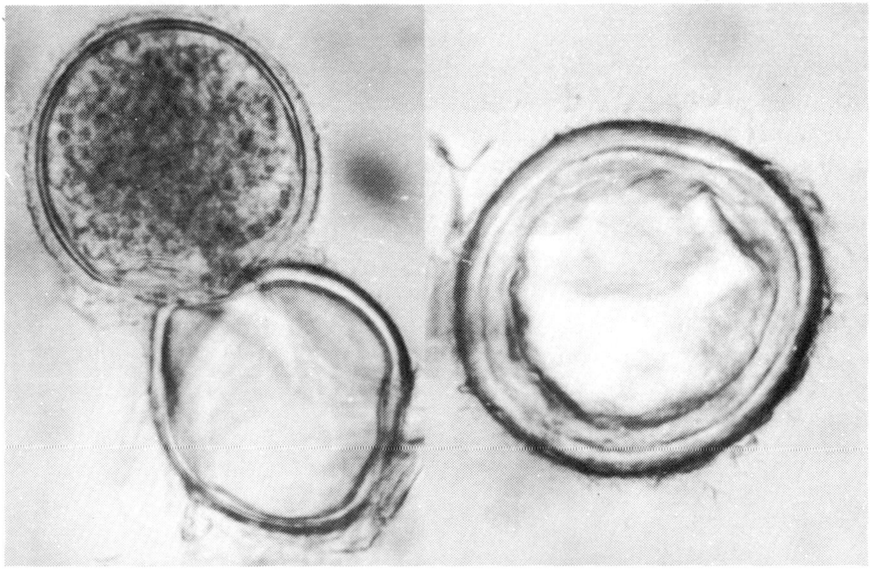

Fig. 5.5. *Protomyces macrosporus*. Left, germination of a resting sporangium or 'chlamydospore', forming a secondary sporangium or 'synascus', which is surrounded by the inner wall of the resting sporangium and which subsequently develops spores. Right, a resting sporangium as found in seed of coriander; the resting sporangium is surrounded by a characteristic thick, three-layered cellulose wall. × 1000

(2. Plectomycetes)

5.3.5.3 EUROTIALES

This class is also called Aspergillales or Plectoascales and comprises about 80 genera including *Eurotium*, perfect states of *Aspergillus*. The ascocarp of many genera within the class is built up by loosely woven hyphae.

Eurotium chevalieri has been encountered in cotton seeds which failed to germinate (Leakey, 1962).

5.3.5.4 ERYSIPHALES

Obligate plant parasites, causing powdery mildews. Mycelium is white, superficial, with haustoria often confined to the epidermal cells. Conidia are formed in basipetal chains (Oidium); also the perithecia (cleistothecia) are external to the host; they have no special opening. In the family Erysiphaceae, the cleistothecium bears various types of appendages and contains one or more asci, according to genus; the ascospores are hyaline. There are few examples of seed contamination, and seed transmission is of little if any importance.

Erysiphe

The perithecium contains more than one ascus, appendages are unbranched or irregularly branched. The conidia are capable of germinating at low humidities, and powdery mildews prevail under dry conditions.

E. cichoriacearum has been encountered on seed of *Zinnia elegans* (Baker and Locke, 1946) but no seed transmission has been established.

E. heraclei (E. umbelliferarum), seed-borne in parsley (Neergaard, 1949AR, 1951AR) and parsnip (Neergaard, 1965AR; Andersen, 1973AR) but transmission has not been proved.

E. pisi, recorded on seed of pea (Crawford, 1927; Alcock, 1931). Uppal, Patal and Kamat (1935) observed the disease in plants grown from untreated seed from infected plants while treated seed developed healthy plants, and they concluded that the pathogen was seed-transmitted; confirmation of this observation is needed.

(3. Pyrenomycetes)

5.3.5.5 HYPOCREALES

The ascocarp is fleshy to leathery, brightly coloured, pale, red, yellow or violet, immersed in or seated on an ostiolate stroma. The order includes saprophytes and plant parasites.

Gibberella

The ascocarps are blue or violet, more or less superficial. The following important seed-borne pathogens are discussed under Fungi Imperfecti I, Subsection 5.3.6.2: *G. avenacea*, imp. state *Fusarium avenaceum*; *G. fujikuroi*, imp. state *F. moniliforme*; *G. intricans*, imp. state *F. equiseti*; *G. gordonia*, imp. state *F. heterosporum*; *G. zeae*, imp. state *F. graminearum*.

Griphosphaeria

The ascocarps are immersed in the host tissue, often covered by a shield-like structure; asci amyloid, staining blue with iodine. *G. nivalis (Micronectriella nivalis)*, imp. state *Fusarium nivale*, see Fungi Imperfecti I, Subsection 5.3.6.2

Nectria

The ascocarps are globose to subglobose, brightly coloured, at maturity brownish, with a papillate ostiole; they are developed on the surface of a cushion-shaped stroma. *N. haematococca*, imp. state *Fusarium solani*, see Fungi Imperfecti I, Subsection 5.3.6.2.

5.3.5.6 CLAVICIPITALES

In the family Clavicipitaceae the perithecia develop on a fleshy stroma. The asci are narrowly cylindrical with a thick perforated cap, and the ascospores are thread-like, being scolecospores, breaking up into the individual segments. Most forms are pathogens.

Balansia

This genus resembles *Claviceps* in many respects; greyish, later purple-black sclerotia (stromata) develop in inflorescences of Gramineae. The conidial state, *Ephelis*, is formed as pycnidium-like, cup-formed structures (sometimes called 'sporodochia'), and is usually referred to Excipulaceae of Sphaeropsidales. Perithecia develop in stipitate spherical stromata. The disease cycle, including mycelium in embryo and systemic infection, is similar to that of *Ustilago nuda*, see Section 14.1, but perennation may occur in perennial grasses. Diehl (1950) monographed the genus.

B. andropogonis, imp. state *Ephelis pallida*, has been reported from *Aristida* sp. (Patel, Gokhale and Kulkarni, 1951), *Chrysopogon aciculatus* (Sydow, Sydow and Butler, 1911; Govindu and Thirumalachar, 1963), *Eragrostis tenuifolia*, *Paspalum scrobiculatum*, *Pennisetum hohenackeri* and *Sorghum halepense* (Govindu and Thirumalachar, 1963).

Distribution: Asia (India, the Philippines).

B. claviceps has been encountered on *Cyrtococcum oxyphyllum* (Ramakrishnan, Srinivasan and Sundaram, 1953; Govindu and Thirumalachar, 1963).

Distribution: Central and South America, India.

B. hemicrypta, black choke, occurs in three-arm grass, *Aristida fendleriana* and other spp. of *Aristida* (Diehl, 1950).

Distribution: North and South America.

B. oryzae, imp. state *Ephelis oryzae*, udbatta disease or sterility disease of rice. The compact sclerotium covers and includes the panicle and rachis of rice (Sydow, 1914; Diehl, 1950; Narasimhan and Thirumalachar, 1943; Mohanty, 1964; Govindu, 1969). Other grass hosts include *Arthraxon ciliaris* var. *coloratus, Cynodon dactylon, Eragrostis tenuifolia*, *Imparata cylindrica*, *Isachne elegans*, *Pennisetum* sp., *Saccolepis indica* and *Secale cereale* (Venkatarayan, 1937; Govindu, 1969). The disease has become serious in recent years in some places in India. Hot water seed treatment is recommended.

Distribution: North and South America, Asia (India, China).

B. sclerotica has been found on *Bothriocohla pertusa*, *Brachiaria distachya*, *Dichantium annulatum*, *Heteropogon contortus*, *Paspalidium flavidum*, *Themeda tremula* and *Urochloa reptans* (Ramakrishnan, Srinivasan and Sundarum, 1953; Govindu and Thirumalachar, 1963). *B.*

sclerotica var. *deformans*, on *Cymbopogon caesius* and *C. coloratus* (Govindu and Thirumalachar, 1963).

Distribution: India.

In India species of *Ephelis* infects inflorescences of Johnson grass, *Sorghum halepense* (Govindu and Thirumalachar, 1954) causing serious losses.

Claviceps

About 25 species; ovary parasitised in 4 species of Juncaceae, 17 species of Cyperaceae and 498 species of Gramineae (Brady, 1962), cf. Section 9.1. *Claviceps* produces purple sclerotia forming stalked stromatic heads in which the perithecia develop; in young ovary infections unicellular conidia, termed *Sphacelia*, are abundantly produced in a sticky, nectar-like secretion called honeydew, attractive to insects; some economically important species are discussed below.

C. gigantea, ergot of maize (in Mexico: diente de caballo, horse's tooth); comma-shaped, 8×5 cm large sclerotia that easily can be removed from seed (Fuentes *et al.*, 1962, 1964). Confined to Mexico; 40 per cent of the ears may be infected in valleys with high humidity.

C. microcephala, related to, or rather a synonym of, *C. purpurea*, by some authors regarded as a separate species, but not by Brady (1962); produces small sclerotia, e.g. in sorghum, millet (*Panicum*), *Pennisetum typhoides*, maize (Futrell and Webster, 1966; Siddiqui and Khan, 1973).

C. paspali, ergot, sugary disease, prevalent in wild and cultivated forms of *Paspalum scrobiculatum* in India (Ramakrishnan, 1963), and occurs on different species of bahia grass and dallis grass, used as forage grasses, *Paspalum* spp., such as *P. notatum*, *P. dilatatum*, *P. distichum* and *P. urvillei* (Anonymous, 1957). Ingestion of sclerotia causes poisoning and gangrene in animals, see Subsection 8.3.1.

Distribution: North, Central and South America, South Europe, Africa, Asia, Australia, New Zealand.

CMI Distribution Map No. 90.

C. phalaridis in canary grasses, *Phalaris* spp., in Australia commonly found in seed samples of *Phalaris tuberosa* (Walker, 1958, 1959, 1970). Introduced to California in seeds of harding grass, *Phalaris aquatica* (Van Kleeck and Dobbins, 1975). Life-cycle is discussed in Section 14.8.

C. purpurea, ergot of grasses and cereals, parasitising the ovary which is transformed to a sclerotium, protruding conspicuously out of the spike in most of the hosts. Several hundred species of Gramineae, Cyperaceae and Juncaceae are susceptible (Brady, 1962). Cross-fertilising hosts such as rye, the classical host, as well as self-sterile hosts, such as some new cultivars of wheat, are particularly susceptible (see Section 11.4). Some important hosts are included in the Annotated List by Noble and Richardson (1968). Some major hosts are included in Table 9.1. Life-cycle is discussed in Section 14.8, see Fig. 14.8; see also Section 9.1. The sclerotia contain a number of poisonous alkaloids causing ergotism in animals and man, discussed in Subsection 8.5.1.

Distribution: World-wide.

Epichloë

Parasites on the stems of grasses which become coated by a bright-coloured stroma, surrounding sheaths and culms of grasses. Ovoid conidia develop, later perithecia immersed in the stroma. Ascospores are hyaline, filiform.

E. typhina, choke of pasture grasses and other grasses, including *Dactylis*, *Holcus*, *Phleum*, *Poa*. In most grasses flowering is suppressed, and seed transmission prevented, but in *Festuca rubra* infections can be carried over in the seed, see Section 13.1. In addition to this host, the fungus is seed-borne also in *Agrostis* spp. and *Poa* spp. (Mühle and Frauenstein, 1970).

Distribution: North America, including Greenland, Europe, Asia.

5.3.5.7 SPHAERIALES

Most of about 5,000 species included in this order are saprophytes, but some are important plant pathogens. Characteristic by the dark-coloured to black, carbonaceous perithecia (cf. Hypocreales) which may occur separately or embedded in a stroma. The ascospores are often expelled violently through an apical pore or a rupture of the ascus.

Chaetomium

Perithecia develop separately on the surface of the substrate, covered with long, often branched hairs and provided with a tuft of long, dark, stiff hairs around the ostiole. The black ascospores escape in a mass of slime. More than 80 species are known. Several species, identified and unidentified, have been encountered on seed of many hosts. Some species exercise antibiotic effects on pathogenic fungi (see Section 16.9).

C. globosum, Fig. 5.6, Mathur, Ram Nath and Mathur (1973) found 13 seed samples of pearl millet, *Pennisetum typhoides*, infected, and the pathogen was shown to be transmitted producing brown leaf spots and blight.

Different species of *Chaetomium* are common contaminants of seeds. Malone and Muskett (1964) described the following species encountered on oat seeds in Northern Ireland: *C. atrosporum, C. bostrychoides, C. crispatum, C. elatum* (frequent), *C. funicola, C. globosum* (frequent), *C. indicum, C. murorum, C. olivaceum, C. spinosum* and *C. tortile*, and Vaidehi (1973) the following from rice seeds in India: *C. aureum, C. bostrychoides, C. caprinum, C. convolutum,* and *C. mollicellum.*

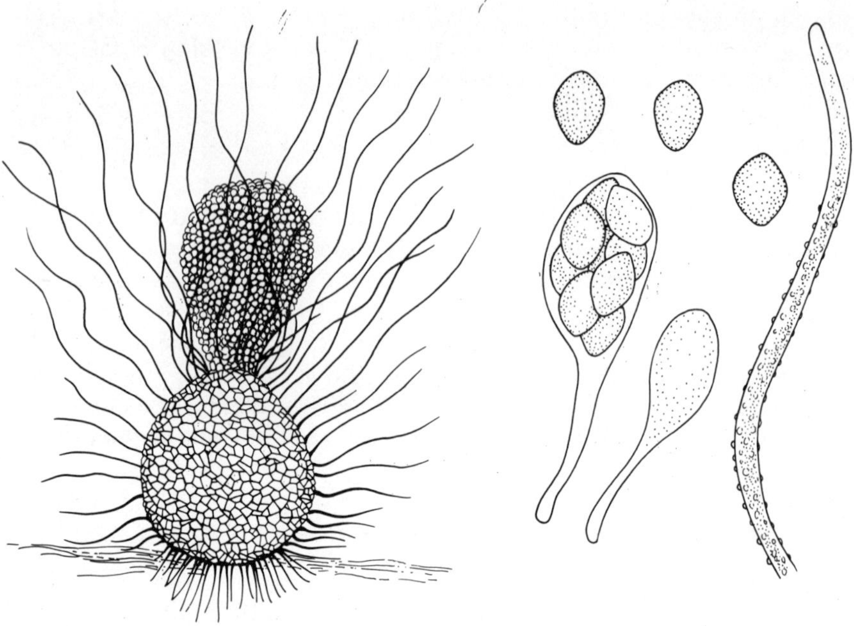

Fig.5.6. *Chaetomium globosum*, perithecium, × 100, asci and ascospores, × 1,000. (Courtesy of J.A. Von Arx)

Diaporthe

Perithecia are formed in a stroma, the ascospores are ellipsoid, fusoid, two-celled and the imp. state is *Phomopsis*.

D. phaseolorum, pod and stem blight of lima bean, occurs in seeds of lima bean, *Phaseolus lunatus*, to which the pathogen is confined (Luttrell, 1947). *D. phaseolorum* var. *sojae* has a wider host range including soybean, cowpea, Phaseolus bean, lima bean, lupin, pepper (*Capsicum*) and others, in which it may infect fruits and seeds (Luttrell, 1947). Usually found in the *Phomopsis* state on seeds. *D. phaseolorum* var. *caulivora* in soybean mainly as internal infections (Peterson and Strelecki, 1965; Nicholson, Dhingra and Sinclair, 1972).

Distribution: North and South America, Africa, Asia.

D. vexans, fruit rot of eggplant, imp. state *Phomopsis vexans*, see Fungi Imperfecti I, Subsection 5.3.6.3.

Gaeumannomyces

Ascocarps are thick-walled, with a short cylindrical beak; ascospores filiform.

G. graminis (Ophiobolus graminis), imp. state *Cephalosporium maydis* (invalid name), take-all disease of wheat, late blight or slow wilt of maize. The pathogen is predominantly soil- or plant-residue-borne. Brittlebank (1921) suggested that infected debris may accompany wheat seed but transmission in this host has not been established. In maize, the imperfect state may be seed-borne (Samra, Sabet and Hingorani, 1963). Principal cause of the stalk rot complex of maize, important in Egypt where 25 per cent of stands may be infected; hot water treatment recommended (Sabet *et al.*, 1962; Samra *et al.*, 1972).

Distribution: Under the name *C. maydis* the pathogen has been regarded as possibly restricted to Egypt and India (Payak *et al.*, 1970) and therefore as an important quarantine object. However, *G. graminis* is widely distributed: world-wide, although scattered in South America, South Africa and Asia; it should still be considered for quarantine.

Gnomonia

There are several important plant pathogens in this genus. The perithecia are embedded in the substrate, not in a stroma; they are provided with a neck that projects prominently. The imperfect states are in some species *Colletotrichum (Gloeosporium)*, *Marssonina* and *Ascochyta*. *G. errabunda*, imp. state *Discula*, and *G. leptostyla*, imp. state *Marssonina juglandis*, see Fungi Imperfecti I, Subsection 5.3.6.3.

Glomerella

This genus is similar to *Gnomonia* but the perithecia are usually embedded in a stroma or they may occur separately; the ascospores are one-celled, hyaline. The imperfect state is always *Colletotrichum (Gloeosporium)*.

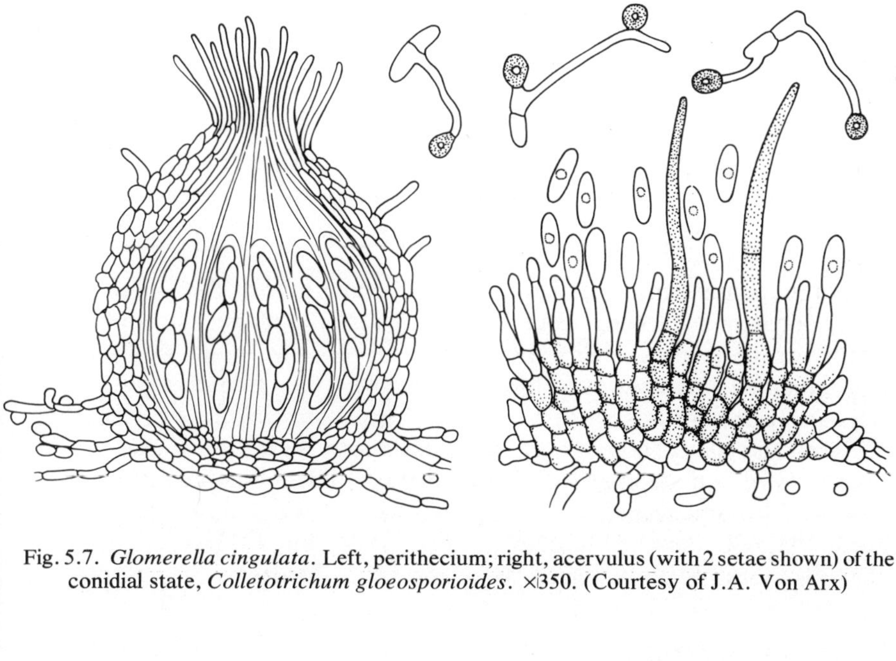

Fig. 5.7. *Glomerella cingulata*. Left, perithecium; right, acervulus (with 2 setae shown) of the conidial state, *Colletotrichum gloeosporioides*. ×350. (Courtesy of J.A. Von Arx)

G. cingulata, Fig. 5.7, anthracnose, occurring in a wide range of hosts, imp. state *Colletotrichum*, described under different names, now regarded as synonyms of *Colletotrichum gloeosporioides*, see Fungi Imperfecti I, Subsection 5.3.6.3.

Other species described under Fungi Imperfecti, Subsection 5.3.6.3: *G. glycines*, soybean anthracnose, imp. state *C. dematium* f.sp. *truncatum* (syn. *G. glycines*); *G. gossypii*, seedling blight, boll rot of cotton, imp. state *Colletotrichum gossypii*; *G. graminicola*, red rot, midrib spot of sorghum, imp. state *Colletotrichum graminicola* (syn. *C. falcatum*).

Khuskia

Perithecia are aggregated, globose or subglobose, with a papillate ostiole erumpent. Ascospores hyaline, granular, curved, inequilateral, one- to two-celled. *K. oryzae*, imp. state *Nigrospora oryzae*, see Fungi Imperfecti I, Subsection 5.3.6.2.

(4. Discomycetes)

5.3.5.8 PHACIDIALES

The fruit bodies of Discomycetes, apothecia, are in this order immersed in the substrate; they are leathery, waxy, at maturing opening by irregular tearing or with a longitudinal split. Saprophytes and plant parasitcs.

Lophodermium

Black elliptic apothecia, rupturing longitudinally at maturity. *L. pinastri*, pine needle cast, imp. state *Leptostroma pinastri*, see Fungi Imperfecti I, Subsection 5.3.6.3.

5.3.5.9 HELOTIALES

Cup- or disc-shaped apothecia, asci inoperculate, i.e. opening by an irregular apical split to discharge the spores; the typical cup fungi.

Ciboria

Apothecia are stipitate, cup- or funnel-shaped, later flat, often large, usually brownish, arising from a more or less developed stroma; ascospores hyaline, one-celled. There are no macroconidia but globose microconidia. Parasites of inflorescences and fruits. *Stromatinia*, one of two subgenera, is characterised by stromatisation of fruits or seeds.

C. alni (Sclerotinia alni). Fruits of *Alnus incana*, *A. glutinosa*, *A. oregona* and *A. viridis* are stromatised (Rostrup, 1897; Kienholz and Cash, 1937; Ferdinandsen and Jørgensen, 1939; Buchwald, 1947; Lakon, 1950; Neergaard, 1956AR, 1964AR, Urošević, 1957; Andersen, 1973AR).

Distribution: North America. Widespread in Europe.

C. betulae. The fruits of *Betula papyrifera*, *B. pendula* and *B. pubescens* are stromatised; common in commercial seeds (Buchwald, 1947; Lacon, 1950; Urošević, 1958).

Distribution: North America, Europe, Asia.

C. batschiana (Sclerotinia pseudotuberosa). The fruits of *Quercus petraea* and *Q. robur* are stromatised. Rather common (Ferdinandsen and Jørgensen, 1939; Buchwald, 1947).

Distribution: Europe.

C. granigena stromatises the caryopsis of brome grasses, *Bromus* spp. (Buchwald, 1949).

C. rufo-fusca, sometimes occurs on seeds of *Abies alba* but is more common on the cone scales which are stromatised (Buchwald, Klinge and Toft, 1961).

Distribution: Europe (Denmark, Germany) and possibly U.S.A.

Gloeotinia

Apothecia are fleshy, cupulate, stalked, arising from stroma which is undifferentiated in being composed of intertwining hyphae only. One-celled macroconidia and spermatia (microconidia) are formed.

G. temulenta (Phialea temulenta), blind seed disease, invades the caryopsis of Gramineae, and when the scutellum is infected the seed usually fails to germinate; heavily infected seed is usually opaque when viewed in a diaphanoscope, slightly infected or healthy are translucent (Wilson, Noble and Gray, 1945). Host genera listed by Noble and Richardson (1968): *Agrostis*, *Aira*, *Alopecurus*, *Bromus*, *Cynosurus*, *Dactylis*, *Danthonia*, *Deschampsia*, *Festuca*, *Glyceria*, *Holcus*, *Hordeum*, *Lolium*, *Phleum*, *Poa* and *Secale*; to which may be added *Phalaris arundinacea* (O'Rourke, 1968). Serious in some places in *Lolium*, particular *L. perenne*, originally described from New Zealand (Hyde, 1932, 1938*a*, 1938*b*). The disease has been serious in seed crops in Oregon, U.S.A., but controlled (Hardison, 1963); in New Zealand, Britain and recently in the Netherlands (De Tempe, 1966, 1971). The life-cycle is discussed in Section 14.8, see also Section 11.5.

Distribution: North America (Oregon), Europe, Australia, New Zealand.

CMI Distribution Map No. 348.

Sclerotinia

Apothecia are stalked arising from stromata developed in the host tissue. The stroma may be in the form of a sclerotium or it may have a diffuse, indefinite form. Macroconidia are usually formed; and according to some authorities the genus should be divided into distinct genera and subgenera, *Botryotinia* for forms with *Botrytis*, *Monilinia* with forms having *Monilia* as imperfect state. The following species are discussed under

Fungi Imperfecti I, Subsection 5.3.6.2: *S. fuckeliana*, imp. state *Botrytis cinerea*; *S. laxa*, imp. state *Monilia laxa*; *S. ricini*, imp. state *Botrytis ricini*.

S. sclerotiorum (Whetzelinia sclerotiorum), watery soft rot, white blight. This serious plant pathogen is quite plurivorous, and the fungus may accompany seeds in two ways: as easily detectable sclerotia, Fig. 19.9, or as mycelial infection within the seed as recorded for *Brassica* (Neergaard, 1958*a*). Seed contamination or seed transmission has been reviewed for the following hosts by Noble and Richardson (1968) and Schlösser (1968): *Brassica* spp. (also mycelial infection), *Capsicum*, *Carthamus tinctorius* (mycelial infection — Sackston, 1960), *Carum carvi* (also as mycelial infection), *Centaurea cyanus*, *Cheiranthus*, *Cynara scolymus*, *Chrysanthemum*, *Cucurbita*, *Dahlia variabilis*, *Daucus carota*, *Glycine max*, *Helianthus annuus* (also as mycelial infection), *Hibiscus sabdariffa*, *Hordeum*, *Linum usitatissimum*, *Lupinus*, *Medicago sativa*, *Melilotus*, *Phaseolus* spp. (also as mycelial infection), *Pisum* (also as mycelial infection), *Quercus*, *Raphanus sativus*, *Scorzonera hispanica*, *Sinapis alba*, *Solanum melongena*, *Trifolium incarnatum*, *T. pratense* (both also as mycelial infection), *Vicia villosa* (also as mycelial infection), *Zea mays*. Hosts reported from Denmark (Andersen, 1974 — index) are: *Cichorium endivia*, *Cosmos bipinnatus*, *Cotonoaster dielsianus*, *Lupinus polyphyllus*, *Matthiola annua*, *Pastinaca sativa*, *Petroselinum sativum*, *Pisum arvense*, *Pseudotsuga taxifolia*, *Sinapis alba*, *Spinacia oleracea*, *Tagetes erecta*, *Trifolium pratense*, *Vicia faba*, *Zinnia elegans*. Moreover, soybean, also as mycelial infection (Nicholson, Dhingra and Sinclair, 1972, 1973) and French bean as mycelial infection (Steadman, 1975). Testing procedure for sunflower has been worked out in Roumania (Tircomnicu and Iliescu, 1973). The disease cycle is described in Section 14.6 and further dealt with in Subsection 1.3.3, and economic importance in Subsection 1.2.2, bean, and in Subsection 1.2.4, cabbage.

Distribution: ubiquitous.

S. trifoliorum, clover rot. Soil transmission by mycelial growth and air dissemination of ascospores are predominant, but sclerotia may be found in seed of red clover, and mycelial infection of seed has also been established like *S. sclerotiorum*; has no conidial state (Doyer, 1938; Justham and Ogilvie, 1950; Brett, *fide* Sampson and Western, 1954; Andersen, 1970AR). Dillon Weston *et al.* (1946) pointed out that sclerotia may be produced on infected plants and be harvested and threshed with the seed. The disease is one of the most serious in herbage legumes, some fields may be completely ruined under cool humid conditions as in Northern Europe.

Distribution: North and Central America, South America (Chile), Europe, Asia (scattered), Australia, New Zealand.

CMI Distribution Map No. 274.

5.3.5.10 MYRIANGIALES

Globose asci are developed singly in stromatic locules, or in locules in a gelatinous mass.

Elsinoe

Plant parasites producing ascostromata in or on the epidermis, as well as acervuli or sporodochia of the imperfect state, *Sphaceloma*.

E. phaseoli, scab of beans. Circumstantial evidence that the fungus has been introduced with lima bean seed to the U.S.A. (Sitterly and Epps, 1958) and to Rhodesia (Bates, in Staples, 1958). At the Institute of Seed Pathology, Copenhagen, the fungus produced the scab on French bean in growing-on tests of four seed samples from Brazil (Lasca *et al.*, 1975, in preparation).

CMI Distribution Map No. 194.

5.3.5.11 PLEOSPORALES

This large group of Ascomycetes includes some very important genera of seed-borne pathogens, such as *Cochliobolus*, *Pleospora* and *Pyrenophora*. The development of the perithecia, by many mycologists named more precisely pseudothecia, is fundamentally different from that of typical perithecia; the asci arise in large spherical ostiolate locules in stromata and are bitunicate (i.e. with two separable walls); the stromata are called ascostromata, characteristic of Loculoascomycetes.

Cochliobolus

Pseudothecia without setae but often covered with conidiophores; ascospores filiform, more or less spirally coiled within ascus. Species discussed under Fungi Imperfecti I, Subsection 5.3.6.2: *C. geniculatus*, imp. state *Curvularia geniculata*; *C. heterostrophus*, southern leaf spot or blight of maize, imp. state *Drechslera maydis*; *C. lunatus*, imp. state *Curvularia lunata*; *C. miyabeanus*, brown spot of rice, imp. state *Drechslera oryzae*; *C. sativus*, foot rot, black point, imp. state *Drechslera sorokiniana*; *C. setariae*, seed rot, imp. state *Drechslera setariae*; *C. victoriae*, Victoria blight, imp. state *Drechslera victoriae*.

Didymella

Only one ascus-bearing cavity in the globose stroma, pseudothecia being deeply immersed in the substrate, only erumpent at the apex; ascospores hyaline. Conidial states *Discella* (syn. *Diplodina*).

D. bryoniae (Mycosphaerella melonis, M. citrullina), imp. state *Phoma cucurbitacearum*, syn. *Phyllosticta (Discella?) citrullina*, leaf spot, black rot, gummy stem blight of cucumber and watermelon. Seed-transmitted in

Citrullus vulgaris (Parris, 1947; Rankin, 1954; Andersen and Broe, 1974AR), in *Cucumis sativus* (Brown, Howard and Knight, 1970; Andersen, 1971AR*b*, 1973AR; Andersen and Broe, 1974AR; Lee *et al.*, in preparation), and in *Cucurbita pepo* (Lee *et al.*, in preparation).

Distribution: North and South America, Europe, Asia.

CMI Distribution Map No. 450.

D. lycopersici, stem canker or stem rot of tomato, imp. state *Phoma lycopersici (Ascochyta lycopersici)*. The fungus is infrequently detected in seed; it is transmitted to seedlings or introduced to soil, especially sterilised soil for glasshouse crops (Schoevers, 1929; Alcock, 1931; Doyer, 1938; Ogilvie, 1945; Van Koot and Brons, 1949; Williams, Sheard and Read, 1953; Fisher, 1954; Phillips, 1955, 1956; Knight and Keyworth, 1960; Derbyshire, 1960; Verhoeff, 1963). The disease has been destructive in Northern Europe but has decreased in recent years. Hot water treatment has been used. The pathogen should be considered as an important quarantine object.

Distribution: In most of Europe, and in scattered limited regions in North and Central America, Africa, Asia, New Zealand.

CMI Distribution Map No. 324.

Leptosphaeria

Spherical or conical, ostiolate pseudothecia immersed in host tissue, mostly in stems and leaves; ascospores fusoid to scolecosporous, septate. Saprophytes, some parasites. Three species are discussed under Fungi Imperfecti I, Subsection 5.3.6.2: *L. avenaria*, black stem, leaf blotch, imp. state *Septoria avenae*; *L. maculans*, black rot, dry rot of crucifers, imp. state *Phoma lingam*; *L. nodorum*, glume blotch, imp. state *Septoria nodorum*.

Physalospora

Pseudothecia are globose or setose, with a fleshy wall; ascospores are one-celled, more than 20 μ long, hyaline or bright, with granular or milky content. *P. rhodina*, imp. state *Diplodia gossypina* and *P. zeicola*, imp. state *Diplodia frumenti*, see Fungi Imperfecti, Subsection 5.3.6.3.

Pleospora

Black globose, sometimes flattened, ostiolate pseudothecia initially embedded in the host, later superficial; ascospores muriform. Three species are discussed under Fungi Imperfecti I, Subsections 5.3.6.2 and 5.3.6.3: *P. björlingii*, black leg of beet, imp. state *Phoma betae*; *P. herbarum*, see Fig. 5.8, imp. state *Stemphylium botryosum*; *P. calvescens*, imp. state *Drechslera papaveris*.

Pyrenophora

Closely related to *Pleospora* but the pseudothecia are hairy, ascospores muriform. The following species are discussed under Fungi Imperfecti I, Subsection 5.3.6.2: *P. avenae*, leaf spot, imp. state *Drechslera avenae*; *P. bromi*, brown leaf spot, imp. state *Drechslera bromi*; *P. graminea*, leaf stripe of barley, imp. state *Drechslera graminea*; *P. semeniperda*, sterility disease, imp. state *Drechslera verticillata*; *P. teres*, net blotch of barley, imp. state *Drechslera teres*; and *P. tritici-repentis*, imp. state *Drechslera tritici-repentis*.

Trichometasphaeria

Pseudothecia are perithecium-like, ostiolate, short setae surrounding the pore; ascospores are phragmosporous. For *T. turcica*, imp. state *Drechslera turcica*, see Fungi Imperfecti I, Subsection 5.3.6.2.

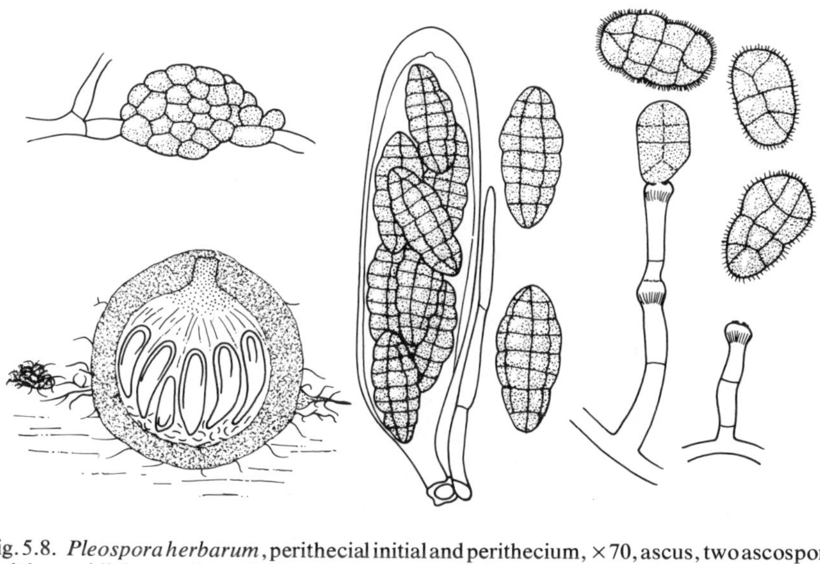

Fig. 5.8. *Pleospora herbarum*, perithecial initial and perithecium, ×70, ascus, two ascospores, and the conidial state, *Stemphylium botryosum*, with three separate conidia. ×500. (Courtesy of J.A. Von Arx)

5.3.5.12 DOTHIDEALES (*Dothiorales*)

Pseudothecial cavities, usually globose, are formed within massive stromata, which are immersed within or seated upon the host. The pseudothecia lack definite walls and are usually opened wide at maturity. Most of the species are leaf parasites in tropical and subtropical regions, but some occur in the temperate zones.

Guignardia

Globose ostiolate pseudothecia deeply immersed in the host tissue; ascospores nonseptate.

G. fulvida, stem break, browning of flax, imp. state *Aureobasidium lini*, is discussed under Fungi Imperfecti I, Subsection 5.3.6.2.

G. robiniae is harboured as dormant mycelium in the seed of *Robinia pseudacacia*, and seedlings become infected, most severely at low temperatures and in wet soil. Boiling water or hot water at 70°C for 5 min. improves germination but does not kill the fungus, whereas organomercurial combined with hot water is effective (Sato and Shôji, 1960).

Microcyclus (Melanopsammopsis)

Black spherical pseudothecia embedded in the erumpent stroma which tapers basally to a centrally attached foot; ascospores two-celled, hyaline. Two imperfect states: a pycnidial state, *Aposphaeria* (similar to *Phoma*; superficial pycnidia), and a hyphomycetous state, *Fusicladium* (1–3-celled, usually two-celled, brown conidia, single or in short chains).

M. ulei (Dothidella ulei, Melanopsammopsis ulei), imp. states (1) *Aposphaeria ulei* and (2) *Fusicladium macrosporum*, South American leaf blight of rubber and para rubber. Although actual seed transmission in *Hevea brasiliensis* has not been established, severe quarantine precautions include also seed. Importation of seed to Africa is prohibited but import permit may be granted on conditions, including especially disinfection of seed and container according to instructions; imported material is subject to post-entry cultivation (IAPSC, 1962; Caresche and Rayner, 1969). The pathogen causes blight and more or less complete defoliation of trees, resulting in serious reduction of yields.

Distribution: Central and South America, between latitudes 18° north and 24° south.

CMI Distribution Map No. 27.

Mycosphaerella

Ostiolate pseudothecia are small, seated on or immersed in a stroma; ascospores are two-celled; the conidial states are: *Ramularia*, *Cercospora*, *Phyllosticta*, *Ascochyta*, *Septoria* and others.

M. brassicicola, black ringspot of crucifers, has been isolated from cabbage seed, and seed taken from beneath lesions on the pods produced seedlings with lesions on the cotyledons; the disease was also established in field experiments from infected seed (Huber and Gould, 1949). Circumstantial evidence that the pathogen was introduced with seeds to Saskatchewan, Canada (Vanterpool, 1960).

Distribution: North and South America, Europe, Africa, Asia, Australia, New Zealand.

CMI Distribution Map No. 189.

Species under Fungi Imperfecti I, Subsections 5.3.6.2 and 5.3.6.3:

M. arachidis, tikka disease, brown spot of groundnut, imp. state *Cercospora arachidicola*; *M. berkeleyi*, tikka disease of groundnut, imp. state *Cercospora personata*; *M. davisii*, imp. state *Cercospora davisii*; *M. linorum*, pasmo of flax, imp. state *Septoria linicola*; *M. phaseolicola*, frog eye leaf spot of soybean, imp. state *Cercospora sojina*; *M. pinodes*, foot rot of pea, imp. state *Ascochyta pinodes*; and *M. rabiei (Didymella rabiei)*, anthracnose of chick-pea, imp. state *Ascochyta rabiei*.

Sphaerulina

Pseudothecia are separate, immersed in the substrate; asci are fasciculate and ascospores phragmosporous, hyaline. *S. oryzina*, imp. state *Cercospora oryzae*, see Fungi Imperfecti I, Subsection 5.3.6.2.

5.3.6 Fungi Imperfecti (Deuteromycetina) I

(Related or assumed to be related to Ascomycetes)
In principle, classification of fungi is based on the morphology of the sexual state and is thus believed to provide a 'natural' system, i.e. to reflect the phylogenetic evolution of the 'Kingdom Fungi' (Ainsworth, 1971) or 'Subkingdom Fungi' or 'Division Fungi' of the Plant Kingdom.

Fungi Imperfecti is an artificial 'Subdivision' (Ainsworth, 1971) or 'Class', based on the asexual state of those fungi which lack or apparently lack a sexual cycle and therefore cannot be conveniently placed in the 'natural' classification system. When the sexual state is discovered, both states, the sexual or perfect and the asexual or imperfect states, are taxonomically transferred to the group to which the perfect state belongs.

Most Fungi Imperfecti are asexual states of Ascomycetes; a small minority is connected with Basidiomycetes. By convention, the asexual states of Zygomycetes (with some exceptions) and Oomycetes are not placed in Fungi Imperfecti.

In a generally accepted broader sense, Fungi Imperfecti comprise any asexual reproductive forms of the mentioned related groups of fungi, including those having a known perfect state.

Most seed-borne fungi are either Fungi Imperfecti *sensu stricto* or Ascomycetes that occur on seed in their asexual state. Consequently, these fungi are usually identified on seed in the imperfect state. Although strictly botanically it is incorrect, in this survey all related fungi that occur on seeds in an asexual state are therefore with a few traditional exceptions discussed below under Fungi Imperfecti.

To emphasise the connection with their perfect states, Fungi Imperfecti related to Ascomycetes have been treated in this subsection as Fungi Imperfecti I, separately from those related to Basidiomycetes which have been treated in Subsection 5.3.8, Fungi Imperfecti II.

The *classification* of Fungi Imperfecti is based on the morphology of the asexual reproductive structures, and the principal distinction is made between presence or absence of special fruit bodies. In Blastomycetes reproduction occurs by budding of individual cells without any distinct differentiation of these. In Hyphomycetes conidiophores and conidia (conidium, any asexual spore except sporangiospore) are produced loosely from the ordinary hyphae. In Coelomycetes conidiophores are clustered into fruit bodies, usually built up by special mycelial tissues; they are either of the type of flat, open aggregations of closely packed conidiophores, the *acervulus* (Melanconiales), or in the form of a usually globose or flask-shaped body, the *pycnidium* (Sphaeropsidales).

Classification of the form genera within Fungi Imperfecti is based on four further principal criteria: (1) conidial type, according to Saccardo (1882–1925) (Table 5.3); (2) arrangement of conidia on the hypha bearing them which is usually differentiated as a conidiophore; (3) colour of conidia; and (4) type of conidiogenesis, especially as defined by type of the conidium-producing cell.

Table 5.3. Descriptive names of conidia and spores, according to Saccardo

			Names of conidia	
Shape	*Septation*	*Form designation*	*Hyaline*	*Coloured*
Globose, ellipsoid, ovoid, allantoid, cylindrical, lunate, falcate, obclavate, and of many other shapes	0 cross septum (1-celled)	Amerosporae	Hyalosporae	Phaeosporae
	one cross septum (2-celled)	Dimerosporae	Hyalodidymae	Phaeodidymae
	2–several cross septa (multicellular)	Phagmosporae	Hyalophragmiae	Phaeophagmiae
	Cross and longitudinal septa (muriform)	Dictyosporae	Hyalodictyae	Phaeodictyae
Filiform	Aseptate or septate	Scolecosporae	Hyaloscolecosporae	Phaeoscolecosporae

Colour and cell number of conidia (and spores) are in modern taxonomy much less strictly used as criteria for classification, emphasis being laid on conidiogenesis (and sporogenesis in perfect states).

Based on the mechanism of conidial production distinction is made between the following main types of conidia (between which there are intermediate forms):

(1) *Blastoconidia* (blastospores) arise as a bud from the wall of the conidiophore; they are typically sprouting cells. Examples: Blastomycetes; *Cercospora*.

(2) *Poroconidia* (porospores) arise from a protusion of protoplasm through a pore in the wall of the conidiophore. Examples: *Alternaria*, Fig. 5.9; *Drechslera*. Poroconidia may be regarded as blastoconidia borne through pores of thick-walled conidiophores.

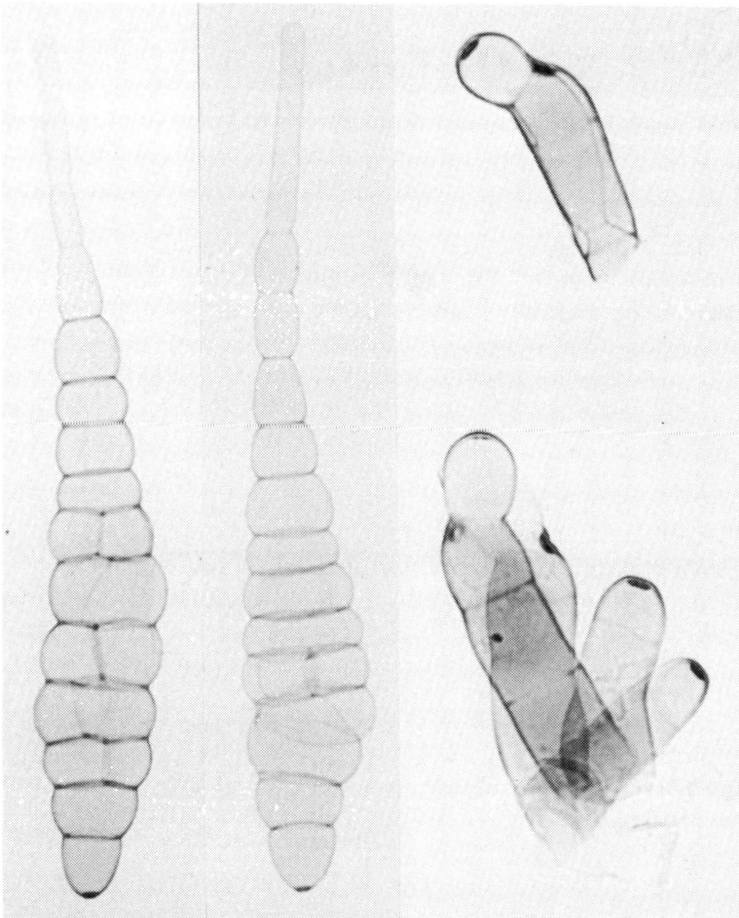

Fig. 5.9. Conidia (poroconidia) and conidiophores of *Alternaria brassicae,* showing slightly sunken scars; conidia are rounded off at the base and easily detached. ×500. (Courtesy of M. B. Ellis; photo D. W. Fry)

(3) *Phialoconidia* (phialospores) are formed either within the flask- or bottle-like termination of the conidiophore, called a phialide, or an expanded conidiophore tip being separated by a septum. Examples: *Acremonium*, see Fig. 5.10, *Aspergillus*, *Penicillium*.

(4) *Arthroconidia* or *thalloconidia* (thallospores) are formed by fragmentation of hyphae, separation at septa or by fracture across cell walls; they are fission cells; hence, it is an asexual spore having no conidiophore. Examples: *Sporotrichum*, not discussed in this survey.

(5) *Chlamydospores* ('chlamydoconidia' would be preferable). A thick-walled intercellary or terminal spore type. Example: *Fusarium*.

Some terminal chlamydospores are called aleurospores, not dealt with in this book. *Dictyochlamydospores*, as found in some species of *Phoma*, have transverse and longitudinal septa.

Another distinction between types of conidia pertains to the mode of spread: (1) dry conidia (in Hyphomycetes, *Xerosporae*) which are separated without slime, and (2) slimy conidia (in Hyphomycetes, *Gloiosporae*) which are held together in slimy drops or (when forced out of pycnidia) in a tendril-like mass, or cirrus. An estimated 75 per cent of Hyphomycetes have dry, 'dusty' conidia, which are spread by wind or air-currents, the remaining having slimy conidia being disseminated by water splashes or on animals. All Coelomycetes have slimy conidia.

Conidia may arise solitary, or be catenate, formed in simple or branched chains. The chain may be *acropetal*, the apical conidium, the one at the tip, being the youngest, as in *Alternaria*, or it may be *basipetal*, the apical conidium being the oldest, as in *Coniosporium*, not discussed in this survey, and in *Aspergillus*.

For identification to genus the following handbooks dealing with larger groups of Fungi Imperfecti, are useful: Barnett and Hunter (1972): *Illustrated Genera of Imperfect Fungi* (3 ed.); Barron (1968): *The Genera of Hyphomycetes from Soil*; Ellis (1971): *Dematiaceous Hyphomycetes*; Grove (1935, 1937, reprints 1967): *British Stem- and Leaf-Fungi* (Coelomytes); Kendrick and Carmichael (1973): *Hyphomycetes*; and Sutton (1973): *Coelomycetes*, both in Ainsworth, Sparrow and Sussman, eds. (1973): *The Fungi. An Advanced Treatise*. Vol. IV A; and Subramanian (1971): *Hyphomycetes*. Further references, dealing with special groups, are given below.

5.3.6.1 BLASTOMYCETES OR ASPOROGENOUS YEASTS

One-celled fungi characteristic of having asexual reproduction by budding, transverse division, or both. Many forms are believed to have lost their ability to form ascospores, as in Saccharomycetes. The process of budding may result in chains of cells which may branch. Some asporogenous yeasts are important human and animal pathogens (*Candida, Cryptococcus*), very few are plant pathogens.

Rhodotorula

The cells are round to elongated, and may by budding form short 'pseudohyphae'. They produce pink-red or orange colonies; some are frequent laboratory contaminants.

R. glutinis var. *rubescens* has been encountered in seeds of Ladino white clover, *Trifolium repens*. Infected seeds are shrivelled, pitted, generally non-viable, and the fungus is harboured in the seed coat. The pathogen produces a leaf spot similar to bacterial leaf spot (Kilpatrick,

1959). The disease may occur where humidity is high over long periods.
Distribution: U.S.A. (New Hampshire).

5.3.6.2 HYPHOMYCETES

Conidia are produced on loose hyphae, not in fruit bodies nor in stromata, but conidiophores are in some genera aggregated in sporodochia as in *Fusarium*. Conidiophores are single or branched, conidia are borne singly, in groups, or in chains. A synopsis of seed-borne genera is given in Table 5.4.

Mucedinaceae

Conidiophores free, conidia and hyphae colourless or in pale or bright colours.

Table 5.4. Hyphomycetes. Classification of the surveyed seed-borne genera

Family	*Conidial type*	*Form-Genus*
Mucedinaceae	Hyalosporae	Acremonium
		Acrocylindrium
		Aspergillus
		Botrytis
		Geniculodendron
		Hymenula
		Monilia
		Penicillium
		Verticillium
	Hyalodidymae	Rhynchosporium
		Trichothecium
	Hyalophragmiae	Pleiochaeta
Dematiaceae		Ramularia
	Phaeosporae	Aureobasidium
		Chalara
		Nigrospora
		Ustilaginoidea
	Phaeodidymae	Cercosporidium
		Cladosporium
		Fusicladium
	Phaeophragmiae	Acroconidiella
		Cercospora
		Corynespora
		Curvularia
		Drechslera
		Isariopsis
		Mycocentrospora
		Phaeoisariopsis
		Pyricularia
		Trichoconis
	Phaeodictyae	Alternaria
		Stemphylium
		Ulocladium
Tuberculariaceae	Hyalophragmiae	Fusarium
		Gloeocercospora
		Ramulispora
	Phaeospora	Myrothecium
	Phaeodictyae	Epicoccum

(1. Hyalosporae)

Acremonium (Cephalosporium)

Conidia are formed in sticky balls, Fig. 5.10, or rarely in chains from phialides arising from the vegetative hyphae (Gams, 1971).

A. strictum (*Cephalosporium acremonium* auct.), Fig. 5.10, and/or the closely related *A. kiliense*, kernel rot and black bundle disease of maize (cf. *Gaeumannomyces graminis*) established as seed-transmitted in the U.S.A. (Reddy and Holbert, 1924), commonly present in maize seed in Egypt (Sabet *et al.*, 1962) and India (Dhanraj and Mathur, 1965; Singh, Chaube and Singh, 1971), causing discoloration of the kernels, sometimes kernel rot. The pathogen may invade the vascular bundles and causes stalk rot. *A. strictum* is commonly isolated from soil.

Distribution: North America, Europe, North Africa, India.

A. zeae, corn stalk rot. Seed-borne in maize but of little importance in field (Sumner, 1967); has also been encountered in the caryopsis of *Sorghum vulgare* (Gams, 1971). Economic significance has not been established, as so far the species has not been well distinguished in literature from *A. kiliense*, cf. above.

Distribution: North America, Europe.

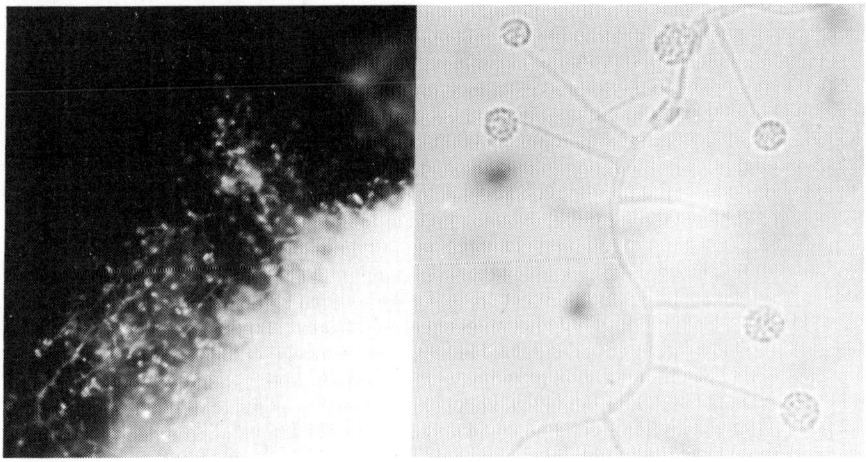

Fig. 5.10. *Acremonium strictum* on maize. Left, habit characters; right, phialides with drops of phialoconidia. ×100

Acrocylindrium

Closely related to *Verticillium*, by some mycologists regarded as identical with this genus.

A. oryzae, sheath rot, may seriously infect young ears of rice, whereas grains are rarely affected (Tasugi and Ikeda, 1956; Chen and Chien, 1964).

Distribution: Asia (Japan, Taiwan).

Aspergillus
Black moulds

Conidiophores are long, erect, terminating in a bulbous head, the vesicle, bearing numerous sterigmata, one or two layers of these may be reproduced. Globose phialoconidia are formed in basipetal chains. About 50 species, widespread, several species, including *A. flavus*, Fig. 5.11, *A. fumigatus*, *A. glaucus* and *A. niger* are important pathogens on stored grains and other seeds, see Chapter 7, and important producers of mycotoxins, see Chapter 8.

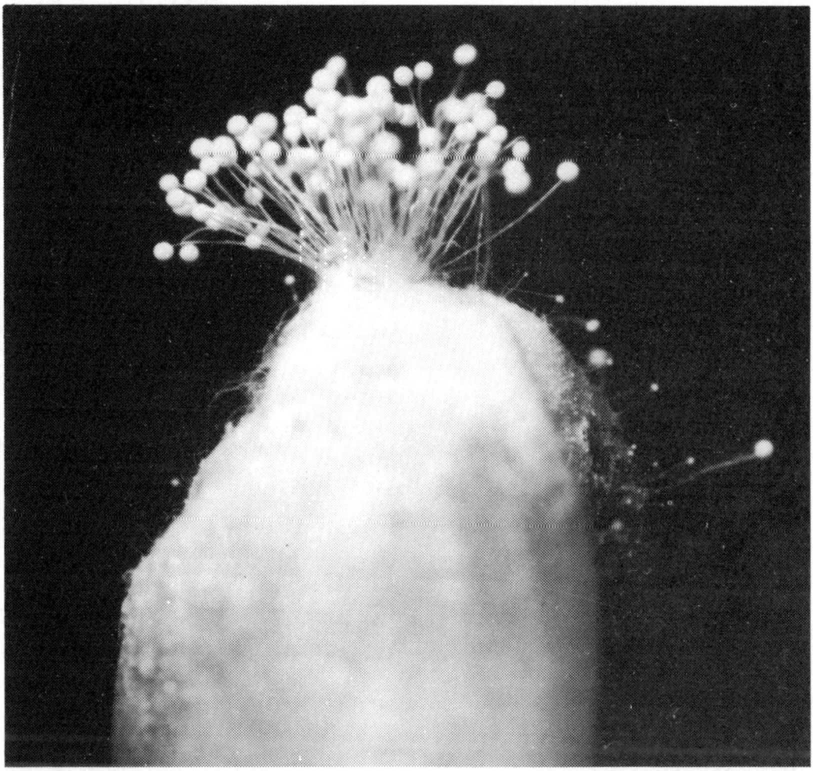

Fig. 5.11. *Aspergillus flavus* on wheat. (Courtesy of Clyde M. Christensen)

Botrytis
Grey mould

Blastoconidia in racemose arrangement, grouped like grapes. Many serious pathogens.

B. allii, common in seed of onion, *Allium cepa*, and chives, *A. schoenoprasum*; occurs also in leek, *A. porrum* (Andersen, 1974 — index). Seed crops in Israel sometimes badly infected; losses of seed may reach 80 per cent (Netzer and Dishon, 1966). The fungus causes seedling damping-off and storage rot.

Distribution: North America, Europe, Africa, East Asia, Australia, New Zealand.

CMI Distribution Map No. 169.

B. anthophila, anther mould of red clover, common in seed, but overlooked in the field (Doyer, 1938; Aleshina, 1957; Ramazanova, 1958, 1960; Salonen, 1960; Jost, Volken and Kern, 1964; Paulech, 1965; Bennum, 1972). Life-cycle, see Section 14.3.

Distribution: North America (Washington State, Oregon), Europe, Africa, Asia, Australia, New Zealand.

CMI Distribution Map No. 167.

B. cinerea, common grey mould, perfect state *Sclerotinia fuckeliana* (only known in the laboratory), a ubiquitous plant parasite, having a wide host range, often seriously attacking seed crops, such as flower seed crops. It is seed-borne in many hosts, of which Noble and Richardson have listed (with references) 44 genera. It often reduces seed germination and produces abnormal seedlings. Longevity in seed is limited to about a year under ordinary storage conditions, see Subsection 20.2.2.

B. fabae, chocolate spot of broad bean and vetches, has been proved to be seed-transmitted in *Vicia faba* and *V. villosa* (Noble and Richardson, 1968). Controlled by hot water treatment (Bikmukhametova, 1963).

Distribution: South America, Europe, Africa, Asia, Australia, New Zealand.

CMI Distribution Map No. 162.

B. ricini, grey mould of castor bean, perfect state *Sclerotinia ricini*, often infects the capsules ('capsule mould') and the pathogen is internally borne or present as sclerotia attached to seed coats and hulls (Godfrey, 1923). Cultivars with compact inflorescences and dwarf internodes more susceptible than those with interspersed, staminate flowers (Thomas and Orellana, 1963).

Geniculodendron

One-celled, pyriform blastoconidia are produced from sympodial conidiogenous cells which are formed singly or verticillately on conidiophores.

G. pyriforme, psychrophilic fungus of coniferous seed. Borne in the endosperm and embryo of imported seed of Sitkaspruce (*Picea sitchensis*) in England (Salt, 1974), encountered in ungerminated seeds of *Picea glauca*, *Pinus resinosus*, *P. strobus* and *P. sylvestris* in Canada (Epners, 1964). 'Stored seed usually contains only 0–5 per cent of infected seed, but this is sufficient to cause patchy emergence in nursery seed beds in cool springs, and substantial losses, where seed is sown earlier than the normal time of early mid-March', in England (Salt, 1968). The fungus attacks only the seed. Favoured by stratification of seed. For testing procedure, consult Salt (1974). Should be considered for quarantine.

Distribution: North America, United Kingdom.

Hymenula

Phialides are procumbent, phialoconidia do not form heads but conglomerate in larger masses (Gams, 1971).

H. cerealis (Cephalosporium gramineum), 'Cephalosporium' stripe disease, occurs in variable frequency in seeds of barley (Lutey, 1962), oats (Kiewnick, 1963), and wheat (Bruehl, 1957). Infected seeds may introduce this important xylem-invading pathogen to new areas.

Distribution: North America, Europe, South Africa, Asia (Japan).

Monilia

Perfect state *Sclerotinia (Monilinia)* or *Neurospora*. Conidia formed as acropetal chains of one-celled blastoconidia.

M. laxa, perfect state *Sclerotinia laxa*, may occur in the nut of hazelnut, *Corylus avellana* (Lovisolo, 1951), but this is of doubtful significance for transmission.

M. roreri, Monilia rot, grey pod rot of cacao, *Theobroma cacao*, the seed is infected and becomes decayed, in older pods the mycelium grows around the seed (Wellman, 1954, 1972; Weber, 1973). Only the pods are infected. Quarantine for preserved pollen must be considered. Spores may be transmitted with pollen that has been freeze-dried at −50°C — the spores may germinate still after three months (Instituto Nacional de Investigaciones Agropecurias, 1965). Important quarantine object in Africa.

Distribution: Tropical Central and South America.

CMI Distribution Map No. 13.

Penicillium

Seed rot, blue eye

Perfect states (*Carpenteles*, *Talaromyces*) probably do not occur on seed. Phialoconidia are formed in basipetal chains from phialides which are grouped in broom-like whorls, the 'penicillus', and these may be twice or several times branched verticillately, the ultimate branches being the phialides.

Different species, including *P. chrysogenum*, *P. cyclopium* (see Subsection 7.2.2), *P. oxalicum*, *P. patulum*, *P. purpurogenum* var. *rubrisclerotium* occur as storage fungi in grain, such as maize (Noble and Richardson, 1968, with references); moreover, as important toxin producers, *P. citrinum*, *P. islandicum*, *P. rubrum* and *P. viridicatum* (see Subsection 7.5.5).

P. oxalicum, isolated from seedlings of *Sorghum vulgare* var. *subglabrescens*, cv. group 'Milos', grown from untreated seed sown in sterilised soil (Tarr, 1962). The pathogen causes typical seedling wilt. Similar symptoms in maize, in which the fungus also occurs in seed.

Distribution: Probably ubiquitous.

Verticillium

Conidia formed from phialides on conidiophores with branches arranged in whorls. The conidia often held together in slimy, globose to oval drops. Review of Verticillium diseases by Pegg (1974).

V. alboatrum, Verticillium wilt, 'alkaline soil wilt', has been established in seed of many species of the very wide host range, the following species or genera of hosts listed, with references, by Noble and Richardson (1968): *Allium*, *Apium graveolens*, *Carthamus tinctorius*, *Gossypium*, *Helianthus annuus*, *Lactuca sativa*, *Linum usitatissimum*, *Medicago*, *Senecio vulgaris*, *Solanum melongena*, *Trifolium hybridum*. The pathogen is predominantly soil-borne but is not infrequently encountered in seed of many hosts, and seed transmission is probably generally underestimated. There is inconsistency in the use of species names, *V. alboatrum* (without microsclerotia) sometimes used for *V. dahliae* (with microsclerotia). A major wilt disease of the world, predominant in cooler temperate regions in alkaline soils, in contrast to Fusarium wilt, see *Fusarium oxysporum*.

Distribution: Probably world-wide.

CMI Distribution Map No. 365.

V. dahliae, morphology as for *V. alboatrum*, but distinguished from this in producing microsclerotia, Fig. 13.4; it is undoubtedly seed-borne in as many hosts as the latter species; for additional hosts, including tomato, see Noble and Richardson (1968), and on importance in cotton seed, Karaca, Ceylan and Karćílioǧlu (1973) and Shtok and Idessis (1974), safflower seed, Goethals (1971) and Klisiewicz (1974, 1975), and seed of *Xanthium pungens*, Evans (1968).

Distribution: Widespread

CMI Distribution Map No. 366.

(2. Hyalodidymae)

Rhynchosporium

Conidia are produced directly from vegetative hyphae and are fusoid or irregularly fusoid, two-celled.

R. secalis, scald or leaf blotch of barley, has been consistently encountered in seed of barley (Skoropad, 1959; Habgood, 1971; Kay and Owen, 1973) but the fungus is undoubtedly carried over predominantly on infected plant debris (Shipton, Boyd and Ali, 1974, review). Infections of seedlings depend on the growth rate of the plumule, at higher temperatures (22–27°C) seedlings escape infection. Scald has attracted increasing interest in recent years because of the growing importance of the disease due to changed agricultural practices. Organic mercurials are effective for seed treatment; the pathogen has a wide host range within Gramineae.

Distribution: N. and S. America, Europe, Africa, Australia, New Zealand.

CMI Distribution Map No. 383.

Trichothecium (Cephalothecium)

Conidiophores are erect, tapering, conidia are produced on single conidiophores in clusters formed by basipetal, irregular, fragile chains.

T. roseum (Cephalothecium roseum), a common saprophyte on seeds (Doyer, 1938; Malone and Muskett, 1964). The fungus is a weak pathogen, frequent cause of rot in stored apples, and it seems sometimes to be weakly pathogenic to seedlings, e.g. of *Zinnia elegans*, as observed in seed testing on blotter (Neergaard, 1958*c*); in forest tree seed, e.g. *Betula*, *Fagus sylvatica*, *Fraxinus excelsior*, *Euonymus*, *Pinus sylvestris*, *Picea excelsa* and *Quercus* it may cause seed rot (Urošević, 1961*b*); moreover, the fungus has been encountered in seeds of *Abies alba*, *A. cephalonica, Cedrus deodara, Cytisus alpinum, Fagus sylvatica* and *Tilia cordata* (Neergaard, 1956AR, 1958AR*a*, 1958AR*b*, 1962AR).

Distribution: Ubiquitous.

(3. Hyalophragmiae)

Pleiochaeta

Conidia bearing several appendages on one of the two apical cells; these are lighter than the other cells of the conidium.

P. setosa (Ceratophorum setosum), lupin leaf spot, brown spot, in seeds of *Lupinus angustifolius* and *L. albus* (Du Plessis and Truter, 1953; Ostazeski, 1960). The fungus produces distinct discoloration of the seed. Hot water treatment eliminates the pathogen.

Distribution: North America, Europe, South Africa, East Asia, Australia.

CMI Distribution Map No. 243.

Ramularia

Blastoconidia are developed in acropetal chains on simple or branched conidiophores. Common leaf spot pathogens, especially in temperate zones.

R. beticola, leaf spot of beet, is regarded as seed-borne (McKay, 1952), and isolated from seed clusters by Von Plotho (1953), but seed transmission seems never to have been established. Leaf spot disease of cooler climates, less severe than the one produced by *Cercospora beticola* in warmer climates.

Distribution: North America, Europe.

R. cyclaminicola, cyclamen wilt and leaf spot, seed-borne in the seed coat of cyclamen, according to Jacques (1942). Pathogenesis described by Baker, Dimock and Davis (1950). Considered as a quarantine object.

Distribution: North America.

Dematiaceae

Conidiophores free, conidia (blastoconidia) and hyphae more or less brown.

(1. Phaeosporae)

Aureobasidium (Pullularia)

The major morphological feature is production of thick-walled chlamydospore-like cells, from which hyaline, oval conidia are produced on short phialidic papillae, subsequently secondary conidia are formed by yeast-like budding of primary conidia; the conidia are aggregated in slimy masses.

A. lini (A. pullulans var. *lini, Polyspora lini)*, perfect state *Guignardia fulvida*, stem break, browning of flax. Commonly seed-borne, infection usually superficial (Wallen and Skolko, 1951; Hoffmann, 1954; Schrödter and Hoffmann, 1961; Malone and Muskett, 1964; Henry and Ellis, 1971). The perfect state described by Sanderson (1965). Serious disease in flax causing fracture at the stem base (stem break), later browning; seeds may be affected and killed.

Distribution: North America, Europe, Asia, Australia, New Zealand.

CMI Distribution Map No. 88.

A. pullulans (Pullularia pullulans), ubiquitous saprophyte, rather common contaminants of seed (Malone and Muskett, 1964). In ryegrass conidia may easily be confused with those of *Gloeotinia temulenta*!

Chalara

Phialoconidia from erect, unbranched phialidic conidiophores.

C. quercina, oak wilt, perfect state *Ceratocystis fagacearum*. Considered as a quarantine object for import of seeds of chestnut, *Castanea molissima*, and oaks, *Quercus* spp., in Italy (Anonymous, 1955), New Zealand (Pl.Prot.Service, New Zealand, personal communication, 1966), Czechoslovakia (FAO Pl.Prot.Bull. 8:72) and Turkey (FAO Pl.Prot.Bull. 10:93).

Distribution: North America.

CMI Distribution Map No. 254.

Kabatiella

The genus is by some mycologists regarded as a synonym of *Aureobasidium*, by others the genus is referred to Melanconiales, a view followed in this book, see below, Subsection 5.3.6.3.

Nigrospora

Depressed globose, opaque and shining black conidia are produced from short conidiophores.

N. oryzae, perfect state *Khuskia oryzae*. A saprophyte or very weak parasite producing grain spots of rice, sorghum and maize (Padwick, 1950; Kulik, 1963; Focke, 1964; Mohamed and Fathi, 1965).

Distribution: Widespread, probably in all parts of the world, particularly in the tropics.

Ustilaginoidea

Conidia solitary, usually spherical, olive green, verruculose, produced from minute sterigmata.

U. virens. Parasitises the ovary of rice which is transformed to a sclerotium covered by a dark olive green mass of conidia. This fructification resembles smut ('false smut'). The perfect state, established by Hashioka, Yoshino and Yamamoto (1951), seems to be at least closely related to *Claviceps*, see review by Ou (1972); Ikegami (1960) successfully inoculated plants with ascospores as well as conidia. The disease usually causes little damage but has in recent years become serious in some places in India.

Distribution: North, Central and South America, Europe (Italy), Africa, Asia, Australia.

CMI Distribution Map No. 347.

(2. Phaeodidymae)

Cercosporidium

Conidiophores simple, geniculate, fasciculate, arising from a stroma, conidial scars thickened and conspicuous, conidia clavate, cylindrical, obclavate or fusiform, with a conspicuous thickened hilum (Ellis, 1971).

C. graminis (Passalora graminis). Encountered on seed of timothy, *Phleum pratense* (Groves, *fide* Noble and Richardson, 1968). The fungus has a wide host range in Gramineae.

Distribution: North and South America, Europe.

C. punctum (Marssonina kirchneri, Passalora kirchneri; pycnidial state *Phoma anethi)*, leaf and stem spot or scurf of dill. Frequent in seed of dill, *Anethum graveolens*, grown in Denmark and Sweden, and also encountered in seed of root parsley, *Petroselinum crispum* var. *radicosum* (Neergaard 1950AR, 1951AR; Andersen, 1974 — index). Seed viability may be substantially reduced (Frauenstein, 1968). The disease is destructive in seed crops of dill in Scandinavia, favoured by humid growing conditions, see Subsection 19.6.6.

Distribution: North America, Europe, Africa, Asia.

Cladosporium

Ubiquitous saprophytes, some species are plant parasites; conidia (blastoconidia) solitary or in chains often being branched, produced on straight or flexuous, unbranched, or apically short-branched conidiophores.

C. cladosporioides, and less frequently *C. herbarum*, are common saprophytes on dead plant parts and on seeds, particularly after wet seasons (Malone and Muskett, 1964). A form of *C. cladosporioides* has been described as pathogenic on pea, including the seeds (Snyder, 1934), later observed in Rhodesia (Bates, 1955).

Distribution: Ubiquitous.

C. cucumerinum has been suggested to be seed-borne in cucumber, melon and squash, and seed treatment, using mercuric chloride, has been suggested (Ellis, 1953; Sherf, 1964).

Distribution: North and Central America, Europe, Africa, Asia.

CMI Distribution Map No. 310.

(3. Phaeophragmiae)

Acroconidiella

Conidia (poroconidia) are solitary, ellipsoid, constricted at septa, echinulate, developed on simple or branched, often geniculate conidiophores.

A. eschscholtziae (Heterosporium eschscholtziae), leaf blight of Californian poppy. Frequent in seed of *Eschscholtzia californica*, hot water seed treatment used in California (Davis, 1952; Andersen, 1974 — index). Mode of seed infection, see Subsection 12.2.4.

Distribution: North America, Europe.

A. tropaeoli (Heterosporium tropaeoli), leaf spot, stem canker, is a serious pathogen in seed fields of nasturtium, *Tropaeolum majus*, in California (Baker and Davis, 1950*b*), where it is commonly seed-transmitted, cf. Fig. 13.12. Controlled by hot water seed treatment, cf. Subsection 19.4.2.

Distribution: North and South America, Africa, Asia, Australia.

Cercospora

Conidiophores erect, sometimes geniculate, single or in fascicles; conidia (blastoconidia) solitary or acrogenous, obclavate, pale, smooth. This is a large genus with more than 1,000 species, many are important plant pathogens, producing leaf spot, predominantly in subtropical and tropical zones; perfect states are *Mycosphaerella*. For identification, see the monograph by Chupp (1953), and Vasudeva (1963*b*).

C. arachidicola, tikka disease of groundnut, perfect state *Mycosphaerella arachidis*, may occur in the seed but plant debris and volunteers are more important in carrying over the pathogen (Hemingway, 1954; Gibbons, 1966). A most serious pathogen, see Subsection 1.2.2.

Distribution: North and South America, Africa, Asia, Australia.

CMI Distribution Map No. 166.

C. beticola, Cercospora leaf spot of beets, common in the seed of *Beta vulgaris*, in regions where the fungus predominates, particularly in warmer temperate and subtropical climates, and seed transmission is important in some regions, perhaps predominant in temperate regions, e.g. Austria (Wenzl, 1959*a*, 1959*b*; Andersen, 1974 — index), less in warmer regions, e.g. Turkey, see Subsection 1.3.3. In cooler regions, such as Denmark, the disease is a rarity. Seed treatment with organomercurials not fully effective.

Distribution: Wherever beet is grown.

CMI Distribution Map No. 96.

C. capsici and *C. capsicicola (Cladosporium capsici)*, Cercospora leaf spot of chilli and pepper, reported as seed-borne in *Capsicum annuum* (Miller and Taylor, 1952; Doolittle, 1953).

Distribution of *C. capsici*: In tropical zones of North, Central and South America, Africa, Asia (India).

CMI Distribution Map No. 248.

C. carotae, Cercospora blight of carrot, in seed of carrot (Thomas, 1943; Saponaro, 1964).

Distribution: North and South America, Europe, Asia, Australia.

C. carthami, Cercospora leaf spot of safflower, *Carthamus tinctorius*, may be quite destructive under high humidity, on leaves as well as on the bracts, and the entire capitulum may be involved (Sundararaman and Ramakrishnan, 1928). Apparently introduced with seed into Israel (Ashri, 1961).

Distribution: Europe (Israel), Asia.

C. corchori, Cercospora leaf spot of jute, *Corchorus capsularis*, infects capsules and seeds (Chowdhury, 1948; Mukherji and Ghosh, 1960), infection percentage up to 45 recorded, and hot water treatment, 70°C for 10 minutes, effective (Odukoya, Mathur and Neergaard, unpublished).

Distribution: North America, South America (Brazil), Africa (Nigeria), Asia (India).

C. davisii, summer black stem and leaf spot of sweet clover, perfect state *Mycosphaerella davisii*, recorded on seed (Latch and Hanson, 1960) but transmission probably not established.

Distribution: North America.

C. insulana, Cercospora leaf spot, frog eye. Seed transmission in annual statice, *Limonium bonduelli*, demonstrated in Florida, U.S.A. (Engelhard, 1975).

Distribution: North and Central America, Europe.

C. kikuchii, purple blotch, purple seed stain of soybean, see Colour Plate

IV, commonly seed-borne in soybean, producing a bright purple stain although infected seed may remain unstained (Kilpatrick, 1957; Kulik, 1957; Jones, 1959; Crane and Crittenden, 1966; Nene and Srivastava, 1971; Laviolette and Athow, 1972; Agarwal and Joshi, 1972), particularly invading the hourglass and parenchyma cell layers at the hilum region (Ilyas *et al.*, 1975); and in guar or cluster bean, *Cyamopsis tetragonoloba*. Infected seeds produce diseased seedlings, the stem is girdled and the plant dies (Weber, 1973).

Distribution: North America, Europe, Asia.

C. nicotianae, frog eye, green spot of tobacco, appears in the clean nursery beds when infected or contaminated seeds are used (Park and Fernando, 1937). Hopkins (1956) indicates that infection, leading to heavy losses, can be traced to planting out infected seedlings but he does not mention seed as source of infection but recommends hygienic precautions related to the seed bed. Destructive under humid conditions.

Distribution: North, Central and South America, Europe, Africa, Asia, Australia, probably in all tobacco growing areas.

CMI Distribution Map No. 172.

C. oryzae, narrow leaf spot of rice, perfect state *Sphaerulina oryzina*, produces a glume spot, seed treatment with organomercurials has been found effective (Cherewick, 1954).

Distribution: North America, Asia, in all countries where rice is grown intensively (Chupp, 1953).

CMI Distribution Map No. 71.

C. personata, blotch, tikka disease of groundnut, perfect state *Mycosphaerella berkeleyi*; seed may be infected but seed transmission is of little importance because plant debris and volunteer plants act as efficient reservoirs (Hemingway, 1954). This species occurs often associated with another, *Cercospora arachidis*, both causing a serious leaf spot disease.

Distribution: Probably wherever groundnut is grown.

CMI Distribution Map No. 152.

C. ricinella, white leaf spot of castor bean, *Ricinus communis*, was found by Singh (1948) in seed for export from India. The disease is common wherever castor bean is grown.

Distribution: N. and S. America, Europe, Africa, Asia, Australia.

C. sesami, seed-borne in *Sesamum indicum* as established by Thomas (1950) and Mathur and Kabeere (1975); detected on sesame seed and subsequently on plants grown from this seed, introduced into the U.S.A. from China (Leppik, 1962*a*).

Distribution: North America, Africa, Asia, probably in most areas where sesame is grown.

C. sojina, frog eye leaf spot of soybean, perfect state *Mycosphaerella phaseolicola*, produces grey brown or 'papillate' discoloration in seed of soybean, markedly different from discoloration due to *C. kikuchii* (Sherwin and Kreitlow, 1952). Seed treatment with organomercurials or thiram recommended (Mikhaïlenko, 1965). Leaf spot sometimes leading to shedding of the leaves.

Distribution: Widespread, North America, Europe (U.S.S.R.), Asia, Australia.

C. sorghi, grey leaf spot of sorghum, has been encountered in seeds of sorghum during seed health testing at the Institute of Seed Pathology in Copenhagen, but transmission has not been established. The disease can be severe under certain climatic conditions. Maize is also attacked.

Distribution: North, Central and South America, Africa, Asia.

CMI Distribution Map No. 338.

C. traversiana, leaf spot in fenugreek, *Trigonella foenum-graecum*, cause of a destructive systemic leaf spot disease; the fungus has been encountered and intercepted in the U.S.A., harboured in the seed coat of seed introduced from India and Ethiopia (Leppik, 1959, 1960).

Distribution: (World survey based on seed samples, by Leppik, 1959, 1960); North and South America, Europe, Africa, Asia.

C. zebrina, summer black stem and leaf spot of medick and clover, listed as seed-borne by Noble and Richardson (1968), but recent records lacking; seed treatment advised.

Distribution: North America, Europe, Asia.

Corynespora

Conidiophores are erect with successive cylindrical proliferations, and produce obclavate, pyriform, cylindrical conidia (poroconidia), solitary or in short acropetal chains; the conidia have a dark basal hilum or truncate base.

C. cassiicola, Fig. 5.12, target spot of soybean, Corynespora blight of sesame, seed-borne in these two hosts (Liu, 1948; Stone and Jones, 1960; Mathur and Kabeere, 1975), soybean and probably also in cowpea (Wei, 1950). Very common in the tropics on a wide host range, producing leaf spot diseases, and occurring on fruits; sometimes destructive on sesame and some of the other hosts.

Distribution: Widespread, N. and S. America, Europe, Africa, Asia, Australia.

Curvularia

Conidiophores are usually erect, straight, often geniculate; the conidia are produced singly at the tip, and by successive growth of the apex of the conidiophore, on the sides, so that the conidia form a cluster arranged spirally around this portion of the conidiophore. The conidia (poroconidia) are curved, clavate, ellipsoid or pyriform, with three or more transverse septa, in some species with a protuberant hilum. Saprophytes, some species weak plant pathogens. Benoit and Mathur (1970) have monographed seed-borne species of *Curvularia* and described their

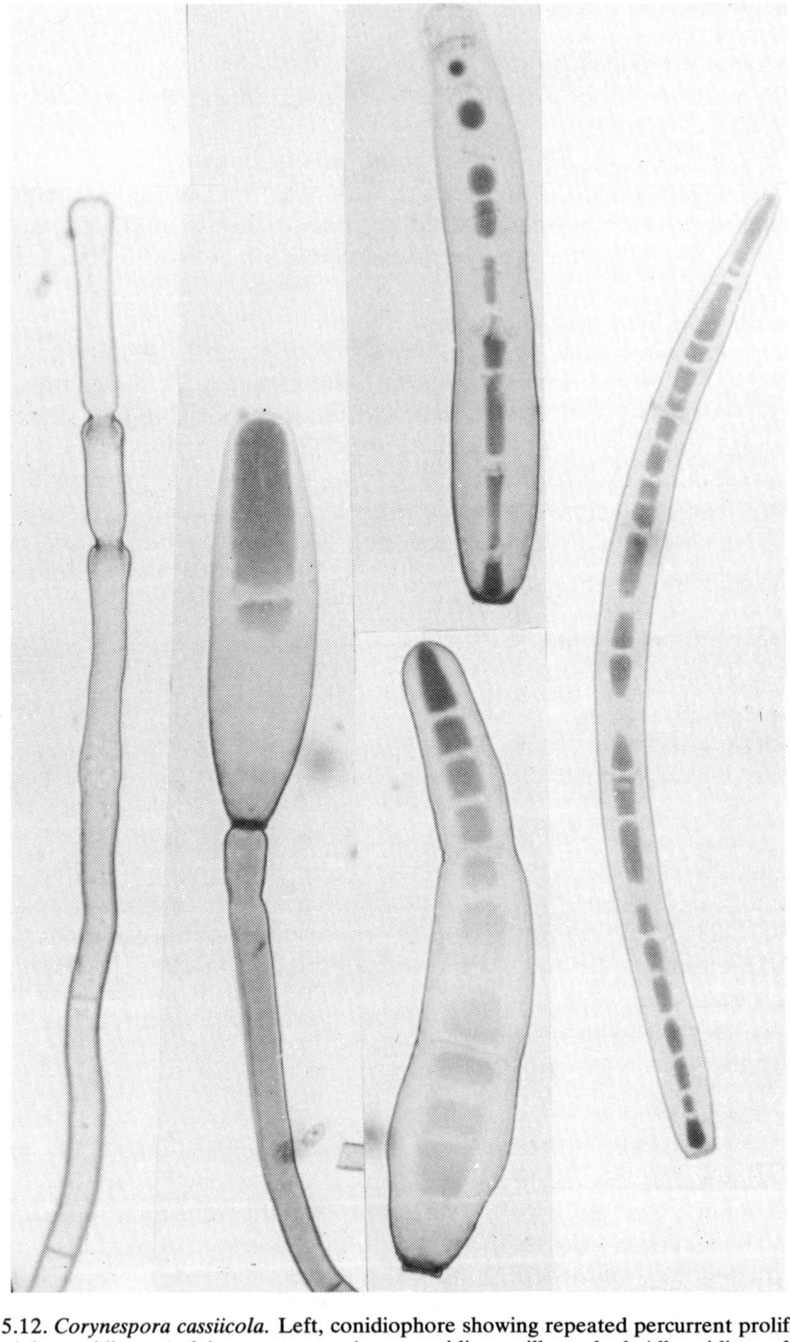

Fig. 5.12. *Corynespora cassiicola.* Left, conidiophore showing repeated percurrent proliferation; right, conidia, one of these a young primary conidium, still attached. All conidia are from the same colony. ×800. (Courtesy of M. B. Ellis; photo D. W. Fry)

habit characters on seed, and reviewed their occurrence as pathogens mainly on cereals; the following species are included: *C. clavata*, *C. cymbopogonis*, *C. eragrostidis*, *C. geniculata*, *C. inaequalis*, *C. intermedia*, *C. lunata*, *C. oryzae*, *C. pallescens*, *C. siddiquii*, *C. trifolii* and *C. uncinata*, all recorded on rice seed.

Drechslera

(Syn. *Helminthosporium* pro parte)

Conidiophores are septate, geniculate, and conidia (poroconidia) cylindrical rounded at the ends, usually straight or slightly curved, fusoid or obclavate, borne singly and acrogenously from the tip of the conidiophore or from its successive growing points. Two subgenera or sections are distinguished, *Bipolaris*, germinating from the end cells, and *Drechslera* sensu stricto, germinating from any cell of the conidium. Most species occur on Gramineae, many as serious parasites, cf. Section 9.1. For identification, consult Ellis (1971) and Chidambaram, Mathur and Neergaard (1973), and cf. Drechsler (1923) and Luttrell (1951) on graminicolous species; see also Ammon (1962).

(The generic name *Helminthosporium* is only validly used for fungi congeneric with the commonly known causal organism of silver scurf of potato, *Helminthosporium solani*, syn. *Spondylocladium atrovirens*).

D. avenae, Fig. 5.13, leaf spot and seedling blight, or leaf blotch of oats, perfect state *Pyrenophora avenae*. Commonly transmitted from oat seeds (Dennis, 1933; Moore, 1945; Malone, 1962; De Tempe, 1964; Malone and Muskett, 1964; Zwatz, 1965; Soteros and Sheridan, 1972). A major pathogen of oats. For mechanism of seed transmission, see Section 14.3. Seed treatment by mercurial compounds was effective until mercury resistant strains of the fungus developed, see Section 22.6.

Distribution: Probably present wherever oats are growing under moist, cool conditions.

CMI Distribution Map No. 105.

D. bromi, brown leaf spot of brome grasses, perfect state *Pyrenophora bromi*. Seed-transmitted in smooth brome, *Bromus inermis*, as reported by Kreitlow (1953) but in North America transmission by ascospores from perithecia in overwintering plant residues is predominant.

Distribution: North America, Europe, Australia.

D. catenaria, leaf spot and blight of grasses. Very frequent in seeds of many grasses, including *Agrostis stolonifera*, *A. tenuis*, *Dactylis glomerata*, *Festuca pratensis*, *F. rubra*, *Lolium multiflorum*, *L. perenne*, *Phleum pratense*, *Poa annua*, *P. pratensis* and *P. trivialis* (Andersen, 1974 — index), and in *Buchloë dactyloides* (Meehan, 1947*b*).

Distribution: North America, Europe.

D. dictyoides, perfect state *Pyrenophora dictyoides*, net blotch of grasses. On seeds of grasses, including *Bromus inermis*, *Dactylis glomerata*,

Festuca pratensis, F. rubra; *D. dictyoides* f.sp. *perenne* on seeds of *Lolium multiflorum* and *L. perenne* (Bravermann and Graham, 1960; Mäkelä, 1972*a*; Andersen, 1974 — index). Causes net blotch in different grasses, perhaps all of the hosts mentioned.

Distribution: North America, Europe.

D. graminea, Fig. 5.13, leaf stripe disease of barley, perfect state *Pyrenophora graminea*. Early recognised as seed-transmitted (Ravn, 1900), frequent in barley seed (Voitova, 1963; De Tempe, 1964; Noble and Richardson, 1968; Halkilahti, 1971), occurring as infection in the pericarp

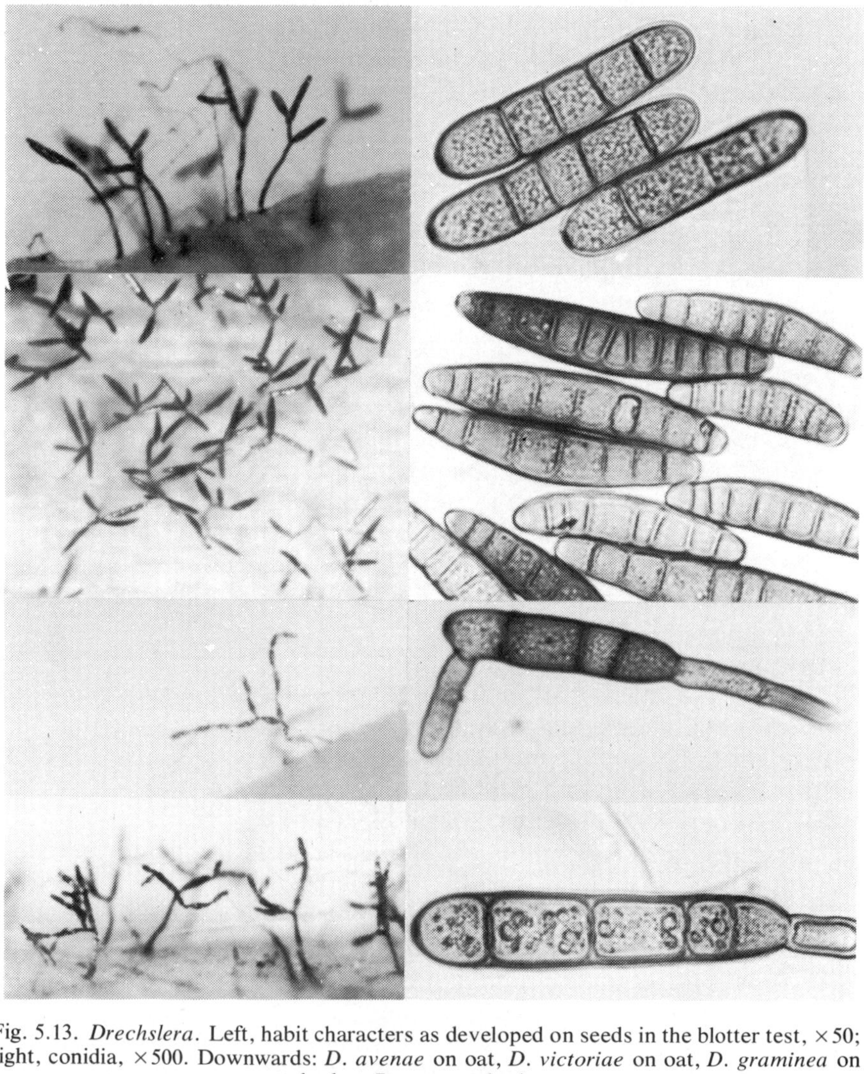

Fig. 5.13. *Drechslera*. Left, habit characters as developed on seeds in the blotter test, ×50; right, conidia, ×500. Downwards: *D. avenae* on oat, *D. victoriae* on oat, *D. graminea* on barley, *D. teres* on barley

leading to systemic infection and loss of the plant, or as contamination, infection then depending on soil conditions; for disease cycle see Section 14.3, and see Figs 13.14, 14.2(9) and 14.5; on economic importance, see Subsection 1.2.1, barley.

Distribution: Cosmopolitan.

D. maydis, Fig. 5.14, southern leaf spot or blight of maize, perfect state *Cochliobolus heterostrophus*. Ullstrup (1941) established two pathogenic races, both seed-borne, and later the importance of seed transmission has been substantiated as relevant to spread of a virulent race on Texas male-sterile (Tms) cultivars (Moore, 1970; Crosier and Braverman, 1971;

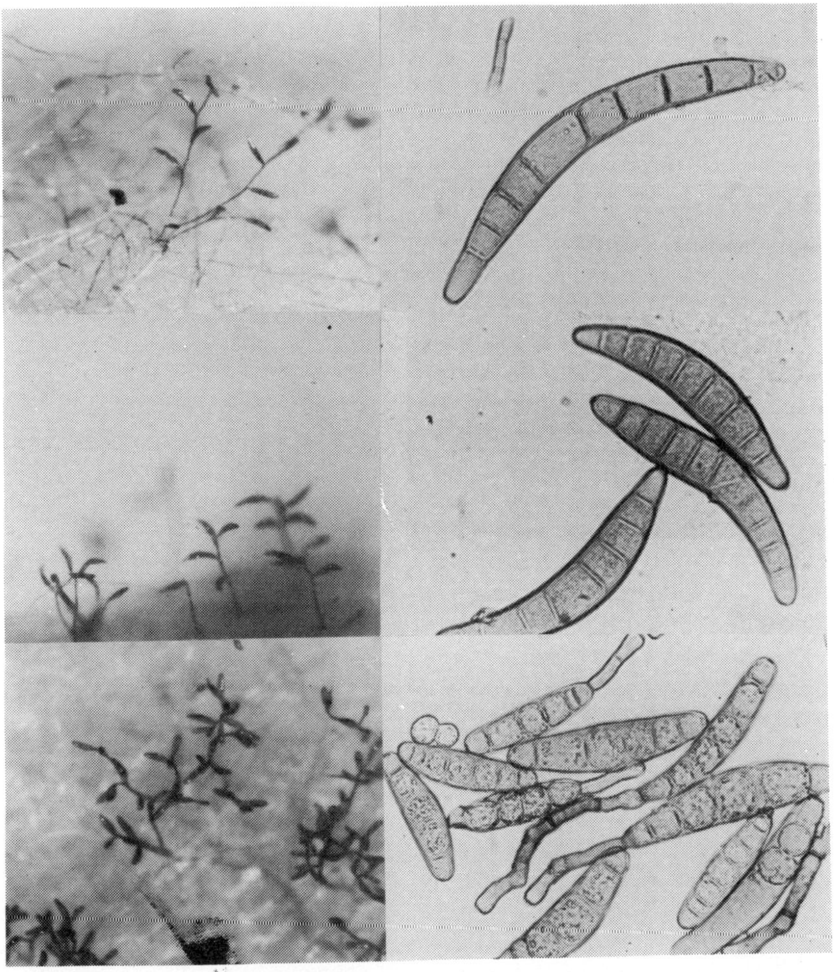

Fig. 5.14. *Drechslera*. Left, habit characters as developed on seeds in the blotter test, ×50; right, conidia, ×500. Downwards: *D. maydis* on maize, *D. oryzae* on rice, *D. sorghicola* on sorghum

Tuite and Caldwell, 1971; White and Ellett, 1971; Boothroyd, 1971; Kommedahl and Lang, 1971; Kommedahl, Lang and Blanchard, 1971; Warmke and Schenck, 1971; Morgan, 1971; Penčić and Smiljaković, 1971; Scott, 1971; Kulik, 1971; Singh, Mathur and Neergaard, 1974*a*, 1974*b*). The pathogen caused a devastating epidemic on susceptible cultivars in the U.S.A. in the summer of 1970. Hosts other than maize may be attacked, and the fungus has been encountered in seeds of *Citrullus lanatus*, *Lactuca sativa*, *Paspalum scorbiculatum*, *Phaseolus aureus*, *P. mungo*, *Sorghum vulgare* and *Vigna sinensis* (Fatima, Mathur and Neergaard, 1974) and of *Pennisetum typhoides* (Mathur, Ram Nath and Mathur, 1973). Testing procedures evaluated (Singh, Mathur and Neergaard, 1974*a*). For economic significance, see Subsection 1.2.1, maize. Interception in India of race T on seed of *Sorghum vulgare* imported from the U.S.A. (Ram Nath *et al*., 1973).

Distribution: North and South America, Europe, Asia, Australia.

CMI Distribution Map No. 346.

D. oryzae, Fig. 5.14 and Colour Plate IV, brown spot disease of rice, perfect state *Cochliobolus miyabeanus*. Very frequent in seed of rice (Chidambaram, Mathur and Neergaard, 1973; Aulakh, Mathur and Neergaard, 1974*a*). Important in producing 'abnormal seedlings' in laboratory testing of rice for germination (Guerrero, Mathur and Neergaard, 1972). Laboratory counts and field incidence compared (Aulakh, Mathur and Neergaard, 1974*b*). A major pathogen of rice, responsible for heavy losses, particularly under poor farming conditions, see Subsection 1.2.1. Trace infections have been observed in seeds of wheat and sorghum. Economic importance, see Subsection 1.2.1, rice.

Distribution: Probably wherever rice is grown.

CMI Distribution Map No. 92.

D. papaveris (Helminthosporium papaveris, Dendryphion penicillatum), perfect state *Pleospora papaveracea (P. calvescens)*, leaf and stem spot of poppies. Frequent in seeds of *Papaver*, including *P. alpinum*, *P. bracteatum*, *P. glaucum*, *P. mursellii*, *P. nudicaule*, *P. orientale*, *P. rhoeas*, *P. somniferum*, *P. umbrosum* (Cristoff, 1930; Barbacka, 1936; Neergaard, 1937AR–1951AR; Reinmuth, 1943; Meffert, 1950; Radulescu, Perceă and Docea, 1961; Radulescu and Persecă, 1964; Andersen, 1974 — index). The disease is destructive in opium poppy crops in southern Europe, reducing yields substantially. Difficult to eradicate infection by chemical seed treatment.

Distribution: North America, Europe, South Africa, Asia.

D. phlei (Helminthosporium dictyoides var. *phlei),* leaf streak disease of timothy. Common in seed of timothy, *Phleum pratense*, and also recorded in *Poa pratensis* and *P. trivialis* (Andersen, 1974 — index).

Distribution: North America, Europe.

D. poae (syn. *D. vagans*), melting out, leaf spot. Rather common in seeds of *Poa pratensis* and *P. trivialis* (Andersen, 1974 — index).

Distribution: North America, Europe.

D. rostrata, Fig. 5.15, leaf spot, is rather frequent on seeds of a wide range of Gramineae, including (acc. to Chidambaram, Mathur and Neergaard, 1973): *Brachiaria ruziziensis*, *Cenchrus ciliaris, Chloris gayana, Eleusine coracana*, *Luffa acutangula*, *Oryza sativa*, *Pennisetum glaucum*, *Setaria italica*, *Sorghum bicolor*, *S. vulgare*, *Triticum aestivum* and *Zea mays*. Moreover in *Axonopus* sp. (Meehan, 1947*a*), *Buchloë dactyloides* (Meehan, 1947*b*), *Cynodon dactylon* (Meehan, 1947*a*), *Paspalum* sp. (Meehan, 1947*a*; Young, Lefebvre and Johnson, 1947), *Phleum*

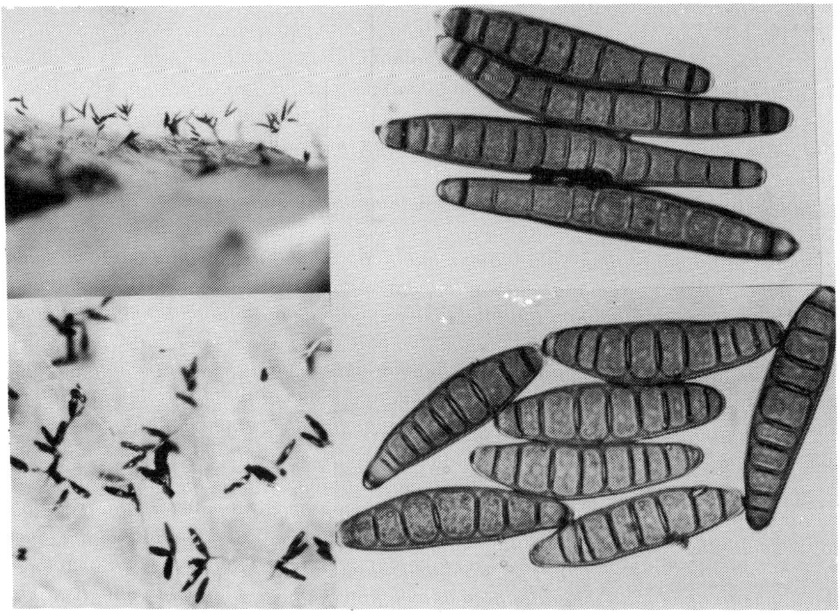

Fig. 5.15. *Drechslera*. Left, habit characters as developed on seeds in the blotter test, ×50; right, conidia, ×500. Downwards: *D. rostrata* on pearl millet, *D. sorokiniana* on sorghum

pratense (Noble and Richardson, 1968), *Poa pratensis* (Andersen, 1961AR) and *Poa* spp. (Noble and Richardson, 1968). The fungus has been found in seeds of hosts other than Gramineae, e.g. *Beta vulgaris*, *Calendula officinalis*, *Phaseolus aureus*, *P. mungo*, *Sesamum indicum* and *Vigna sinensis* (Chidambaram, Mathur and Neergaard, 1973). A mild pathogen, producing leaf spots in Gramineae; *D. longirostrata* is closely related to *D. rostrata*, or possibly a synonym. It occurs in seeds of many hosts, particularly of Gramineae (Ram Nath *et al.*, 1970).

Distribution: North America, Europe, Africa, Asia, mostly in tropical and subtropical regions.

D. sacchari, eye spot and seedling blight of sugarcanes. Found to be seed-transmitted in Jamaica (Loveless and Smith, 1956); seedlings are killed, surviving seedlings are rendered susceptible by accidental wounding during transplanting.

Distribution: North, Central and South America, Europe (Italy), Africa, Australia, widespread where sugarcanes are grown.

CMI Distribution Map No. 349.

D. sesami, leaf blotch or blight or stem rot of sesame. The capsules are attacked and the seeds infected or contaminated (Poole, 1956).

Distribution: North America, Asia.

D. setariae, seed rot, perfect state *Cochliobolus setariae*. In seeds of *Setaria italica* (Grewal and Pal, 1965).

Distribution: North America, Europe, Asia.

D. siccans, foot rot and leaf spot. Rather infrequent on seeds of different Gramineae such as *Avena sativa*, *Bromus unioloides*, *Festuca arundinacea*, *F. pratensis*, *Hordeum vulgare* and *Lolium perenne* (Chidambaram, Mathur and Neergaard, 1973), and *Dactylis glomerata*, *Festuca ovina*, *F. pratensis*, *F. rubra*, *Hordeum vulgare*, *Lolium multiflorum*, *L. perenne*, *Phleum pratense*, *Poa pratensis* and *P. trivialis* (Mäkelä, 1972*a*; Andersen, 1974 — index).

Distribution: North America, Europe, Asia.

D. sorghicola, Fig. 5.14, target spot of sorghum. Encountered on seeds of *Sorghum vulgare* (Chidambaram, Mathur and Neergaard, 1973; Mathur, Mathur and Neergaard, 1975).

Distribution: North America, Europe, Africa, Asia, Australia.

D. sorokiniana (syn. *Helminthosporium sativum*), Fig. 5.15, seedling blight, foot rot, ear blight, perfect state *Cochliobolus sativus*. Very frequent in low percentages in seeds of many hosts, particularly in those of Gramineae, and largely seed-transmitted. Seed infection has been reported in the following hosts (Greaney and Machacek, 1946; De Tempe, 1970*a*; Kolk, 1970; Mäkelä, 1972*a*; Chidambaram, Mathur and Neergaard, 1973; Andersen, 1974 — index): *Acacia farnesiana*, *Agropyron elongatum*, *Anethum graveolens*, *Avena sativa* (frequent), *Beta vulgaris*, *Brachiaria ruziziensis*, *Brassica napus*, *B. oleracea*, *Bromus inermis*, *B. unioloides*, *Calendula officinalis*, *Capsicum annuum*, *Centaurea imperialis*, *Chrysanthemum carinatum*, *C. cinerariifolium*, *Coriandrum sativum*, *Cucurbita pepo*, *Cuminum cyminum*, *Cynodon dactylon*, *Dactylis glomerata* (frequent), *Daucus carota*, *Dianthus caryophyllus*, *Dimorphotheca aurantiaca*, *Eruca sativa*, *Eschscholtzia californica*, *Festuca ovina*, *F. pratensis*, *F. rubra* (frequent), *Glycine max*, *Helianthus annuus, Hordeum vulgare* (very frequent), *Lactuca sativa, Lagurus ovatus*, *Larix decidua*, *Lathyrus odoratus*, *Lolium multiflorum*, *Lolium perenne*, *Oryza sativa*, *Papaver somniferum*, *Pastinaca sativa*, *Pennisetum glaucum*, *Petroselinum sativum*, *Petunia hybrida*, *Phaseolus* *vulgaris*, *Phleum pratense*, *Phlox drummondii*, *Poa pratensis*, *Raphanus sativus*, *Secale cereale*, *Sesamum indicum*, *Setaria italica*, *Sorghum vulgare*, *Spinacia oleracea*, *Tagetes erecta*, *Trifolium repens*, *Triticum aestivum, Vigna sinensis* and *Zinnia elegans*. Mode of seed transmission has been studied by Mead (1942*a*), Ammon (1962), Tyulina (1971), and Jørgensen (1974) in barley. For impact of the time of infection during seed development, as related to weather conditions, see Section 11.3. The pathogen is common and particularly important in cereals, see Subsection 1.2.1, under wheat and barley.

Distribution: Widespread, North, Central and South America, Europe, Africa, Asia, Australia.

CMI Distribution Map No. 322.

D. teres, Fig. 5.13, net blotch of barley, perfect state *Pyrenophora teres*. Very commonly seed-transmitted in barley (Ravn, 1900, a monograph; Machacek and Wallace, 1942; Kenneth, 1962; De Tempe, 1964; Piening, 1968; Andersen, 1974 — index). Economic importance, see Subsection 1.2.1, under barley. A pathogenic variant, *D. teres* f.sp. *maculata*, producing spot blotch has recently been described (Smedegård-Petersen, 1971).

Distribution: North and South America, Europe, Africa, Asia, Australia.

CMI Distribution Map No. 364.

D. turcica, northern leaf blight of maize, black ear rot, crown rot, perfect state *Trichometasphaeria turcica*. Encountered frequently in seeds of *Sorghum bicolor* and *S. vulgare* (Chidambaram, Mathur and Neergaard, 1973) and can survive for two years in the seed (*fide* Noble and Richardson, 1968). Seems not to have been reported from seeds of maize.

Distribution: Widespread in tropical and subtropical areas of North, Central and South America, Europe, Asia and Australia.

CMI Distribution Map No. 257.

D. verticillata (syn. *Helminthosporium cyclops*), sterility disease, perfect state *Pyrenophora seminiperda*. In seeds of *Bromus* spp. (Cooke and Shaw, 1952), different Gramineae (Shoemaker, 1966), and *Triticum sativum* (Wallace, 1959).

Distribution: North America, Africa, Australia.

D. victoriae, Fig. 5.13, Victoria blight of oats, perfect state *Cochliobolus victoriae*. In seeds of *Agropyron* sp. (Meehan, 1947*a*), *Avena sativa* (Tveit, 1956; Jacquet, Leben and Arny, 1957; Ivanoff, 1959), *Dactylis glomerata* and *Festuca rubra* (Meehan, 1947*a*), *Setaria viridis* (Meehan, 1947*b*), and *Triticum aestivum* (Chidambaram, Mathur and Neergaard, 1973). The pathogen is destructive in oats with Victoria parentage, bred for crown rust resistance, see Subsection 1.2.1, under oats. Should be regarded as an important quarantine object for many developing countries.

Distribution: North and South America, Europe, Australia.

CMI Distribution Map No. 267.

Other species of less importance may be encountered in seeds of Gramineae, some of them very frequently. The following may be mentioned: *D. avenacea* in wheat (Howell and Noble, 1963); *D. bicolor* in different hosts (Chidambaram, Mathur and Neergaard, 1973); *D. buchloës* in barley (Crosier, 1961); *D. cynodontis* in *Agropyron elongatum*, *Hordeum vulgare* and *Triticum aestivum* (Chidambaram, Mathur and Neergaard, 1973); *D. dactylidis*, common in *Dactylis glomerata* (Tulloch and Leach, 1972; Mäkelä, 1972*a*); *D. gigantea* in *Bromus* sp. (Mühle, 1953), *Phleum pratense* and *Poa* spp. (Noble and Richardson, 1968); *D. halodes* in rice (Noble and Richardson, 1968) and maize (Kerr, in Noble and Richardson, 1968); *D. hawaiiensis* in barley, millet, oats, rice, sorghum, wheat and many non-graminaceous hosts (Chidambaram, Mathur and Neergaard, 1973); *D. tetramera*, common in seed of many hosts, 67 species listed by Chidambaram, Mathur and Neergaard (1973); *D. tritici-repentis* (syn. *D. tritici-vulgaris*), perfect state *Pyrenophora tritici-repentis* (Duff, 1954; Vanterpool, 1963).

Heterosporium

In recent years the position of this genus has become doubtful, being considered as a synonym of *Cladosporium* (De Vries, 1952; Hughes, 1958), a view accepted by Ellis (1971). *H. eschscholtziae* and *H. tropaeoli* are discussed under *Acroconidiella*.

Mycocentrospora (Centrospora)

Conidiophores are straight or flexuous, geniculate, conidia solitary and acropleurogenous, usually with a basal obclavate, one- to multi-septate appendage.

M. acerina (Centrospora acerina), halo blight of pansy. In seeds of caraway, *Carum carvi* (Westerdijk and Van Luijk, 1924) and *Viola tricolor* (Neergaard, 1943; Gill and Ratcliff, 1967; Gill, 1971). A serious, usually soil-borne plant pathogen having a wide host range.

Distribution: North America, Europe, New Zealand.

Phaeoisariopsis

Conidiophores are straight or slightly flexuous, loosely grouped in synnemata; conidia are dry, borne solitary, obclavate or cylindrical, with three or more transverse septa.

P. griseola (Isariopsis griseola), angular leaf spot of Phaseolus bean. Produces a red-brown discoloration of bean seed from which the pathogen may be transmitted to the seedlings (Orozco-Sarria and Cardova-Alvarez, 1959).

Distribution: Cosmopolitan, North and South America, Europe, Africa, Asia, Australia.

CMI Distribution Map No. 328.

Pyricularia

Conidia (blastoconidia) produced solitary on the conidiophore in a sympodial fashion, pyriform to obclavate, two- or three-septate, hyaline to pale olive-coloured.

P. oryzae, blast of rice, frequently seed-borne in all rice-producing countries (for references, see Noble and Richardson, 1968), surveys based on infected seed samples made for Nigeria (Komolafe *et al.*, unpublished; Awoderu and Onuarah, 1974), Nepal (Shrestha, Mathur and Neergaard, in press), Thailand. Transmission from untreated seed in the U.S.A. (Lamey, 1970). The principal, most destructive disease of rice; heavy infections on the panicles are often detrimental to yield (Ou, 1972). Numerous pathogenic races, see Section 17.2, imposing quarantine provisions against their international spread. Seed treatment by mercurial compounds recommended.

Distribution: World-wide in rice-growing regions.

CMI Distribution Map No. 51.

Trichoconis

The species discussed below as *T. padwickii* does not in fact belong to *Trichoconis*, type species *T. caudata*, being different from this in conidiogenesis; it has therefore been provisionally referred to *Alternaria* by Ellis (1971). This transfer, however, is not acceptable because the conidia are phragmosporous, not dictyosporous. During the preparation of this book a new genus *Trichoniella* has been erected with one species, *T. padwickii* (Jain, 1975). *T. padwickii* has been maintained here provisionally in its original name.

T. padwickii (Alternaria padwickii), stackburn disease, seedling blight of rice, see Colour Plate IV. Very common in rice seed (Mathur and Neergaard, 1970; Mathur, Mallya and Neergaard, 1972; Cheeran and Raj, 1972), about 75 per cent of about 400 seed samples having been found infected. Pre- and post-emergence killing of infected seedlings.

Distribution: South America (Surinam), Europe, Africa, Asia, Australia.

CMI Distribution Map No. 314.

(4. Phaeodictyae)

Alternaria

Conidiophores occur solitary or in fascicles, and are usually erect, simple or branched; conidia (poroconidia) catenate or solitary, ovoid, cylindrical or obclavate, rounded at the base, smooth, echinulate or ver-

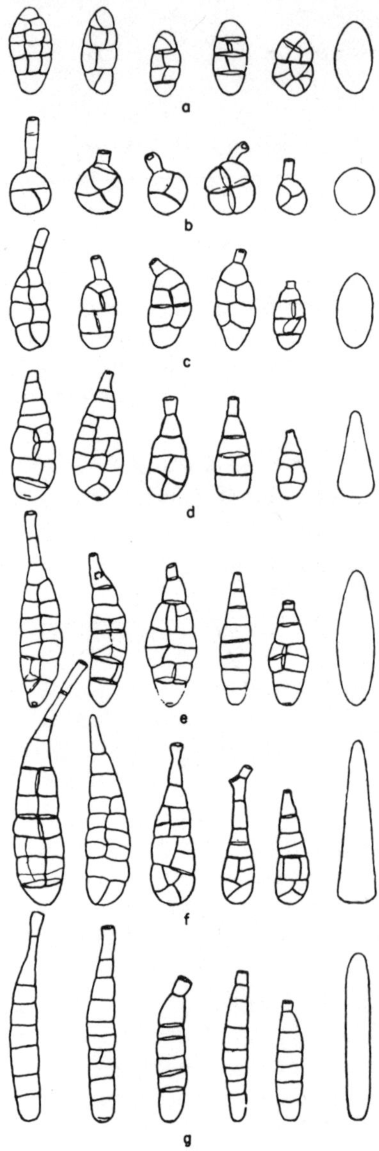

Fig. 5.16. *Alternaria tenuis* auct. Polymorphy of conidia: Form types—(a) 'beakless oval or ball'; (b) 'ball'; (c) 'short oval'; (d) 'short cone'; (e) 'long oval'; (f) 'long cone'; (g) 'cylinder'. ×450. (Neergaard, 1945)

rucose, and usually provided with a beak which may be short and thick, or elongated and tapering or filiform; the conidial body has transverse, longitudinal and sometimes oblique septa. Many species are serious leaf parasites. For identification, consult Groves and Skolko (1944); Neergaard (1945); Joly (1964) and Ellis (1971).

A. alternata (Fr.) Keissler, syn. *A. tenuis* Nees, is as explained by Simmons (1967) the correct name of the type species of *Alternaria*. However, *A. 'tenuis'* is the commonly used name for a collective species of *Alternaria*, generally characterised by long conidial chains and markedly polymorph conidia, Fig. 5.16. Members of this collective species which may be difficult or almost impossible to distinguish in their conidial states, belong nevertheless in their perfect state to different species of *Chlatrospora* and *Pleospora* (Simmons, 1952). *A. alternata* as a species having been well defined from authentic herbarium material, should not be adopted as name for the loosely defined group species for which it is preferable still to use *A. tenuis* auct., as defined by Wiltshire (1933) and further substantiated by Neergaard (1945). *A. alternata,* which may be regarded as a member of the group species *A. 'tenuis'*, needs further investigation based on living material that offers sufficient evidence for identification with the type species.

A. brassicae, Fig. 5.9 and Fig. 5.17(12), grey leaf spot. Very common, although less common and less severe in seeds of cabbage and cauliflower, *Brassica oleracea*, in humid temperate regions and common also in other cruciferous seeds, including *Brassica campestris*, *B. napus*, *B. nigra, Iberis amara, I. coronaria, I. umbellata, Lepidium sativum, Matthiola incana*, *Raphanus sativus*, *Sinapis alba*, *S. nigra* (Groves and Skolko, 1944; Neergaard, 1935AR–1951AR, 1945; Andersen, 1974 — index; Richardson, 1970). Important pathogen in seed crops in temperate regions.

Distribution: World-wide.

CMI Distribution Map No. 353.

A. brassicicola, Fig. 5.17(4), black spot of cabbage and other crucifers. Very common in high infection percentages on seeds of cabbage and cauliflower, *Brassica oleracea*, in humid temperate regions, less severe in other cruciferous seeds (Nielsen, 1933; Neergaard, 1935AR–1954AR, 1945; Andersen, 1974 — index; Schimmer, 1953; Richardson, 1970; Petrie, 1974). Common in seeds of most other cruciferous crops, including *Aubrietia hybrida*, *Brassica campestris*, *B. napus*, *B. niger*, *Cheiranthus cheiri*, *Crambe maritima*, *Eruca sativa*, *Iberis amara*, *I. coronaria*, *I. umbellata*, *Matthiola incana*, *Lepidium sativum*, *Raphanus sativus*, *Sinapis alba*, *S. nigra* (Neergaard 1936AR–1951AR, Andersen, 1974 — index). A major pathogen on cruciferous seed crops in temperate regions, substantially lowering yields quantitatively and qualitatively. Disease cycle is discussed in Section 14.4, impact of weather conditions in relation

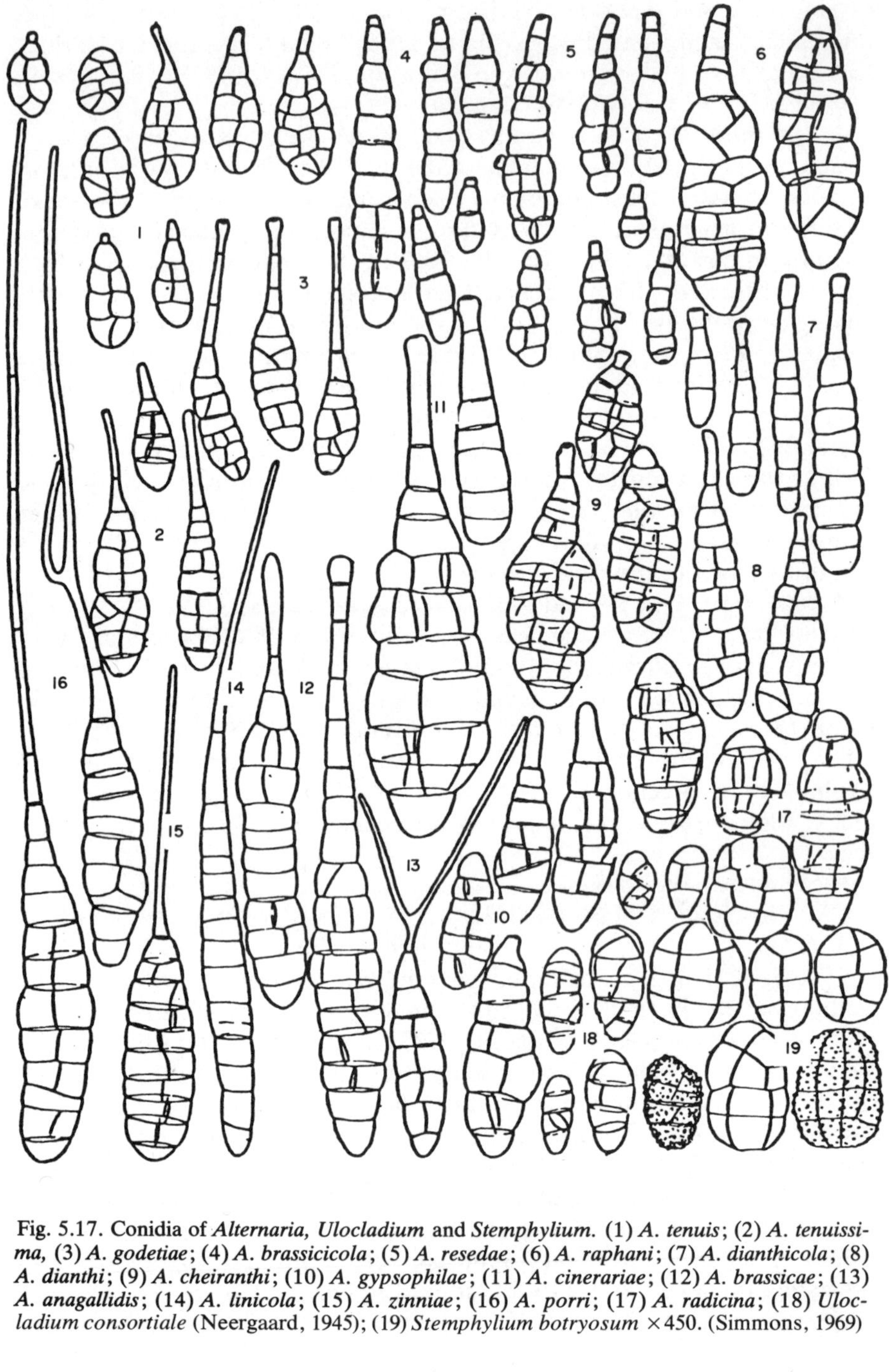

Fig. 5.17. Conidia of *Alternaria, Ulocladium* and *Stemphylium*. (1) *A. tenuis*; (2) *A. tenuissima,* (3) *A. godetiae*; (4) *A. brassicicola*; (5) *A. resedae*; (6) *A. raphani*; (7) *A. dianthicola*; (8) *A. dianthi*; (9) *A. cheiranthi*; (10) *A. gypsophilae*; (11) *A. cinerariae*; (12) *A. brassicae*; (13) *A. anagallidis*; (14) *A. linicola*; (15) *A. zinniae*; (16) *A. porri*; (17) *A. radicina*; (18) *Ulocladium consortiale* (Neergaard, 1945); (19) *Stemphylium botryosum* ×450. (Simmons, 1969)

to infection time, Section 11.3, and economic significance, Subsection 1.2.4, cabbage.

Distribution: North and South America, Europe, Africa, Asia, probably world-wide.

A. cheiranthi, Fig. 5.17(9), black mould of wallflower. Common in seed of *Cheiranthus cheiri* (Neergaard, 1936AR–1951AR, 1945; Andersen, 1974 — index). Sometimes considerable damage in North Europe.

Distribution: Europe.

A. chrysanthemi. Rather common on seeds of *Chrysanthemum carinatum* and *C. maximum* (Neergaard, 1948AR–1951AR; Schmidt, 1958; Andersen, 1974 — index). For identification of the species see Simmons (1965).

Distribution: North America, Europe, India.

A. cichorii. Common in seeds of *Cichorium endivia* and *C. intybus* (Andersen, 1974 — index). Leaf spot on chicory and endive.

Distribution: North and South America, Europe, Asia.

A. crassa, leaf and pod blight of Datura. Seed-transmitted in *Datura metel* var. *muricata* (Halfon-Meiri, 1973). Complete control by seed treatment was not obtained by organomercury dust or by the thiram soak method.

Distribution: North and Central America, Europe, Africa, Asia.

A. cucumerina (A. nigrescens, A. brassicae var. *nigrescens),* leaf blight of cucurbits. In seeds of cucumber (Munn, 1921).

Distribution: North and South America, Europe, Africa, Asia, Australia, New Zealand.

A. cyamopsidis (A. cucumerina var. *cyamopsidis),* Alternaria leaf spot of guar. Seed-borne in guar, *Cyamopsis tetragonoloba* (Sowell, 1965).

Distribution: North America.

A. dianthi, Fig. 5.17(8), leaf spot, leaf blight of carnation. Seed-borne, although rather infrequently, in *Dianthus barbatus*, *D. caryophyllus* and *D. chinensis* (Crosier and Heit, 1948; Andersen, 1974 — index).

Distribution: North America, Europe.

A. dianthicola, Fig. 5.17(7), black mould, Alternaria petal blight of carnation. Common in seeds of *Dianthus barbatus, D. caryophyllus, D. caryophyllus* var. *heddewigii, D. chinensis, D. plumarius* (Neergaard, 1945; Andersen, 1974 — index; Baker and Davis, 1950*c*). Severe in some regions and sometimes under glass.

Distribution: North America, Europe.

A. godetiae, Fig. 5.17(3). Low percentages of infection in seeds of *Clarkia amoena* subsp. *whitneyi (Godetia hybrida)* (Neergaard, 1945; Moore and Moore, 1949). In Denmark the pathogen caused destructive attacks in 1937–1941 but has since then lost economic significance, possibly because of natural selection of resistant genotypes in susceptible cultivars like Kelvedon Glory and Sybil Sherwood. In England serious in 1947.

Distribution: Europe.

A. lini, Alternaria blight of flax. A small-spored species (Dey, 1933), resembling *A. brassicicola*, seed-borne in linseed (Arya and Prasada, 1953; Leduc, 1958). Serious crop losses in India.

Distribution: Asia (India), Europe.

A. linicola, Fig. 5.17(14), Alternaria blight of flax. Rather common in seed of *Linum grandiflorum* and *L. usitatissimum* (Groves and Skolko, 1944; Neergaard, 1945; Moore, 1946*b*; Leduc, 1958; Andersen, 1974 — index). A weak pathogen, prevailing under adverse crop growing conditions.

Distribution: North America, Europe.

A. longipes, brown spot of tobacco. Seed-transmitted in tobacco (Gulyas, 1930; Hopkins, 1956; Heursel, 1961). Severe epidemics in some tobacco growing areas, as in East Africa.

Distribution: North and South America, Europe, Africa, Asia.

CMI Distribution Map No. 63.

A. porri, Fig. 5.17(16), comprises several formae speciales (Neergaard, 1945) including *A. porri* f.sp. *porri*, *A. porri* f.sp. *dauci*, *A. porri* f.sp. *solani*.

A. porri, f.sp. unidentified, has been encountered on seeds of *Alyssum benthami* (Andersen, 1969AR), *Bellis perennis* (Neergaard, 1964AR, 1965AR), *Celosus cristata* (Andersen, 1953AR), *Dianthus caryophyllus* and *D. caryophyllus* var. *heddewigii* (Neergaard, 1964AR), *Helichrysum bracteatum* (Neergaard, 1966AR), *Scabiosa atropurpurea* (Andersen, 1973), *Tragopogon porrifolius* (Andersen, 1973AR) and *Thymus vulgaris* (Neergaard, 1967AR).

A. porri f.sp. *calendulae* f.sp.nov., leaf spot in marigold. Seed-borne in *Calendula officinalis* (Neergaard, 1945; Baker and Davis, 1950*c*; Andersen and Broe, 1974AR). Pathogenic to marigold and lettuce (Baker and Davis, 1950*c*), and probably to other Compositae, while marigold is not infected by *A. porri* f.sp. *porri*, *A. porri* f.sp. *dauci* or *A. porri* f.sp. *solani*.

Distribution: North America, Europe.

A. porri f.sp. *dauci (A. dauci)*, carrot leaf blight, widely seed-transmitted in carrot (Neergaard, 1945; Groves and Skolko, 1944; Hewett, 1964; Maude, 1966*b*; Andersen, 1974 — index), and encountered in seeds of celery, dill and parsley (Andersen, 1974 — index). Considerable losses, also in seed crops.

Distribution: North and South America, Europe, Africa, Asia, Australia.

A. porri f.sp. *lallemantiae,* leaf spot, dry spot of *Lallemantia*. Seed-borne in *Lallemantia ibirica* (Schmidt, 1954; Pietkiwicz, 1958; Todorova, 1960). Seed treatment ineffective.

Distribution: Europe.

A. porri f.sp. *porri,* purple blotch of onion and leek. In seeds of onion (Neergaard, 1945; Andersen, 1952AR) and leek (Andersen, 1952AR; Neergaard, 1964AR). Sometimes causes considerable losses in the U.S.A., Italy and Denmark.

Distribution: North, Central and South America, Europe, Africa, Asia, Australia, New Zealand.

CMI Distribution Map No. 350.

A. porri f.sp. *solani (A. solani),* early blight of potato and tomato. Seed-transmitted in *Ageratum houstonianum* (Neergaard, 1945, 1956AR, Andersen, 1974 — index), *Capsicum annuum* (Neergaard, 1966AR), and in tomato as seed coat infection (Massee, 1914; McWhorter, 1927; Van Haltern, 1935; Miller and Crosier, 1936; Samson, 1939; Moore, Thomas and Vaughan, 1943; Hodosy, 1966). Seed transmission in tomato considered important in the U.S.A. A major pathogen, particularly in potato and tomato; for economic significance, see Subsection 1.2.4, tomato.

Distribution: World-wide.

CMI Distribution Map No. 89.

A. radicina (Stemphylium radicinum, Pseudostemphylium radicinum), Fig. 5.17(17), black root rot of carrot. Commonly seed-transmitted in carrot (Doyer, 1938; Neergaard, 1937, 1945; Groves and Skolko, 1944; De Tempe, 1962; Lange-De La Camp, 1966; Maude, 1966*b*; Andersen, 1974 — index), also common in celery and celeriac, dill and fennel, and occasionally found in seed of parsnip (Andersen, 1974 — index). For economic importance, see Subsection 1.2.4, carrot. As a variety *A. radicina* var. *petroselini* may be regarded a form of the fungus commonly encountered in seed of parsley (Neergaard, 1945; Andersen, 1974 — index).

Distribution: North America, Africa (Nigeria), Europe, Asia.

A. raphani (A. matthiolae), Fig. 5.17(6), Alternaria leaf spot of radish and stock. The fungus produces chlamydospores abundantly. In North Europe frequent in seeds of radish and garden stock, *Matthiola incana* and *M. bicornis* (Groves and Skolko, 1944; Neergaard, 1945; Atkinson, 1950; Andersen, 1974 — index), also sometimes in seeds of cabbage (Andersen, 1974 — index), *Cheiranthus cheiri*, *Iberis amara*, *I. umbellata* and *Sinapis alba* (Andersen, 1974 — index). Depreciation of seed quality, particularly in radish (Neergaard, 1945).

Distribution: North America, Europe, Asia (India).

A. sesami, sesame blight. Seed transmission important, especially for international spread of the pathogen (Göbelez, 1960; Leppik and Sowell, 1964; Leppik, 1968; Mathur and Kabeere, 1975). A serious pathogen, causing quantitative and qualitative losses.

Distribution (world mapping based on seed samples, cf. Leppik and Sowell, 1964): North and South America (scattered areas), South Europe, Africa (scattered), Asia.

CMI Distribution Map No. 410.

A. tenuis (on nomenclature, cf. *A. alternata*), Figs 5.16 and 5.17(1). A common saprophyte or weak parasite, occurring on hundreds of host plants. Neergaard (1945) listed 117 plant species in which the fungus has been recorded on seeds, with indication of frequency for some of the hosts; Groves and Skolko (1944) provide a similar list; the fungus usually occurs as seed coat infection, cf. discussion in Section 10.5. Although profuse infection may lower the quality of some kinds of seed, sometimes the main trouble caused by the fungus as a contaminant is that in seed health testing this collective species often is confused with pathogenic species of *Alternaria*. In identification, attention should be paid to the polymorphy of the conidia (Fig. 5.16) by which it can be distinguished from the less polymorph conidia of most pathogenic species.

Distribution: Ubiquitous.

A. triticina, Alternaria leaf blotch of wheat. Seed-transmitted in wheat (Prasada and Prabhu, 1963; Prabhu and Prasada, 1966, 1967; Kumar and Arya, 1973*a*, 1973*b*). Serious on Mexican cultivars and their derivatives, see Subsection 1.2.1, wheat.

Distribution: India.

A. zinniae, Fig. 5.17(15), leaf and flower blight of zinnia and other ornamentals, especially in Compositae. More or less frequent in seeds of *Acacia farnesiana*, *Ageratum houstonianum*, *Bellis perennis*, *Calendula officinalis*, *Calliopsis bicolor*, *Callistephus chinensis*, *Centaurea cyanus*, *Chrysanthemum carinatum*, *Clarkia elegans*, *Cosmos bipinnatus*, *Gaillardia picta*, *Gentiana acaulis*, *Gerbera jamesonii*, *Helianthus annuus*, *Impatiens* sp., *Lactuca sativa*, *Papaver nudicaule*, *Reseda odorata*, *Senecio cruentus*, *Tagetes erecta*, *Viola tricolor*, and very frequent and in high infection percentages in *Zinnia elegans* and *Z. haageana* (Neergaard, 1942AR–1951AR, 1945; Baker, 1956; McDonald and Martens, 1963; Andersen, 1974 — index). Seed treatment, see Subsection 21.3.1 (5) and Table 22.3.

Distribution: North America, Europe.

Several other species of *Alternaria* are seed-borne, including *A. anagallidis* var. *linariae*, cf. Fig. 5.17(13), on *Antirrhinum majus* (Neergaard, 1945; Andersen, 1974 — index) and *Linaria maroccana* (Neergaard, 1945), *A. cinerariae (A. senecionis)*, Fig. 5.17(11), frequently in *Senecio cruentus* (Neergaard, 1945; Andersen, 1974 — index), *A. gypsophilae*, Fig. 5.17(10), on *Dianthus caryophyllus* (Neergaard, 1945) and frequently *Gypsophila elegans* (Neergaard, 1945; Andersen, 1974 — index), *A. resedae*, Fig. 5.17(5), frequently on *Reseda odorata* (Neergaard, 1945; Andersen, 1974 — index), and *A. tagetae* (Andersen, 1974 — index).

Stemphylium

Conidiophores are usually straight or somewhat curved, unbranched or occasionally branched, apical conidiogenous cell with a single pore, several successive conidiogenous cells may be produced by successive conidiophore growth through apical pores — or leaving such pores as secondary phenomena; conidia (probably blastoconidia) solitary, rounded at the ends, package-formed, obclavate or broadly ellipsoidal, in some species with a conical apex, verrucose or echinulate, often constricted at one or more septa, typically at the middle septum. For identification see Graham and Zeiders (1960), Simmons (1967, 1969) and Ellis (1971).

S. botryosum, perfect state *Pleospora herbarum*, Fig. 5.8 and Fig. 5.17 (19). This ubiquitous saprophyte or weak pathogen occurs commonly but usually in low percentages in seeds of many hosts. Neergaard (1945) listed 34 such hosts, Noble and Richardson (1968) 14 genera, partly different from those included in the former list. However, some of the cases reported may have included other, morphologically closely related species, which by further studies may be confirmed to be seed-borne as well, such as *S. globuliferum* (perfect state *Pleospora* sp.); *S. majusculum* (perfect state *Pleospora* sp.); and particularly *S. vesicarium* (perfect state *Pleospora* sp.), common in onion and certainly also in seeds of this host (cf. Simmons, 1969). In bean seeds the fungus is known as cause of 'red nose' (Doyer, 1938).

Distribution: Ubiquitous.

S. callistephi, Stemphylium leaf spot of China aster. Natural seed transmission in *Callistephus chinensis* experimentally confirmed (Baker and Davis, 1950*a*).

Distribution: North America.

S. loti, leaf spot of bird's foot trefoil. Causes shrivelling and discoloration of the seeds; young seedlings may be killed by the fungus growing from the cotyledons into the stem (Graham, 1953*b*).

Distribution: North America.

S. sarciniforme, leaf spot. Groves and Skolko (1944) have isolated this species from clover seed in Canada; this is perhaps the only case of seed infection known, while *S. botryosum* is rather common in clover seed.

Distribution: North and South America, Europe, Africa (Morocco, Rhodesia), Asia, Australia.

CMI Distribution Map No. 139.

S. solani, grey leaf spot. Seed treatment advised against seed-borne infection in tomato (Linn and Wright, 1951).

Distribution: North, Central and South America, Europe, Africa, Asia, Australia.

CMI Distribution Map No. 333.

S. trifolii. Encountered in seeds of *Trifolium repens* (Kilpatrick, 1958).

Distribution: North America.

Ulocladium (Pseudostemphylium)

Conidiophores are straight or flexuous, often geniculate, unbranched or branched; conidia solitary, in some species secondary conidia are formed on a short secondary conidiophore; by successive growth of the tip of the conidiophore conidia are produced successively from below each terminally produced conidium, resulting in botryose clusters of conidia; the conidia are broadly ellipsoidal, clavate or pyriform, but typically not obclavate, Fig. 5.18; they are smooth or verrucose. Occur commonly on dead plants, paper and similar cellulose materials. For identification consult Simmons (1967).

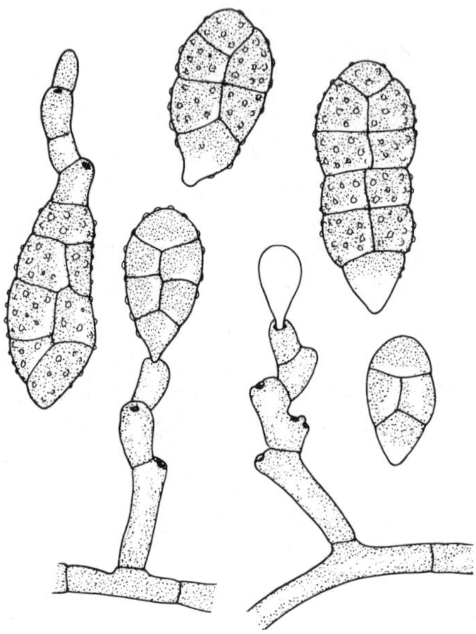

Fig. 5.18. *Ulocladium consortiale,* conidiophores and conidia. ×1,000. (Courtesy of J. A. Von Arx)

In seeds different species of *Ulocladium* are encountered in low percentages as saprophytes, including particularly *U. consortiale (Stemphylium consortiale, S. ilicis)*, Figs 5.17(18) and Fig. 5.18. Groves and Skolko (1944) recorded this species on seeds of many hosts, and Neergaard (1945) listed 45 species of plants in which it was recorded on seed. Some of these reported cases have probably been other species, including *U. atrum*.

Tuberculariaceae

Conidiophores usually short, aggregated in more or less cushion-shaped sporodochia, which are often waxy or gelatinous.

(1. Hyalophragmiae)

Fusarium

Many species within this genus are major plant pathogens, and many are seed-borne. The main criteria are based on morphology of conidia, in seed health testing habit characters of colonies on seeds are important for identification. There are three kinds of conidia: macroconidia (phialoconidia), microconidia (phialoconidia), and chlamydoconidia (chlamydospores). Conidial masses appear pale or brightly coloured, and slimy.

Though the conidiogenous cell in *Fusarium* is regarded as a phialide (Mason, 1933), in some species (Section Arthrosporiella) blastoconidia are developed. The conidiophores are simple or branched, each branch ending in one phialide, or in some species in two or more phialides (called polyphialides) as in *F. moniliforme* and *F. semitectum*. The morphology of macroconidia, the form of the basal cell being important, as well as that of the conidiogenous cell are fundamental diagnostic characters. Presence and form of microconidia, which usually are one-celled, and chlamydoconidia, which can be terminal, or intercalary, often formed in chains, are additional diagnostic characters of importance. Wollenweber and Reinking (1935) classified *Fusarium* in 16 Sections; Booth (1971) reduced the number of Sections to 12. In the following survey of seed-borne species the name of the Section concerned, according to Booth, is indicated for each species.

For identification of species, consult Booth (1971), cf. previous important monographs or detailed studies, particularly Wollenweber and Reinking (1935), Gordon (1952, 1954, 1960) and (in Russian) Bilai (1955). For identification in seed health testing, the descriptions based on habit characters as seen on seeds, given by Ram Nath, Neergaard and Mathur (1970), are useful. The total number of described species has been about a thousand but the number of species considered valid is now reduced to about fifty–hundred. See Colour Plates I, II and IV.

The perfect states of *Fusarium* are without exception Ascomycetes of the family Hypocreaceae (*Calonectria*, *Gibberella*, *Griphosphaeria*, *Micronectriella* and *Nectria*).

F. avenaceum, Fig. 5.19(5), Section Arthrosporiella, perfect state *Gibberella avenacea*. This species has a very wide host range, and is encountered rather frequently in seeds of many hosts. Noble and Richardson (1968) listed 28 genera of host plants, in which the fungus has been recorded on seeds, Ram Nath, Neergaard and Mathur (1970) and the Andersen (1974) index have listed 69 species of 55 genera. The pathogen may sometimes be serious but is regarded as less so in cereals than are certain other species of *Fusarium*, cf. Subsection 1.2.1, wheat.

Distribution: World-wide.

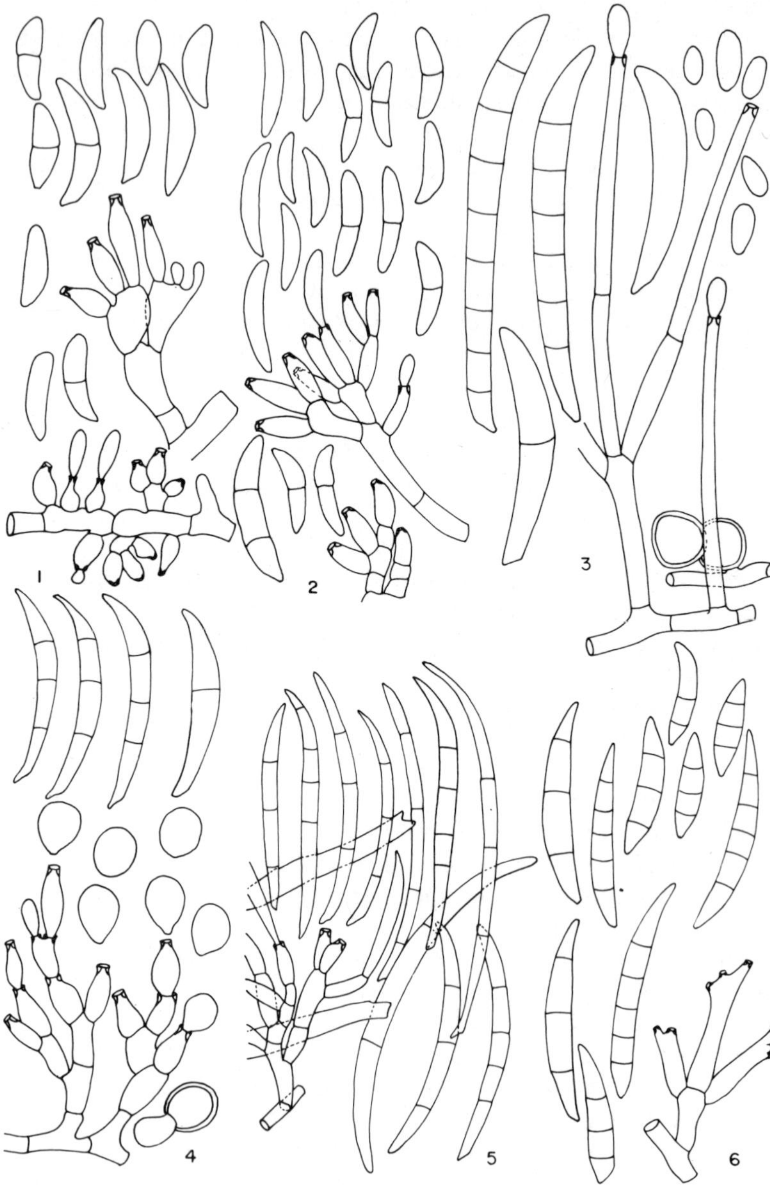

Fig. 5.19. Conidia and conidiophores of (1) *Fusarium dimerum*; (2) *F. nivale*; (3) *F. solani*; (4) *F. poae*; (5) *F. avenaceum*; (6) *F. semitectum*. Microconidia are present in (3) and (4). ×700, except (5) which is ×650. (Courtesy of C. Booth)

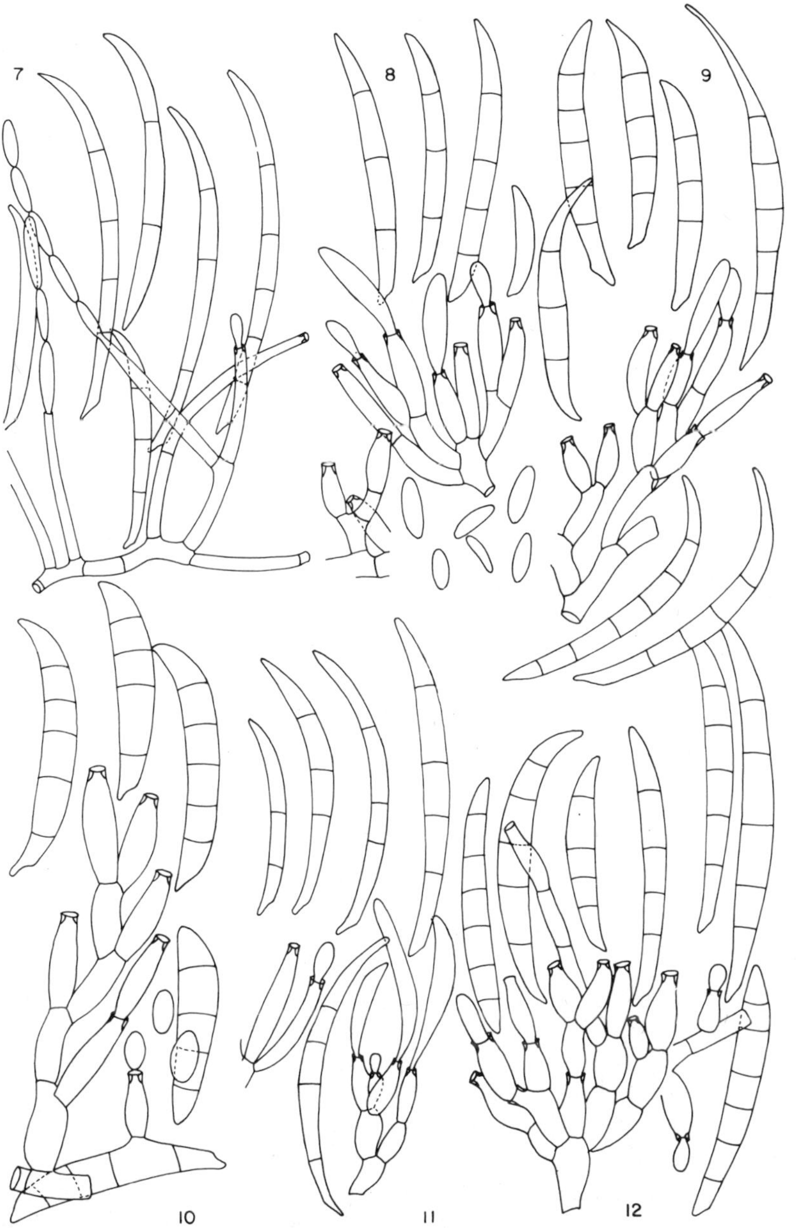

Fig. 5.20. Conidia and conidiophores of (7) *Fusarium moniliforme*; (8) *F. oxysporum*; (9) *F. equiseti*; (10) *F. culmorum*; (11) *F. heterosporum*; (12) *F. graminearum*. Microconidia are present in (7), (8) and (10). ×700. (Courtesy of C. Booth)

F. culmorum, Fig. 5.20(4), Section Discolor. A serious pathogen of cereals, particularly on wheat, but has a wide range of hosts also outside of Gramineae. Noble and Richardson (1968) have listed the fungus as seed-borne on 11 plant genera, Ram Nath, Neergaard and Mathur (1970) on 23 species of 20 genera, while the Andersen (1974) index listed 48 species of 42 genera. Technique of detection in wheat is discussed by Malalasekara and Colhoun (1969). Localisation in seed, see Sections 11.5 and 13.5.

Distribution: North America, South America (scattered areas), Europe, East Africa, Asia (India), Australia, New Zealand.

CMI Distribution Map No. 440.

F. dimerum, Fig. 5.19 (1), Section Arachnites. A saprophyte reported from a few genera of plants but predominantly from rice. Frequent on seed of this host (Ram Nath, Neergaard and Mathur, 1970).

Distribution: Europe, Asia, Australia.

F. equiseti (F. scirpi), Fig. 5.20(3), Section Gibbosum, perfect state *Gibberella intricans*. This species is prevalent in warm temperate and subtropical areas on many hosts and may cause stem and root rot in cereals but is usually not serious. It has been found fairly frequently in seeds of many hosts, including *Abelmoschus esculentus*, *Capsicum annuum*, *Pennisetum glaucum*, *Phaseolus aureus*, *Oryza sativa*, *Sorghum vulgare* and *Zea mays* (Ram Nath, Neergaard and Mathur, 1970), and has been reported on further 5 hosts, according to Andersen (1974) index, while 7 genera are included in the Annotated List (Noble and Richardson, 1968).

Distribution: North America, Europe, Africa, Asia.

F. fusarioides (F. chlamydosporum), Section Arthrosporiella. Pathogenic on different hosts, and encountered commonly on seeds of *Pennisetum glaucum* and *Sorghum vulgare*, rarely on *Oryza sativa*.

Distribution: Europe, Asia.

F. graminearum (*F. roseum* pro parte), Fig. 5.20(6), Section Discolor, perfect state *Gibberella zeae*, important pathogen in cereals, causing seedling blight, crown rot, head and kernel blight or scab. On many gramineous hosts but also in a wider host range, pathogenicity between other hosts and Gramineae having been confirmed by cross-inoculations. Noble and Richardson (1968) have listed 8 genera of hosts carrying the pathogen on seed, Ram Nath, Neergaard and Mathur (1970) 15 species (13 genera), and the Andersen (1974) index 11 species of 10 genera, all except one being Gramineae. Influence of soil temperature and host on development of disease, see Subsection 15.2.2, and about the impact of time of infection in wheat, see Section 11.4. Economic importance is discussed in Subsection 1.2.1, wheat and maize.

Distribution: North and South America, Europe, Africa, Asia, Australia, New Zealand.

F. heterosporum (F. reticulatum), Fig. 5.20(5), Section Discolor, perfect state *Gibberella gordonia*. Causes head blight in cereals and other Gramineae, but may occur on hosts outside this family. Noble and Richardson (1968) indicate its presence on seeds of rice and *Poa* spp.

Distribution: North America, Europe, Africa, Asia.

F. moniliforme, Fig. 5.20(1), Section Liseola, Gibberella ear rot, bakanae disease, perfect state *Gibberella fujikuroi*. A serious pathogen in cereals but has wide host range including 31 plant families in which it causes seedling blight, scorch, foot rot, stunting and hypertrophy (Booth, 1971). Seed-borne in at least 10 plant families according to the Annotated List (Noble and Richardson, 1968), recorded in seeds of the following species by Ram Nath, Neergaard and Mathur (1970) and according to the Andersen (1974) index: *Abelmoschus esculentus*, *Capsicum annuum*, *Coriandrum sativum*, *Cuminum cyminum*, *Dahlia variabilis*, *Dalbergia sissoo, Glycine max, Gossypium* sp., *Helianthus annuus, Lycopersicon esculentum*, *Luffa acutangula*, *Oryza sativa*, *Pennisetum glaucum*, *Phaseolus aureus*, *Rudbeckia hybrida*, *Sorghum vulgare*, *Tagetes erecta*, *T. patula*, *Zea mays* (see also Rintelen, 1967), *Zinnia elegans* and *Z. haageana*; perithecia rather frequently encountered on rice seed. For economic significance, see Subsection 1.2.1, rice, maize and sorghum; also important on cotton.

Distribution: North, Central and South America, Europe, Africa, Asia, Australia, New Zealand.

CMI Distribution Map No.102.

F. moniliforme var. *subglutinans,* perfect state *Gibberella fujikuroi* var. *subglutinans*. Often misidentified as *F. moniliforme* but readily separated from this by the presence of polyphialides producing microconidia in heads in contrast to the main species producing microconidial chains. Important pathogen of cereals but occurs in a wide range of hosts other than Gramineae. Encountered frequently in seeds of maize.

Distribution: North and South America, Europe, Africa, Asia, Australia.

CMI Distribution Map No. 191.

F. nivale, Fig. 5.19(2), Section Arachnites, snow mould, brown foot rot, perfect state *Griphosphaeria nivalis (Micronectriella nivalis, Calonectria nivalis)*. This major pathogen of Gramineae, particularly of cereals in cool temperate regions, is common on seeds of these hosts, cf. Noble and Richardson (1968), Ram Nath, Neergaard and Mathur (1970), De Tempe (1970*b*), Schmidt (1971AR, 1972AR), the Andersen (1974) index, and Kolk and Olsson (1975), indicating seed infection of *Avena sativa*, *Dactylis glomerata*, *Festuca pratensis*, *F. rubra*, *Hordeum distichon, H. vulgare, Lolium multiflorum, L. perenne, Oryza sativa, Secale cereale* and *Triticum aestivum*. On economic importance see Subsection 1.2.1, wheat, and on epidemiological implications of seed transmission and air dissemination see Subsection 1.3.3.

Distribution: North and Central America, Europe, Asia, Australia, New Zealand.

CMI Distribution Map No. 432.

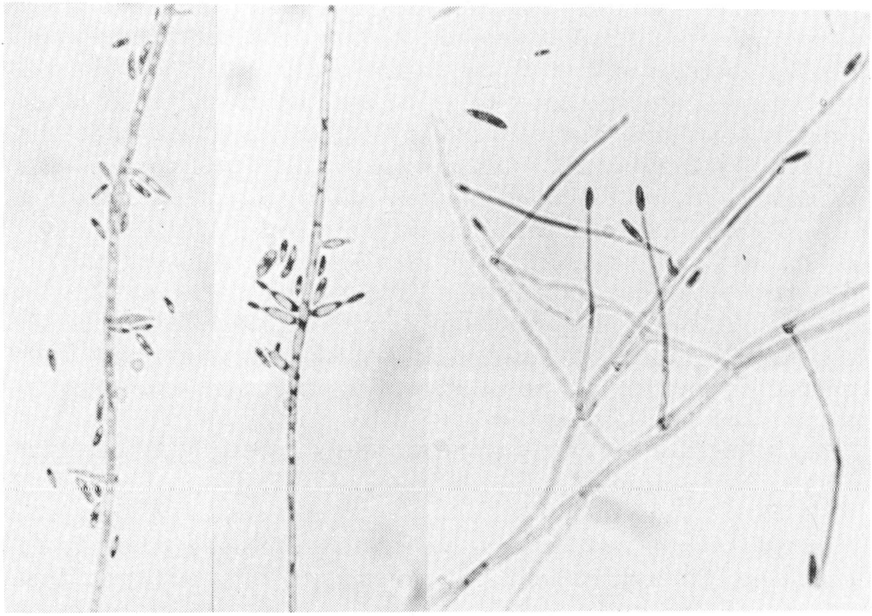

Fig. 5.21. Conidiogenesis of microconidia in, left, *Fusarium oxysporum,* right, *F. solani.* Note the difference in length of conidiophores between the two species, an important diagnostic character. ×250. (Courtesy of C. Booth)

F. oxysporum, Figs 5.20(2) and 5.21, Section Elegans, Fusarium wilt of many hosts. The most important plant pathogenic species of *Fusarium*, having a wide host range, and including numerous *formae speciales*, some of which contain two or several pathogenic races causing devastating wilt diseases (cf. Section 17.2), and many of which are seed-borne as listed (with or without reference to *formae speciales*) by Noble and Richardson (1968) and the Andersen (1974) index for the following hosts: *Allium*, *Asparagus officinalis, Beta vulgaris, Bromus, Callistephus chinensis, Cannabis sativa, Citrullus vulgaris, Cucumis sativus, Glycine max, Gossypium, Lens culinaris, Linum grandiflorum, L. usitatissimum, Lupinus luteus, Lycopersicon esculentum* (see also Staunton, 1971), *Matthiola incana*, *Medicago sativa*, *Oryza sativa*, *Phaseolus vulgaris*, *Pisum sativum*,

Pseudotsuga menziesii, *Solanum melongena*, *Sorghum vulgare*, *Spinacia oleracea*, *Tagetes*, *Trifolium pratense*, *T. repens*, *Vicia faba*, *Vigna* and *Zea mays* (on this host, see also Rintelen, 1967). Also in *Cuminum cyminum* (Singh, Choudhary and Patel, 1972), *Cyclamen persicum* (Tompkins and Snyder, 1972), *Elaeis guineensis* (Locke and Colhoun, 1973), *Gossypium* (Gubanov and Sabirov, 1972; Bakry and Rizk, 1967), *Glycine max* (Saharan and Gupta, 1972). Economic significance is discussed in Subsection 1.2.4, tomato, Subsection 1.2.5, cotton, and Subsection 1.3.2. *Formae speciales* and races are listed in Section 17.2. On disease cycle of *F. oxysporum* f.sp. *callistephi*, see Section 14.7.

Distribution: World-wide, but pathogenic races may have limited distribution, defined by host range.

CMI Distribution Maps No. 32 (*Linum*) and No. 362 (*Gossypium*).

F. poae, Fig. 5.19(4), Section Sporotrichiella, has a wide host range, predominantly belonging to Gramineae. Encountered in seeds of *Avena sativa*, *Bromus*, *Dianthus*, *Festuca*, *Glycine max*, *Hordeum vulgare*, *Poa pratensis, Triticum aestivum* and *Zea mays* (Noble and Richardson, 1968; Schmidt, 1971AR, 1972AR; Andersen, 1971AR*b*; Andersen and Broe, 1974AR). Ram Nath, Neergaard and Mathur (1970) listed 24 plant species reported from literature.

Distribution: Common in temperate regions, North and South America, Europe, Africa, Australia, New Zealand.

F. semitectum, Fig. 5.19(6), Section Arthrosporiella, is predominantly a secondary pathogenic invader of plant tissues. Rather common on seeds of many hosts including *Abelmoschus esculentus*, *Coriandrum sativum*, *Cuminum cyminum*, *Desmodium intortum*, *Luffa acutangula*, *Pennisetum glaucum*, *Oryza sativa*, *Phaseolus aureus*, *Sorghum vulgare* and *Zea mays*. Further 14 plant species are listed by Ram Nath, Neergaard and Mathur (1970). Seed-borne infection in soybean controlled by seed treatment (Saharan and Gupta, 1974). Prevailing in tropical and subtropical regions.

Distribution: North America, Europe, Africa, Asia.

F. solani, Figs 5.19(3) and 5.21, Section Martiella, perfect state *Nectria haematococca*. This ubiquitous pathogen occurs on seed of a wide range of hosts, including, according to Ram Nath, Neergaard and Mathur (1970): *Capsicum annuum*, *Coriandrum sativum*, *Cuminum cyminum*, *Dalbergia sissoo, Foeniculum vulgare, Glycine max, Linum usitatissimum*, *Lycopersicon esculentum*, *Oryza sativa*, *Phaseolus vulgaris* and *Sorghum vulgare*, and further 16 plant species reported by other authors. Noble and Richardson (1968) listed: *Allium, Arachis hypogaea, Cucurbita, Gossypium*, *Hibiscus esculentus* (see also Ribeiro *et al.*, 1971), *Trifolium pratense* and *Zea mays*. The pathogen causes foot rot in many hosts such as broad bean, Phaseolus bean and pea.

Distribution: World-wide.

Gloeocercospora

The conidiophores develop in sporodochia in which filiform, septate blastoconidia are produced in a slimy matrix (Deighton, 1971).

G. sorghi, zonate leaf spot, is seed-borne in millets, the hosts not specified to species (Leukel and Martin, 1953) and in *Sorghum vulgare* (Ciccarone, 1949; Bain, 1950; Mathur, Mathur and Neergaard, 1975); seed treatment is recommended in the U.S.A.

Distribution: North and South America, Africa, Asia, Australia.

CMI Distribution Map No. 339.

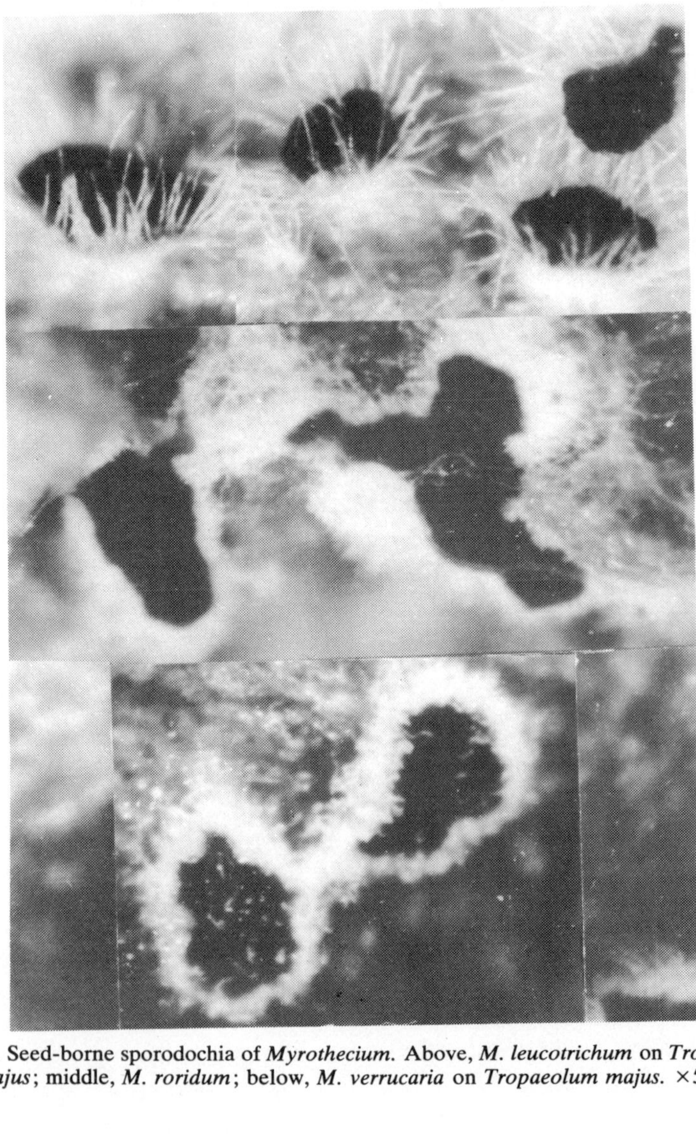

Fig. 5.22. Seed-borne sporodochia of *Myrothecium.* Above, *M. leucotrichum* on *Tropaeolum majus*; middle, *M. roridum*; below, *M. verrucaria* on *Tropaeolum majus.* ×50

Ramulispora

Simple or branched conidiophores, which are developed from stromata, produce filiform conidia possessing lateral branches; the conidia are held together in gelatinous aggregates.

R. sorghi (R. andropogonis), sooty stripe, is considered to be seed-borne (Bain, 1942; Tarr, 1962) but seed transmission seems not to have been established. Important pathogen on sorghum.

Distribution: North and South America, Africa, Asia.

(2. Phaeosporae)

Myrothecium

Sporodochia shallow cups, with a white margin of contorted hyphae, with or without setae. Phialoconidia are produced on branched conidiophores, forming viscid masses.

M. leucotrichum (M. jollymannii), occurs in low percentages in cowpea seed, and has been proved to be quite virulent in this and other hosts (Nguyen, Mathur and Neergaard, 1973). See Fig. 5.22.

Distribution: Africa, Asia (India).

M. roridum has been recorded in low percentages in seeds of cotton (Cognée and Bird, 1965, see also Bagga, 1970), *Dahlia variabilis*, and nasturtium, *Tropaeolum majus* (Nguyen, Mathur and Neergaard, 1973), and mungbean, *Phaseolus aureus* (Ram Nath, Mathur and Neergaard, 1970). See Fig. 5.22.

Distribution: North America, Europe, Africa, Asia (India).

CMI Distribution Map No. 458.

M. verrucaria has been found in low percentages in seeds of rice and soybean (Nguyen, Mathur and Neergaard, 1973). A mild pathogen. See Fig. 5.22.

Distribution: North America, Europe (?), Africa (Egypt), Asia (India).

(3. Phaeodictyae)

Epicoccum

Conidia are brown, subspherical or pyriform, verrucose or echinulate, muriform, produced singly on short stout conidiophores in sporodochia appearing black. One species is recognised, *E. purpurascens*. Another, *E. andropogonis*, maintained by some mycologists, is by others (e.g. Ellis, 1971) classified under its original name, *Cerebella andropogonis*; it occurs sometimes on ergots.

E. purpurascens (*E. neglectum*, *E. nigrum*, *E. oryzae*, and about 50 other synonyms), Fig. 5.23, is a common contaminant on all kinds of seed, appearing as black sporodochia, often producing a conspicuous red pigment forming local coloration around these (Neergaard and Saad,

1962; Malone and Muskett, 1964; Limonard, 1968; Kang, Neergaard and Mathur, 1972; Agarwal, Mathur and Neergaard, 1972). A common saprophyte and secondary invader on all sorts of plants.

Distribution: Cosmopolitan.

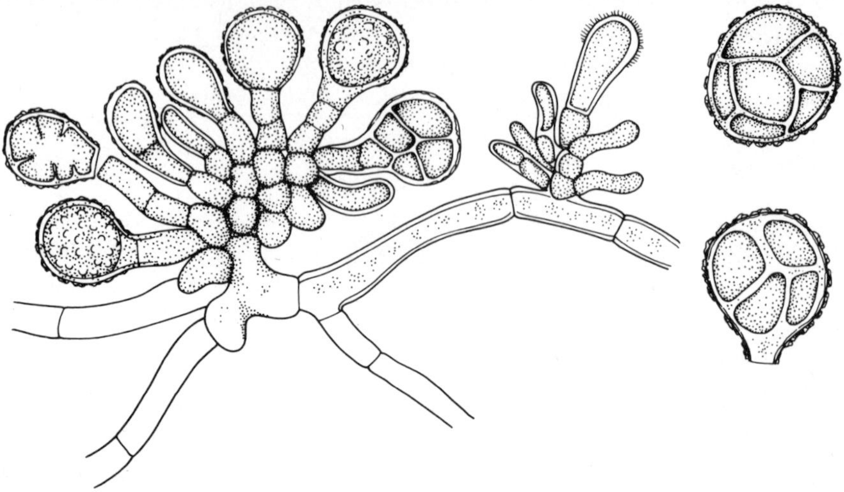

Fig. 5.23. *Epicoccum purpurascens,* conidiogenous parts and conidia, × 1,000. (Courtesy of J. A. Von Arx)

5.3.6.3 COELOMYCETES

This class of Fungi Imperfecti comprises forms in which conidia are developed within a pseudoparenchymatous cavity formed by hyphae, and which may form a (1) more or less open structure, called acervulus (order Melanconiales) or (2) closed, frequently flask-like structure, called pycnidium (order Sphaeropsidales).

Melanconiales (cf. Table 5.5)

The open, pseudoparenchymatic fruit-body is an acervulus, which may be epidermal, subepidermal or subcuticular (Von Arx, 1957*a*, 1957*b*). There is one family, Melanconiaceae. Saprophytes or plant pathogens.

(1. Hyalosporae)

Colletotrichum (Gloeosporium)

Acervula are typically provided with setae; conidia (phialoconidia) are cylindrical, ellipsoidal or falcate, developed from phialides in basipetal succession from phialides. Germ tubes from conidia form appressoria. The original total number of about 800 described species of *Colletotrichum*, and (partly) *Gloeosporium*, has now been reduced to a dozen of species considered valid (Von Arx, 1957*a*, 1957*b*, 1970*a*).

Table 5.5. Coelomycetes. Classification of the surveyed seed-borne genera

Order	*Family*	*Conidial type*	*Genus*
Melanconiales	Melanconiaceae	Hyalosporae	Colletotrichum (Gloeosporium) Discula Kabatiella Sphaceloma
		Hyalodidymae	Marssonina
		Phaeophragmiae	Pestalotiopsis Truncatella
Sphaeropsidales (Phomales)	Sphaerioidaceae (Phomaceae)	Hyalosporae	Aposphaeria Macrophoma Macrophomina Phoma Phomopsis Phyllosticta Plenodomus Selenophoma
		Phaeosporae	Coniothyrium
		Hyalodidymae	Ascochyta
		Phaeodidymae	Botryodiplodia Diplodia
		Hyalophragmiae	Stagonospora
		Hyaloscolecosporae	Dilophospora Septoria
	Leptostromataceae	Hyalosporae	Leptostroma
	Excipulaceae	Hyalodidymae	Discella (Diplodina)
		Scolecosporae	Ephelis

C. acutatum, Fig. 5.25, has been encountered in seed of *Capsicum annuum* (Lee, 1973, in preparation) and *Zinnia elegans* (Kulshrestha, Mathur and Neergaard, in preparation). The fungus causes fruit rot in eggfruit, tomato, papaver, strawberry and others (Simmonds, 1965).

Distribution: Europe (Italy), Asia (Turkey), Australia.

C. dematium (C. circinans, C. curvatum, C. glycines, C. indicum, C. spinaciae, C. truncatum, a total of about 100 synonyms, *fide* Von Arx, 1957*a*, 1957*b*), Fig. 5.24. A polyphagous saprophyte, three pathogenic *formae speciales* having been distinguished: *C. dematium* f.sp. *truncatum* on legumes, seed-borne in *Cyamopsis tetragonoloba* (Sowell, 1965), *Desmodium* spp. (Wells and Forbes, 1963), *Glycine max* (Tiffani, 1951; Kulshrestha, Mathur and Neergaard, in preparation) and *Phaseolus lunatus* (Chambers and Andes, 1962; Toler and Wester, 1966); *C. dematium* f.sp. *circinans* in *Allium* sp. (Neergaard, 1940AR, 1941AR, 1951AR; Behr, 1963) and *C. dematium* f.sp. *spinaciae*, frequent in seeds of *Spinacia oleracea* (Doyer, 1938; Neergaard, 1939AR, 1951AR; Baker, 1954; Von Arx, 1970; Andersen, 1974 — index; Buchwald, 1974). The fungus has been recorded (under different names) on *Crotalaria juncea* (Whiteside and Herd, 1966), *Gossypium* (Bhide, Desai and Rane, 1957; Leakey, 1962; Sokolova and Sokolova, 1964). Also in *Abelmoschus esculentus*, *Capsicum annuum*, *C. frutescens*, *Dolichos biflorus*, *Hibiscus cannabinus*, *Oryza sativa*, *Tropaeolum majus*, *Vigna sinensis* (Kulshrestha, Mathur and Neergaard, in preparation). Saprophyte, or cause of leaf spot or anthracnose in many hosts. Disease cycle in soybean is discussed in Section 14.1.

Distribution: World-wide.

C. fuscum, foxglove anthracnose. Frequently transmitted from seed in the U.S.A. (Thomas, 1951), seed-borne and serious on *Digitalis lanata* in Czechoslovakia (Jechova, 1965).

Distribution: North America, Europe.

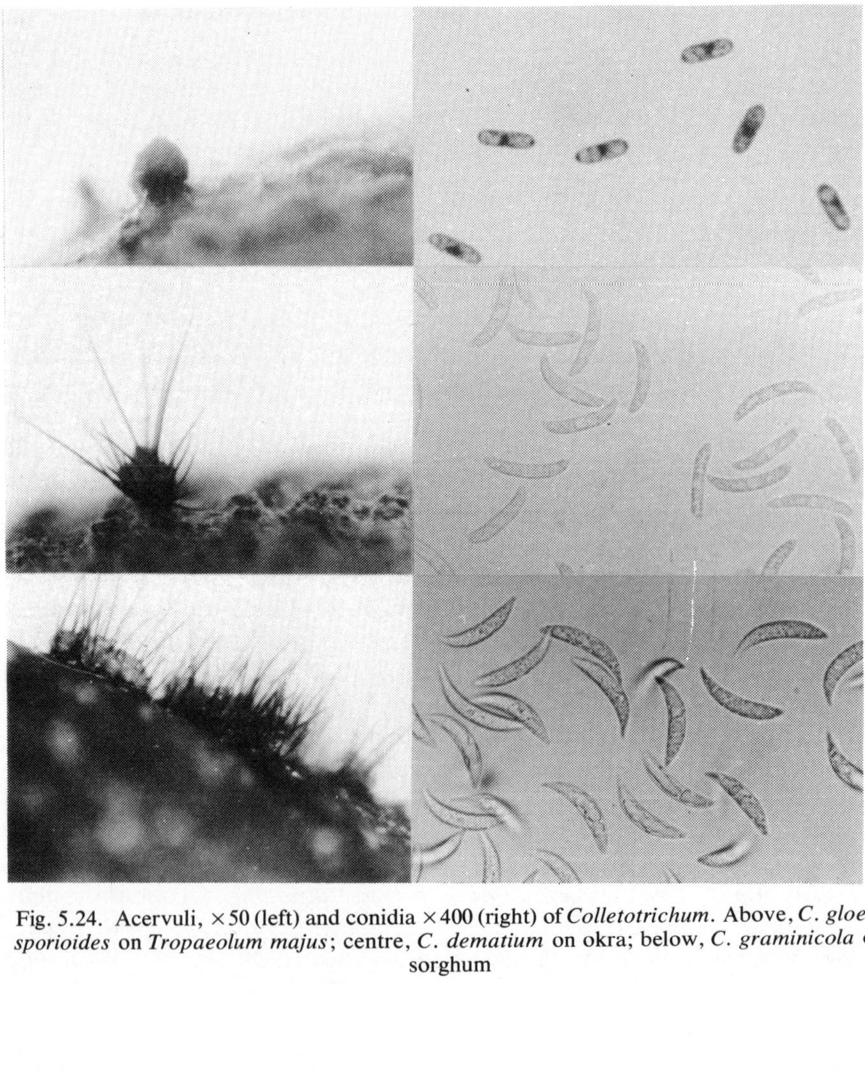

Fig. 5.24. Acervuli, ×50 (left) and conidia ×400 (right) of *Colletotrichum*. Above, *C. gloeosporioides* on *Tropaeolum majus*; centre, *C. dematium* on okra; below, *C. graminicola* on sorghum

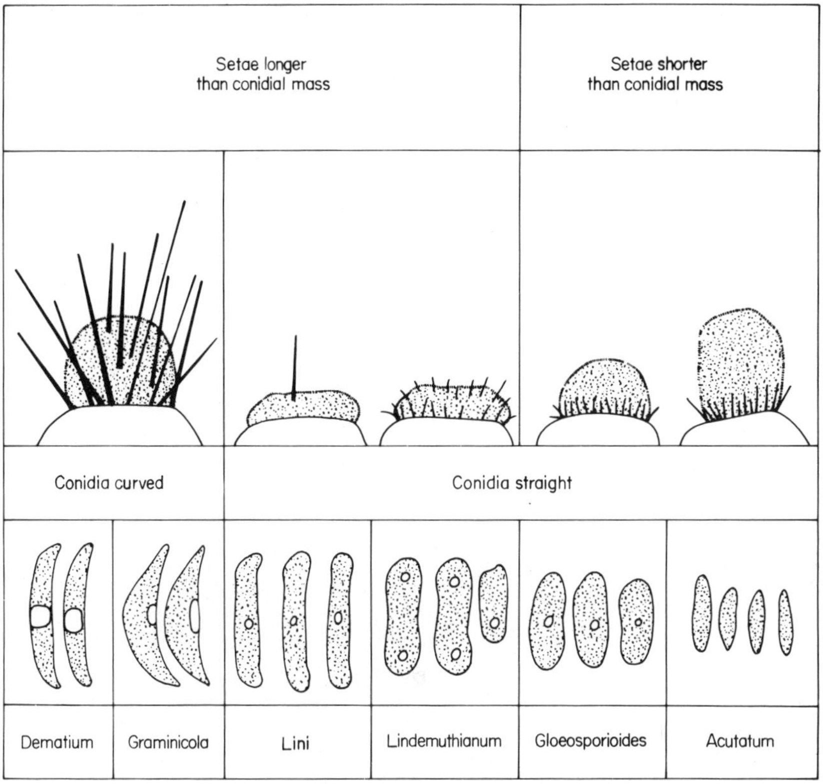

Fig. 5.25. Diagrammatic key for identification of seed-borne species of *Colletotrichum*, based on characteristics of acervuli and conidia. (Kulshrestha, Mathur and Neergaard, in preparation)

C. gloeosporioides (*C. antirrhini, C. cajani, C. capsici, C. destructivum, C. hibisci, C. higginsianum, C. melongenae, C. piperatum, C. tabacum, C. violaetricoloris,* about 750 synonyms for the conidial and ascigerous states, cf. Von Arx, 1970*a*), Fig. 5.24 and Fig. 5.7, perfect state *Glomerella cingulata*. In seeds of *Acacia decurrens* var. *dealbata* (Hashimoto, 1968), *Antirrhinum majus* (Andersen and Broe, 1974AR), *Bixa orellana* (Peregrine, 1970), *Cajanus cajan* (Tucker, 1927), *Capsicum* spp. (Smith and Crossan, 1958; Grover and Bansal, 1971), *Corchorus* (see Noble and Richardson, 1968), *Coriandrum sativum* (Sehgal, Desphande and Srivastava, 1965), *Hibiscus cannabinus* (see Noble and Richardson, 1968), *Raphanus sativus* (Neergaard, 1951AR; Andersen and Lundsten, 1968AR; Andersen, 1971AR*a*), *Nicotiana tabacum* (Bates, 1958), *Ornithopus sativus* (Shulga, 1964; Neofitova, 1964), *Sorghum vulgare* (Kulshrestha, Mathur and Neergaard, in preparation), *Trifolium pratense* (Butler, 1953*b*) and *Tropaeolum majus* (Kulshrestha, Mathur and Neergaard, in preparation). Under the name of the perfect state Noble and Richardson (1968) listed 12 genera, 9 of which are not mentioned above. This is the most widespread species within the genus; in many hosts it is a serious pathogen.

Distribution: World-wide.

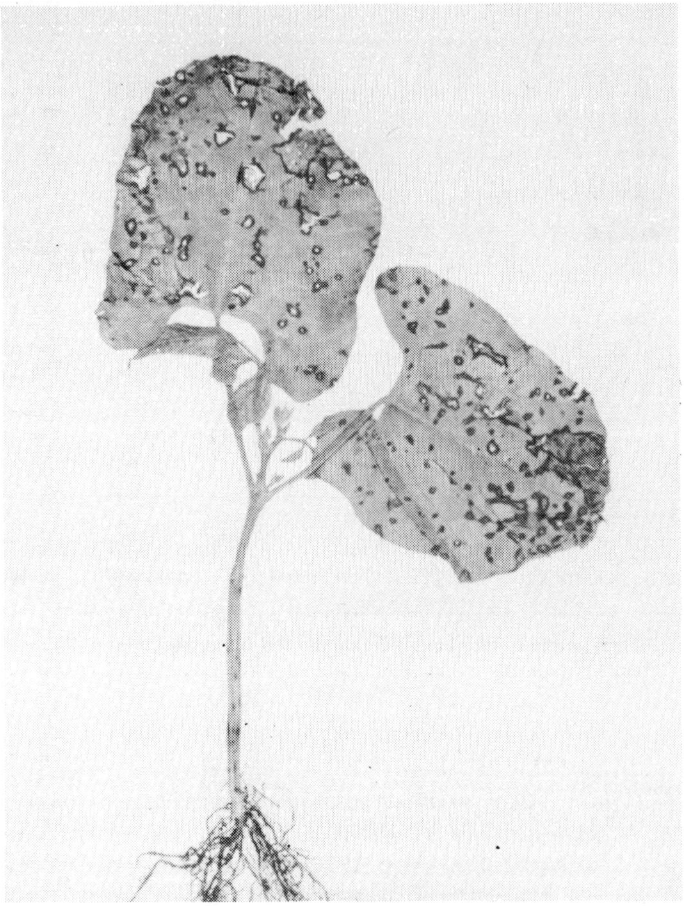

Fig. 5.26. Bean anthracnose, *Colletotrichum lindemuthianum*. (Courtesy of E. Hellmers)

C. gossypii, pink boll rot, cotton anthracnose, perfect state *Glomerella gossypii*. Seed transmission in cotton, *Gossypium* spp. (Ray and McLaughlin, 1942; Arndt, 1953, 1956; Smith, 1953*b*; Cauquil, 1960; Bagga, 1970; Fröhlich, 1974). In the U.S.S.R. all imported cotton seed is examined and sown in quarantine (Sokolova and Sokolova, 1964). Seed treatment in captan recommended in the U.S.A. Predominant in humid regions, causing damping-off and boll rot.

Distribution: In scattered areas in North, Central and South America, Europe, Africa, Asia, Australia.

CMI Distribution Map No. 317.

C. graminicola (about 30 synonyms according to Von Arx, 1957*a*, 1957*b*), Fig. 5.24, perfect state *Glomerella graminicola* (cf. Politis, 1975). Saprophytic and parasitic in many Gramineae. In seeds of *Dactylis glomerata*, *Paspalum* spp., *Phleum pratense*, *Poa* sp., *Sorghum halepense, S. nitidum, S. sudanense, S. vulgare* and *Zea mays* (Noble and Richardson, 1968; Chohan, 1967; Warren and Nicholson, 1975). In some areas a serious pathogen of cereals.

Distribution: North and South America, Europe, Africa, Asia.

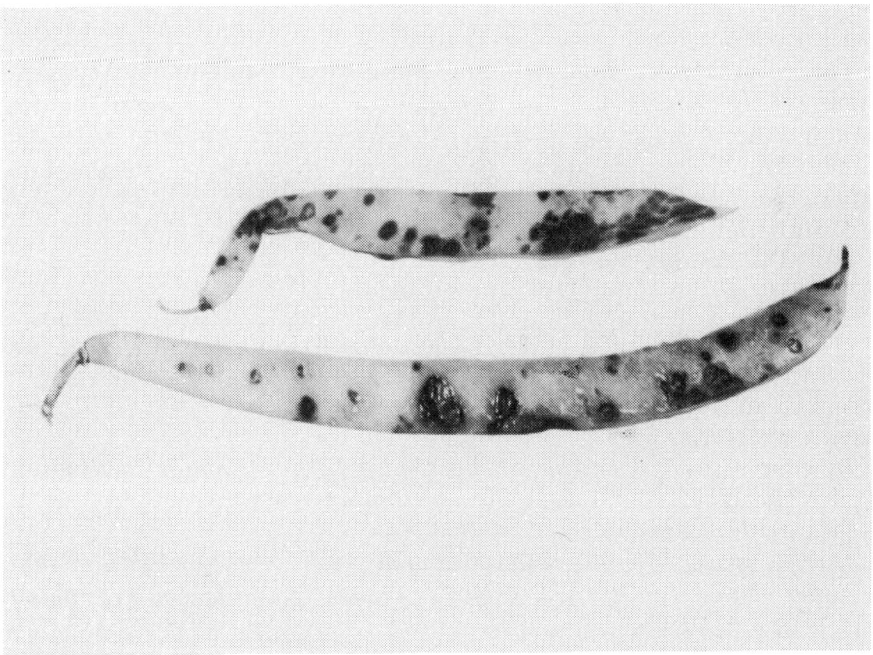

Fig. 5.27. Bean anthracnose, *Colletotrichum lindemuthianum*. (Courtesy of E. Hellmers)

C. lindemuthianum, bean anthracnose, morphologically closely related to *C. gloeosporioides*, perfect state recently described, *Glomerella cingulata* f.sp. *phaseoli* (Kimati and Galli, 1970). First fungus that was shown to be seed-transmitted (Frank, 1883; Whetzel, 1906). There is an extensive literature on seed transmission, cf. Noble and Richardson (1968). Schaffnit and Böning (1925) published a monograph on the fungus. Many pathogenic races, see Section 17.2. On economic significance, see Subsection 1.2.2, bean. Figs 5.26 and Fig. 5.27.

Distribution: World-wide, wherever Phaseolus bean is grown.

CMI Distribution Map No. 177.

C. lini (C. linicola), flax anthracnose, stem canker. Commonly seed-transmitted in flax (Doyer, 1938; Malone and Muskett, 1964; Strel'skaya, 1958; Pietkiewicz and Czyzewska, 1960; Kulshrestha, Mathur and Neergaard, in preparation). A destructive pathogen.

Distribution: North and South America, Europe, Africa, Asia.

CMI Distribution Map No. 159.

C. malvarum, anthracnose, leaf scorch. The conidia cannot be distinguished from those of *C. gloeosporioides* but a perfect state has not been found, hence the original name has been maintained (Von Arx, 1957*a*, 1957*b*). Seed transmission established in Germany (Pape, 1964). Pathogen on different Malvaceae.

Distribution: North America, Europe.

C. orbiculare (C. lagenarium), anthracnose. Morphologically very close to *C. gloeosporioides* but perfect state has not been identified with certainty. In seeds of watermelon, *Citrullus vulgaris* (Rankin, 1954), and cucumber, *Cucumis sativus* (Noble and Richardson, 1968). Two pathogenic races on watermelon.

Distribution: World-wide.

CMI Distribution Map No. 313.

Discula

Acervuli are intra- and subepidermal, variable being flat or pycnidium-like, the crowded conidiophores are short or fairly long, conidia borne in basipetal succession, ellipsoidal to cylindrical; their germ tubes do not form appressoria.

D. campestris (Gloeosporium acericola, G. quercinum). In fruits of *Acer* (Prisyaznuk, 1949) and *Quercus* (Urošević, 1961*a*).

Distribution: Europe.

Discula state of *Gnomonia errabunda* (probable synonym: *Gloeosporium ulmicola*). On seed of *Ulmus pumila* in Roumania (Georgescu and Petrescu, 1954).

Distribution: North America, Europe.

Kabatiella (Polyspora)

Acervuli are hyaline, conidia ovoid to pyriform, produced in cluster of 2–8 almost simultaneously from polyphialidic conidiogenous cells.

K. caulivora (Gloeosporium caulivorum), scorch, northern anthracnose. The conidia can be carried on seed in a viable condition for at least 2–3 years (Sampson, 1928*b*; Leach, 1962*c*), and seed transmission has been established in *Trifolium incarnatum* and *T. pratense* (Minyaeva, 1951; Azbukina, 1956; Cole and Couch, 1958; Leach, 1962*c*). The disease is sometimes destructive, in Denmark causing failure of seed crops. Seed treatment effective.

Distribution: North America, Europe, Morocco, Japan, Australia, New Zealand.

CMI Distribution Map No. 351.

Sphaceloma

Acervuli are developed within epidermis and are flat, often very small, bursting in a 'hyphomycetous way' (Von Arx, 1970*a*). Conidia (phialoconidia) are formed acrogenously, sometimes pleurogenously, and are ellipsoidal to ovoid. Members of this genus cause scab, anthracnose or necrotic spots on leaves, stems and fruits. Perfect state: *Elsinoë*.

S. violae, scab of pansy and violet. Seed capsules are attacked (Pirone, 1970). Seed treatment advised (Pape, 1964).

Distribution: North America, Australia.

(2. Hyalodidymae)

Marssonina

Acervuli subcuticular, intra- or subepidermal, conidia long ovoid, often curved.

M. juglandis, blotch, anthracnose of walnut, perfect state *Gnomonia leptostyla*. Perithecia may be found in fallen nuts of *Juglans regia* (Miller, 1953).

Distribution: North and South America, Europe, Africa, Asia.

CMI Distribution Map No. 384.

(3. Phaeophragmiae)

Pestalotiopsis

Acervuli without setae, conidia (blastoconidia) five-celled, three central cells being pigmented, end cells hyaline, with two or more apical simple or branched appendages; there may be one basal, simple or branched appendage. Guba (1961) maintains in his monograph almost 200 species in *Pestalotia*, section Quinqueloculatae.

P. funerea. In seeds of *Libocedrus decurrens*, *Pinus pinaster* and *Pseudotsuga menziesii* (Batko, in Holmes and Buszewicz, 1955). A common pathogen on conifers.

Distribution: North America, Europe.

P. gracilis. In seed of *Clarkia amoena* subsp. *whitneyi* (syn. *Godetia whitneyi*) in Egypt (Michail, 1964).

Distribution: Africa.

Truncatella

Acervuli without setae, conidia (blastoconidia) four-celled, the two median cells being pigmented, end cells hyaline, one to four apical simple or branched appendages, with or without a basal unbranched appendage. Guba (1961) maintains many species in *Pestalotia*, section Quadriloculatae.

T. hartigii. In seeds of *Abies* sp. and *Nothophagus* sp. (Batko, in Holmes and Buszewicz, 1957).

Sphaeropsidales (Phomales) (cf. Table 5.5)

The fruit-body is a globose or more or less flask-like pycnidium. Most of the seed-borne species discussed below belong to the family Sphaerioidaceae (Phomaceae) which produce dark, often globose, hard-textured pycnidia. One genus, *Leptostroma*, is classified under Leptostromataceae which have pycnidia with a poorly developed basal portion.

Sphaerioidaceae (Phomaceae)

(1. Hyalosporae)

Aposphaeria

See *Microcyclus* (Subsection 5.3.5.12).

Macrophoma

Pycnidia are similar to those of *Phoma*, but the conidia longer than in this genus, up to more than 15 μ long, cylindrical to ellipsoid. Saprophytes and parasites in leaves and often on stems. Closely related to or perhaps to be classified as *Botryodiplodia*.

M. araucariae. Several seed consignments have been intercepted by the quarantine service of the United States Department of Agriculture, including seeds of *Araucaria excelsa* imported from New Zealand and from Norfolk Islands as well as seeds of *A. cookei* from New Zealand (Girard, 1969, 1971, 1973, 1974).

Distribution: Australia, New Zealand.

Macrophomina

Pycnidium is provided with a pore, conidia are elongated, cylindrical to ellipsoidal.

M. phaseolina (M. phaseoli; Rhizoctonia bataticola; Sclerotium bataticola), Fig. 5.28, charcoal rot, stem rot, ashy stem blight. This pathogen, very common in the tropics, is seed-borne in a considerable range of hosts including *Arachis hypogaea, Corchorus* spp., *Glycine max, Gossypium* spp., *Phaseolus lunatus*, *P. vulgaris* (see also Watanabe, 1972), *Sesamum indicum*, *Setaria italica*, *Sorghum vulgare* (possibly), *Trifolium subterraneum* and *Vigna unguiculata* (Noble and Richardson,

1968, with many references). In addition to these hosts, the fungus has been found on seeds of the following four hosts (Chidambaram and Mathur, 1975): *Hibiscus esculentus*, *Luffa acutangula*, *Phaseolus aureus* and *Tagetes* spp.; also in *Helianthus annuus* (Fakir, 1975, in preparation). A serious pathogen causing root and stem rot in a wide range of hosts. Sclerotia survive in soil for a long period but seed transmission may be important for introduction into new areas.

Distribution: Widespread throughout the tropics.

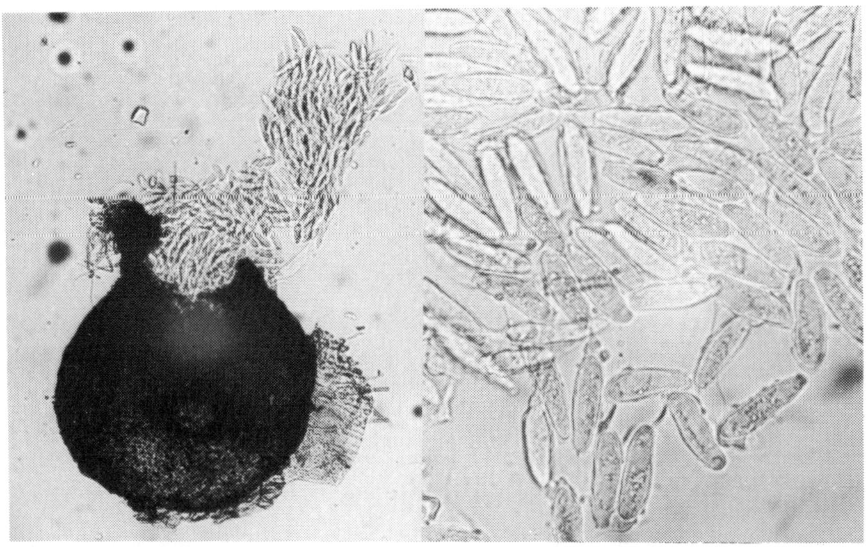

Fig. 5.28. *Macrophomina phaseolina* from *Phaseolus mungo*. Left, pycnidium, ×100; right, conidia, ×500

Phoma

Pycnidia are globular or lens-shaped, mostly superficially formed, simple or compound, and provided with a distinct somewhat protruding ostiole. The wall of pycnidium is pseudoparenchymatous, consisting of a few cell layers. The conidia are typically one-celled but may sometimes secondarily have two or more cells, they are ellipsoidal to cylindrical, formed on very short, tap-like conidiogenous cells, the first conidium endogenously, the following by successive 'budding' from the same conidiogenous cell (Boerema, 1964); they are, however, according to modern terminology (Kendrick, 1971) regarded as phialoconidia. The type species is *Phoma herbarum*, Fig. 5.30, a common saprophyte which is also encountered in seed of different kinds. About 2,000 species have been described; many of these are based, by tradition, merely on the host, and will probably by revision of the genus be referred to synonymy, e.g. species such as *Phoma bellidis* and *P. nemophilae*, listed below. Many species described as *Phyllosticta* belong to *Phoma*. The genus *Peyronellaea*, separated from *Phoma* on the basis of dictyochlamydospore-formation, which is an unstable character, is not accepted here (cf. Boerema, Dorenbosch and Kesteren, 1965, 1968).

P. anethi, see *Cercosporidium punctum*.

P. apiicola, celery root rot, celery scab. Occurs in seed, and may then attack seedlings (Goossens, 1928; Brett, 1939; Hewett, 1971; Andersen, 1974 — index). *Phoma* sp., sometimes *Phoma herbarum*, generally found on celery seed in germinator, is non-parasitic. Serious in areas where the host is constantly grown.

Distribution: North America, Europe, New Zealand.

P. aubrietiae. Seed borne in *Aubrietia* (Boerema and Valckx, 1970).

Distribution: Europe.

P. bellidis, common on seeds of *Bellis perennis* in which it may produce damping-off (Neergaard, 1945AR–1951AR; Andersen, 1974 — index).

Distribution: Europe.

P. betae (Phyllosticta spinaciae), Fig. 5.29, black leg, damping-off, perfect state *Pleospora björlingii*. Very common in seeds of *Beta vulgaris* and *Spinacia oleracea* (Doyer, 1938; McKay, 1952; Gates and Hull, 1954; Byford, 1963, 1972; Pozhar and Slyshko, 1962; Mangan, 1971, 1974; Kietreiber, 1970AR, 1971AR, 1973AR, 1974AR; Andersen, 1974—index). Seed transmission frequent and important, particularly in moist climates, cf. Section 11.3.

Distribution: Wherever beets are grown.

CMI Distribution Map No. 427.

P. destructiva, Phoma rot, fruit and stem rot. Rather common in seeds of *Capsicum annuum* (Andersen, 1974 — index), sometimes in seeds of *Lycopersicon esculentum* (Galli and Tokechi, 1961).

Distribution: North and South America, Europe.

P. eucalyptica. Intercepted in the U.S.A. by the quarantine service of the United States Department of Agriculture on seed of a considerable range of *Eucalyptus* spp. (17 species listed in 1971, 8 species in 1972), imported from Australia and New Zealand (Girard, 1973, 1974).

Distribution: Australia, New Zealand.

P. exigua (*Ascochyta phaseolorum*, cf. Boerema, 1972), leaf and pod spot of legumes, in seeds of *Phaseolus vulgaris* (Sneep, 1945; Andersen, 1972AR, 1973AR, 1974AR) and of *Vigna* sp. (Whiteside and Herd, 1966).

Distribution: North America, Europe, Africa, Australia.

P. exigua var. *linicola* (*Ascochyta linicola,* cf. Boerema and Höweler, 1967), foot rot. Rather common in seeds of flax, *Linum usitatissimum* (Muskett and Colhoun, 1947; Reeder and Vanterpool, 1953; Neergaard, 1956AR, 1958AR; Malone and Muskett, 1964). Serious in flax, often deep-seated in the seed and difficult to control by seed treatment.

Distribution: North America, Europe.

P. glumarum (Phyllosticta glumarum, Phyllosticta glumicola, Phoma indianensis), brown-bordered leaf and glume spot of rice. Rather frequent in seeds of rice (Padwick, 1950; Roy, 1966; Ou, 1972; observed at the Institute of Seed Pathology for Developing Countries, unpublished). Infected kernels develop poorly.

Distribution: North and South America, Africa, Asia.

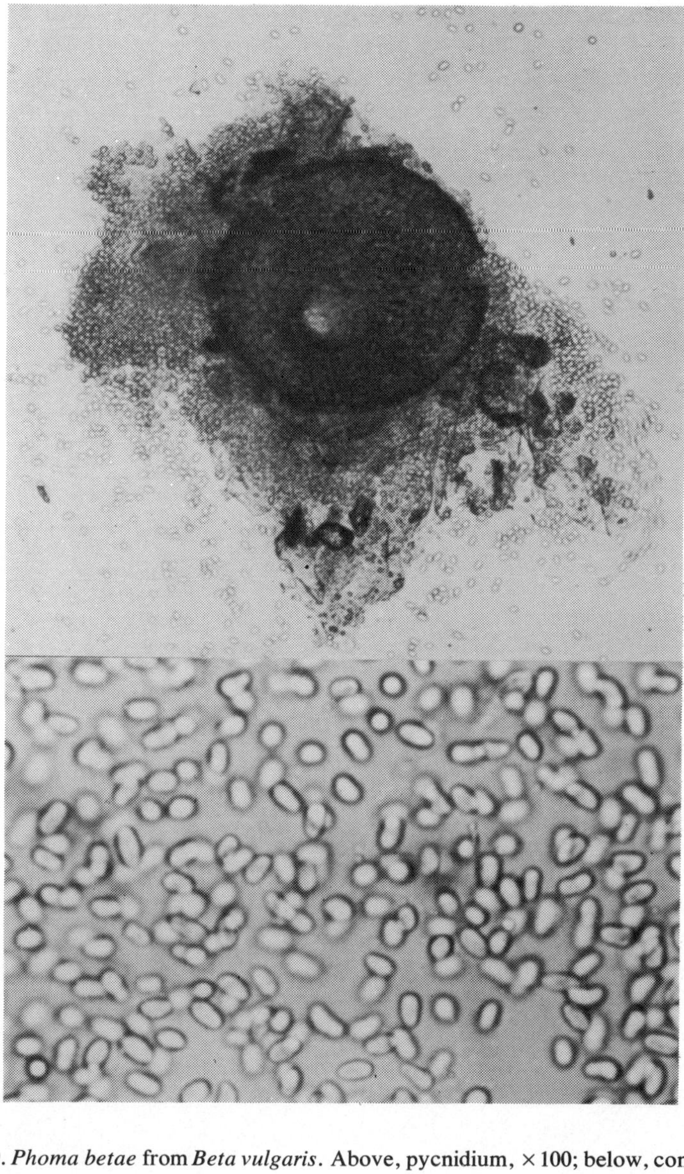

Fig. 5.29. *Phoma betae* from *Beta vulgaris*. Above, pycnidium, ×100; below, conidia, ×500

P. herbarum, Fig. 5.30. A rather common saprophyte in many kinds of seed.

P. lingam (Plenodomus lingam), Fig. 5.31, black leg of crucifers, perfect state *Leptosphaeria maculans*. Seed-transmitted and rather frequent in seeds of cabbage and other crucifers but usually in very low percentages (Walker, 1922, 1923, 1934; Neergaard, 1937AR–1951AR, 1969; Bontea, 1955; Darpoux, Louvet and Ponchet, 1957; Lloyd, 1959*a*; Allen and Smith, 1961; Jacobsen and Williams, 1971; Giessmann and Daebeler, 1973; Andersen, 1974 — index). Discussion on the role of the perfect state in air dissemination of inoculum in Subsection 1.3.3. A major pathogen of crucifers, cf. Walker (1969), and see Subsection 1.2.4, cabbage. Hot water treatment recommended but not fully effective, but see Table 22.3.

Distribution: North and Central America, Europe, Africa, Asia, Australia, New Zealand.

CMI Distribution Map No. 73.

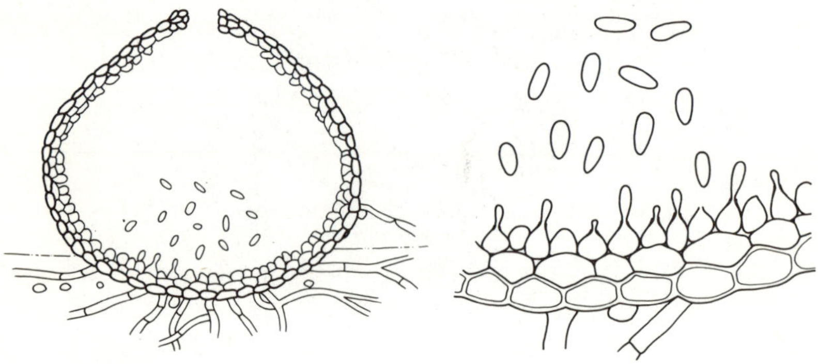

Fig. 5.30. *Phoma herbarum*, pycnidium, ×250, and part of the pycnidial wall with conidiogenous cells and conidia, ×1,000. (Courtesy of J. A. Von Arx)

P. medicaginis (Ascochyta imperfecta), spring black stem. Rather common in seeds of *Medicago lupulina* (Johansen, 1948; Andersen, 1974 — index), *M. sativa* (Cormack, 1945; Kernkamp and Hemerick, 1953; Mead, 1953; Mead and Cormack, 1961; Weltzien, 1957*b*; Csaplińska, 1967; Marasas and Bredell, 1973; Andersen, 1974 — index), *Trifolium pratense* (Andersen, 1974 — index), *T. repens* (Andersen, 1974 — index). The pathogen causes serious losses, particularly in seed crops and herbage production of lucerne.

Distribution: North America, Europe.

P. medicaginis var. *pinodella (Ascochyta pinodella),* foot rot, collar rot. Rather common but less common than *Ascochyta pisi*, in seed of pea, *Pisum arvense* and *P. sativum* (Jones, 1927; Noll, 1940; Sörgel and Unger, 1954; Wallen, 1962, 1965; Bajan, 1968; Golubev, 1971; Anselme and Champion, 1971), also encountered in seed of *Trifolium hybridum*

PLATES I and II

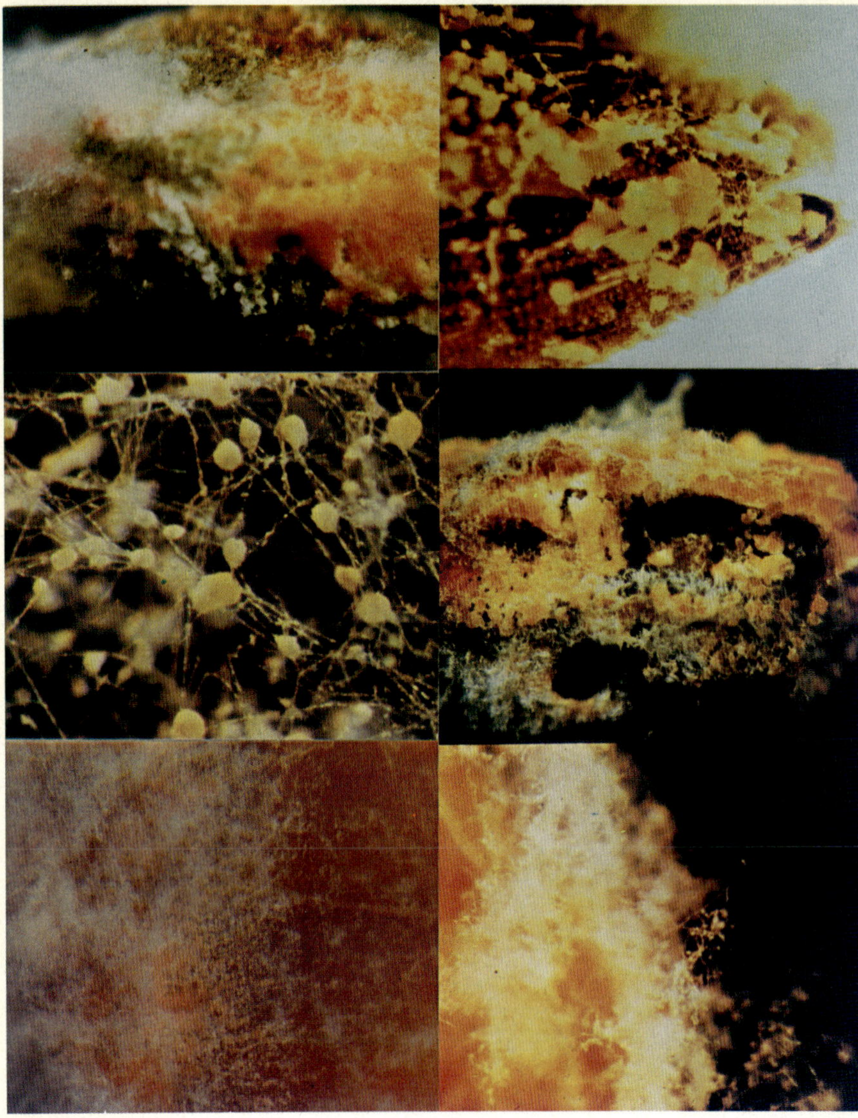

Plate I. Habit characters of *Fusarium* on seeds after 7 days of incubation in the blotter test (cf. Section 28.1). Left, from above: *Fusarium avenaceum* on barley, ×25; *F. culmorum* on barley, ×50; *F. moniliforme* on maize, ×25. Right, from above: *Fusarium dimerum* on rice, ×50; *F. equiseti* on *Phaseolus aureus*, ×25; *F. moniliforme* on maize, ×50. The habit characters, typical of the species, persist from host to host

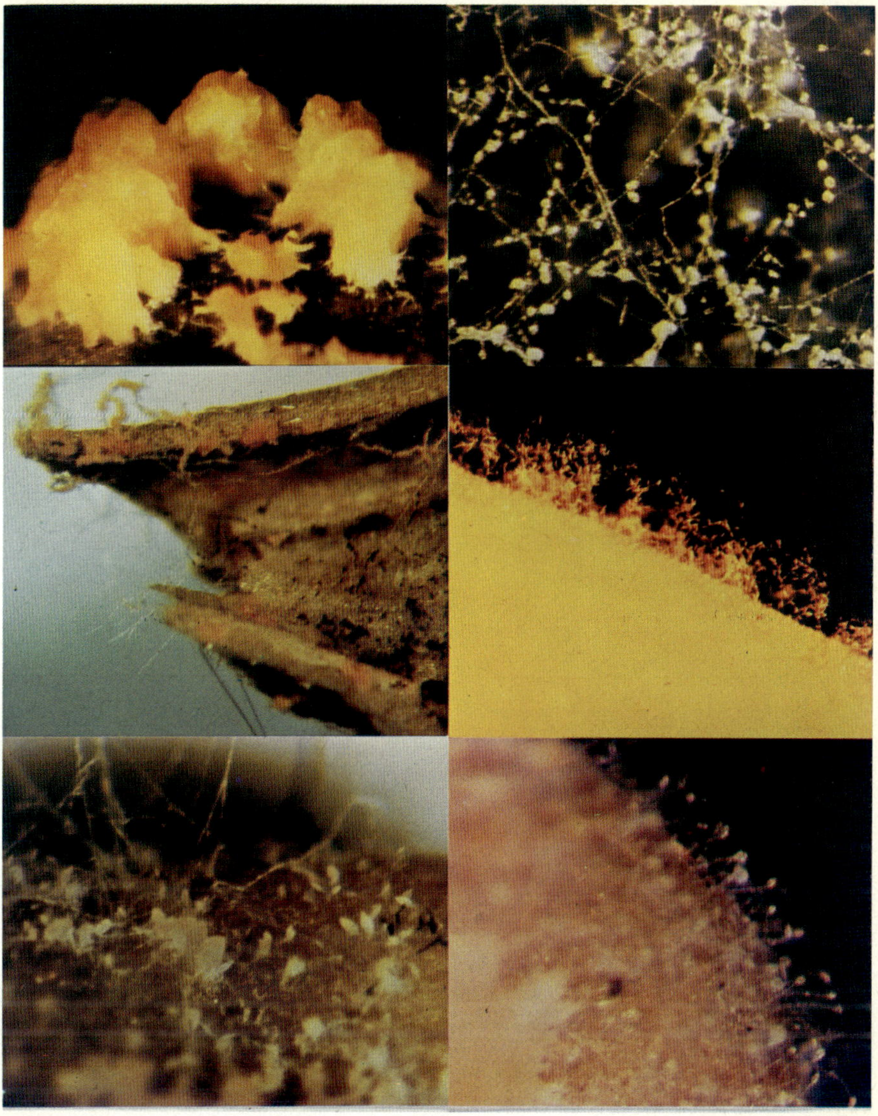

Plate II. Habit characters of *Fusarium* on seeds after 7 days of incubation in the blotter test (cf. Section 28.1). Left, from above: *Fusarium longipes* on pearl millet, ×50; *F. nivale* on barley, ×50; *F. oxysporum* on soybean, ×50. Right, from above: *Fusarium poae* on barley, ×50; *F. semitectum* on rice, ×25; *F. solani* on jute, ×50. The habit characters, typical of the species, persist from host to host

(Leach, 1960) and *Lathyrus odoratus* (Noble, in Noble and Richardson, 1968). For assessment of seed-borne inoculum, see Sections 32.4, 32.5 and 32.6, and for seed treatment, Table 22.3.

Distribution: North America, Europe.

P. nemophilae. Common in seeds of '*Nemophila menziesii*', including *N. atomaria* and *N. insignis* (Neergaard, 1936AR–1951AR; Crosier and Heit, 1948; Andersen, 1974 — index). On pathogenicity, see Neergaard (1938AR).

Distribution: North America, Europe.

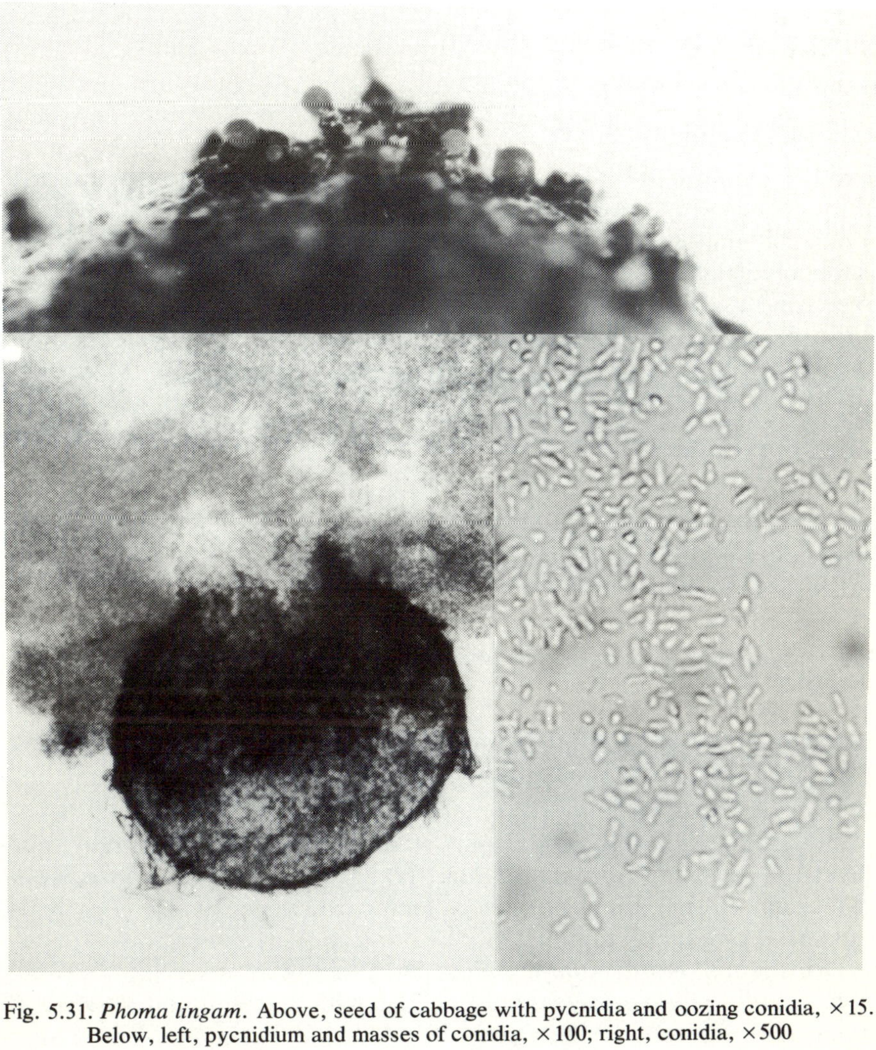

Fig. 5.31. *Phoma lingam*. Above, seed of cabbage with pycnidia and oozing conidia, ×15. Below, left, pycnidium and masses of conidia, ×100; right, conidia, ×500

P. rostrupii (P. sanguinolenta), Phoma root rot of carrot. Infrequently found in seed of carrot (Jørgensen, 1934; Neergaard, unpublished; Osnitskaya, 1954; Andersen, 1973AR, 1974AR; Buchwald, 1974). Distinguished by the deep red conidial exudate from other (saprophytic) spp. of *Phoma* encountered in carrot seed. Occasionally serious on carrot in storage, cf. Subsection 19.6.5.

Distribution: Europe.

P. sabdariffae, blight of kenaf. Encountered in seed of *Hibiscus cannabinus* in Philippines, and controlled by seed treatment, 0.1 per cent mercuric chloride (Quiniones and Orillo, 1952).

Distribution: Central America, Asia.

P. trachelii. Common in seed of *Campanula medium* (Neergaard, 1941AR–1945AR, 1949AR–1951AR; Andersen, 1974 — index).

Distribution: Europe.

P. valerianellae. Frequent in seeds of corn salad, *Valerianella locusta* var. *oleracea* and of other species of *Valerianella*, involving *V. auriculata*, *V. carinata*, *V. coronata*, *V. dentata* and *V. rimosa*, and has also been encountered in seeds of *Fedia cornucopiae* (Boerema and De Jong, 1968; Limonard, 1968). Other authors have reported incidences of undoubtedly this species in seeds of *Valerianella locusta* (Doyer, 1938; Neergaard, 1950AR, 1951AR, 1958AR, 1964AR, 1966AR; Andersen, 1952AR, 1970AR, 1971AR; Rohloff, 1968*a*). The pathogen causes damping-off and other damage.

Distribution: Europe.

Phomopsis

The fruit body is developed in a thick-walled, conic stroma and is provided with a single ostiole. The conidiophore is multibranched, the conidiogenous cell phialidic. Typically two types of conidia are produced, fusiform, often biguttulate *alpha-conidia*, and filiform, curved, s-formed, or hooked at the upper end, *beta-conidia*. The perfect state is *Diaporthe*.

P. endogena produces mummification of chestnuts, *Castanea* (Ciferri, 1950).

Distribution: Europe (Italy).

P. heveae, stem die-back, has been intercepted in India on rubber seeds from Malaysia (Srivastava, 1956).

Distribution: Asia.

P. leptostromiformis, Phomopsis stem blight of *Lupinus luteus*, may persist in seeds for more than 3 years (Ostazeski and Wells, 1960; Strukchinskas, 1971). Serious in yellow lupin; causes poisoning in animals, lupinosis, see Subsection 8.3.2.

Distribution: North America, Europe, Africa, Australia, New Zealand.

P. occulta (Phoma occulta Sacc.*)* has been encountered in seeds of *Juniperus virginiana* (Neergaard, 1958AR) and *Larix leptolepis* (Neergaard, 1956AR). Common on fallen cones of different conifers, occasionally parasitic on seedlings.

Distribution: Europe.

P. sojae, pod and stem blight of soybean, perfect state *Diaporthe phaseolorum* var. *sojae*. Very frequently seed-borne in soybean in humid climates (Liu, 1948; Kurata, 1960; Nicholson, Dhingra and Sinclair, 1972; Ellis *et al.*, 1974), and lima bean, *Phaseolus lunatus* (Zaumeyer and Thomas, 1950).

Distribution: North America, Asia.

P. vexans, fruit rot of eggplant, perfect state *Diaporthe vexans*. Seed-transmitted in eggplant (Porter, 1943; Felix, Lallmahomed and Orieux, 1965), hot water treatment effective.

Distribution: North, Central and South America, Africa, Asia, Australia.

CMI Distribution Map No. 329.

Different species of *Phomopsis* have been intercepted in the U.S.A. by the USDA quarantine service, on various kinds of forest seed. These include: *Phomopsis cocoes* on palm seed, including *Howea belmoreana*, *H. forsteriana* and *Hedyscepe canterburyana*, imported from New Zealand; *P. conorum* on *Pseudotsuga taxifolia* from Argentina; *P. imperialis* on *Paulownia libicina* from the United Kingdom, *P. palmicola* on *Howea forsteriana* from Australia and New Zealand, and *P. pandani* from *Pandanus* sp., screw pine, from Okinawa (Girard, 1969, 1971, 1973, 1974).

Phyllosticta

The genus is under revision. Recent revisions consider *Phyllosticta cruenta (P. convalariae* Pers.) as type species (Donk, 1968; Van Der Aa, 1973) and this has been confirmed by Punithalingam (1974) who examined type material. Traditionally the distinction between *Phoma* and *Phyllosticta* has been based on the very arbitrary, now obsolete criterion that *Phoma* is parasitic on stem and roots while *Phyllosticta* spp. are leaf parasites. More than 1,500 species of *Phyllosticta* have been described. *Phyllostictina* is regarded as a synonym by some authors. For the species discussed below, the name has been adopted merely as currently used in literature.

P. antirrhini, leaf and stem blight, stem rot of snapdragon. Frequently encountered on seed of *Antirrhinum majus* (Buddin and Wakefield, 1924; Harrar and Miller, 1938; Neergaard, 1941AR, 1948AR, 1951AR; Rosser, 1961; Harman, Heit and Braverman, 1971; Andersen, 1974 — index). Sometimes quite destructive.

Distribution: North America, Europe.

Plenodomus

Fruit bodies are developed as a sclerotium-like stroma, in which a cavity is developed, and at maturity an opening, a pore (not an ostiole as in *Phoma*) is formed, around which an often protruded papilla is formed. The type species is *P. lingam*. Conidia are cylindrical, produced from the conidiophores by successive budding. Von Arx (1970*b*) prefers to regard *Plenodomus* as a section of *Phoma*.

P. lingam, see *Phoma lingam*.

Selenophoma (Lunospora)

Pycnidia are spherical, compact, ostiolate, erumpent, conidia falcate, lunate, or boomerang-shaped. Pathogens producing leaf spot diseases in Gramineae (see Sprague, 1950).

S. bromigena, halo spot. Recorded on seeds of *Agropyron* sp. (Groves and Elliot, *fide* Noble and Richardson, 1968) and *Bromus inermis* (Allison, 1945; Smith, 1970*a*).

Distribution: North America, Europe (Hungary).

S. donacis (Septoria donacis, S. oxyspora, Lunospora oxyspora), eye spot, halo spot. The fungus has been encountered in seeds of *Agropyron* sp. (Groves and Elliott, *fide* Noble and Richardson, 1968) and *Festuca* sp. (Groves, *fide* Noble and Richardson, 1968). It occurs on a considerable range of genera within Gramineae, including oats, rye and wheat.

Distribution: North America, Europe.

S. donacis var. *stomaticola (Phyllosticta stomaticola, Septoria culmifida),* eye spot, halo spot. The pathogen has been reported from seeds of *Dactylis glomerata* (Noble, in Noble and Richardson, 1968), *Hordeum vulgare* (McKay, 1946; Noble, in Noble and Richardson, 1968), and of *Triticum aestivum* (Crosier, 1963). Most important on cocksfoot, *Dactylis*, causing serious damage in wet weather. On barley believed to have been introduced into Scotland with seed from Denmark. The pathogen has a wide host range of Gramineae.

Distribution: North America (including Greenland), Europe, Algeria.

(2. Phaeosporae)

Coniothyrium

Pycnidia are immersed, subepidermic, then erumpent; conidiophores produce many conidia successively from a simple flask-shaped conidiogenous cell; conidia are pigmented, typically aseptate but may be one-septate, they are cylindrical to oval, thick-walled.

C. hellebori, black spot, has been intercepted in the U.S.A. on seed of larkspur, *Helleborus niger*, imported from Italy (Limber, 1962). A serious pathogen of Christmas-rose.

Distribution: North America, Europe.

C. olivaceum, seedling blight and leaf spot of flax. In Canada Vanterpool (1947) found this pathogen in 12 out of 32 seed samples of flax. In Denmark encountered on seeds of *Acer negundo* (Andersen and Neergaard, 1956AR).

Distribution: North and South America, Europe.

(3. Hyalodidymae)

Ascochyta

Thin-walled glabrous pycnidia which at maturity typically develop a pore (not an ostiole as in *Phoma*). Conidiogenesis not fully established but Brewer and Boerema (1965) believe that conidia are produced by septation from the conidiogenous cell, not by budding. About 500 species have been described but many are synonyms.

A. chenopodii (Diplodina chenopodii, Ascochyta atriplicis). Encountered in seed samples of *Chenopodium quinoa* from Bolivia where the host is grown as a crop plant (current investigations at the Institute of Seed Pathology, Copenhagen). Seedlings became infected in blotter tests. Economic significance not established.

Distribution: South America (Bolivia), rather common on Chenopodiaceae in northern Europe, Asia (Siberia).

A. chrysanthemi, ray blight, black stem rot of chrysanthemum, perfect state *Mycosphaerella ligulicola*. Particularly serious in cuttings and is regarded an important quarantine object. For this reason it is of interest that the fungus also has been encountered in seed of *Chrysanthemum carinatum* (Neergaard, unpublished; Andersen, 1968AR, 1969AR, 1972AR).

Distribution: North America, Europe, Africa, New Zealand.

A. fabae, leaf and pod spot of broad bean. Seed transmission established in England (Hewett, 1966, 1973). Also encountered on seed of *Vicia faba* in Denmark (Andersen, 1968AR, 1970AR, 1973AR). Discussion on rate of seed transmission and rate of increase in Section 32.2.

Distribution: N. and S. America, Europe, North Africa, Australasia.

A. gossypii, Ascochyta blight, wet weather canker of cotton. Regarded as internally borne in seed (Smith, 1950, 1953*b*). Seed treatment recommended.

Distribution: North and South America, Europe, North Africa, Australasia.

CMI Distribution Map No. 259.

A. pinodella, see *Phoma medicaginis* var. *pinodella*.

A. pinodes, Fig. 5.32, foot rot, blight of pea, perfect state *Mycosphaerella pinodes*. Seed-transmitted usually from low percentages of infection (Jones, 1927; Doyer, 1938; Kerling, 1949; Sörgel and Unger, 1954; Skolko, Groves and Wallen, 1954; Walker, 1961; Kennedy, 1961; Marshall, 1962; Anselme and Champion, 1962; Matthews, 1964; Wallen, 1965; Maude, 1966*b*). About disease cycle, see Section 14.2. More serious than the other members of the 'Ascochyta-complex' of peas, see Sections 32.4, 32.5 and 32.6. Rates of seed transmission and increase, see Section 32.2.

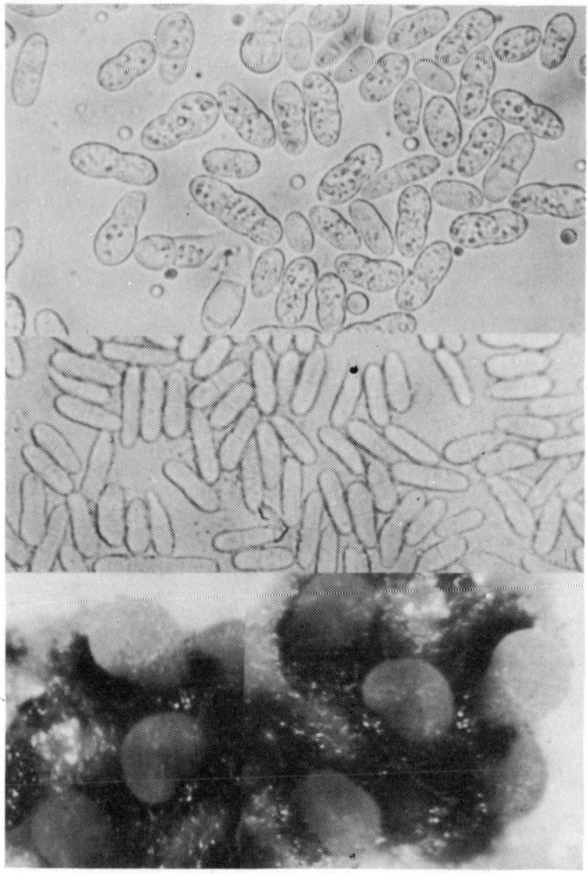

Fig. 5.32. Above, conidia of *A. pinodes* and *A. rabiei*, all ×500; below, pycnidia of *A. rabiei* covered by oozing conidia, ×50

Distribution: North and South America, Europe, Africa, East Asia, Australia, New Zealand.

CMI Distribution Map No. 316.

A. pisi, Fig. 5.33, leaf and pod spot. Very common in seeds of pea grown in moist climates, cf. Table 32.8 (Krüger, 1895; Jørgensen, 1934; Doyer, 1938; Skolko, Groves and Wallen, 1954; Neergaard, 1937AR–1951AR; Kaiser, 1972*a*; Andersen, 1974 — index). See comment under *A. pinodes*. Disease cycle is described in Section 14.2.

Distribution: World-wide.

CMI Distribution Map No. 273.

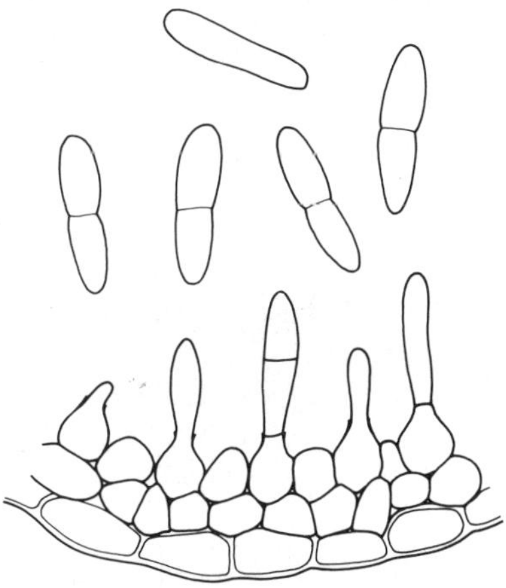

Fig. 5.33. *Ascochyta pisi*, parts of the pycnidial wall with conidiogenous cells and phialoconidia, ×1,000. (Courtesy of J. A. Von Arx)

A. rabiei (Phyllostica rabiei), Fig. 5.32, Ascochyta blight of chick pea, perfect state *Didymella rabiei (Mycosphaerella rabiei)*. Commonly seed-transmitted in chickpea, *Cicer arietinum* (Luthra, Bedi and Singh, 1932; Sattar, 1933; Zachos, 1951; Lukashevich, 1958; Khachatryan, 1961; Zachos, Panagopoulos and Makris, 1963; Solel and Kostrinski, 1964; Scharif, Niemann and Ghanea, 1967; Halfon-Meiri, 1970; Maden *et al.*, 1975). For economic importance, see Subsection 1.2.2. See also Section 19.9 on control. A major pathogen of the host.

Distribution: Europe, Africa, Asia.

CMI Distribution Map No. 151.

Diplodina

Diplodina (type species *D. salicis*) is now regarded as a synonym of *Discella* (Boerema, 1970). *D. lycopersici* is discussed under its perfect state *Didymella lycopersici*.

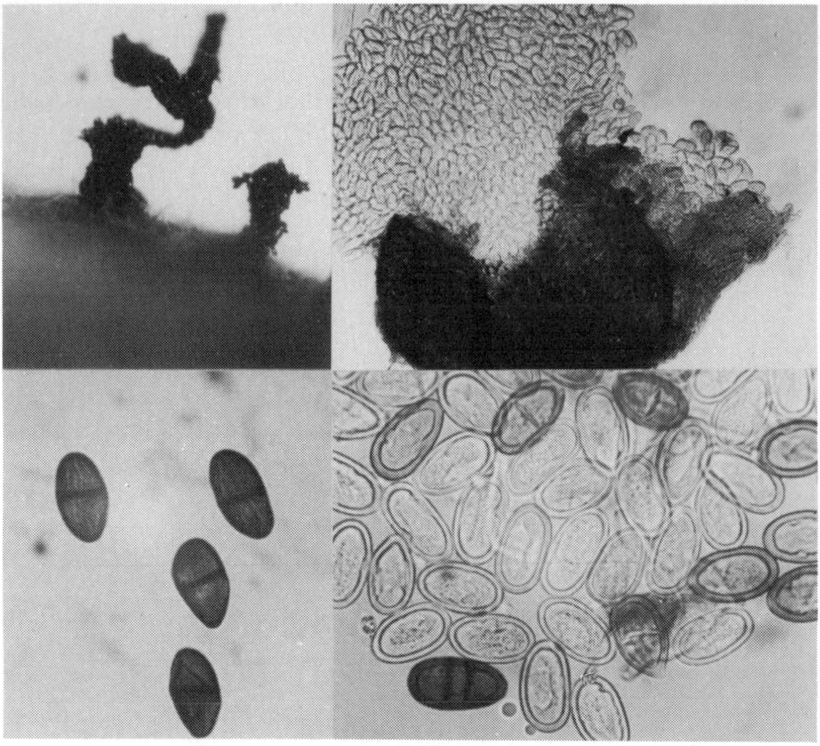

Fig. 5.34. *Botryodiplodia.* Above, left, pycnidia with conidial ooze, ×15; right, pycnidium with immature conidia, ×100. Below left, mature conidia, ×500; right, immature conidia, and one mature conidium, ×500

(4. Phaeodidymae)

Botryodiplodia (Lasiodiplodia)

Pycnidia are thick-walled and aggregated in botryose clusters, immersed in a stroma, erumpent, often with a distinct papillate ostiole; conidia are ellipsoid, one-septate, pigmented with longitudinal striations. The mature conidia are uncoloured and not yet septate, see Fig. 5.34.

B. theobromae (syn. *Diplodia natalensis*), die-back and charcoal pod rot of cacao; some authors have suggested *Physalospora rhodina* as perfect state; encountered in a range of hosts, in some possibly as a saprophyte, but in others definitely as a pathogen, especially in cacao. Recorded in seeds of *Arachis hypogaea* (Gilman, 1965), *Gossypium* (Anonymous, 1951), *Hevea brasiliensis* (Srivastava, 1964), five *Musa* spp., including *M. acuminata*, *M. balbisiana*, *M. gracilis*, *M. reflexa* and *M. textilis*, with seed coat infection, from Panama, Honduras and

Malaysia (Goos, Cox and Stotzky, 1961), *Strophanthus intermedius* (Calpouzos, 1956), and *Theobroma cacao* (Wellman, 1954). The pathogen causes fruit rot in banana and cacao, and a die-back in rubber, *Hevea brasiliensis*. It may be severe in poorly managed crops.

Distribution: Widespread in tropical zones of North, Central and South America, Africa, Asia.

Diplodia

Pycnidia are erumpent, subglobose, usually papillate, thick-walled; conidia are one-septate, pigmented, often slowly maturing. About 700 species described, including some important plant pathogens.

D. corchori, black band of jute. Infected seeds of *Corchorus* spp., including *C. capsularis* and *C. olitorius*, recorded by Shaw (Shaw, 1921; Imam, Fazli and Ahmed, 1960). Causes wilting due to rotting around the stem just above the soil line.

Distribution: Asia (India, Bangladesh).

D. frumenti, ear rot of maize, perfect state *Physalospora zeicola*, can be borne in the seed of maize (Eddins, 1930). The economic importance of this pathogen has not been established.

Distribution: North America.

D. gossypina, Diplodia rot, black boll rot of cotton, perfect state *Physalospora rhodina*. The fungus infects cotton seed (Crawford, 1923; Crosier, 1944; Roncadori, McCarter and Crawford, 1971). Lint of infected bolls become matted, smutty, worthless, bolls fail to open.

Distribution: North and Central America, Africa.

D. macrospora, dry rot, stalk rot of maize. Seems to have been less encountered in maize seed than *D. maydis*, cf. Noble and Richardson (1968) and is less aggressive.

Distribution: North and South America, Europe (Roumania), Africa, Asia, Australia.

CMI Distribution Map No. 227.

D. maydis (D. zeae), Fig. 5.35 and Colour Plate IV, stalk rot of maize, dry ear rot of maize. Commonly seed-transmitted and causing seed rot in maize (Hoppe, 1938, 1939, 1941, 1942; Crosier, 1946; Dickson, 1956; Leont'eva and Gerasimov, 1957; Nemlienko and Grisenko, 1962; Kapoor, 1964). The most serious member of the Diplodia complex of maize, prevalent in cooler regions.

Distribution: N. and S. America, Europe, Africa, Asia, Australia.

D. palmicola. Old records on the epicarp of *Cocos nucifera* (Fries, *fide* Noble and Richardson, 1968; Von Thümen, 1871, *fide* Noble and Richardson, 1968), cf. Wollenweber and Hochapfel (1943). Causes fruit rot in apple and citrus fruits.

Distribution: Europe, Asia (Indonesia, Sri Lanka), Oceania (Polynesia).

D. pinea, seedling collar rot. Recorded from seeds of *Pinus nigra* in the U.S.A. (USDA, Forest Service, 1948). The pathogen is cause of a sometimes serious needle cast disease in *Pinus*.

Distribution: North and South America, Europe, Africa, Australia. CMI Distribution Map No. 459.

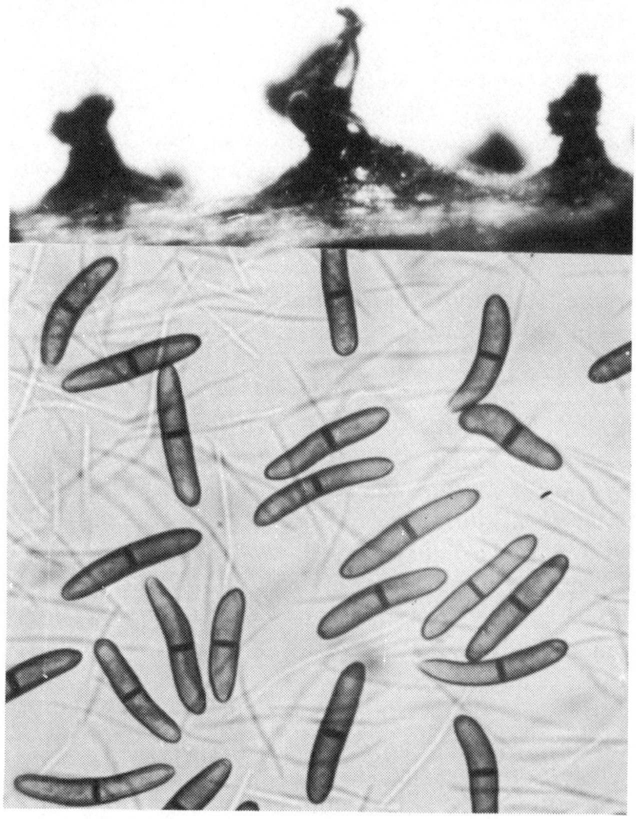

Fig. 5.35. *Diplodia maydis*. Above, pycnidia developed on a maize kernel, ×15; below, two kinds of conidia, two-celled phaeoconidia, and hyaline scolecoconidia, ×500

(5. Hyalophragmiae)

Stagonospora

Pycnidia are thin-walled, immersed, globose to lens-shaped; conidia (blastoconidia) cylindrical, fusiform, truncate at the base, multiseptate. About 200 species are described.

S. meliloti, leaf spot, root and crown rot of sweet clovers, may be carried on seed of *Melilotus* (Jones, 1939).

Distribution: North America.

(6. Hyaloscolecosporae)

Dilophospora

Fruit bodies formed within a black stroma, pierced by a pore; conidia (phialoconidia) cylindrical, one- to three-septate, furnished with several apical and basal branched pluricellular appendages and produced from ampulliform phialides.

D. alopecuri, twist of grasses, plumed spore disease. Seed-borne in *Agrostis spica-venti* (Becker, 1955), *A. tenuis* (Protsenko, 1957), *Avena sativa* (Kotte, 1934), *Dactylis glomerata* (Mühle, 1953), *Festuca duriuscula* (Noble, unpublished), *Secale cereale* (Kotte, 1934; Becker, 1955), *Triticum aestivum* (Becker, 1955). The fungus often occurs associated with *Anguina*, including *A. tritici*, often on the nematode galls, but a pathogenic interrelationship has not been proved.

Distribution: North America, Europe.

Septoria

Pycnidia are glabrous, thin-walled, immersed, sometimes papillate; conidia (blastoconidia) filiform, multiseptate, produced from ampulliform or elongated conidiogenous cells. 1,000–2,000 species described, many destructive leaf and stem pathogens.

S. apiicola (S. apii, S. apii-graveolentis), late blight of celery. Very frequent in seeds of celery and celeriac (Klebahn, 1913; Krout, 1921; Jørgensen, 1934; Doyer, 1938; Neergaard, 1940AR–1951AR; Cook, 1941; Bant and Storey, 1952; Hewett, 1962; Sheridan, 1963, 1966; Maude, 1964; Berger, 1971; Andersen, 1974 — index). A destructive leaf pathogen causing heavy losses, see Subsection 1.2.4, disease cycle is discussed in Section 14.4.

Distribution: North and South America, Europe, Africa, Asia.

S. avenae, leaf blotch, stem breaking, black stem, perfect state *Leptosphaeria avenaria*. Rather commonly carried on seed of oats but seed transmission seems to be very limited (Crosier, 1954; Noble and Montgomerie, 1954; Hooker, 1957; Clark and Zillinsky, 1960; Chambers, 1963). Sometimes very injurious.

Distribution: North and South America, Europe, Africa, Asia.

CMI Distribution Map No. 323.

S. cyclaminis, Septoria leaf spot of cyclamen. Infected plants of *Cyclamen persicum* were grown in two nurseries in California, U.S.A., from seed imported from Europe; the disease was new for the country (Raabe, 1971).

Distribution: Europe, California.

S. digitalis. Encountered in seed on *Digitalis lanata* in Czechoslovakia (Jechová, 1964). The pathogen caused considerable loss of yield.

Distribution: Europe.

S. drummondii. Found seed-transmitted in *Phlox drummondii*, imported from Europe to New Zealand (Wenham, 1958).

Distribution: Widely distributed, North America, Europe.

S. glycines, brown spot of soybean. Seed transmission studied by MacNeill and Zalasky (1957). The pathogen acts synergistically with *Pseudomonas glycinea*, see Section 16.10.

Distribution: North America, South America (Colombia), Europe (Roumania), East Asia.

CMI Distribution Map No. 361.

S. lactucae, leaf spot of lettuce. Carried as pycnidia on seeds (Moore, 1940; Smith, 1961; O'Reilly, personal communication, 1969). Serious in Australia, see Subsection 19.2.4.

Distribution: North America, Europe, Asia (India), Australia.

CMI Distribution Map No. 485.

S. linicola, pasmo, rust blotch of flax, perfect state *Mycosphaerella linorum*. This important pathogen may occur in the seed coat as well as in plant debris mixed with the seed (Loughnane, McKay and Lafferty, 1946; Muskett and Colhoun, 1947; Anselme, 1952; Christensen, Henderson and Aragaki, 1953; Cruickshank, 1954; Chambers, 1963; Sanderson, 1963; Malone and Muskett, 1964). A major pathogen of oil flax, see Subsection 1.2.3. Mechanism of seed infection is discussed in Section 11.5 and Subsection 12.2.4.

Distribution: North and South America, Europe, Africa, Asia (U.S.S.R.), Australia, New Zealand.

CMI Distribution Map No. 18.

S. lycopersici, tomato leaf spot. Although suspected as occasionally seed-borne by some authors (seed treatment has been recommended), occurrence in seed of tomato as well as seed transmission have not been established (see Landaeta, 1952; Merlo and Perera, 1964; Rizinski, 1966).

Distribution: World-wide.

CMI Distribution Map No. 108.

S. nodorum, Fig. 5.36, glume blotch, perfect state *Leptosphaeria nodorum*. Frequently transmitted, often from high percentages of seed infection in its principal host, wheat (Machacek, 1945; Mead, Broadfoot and Machacek, 1950; Machacek and Wallace, 1952; Kietreiber, 1961, 1962, 1964, 1966; Hewett, 1965, 1975; Shipton, 1966*b*; Ponchet, 1966; Baker, 1971; Schmidt, 1971AR, 1972AR; Cooke and Fozzard, 1973). Seed infection in different cultivars of oats reported from South Africa (Von Wechmar, 1965); Hewett (1975) found close correlation between infection percentage recorded in agar plate tests and percentages of diseased seedlings in the field. Also seed-borne in barley (Holmes and Colhoun, 1973). Influence of time of infection and of rainfall on epidemics and establishment of kernel infection of wheat are discussed in Sections 11.2 and 11.3. A serious pathogen of wheat in humid temperate climates, see Subsection

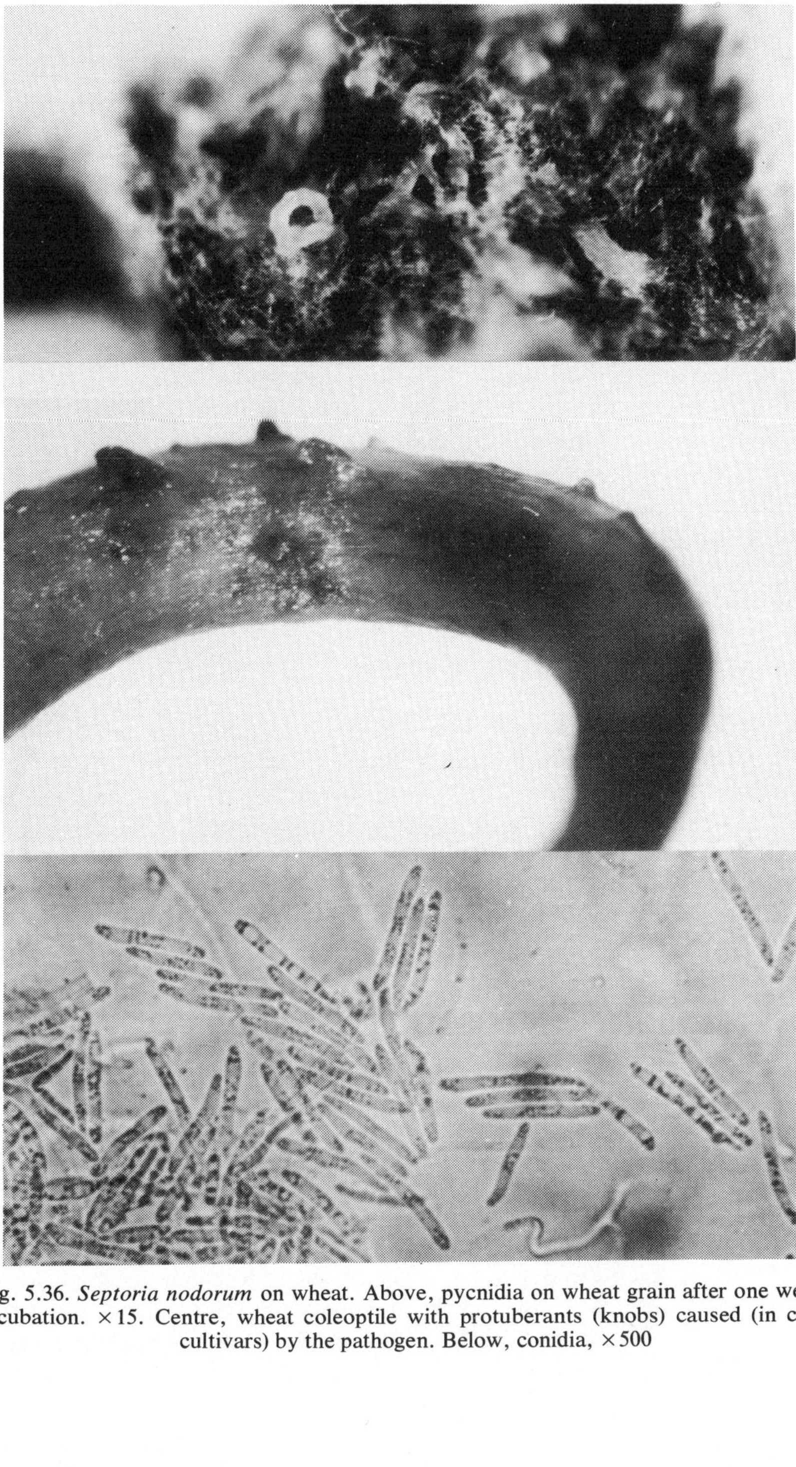

Fig. 5.36. *Septoria nodorum* on wheat. Above, pycnidia on wheat grain after one week of incubation. ×15. Centre, wheat coleoptile with protuberants (knobs) caused (in certain cultivars) by the pathogen. Below, conidia, ×500

1.2.1, wheat, cf. on mechanism of yield reduction, Section 32.4.

Distribution: North, Central and South America, Europe, East and South Africa, Asia, Australia, New Zealand.

CMI Distribution Map No. 283.

S. passerinii, leaf spot of barley. Occasionally seed-borne in barley (Noble, in Noble and Richardson, 1968), causes kernel shrinkage increasing cleaning losses (Green and Bendelow, 1961).

Distribution: North America, Europe, Africa, Australia.

S. petroselini, parsley leaf spot. Very frequent on seed of parsley (Grove, 1935; Doyer, 1938; Neergaard, 1940AR–1951AR; Moore, 1946*a*; Andersen, 1974 — index). Widespread on parsley but less harmful than Septoria late blight of celery.

Distribution: Wherever parsley is grown.

S. pisi, leaf blotch, Septoria blotch, Septoria blight. Sometimes carried in the seed (Brien *et al.*, 1955; Chupp and Sherf, 1960).

Distribution: In scattered localities in North and South America, Europe, South Africa, New Zealand.

S. secalis, leaf blotch. Occasionally encountered in seed of rye (Greaney and Machacek, 1946; Anonymous, 1953).

Distribution: North America, Europe.

S. tageticola, leaf spot of marigold. Borne as pycnidia on seeds of *Tagetes erecta*.

Distribution: North America (Florida).

S. tritici, speckled leaf blotch, perfect state *Mycosphaerella* sp. The pathogen seems to be present with wheat seed mainly on debris mixed with the grain (Noble, in Noble and Richardson, 1968; Pollack, 1945; Gattani and Abdulla, 1959; Jovićević, 1960; Andersen, 1972AR). Recently Brokenshire (1975) studied an agar plate technique for detection on seed. Increasing importance in recent years due to susceptibility of new cultivars.

Distribution: North, Central and South America, Europe, Africa, Asia, Australia, New Zealand.

CMI Distribution Map No. 397.

Lepstromataceae
(Hyalosporae)

Leptostroma

Pycnidia are shield-shaped, elongated, black with a poorly developed basal portion and with slit-like ostioles or fissures; conidia ovoid or allantoid.

L. pinastri, needle cast, leaf cast of Scotch pine, perfect state *Lophodermium pinastri*. May occur on fragments of needle mixed with seed. Noble (in Noble and Richardson, 1968) found in Scotland such concomitant contamination in seed of *Pinus nigra* var. *poiretiana* intended for export.

Distribution: North America, Europe, North Africa and Rhodesia, Asia, Australia, New Zealand.

CMI Distribution Map No. 371.

Excipulaceae
(Hyalodidymae)

Discella (Diplodina)

Pycnidia separate, acervulus-like; conidiophores elongate, branched, septate, forming phialoconidia on small lateral branches; conidia oval, two-celled. See *Didymella*, Subsection 5.3.5.11.

(Scolecosporae)

Ephelis

See *Balansia*, Subsection 5.3.5.6.

5.3.7 Basidiomycetes (Basidiomycetina)

The sexual spores, basidiospores, are produced externally on a single cell, a club-shaped or tubular basidium, typically four on each. There are two subclasses: Hymenomycetes produce basidia in an often conspicuous fruiting body, a basidiocarp, such as mushrooms and puff-balls, whereas in Teliomycetes basidia are more loosely organised in crusts or columns, or they develop individually from septate resting spores or teliospores, which often form powdery sori.

Clamp-connections are present in many Basidiomycetes. A clamp is a hyphal outgrowth which makes a connection between two cells when a septum is being formed with conjugate nuclear division; the clamp fuses with the lower cell. In non-sporulating fungi clamps may be a helpful diagnostic character. Clamps do not occur in rust fungi.

(1. Teliomycetes)

5.3.7.1 UREDINALES (RUST FUNGI, OR RUSTS)

'Obligate' plant parasites, producing various kinds of spore pustule, and many of which complete their life-cycle in two different unrelated host plants, such as black stem rust, *Puccinia graminis*, grasses and cereals being one set of alternate hosts, and different species of *Berberis* the other. A complete life-cycle includes five kinds of spores, traditionally numbered as follows:

0. *Pycniospores* (spermatia), produced in pycnia (which resemble pycnidia of Sphaeropsidales).

I. *Aeciospores* (aecidiospores), produced in aecia (aecidia, cluster cups). These spores normally result from fertilisation originating from the pycniospores.

II. *Urediniospores* (uredospores, summer spores), produced in uredinia (uredosori). These are conidia producing new uredinia, or state III.

III. *Teliospores* (teleutospores, winter spores), produced in telia (teleutosori). These are resting spores which on germination produce basidia.

IV. *Basidiospores* (sporidia), produced on basidia (promycelia).

Rust fungi which pass through states 0 and I in one alternate host, and through II and III in the other are *heteroecious*, while rust fungi which complete their life-cycle on one host (or collateral hosts) are *autoecious*. Rust fungi which have more forms than states III, IV and perhaps 0 are referred to as *long-cycle* or *macrocyclic* (e.g. *Puccinia graminis*), while those possessing only III and IV (and some 0) are called *short-cycle* or *microcyclic* (e.g. *Puccinia malvacearum*).

Puccinia

This genus includes heteroecious or autoecious rusts attacking many families of angiosperms. Teliospores are two-celled.

P. allii (P. porri), onion rust. May occur in seeds of *Allium* spp. (Alcock, 1931; Campacci, 1950). Significance of seed transmission not established. Autoecious. Important in some places.

Distribution: Europe, Asia, New Zealand.

CMI Distribution Map No. 400.

P. antirrhini, snapdragon rust. 'Seed-packet-borne', particularly as uredospores, in packets from which at sowing adjacent plants, which are beyond the resistant seedling stage, may be infected, cf. Section 13.8 and Section 24.1, example 8, and Section 24.2, example 1; see also Baker (1952). Introduction into South Africa in 1939 (*fide* Wilson and Henderson, 1966), into Rhodesia 1940 (Hopkins, see Section 13.8), and into Australia (Walker, 1954). The rust is considered heteroecious although the alternate host is unknown. The pathogen is sometimes destructive in North America, decreasing seed production.

Distribution: North and Central America, Europe, Africa, Middle East, Australia, New Zealand.

CMI Distribution Map No. 40.

P. carthami, safflower rust. Seed-transmitted as teliospores on seed of *Carthamus tinctorius*, transmission being favoured by lower temperatures (Calvert and Thomas, 1954). Unsuccessful seed treatment reported (Darpoux, 1948; Thomas, 1952; Schuster and Christiansen, 1952). Seed treatment recommended in Mexico (Gallegos and Rodriguez, 1967) and in Iran (Behboudi, 1968), volatile mercury compounds effective when treated seed is stored for about two months (Thomas, 1959*b*), also maneb effective (Thomas, 1971). Pathogenic races, see Section 17.2. (4.2).

Distribution: North and Central America, Europe, East Africa, Asia, New Zealand.

P. heterospora may occur as teliospores on seed bracts and seeds of hollyhock, *Althaea* spp. (Taubenhaus and Ezekiel, 1933).

Distribution: North and Central America.

P. malvacearum, hollyhock rust. May be transmitted from debris mixed with seed (Baker, 1952; Wilson and Henderson, 1966) or as teliospores on the seeds or bracts (Taubenhaus, 1911). The rust is microcyclic. Very common on a wide host range of Malvaceae.

Distribution: North and Central America, Europe, Africa, Asia, Australia.

Sometimes heteroecious rust fungi may occur, even profusely, as sori on seeds, such as *Puccinia graminis*, black stem rust, in wheat (Hungerford, 1920; Nattrass, *fide* Noble and Richardson, 1968, and see Chapter 14, Introduction) and *P. striiformis (P. glumarum)*, yellow rust in wheat (Hermansen, 1960; Naumova, 1960; Vardosanidze and Stradze, 1962; Savosta, 1964). However, seed transmission has not been proved although suspected, and is not very likely. Uredospores are too shortlived to ensure infection after emergence of seedlings, and basidiospores, produced from seed-borne teliospores, have little chance of ensuring infection of an alternate host, although the possibility cannot be excluded.

Uromyces

Autoecious and heteroecious rust fungi; life-cycles and morphology similar to those of *Puccinia*; teliospores are one-celled.

U. betae, beet rust. Both uredo- and teliospores are frequent, and often abundantly present on seed clusters of *Beta vulgaris*, e.g. sugarbeet and garden beet (Alcock, 1931; Neergaard, 1951AR; Andersen, 1974 — index). First record of the disease in Canada was on sugarbeet raised from seed imported from Europe, and grown in two isolated localities; in this incidence the seed carried large numbers of uredo- and teliospores (Newton and Peturson, 1943). The disease is rarely serious.

Distribution: North and South America, Europe, Africa, Australasia.

U. lespedezae-procumbentis, bush clover rust, has been intercepted in the U.S.A. on seed from Japan (Anonymous, 1942).

Distribution: North America, Asia.

U. viciae-fabae (U. fabae), lentil rust, has been encountered on debris mixed with seed of lentil, *Lens culinaris* (Prasada and Verma, 1948). Autoecious.

Distribution: Almost world-wide.

CMI Distribution Map No. 200.

Melampsora

Heteroecious, a few autoecious, rusts occurring in phanerogams. Teliospores are usually one-celled; their sori form crusts of a single layer of spores.

M. lini, flax rust. Teliospores may be present abundantly in debris mixed with seed of *Linum usitatissimum*, and teleutospores may be attached to the seed (Muskett, 1950). In India Vasudeva (1963*a*) found that teliospores in seed which had been stored on the hills could freshly infect local crops in other localities whence the disease spread into the plains. Seed treatment recommended (Anselme, 1956). The existence of many pathogenic races, see Section 17.2 (4.2), substantiates the feared risk of international spread of these with seed and suggests quarantine precautions.

Distribution: World-wide, although in scattered regions in South America and Africa.

CMI Distribution Map No. 68.

5.3.7.2 USTILAGINALES (SMUT FUNGI, OR SMUTS)

Parasites on flowering plants, especially Gramineae and Cyperaceae, primarily being limited to the ovary or inflorescence, see Section 9.1, but others attacking anthers, or leaves and stem. Resting spores (chlamydospores) or teliospores are formed in sori which after maturation usually are uncovered appearing as a dark powder. By germination these form a basidium or promycelium from which basidiospores or sporidia are produced. The germ tubes from these are parasitic forming the critical phase in establishment of infection, see Section 17.1, Subsection 12.2.3, and Section 14.5. There are primarily two infection types in seed-borne smuts: (1) *seedling infection,* brought about only in young seedlings, the overwintering inoculum being resting spores ('covered' smuts), and (2) *embryo infection,* brought about during flowering of the mother crop ('loose' smuts), infected grains being the overwintering inoculum; both are systemic. Some seed-borne smuts produce (3) *local infections.* Mature smut spores of the 'covered' smut type are very long-lived, see Subsection 20.2.2, while those of the 'loose' smut type are short-lived. Two families are considered, Ustilaginaceae and Tilletiaceae.

Reviews and bibliographies on smuts have been published by Liro (1924); Viennot-Bourgin (1949); Fischer (1951*a*, 1953); Mundkur and Thirumalachar (1952); Zundel (1953); Fischer and Holton (1957); Savulescu (1957*a*); Niemann (1962) and Schuhmann (1962).

Ustilaginaceae

Basidiospores are produced on the sides of a four-celled promycelium. In some species the promycelium grows directly into infective hyphae. Most species depend on seed transmission but soil transmission may be important in some, particularly in those occurring in warmer climates.

Sphacelotheca

Teliospores are produced within a false membrane of sterile hyphae forming a peridium, sori develop around a central columella composed mostly of vascular remains of host tissue. Most species occur in Gramineae. The following are of major importance in millets and sorghum, one species also in maize.

S. cruenta (*S. holci,* now regarded as a 'pathogenic race', or rather *forma specialis*), loose kernel smut, occurs in a range of *Sorghum* species, including *S. bicolor, S. caffrorum, S. caudatum, S. halepense, S. saccharatum, S. sudanense*, and in the main host, *S. vulgare* (Faris and Reed, 1925; Johnston, Lefebvre and Hansing, 1938; Melchers and Hansing, 1943; Leukel and Martin, 1950; Leukel, Martin and Lefebvre, 1960; Niemann, 1962; Tarr, 1962; Ramakrishnan, 1963). Seedling infection. The false membrane of the smut galls ruptures earlier in this species than in *S. sorghi*. Economic importance is less than that of *S. sorghi*. Height of infected plants may be considerably reduced. Pathogenic races, see Section 17.2. Soil transmission of little significance. Seed treatment as for *S. sorghi*.

Distribution: North and Central America, South America (scattered areas), Europe, Africa, Asia.

CMI Distribution Map No. 408.

S. destruens (S. panici-miliaceae), head smut of millet, predominates in *Panicum miliaceum* and occurs moreover in *P. carthaginense, P. crus-galli, P. dichotomiflorum, P. spinescens* and *Setaria italica* (Buchheim, 1930; Martin, 1943; Richter and Müller, 1943; Dillon Weston and Schofield, 1952; Todorova, 1953; Fedoseyeva, 1957; Lyu, 1957; Maslavskii, 1958, 1959; Kühnel, 1961; Niemann, 1962; Ramakrishnan, 1963). Seedling infection. Very important on millet in south and eastern Europe. Best seed treatment so far by organo-mercury compounds.

Distribution: North America, South America (Argentina), Europe, Africa, Asia, Australia.

CMI Distribution Map No. 219.

S. reiliana (Sorosporium reilianum, S. holci-sorghi), head smut of sorghum. The host range includes *Andropogon arundinacea, A. distachyos, Euchlaena mexicana, Sorghum bicolor* var. *charisianum, S. caffrorum* var. *albofuscum, S. caudatum, S. halepense, S. sudanense, S. verticilliforme, S. vulgare* and *Zea mays* (Potter, 1914; Dana and Zundel, 1920*a*, 1920*b*; Christensen, 1926; Geschele, 1927; Bressman and Barss, 1933; Ling, 1945; Kispatic, 1948; Jacks and Graham, 1955; Stewart and Reyes, 1958; Kühnel, 1959*b*; Niemann, 1962; Tarr, 1962; Wilson and Frederiksen, 1970; Harris, Fisher and Sowell, 1971). Seedling infection. Soil transmission is predominant. Important in the U.S.A. Sensitive to temperature, Subsection 15.2.1. Pathogenic races, see Section 17.2 (4.1).

Distribution: World-wide.

CMI Distribution Map No. 69.

S. sorghi, covered kernel smut, grain smut, attacks species of *Sorghum*, including *S. caffrorum*, *S. caudatum*, *S. dochna*, *S. gambicum*, *S. guineense*, *S. halepense*, *S. margaritiferum*, *S. membranaceum*, *S. notabile*, *S. saccharatum*, *S. sudanense* and *S. vulgare* (Kulkarni, 1922, 1944; Reed and Faris, 1924; Melchers, Ficke and Johnston, 1932; Melchers, 1933; Melchers and Hansing, 1938, 1943; Goidanich and Scardovi, 1946; Porter, 1954, 1955*b*; Tarr, 1954, 1962; Webster and Leukel, 1959; Zogg and Lanini, 1957; El-Helaly and Ibrahim, 1957; Hsi, 1958; Krüger and Gorter, 1960; Niemann, 1962; Markov, 1964; Zakari and Geraldo, 1965; Mathur and Dalela, 1971). Seedling infection. Seed transmission predominant. Most serious pathogen in sorghum, see Subsection 1.2.1. Several pathogenic races, Section 17.2 (4.1). Seed treatment by organo-mercury compounds, Table 22.3.

Distribution: World-wide.

CMI Distribution Map No. 220.

Other species of *Sphacelotheca* of minor importance occurring on Gramineae are *S. cordobensis* on *Paspalum* (Fischer, 1953); *S. erianthi* on *Saccharum* (Mundkur and Thirumalachar, 1952); *S. macrospora* on *Saccharum* (Yen and Wang, 1955); *S. montaniensis* on *Muhlenbergia* (Fischer, 1953); *S. occidentalis* on *Andropogon gerardi* (Dunleavy, 1956); *S. pamparum* on *Pennisetum* and *Setaria* (Fischer, 1953); and *S. sorghicola* on *Sorghum vulgare* (Niemann, 1962).

Thecaphora

Sporeballs are composed by few to many, more or less permanently, very firmly united teliospores.

T. deformans, seed smut of legumes, produces sori in seeds of *Lathyrus pratensis* (Ainsworth and Sampson, 1950), and of species of *Astragalus*, *Desmodium*, *Homalobus*, *Hosackia*, *Lohis*, *Lupinus*, *Medicago*, *Meibomia*, *Melilotus*, *Orobus*, *Trifolium*, *Vicia* (Niemann, 1962) and *Ulex* (Brett, 1966). There seems to be no detailed information about life-cycle or proof of seed transmission. Possibly the pathogen produces local infections.

T. seminis-convolvuli occurs as sori in seeds of *Convolvulus arvensis* and *Calystegia sepium* (Ainsworth and Sampson, 1950). No information about life-cycle or on establishment of seed transmission.

Tolyposporium

Sori develop in florescence, usually ovaries of Gramineae, teliospores are firmly united in irregularly formed sporeballs.

T. bullatum attacks *Panicum crus-galli*, *P. erectum*, *P. frumentaceum*, *Pennisetum lonis*, *P. typhoides* and *Setaria* sp. (Corbetta, 1954; Niemann,

1962). Mode of infection is unknown but local infections of ovaries are assumed.

T. ehrenbergii (T. filiferum), long smut of sorghum, has been encountered in *Sorghum bicolor* var. *technicum*, *S. halepense*, *S. purpureo-sericeum*, *S. sudanense* and *S. vulgare* (Tarr, 1962; Niemann, 1962; Ramakrishnan, 1963; Bamdadian, 1968; Doshimov, 1969). Important in some countries of Africa. Life-cycle is not fully elucidated, probably air dissemination and local infections occur rather than seedling infection (Tarr, 1962).

Distribution: Africa, Asia.

CMI Distribution Map No. 377.

T. hessii, seed smut, has been reported on seed of pipewood, *Eriocaulon lanatum*, in Angola (Müller, 1955).

T. penicillariae, smut of bajri, occurs on *Pennisetum glaucum* and *P. typhoides*. Seed transmission doubtful. Fischer and Holton (1957) discuss this smut as an example of local infection, inoculations of seeds or seedlings having been unsuccessful, and Ramakrishnan (1963) refers to conflicting results from seed treatment. Wells, Burton and Ourecky (1963) observed the fungus for the first time on *Pennisetum glaucum* in the U.S.A., and suggested that the pathogen might have been introduced on seed. Intercepted in 1972 in the U.S.A. by the USDA plant quarantine service on seed of *Pennisetum typhoides*, imported from Nigeria and South Africa (Girard, 1974).

Distribution: North America (Georgia), Africa, Asia (India, Iran and Pakistan).

CMI Distribution Map No. 150.

Ustilago

The largest genus amongst the smuts, perhaps 300–600 accepted species, parasitising especially on Gramineae. Sori of dark powdery spore masses are covered for some time by a membrane of host tissues; teliospores are individual, one-celled, germinating by a septate promycelium.

U. avenae, loose smut of oats, occurs, in addition to its principal host, *Avena sativa*, also on *A. abyssinica*, *A. algeriensis*, *A. barbata*, *A. byzantina*, *A. fatua*, *A. longissima*, *A. ludoviciana*, *A. nuda*, *A. orientalis*, *A. pilosa*, *A. pubescens*, *A. sterilis*, *A. strigosa* and *A. tartarica*, and on *Arrhenatherum elatius* (Batholomew and Jones, 1923; Gassner, 1924; Zade, 1924; Diehl, 1925; Rösch, 1926; Gage, 1927; Johnston, 1927; Tamme, 1927; Haarring, 1930; Western, 1936; McKay, 1936; Kitunen, 1937; Sampson and Western, 1938; Halisky, 1956; Leukel and Earhart, 1956; Rusch, 1957; Hille, 1958; Zemanek, 1960; Narain and Luke, 1963; Thiede, 1963; Mills, 1966). The fungus is similar to *U. nigra* in barley which by Fischer (1953) is regarded as identical with *U. avenae*. Seedling infection. Many pathogenic races, see Section 17.2 (4.1). Economic importance,

Subsection 1.2.1. Seed treatment, Table 22.3.

Distribution: World-wide where oats are grown.

CMI Distribution Map No. 238.

U. bullata, head smut, loose smut of grasses. A collective species comprising three previously separated species, now regarded as *formae speciales* (Fischer, 1937): '*U. bullata*'s.str. occurs on 5 different species of *Agropyron*; '*U. bromivora*' occurs on 23 different species of *Bromus*, and on *Brachypodium ramosum*; and finally '*U. lorenziana*' on *Elymus* spp., *Festuca idahoensis* and *F. viridula*, *Hordeum compressum*, *H. jubatum*, *H. maritimum* and *H. nodosum*, and *Silarion hystrix* (Fischer, 1953). Seedling infection, but the fungus may persist in perennial grasses (Niemann, 1962). In some regions the disease is serious on fodder grasses. Seed treatment by mercury compounds effective.

Distribution: World-wide, predominating in semi-arid areas.

U. crameri, millet kernel smut. In addition to its principal host, *Setaria italica*, the pathogen occurs on *S. ambigua*, *S. germanica* and *S. viridis* (Wang, 1936, 1943, 1944; Ling, 1945; Ainsworth and Sampson, 1950; Fischer, 1953; Kühnel, 1961; Niemann, 1962; Leppik, 1962*a*; Ramakrishnan, 1963; Popescu, 1970). Seedling infection. In some places, e.g. China, a major pathogen of millet. Pathogenic races, Section 17.2 (4.1). Seed treatment by mercury compounds effective.

Distribution: North and South America (Argentina), Europe, Africa, Asia, Australia.

CMI Distribution Map No. 239.

U. hordei (U. kolleri, U. levis), covered smut of barley and oats. The form on barley occurs on the following hosts: *Hordeum distichon, H. hexastichon, H. hibernum, H. trifurcatum* and *H. vulgare*, and has also been encountered on *Elymus glaucus* and *Secale cereale* (Appel and Gassner, 1907*a*; Faris, 1924*a*; Rump, 1926; Taylor and Zehner, 1931; Ruttle, 1934; Tapke, 1938, 1948, 1952; Fischer, 1939, 1951*b*; Greaney and Machacek, 1946; Ainsworth and Sampson, 1950; Porter, 1956; Wells, 1958; Gray, 1958; Chatterjee, 1958; Niemann, 1961*a*, 1962; Schafer, Dickson and Shands, 1962*a*; Ponirovskii, 1962). Seedling infection. Many pathogenic races, see Section 17.2 (4.1). Disease cycle is discussed in Section 14.5, the influence of various edaphic factors on development of the disease, in Subsections 15.2.1, 15.2.2 and 15.2.3, and economic importance briefly in Subsection 1.2.1, barley. Seed treatment, Table 22.3.

The form on oats has the following species of *Avena* as hosts: *A. algeriensis*, *A. barbata*, *A. byzantina*, *A. fatua*, *A. nuda*, *A. orientalis*, *A. sativa, A. sterilis, A. strigosa* and *A. tartarica* (Heald, 1919; Reed and Faris, 1924; Sampson, 1928*a*, 1929; Kingsley, 1933; Mundkur, 1934; Brandwein, 1940; Cherewick, 1944, 1965; Greaney and Machacek, 1946; Leblond, 1948; Ainsworth and Sampson, 1950; Gray, 1954; Jacks and

Cruickshank, 1956; Niemann, 1962). Seedling infection. Many pathogenic races, see Section 17.2 (4.1). Influence of soil moisture on development of the disease, Subsection 15.2.1. Economic importance, Subsection 1.2.1, oats. Widespread, but less frequent than *U. avenae*, cf. Subsection 32.1.3 about decline of prevalence in the U.S.A. Seed treatment, Table 22.3.

Distribution: World-wide where the hosts are grown.

U. maydis (U. zeae), corn smut, maize smut, blister smut, see Fig. 5.37. Seed transmission limited, soil transmission predominant; in *Zea mays* (Borzini, 1935; Itzerott, 1938; Middendorf, 1958; Beke and Marton, 1958; Niemann, 1962; Christensen, 1963*a*; Drăcea, 1970; Batiashvili and Natsvlishvili, 1972). Most infections are local, although young seedlings may be invaded systemically; meristematic tissues of above-ground parts are susceptible to direct penetration. Seed transmission primarily of importance in spreading the pathogen to new areas. Seed treatment by organomercurials effective against seed-borne inoculum, but not sufficient against soil-borne.

Distribution: World-wide, except Australia and New Zealand.
CMI Distribution Map No. 93.

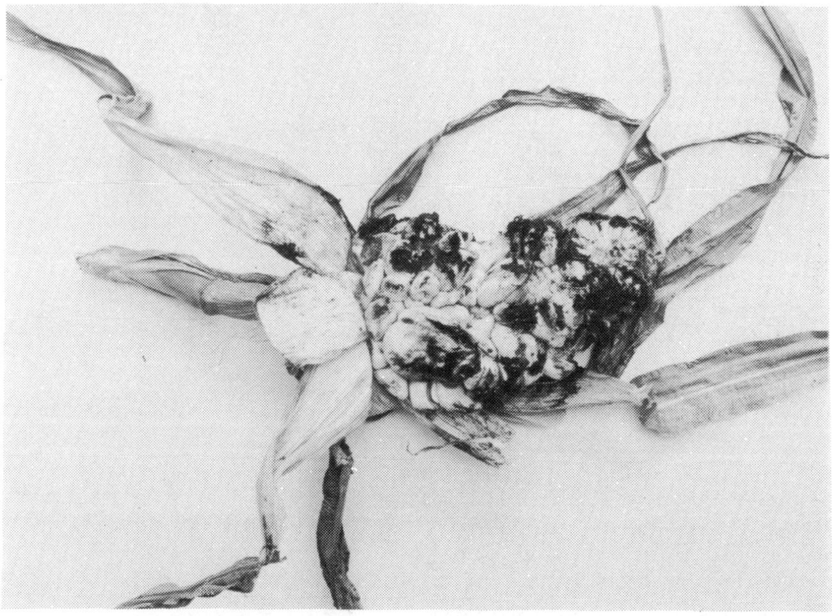

Fig. 5.37. Maize smut, *Ustilago maydis,* affecting the kernels of the ear

U. nigra (by some authors, e.g. Fischer, 1943, considered as a synonym of *U. avenae*), semi-loose smut, false loose smut of barley, black loose smut. The hosts are *Hordeum distichon, H. hexastichon, H. pusillum*

and *H. vulgare* (Johnson, 1914; Tisdale and Tapke, 1924; Leukel, 1936; Tapke, 1943, 1951; Lachance, 1945–1947; Greaney and Machacek, 1946; Ainsworth and Sampson, 1950; Podhradszky and Király, 1954; Savulescu, 1957*b*; Pedersen, 1957; Niemann, 1962; Hirschhoorn, 1962; Dunenko,

Fig. 5.38. Spikes of barley attacked by loose smut, *Ustilago nuda*. (Courtesy of Department of Agriculture and Fisheries for Scotland, Edinburgh)

1974). Seedling infection. Many pathogenic races, Section 17.2 (4.1). *U. nigra* is an intermediate form between *U. nuda* and *U. hordei* and can hybridise with these species. The two loose smuts may be confused on barley but the smutted spikes appear 10–12 days later in *U. nigra* than in *U. nuda*, the spores are darker and less 'loose'. Identification can readily be made by spore germination test. In some areas more harmful than *U. nuda*. Seed treatment as for covered smut.

Distribution: North America, Europe, Asia.

U. nuda (U. nuda f.sp. *hordei)*, loose smut of barley, brown loose smut, see Fig. 5.38. Occurs on all grown species of barley, including *Hordeum distichon, H. hexastichon, H. hibernum, H. murinum, H. stendeli, H. trifurcatum* and *H. vulgare* (Appel and Riehm, 1911; Liro, 1924; Feistritzer, 1931; Winkelmann, 1932, 1955; Leukel, 1936; Honecker, 1942; Fischer, 1943, 1953; Semeniuk and Ross, 1942; Simmonds, 1946; Ainsworth and Sampson, 1950; Popp, 1951, 1955, 1958*a*, 1958*b*, 1959; Tyner, 1951; Oertel, 1955; Porter, 1955*a*; Niemann, 1955, 1956*b*, 1957, 1958, 1961*a*, 1961*b*, 1962; Wagner, 1956; Marshall, 1959; Malik and Batts, 1960*a*, 1960*b*, 1960*c*; Kavanagh, 1961, 1964; Grainger, 1962; Russell, 1962; Mumford and Rasmusson, 1963; Edgington and Reinbergs, 1966; Kozera, 1968; Kietreiber, 1968AR–1971AR, 1973AR–1974AR). Embryo infection. Mode of infection is described in Subsection 12.2.3, impact of time of infection in Section 11.4, location of infection Section 13.2 and disease cycle in Section 14.1. Many pathogenic races, see Section 17.2 (4.1). Differentiation between *U. nuda* and

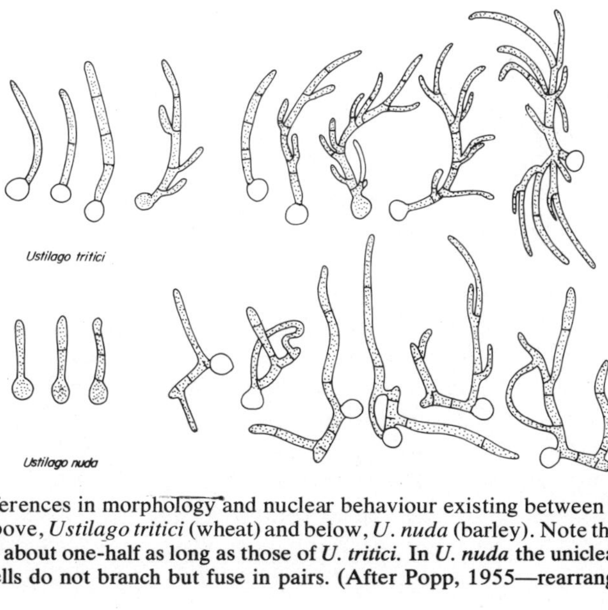

Fig. 5.39. Differences in morphology and nuclear behaviour existing between the two loose smut fungi. Above, *Ustilago tritici* (wheat) and below, *U. nuda* (barley). Note that promycelia of *U. nuda* are about one-half as long as those of *U. tritici*. In *U. nuda* the unicleate promycelial cells do not branch but fuse in pairs. (After Popp, 1955—rearranged)

U. tritici is based on teliospore germination, see Fig. 5.39 (Popp, 1951, cf. also Zambettakis, 1973). Economic importance is discussed in Subsection 1.2.1, barley. There is an extensive literature on seed treatment, hot water treatment being the classical procedure, see Section 21.1 and Subsection 21.3.1, now often replaced by treatment with systemic fungicides, see Section 21.5 (14) and Table 22.3. Testing for seed infection by the embryo count procedure is widely used, see Section 27.4; there is close positive correlation between laboratory results and field incidence.

Distribution: World-wide.

CMI Distribution Map No. 368.

U. striiformis (U. salvei), stripe smut of grasses. This species has a wide host range within Gramineae, including, *fide* Zundel (1953), 68 species of 26 genera, e.g. species of *Agropyron*, *Avena*, *Bromus*, *Dactylis*, *Elymus*, *Festuca*, *Holcus*, *Lolium*, *Phleum*, e.g. *P. pratense*, *Poa*, e.g. *Poa pratensis*, and *Setaria italica* (see Niemann, 1962); also on *Agrostis palustris* (Hardison, 1968). Important in timothy (Kreitlow, 1947). The fungus occurs in perennial grasses, often persists for a considerable period in plant residues and soil but may be transmitted by seeds causing seedling infection (Liro, 1924; Fischer, 1940; Kreitlow, 1947; Ainsworth and Sampson, 1950; Thirumalachar and Dickson, 1953). Many pathogenic races, see Section 17.2 (4.1). The pathogen causes considerable losses in hay and seed production.

Distribution: North America, Europe, Asia, Australia.

U. tritici (U. nuda f.sp. *tritici),* loose smut of wheat. This smut occurs in all grown wheats, including *Triticum dicoccum*, *T. durum*, *T. monococcum, T. polonicum, T. spelta, T. turgidum* and *T. aestivum* (Lang, 1910; Appel and Riehm, 1911; Liro, 1924; Tiemann, 1925; Kourssanow, 1928; Piekenbrock, 1927; Vanderwalle, 1942, 1953; Oort, 1940*a*, 1940*b*, 1944; Pal and Mundkur, 1945; Comptom and Caldwell, 1946; Ross, Semeniuk, Taylor and Jenkins, 1948; Ainsworth and Sampson, 1950; Popp, 1951, 1959; Zundel, 1953; Ohms and Bever, 1955, 1956; Batts, 1955*a*, 1955*b*, 1956, 1959; Geshele, 1956, 1959; Batts and Jeater, 1958*a*, 1958*b*; Rod, 1958–1959; Savulescu and Stãnescu, 1959; Mathur and Hansing, 1960; Mantle, 1961; Mathur, 1961; Niemann, 1962; Hsu and Schafer, 1962; Bondar, 1963; Gorter, 1964; Doling, 1965*a*, 1965*b*; Gothwal, 1972*a*, 1972*b*). Embryo infection. Mode of infection is described in Subsection 12.2.3, and in Section 9.1, influence of time of infection in relation to flowering of wheat, Section 11.4, importance of location of infection within the embryo, Section 13.2; disease cycle is the same as that of loose smut in barley, see Section 14.1; differentiation between *U. tritici* and *U. nuda*, see above under *U. nuda*, and see Fig. 5.39. Economic importance, see Subsection 1.2.2, wheat. Seed treatment as for *U. nuda*. Testing for seed infection by the embryo count method is more difficult for wheat than for barley, see Section 27.4. *U. vavilovi* is closely related, infects

Secale cereale as well as wheat (Jaczewski, 1925; Shands and Schaller, 1946; Savulescu and Savulescu, 1955).

Distribution: World-wide.

CMI Distribution Map No. 368.

Other seed-borne smuts of varying importance are *U. compacta* on *Lolium multiflorum* (Fischer, 1953); *U. cynodontis* in *Cynodon dactylon*, seedling infection (McCain and Halisky, 1962); *U. heterogena (U. togata),* bunt on *Panicum dichotomiflorum* and *P. virgatum* (Fischer, 1953); *U. kenjiana* on *Sorghum vulgare* (Niemann, 1962); *U. mulfordiana*, head smut, on *Festuca* spp. including *F. octoflora* and *F. pacifica* (Fischer, 1953); *U. neglecta*, kernel smut of *Setaria*, on *Setaria glauca* and *S. viridis*; *U. paradoxa* on *Panicum frumentaceum (Echinochloa frumentacea)* (Sharma, 1963); *U. sphaerocarpa*, head smut, on *Festuca amplissima* (Fischer, 1953); *U. syntherismae* on *Digitaria sanguinalis* (Gaskin, 1958); *U. tragopogonis-pratensis* on *Tragopogon porrifolius* (Liro, 1924); *U. trebouxii* on many Gramineae including *Panicum virgatum*, *Poa pratensis* and *P. secunda* (Fischer, 1953; Greene, 1966); and *U. turcomanica* on *Agropyron orientale* and *A. squarrosum*, seedling infection (Fischer and Meiners, 1952).

Tilletiaceae

Basidiospores or sporidia are produced in a whorl at the end of a usually one-celled, in some one-septate, promycelium. From the promycelium primary conidia or sporidia are formed, then after conjugation with one another these produce secondary conidia or sporidia (Fig. 14.6) which are forcibly ejected (ballistospores). Finally, from the secondary sporidia infective hyphae are produced. In *Tilletia* and *Urocystis* seedling infection is common, but there are many exceptions, including local infection of ovaries in the former genus and shoot infection in the latter.

Tilletia

Powdery sori of teliospores are formed in ovaries. The teliospores usually have a reticulate surface, see Fig. 14.6, there is a considerable variation in spore morphology, and different graminicolous species may cross; hence speciation is difficult and subject to some controversial views amongst specialists.

T. barclayana (Neovossia barclayana, T. horrida), black smut, covered smut, kernel smut of rice. Occurs in rice as sori within the glumes affecting only a few ovaries in each panicle (Rutgers, 1914; Teng, 1931; Reyes, 1933, 1939; Padwick, 1950; Tullis and Johnson, 1952; Templeton, Johnston and Henry, 1961; Templeton, 1963; Chauhan and Verma, 1964; Ou, 1972). Local infection of opening flowers, not embryo infection or systemic seedling infection (Chowdhury, 1946*b*). Each infected ovary is a separate local infection. Soil transmission predominant, seed transmission probably very limited. Usually not economically important. Seed treatment by mercury may be considered to prevent introduction to new regions.

Distribution: In scattered regions, North America, South America, Africa (Sierra Leone), South-East Asia (including India, etc., and Japan).

CMI Distribution Map No. 75.

T. caries (T. tritici, T. secalis), see Figs 5.40 and 5.41, bunt, rough-spored bunt, stinking smut of wheat and grasses. Host range includes the cultivated species of wheat, *Triticum aestivum*, *T. dicoccum*, *T. durum* and *T. spelta*, and many other Gramineae, including *Aegilops cylindrica*, *Agropyron cristatum*, *A. subsecundum*, *A. trachycaulum*, *Bromus ciliatus*, *B. japonicus*, *B. marginatus*, *B. rigidus*, *B. sterilis*, *B. tectorum*, *Festuca octoflora*, *Lolium multiflorum*, *L. perenne*, *Poa reflexa* and *Secale cereale* (Kühn, 1874*a*; Barrus, 1916; Heald, 1921; Hungerford, 1922; Heald and Boyle, 1923; Faris, 1924*b*; Woolman and Humphrey, 1924; Caspar, 1926; Rabien, 1927; Gentner, 1929; Woolman, 1930; Giljarouskij and Zak, 1930; Bonne, 1931; Bressman, 1932; Feucht, 1932; Gaudineau, 1932; Mourashkinsky, 1932; Holton and Heald, 1936, 1941; Fischer, 1936; Gassner, 1938*a*, 1938*b*, 1938*c*, 1944; Churchward, 1940; Stevens, 1940; Rodenhiser and Taylor, 1940*a*, 1940*b*; Cherewick, 1944; Popp, 1947; Kedding, 1952*a*; Holton and Purdy, 1954; Zscheile, 1956; Weltzien, 1957*a*; Purdy and Kendrick, 1957, 1963; Kühnel, 1958, 1959*a*, 1960; Hansen, 1959; Kendrick and Holton, 1960; Marshall, 1960; Glaeser, 1961; Witkowski and Witkowska, 1962; Schuhmann, 1962; Swinburne, 1963; Kuiper, 1965; Dumitras, 1970; Özkan, 1971; Kietreiber, 1970AR–1971AR, 1973AR–1974AR).

Seedling infection. Many pathogenic races, see Section 17.2 (4.1), life-cycle is discussed in Section 14.5, influence of edaphic factors on establishment of infection in Subsections 15.2.1, 15.2.2 and 15.2.3, of light intensity in Subsection 15.2.4, on influence of photoperiodism on susceptibility of host in Subsection 15.2.5; moreover, on inoculum potential in Section 17.3, factors of host resistance in Subsection 18.2.1 and Section 18.4, influence of manuring is dealt with in Subsection 19.6.8, and of sowing date and depth of sowing in Subsection 19.6.2. Testing procedures for detection of seed-borne inoculum are described in Section 27.2, and tolerances to be considered for seed health certification schemes in Section 32.6; this smut is the economically most important of all smuts, see Subsection 1.2.1, wheat. For seed treatment a range of chemicals may be used, see Subsection 22.1.4, Section 22.6 and Table 22.3.

Distribution: World-wide where wheat is grown.

CMI Distribution Map No. 294.

T. contraversa (T. brevifaciens), dwarf bunt, occurs on a considerable range of hosts: *Aegilops cylindrica, A. triuncialis, A. umbellata, Agropy-*

Fig. 5.40. Spikes of wheat attacked by bunt or stinking smut, *Tilletia caries*. Left, smut balls cut through; right, a healthy spike. (Courtesy of Knud E. Hansen; photo J. O. Thögersen)

ron amurense, *A. cristatum*, *A. dasystachyum*, *A. desertorum*, *A. diazii*, *A. elongatum*, *A. inerme*, *A. intermedium*, *A. litorale*, *A. orientale*, *A. repens* (type host), *A. rigidum*, *A. riparium*, *A. sibiricum*, *A. subsecundum*, *A. trachycaulum*, *A. trichophorum*, *A. triticeum*, *Alopecurus myosuroides*, *Arrhenatherum elatius*, *Bromus erectus*, *B. marginatus*, *Dactylis glomerata*, *Elymus aralensis*, *E. canadensis*, *E. crinitus*, *E. glaucus*, *E. sibiricus*, *E. triticoides*, *Festuca idahoensis*, *F. pratensis*, *F. rubra*, *Hordeum bulbosum*, *H. leporimum*, *H. vulgare*, *Koeleria cristata*, *Lolium multiflorum*, *L. perenne*, *L. remotum*, *Phalaris arundinacea*, *Secale cereale*, *Triticum aestivum*, *T. dicoccum*, *T. glaucum*, *T. spelta*, *T. villosum* (Kühn, 1874*b*; Young, 1935; Bamberg, 1941; Bamberg *et al*., 1947; Tyler, 1945, 1953, 1958; Fischer, 1952; Warmbrunn, 1952, 1953, 1954; Hardison, 1954; Conners, 1954; Niemann, 1954; Riehm, 1954, Aebi, 1955, 1957; Duran and Fischer, 1956; Fischer and Duran, 1956; Holton, Kendrick and Meiners, 1956; Jensen and Tyler, 1956; Meiners and Hardison, 1957; Purdy, 1957*b*; Fischer and Holton, 1957; Baylis, 1958; Böning, 1958, 1959; Kanzler, 1958; Todorova, 1958; Tyler and Jensen, 1958; Graham, 1959; Zogg, 1959; Schuhmann, 1962; Savulescu *et al*., 1963; Zakharova, 1963; Wagner, 1964; Hoffmann, Kendrick and Holton, 1965; Dumitras, 1970; Özkan, 1971; Wallen and Ednie, 1972; Johnsson, 1973; Kietreiber, 1970AR–1971AR, 1973AR–1974AR). For other hosts see Zakharova (1963). Seedling infection. Many pathogenic races, see Section 17.2 (4.1); life-cycle is similar to that of *T. caries* but soil transmission is important, as teliospores may persist for years. Growth of infected plants is much reduced, down to between one-third and one-fifth of the normal size. Economic importance is considerable, see Subsection 1.2.1, wheat. Seed treatment as for common bunts, see Table 22.3. Important quarantine object.

Distribution: North America, South America (Argentina, Uruguay), Europe, Africa (Libya, possibly Sudan), West Asia.

CMI Distribution Map No. 297.

T. foetida (T. laevis, T. foetens), see Fig. 5.41, smooth-spored bunt, stinking smut. Host range of this species is considered the same as for *T. caries*, and in literature both species are often treated together. Differences in pathogenicity in the two species are often more pronounced between pathogenic races within one of the species than it is between the two species. On pathogenic races, see Section 17.2 (4.1). For references, see *T. caries* (further references specifically on *T. foetida*: Hanna, Vickery and Pucher, 1932; Radulescu, 1938; Apostolides, 1940; Rodenhiser and Holton, 1945, 1953; Holton, 1947; Berend, 1952, 1957; Urries, 1958; Mathur and Misra, 1961; Todorova, 1962; Naqvi, Bakshi and Joshi, 1963; Kunowskii and Panaĭotova, 1971; Tusa and Pîrvu, 1973). Life-cycle (see Section 9.1 on ovary infection), influence of edaphic and other external factors, economic importance and control are as for *T. caries*.

Distribution: World-wide, by and large as for *T. caries*.

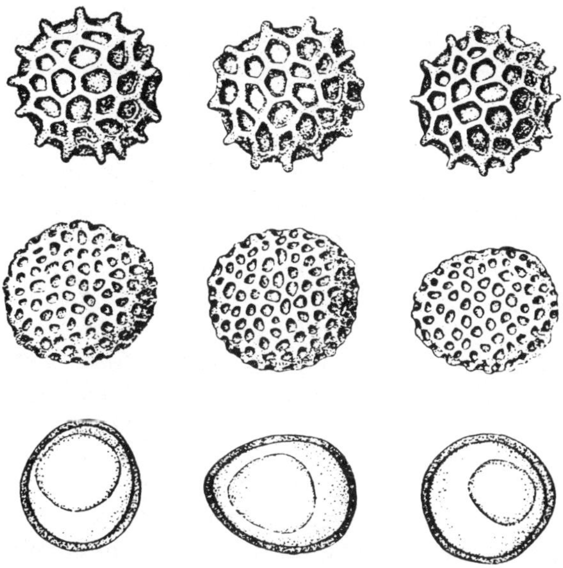

Fig. 5.41. Teliospores of three species of *Tilletia* attacking wheat. Above, *T. caries*; centre, *T. intermedia*; below, *T. foetida*. ×1,000. (After Savulescu, 1944)

CMI Distribution Map No. 295.

T. intermedia, see Fig. 5.41, is regarded as an intermediate form between *T. caries* and *T. foetida* and is of little economic importance.

T. pancicii, occurring on barley, is rather to be regarded as *T. contraversa* (Duran and Fischer, 1956).

T. triticoides is regarded as identical with *T. caries* (Schuhmann, 1962).

Other seed-borne species of probably less economic importance are: *T. buchloeana*, kernel smut of buffalo grass, *Buchloë dactyloides* (Fischer, 1953); *T. decipiens*, bunt of bent grasses, on *Agrostis* spp. (Fischer, 1953; Radulescu, 1962); *T. elymi*, kernel smut on blue wild rye, *Elymus glaucus* (Fischer, 1953); *T. fusca (T. guyotiana)*, kernel smut, on *Bromus* spp., *Festuca* spp., and *Poa* spp. (Fischer, 1953); *T. holci* on creeping soft grass, *Holcus mollis* (Das, 1948), *T. lolii* on darnel, *Lolium temulentum* (Ainsworth and Sampson, 1950); *T. pallida* on *Agrostis canina* (Conners and Savile, 1952); *T. paradoxa*, bunt, on timothy, *Phleum pratense* (Jaczewski, 1926); *T. setariae* on *Setaria lutescens* (Ling, 1945); *T. setariicola*, smut of *Setaria intermedia* (Pavgi and Thirumalachar, 1952); *T. texana* on *Festuca* spp., and on *Hordeum brachyantherum* and *H. pusillum* (Fischer, 1953).

Urocystis

Sori of powdery spore masses occur in different vegetative parts of the host, only few occur in reproductive organs. Few to many teliospores are firmly united in balls which are partially or completely surrounded by a layer of sterile cells. By germination the teliospore forms a basidium or promycelium, at the end of which a whorl of primary sporidia are formed; they fuse in pairs; in some species instead of sporidia, mycelial branches appear, as in *U. cepulae*. In other species secondary sporidia are formed. Parasites in Gramineae, Ranunculaceae and Liliaceae.

U. agropyri (U. tritici), flag smut or stripe smut of wheat. In addition to the principal host, *Triticum aestivum*, a considerable number of species of the following genera may be attacked: *Agropyron*, *Agrostis*, *Alopecurus*, *Amphibromus, Arctagrostis, Avena, Briza, Bromus, Calamagrostis, Carex, Dactylis, Deschampsia, Elymus, Festuca, Glyceria, Hesperochloa, Hordeum*, *Koeleria*, *Lolium*, *Melica*, *Phleum*, *Poa*, *Seslena*, *Sitanion* and *Trisetum* (Zundel, 1953; Fischer, 1953; Schuhmann, 1962, Purdy, 1963*b*, 1965; Duran, 1968; Hodges, 1970). Seedling infection takes place before the seedlings are more than 4 mm long and become resistant. Many pathogenic races, see Section 17.2 (4.1), influence of edaphic factors on establishment of infection, see Subsections 15.2.1 and 15.2.2, optimum temperature for infection relatively high, about 19–23°C. Of economic importance only in warm climates, resistant cultivars in the U.S.A. Seed treatment, see Table 22.3.

Distribution: In all parts of the world, although scattered.

CMI Distribution Map No. 80.

U. occulta, flag or stalk smut of rye. Occurs on *Secale cereale*, sometimes on *S. montanum* (Hecke, 1907; Ravn, 1912; Lemcke, 1919; Müller and Molz, 1922; Stakman and Levine, 1916; Biraghi, 1933; Ling, 1940; Taylor and Dillon Weston, 1945; Zhukova, 1951; Schuhmann, 1962; Shembel', 1970). Seedling infection from seed-borne spores, or infection of young side shoots. Pathogenic races are not known. Economic importance in some regions of Europe. Seed treatment with mercury compounds.

Distribution: North and South America, Europe (widespread), Asia, Australia.

Other seed-borne species of *Urocystis* of less importance are: *U. arrhenatheri* on *Arrhenatherum* sp. (Savulescu, 1951) and *U. violae* on *Viola odorata* (Pape, 1925).

(2. Hymenomycetes)

Fruit body usually well developed, shape and size vary greatly; the subclass includes such forms as mushrooms, puffballs, conks. Clamp connections may be present.

5.3.7.3 EXOBASIDIALES

A small order, with a single family, Exobasidiaceae, of leaf parasites on phanerogams, mostly belonging to Ericaceae, and often causing swelling of infected tissues. The basidia are produced directly on hyphae between the epidermal cells; they pierce the cuticle, forming externally four to eight spores. Genus considered: *Exobasidium*.

E. vexans, blister blight of tea. In India Srivastava (*fide* Noble and Richardson, 1968) found circumstantial evidence of seed transmission. Only tea, *Thea assamica* and *T. sinensis*, is infected. Yield losses may go up to 60 per cent as in South India; in some regions of South East Asia it causes annual average losses of 10 per cent or more. Very important quarantine object.

Distribution: Limited to South East Asia, including India, China and Japan. Wellman (1972) indicates that the fungus can be found in Southern United States, and warns against introduction from there to the tropics. The pathogen is not listed in USDA Index of Plant Diseases in the U.S.A., 1960.

CMI Distribution Map No. 45.

5.3.7.4 APHYLLOPHORALES

(Polyporales)

Most forms have a large, conspicuous fruiting body, basidiocarp, in which the spore-bearing structure may be flattened, as in Cortiaceae, represented below by *Thanatephorus*; club-like or tooth-like; or the hymenium may form pores as in Polyporaceae, including *Polyporus* and *Fomes*, or lamellae (gills) as in Schizophyllaceae, represented below by *Schizophyllum*. In poroid and lamellate forms the fruit bodies are tough, papery, leathery, or woody, not fleshy and putrescent as in Agaricales.

Fomes

Fruit body somewhat hoof-shaped, woody, perennial, viable for many years; a new layer of tubes is formed each year, and only recently formed layers of tubes sporulate.

F. annosus (Fomitopsis annosus, Polyporus annosus, Trametes radiciperda), root and heart rot or butt rot of trees. In England this serious pathogen of forest trees has been encountered in laboratory seed germination test of *Abies nobilis* (Batko, in Holmes and Buszewicz, 1957), suggesting the possibility that the fungus could be introduced to nurseries with seed. The parasite decomposes lignins and celluloses in conifers and broad leaved trees, resulting in a redbrown heart rot.

Distribution: North America, Europe (common in the northern and central regions), Asia (India, Japan).

Thanatephorus

Hymenium present as a thin film of mycelium on the substrate, bearing short broad basidia; each produces four to eight basidiospores. Imperfect state *Rhizoctonia*.

T. cucumeris (Corticium vagum var. *solani, Pellicularia filamentosa),* Rhizoctonia fungus, Rhizoctonia disease, black scurf, imperfect state *Rhizoctonia solani*, see Subsection 5.3.8.2.

Schizophyllum

Fruit body is thin, leathery, laterally attached; the gills are longitudinally split along the edge, hence the name, *schizo* (from Greek *schizein*, to cleave, split).

Only one species is common, *S. commune (S. alneum)*. This fungus produces a rapid deterioration of stored cones of *Pseudotsuga menziesii*, douglas fir, in Washington State, U.S.A., associated with loss of seed viability (Shea and Rediske, 1964). In the reported case the fungus developed dense white mats on cones stored in sacks at high relative humidity and at temperature of 20°C. Also stored acorns, *Quercus*, are attacked (Uroševič, 1957), and in Malaysia the fungus has been encountered on the surface of imported seed of oil palm, *Elaeis guineensis* (Williams, 1963). The fungus is a common cosmopolite on dead parts of trees but may attack some trees as an active wood destroyer (white rot). In sugarcane it causes a wound rot.

Distribution: World-wide, mainly in equatorial regions of America, Africa, Asia; a common saprophyte in northern Europe.

5.3.7.5 AGARICALES

Fruit bodies are usually large and conspicuous, largely fleshy, commonly called mushrooms or toadstools. Basidia are borne on the surface of gills or lamellae (Agaricaceae) or inside tubes (Boletaceae). The two genera considered below belong to Agaricaceae.

Coprinus

Inky-cap mushrooms

After spore discharge the gills deliquesce by autodigestion into a black, inky liquid (that may be used for writing). Many species occur on dung, some on wood, all species are saprophytes.

C. cinereus (C. lagopus), or possibly *C. macrocephalus* (see Noble and Richardson, 1968), has been found in a considerable number of seed samples tested in the Copenhagen germinator, occurring as conspicuous mushrooms on seeds and blotter. The fungus, which obviously is always the same species, has been encountered in seeds of *Anethum graveolens*, *Chrysanthemum carinatum*, *Daucus carota*, *Pastinaca sativa*, *Petroselinum crispum, Spinacia oleracea* (relatively frequent), *Tragopogon porrifolius* and *Zinnia elegans* (Neergaard, 1942AR, 1945AR*a*, 1945AR*b*, 1948AR, 1949AR, 1950AR, 1951AR) and in *Beta vulgaris*, garden beet, and *Tetragonia expansa*, New Zealand spinach (Buller and Newton, 1927; Crosier, Patrick, Heit and McSwain, 1949). The startling presence of the mushroom in seed is of no consequence to its quality but may be regarded as an interesting curiosity.

Distribution: Probably world-wide.

Marasmius

The fruit body is thin-fleshy, with a hard, cartilaginous stalk or stripe; spores are colourless. The 'mushrooms' do not deteriorate rapidly but persist for long in a dry shrivelled condition (Greek, *marasmos* withering). Widely known is *M. oreades*, fairy ring mushroom, common in lawns.

M. perniciosus (Crinipellis perniciosus), witches' broom of cacao. This most important and very destructive pathogen of cacao, which attacks all species of *Theobroma*, can infect developing fruits and seeds and may possibly persist in these. It is an important quarantine object in Africa where the disease is unknown (Inter-African Phytosanitary Commission, 1962; Caresche and Rayner, 1969). Seeds imported into the Dominican Republic in 1974 and 1975 from Trinidad, developed up to 70 per cent infected seedlings. All consignments and seedlings were destroyed. The pathogen is not present in the Dominican Republic (Manuel Mendez, personal communication, 1976).

Distribution: Common in some countries of South America.

5.3.8 Fungi Imperfecti II

(Related to Basidiomycetes)

5.3.8.1 BLASTOMYCETES

Characteristics of this class are given in Subsection 5.3.6.1. The genus below, *Itersonilia*, is provisionally referred to the order Sporobolomycetales which reproduce by forcibly abjected spores, ballistospores (as in Basidiomycetes).

Itersonilia

Mycelium with clamp connection at every septum; chlamydospores may be formed; no budding cells are produced; ballistospores are kidney-shaped.

I. pastinacae, parsnip canker, has been encountered in parsnip seed crops in England and was recorded in 6 of 50 commercial seed samples (Channon, 1967). Seed infection is largely superficial although sometimes internal (Channon, 1969), but the fungus may also be present as concomitant contamination, carried in debris (Smith, 1966). Cause of root canker and petal blight. Seed treatment by thiram soak, 30°C for 24 hours, effective.

Distribution: Europe, Australia.

5.3.8.2 HYPHOMYCETES

[Agonomycetales (Mycelia sterilia)]

Rhizoctonia

Septate, brown mycelium, often bearing clamps, but producing no conidia (*Mycelia sterilia*). Some forms develop superficial, irregular, scale-like black sclerotia. The morphology of the mycelium varies considerably being slender, hyaline when growing within plant tissues, while older hyphae are short-celled, abundantly branched, stout, often constricted at septa.

R. bataticola, see *Macrophomina phaseolina.*

R. leguminicola, blackpatch of red clover and other legumes. The hyphae may be conspicuously visible as coarse mycelium on the surface of seed, or they occur as internal seed infection in legumes, including soybean, *Glycine max* (Weimer, 1950), giant trefoil, *Lotus uliginosus* (Wells, 1953) and red clover, *Trifolium pratense* (Gough and Elliott, 1956), and undoubtedly many other leguminous hosts. In some regions in the U.S.A. blackpatch is a major disease in red clover seed crops, as great as 50 per cent weight loss of seed yields may occur in heavily infected crops, and infected seed may fail to germinate.

Distribution: U.S.A.

R. solani, common Rhizoctonia fungus, see Fig. 5.42, perfect state *Thanatephorus cucumeris*. This ubiquitous plant pathogen, well known as soil-borne, is commonly seed-borne. In Denmark the fungus has been recorded in seeds of the following hosts: *Brassica campestris, B. napus, B. oleracea, B. rapus, Cichorium endivia, C. intybus, Dianthus caryophyllus, Helianthus annuus, Humulus japonicus, Iberis amara, Impatiens balsamina, Lactuca sativa, Pisum sativum, Raphanus sativus, Sinapis alba,*

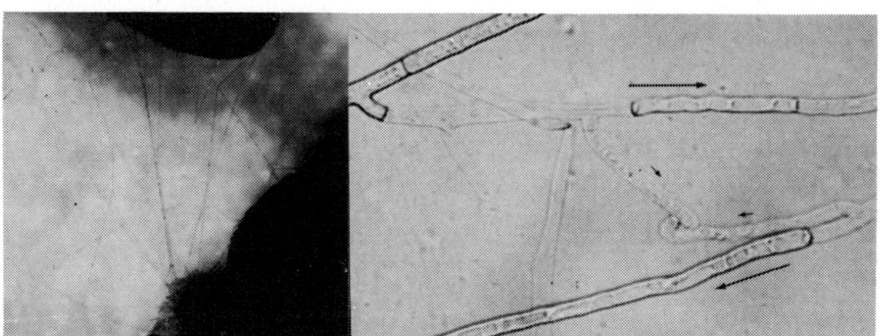

Fig. 5.42. *Rhizoctonia solani,* isolates from *Corchorus.* Left, mycelial threads as seen in a blotter seed health test. Right, anastomosis between an isolate from bean seed and an authentic isolate of Anastomosis Group No. 4, resulting in characteristic vacuolation and ultimate killing of the anastomosed cells, ×300. On pathogenic specialisation of this fungus, see Section 17.2 (4.3)

Spinacia oleracea, *Trifolium pratense*, *T. repens*, *Ulmus campestris*, *Vicia faba*, *Viola tricolor*, *Zinnia elegans* (Neergaard, 1940AR–1942AR, 1948AR, 1950AR, 1951AR; Andersen, 1974 — index). Noble and Richardson (1968) have references to seed infection in 37 plant genera; Leach and Pierpoint (1956, 1958) demonstrated seed transmission in *Phaseolus vulgaris*, *P. lunatus* and *Agrostis tenuis*, and Onesirosan (1975) that it is seed-borne in cowpea, *Vigna unguiculata*. Baker (1947) studied in detail seed transmission in relation to control of damping-off of egg-plant, pepper and tomato. The pathogen causes a range of diseases in herbaceous and woody plants, including damping-off, stem canker, root rot, foliage blight or leaf spot, and decay of stored products. The fungus can be troublesome in seed-testing laboratories, where it spreads rapidly in germinators causing damping-off of seedlings in the tests (Neergaard, 1942AR). Selective pathogenicity amongst strains of the fungus has been established, see Section 17.2 (4.3). Chemical seed treatment often inadequate, hot water treatment and particularly aerated steam treatment more promising, see Subsection 21.3.1 (5).

Distribution: World-wide.

Sclerotium

A form genus producing distinct more or less regularly shaped sclerotia, often with a brown or black outer layer; conidia are absent or unknown. The sclerotia are variable in size, shape and colour, usually connected by mycelial threads.

S. bataticola, see *Macrophomina phaseoli*.

S. cepivorum white rot of onion. Seed transmission doubtful; suggested seed treatments are directed rather against soil-borne inoculum (Croxall, Sidwell and Jenkins, 1953; Hickman and Coley-Smith, 1960). However, presence of sclerotia in poorly cleaned seed lots may be assumed as a possibility.

Distribution: North and South America (scattered areas), Europe (widespread), Africa (scattered), Asia, Australia, New Zealand.

CMI Distribution Map No. 331.

S. rolfsii, perfect state *Corticium (Pellicularia) rolfsii*. The fungus has been established in seeds of groundnut, *Arachis hypogaea* (Clinton, 1957), lettuce, *Lactuca sativa* (Silveira, 1947), rice, *Oryza sativa* (De Tempe, *fide* Noble and Richardson, 1968), French bean, *Phaseolus vulgaris* (Singh and Mathur, 1974) and wheat, *Triticum aestivum* (Epps, Patterson and Freeman, 1951). Serious in many crops in the low-country tropics, especially in moist climates, and primarily as a root parasite. Clamp connections used for identification in seed health testing (Naumova, 1972).

Distribution: World-wide, in warm climates.

CMI Distribution Map No. 311.

6 Seed-borne Nematodes

INTRODUCTION

Nematodes, or eelworms, are long thin roundworms. Most nematodes are from 0.3 to 1 mm in length, some up to 4 mm, and certain species of *Longidorus* may range even up to 10 mm; they range in breadth from 15 to 35 μ; the body usually tapers at each end and is more or less transparent; the fertilised females of some species become swollen and sedentary, having a sphaeroid or pear- or lemon-shaped body which becomes an egg-carrying case or cyst (Fig. 6.1). The nematodes are unsegmented but covered by a colourless, usually annulated cuticle beneath which are two

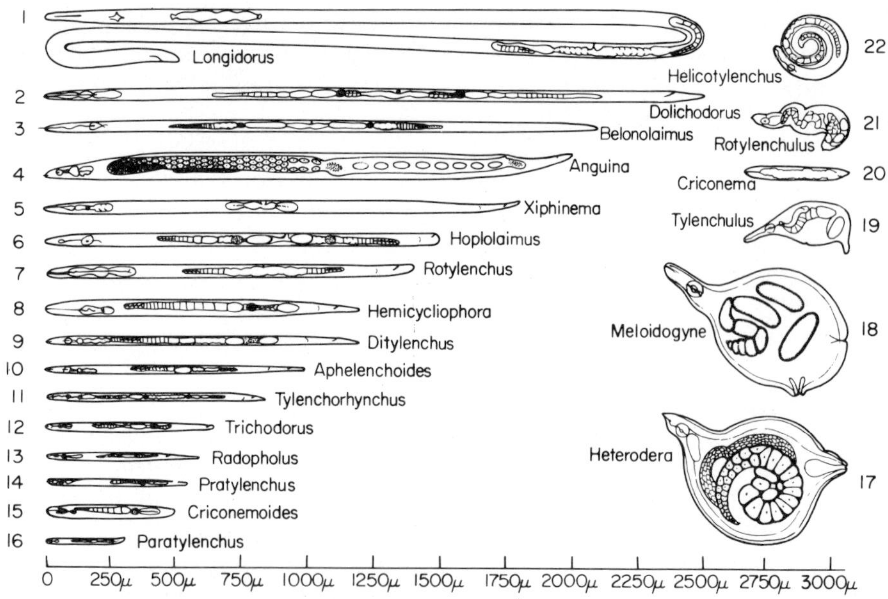

Fig. 6.1. Morphology and relative size of the most important plant parasitic nematodes. (Agrios, 1969) The following genera contain seed-borne species: *Anguina, Ditylenchus,* ***Aphelenchoides*** and ***Heterodera.*** Three genera are important vectors of seed-borne viruses: ***Longidorus, Xiphinema*** and ***Trichodorus.*** Species of some genera may enter a synergistic relationship with seed-borne bacteria and/or fungi: *Belonolaimus, Anguina, Hoplolaimus, Rotylenchus, Ditylenchus, Aphelenchoides, Tylenchorhynchus, Pratylenchus, Rotylenchulus, Meloidogyne* and *Heterodera*

layers, a subcuticular one over a muscular layer. The mouth is provided with papillae leading to a buccal cavity; this leads to the oesophagus, a muscular tube that leads to the intestine which is a simple tube without muscles. All known plant parasitic nematodes possess in all the feeding stages a hollow buccal stylet or mouth spear which is extruded from the buccal cavity to puncture plant cells. Saprozoic nematodes do not have a spear (Fig. 6.2).

About 9,000 species of nematodes are known. Many of these live in soil or water, some 5,000 species having been recorded from soil (Peacock, 1957). Saprozoic soil-inhabiting nematodes live in decaying matter and may, naturally, contaminate seeds by means of infested soil. About 100 species of soil nematodes are known to attack plants, and a total of several hundred species are plant parasites. Depending on the host–parasite association, plant parasitic nematodes may be sedentary or migratory, and ectoparasitic or endoparasitic in their feeding habits.

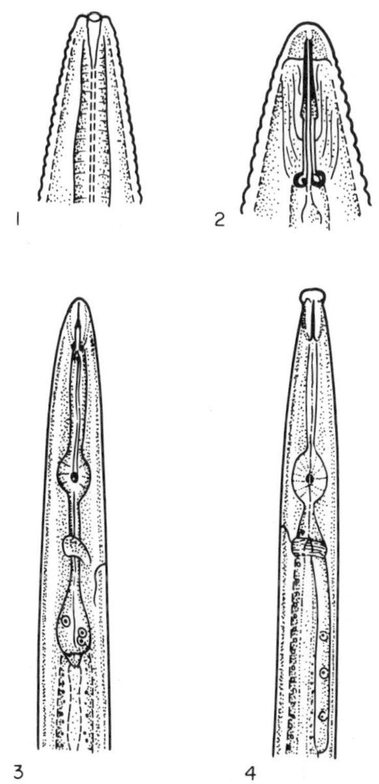

Fig. 6.2. Head and oesophagus of two types of nematode. (1) Head end of a non-spear-bearing free-living saprozoic nematode; (2) head end of a typical plant parasitic nematode (Tylenchida) provided with a mouth spear; (3) oesophagus of *Tylenchorhynchus*; (4) oesophagus of *Aphelenchoides*. Note mouth spear present in both these two plant parasitic nematodes. (Siddiqi, from Caresche *et al.*, 1969)

The life history of most plant parasitic nematodes is similar: eggs hatch into larvae which pass through four stages each of which is terminated by a moult. The first moult usually occurs in the egg. After the final moult the nematodes differentiate into adult males and females. Some of these stages are very resistant to drought. The second stage of *Ditylenchus dipsaci* and *Anguina tritici* can withstand extreme dry conditions for years and thus account for seed transmission of these two species. In the cyst-forming nematodes of *Heterodera* the unhatched eggs or the larvae in the cysts are drought-resistant.

Some families of nematodes are important seed-borne pathogens, other families of plant parasitic nematodes are vectors of seed-borne viruses, and still others may enter a synergistic relationship with pathogenic seed-borne fungi and bacteria.

Seed-borne. In the family Tylenchidae two genera contain seed-borne nematodes: *Anguina*, seed-gall nematodes, and *Ditylenchus*, stem and bulb nematodes. In the family Aphelenchoididae the following two genera: *Aphelenchoides,* bud and leaf nematodes, and *Rhadinaphelenchus.* In the family Heteroderidae species of one genus may be transmitted by seed: *Heterodera*, cyst nematodes.

Vectors. Three genera of seed-borne nematodes are vectors of seed-borne and other viruses (see Section 16.8), namely two of the family Tylencholamidae transmitting polyhedric viruses: *Longidorus,* needle nematodes, and *Xiphinema*, dagger nematodes, and one genus of the family Trichochoridae transmitting rod-shaped viruses: *Trichodorus,* stubby root nematodes.

Biotic relationships. Finally, species which may be associated with other plant pathogens may exercise a synergistic effect aggravating the disease complex produced by the two associated pathogens.

In the family Tylenchidae this holds true for *Tylenchorhynchus*, stunt nematodes (Section 16.7) as well as for species of the above genera *Anguina* (Section 16.6) and *Ditylenchus* (Sections 16.6 and 16.7).

In the family Hoplolaimidae four genera may have synergistic effects on seed-borne pathogenic fungi (Section 16.7): *Hoplolaimus,* lance nematodes; *Pratylenchus,* lesion or meadow nematodes; *Rotylenchulus,* reniform nematodes; *Belonolaimus*, sting nematodes.

In the family Heteroderidae one genus is of particular importance in enhancing diseases caused by various seed-borne bacteria and fungi (Sections 16.6 and 16.7): *Meloidogyne,* root knot nematodes.

Moreover, species of the genus *Heterodera*, listed above, may aggravate *Fusarium oxysporum* wilts (Section 16.7).

6.1 SURVEY OF SEED-BORNE NEMATODES

Nematodes may be carried with seed in three different ways. They may be borne in the seed itself as larvae harboured beneath the hull of cereals and grasses or in minor cavities of seed, in particular in the region of hilum (*Aphelenchoides*, *Ditylenchus*) or in lesions in seed (*Ditylenchus*). Secondly, they may be carried with the seed in plant debris from infected mother plants (*Ditylenchus*). Thirdly, nematode galls may be mixed with the seed (*Anguina*) or nematode cysts may be present in peds of soil mixed with the seed (*Heterodera*).

Saprozoic nematodes are commonly encountered on seeds. Goffart (1930) surveyed the seed-borne fauna of nematodes of more than 50 plant species and recorded several species of saprozoic soil nematodes belonging to the families Rhabditidae and Cephalobidae which often occur in great numbers in decomposing plant material or soil. The most frequent species was *Cephalobus rigidus* which he found on seeds of about 35 plant species.

Species of four or five genera of plant parasitic nematodes are seed-transmitted. The genera are listed in alphabetical order.

6.1.1 Anguina

Seed gall nematodes

Gall-forming endoparasites of seeds, stems and leaves of cereals, grasses and other plants. During part of their life they may feed as ectoparasites. In the seed-borne species galls are formed by transformation of ovaries. Four species are seed-borne.

Anguina agropyronifloris

The nematode causes seed galls to develop in the caryopsis of western wheatgrass, *Agropyron smithii* (Lakon, 1953; Norton and Sass, 1966) and *Trisetum flavescens* (Lakon, 1953).

The nematodes are ecto- and endoparasitic in the root — endoparasitic in the scutellar node and ectoparasitic between the leaf sheaths. The undeveloped fertile ovary is invaded, before any stamen or pistil differentiation occurs.

Distribution: U.S.A., notably in Kansas, Montana and South Dakota.

Anguina agrostis

Seed gall, bent grass nematode

The nematode produces seed galls contaminating seeds of many grasses, *Agrostis* spp., including *A. alba*, *A. canina*, *A. exarata*, *A. palustris*, *A. stolonifera*, *A. tenuis* and *A. vulgaris* (T. Goodey, 1930; Courtney and Howell, 1952; Hardison and Jensen, 1953; Apt, Austenson and Courtney, 1960; Hague, 1963), *Arctagrostis agrostis* (T. Goodey, 1952),

Arrhenatherum elatius (Dorph-Petersen, 1903AR, 1921AR), *Dactylis glomerata*, very common in Danish seed, recorded consistently during 75 years (O. Rostrup, 1901AR; Dorph-Petersen, 1903–1936AR; Stahl, 1937–1962AR; S. Rostrup, Bovien and Thomsen, 1940; Hardison and Jensen, 1953, observed short and thick galls caused by *Anguina* sp., whereas those incited by *A. agrostis* are long and thin), *Dupontia fisheri* (T. Goodey, 1952), *Festuca duriuscula*, often recorded in Danish seed during recent 75 years (O. Rostrup, 1901AR; Marcinovski, 1910; Dorph-Petersen, 1903AR, 1905–1906AR, 1913–1919AR, 1924–1927AR, 1933–1935AR; S. Rostrup, Bovien and Thomsen, 1940; Stahl, 1953–1955AR, 1957–1959AR, 1962–1963AR*b*), *Festuca ovina* (Marcinovski, 1910), *Festuca rubra* (Hardison, 1953; Hardison and Jensen, 1953); Courtney (1958, *fide* Thorne, 1961) observed that the galls of fescue appear much the same as normal seeds and that they can be identified as galls only by dissecting the palea from each seed; however, the galls are much lighter than healthy seed and can be mechanically separated; *Holcus lanatus*, often in Danish seed (O. Rostrup, 1901AR; Dorph-Petersen, 1903–1930AR), *Koerulea glauca* (Marcinovski, 1910), *Phalaris* spp. (Filipjev and Schuurmans Stekhoven, 1941; Goffart, 1951), *Phleum boehemeri* (Marcinovski, 1910), *Phleum pratense* (Goffart, 1951), *Poa alpina* (Marcinovski, 1910), *Poa annua* (Marcinovski, 1910), and *Trisetum flavescens* (Dorph-Petersen, 1903AR, 1904AR, 1906AR).

When the second-stage larvae have been released from galls they migrate to the growing points of the young plants where they probably feed ectoparasitically. The larvae enter developing ovaries and produce long spindle-shaped purple galls (Fig. 6.3). The galls are toxic to livestock. Cattle, sheep and horses suffer from nervous disorder, including falling, lack of co-ordination, and the disease condition may result in the death of animals.

Distribution: North America, Europe, New Zealand.

Anguina spermophaga

True seeds of sugarcane which were brought from Turkistan, U.S.S.R. to a quarantine greenhouse in Virginia, U.S.A., harboured this pathogen and were destroyed under quarantine but may be expected to reappear in any tropical region (Steiner, 1937; Thorne, 1961; cf. also Wellman, 1972).

Anguina tritici

Ear cockle, nematode galls, purples, peppercorn, spikelet nematode

The seed galls (see Fig. 19.8) have been recorded in *Avena sativa* (Leukel, 1924; Thorne, 1961), *Secale cereale* (Leukel, 1929), sometimes associated with *Dilophospora alopecuri* as observed in Germany (Becker, 1955), *Triticum aestivum* (Goodey and Hooper, 1958; Bloom, 1963; Marshall, 1960) in which it is widespread in many wheat-growing areas where efficient seed cleaning is not carried out; often associated with *Dilopho-*

spora alopecuri and *Corynebacterium tritici*. In most of Europe the pathogen has disappeared from wheat seed, due to efficient improvement of seed cleaning during the last decades. See Table 32.8.

The life-cycle is dealt with in Section 14.8. Predominant in cool, moist climates, losses may sometimes go up to about 70 per cent.

Distribution: North and South America, Europe, although now infrequent, North Africa, Asia, Australia, New Zealand; often serious in tropical Far and Middle East.

6.1.2 Aphelenchoides

Bud and leaf nematodes

Saprozoic and parasitic species. The slender body often very small, usually below 1 mm; however, one of the two seed-borne species, *A. ritzemabosi* is often more than 1 mm long. Tail conoid to blunt, often mucronate but never filiform.

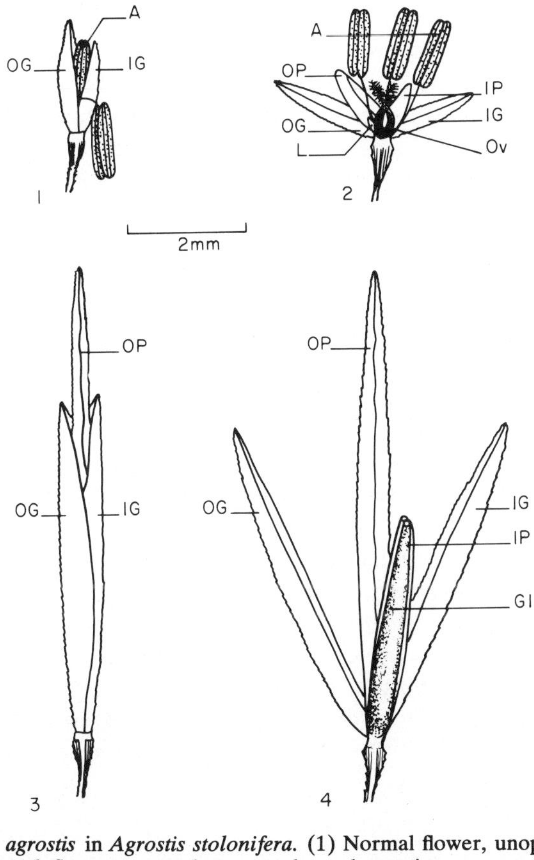

Fig. 6.3. *Anguina agrostis* in *Agrostis stolonifera*. (1) Normal flower, unopened, one anther extruded; (2) normal flower, opened out to show the various parts; (3) galled flower, unopened; (4) galled flower, opened to show the parts. A, anthers; Gl, gall; IG, inner glume; IP, inner palea; OG, outer glume; OP, outer palea; Ov, ovary. (T. Goodey, 1930)

Aphelenchoides besseyi
(syn. *A. oryzae*)
White tip nematode, summer crimp nematode of strawberries

The nematodes are harboured beneath the hull of rice grains where they may remain viable from season to season (Cralley, 1949), even for two or three years (Yoshii and Yamamoto, 1950; Todd, 1953; Todd and Atkins, 1958). Only an average of five or six nematodes are present in each seed (Thorne, 1961). The nematodes may be found remaining on the seeds during their germination (Goto and Fukatsu, 1952). In a survey organised by the Inter-African Phytosanitary Commission many seed lots were found to be infested (Hooper and Merny, 1966). Cralley (1952) established in the U.S.A. that seed transmission prevails, whereas the nematode does not overwinter in soil. Hot water seed treatment or methyl bromide fumigation is recommended (Todd and Atkins, 1959).

The nematode is an ectoparasite, at first living between young folded leaves, then entering the spikelets and developing within the glumes as the grains mature. The prominent symptom is the 'white tip' of the leaves which become chlorotic or white, later brownish, for a distance of up to 5 cm from the tip, while the lower parts of infected leaves are darker green than normal. Infected plants are stunted, produce few panicles; yield losses in susceptible cultivars may go up to 50 per cent. The white tip symptom usually appears when the plants begin to elongate, although infected plants may be symptomless. Resistant cultivars rarely show foliar or panicle symptoms but yields may be reduced (Atkins and Todd, 1959). For further details consult Ou (1972).

Distribution: North America, Europe (U.S.S.R., Italy), Africa, Asia, Australia.

Aphelenchoides ritzemabosi
Chrysanthemum foliar nematode, chrysanthemum eelworm

This widespread nematode has been reported to be seed-transmitted in *Callistephus chinensis* (E. B. Brown, 1956).

Endoparasitic; it is most frequent in leaves but also in buds and meristems when the nematode overwinters in temperate zones. Eggs and all four larval stages are produced inside the leaf. In chrysanthemum infection results in brown to black leaf spots, which are sharply confined to areas between main leaf veins; later the entire leaf may be killed. The first stems may become dwarfed and the plant forced to produce other stems. Infections of buds may lead to stunting or distortion of plants, or dwarfing and death of shoots. Other hosts include *Dahlia*, *Delphinium*, *Phlox*, *Zinnia elegans*, occasionally tobacco and strawberry.

Distribution: North and Central America, Europe, widely distributed in temperate areas, and introduced to upland areas in India and some East African countries.

6.1.3 Ditylenchus

Stem and bulb nematodes

Among about thirty species, some of which are obligate plant parasites, others soil inhabitants, there are two seed-borne species.

Ditylenchus angustus

Rice stem nematode

Ufra or Dak pora (Bangladesh and India)

The nematode is found as small cottony masses of larvae of the fourth stage beneath the glumes (Seth, 1939; Swarup, personal communication, 1967), as established in Burma and India but not in Thailand (Hashioka, 1963). Hot water seed treatment is recommended (Anonymous, 1943).

The nematode is an obligate ectoparasite. The seedlings are infected when they are only a few days old; the nematodes occur at the base of the peduncle, the stem above the node and within the glumes. Infection leads to chlorosis, later withering of leaves, or leaves may become twisted. The panicles may become twisted or otherwise distorted, spikelets may be empty or only partly filled. The nematode is restricted to certain tropical areas, when the humidity is high during the growing and ripening season. Yield losses may be considerable — up to 90 per cent in Thailand.

Distribution: Asia, Egypt, Madagascar.

Ditylenchus dipsaci

Stem and bulb nematode, teasel nematode; onion blast; tulip root, segging (and other names of the disease in individual hosts)

Reported as seed-borne in the following hosts, in some of which seed transmission is very important: *Allium ascalonicum* (Hague, 1968), *A. cepa* (Ritzema Bos, 1888–1892; T. Goodey, 1945*a*, 1945*b*, who observed nematodes near the hilum; Southey, 1965, who found them below the seed coat; Seinhorst and Koert, 1971, see Section 17.3). *Avena sativa* (T. Goodey, 1939, transmission 'negligible' from the practical point of view compared with soil infestation; Robertson, 1955; J. B. Goodey, 1965, negative statement on seed transmission in oats), *Beta vulgaris* (Dunning, 1956, seed-borne but importance of seed transmission doubtful), *Crepis* sp., syn. *Barkhausia taraxacifolia* (Maige, 1906), *Dipsacus fullonum,* the type host of the nematode (Kühn, 1857; J. B. Goodey, 1949; Courtney, 1952; J. L. Jones, 1955), *Hypochoeris radicata* (Hodson, 1929; Goodfrey, 1931), *Medicago sativa*, mainly with plant debris with the seed (Goodey, 1950; Hawn, 1963, who reports transmission of *Corynebacterium insidiosum* by the seed-borne nematode; Thorne, 1961, cf. Section 16.6; Nielsen, 1964), *Plantago lanceolata* (Hodson, 1929), *P. maritima* (Hodson, 1929), *Trifolium pratense,* mainly with plant debris with the seed (Lester and Large, 1958, on distribution in England; Tinnilä, 1959, 1960; Nielsen, 1964), but also tenaciously adhering to the seed coat (Cobb, 1925), *Vicia faba* (Diercks and Klewitz, 1957, 1962). In Morocco 87 per cent of seed lots of

Vicia faba, comprising 19 tons, tested in 1972 for export, were contaminated (Schreiber, 1974, in litt.), see Fig. 6.4. Fumigation of seed by methyl bromide is recommended, see Section 21.5(11).

The nematode (Fig. 6.5) is endoparasitic, infecting more than 400 species of 40 plant families; many pathogenic races. It is severe in red clover (clover sickness) and lucerne; in these hosts it causes deformed buds, stunted shoots, stems with galls, and many short side stems may give affected plants a tufted appearance; it may be destructive in onions producing softening of the scales, leading to desiccation in dry seasons, soft rot in presence of moisture (onion bloat). In other plants the pathogen produces decay of underground parts, and thickening, dwarfing and distortion of shoots. The saliva of the nematode contains a pectinase that dissolves the middle lamella and causes the cells to separate. It is most serious in cooler, moist climates, and prevalent on heavy soils.

Distribution: World-wide, North and South America, Europe, Africa, Asia, Australia.

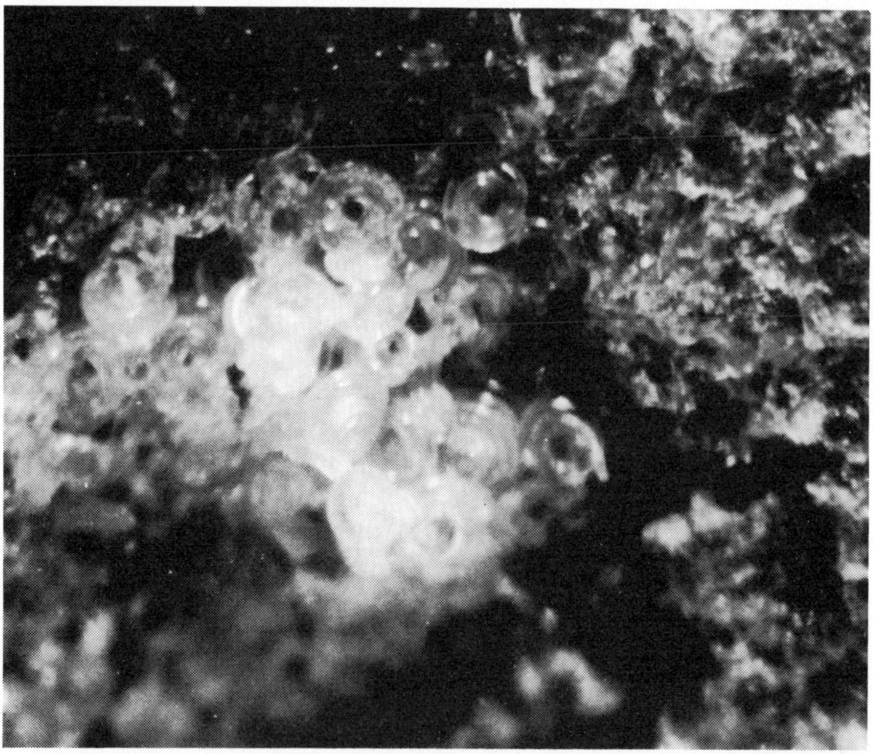

Fig. 6.4. Nematode 'wool' of stem nematode, *Ditylenchus dipsaci*, 'giant race', located below the seed coat of broad bean, *Vicia faba*. Ring-formed curling of individual nematodes is visible. (Courtesy of K. Schlüter)

6.1.4 Heterodera

Cyst nematodes

The female has a tough, white sac-like body which retains most or all of eggs; the body becomes a brown or yellow cyst at maturity. The cysts are typically ovate to spheroid and usually range from 0.4 to 0.8 mm in length, and contain, according to species, from about 10 to more than 600 eggs (Thorne, 1961).

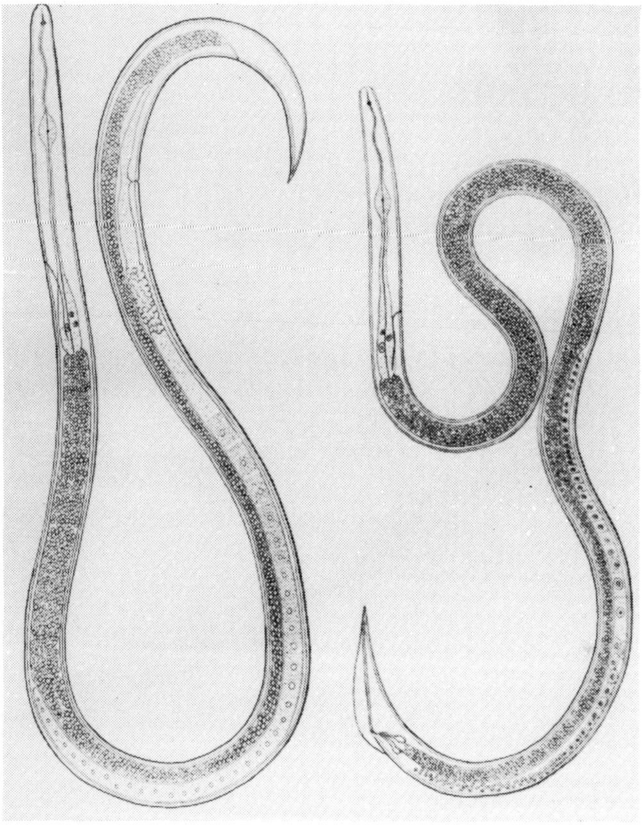

Fig. 6.5 Stem nematode, *Ditylenchus dipsaci*. Left, a female; right, a male ×190. Marcinovski, 1910)

Heterodera glycines

Soybean cyst nematode

Cysts of the nematode may be carried with seeds of soybean, either loose or in small 'peds' of soil, mixed with the seed (Epps, 1968, 1969). Each cyst contains 300–600 eggs which pass through the embryonic stages and reach the second larval stage which lies quiescent in the dry cyst.

The nematode is endoparasitic, causing premature yellowing of leaves and sometimes dwarfing of plants which bear few leaves and few small seeds. Roots of infected plants are weaker, have fewer bacterial nodules than normal but are otherwise normal. In heavily infested fields yields may be reduced up to 75 per cent. For further details, see Agrios (1969).

Distribution: North America, Asia.

Heterodera goettingiana

Pea cyst nematode

May occur in soil clods, mixed with pea seed; intercepted in imported seed in India (Sethi *et al.*, 1972).

Infected plants become stunted, leaves yellow-green and the leaves die when affected early in the season. In addition to pea, species of *Vicia* and *Lupinus* as well as soybean may be attacked.

Distribution: U.S.A., Europe.

Heterodera schachtii

Sugarbeet nematode

The cysts may be carried with sugarbeet as concomitant contamination, loose or in peds of soil; contaminated seed intercepted in India (Sethi *et al.*, 1972). The cysts may contain from less than 10 to more than 600 eggs which pass through the embryonic stages to second stage larvae within the cyst (Thorne, 1961).

The nematode is endoparasitic, mass invasion of larvae, particularly in young root tips, may cause serious reduction of roots, with many dead rootlets, the roots look 'hairy' and carry the cysts of the females, plants become stunted, occurring usually in patches in the field, young seedlings may be killed. The nematode attacks most species of *Beta* and *Brassica* and has a wide host range, including about 100 plant species of a dozen plant families.

Distribution: North America, Europe.

6.1.5 Rhadinaphelenchus

May be described as a very slender nematode, about 1 mm long, and characterised by having quadrangular front portion of the head, and a sclerotised spade-like extension of the male tail.

Rhadinaphelenchus cocophilus

Coconut nematode, red ring disease

Introduction of the pathogen to Africa with infected coconuts for propagation is considered by quarantine authorities to be a serious risk and importation of seeds for planting is therefore prohibited from all areas of America and West India (Anonymous, 1966). Nematodes have been found in the husks of dropped nuts (Fenwick, 1957).

It is endoparasitic in *Cocos nucifera* in which it causes yellowing and browning of leaves, withering, and death of growing point and inflorescence, and death of the tree. A cross-section of the trunk of four- to ten-year-old trees reveals dull red ring at about 2 in from the periphery, 1–2 in wide, and nematodes may be found in adjacent tissues (Thorne, 1961). The main vector is a large beetle, *Rhyncophorus palmarum*, which feeds in palms.

Distribution: Most areas of tropical America, Central America, Venezuela, Brazil.

7 Storage Fungi

7.1 IMPORTANCE

'The people need food, not production statistics, and a crop is not food until it is eaten. A program to reduce storage losses probably could result in a 10–20 per cent increase in available food now in some of the developing countries, and might also assure that whatever increases occur in production in the future would be used for the nourishment of people, not for feeding pests' (Christensen and Kaufmann, 1969).

To continue quoting from Christensen and Kaufmann's animated and informative book, *Grain Storage, The Role of Fungi in Quality Loss*, there are six major types of losses caused by fungi growing in stored grains. They are (1) decrease in germinability, (2) discoloration of part (usually the germ or embryo) or all of the seed or kernel, (3) heating and mustiness, (4) various biochemical changes, (5) production of toxins that if consumed may be injurious to man and to domestic animals, (6) loss in weight.

There are no exact statistics on the world losses of food grains and other seeds due to storage fungi but a rough and probably conservative estimate (FAO) is 5 per cent of all harvested food grains before consumption; in some countries, such as India and certain countries in Africa and South America, it may be 30 per cent of the annual harvest. The fact that particularly some tropical developing countries suffer from these losses makes the problem of storage fungi (and pests) a major one in world agriculture.

7.2 FUNGI INVOLVED

7.2.1 Definition

Distinction is made between two ecological categories of fungi that invade seeds, field fungi and storage fungi (Christensen and Kaufmann, 1965, 1969).

Field fungi are those that invade seeds on the developing plants in the field, often in the maturing seeds and after the crop is cut and swathed but before it is threshed. They may be pathogens or saprophytes. Common field fungi are *Alternaria tenuis*, *Cladosporium herbarum*, *Curvularia* spp., *Epicoccum purpurascens*, *Fusarium* spp., and in some seeds the pathogenic fungi *Fusarium moniliforme* and *Verticillium alboatrum* (Malone and Muskett, 1964). To these may be added a range of other pathogenic fungi (Noble and Richardson, 1968). These may cause discoloration of seeds, reduction of germination and diseases of the seedlings or growing plants. During storage of the seeds their activity is usually arrested because they require high relative humidity for growth — usually above 95 per cent.

Storage fungi are those that grow on stored products. Most of them are able to grow without free water, and on media with a high osmotic pressure. Most of the storage flora are species of *Aspergillus* and *Penicillium* which are active at relative humidities ranging from 70 to 90 per cent. It has been established that storage fungi usually do not invade before harvest (Christensen and Kaufmann, 1969; Christensen, 1971) but they may be found on the seed in very low percentages often below 1 per cent, nevertheless providing for the presence of inoculum of storage fungi (Tuite, 1959, 1961; Qasem and Christensen, 1958). They may be present not only as contamination but as dormant mycelium within the tissues of pericarp or seed coat (Warnock and Preece, 1971), cf. Section 10.5.

7.2.2 Categories of storage fungi

The predominant species of storage fungi belong to *Aspergillus*, Fig 7.1, of which about five are common, according to Christensen and Kaufmann (1969), and *Penicillium*. According to these authors the overwhelmingly predominant species of *Aspergillus* are *Aspergillus restrictus* and members of the *A. repens–A. amstelodami–A. ruber* group. In addition to these, Welling (1974) found *Penicillium cyclopium* to be dominant in Danish stored barley grain. The specific composition of the fungal flora in store is highly dependent on the water content of the seed, and very small variations in the moisture percentage may change the flora substantially, qualitatively as well as quantitatively. Christensen and López (1963) established that the following low limits of moisture content, based on per cent wet weight, permit invasion of seeds when starchy cereal seeds such as wheat, barley and maize are kept at 20–25°C: *Aspergillus halophilicus* and *A. restrictus* 13.2 per cent. *A. candidus* 15.0 per cent, *A. ochraceus* 15.2 per cent, and *A. flavus* 18.0 per cent. The colonies of *A. restrictus* from wheat, kept for 100 days at 20–25°C, were at moisture content of 14.4 per cent numbered 750,000, at 14.6 per cent 2,710,000 and at 14.9 per cent 4,910,000; that is, an increase of 0.2 per cent in the moisture content may lead to a five-fold increase. Five phases in the developing microflora in stored seeds are recognised (Clarke, Niles and Hill, 1967; Clarke *et al.*, 1969). In the first phase field fungi are predominant, such as *Alternaria, Cladosporium* and *Fusarium*; these decrease during the second phase and are replaced in the third by yeast, sometimes accompanied by *Penicillium roqueforti*. In the fourth phase typical storage fungi occur, and by spontaneous heating of the grain the growth of the thermotolerant and thermophilic organisms of the fifth phase is promoted, such as *Absidia* spp. and *Mucor pusillus* at about 30–35°C, *Aspergillus fumigatus* at about 40°C, and *Humicola lanuginosa* and *Micropolyspora faeni* at 50–60°C (Mulinge and Chesters, 1970; Lacey, 1971).

Fig. 7.1. Species of *Aspergillus* grown from maize, rice, wheat and barley. (Courtesy of Clyde M. Christensen)

7.2.3 Other organisms

In addition to storage fungi producing damage, a range of other organisms may develop during storage of seeds, accelerating the process of deterioration. Insects and mites, especially when they breed, increase the moisture content of grain, raise the temperature and carry spores around. After three months of storage, grain at 12.1 per cent moisture, stored at 75 per cent relative humidity, had a moisture content of 14.6–14.8 per cent when not infested with the granary weevil, *Sityphilus granarius*, but when the weevils were present the moisture content arose to a level ranging from 17.6 to 23.0 per cent, increasing with the number of weevils (Agrawal, Christensen and Hodson, 1957). Abundance of grain-infesting mites may produce similar effects (Griffiths, Hodson and Christensen, 1959).

7.3 PREDISPOSING CONDITIONS

Where grain is harvested under humid conditions, as is common in Scandinavia, the moisture content is often 25 per cent or more, and rapid drying is required. It may be difficult to obtain this because the capacity of drying plants is limited and often grain is kept for some time at a too high moisture content, resulting in spontaneous heating and initial invasion of storage fungi (Kaufmann, 1959). This means that even after subsequent drying such grain may have deteriorated when it arrives at the storage site, although damage may not be visible in ordinary inspection. Once storage fungi are established they will continue to develop even under storage conditions that do not permit invasion of sound grain by fungi.

Mechanical damage in seed, cracks, breaks or scratches in the pericarp or seed coat, are conditions that substantially favour invasion of storage fungi. Hand-shelled maize or hand-threshed wheat is about three times as resistant to damage by storage fungi as the average lot of U.S.A. No. 2 grade grain (Qasem and Christensen, 1960). Welling (1968) compared four degrees of threshing injury in barley grain stored under conditions favourable for development of storage fungi and found increasing mycelial growth and damage with increasing degree of injury.

7.4 CONDITIONS DURING STORAGE IN RELATION TO DEVELOPMENT OF DAMAGE

7.4.1 Humidity

Moisture content and temperature are the primary factors determining development of storage fungi in seed. Storage fungi grow at moisture contents in equilibrium with relative humidities ranging from 65–70 to 85–90 per cent. Examples of minimum relative humidity that permits growth of some prevailing storage fungi are: 65 per cent, *Aspergillus halophilicus*; 70–73 per cent, *A. restrictus*, *A. repens*; 80 per cent, *A. candidus*, *A. ochraceus*; 85 per cent, *A. flavus*; 85–95 per cent, various species of *Penicillium* (Christensen, 1973), cf. Section 20.1.

There may be considerable variation of the moisture level within a bulk. Any minor part of the bulk may act as a focus of infection if the moisture content there is high enough to permit invasion of storage fungi. The 'average moisture content' of a seed bulk makes little sense. It is the moisture content in *different* places of a bulk that indicates potential of local infection — only the *range* provides information about where in the bulk critical levels of moisture content may lead to development of fungi and deterioration (Christensen and Kaufmann, 1969): 'It is of small comfort and little use . . . to know that the average moisture content in a given bin of wheat is 13.0 per cent if some of the grain has a moisture content of 16 per cent and is heating and musty, and other portions of the grain have a moisture content of 10.0 per cent'. The problem of unequal distribution of moisture in the grain mass may lurk in bins where no forced aeration system is available.

The moisture requirements of storage fungi in cereal seeds varies within a range from 13 to 18 per cent, but in soybean the limits are 1.5 per cent lower, and in flax seed even 1 per cent lower. Production of perithecia of different species of *Aspergillus* on cereal grains is a safe indication of the moisture level (Christensen and Kaufmann, 1965). Minimum moisture content for production of perithecia of *A. amstelodami*, *A. chevalieri*, *A. repens* and *A. ruber* is 14–15.1 per cent; above 17 per cent perithecia are seldom produced due to predominance of other fungi.

In recent literature a different term for moisture level is preferred to 'moisture content' as criterion for storability of seed — *water activity*. Seed stored in a closed room produces a vapour pressure in equilibrium with the moisture content of the seed (Section 20.1). This pressure is determined by the water activity a_w, of material (Scott, 1957). The relationship is expressed in the equation

$$a_w = \frac{\mathrm{RH}}{100}$$

where a_w is water activity, RH relative humidity of the atmosphere. The definition of a_w is

$$a_w = \frac{p}{p_0}$$

where p is the vapour pressure of the material at a fixed temperature, p_0 the vapour pressure of pure water at the same temperature.

Development of storage fungi depends on the water activity of the product rather than on its relative moisture content. Soybean has a water activity different from wheat or barley which is reflected by the fact that in soybean mould will grow at a moisture content 1.5 per cent lower than in cereal grain. The water activity of healthy undamaged grain of barley, oats, rye and wheat is approximately the same; there are only minor differences unimportant in practice (Sijbring, 1963; Frederiksen and Kjeldsen, 1970*b*). However, mechanical injury and infection by fungi influence the water activity. Frederiksen and Kjeldsen (1970*b*) found that storage fungi increase the water activity in barley while they reduce it in rye.

7.4.2 Temperature

The cardinal temperatures for the growth of most storage fungi are 0–5°C, 30–33°C and 50–55°C. Some species of *Aspergillus* have a high optimum temperature in moist grains — *A. flavus* 35°C and *A. candidus* 35–40°C. At 12–15°C most storage fungi grow very slowly in cereals with moisture content of 15–16 per cent and at 5–8°C they almost cease growing (Christensen and Kaufmann, 1965; Christensen, 1974). At temperatures of 5–10°C sound wheat and maize with a moisture content of 15–16 per cent or even higher did not deteriorate when stored for a year (Papavizas and Christensen, 1958; Qasem and Christensen, 1958); and maize kept for two years at 5°C still germinated 100 per cent and was entirely free of storage fungi (Qasem and Christensen, 1958). It appears from this that low temperatures sometimes can counteract high moisture content in preserving high quality in stored grains. It must be borne in mind, however, that if grains are already invaded even moderately by storage fungi, these can still grow at very low temperatures, even down to freezing point or below; this holds true for some species of *Penicillium* common in seed, provided the relative humidity is high (Christensen, 1973).

Temperatures may be reduced in the grain mass by aeration using relatively low rates of airflow, in temperate regions often enough to maintain a uniform temperature, 8–10°C, throughout the mass (Burges and Burrell, 1964; Kaufmann, 1964*b*).

If grain at a very low temperature, a few degrees C, is taken out of store when the temperature and humidity of the outside air are just moderately high, condensation will occur and the moisture content of the grain will increase rapidly (Christensen and Kaufmann, 1968). For this reason it is usually not desirable to maintain a temperature *below* 8–10°C.

Most of the field fungi are sensitive to high temperatures and usually disappear under such conditions (Lutey and Christensen, 1963). *Alternaria tenuis* and *Rhizopus stolonifer* may, however, develop at temperatures above 40°C (Hellberg and Kolk, 1972).

7.5 HARMFUL EFFECTS

7.5.1 Loss of germination power

One of several causes of decrease in germination percentage of seeds during storage is invasion of the embryos by storage fungi. In storage experiments, by comparing seed completely free of storage fungi with seeds in which such fungi were present, peas kept at 85 per cent relative humidity and 30°C lost germination capacity within six months when invaded by *Aspergillus* spp., *A. flavus*, *A. candidus* and *A. ruber*, whereas uninfected seed maintained a germination percentage of 95 (Fields and King, 1962). Maize, wheat and sorghum retained a germination percentage of 90–95 throughout the experimental period at a moisture content favourable to growth of storage fungi, whereas infected samples lost their germinability within a few weeks or months (Christensen and López, 1963); López established by similar experiments that *Aspergillus flavus* causes mortality in groundnut seed (Christensen, 1973).

7.5.2 Seed discoloration and decay

Storage fungi may cause discoloration of the embryo or of the whole seed (Christensen and Kaufmann, 1969). In the grain trade wheat kernels that have developed dark germs are called 'sick wheat' which, however, is virtually dead wheat — it does not germinate, and usually such kernels are heavily invaded by storage fungi (Christensen, 1955). They are also of poor feeding quality.

7.5.3 Fat acidity

Deterioration of stored grain containing fats or oils may be accompanied by increase in fatty acids, measured in terms of Fat Acidity Value, FAV. The odours and flavours of these fatty acids make partially spoiled fats rancid.

Increase in fatty acids in deteriorating seeds is caused by storage fungi. The biochemical changes involved take place in seeds at moisture level favourable to growth of moulds, and are largely proportional to the level of moisture content and extent of mould growth, as was first shown for soybean (Milner and Geddes, 1946). In seed where mould growth is inhibited, fat acidity remains essentially constant, also at high moisture content levels (Milner, Christensen and Geddes, 1947).

Rancidity is an important quality factor in cotton seed oil for human consumption. Storage fungi gain access to the interior parts of the seed through the chalazal region which provides less protection than the seed coat; the seed may be seriously affected internally without showing any external evidence (Christensen, Olafson and Geddes, 1949).

Also in deteriorating maize, storage fungi incite production of free fatty acids. Nagel and Semeniuk (1947) and Goodman and Christensen (1952) showed that *Aspergillus amstelodami*, *A. candidus*, *A. flavus* and *Penicillium solitum* are all active, but that they produce very different amounts of fatty acids. They also showed that the initial increase in fatty acids, after reaching a peak, is followed by a decrease. Goodman and Christensen demonstrated that the increase is due to fungus lipases acting upon the maize oil, and that the lipases to some extent are adaptive — some isolates produce more lipase when grown on a medium containing maize oil than when grown on a medium in which the carbohydrate source is sucrose. Fatty acid value which has been suggested as a quantitative indicator of deterioration in stored grain, may provide unreliable results for two reasons. Production of fatty acids varies with the species and possibly with the strain of the fungus and, secondly, a fungus may be relatively productive but may itself subsequently consume portions of the fatty acids produced (Christensen and Kaufmann, 1969).

7.5.4 Heating and mustiness

Any moist organic material may heat in storage — hay, manure, feed, compost heaps. Spontaneous heating of composted manure is utilised for the cultivation of mushrooms to make the medium suitable for mycelial growth of the fungus.

Spontaneous heating and associated characteristics of respiration, as in moist stored grains, is a biological process; microbes are responsible for most of the heating up to a higher temperature of 70–75°C; fungi act at lower moisture contents, bacteria at higher. The principal formula for respiration when 1 g molecular weight of hexose sugars are consumed for food is:

$$(180\text{ g})\ C_6H_{12}O_6 + 6O_2 \rightarrow 6CO_2 + (108\text{ g})\ 6H_2O + 677.2\text{ cal}$$

In analysing the factors involved in respiration and heating, the salient problem has been to determine how much carbon dioxide is produced by seed, and how much by fungi. To obtain an unambiguous answer, experimental techniques must exclude the presence of fungi from the seed to be investigated (Christensen and Kaufmann, 1969). At higher temperatures, such as 45°C, moist grain is killed within a few days and thus eliminated as a direct source of respiration, while storage fungi are vigorous and active.

Milner and Geddes (1945, 1946) demonstrated that spontaneous heating in soybean, stored at humidity levels favourable to mould growth, is directly and positively correlated with increase of respiration: this increase was associated with increase in growth of *Aspergillus glaucus* and *A. flavus*.

The impact of storage fungi on the respiration and heating of grain has been substantiated by Carter (1950) and Hummel *et al.* (1954) who worked with wheat.

In moist grain, such as grain harvested by combine under humid conditions and awaiting further processing including drying, heating may occur within hours or a few days (Fig. 7.2). In oats, with a moisture content of 29 per cent, the temperature of grain at 30 cm from the surface went up to 40°C after 54 h under experimental conditions while oxygen fell to 1 per cent and carbon dioxide increased to 30 per cent (cf. Fig. 7.3). This heating was followed by rapid development of fungi including *Aspergillus fumigatus* and *Rhizopus* spp. (Hellberg and Kolk, 1972).

Heating caused by insects is more difficult to control without fumigation. It begins when adult insects congregate and breed rapidly in pockets in the grain bulk, the growing population produces heat and moisture, moulds grow quickly and crusts are formed; in Canada the rusty grain beetle, *Cryptolestes ferrugineus*, the saw-toothed grain beetle, *Oryzaephilus surinamensis*, and the red flour beetle, *Tribolium castaneum*, are the worst causes of heating; red flour beetles can increase 70-fold in a month (Sinha, 1971).

In a grain mass a rise in temperature may develop locally. Grain is a good insulator and provides excellent conditions for development of 'hot spots'; these may approach or reach the final stages of spoilage while most other parts of the bulk may not be affected. Various devices such as temperature detection cables assist in measuring temperature at different places within the grain bulk. A vacuum probe for sampling any portion of grain in any type of bin is available (Christensen and Kaufmann, 1965, 1969).

7.5.5 Toxin production

Many fungi, both field fungi and storage fungi, produce metabolites that are poisonous, sometimes fatal, to man and animals. Production of these

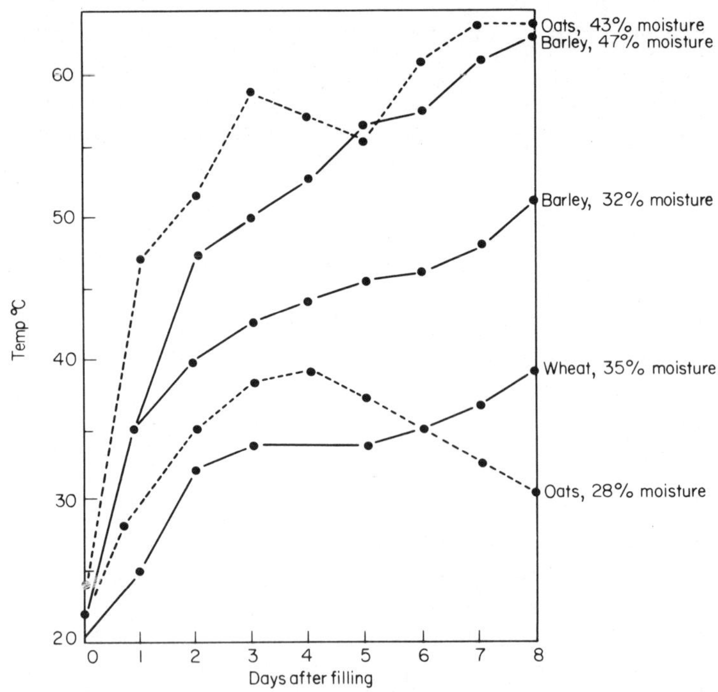

Fig. 7.2. Temperature recorded at spontaneous heating in moist grains of barley, oats and wheat. (Hellberg and Kolk, 1972)

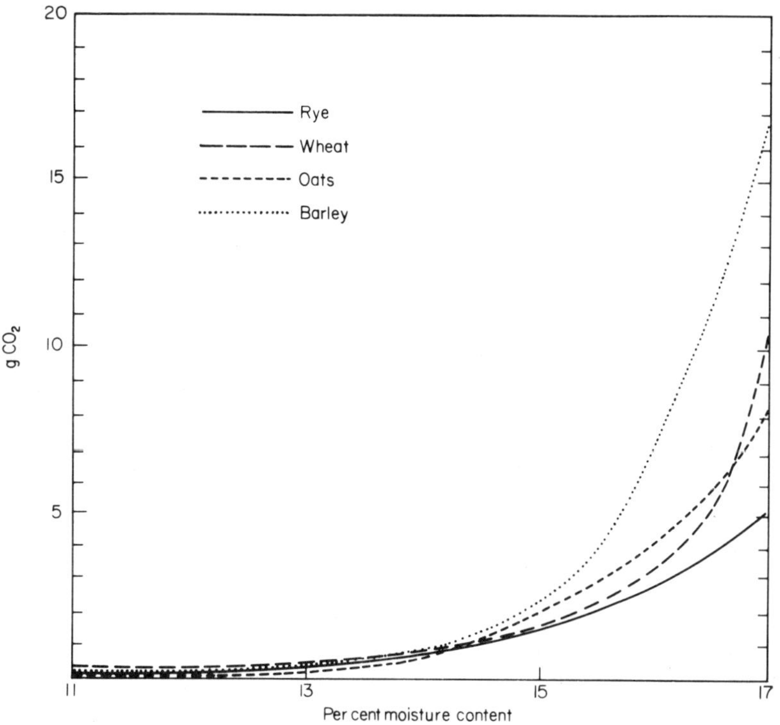

Fig. 7.3. Respiration in grains of barley, oats, rye and wheat at 37.8°C, kept at different levels of moisture content and measured in g CO_2 per 100 kg dry matter per 24 h. (Bailey, from Frederiksen and Kjeldsen, 1970*a*)

mycotoxins depends on species or strain of the fungus and on the ecological conditions for its development, in particular food source, temperature and humidity. Mycotoxins and mycotoxicoses are dealt with in Chapter 8.

Important producers of mycotoxins among storage fungi in different kinds of grain are *Aspergillus flavus* (aflatoxin complex), *A. candidus*, *A. fumigatus*, *A. ochraceus* (ochratoxin complex), *Penicillium islandicum* (islanditoxin), *P. citrinum* (citrinin), *P. rubrum* (rubratoxin) and *P. viridicatum* (hepatotoxin, nephrotoxin), all sharing responsibility for poisonous 'mouldy grain'. It must be noted, however, that grain can be highly toxic without appearing at all 'mouldy'. Some of the conditions leading to toxin production by storage fungi will be briefly discussed, exemplified by *A. flavus*.

Groundnut is the classical medium for production of aflatoxin. The first notorious example was an incidence in England in 1960. 100,000 turkey poults died from feed in which one of the constituents was groundnut invaded by *A. flavus* (Asplin and Carnaghan, 1961). The fungus may produce the toxin on different grains in varying quantities but most strains from tropical and subtropical regions produce aflatoxin (Diener and Davis, 1969). Aflatoxin is produced in the highest amounts in groundnut, wheat and rice but may also be produced in cotton seed, soybeans and maize. In Denmark isolates from various feed samples only produced small amounts in barley and rye, and one from groundnut, known to be productive, produced only low concentrations when grown on barley and rye (Krogh *et al.*, 1966). No Danish isolate of *A. flavus* has been found to produce aflatoxin (Krogh and Hald, 1969).

The minimum requirement for growth of *A. flavus* and for its production of aflatoxin is about 85 per cent relative humidity. Mycelial growth can take place over a wide temperature range from 6–8 to 44–46°C, although aflatoxin is produced only between 13 and 41°C with the optimum about 28°C (Diener and Davis, 1969).

Microbial interaction may influence aflatoxin production. When *Aspergillus niger* or *Rhizoctonia solani* is present, *A. flavus* produces less aflatoxin, and there may be other such microbial interactions (Marth, 1973).

Attempts have been made to determine at what point in the production of groundnut *A. flavus* infection takes place, and the conditions for development of toxicity. The fungus becomes dominant, especially when shells are damaged before harvest; kernels are exposed to infections leading to production of aflatoxin. Delayed harvest and bad post-harvest drying conditions increase toxin production (McDonald, 1968). When in Nigeria the crop is left on the ground for more than a month after normal harvest there is a considerable risk of aflatoxin being present at lifting (McDonald and Harkness, 1968).

7.6 PRECAUTIONS

7.6.1 Harvesting and processing

Storability of seed depends on two primary conditions — adequate moisture content of the seed and intact condition of the pericarp or seed coat. Techniques of weathering of the crop and prompt drying after harvest, techniques of threshing, transportation and safe movement of the bulk (Sections 2.9 and 19.2), all these techniques may have reached a fairly high standard in many technically advanced countries but could be greatly improved in other vast areas of the world. Most farmers in the developing world produce grain in small farms, and millions of them harvest their crops under continuous hot and rainy conditions. The salient point is not so much to improve procedures in technically advanced countries but rather to solve technical problems in countries where post-harvest losses permanently run up into 10, perhaps 20 or 30 per cent, and where in some places mouldy grain is almost a standard for human consumption.

7.6.2 Testing for storability before storage

Wherever there is risk of poor storage it is just as important to predict storability of the crop as it is to forecast diseases in the field. Standard procedures are available for testing storage quality of seeds. Grain bulks of poor storability may still have good market value for immediate use or processing.

Test for moisture content. The moisture level of seed is crucial to its keeping quality, and proper sampling is important. Moisture content may vary greatly within a bulk; a range of 1 to 2 per cent at arrival at storage site is common in grain of cereals, and therefore the highest moisture content in any portion of a given lot must be known. Mixing grain of different moisture content to find an average or to meet a given grade is very risky. The standard of moisture content is based on wet weight. Moisture content is usually determined by means of electric moisture meters; accuracy of the meters may be checked by oven drying (Christensen and Kaufmann, 1969).

Test for injuries. As injuries in pericarp or seed coat provide sites for invasion by storage fungi, it is important to examine it for such injuries before storage. Breaks, cracks and especially very slight mechanical damage can be detected by staining, e.g. in 0.1 per cent aqueous solution of fast green for maize, or in 0.4 per cent aqueous indigo-carmine solution for barley.

Test for contamination. Tests to detect the number and kinds of storage fungi present indicate to what extent invasion may already have taken place; this is essential in any rating for storability. Several procedures may be followed. Pretreated seeds may be plated on an agar medium with high osmotic pressure which is selective for storage fungi, e.g. containing 18 per cent NaCl as used by Christensen and Qasem (1962) to detect *Aspergillus restrictus*. Christensen (1973) recommends T-6 agar as satisfactory for detection of storage fungi; its composition is Difco tomato juice agar, 25 g, Difco powdered agar, 15 g, NaCl, 60 g; distilled water, 900 ml.

A provisional evaluation can be made by direct microscopic examination of seeds for mycelium on the inside of the hulls of barley (Tuite and Christensen, 1955), below the pericarp or on the surface of the embryo of wheat (Hyde, 1950; Christensen, 1955).

7.6.3 Testing for initial deterioration during storage

The behaviour of a seed bulk can be followed in detail by proper checking of the temperature in different places of the bulk, and by drawing samples from different portions of the stored grain at intervals for subsequent individual testing for moisture content and for storage fungi, as well as for discoloration of embryos.

Temperature fluctuations in stored seed may be detected by means of thermocouple cables suspended in the bin. By proper sampling the size and condition in a detected 'hot spot' of the grain can be determined for further action. A simple procedure is to insert a metal pipe deep in the grain bulk and withdraw it after a few minutes. If it is warm to the back of the hand, the grain is heating (Sinha, 1971). Checking grains for heating should be done regularly, at least once every two weeks.

Procedures as mentioned in Subsection 7.6.2 are used for testing samples for moisture content and incidence of fungi.

Test for embryo discoloration. The procedure is simple: removal of the bran layer or pericarp, and examination under stereoscopic microscope or by use of a good hand lens for dark brown discoloration of the embryo. The damaged tissues are usually heavily loaded with storage fungi.

Obviously, stored seed destined for sowing should be tested regularly for germinability; this test is also a good indicator of the general condition of grain for direct consumption or processing.

Test for fat acidity value. Deterioration of stored grain is accompanied by increase in fatty acids (Subsection 7.5.3), and the fat acidity value, FAV, has been suggested as a measure of deterioration. However, as already pointed out, FAV may vary with the species and perhaps even with the strain of storage fungi involved, and fungi may metabolise fatty acids subsisting on them as source of carbon; therefore a peak value for fat acidity may be followed by a decrease. Because of these substantial sources of variation, figures may be misleading as indicators of the condition of a grain lot.

7.6.4 Establishment of proper storage conditions

7.6.4.1 STORAGE SITE

In technically advanced countries special storehouses and grain silos are used for storing grain or seed, but in vast regions of the world more primitive shelters prevail — the crop is kept in the farmhouse or piled under a light shelter in the field. In the most densely populated regions of the world the major part of the world's grain for human consumption is stored under just such primitive conditions, and this fact must not be disregarded.

The following examples of simple storage arrangements are provided by participants of the First and Second Regional Workshops on Seed Pathology for Developing Countries, held in the Philippines 1969 and Ghana 1970, respectively (Neergaard, 1970*d*, 1971).

Storage in the field. This system is practised in many regions of Africa. In Ethiopia sorghum is stored in pits in the ground on a dry place, the walls plastered with animal manure, and the pit covered by dry leaves which are covered by soil. If water comes in, the grain may become mouldy. Also in Ghana pits are commonly used, or a simple platform on which maize is stacked in cobs. In East Africa (Kenya, Tanzania, Uganda) grain and seeds are stored in cribs above the ground on raised platforms, wooden walls may be provided but are never plastered by mud. In Afghanistan, which has a dry climate, underground storage is often used.

Farmhouse storage. In countries with a humid climate, such as Indonesia with a relative humidity of 90–95 per cent, in the rainy season 100 per cent, the traditional system of storage in small farms is to put the grain on shelves in the kitchen above the oven. In South Vietnam grain is stored in cylindrical bamboo frames which are provided with a layer of palm leaves around the sides and on the bottom; moreover a layer of rice hulls is placed on the bottom, and another layer on the top of the grain bulk. The container is placed in the kitchen. In Ghana maize intended for planting is stored in the farmhouse above the fireplace — the smoke presumably disinfects the seed and controls weevils, and the grain does not get mouldy.

Storage huts. In many countries small simple houses are built for storage, and there is a considerable variety of structures, according to local conditions and traditions. In some regions of Ghana small wooden houses are in use; they are placed on a platform made of wood or bamboo or both, and the roof of palm leaves, or banana or plantain leaves. These houses are used for grains, mainly maize, the cobs being piled up. Similar houses are used for unthreshed rice and for unshelled groundnut; in these the roof is thatched with grass to make the house cool [Fig. 7.4(1)]. In Nigeria there are different types. Maize is kept by farmers in a house arranged as shown in Fig. 7.4(2), which is considered by farmers to be the best. Another arrangement is shown in Fig. 7.4(3). Also, in Upper Volta various types of storehuts are used. In the northern-central part the 'Pilougou' type is predominant [Fig. 7.4(4)]; it is built of straw in the form of a reversed cone placed on a stone block, is about 1.7 m high, and is used for millet and sorghum. In the central and southern parts of the country a type called 'Boré' [Fig. 7.4(5) and 7.4(6)] is used. The house is cylindrical, the walls built up with mud, the conical roof thatched with straw; the roof is removed during the dry season and replaced just before the winter season. 'Dian' [Figs 7.4(7) and 7.4(8)] is commonly used in the southern part. The walls are made with mud, and a bottle-shaped container placed in the house on large stones on a wooden grating; the corners of the house are supported by wooden pillars. The container is filled

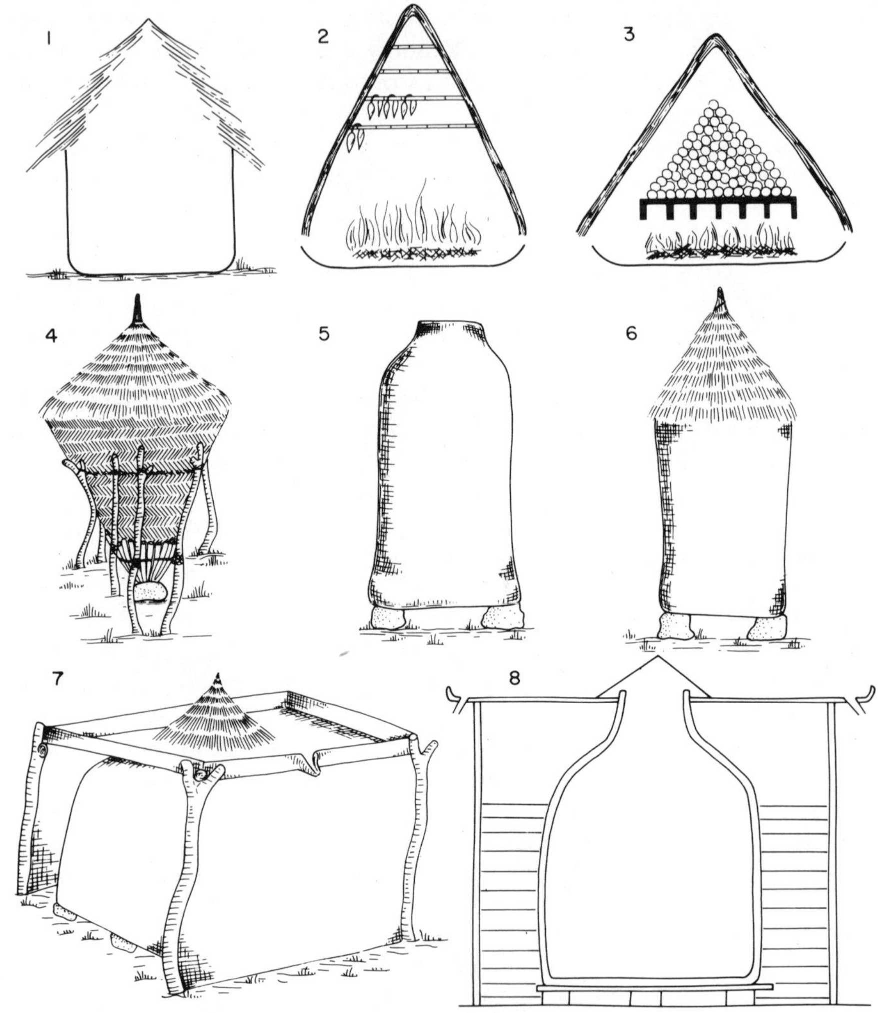

Fig. 7.4. Storehuts. (1) Ordinary type for storing grain and groundnut in Ghana. (2) Storehut used for maize in Nigeria; maize cobs are hanging in pairs on cross bars; fire is made below, to dry. (3) Storehut in Nigeria, thatched by grass; maize is piled up in a pyramid on a platform made by bamboo; below a fireplace in a pit, to improve drying. (4) Storehut, type 'Pilougou', used in the central part of Upper Volta. (5) and (6) Storehut, type, 'Boré', used in the southern and central parts of Upper Volta; the roof removed during the dry season. (7) Storehut, type 'Dian', used in the southern part of Upper Volta. (8) Cross-section shows roof and, around this, floor for drying, at left and right, drain; inside, bottle-shaped storage room around which supporting wood grating; below, stones. (Based on information from participants of the Second Regional Workshop on Seed Pathology for Developing Countries, cf. Neergaard, 1971; Figs (4)–(8), courtesy of Mr. Ouattara Soma)

from the top and emptied through a lateral opening. There are other different types in the country.

Valuable information on airtight storage of dry grain in hot climates has been given by Hyde *et al.* (1973).

7.6.4.2 REGULATION OF STORAGE CONDITIONS

Irrespective of stage of technology a clean storage site is fundamental. Any empty bin, storehouse, store hut, store pit, etc., must be thoroughly cleaned before new grain is stored. Grain debris, droppings from birds and mice, etc., must be removed, and roofs and walls inspected, repaired and made weatherproof. Contact insecticides may be used in empty granaries, such as malathion and pyrethrins-piperonyl butoxide. A dosage of 1.5 g/m^2 has been suggested as workable (Watters, 1974).

Essentially, adequate regulation of storage conditions is to prevent infestation of insects and to keep down humidity and temperature on a safe level. This means, in practice, protection against infestation and provisions for proper drying and aeration.

During recent years forced aeration systems for stored grain have been taken in use. By this precaution both purposes of regulating storage conditions are complied with, reduction of temperature, which in storehouses in Europe and North America may be kept down to a desirable level of 8–10°C, *and* reduction of any rising moisture level by preventing formation of stagnant humid air in the interstitial spaces in the grain mass (Holman, 1957; Johnson, 1957; Delong, 1958; Kaufmann, 1964*a*, 1964*b*; Burges and Burrell, 1964; Bailey, 1967).

Under primitive conditions of agriculture, possibilities of natural ventilation must be considered. As indicated by some of the structures adopted in developing countries, a natural possibility is to pile material for storage on some kind of framework-support through which more or less continuous air passage provides some ventilation. Future experimentation on possibilities of such devices for small farms may prove beneficial for improvements in regions where it is difficult to keep moisture and temperature down to an acceptable level.

If grain heating occurs, the cause should be determined before control measures are applied. If moulds are involved and none of the principal pest insects is present in damp grain pockets, there are probably only few pockets and these should be removed and destroyed. If insects are involved, the infested patches of grain should be carefully separated and burned or buried. Subsequent precautions are different according to local conditions. Often remaining grain should be transferred, and if necessary cleaned, preferably in cold weather, under conditions that are as dry as possible.

8 Impact of Seed-borne and Storage Fungi on Animal and Human Health

8.1 CONDITIONS AND CASES

Many diseases of animals, including man, are caused by seed pathogenic fungi, some by direct effect, others by consumption of foods made toxic by fungi. During the last 10–20 years the number of known mycotoxicoses recognised to be potentially important for the health of man and domestic animals has increased so that there is some improvement in the general understanding of the true impact of these diseases. Such diseases usually develop under characteristic circumstances. These are:

(1) Accumulation of the pathogen in a limited space thus providing high concentrations of the causal agent, as spores dispersed in the air during harvesting or threshing of a contaminated crop, loading of contaminated grain, etc.

(2) Development of increased sensitivity in man and animals by continuous exposure to such high concentrations of a pathogenic agent, in the same way as the well-known allergic response to protracted exposure to air-borne spores, pollen, etc.

(3) Grain, and other food stuffs, containing toxic metabolites which are not detectable by ordinary visual inspection.

(4) Inefficient cleaning of seeds containing toxic fungi in the form of sclerotia such as ergot mixed with grains.

8.2 ALLERGIC DISEASES

Diseases of the respiratory system. Bronchial asthma or 'extrinsic allergic alveolitis' in man may be due to a variety of agents, including pollen, foods and spores of fungi. Respiratory allergies may be caused by smut fungi including *Ustilago maydis*, *Tilletia caries*, *T. foetida* and *Sphacelotheca sorghi*. Wittich and Stakman (1937) described a case of a 48-year-old man who has been suffering from asthma for 22 years. He worked at a station processing grains, in a milling district in Minnesota. A few hours after starting to work, severe asthmatic paroxysms (attacks occurring at intervals) would set in; a microscopic examination of his sputum revealed spores of maize smut, wheat bunt and other smuts. The patient recovered after having been hyposensitised with smut extracts (Wittich, 1939). In further investigations Wittich found that 24 out of 105 patients suffering from respiratory allergy were sensitive to smuts, and most of these lived and worked in milling districts, grain elevators, etc. Similar case reports have been published from the U.S.A. by Harris (1939*a*, 1939*b*) and Waldbott and Ascher (1940), and from Spain Diaz, Lahoz and Recatero (1941) reported two cases of seasonal asthma due to allergy to the wheat bunt organisms. They also reported that spores of the bunt fungi were present in the faeces of persons suffering from digestive allergies.

Other respiratory diseases of agricultural workers have been named 'farmer's lung', and 'thresher's lung', diseases that may be confused. Their activity have been somewhat elucidated in recent years. 'Farmer's lung' is not caused by large fungal spores as previously believed but by spores of aerobic actinomycetes whose diameter is less than 5 μ, often less than 1 μ, present in damp hay (Pepys *et al.*, 1963).

Lacey (1971) studied the microbiology of moist barley storage in unsealed silos in England; he found that concentrations of air-borne spores when mouldy grain was unloaded increased to a maximum of 3,000 times the concentration of spores in the silo when grain was undisturbed. A quarter of the spores were *Aspergillus flavus*; other fungi present were *A. fumigatus* and *Penicillium* spp. *A. fumigatus* can cause allergic response of both the upper and lower respiratory tracts, broncho-pulmonal aspergillosis, in which the fungus is present in the bronchia without invading the tissues (Austwick, 1965). Henderson (1968) reported more than 30 cases of pulmonary allergic aspergillosis. One of these patients died with invasive aspergillosis.

8.3 MYCOTOXICOSES

Chemistry and biosynthesis of mycotoxins produced by plant pathogenic moulds have been reviewed by Mirocha and Christensen (1974*a*). In the following discussions mycotoxins have been grouped according to their ecological origin, into (1) mycotoxins produced by field fungi and (2) mycotoxins produced by storage fungi.

8.3.1 Mycotoxicoses originating from field fungi

Ergotism, in man and animals. Poisoning due to the effect of various alkaloids in the sclerotia of *Claviceps purpurea* is known from ancient time as cause of St. Anthony's Fire of the Middle Ages, then common in

Western Europe (Barger, 1931); even in recent years (1951) fatal cases have been encountered in France among people having used ergot-contaminated flour. The effects of the poison in man ranges from vomiting, lesions on the hands and feet, diarrhoea, impairment of mental function, endocrine imbalance leading to foetal disturbances, agalactia – to the ultimate stages of the disease which occur in two distinct forms, convulsions and peripheral gangrene; severe cases are fatal. Also cattle, sheep, poultry and pigs are sensitive to the poison; it has a contractile effect on the uterus and the circulative system and culminates in gangrene developing as tissue necrosis of the extremities and the end of the tail (Ainsworth and Austwick, 1973; Møller, 1965).

The ergot alkaloids have for long been utilised in medicine in childbirth (see Bové, 1970) and treatment of migraine. In recent years large-scale production of semi-synthetic alkaloids has been based on lysergic acid, gained from *Claviceps paspali*. LSD is a modified form of lysergic acid.

Toxins from *Claviceps paspali* cause damage of the nervous system of domestic animals including horse, mule, pig, sheep and cattle. The disturbance results in trembling and uncoordinated movement, paralysis and death. The condition is known as 'paspalum staggers' (Sarkisov, 1954; Spesivtseva, 1964).

Fusariotoxicoses in man and animals. Several species of *Fusarium* produce different toxins which cause a range of disturbances in man and animals, including nausea and vomiting, by *F. nivale*, haemorrhage (bleeding), bone marrow exhaustion, aleukia by *F. sporotrichioides,* diarrhoea and weight loss by *F. equiseti* (*F. scirpi*) and *F. tricinctum* (Lillehoj, Ciegler and Detroy, 1970).

The oestrogen syndromes, or hyperestrogenism, causing infertility in swine and dairy cattle, are due to the effect of zearalenone called F-2, produced by *Fusarium graminearum*, *F. moniliforme*, *F. oxysporum* and *F. tricinctum* (Mirocha, Christensen and Nelson, 1971; Mirocha and Christensen, 1974*a*). The disturbances are in the form of enlarged uteri and a swollen oedematous vulva in females, shrunken testes in young males, enlarged mammary glands of both sexes, general atrophy of the ovaries and infertility in animals; pig and dairy cattle are affected (Mirocha, Christensen and Nelson, 1967; Mirocha *et al.*, 1968), see Fig. 8.1. Cases of swollen oedematous vulva have been recorded in Danish livestocks of pigs associated with feeding by mouldy grain (Krogh and Hasselager, 1970).

Potentially very dangerous to man is fusariogenin from *Fusarium sporotrichioides*. The toxin is produced in various cereals, in particular millets, and maximum production takes place at 0–5°C. Human beings are very susceptible to this toxin but also horses, cattle and pigs are affected. The disease was detected in Russia under extreme circumstances during

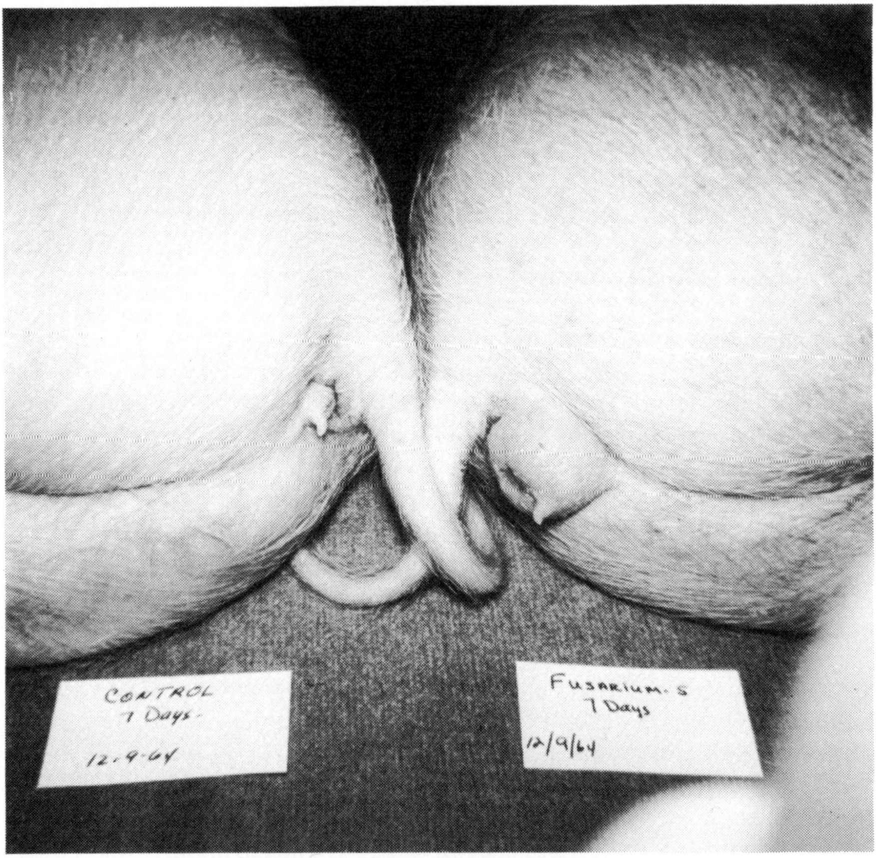

Fig. 8.1. Hyperestrogenism. Left, normal vulva of a control gilt and, right, swollen oedematous vulva of a young gilt produced experimentally by feeding corn invaded by a pure culture of *Fusarium graminearum*. Identical results are obtained when pure F-2 administered to the animals. (Courtesy of C. Mirocha)

World War II. Farmers were prevented from harvesting their crops of grain millet during the autumn of 1943. Because of acute shortage of food, millet was harvested the following spring, after the crop had been covered by snow through winter. The grain was ground up and the flour used for bread. After the foods were consumed, many people became ill and thousands died from a disease called *alimentary toxic aleukia* (ATA) or septic angina. Necrotic ulcers develop in mouth and pharynx, but the disease is characterised by haemorrhages (bleeding) of the skin and mucus membranes, loss of white blood cells, fever and bone marrow exhaustion (Sarkisov, 1954; Spesivtseva, 1964; Bilai, 1965).

Mirocha and Christensen (1974*a*) express the view that toxicity of natural products from *Fusarium* is of greater incidence and perhaps of more importance than aflatoxin.

Pink rot dermatitis in man may be incited by a mycotoxin, containing psoralens, produced by *Sclerotinia sclerotiorum* causing pink rot of celery. In the U.S.A. workers who harvest celery often suffer from blisters on skin that contacts plants infected with the pathogen, while uninfected plants cause no trouble. The toxin contains a photoreactive compound (Scheel *et al.*, 1963). The severe dermatitis develops after exposure of the workers to sunlight or ultraviolet light (Kraybill and Shapiro, 1969). The very wide host range of this plant pathogen suggests that it may be involved in other allergic disorders.

Lupinosis in sheep, cattle and horses. Lupins (*Lupinus* spp.) cause two distinct forms of poisoning in animals, lupine or alkaloid poisoning and lupinosis. The first of these is a nervous disorder caused by the alkaloids present in bitter lupins. The second is an icteric disease (icterus: jaundice) caused by a toxin named hepatotoxin or icterogen. Van Warmelo *et al.* (1970) have shown that this is a mycotoxicosis, caused by a well-known seed-borne pathogen of lupin, *Phomopsis leptostromiformis*. They found that infected pods were more toxic than infected seeds and stalks, Lupinosis has been known in Europe since 1880, originally described by Kühn (1880) who also described the lupin pathogen. Moreover lupinosis is known in Australia, New Zealand and South Africa. The toxin causes in sheep a progressive inappetence and lethargy (apathy) when abnormalities in the blood become pronounced; these appear as an increase in the serum glutamic oxalacetic transminase and total bilirubin levels in the blood, and a severe icterus develops, and immediately before the death the blood sugar levels are higher than normal. Atrophy of the liver with fatty degeneration and cirrhosis (fibrosis with hardening) are constant features. The toxin is different from aflatoxin which produces similar symptoms in animals, including generalised icterus and degeneration of liver.

8.3.2 Mycotoxicoses originating from storage fungi

Aflatoxicosis, in man and animals, is caused by a series of toxins, including the most commonly encountered aflatoxin B_1, B_2, G_1 and G_2, which are produced by *Aspergillus flavus* (from which the name is derived), *A. parasiticus* and other species of the *A. flavus* group. Aflatoxicosis is known to occur in a considerable range of animals, including poultry, especially young turkeys and ducks; cattle, pig, rabbit, dog and monkey; sheep seem to be almost resistant. Applications of aflatoxin to human cell cultures indicate that the toxin damages human cells of liver and other human tissues and that it can initiate development of malignant cell forms in culture (Krogh, 1969*a*). The first recorded incident of animal poisoning that obtained general publicity is known as 'turkey X disease' (Blount, 1961) and occurred in England in 1960 when aflatoxicosis caused heavy losses of turkey poults and other poultry, and the circumstances pointed clearly to the feed as the source of the trouble. The disease developed inappetence, poor growth rate, subcutaneous haemorrhage of the legs, feet and back, a staggering gait and sudden death; livers from the birds were enlarged, pale, firm (Asplin and Carnaghan, 1961; Spensley, 1963). The characteristic effect is liver damage, characterised by proliferation of the bile ducts, fibrosis and necrosis; haemorrhages occur in the intestinal tracts, and degenerative changes in the kidney. After prolonged intake of sublethal amounts malignant liver tumours develop in some animals. Aflatoxin B is regarded as the most active hepatocarcinogen known. In regions of Africa where groundnut, often highly contaminated with aflatoxin, is largely consumed as an important source of protein, the frequency of primary liver cancer is about 100 times higher than in Europe (Wogan, 1966); it has been shown that very small quantities of aflatoxin result in liver cancer in Bantu people in South Africa (Keen, 1971; Martin, Meadley and Keen, 1971). Contaminated food is a potential risk to human health, in particular mouldy grain and other foodstuffs affected by the toxigenic fungi. But also animal products may be contaminated. Ruminants affected by aflatoxin excrete in milk a toxin derivative of it, M-toxin, which is just as harmful as aflatoxin (Wogan, 1966; Schoental, 1967).

The carcinogenic nature of aflatoxin has been substantiated by human tissue culture studies. The toxicant caused suppression in DNA synthesis and mitosis and abnormal production of giant cells (Legator, 1966). The effects of aflatoxin, including acute toxicity and carcinogenic activity, have been extensively studied experimentally in laboratory animals, e.g. rats, guinea pigs, ducklings, dogs, mice (see review by Butler, 1969).

Nephrotoxicoses in animals, in particular in pigs. The nephrotoxic mycotoxins are produced by different species of *Aspergillus* and *Penicillium*, the ochratoxin complex by *A. ochraceus*, citrinin by *P. viridicatum*, *P. citrinum*, *P. expansum* and other species of *Penicillium*; citrinin is found in nature in 'yellowed-rice' invaded by fungi of this group. *P. viridicatum* appears to be the main producer of citrinin in cereals (Krogh and Hasselager, 1968; Krogh, Hasselager and Friis, 1970; Scott *et al.*, 1972). Nephrotoxicosis in pigs, associated with cereal feed naturally contaminated with citrinin and ochratoxin A, has been observed in Denmark (Krogh, Hald and Pedersen, 1973). Mould nephrosis or porcine nephropathy in swine associated with feeding of mouldy grain has been known in Denmark for almost half a century (Larsen, 1928). The effects of the toxins are depression of growth rate and increased water consumption and diaresis. Chronic damage of the kidney in the form of degeneration of the tubular cells, formation of connective tissue, and cysts; the kidneys assume a pale colour and a wood-hard consistency. In humid years cases of mould nephroses in Denmark may reach 6–7 per cent of the pigs at slaughter; in 1969 this was equal to 12 million Danish kroner (Krogh, 1969*b*). A survey in districts in Denmark having high incidence rates revealed that 58 per cent of the cereal samples contained ochratoxin A, 9 per cent contained citrinin (Krogh, Hald and Pedersen, 1973). Krogh *et al.* (1974) found that kidney, liver and muscular tissue of affected bacon pigs at slaughter contained small amounts (67 ppb) of ochratoxin A residues, and as most of the carcases would have passed the meat inspection this possibly may become a public health problem; the dosage mentioned would, however, hardly be significant in production of any known disease.

It seems not to be known whether these nephrotoxins affect other domestic animals but Still *et al.* (1971) suggested that abortion in cows and pigs may be associated with ingestion of foodstuffs containing *Aspergillus ochraceus*.

There are other liver–kidney toxins. The rubratoxins from *Penicillium rubrum* cause liver damage and disseminated haemorrhage in horse, cattle, pig, dog and chicken (Burnside *et al.*, 1957; Forgacs *et al.*, 1958, 1962); luteoskyrin from *Penicillium islandicum* also causes liver damage similar to that caused by aflatoxin and with the ultimate development of malignant liver tumours in various small laboratory animals but not so far encountered in domestic animals (Lillehoj, Ciegler and Detroy, 1970).

Neurotoxicosis. Citreoviridin, a metabolic product of *Penicillium citreo-viride*, causes damage of the central nervous system resulting in progressive paralysis and respiratory failure. Monkey, pigeon and other animals are sensitive (Uraguchi, 1950; Kinosita and Shikata, 1965). Beriberi, described as a thiamine deficiency and predominating in rice-eating countries, has been suggested rather to be a mycotoxicosis (Uraguchi, 1969); there is a symptomatic and toxicological correlation of citreoviridin and cardiac beriberi in animals (Ueno and Ueno, 1972).

8.4 MYCOSES

There are many known fungal infections in man and animals. A great increase in these diseases has been caused by the widespread use of antibacterial antibiotics and sulfonamides and other chemotherapeutic drugs, resulting in lowering of the resistance of patients. This may reflect also on the very small number of diseases in man and animals known to be caused by seed-borne fungi. Probably only aspergillosis may be regarded as important.

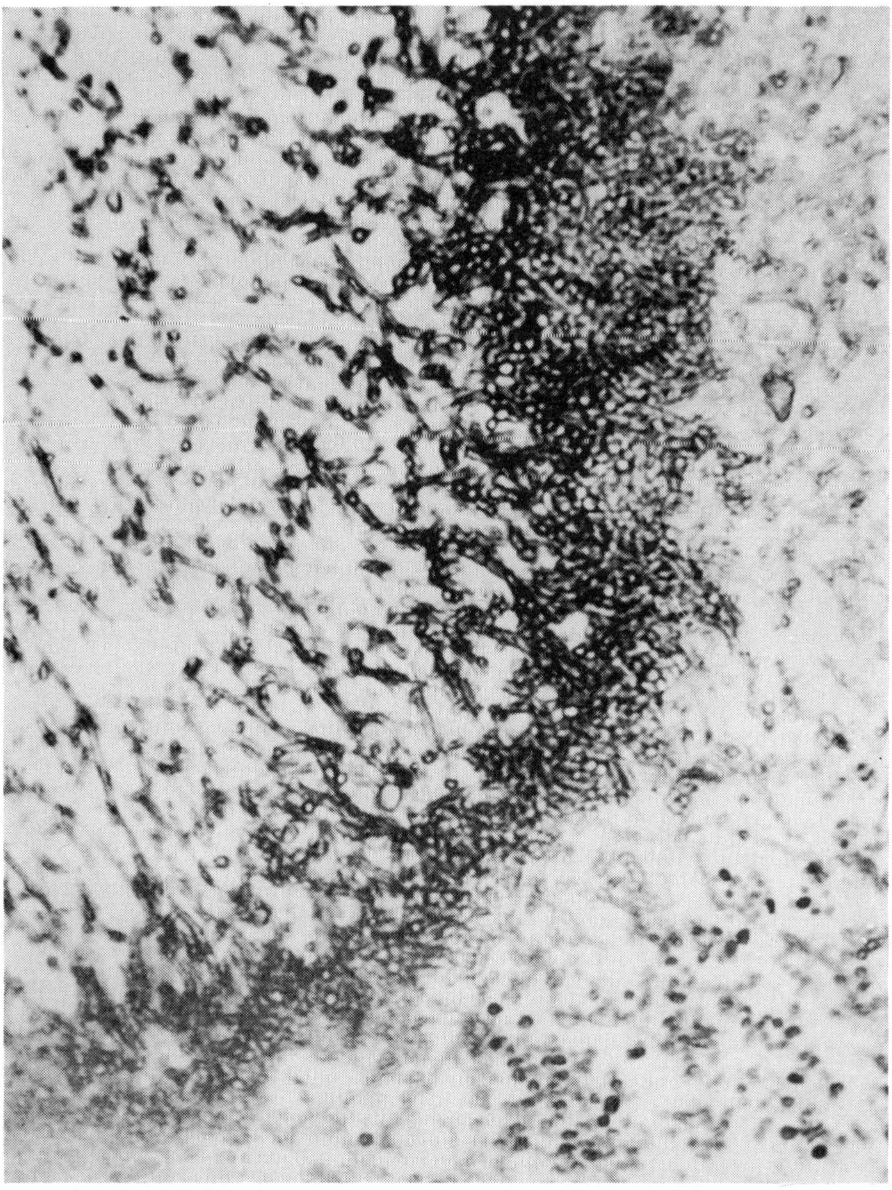

Fig. 8.2. Aspergillosis, section of the lung showing granule with hyphae. Brown-Gram stain. ×380. (After Weed, Baggenstoss and Baugher, from Conant *et al.*, 1971)

Aspergillosis. Any disease in man or animals caused by species of *Aspergillus* is termed aspergillosis. The disease is especially common in birds. In man it is characterised by 'the presence of inflammatory granulomatous lesions in the skin, external ear, nasal tissues, orbit, eye, bronchi or lungs (see Fig 8.2) and occasionally in the nasopharynx, vagina, uterus, heart valves, pleural cavity, mediastinum, bones, brain meninges' (Conant *et al.*, 1971). (Mediastinum, membraneous middle septum, especially between lungs.)

The main cause of infection is *Aspergillus fumigatus*, a common storage fungus in grain. People exposed frequently to massive doses of the spores, such as agricultural workers exposed to dust at harvest or from grain being loaded, are among common patients. Adults are more susceptible than children. Fatalities are now being reported, especially on patients treated with antibacterial antibiotics (Boyd, 1961).

Conant *et al.* (1971) classify in their *Manual of Clinical Mycology* four clinical types of aspergilloses: (1) allergic type, without local lesions, (2) allergic type with lesions in the lungs, (3) saprophytic type, and (4) infectious type. The two first types have been briefly discussed in Section 8.2.

In the *saprophytic type* infection takes place in open cavities in the lung which may have been produced by tuberculosis, carcinoma (cancer in epithelial tissue), bronchiectasis (dilation of the bronchi, local or general), or similiar morbid conditions. *A niger* has been seen in such infections.

The *infectious type* may appear either localised or in a generalised or disseminated form. The first is the most common, in lungs as local lesions in which a mass of living hyphae, usually without sporulation, are established, with some sporulation in cavities. Green, Font and Zimmermann (1969) reported 10 cases of aspergillosis in the orbit (eye-socket) from which infections may extend to the brain. Primary infections of the external ear also occur as well as primary localised infections of the sinuses, conjunctiva, eyelids and intraocular structures have been described, and also infections of fingernails, which were caused by *A. flavus*, *A. nidulans* and *A. glaucus* (Conant *et al.*, 1971). Four of five cases of aspergillosis in man described by Rénon (1897) in his monograph on this disease in man and animals, were in persons who professionally fed pigeons by masticating grain, forcing it directly from the mouth into the beak of the pigeon.

Generalised or disseminated cases are rare but when they occur they are often fatal. Such infections occur predominantly in patients with low resistance due to primary chronic disease but the aspergillus infections may be primary as well, and may develop as meningitis and endocarditis (inflammation of the skin serous membrane lining the cavities of the heart). Most often *A. fumigatus* and *A. niger* are recovered from true infections (Conant *et al.*, 1971).

A disease of the pulmonary tissues of birds is caused by *A. fumigatus*. The disease is rather common among newly hatched fowl, often occurring in epidemic proportions; the lungs are abundantly infected being covered by a felt of mycelium and spores making breathing impossible. The fungus may also infect cattle and sheep causing abortion (Emmons, 1960).

In chickens and turkeys *A. fumigatus* causes a panophthalmitis (general inflammation of the eyeball), and the fungus grows luxuriantly in the vitreous body (Jubb and Kennedy, 1963).

There are examples of occasional 'infections' of man and animals by plant pathogenic fungi, including some seed-borne. *Ustilago maydis,* maize smut, has been isolated from epithelial infection of a Hungarian farm worker in an area where smut was prevalent (Preininger, 1937) and also *Schizophyllum commune* has been claimed as cause of disease on humans (Moore-Landecker, 1972). However, occasional isolations of fungi from lesions in man and animals do not necessarily mean that these fungi are the causal agents; they may be secondary invaders or common laboratory contaminants. Conant *et al.* (1971) have recorded many non-pathogenic fungi as contaminants, including the above *U. maydis* having been frequently encountered. Other of the recorded contaminants in materials from human lesions are *Alternaria* sp., *Aspergillus* sp., *Cephalosporium* sp., *Cladosporium herbarum*, *Drechslera* sp., *Fusarium* sp., *Phoma* sp., *Pleospora* sp., *Trichothecium* sp. (probably *T. roseum*) and *Verticillium* sp.

9 Types of Host in Relation to Types of Pathogen

INTRODUCTION

There are considerable differences between seeds of different kinds in regard to frequency of transmission of any one pathogen, and in the number of different pathogens so transmitted. Seed-borne pathogens and transmission are common in cereals and legumes but relatively infrequent in forestry seed and many flower seeds.

There are several obvious reasons for this. The number of reports on seed transmission of pathogens depends largely on the importance of the crop and, in consequence, the attention given to the seed. Naturally, seed-borne pathogens of cereals have usually been more fully investigated than ornamental annuals, and crops commercially vegetatively propagated have been largely neglected because true seeds of these crops usually do not enter the seed trade.

However, seed transmission of pathogens, both in terms of frequency of transmission and number of different pathogens involved, depends basically on the inherent characters of the host, as well as those of the pathogen. Some of these factors of susceptibility and virulence pertain to the special host–parasite interaction or association that leads to seed transmission.

In some plant families seed transmission of pathogens predominates, in others it is rare. In Gramineae and Leguminosae there are many seed-borne pathogens, and they are frequently transmitted. It is a salient feature of seed transmission that groups of related hosts, such as members of a plant family, transmit groups of related pathogens through seed: smut fungi in Gramineae, viruses or *Colletotrichum* species or certain groups of bacteria in Leguminosae.

The following sections outline a discussion of this phenomenon of 'groupwise' affinity between hosts and pathogens and its impact on seed transmission of disease.

A comprehensive annotated list of seed-borne diseases, arranged according to hosts, has been published by Noble and Richardson (1968).

9.1 GRAMINEAE

In a great variety of hosts belonging to Gramineae there is a considerable range of seed-borne species within the order of Ustilaginales and the genus *Claviceps*. Also a great variety of *Drechslera* species is characteristic of the seed-borne mycoflora of Gramineae. Moreover, seed gall nematodes, members of the genus *Anguina*, are common in Gramineae, having a wide host range within this plant family but apparently no host outside it.

The seed-borne smut fungi, ergot fungi and seed gall nematodes, are specific ovary parasites; they are organotropic parasites, according to Gäumann; the ovary is replaced by reproductive structures of the pathogen. Table 9.1 comprises a survey of the more important hosts and major pathogens within these groups.

All these pathogens invade the ovary before fertilisation, most of them at a very early stage in its development. Barrus (1916) showed that three days prior to emergence of the head of wheat (about Stage 10.0, according to Feeke's scale, cf. Section 11.1), stinking smut, *Tilletia foetida*, was fully established in the ovary, the pistil of the diseased flower was much larger than the healthy pistil and the ovary itself twice as long and somewhat broader than the healthy ovary [Fig. 9.1(1)]. A section of the diseased ovary revealed that while the cells of the pericarp were normal the ovule was already filled with spores, which still had a light-coloured wall and hence passed unobserved to the naked eye when the ovary was crushed. *Ustilago tritici* exercises a similar effect on the rudimentary ovary of wheat, obviously invading it at about the same stage of its development. In contrast to the effect of the bunt fungus, the development of the outer glumes as well as the lemma and palea is completely suppressed already before emergence of the ear, but also in this disease the ovary is increased in size (Fig. 9.2) and filled with the spores of the parasite. Barrus found for *Tilletia foetida* that in many spikelets four or five smut balls are formed. This is because normally unproductive florets are stimulated to produce bunt balls. The successful accomplishment of this stimulation may be attributed to the fact that the parasite invades the floret primordia before elimination of the unproductive florets has taken place. Similar cases of potentially unproductive florets having been induced to form bunt balls have been observed in bunt-infected rye (Gassner and Niemann, 1954) and in dwarf bunt-infected wheat (Fischer and Holton, 1957).

Infection of the ovary of Gramineae by air-borne inoculum of the ergot fungi, including the most common species *Claviceps purpurea*, occurs at a later stage than does the invasion of the above systemic smut fungi from the mother plant into the ovary. Infection by ergot fungi occurs at the beginning of anthesis, still, however, in the unfertilised flower (Section

Table 9.1. Some seed-borne ovary parasites of Gramineae

Host	*Ustilaginales*	*Claviceps*	*Anguina*
Agropyron spp.	Tilletia caries	C. purpurea	
	— contraversa		
	Ustilago bullata		
	— nuda		
	— turcomanica		
Agrostis spp.	Tilletia decipiens	C. purpurea	A. agrostis
	— pallida		
	Urocystis agropyri		
Avena sativa	Ustilago avenae	C. purpurea	A. tritici
	— hordei		
Bromus spp.	Tilletia caries	C. purpurea	
	— contraversa		
	— fusca		
	Ustilago bullata		
Cynodon dactylon	Ustilago cynodontis and other spp.	C. cynodontis	
Dactylis glomerata	Tilletia contraversa	C. purpurea	A. agrostis
	Ustilago sp.		
Elymus spp.	Tilletia contraversa	C. purpurea	
	— elymi		
	Urocystis agropyri		
	Ustilago bullata		
Festuca spp.	Tilletia caries	C. purpurea	A. agrostis
	— contraversa		
	— fusca		
	— texana		
	Urocystis agropyri		
	Ustilago bullata		
	— mulfordiana		
	— sphaerocarpa		
Hordeum vulgare	Tilletia contraversa	C. purpurea	
	— pancicii		
	— texana		
	Urocystis agropyri		
	Ustilago bullata		
	— hordei		
	— nigra		
	— nuda		
Lolium spp.	Tilletia caries	C. purpurea	A. sp.
	— contraversa		
	— lolii		
	Ustilago compacta		
	— nuda		
Oryza sativa	Tilletia barclayana	C. purpurea	A. sp. (?)
Panicum spp.	Neovossia corona	C. maximensis (on P. maximum)	
	Sorosporium syntherismae	C. purpurea	
	Sphacelotheca destruens		
	Tilletia ayresii		
	Tolyposporium bullatum		
	Ustilago crameri		
	— heterogena		

Table 9.1. *continued*

Host	*Ustilaginales*	*Claviceps*	*Anguina*
Paspalum spp.	Sorosporium paspali	C. paspali	
	Sphacelotheca cordobensis	C. purpurea	
	Tilletia rugispora	C. queenslandica	
Pennisetum spp.	Sphacelotheca pamparum	C. microcephala	
	Tolyposporium penicillariae	C. sp. (Sphacelia)	
Phalaris spp.	Ustilago longissima	C. phalaridis	A. agrostis
	— macrospora	C. purpurea	
Poa spp.	Tilletia caries	C. purpurea	A. agrostis
	— contraversa		
	— fusca		
	Urocystis agropyri		
	Ustilago striiformis		
	— trebouxii		
Saccharum spp.	Cintractia pulverulenta	C. purpurea	A. spermophaga
	Sorosporium indicum		
	Sphacelotheca erianthi		
	— macrospora		
	— sacchari		
	Ustilago scitaminea		
Secale cereale	Tilletia caries	C. purpurea	A. tritici
	— contraversa		
	— foetida		
	Urocystis occulta		
	Ustilago nuda		
	— vavilovi		
Setaria spp.	Sphacelotheca destruens	C. purpurea	
	— pamparum		
	Tilletia echinosperma		
	— setariicola		
	Ustilago crameri		
	— neglecta		
Sorghum vulgare and other S. spp.	Sorosporium andropogonis-sorghi	C. purpurea (C. microcephala)	
	Sphacelotheca cruenta	Sphacelia sorghi	
	— reiliana	— spp.	
	— sorghi		
	Tolyposporium ehrenbergii		
	Ustilago kenjiana		
Triticum aestivum and other T. spp.	Neovossia indica	C. purpurea	A. tritici
	Tilletia caries	C. ranunculoides	
	— contraversa		
	— foetida		
	— intermedia		
	Urocystis agropyri		
	Ustilago tritici		
Zea mays	Ustilago maydis	C. gigantea	

11.4). The germ tubes of the ascospores invade the ovary near the base of the ovule, and the mycelium spreads intercellularly in the ovary wall and, subsequently, intracellularly in the ovule (Campbell, 1958). The sclerotium further develops simultaneously with the ripening of the uninfected grains. The various species of *Claviceps* are confined to Gramineae, Cyperaceae and Juncaceae (cf. Brady, 1962).

The nematode seed galls in Gramineae are produced from undifferentiated flower buds, staminate tissue being first invaded, then the carpelled tissue. As in smuts, several galls, up to five, may replace a single kernel, perhaps due to induction of the gall formation in normally unproductive florets in a spikelet; however, no explicit reference to such a relationship known from the smut fungi seems to be available. Seed-gall forming nematodes, *Anguina*, are confined to Gramineae.

There are other specific ovary parasites of Gramineae. *Gloeotinia temulenta,* blind seed disease, infects the ovary of at least 16 genera of Gramineae, *Lolium* being particularly susceptible (cf. Noble and Richardson, 1968, and see Subsection 5.3.5.9). If the parasite invades the ovary up to about one week after fertilisation, the seed is killed. If it is infected after differentiation of the embryo but before the pericarp has dried, all tissues of the young caryopsis are colonised resulting in a moribund, a 'blind seed', looking normal but incapable of germinating, virtually functioning as a sclerotium (Section 11.5).

Biologically interesting, although economically unimportant, is another ovary parasite, *Ustilaginoidea virens*, false smut of rice, one of five species of the genus belonging to *Clavicipitaceae* (imperf. state belonging to *Hyphomycetes*) and parasitising the ovaries of Gramineae in warmer regions. The flower bud is invaded when the staminate and carpelled organs are differentiated. The ovary is destroyed while style, stigmas and anther lobes remain intact within the spore mass (Raychaudhury, 1946). Also the mature caryopsis may be infected.

Drechslera is another genus of fungi showing a remarkable affinity to Gramineae. In the *Annotated List of Seed-Borne Diseases* (Noble and Richardson, 1968) a total of 38 species of *Drechslera*, some of which are indicated under the name of their perfect state, are listed as seed-borne in Gramineae, involving about 115 specific host–pathogen combinations; to this may be added a number of unspecified combinations, Gramineae in general having been indicated as hosts for certain species of *Drechslera*, and in recent years the number of recorded seed-borne species of *Drechslera* in Gramineae has further increased considerably. In contrast, very few species of *Drechslera* attack other plant families; in the Annotated List only four such species have been registered, *D. apii* on *Apium*, *D. sesami* on *Sesamum*, *D. sorokiniana* on *Allium* (but listed for 16 specified genera of Gramineae) and *D. papaveris* on *Papaver*.

The *Drechslera* diseases in Gramineae form a fairly uniform group,

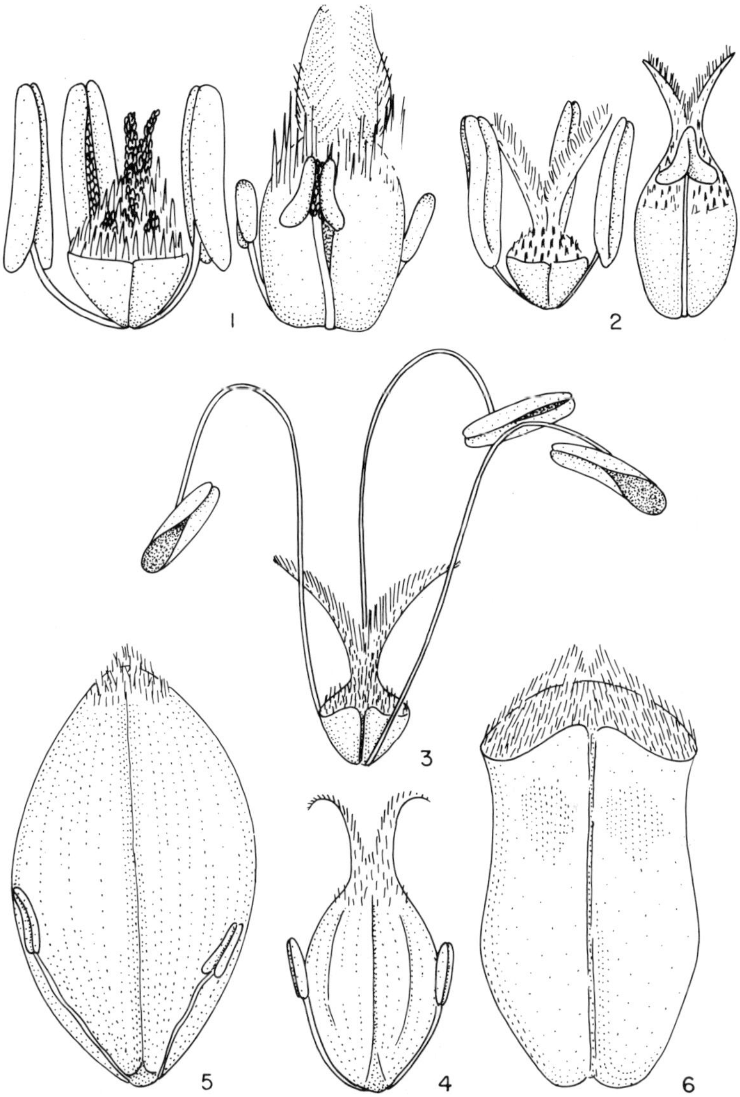

Fig. 9.1. The effect of *Tilletia foetida* on the ovary and caryopsis of wheat. (1) Flowers about three days before head should appear; at the left, from healthy floret, ventral aspect; at the right, from smutted floret, dorsal aspect. (2) Flowers just following emergence of entire head; at the left, from healthy floret, ventral aspect; at the right, from smutted floret, dorsal aspect. (3) Flower from healthy floret soon after stamens are extruded. (4) Flower from smutted floret at about the same period of development as in (3). (5) Smutted kernel and two aborted stamens about the same period of development as in (6). (6) Healthy kernel about six days after pollination. (1) ×75, (2)–(6) ×7.5. (Barrus, 1916)

pathogenesis and symptoms being similar from pathogen to pathogen and host to host in the great majority of combinations. Disregarding the notable exception of the systemic *Drechslera graminea* in barley (Section 14.3), the predominant symptoms of *Drechslera* diseases are seedling blight, various degrees of foot and root rot, lesions of varying severity in the form of oval spots, or blotches on the leaf blades and sheaths, and often elliptical lesions on the culms. The ears may be attacked causing varying degrees of damage to the kernels or seeds, and often developing

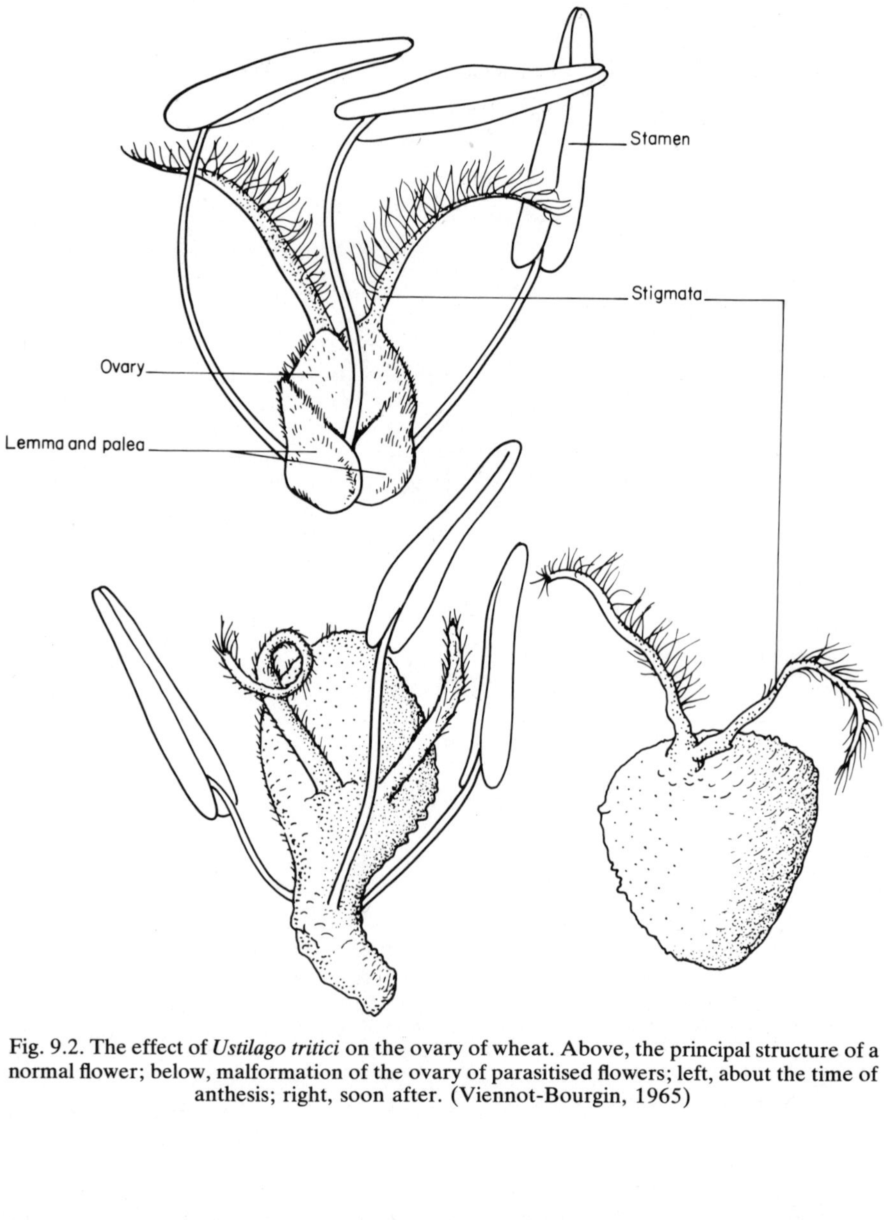

Fig. 9.2. The effect of *Ustilago tritici* on the ovary of wheat. Above, the principal structure of a normal flower; below, malformation of the ovary of parasitised flowers; left, about the time of anthesis; right, soon after. (Viennot-Bourgin, 1965)

black point symptoms in grains (Section 2.5). Long-living dormant mycelium is established in the pericarp, often contributing essentially to accomplishment of transmission.

The fact that all the above pathogens are more or less confined to one family, Gramineae, as well as the fact that the fungi show a considerable variation in comprising many species, and in the smut and ergot fungi many pathogenic races, suggests that these host–parasite relationships, so intimately associated with seed transmission, have a phylogenetically long standing.

9.2 LEGUMINOSAE

Seed-borne diseases are common and severe in Leguminosae; they are more predominant than in almost any other plant family and comparable in importance perhaps only to seed-transmitted diseases in Gramineae. In soybean, for instance, the vast majority of losses from diseases is due to seed-borne pathogens.

There are three groups of commonly occurring pathogens, each of which is biologically fairly uniform; they are viruses, bacteria and a group of fungi that produce 'anthracnose' or 'anthracnose-like' diseases, including in particular *Colletotrichum* and *Glomerella*, as well as *Ascochyta*, *Phoma* and *Phomopsis*.

Many leguminous crops are susceptible to viruses, some harbour a considerable range of them; many are seed-transmitted, and most of them embryo-borne. Table 9.2 surveys the known seed-borne viruses of Leguminosae, arranged according to Hansen's classification (cf. Section 3.3), i.e. on modes of transmission (mechanical or deep; vector transmission, aphids or nematodes being involved in the vector-borne viruses listed) and on shape and size of virus particles (Section 3.4). It appears from the survey that about half of a total of 69 host-virus combinations, namely 35 aphid-borne viruses, are involved while in 14 of the other combinations the viruses are nematode-borne.

There is no explanation as to why seed transmission is so common in this plant family. However, the capability of leguminous hosts to transmit viruses through seed varies exceedingly from species to species, even from cultivar to cultivar; and the extent to which different strains of a virus may be seed-transmitted in the same host may vary considerably (see Section 3.1; and for cultivar-induced variation of seed transmission, Subsection 3.4.2.2, bean common mosaic virus).

Seed-borne bacteria of Leguminosae are surveyed in Table 9.3. They belong to three genera, of which the three species of *Corynebacterium*, confined mainly to the vascular system, and at least five of the seven species or varieties of *Xanthomonas* produce systemic diseases, *X. alfalfae* being an exception in producing parenchymatous infections (the position of *X. vignicola* is unclear). Most *Pseudomonas* spp., on the other hand, produce parenchymatous infections, *P. solanacearum* being exceptionally a systemic pathogen.

Table 9.2. Grouping of seed-borne viruses of Leguminosae

	Virus groups, cf. Subsections of Section 3.4														
	1.1	1.2	1.3	1.4	2.1	2.2	3.1	4.2	5.1	5.2	6.1	9.1	9.2	9.4	
Host	Minglobus	Maphiglobus	Daphiglobus	Genemaglobus	Minflexus	Maphiflexus	Minchorda	Genemapachus	Minlancea	Maphilancea	Maphibrevia	Minvirus	Maphivirus	Dinvirus	*Totals*
Arachis hypogaea			1			1								1	3
Glycine max	1	1		8		1			1				1		13
Lupinus albus						1									1
Lupinus luteus		1				1									2
Medicago sativa											1				1
Melilotus alba						1									1
Phaseolus aureus		1				1									2
Phaseolus coccineus												1			1
Phaseolus mungo	1														1
Phaseolus vulgaris	1	2		2		1			1						7
Pisum sativum						3		1						1	5
Trifolium pratense				1	3	1				1					6
Vicia faba	1				1	2				1			1		6
Vigna catjang		1											1		2
Vigna cylindrica		1													1
Vigna sesquipedalis	1					1									2
Vigna sinensis	1	1		2		2							1		7
Vigna unguiculata	3	1				1	1						1	1	8
Totals	9	9	1	13	4	17	1	1	2	2	1	1	5	3	69

Most of these bacteria become well established in the seed. Undoubtedly all of the species producing systemic infections invade the developing seed through the funiculus and raphe, becoming localised mainly in the subepidermal layers of the seed coat (examples are given in Section 13.4). This site of infection in the seed is extremely favourable for transmission to the seedling. Also a substantial amount of contamination may, however, lead to successful transmission.

The fungi mentioned above, *Colletotrichum*, *Glomerella*, *Ascochyta*, *Phoma*, and *Phomopsis* or *Diaporthe*, belonging to Coelomycetes, all

Table 9.3. Seed-borne bacteria in Leguminosae, cf. Chapter 4

Hosts	*Corynebacterium*	*Pseudomonas*	*Xanthomonas*
Arachis hypogaea		solanacearum	
Glycine max	flaccumfaciens	glycinea tabaci	phaseoli var. sojense
Lablab niger			phaseoli
Lespedeza stipulacea			lespedezae
Medicago sativa	insidiosum		alfalfae
Phaseolus lunatus		syringae	
Phaseolus vulgaris	flaccumfaciens	phaseolicola syringae	phaseoli phaseoli var. fuscans
Pisum sativum		pisi	
Vigna cylindrica		syringae	
Vigna sinensis		syringae	vignicola

Table 9.4. Some seed-borne Coelomycetes of Leguminosae, cf. Chapter 5

Hosts	*Colletotrichum*	*Ascochyta*	*Phoma*	*Phomopsis*	*Diaporthe (Phomopsis)*
Cajanus cajan	gloeosporioides				
Cicer arietinum		rabiei			
Clitoria ternatea	dematium				
Crotalaria juncea	dematium				
Cyamopsis tetragonoloba	dematium			sp.	
Desmodium spp.	dematium				
Glycine max	dematium	sojicola			phaseolorum — v. caulivora — v. sojae
Lablab niger	lindemuthianum				
Lathyrus odoratus	gloeosporioides	lathyri	medicaginis — v. pinodella		
Lupinus spp.	gloeosporioides			leptostromi-formis	phaseolorum — v. sojae
Medicago sativa	trifolii		medicaginis		
Melilotus spp.		caulicola	medicaginis		
Ornithopus sativus	trifolii				
Phaseolus lunatus	lindemuthianum dematium	boltshauseri	exigua		phaseolorum — v. sojae
Phaseolus vulgaris	lindemuthianum dematium	boltshauseri	exigua		phaseolorum — v. sojae
Pisum arvense, sativum	pisi	pinodes pisi	medicaginis — v. pinodella		
Sesbania grandiflora	gloeosporioides				
Trifolium spp.	trifolii		medicaginis — v. pinodella		
Vicia faba	lindemuthianum	fabae			
Vicia spp.	villosum	pisi			
Vigna spp.			exigua		

characteristic of producing deeply penetrating lesions in stems and pods ('anthracnose', 'canker') as well as different degrees of leaf blight and foot rot, form a fairly uniform group within the diseases of Leguminosae, both in terms of general pathogenesis and in means of seed transmission. They are surveyed in Table 9.4. The seed often becomes infected through pods, and infections involves the seed coat and usually also considerable parts of the cotyledons, thus being embryo-borne. The site of infection in the fleshy cotyledons, which form a rich nutritious base for the seedling as well as for pathogens, may essentially account for the important role of seed transmission of these pathogens.

9.3 OTHER PLANT FAMILIES

In Cruciferae the *Alternaria* diseases are comparable in many respects to the *Drechslera* diseases of Gramineae. In both genera there is a range of species confined to one family of host plants, all producing parenchymatous infections affecting seedlings, stems, leaves and fruits, and the pathogens establish themselves as a tenacious dormant mycelium in the seed, *Alternaria* in the seed coat, *Drechslera* in the pericarp of their hosts; from these sites transmission of the pathogen occurs, although both genera of fungi are also largely present as long-living conidia contaminating the seed. *Alternaria brassicae, A. brassicicola* and *A. raphani* infect and are seed-transmitted in many cruciferous crops, including cabbage, radish and flower crops such as *Aubrietia*, *Cheiranthus*, *Iberis* and *Matthiola*, and may occur in any cruciferous crop or weed. *A. cheiranthi* infects *Cheiranthus cheiri* and may infect other crucifers.

Another example of affinity between groups of hosts and seed-borne pathogens is that of stonefruits (Rosaceae) and viruses. In *Prunus avium*, *P. cerasus, P. mahaleb, P. pennsylvanica* and *P. persica* Prunus necrotic ringspot virus is commonly seed-transmitted, also in *P. cerasus* and *P. mahaleb* prune dwarf virus, and in *P. avium* cherry necrotic rusty mottle virus. These viruses are also commonly pollen-borne. Among other members of Rosaceae, *Rubus idaeus*, red raspberry, transmits six viruses through seed, among which is Prunus necrotic ringspot virus; and some other species of *Rubus* even transmit a further two viruses.

10 Morphology and Anatomy of Seeds in Relation to Transmission of Pathogens

Seeds are both vehicles and suscepts of pathogens and they may carry these as contaminations or as infections. Inoculation and establishment of infection of the host depend on (1) conditions favouring exposure of the seed to inoculum, (2) susceptibility of the successive stages in the development from ovule to mature seed, and (3) the microenvironment of the developing seed provided by the mother plant, such as the enclosure formed by fruits or by foliar envelopes of bracts.

There is great variation in morphological and anatomical details between seeds, particularly their enclosing structures, details that highly pertain to the physical and physiological access of pathogens to seeds.

10.1 FRUIT STRUCTURES AND BRACT ENVELOPES AS MICROENVIRONMENT OF DEVELOPING SEEDS

Fruits may be classified according to general structure of the gynoecium, texture of the pericarp, and mode of seed dispersal. In our discussion, however, attention will be concentrated on the properties of fruits in providing the physical microenvironment of developing and maturing seeds.

The fruit wall and the cuticular covering of the fruit surface provide protection to seeds against external agencies including microorganisms. However, the microclimate in enclosures formed by fruits, and sometimes by enveloping bracts, has been little studied although it may be important to establishment of infection of seed. To be considered are dry fruits, which may be dehiscent or indehiscent, and fleshy fruits.

The cavity of young fruits of the dehiscent type, such as pods of legumes, siliquas of crucifers, and capsules of the poppy family, may act as a moist chamber providing favourable conditions for spread and establishment of fungi and bacteria within the fruit, such as *Colletotrichum lindemuthianum*, *Xanthomonas phaseoli* and *Pseudomonas phaseolicola* in bean, *Ascochyta pisi* in pea, *Ascochyta rabiei* in chickpea, *Alternaria brassicicola* in cabbage, *Aureobasidium lini*, *Colletotrichum lini* and *Sep-*

toria linicola in flax (Subsection 12.2.4), *Drechslera papaveris* in poppy, *Acroconidiella eschscholtziae* in *Eschscholtzia californica* (Section 11.5), all of which may produce severe lesions on seeds, and some of the fungi profuse sporulation, within the fruits. The fruit wall may form the nutritional basis for invasion lesions in seeds such as those produced by *Colletotrichum lindemuthianum* in bean and *Ascochyta pisi* in pea, which are often developed directly beneath the external lesions on the pods (Walker, 1969). There is great need for more exact studies on the influence of the physical conditions of hollow enclosures of fruits on establishment of seed infections.

The consistency of *fleshy* fruits such as tomato, cucumber, squash and melon would be expected to provide favourable conditions for establishment of fruit invading pathogens in the seed but seeds of these crop plants are usually remarkably free from fungi. However, the pulp of tomato provides excellent conditions for sealing of the gelatinous matrix, which is often infected by tobacco mosaic virus, on to the hairy seeds (Section 13.6). In avocado pear the mature fruits, lying on the ground, are invaded by *Phytophthora cinnamomi* which then may be established in the embryo and thus transmitted (Section 13.2). As comparable cases may be considered, in tomato, *Phytophthora infestans* (Boyd, 1935) and *Pythium aphanidermatum* (Gattani and Kaul, 1951). The constant humid environment provided by such fruits may be responsible for successful infection of the seed. Oomycetes are hydrophilic fungi and, certainly because of this, only few of them are transmitted by seed, and some of these extremely rarely.

Most types of *indehiscent* fruits, which never split open at maturity, are one-seeded and regarded as functional seeds; many are 'commercial' seeds such as the achene of Compositae and Polygonaceae, mericarp of Malvaceae, caryopsis of Gramineae, mericarp or cremocarp of Umbelliferae, and the utricle of Chenopodiaceae. The fruit coat in these is usually wholly or partly unified with the seed coat and is functionally a seed coat.

Bract envelopes are predominant in Gramineae, in which the glumes envelop, wholly or partly, the caryopsis thus forming a protecting enclosure around the developing seed (Fig. 10.1). The role of this cover in relation to diseases is dual in that (1) it may protect the flower and developing seed from inoculum coming from outside, and, conversely, (2) it may provide favourable incubation conditions for pathogens that have invaded the enclosure, or protect established inoculum.

'Closed flowers' of barley are protected by the glumes from inoculation by loose smut, *Ustilago nuda*, and ergot, *Claviceps purpurea*, thus escaping infection (Section 11.3; Subsection 18.2.1); in some maize cultivars tight husks protect the cob from infection by maize smut, *Ustilago maydis* (Subsection 18.2.1). In most cases, however, the glumes being dead tissues favour establishment of infection or contamination or preserve for a period of time inoculum produced by systemic infection, such as in covered smuts of barley, oats and sorghum, or, in rice, an ectoparasite, the white tip nematode, *Aphelenchoides besseyi* (Section 13.7), and the Ufra disease nematode, *Ditylenchus angustus* (Section 13.8). In oats the teliospores of *Ustilago avenae*, black loose smut, are disseminated during the flowering, partly during the threshing, and may be carried beneath the lemma and palea (Section 14.3). The glumes may themselves harbour pathogens; oospores of *Sclerospora sorghi*, downy mildew of sorghum, are transmitted in the glumes (Section 13.7). In some cultivars of cotton the enveloping bracts provide incubation conditions for *Physalospora rhodina*, grey ear rot, in contrast to bractless cultivars (Subsection 18.2.1).

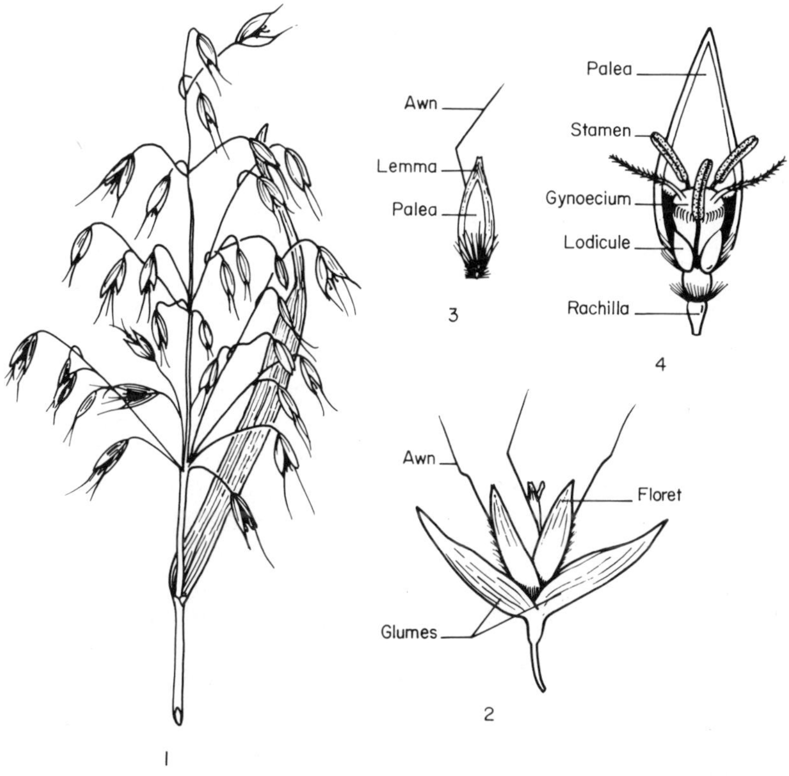

Fig. 10.1. Diagram of inflorescence and flower of oats (*Avena sativa*). (1) Inflorescence; (2) spikelet; (3) floret; (4) opened flower. (Weier, Stocking and Barbour, 1970)

10.2 EMBRYOLOGY AND DEVELOPMENT OF SEED STRUCTURES

In a general discussion on the ontogenesis of seed, the development and structure of the ovule before pollination and fertilisation is vital to understanding of the mechanism of blossom infection and resistance to blossom infection. In our discussion only phanerogams will be considered.

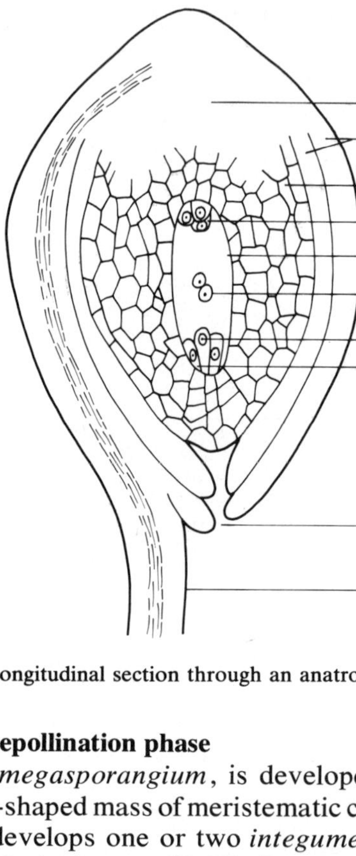

Fig. 10.2. Longitudinal section through an anatropous ovule—cf. Fig. 10.4(3)

10.2.1 The prepollination phase

The *ovule*, or *megasporangium*, is developed from the *placenta* of the ovary. A dome-shaped mass of meristematic cells differentiates as an ovule initial, which develops one or two *integuments* and is connected to the placenta by the *funiculus* or stalk. The integuments do not cover the *nucellus* fully but remain open at one end, forming a narrow canal, called the *micropyle* (Fig. 10.2).

The young nucellus, or central meristematic tissue, bears in the hypodermal layer one or more female archesporial cells. Usually a single *megaspore* mother cell, which is diploid as are the other cells of the nucellus, develops further. It increases in size and undergoes a meiotic division forming four megaspores of which the chalazal megaspore is usually functional. This is the first cell of the *female gametophyte* leading to the formation of the embryo sac (Fig. 10.3). Its nucleus divides mitotically and the resulting nuclei undergo polarisation [Fig. 10.3(2)]. Two consecutive mitotic divisions form eight nuclei [Fig. 10.3(3) and 10.3(4)]. Three of these at the chalazal end form finally the *antipodal cells*, while three nuclei of the micropylar end organise into the egg apparatus comprising the *egg* or *ovum*, and two *synergids*. The two remaining nuclei, the polar nuclei, move towards the centre, fuse forming the *secondary nucleus* which thus is diploid and is important in the fertilisation process.

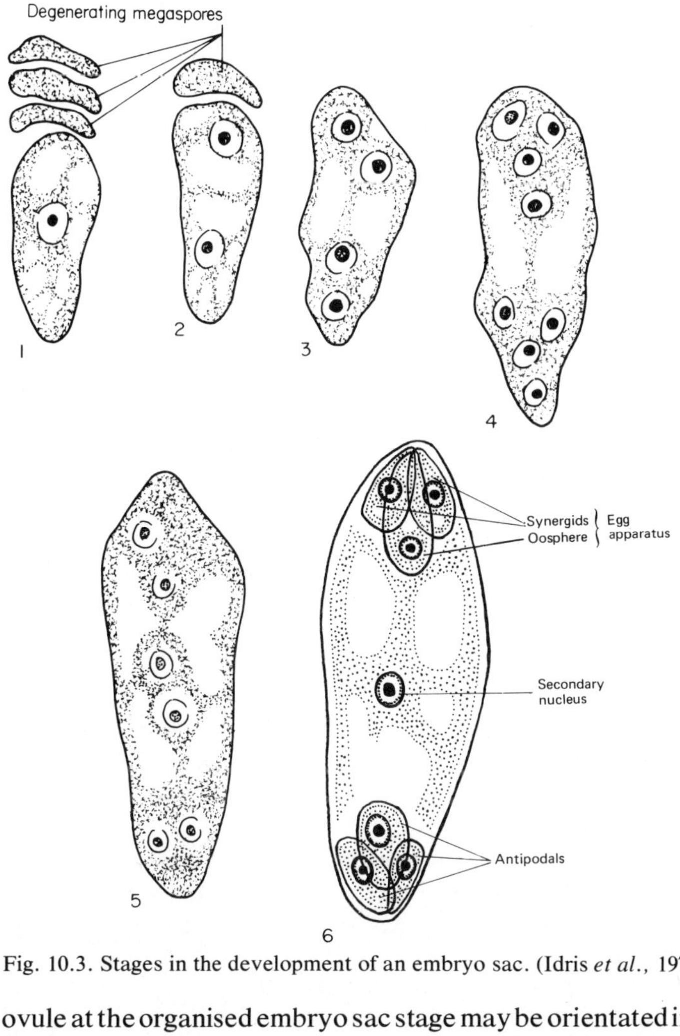

Fig. 10.3. Stages in the development of an embryo sac. (Idris *et al.*, 1972, adapted)

The ovule at the organised embryo sac stage may be orientated in different ways, the positions of the hilum and micropyle thus being defined. Distinction is made between the following common types of ovules, shown in Fig. 10.4:

Orthotropous [Fig. 10.4(1)]: The body of the ovule straight, micropyle, chalaza and funiculus lie on the same longitudinal axis, the micropyle on the top (Polygonaceae, e.g. buckwheat, rhubarb).

Hemianatropous [Fig. 10.4(2)]: The body of the ovule half inverted but straight, the longitudinal axis between micropyle and chalaza lying on a right-angle to the funiculus (Ranunculaceae, e.g. *Ranunculus*).

Anatropous [Fig. 10.4(3); Fig. 10.2]: The body of the ovule becomes inverted so that the micropyle and base of funiculus are adjacent and parallel but the micropyle and chalaza lie on the same longitudinal axis; the integuments remain adnate to the funiculus thus being the raphe. (The most common type, examples being Gramineae, Compositae.)

Campylotropous [Fig. 10.4(4)]: The body of the ovule is transversely placed as in the hemianatropous type, but it shows slight curvature in the front part affecting the embryo sac also (Leguminosae, Resedaceae).

Amphitropous [Fig. 10.4(5)]: The body of the ovule including the embryo sac is strongly bent by unilateral growth, the longitudinal axis between funiculus, chalaza and micropyle being horseshoe-shaped (Centrospermales).

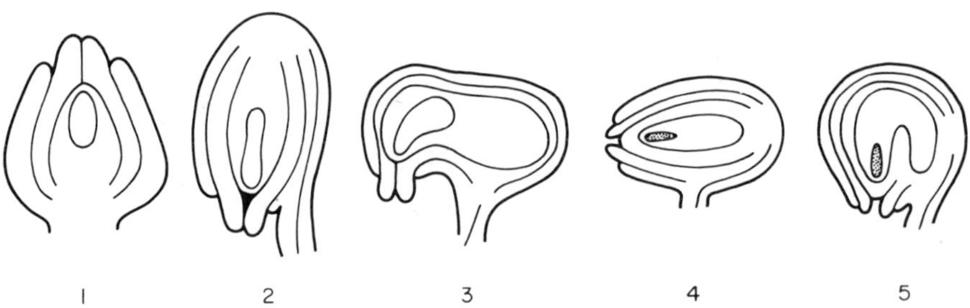

Fig. 10.4. Types of ovules as seen in vertical longitudinal section. (1) Orthotropous; (2) hemianatropous; (3) anatropous; (4) campylotropous; (5) amphitropous. (Adapted, after Prantl; from Maheshwari, 1950)

These ovules reflect the location of micropyle and hilum in the developing and consequently in the mature seed (Fig. 10.5). The differences in structure may possibly have an impact on the preferential avenues of invasion of pathogens into the seed.

Seed transmission of some embryo-borne viruses in some hosts depends on synchronisation of their invasion into the ovule with the initial stage of the prepollination phase. Generally the virus must be established in the mother plants some time before flowering, as shown for the viruses of tobacco ringspot in soybean, bean common mosaic in Phaseolus bean, cowpea mosaic in cowpea, barley stripe mosaic in barley, and lettuce mosaic in lettuce (Section 11.7). Bean common mosaic virus in Phaseolus bean cv. Beka becomes established when it invades the ovule in the mono-nucleate stage; if invasion takes place in the eight-nucleate stage, the ovule is killed; beyond this critical period of necrosis the virus will not be established in the ovule or seed primordium and, hence, there will be no seed transmission (Schippers, 1963, see Section 13.1).

Among fungi *Claviceps*, including *C. purpurea*, *C. paspali* and *C. gigantea*, exemplifies pathogens that can infect the ovulum only in its prepollination, or at least prefertilisation phase. *C. paspali* infects *Paspalum notatum* only before pollination and fertilisation; the ovule is immune only 30 minutes after pollination which suggests that it becomes so already during the phase between pollination and fertilisation (Section 11.3).

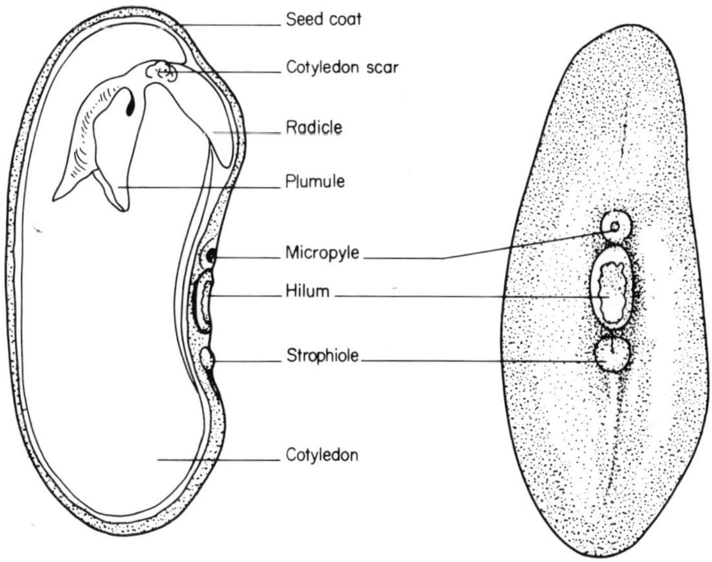

Fig. 10.5. Bean seed. Left, one of the two cotyledons removed; right, bean seed, showing micropyle, hilum and strophiole. (Kozlowski and Gunn, 1972)

10.2.2 The phase between pollination and fertilisation

The length of the period between pollination and fertilisation varies greatly among plant species, from less than an hour as in barley to about a year as in oaks (Section 11.4).

Figure 10.6 indicates the names of component structures of a flower and shows the course of the pollen tube from stigma to ovule. Pollen grains germinate on the stigma and the pollen tube grows through the style. Usually it enters the nucellus through the micropylar end but in some plant species it penetrates through the chalazal end. However, in such cases the pollen tube continues to grow through the nucellar tissue reaching the micropylar end from where it enters the embryo sac. In the embryo sac it releases two male gametes. One of these fuses with the egg nucleus to form the *zygote*, the other fuses with the secondary (diploid) nucleus to form the (triploid) primary *endosperm nucleus* thus accomplishing the *double fertilisation*.

As the pollen tube grows through the style it secretes enzymes such as pectinase (Schoch-Bodmer and Huber, 1947) that soften the middle lamellae of stylar cells, the cell walls becoming mucilaginous. Chemotropic factors, originating in the ovule and the transmitting tissue in the style, are believed to determine the direction of growth of the pollen tube (Welk, Millington and Rosen, 1965). It is difficult to determine the immediate effect of pollination before fertilisation, and it remains to be discovered whether factors which are chemotropic to pollen tubes may have an effect on the establishment of viruses and fungi, and how long stigmata are receptive to pollen after pollination but before fertilisation, and the impact of their condition in this phase on invasion by pathogens.

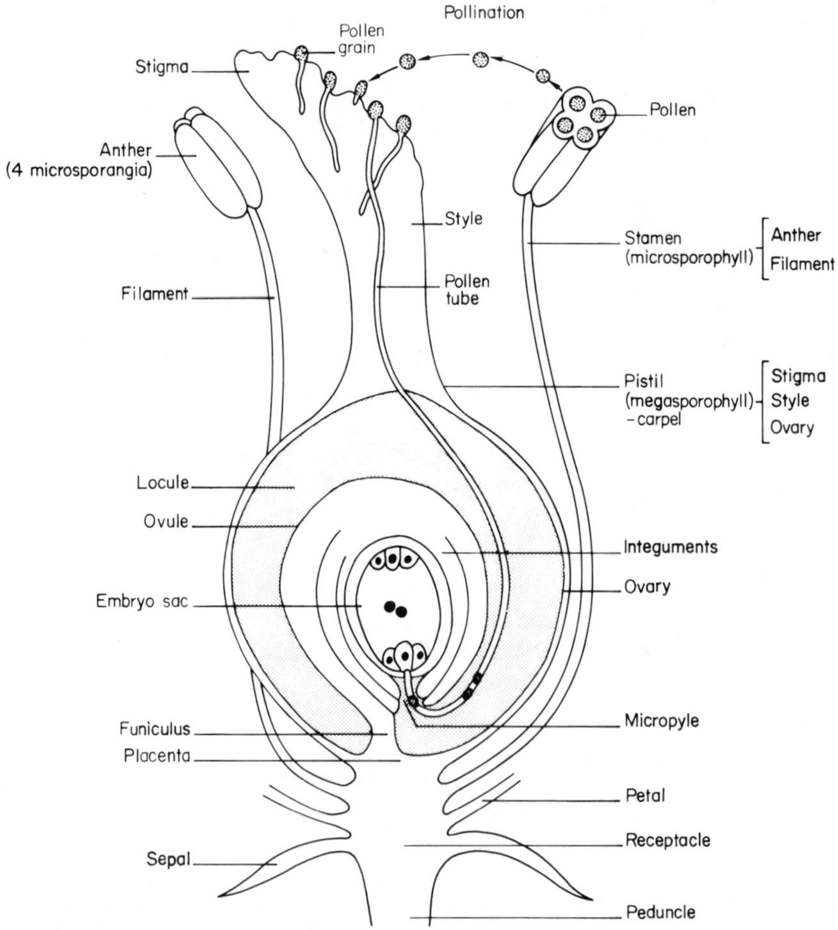

Fig. 10.6. Diagram of median longitudinal section of a flower; component structures labelled. (Bold, 1967)

Many embryo-borne viruses may be transmitted through pollen. This means that the ovule may become infected and a compatible relationship between ovule and virus established when the ovule is ready for fertilisation. This holds true for the viruses of bean common mosaic in Phaseolus bean, bean yellow mosaic in *Melilotus albus*, cucumber mosaic in chickweed, *Stellaria media*, several stone fruit viruses in their hosts, and a number of other viruses (Subsection 12.2.1.1 and Section 13.1). Some viruses may escape from ovules, infected through pollen transmission, and invade the mother plant, see Section 13.1.

There have been few, if any, investigations on invasion of ovules by fungi, in which the time of infection has been related to the precise phase of the ovule; undoubtedly many fungi capable of infecting the unfertilised ovule cause sterilisation (cf. Sections 11.5 and 13.1). On the other hand, in experiments on flower infection with loose smuts of barley and wheat, *Ustilago nuda* and *U. tritici*, highly successful infection has been obtained by inoculation prior to pollination and still 1–2 days after pollination; this indicates susceptibility of the ovule in the phases prior to pollination, between pollination and fertilisation, as well as immediately after fertilisation has been accomplished (Section 11.4).

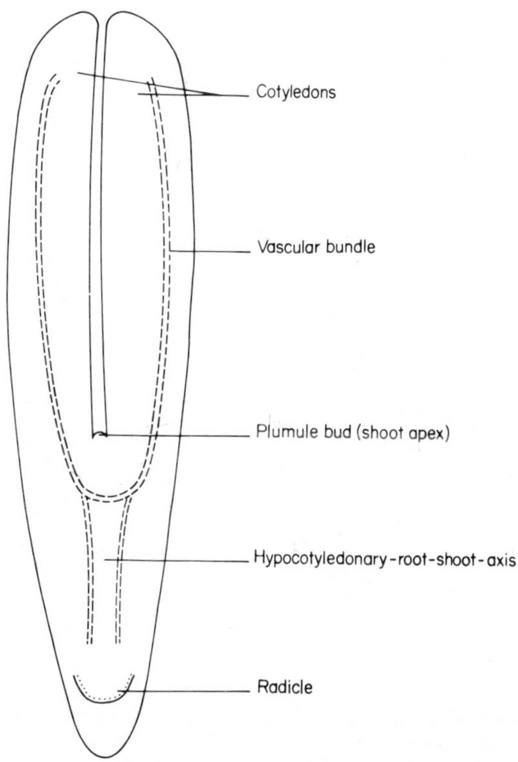

Fig. 10.7. Diagram of a typical dicotyledon embryo

10.2.3 The postfertilisation phase

In the majority of plants the development of endosperm precedes the development of the embryo, rarely it may develop simultaneously and only in a few cases the endosperm development is suppressed. Detailed information on development and types of endosperm is given by Maheshwari (1950). In the majority of crop plants the endosperm is already formed before the embryo starts development, thus providing nourishment for its growth. In seeds the endosperm may persist or may be completely used up by the growing embryo. Even when present the relative amount of endosperm in comparison to other parts of the seed, particularly the embryo, varies very much (Martin, 1946; see next section).

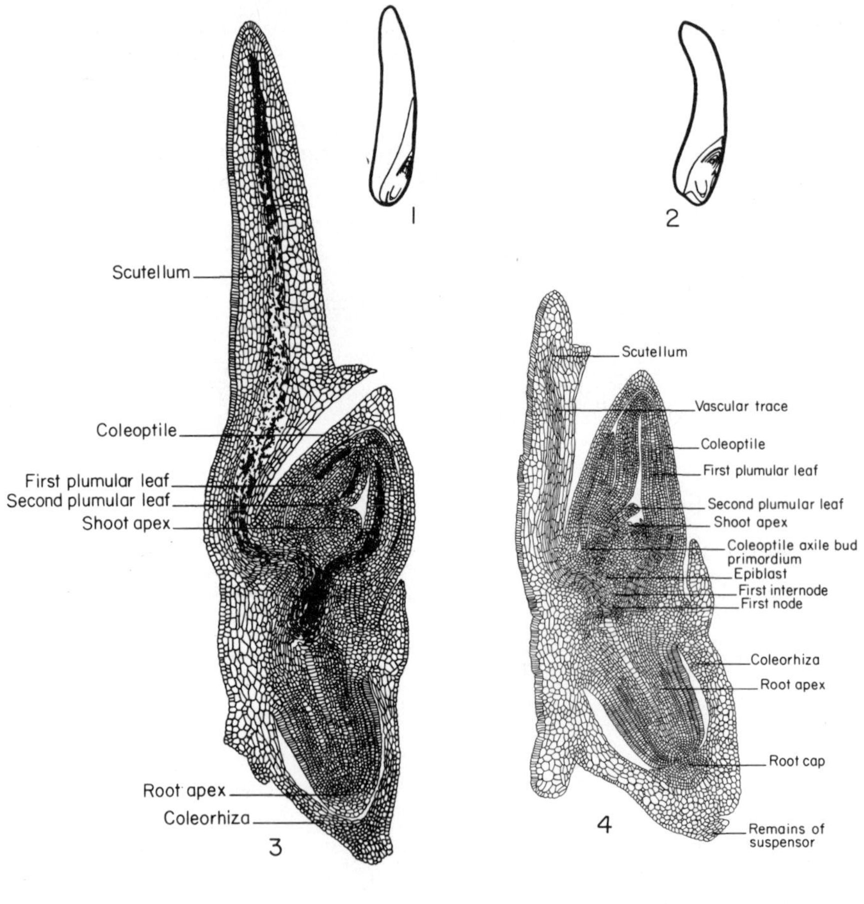

Fig. 10.8. Mature embryos of monocotyledons. (1) Diagram of median longitudinal-side section of oat caryopsis showing location of embryo; (2) median longitudinal-side section of oat embryo; (3) median longitudinal-side section of caryopsis, showing location of wheat embryo; (4) median longitudinal-side section of wheat embryo. (After Avery, 1930; redrawn by courtesy of Dalbir Singh)

The zygote starts developing by a series of transverse and vertical divisions forming the proembryo. These early stages of embryo development do not show any fundamental differences between monocotyledons and dicotyledons. There is a considerable literature available on angiosperm embryogeny (Maheshwari, 1950; Johansen, 1950; Wardlaw, 1955). Subsequent stages in the embryo development of the monocotyledons and dicotyledons lead to two different lines of differentiation. In dicotyledons the embryo becomes globular, heart-shaped before attaining maturity. Throughout the development it maintains bilateral symmetry and in the mature embryo the plumule bud occupies the apical position between the cotyledons (Fig. 10.7). On the other hand, in monocotyledons, after the globular stage the embryo undergoes unilateral growth which is continued up to maturity. The plumule bud apparently occupies a lateral position and the single cotyledon, or *scutellum*, is terminal (Fig. 10.8).

In dicotyledons the plumule bud is in a protected position between the cotyledons, while in monocotyledons it is in direct contact with the adjoining tissue of the endosperm. However, the graminaceous embryo is more specialised with several additional accessory structures, of which particularly the *coleoptile* and the *coleorhiza* are important. The former protects the plumule, the latter the radicle.

The comparative development of the caryopsis, endosperm and embryo, as exemplified by barley, is illustrated in Fig. 10.9, with reference to the sequential development of the caryopsis shown in Fig. 10.10. In the four stages before maturity shown the embryo grows slowly [Fig. 10.9(1)–(4)], while the endosperm develops fast passing through distinct phases — the watery, the milky and later, after the stages shown, the starchy (dough) phase. The elongation of the caryopsis and the endosperm attain almost their full length within about a week.

Most viruses are not capable of invading the seed primordium after fertilisation (Section 11.7) but there are exceptions. Barley stripe mosaic virus may invade the embryo of barley directly even in the hard-dough stage of maturity (Eslich and Afanasiev, 1955; Crowley, 1959), and bean southern mosaic virus may invade the embryo of Phaseolus bean (Crowley, 1959). The mechanism of this invasion is unknown.

The invasion of fungi at different stages in the development of the seed primordium and maturing seed is, naturally, of great importance to the fate of the seed, which often is killed at the early stages (cf. Section 2.5, on seed abortion, and Section 11.5), and to the ultimate location of inoculum within the tissues of mature seed. If invasion takes place when the seed is nearing full maturity, only the seed coat, or pericarp functioning as seed coat, is invaded, often only the epidermal layers (Sections 13.4 and 13.5). At later stages of the seed primordium, or those of the primordium of for instance the caryopsis, the morphological 'barriers' within the seed may hinder penetration of hyphae into deeper layers. In the wheat caryopsis, such 'barriers' are the endocarp, and still deeper, the layers of the integuments; at early invasions hyphae are not arrested by these 'first' and 'second' morphological barriers, and the endosperm, including regions surrounding the embryo, may be invaded, e.g. by *Fusarium culmorum* (Section 13.5, cf. also Section 11.5). In true seeds, such as those of flax, sesame and Californian poppy, *Eschscholtzia californica*, the hyphae are arrested in mature seeds by the outer integument, which in the last host is replaced by the crystal layer — the inner epidermis of the outer integument. In maturing flax seed the hyphae of *Aureobasidium lini* may penetrate into the fibrous layers but are arrested by the cross-cells below

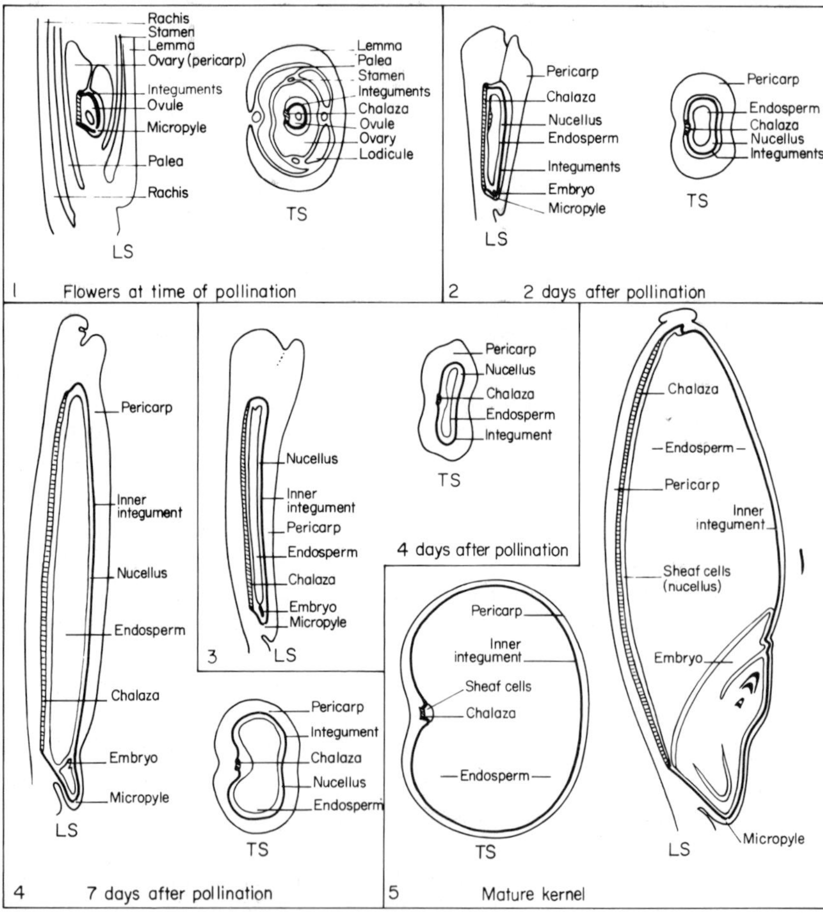

Fig. 10.9. Diagrams of longitudinal and transverse sections of kernels of barley cv. Wisconsin Ped. 51, at five stages of development (palea and lemma have been omitted, except in (1), of flowers fixed at time of pollination). (2) Two days after pollination. (3) Four days after pollination. (4) Seven days after pollination. (5) Mature kernel. LS = longitudinal section, TS = transverse section. (Adapted from Tharp, 1935)

these (Section 11.5); both these layers constitute the inner integument. In sesame seed the inner cuticle of the seed coat and the outer thick cuticle of the endosperm appear to be the barriers for deeper penetration of invading hyphae of *Alternaria sesami* (Fig. 10.11). If infection occurs after these cuticles have been established the mycelium remains localised outside these barriers (Singh, unpublished).

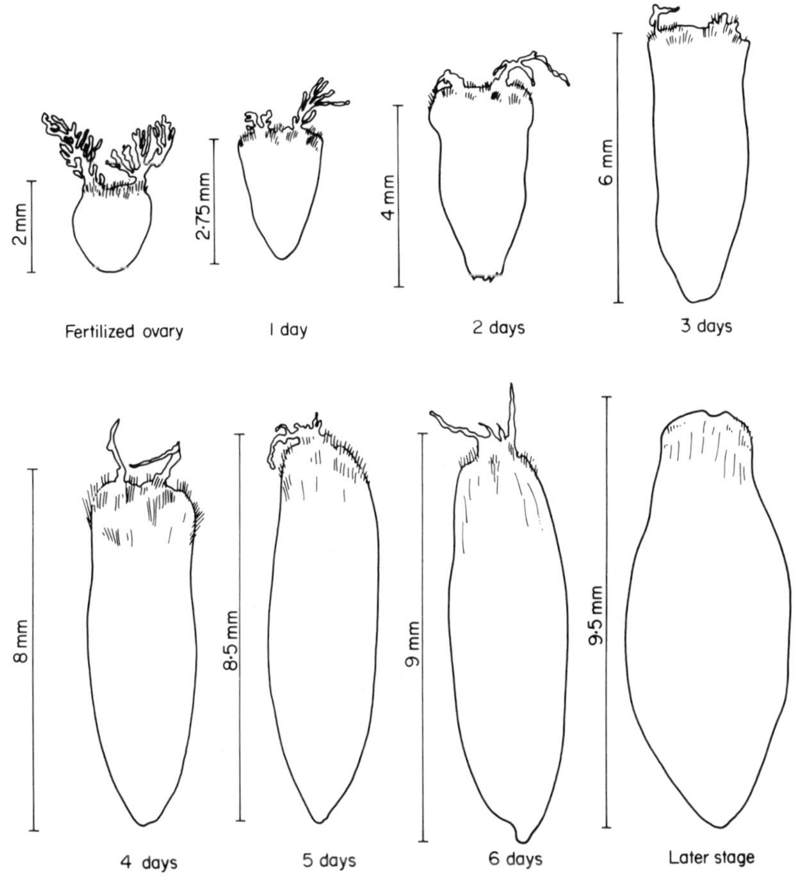

Fig. 10.10. Development of the caryopsis of barley cv. Hannchen. (Drawn from photos in Harlan, 1920). Cf. development shown in Fig. 10.9

10.3 GROSS ARCHITECTURE OF SEED STRUCTURES

Seeds vary exceedingly in size from the minute seeds of orchids to the massive structures of coconuts. Evaluation of seed size in relation to mechanism of transmission of pathogens would only be valid if seeds of different size within the same family, genus or species of hosts were compared.

The outer structure of seeds, such as appendages of various kinds, and sculpture of seed coat, or sculpture of the fruit coat when functioning as a seed coat, are often important for establishment of infection or in providing conditions for anchorage of contamination.

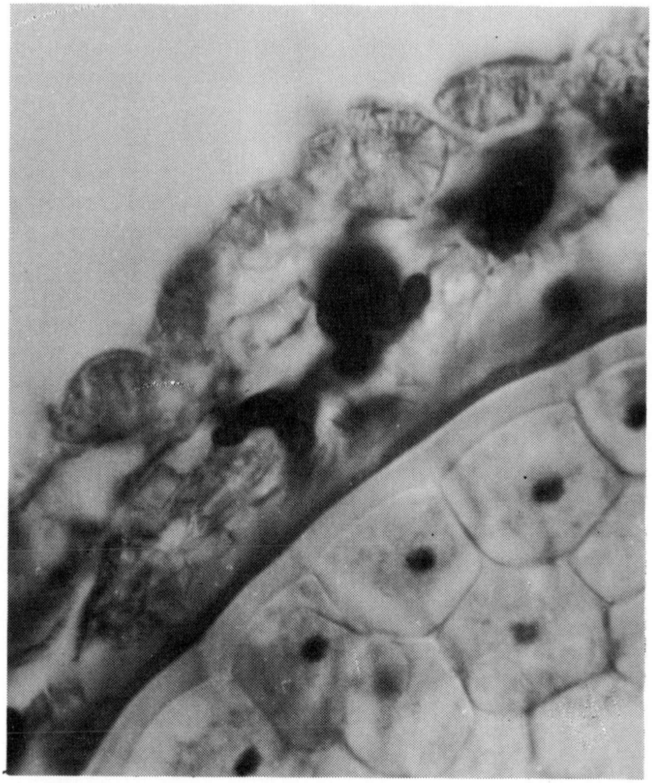

Fig. 10.11. Hyphae of *Alternaria sesami* in the seed coat of sesame, *Sesamum indicum*, arrested by the two layers of the inner cuticle of the seed coat (above) and the outer thick cuticle of the endosperm (below). Longitudinal section. Cf. Fig. 10.18

The bracts in beet seeds are susceptible to infection by *Phoma betae* and *Cercospora beticola*, and the glumes of cereals such as barley, rice, sorghum and oats, assist efficiently in harbouring infections of major pathogens within groups such as *Drechslera*, smuts and downy mildews (Section 13.7).

Seeds with different types of surface structure retain fungicide dusts at very different rates (Subsection 22.1.2; cf. Fig. 22.2 showing seed surface of seven kinds of seed). This suggests that the rate of adherence of pathogen propagules, such as spores, also is considerably influenced by the properties of the seed surface. It is well known that smut spores, e.g. of bunt, are carried predominantly on the ventral side of wheat kernels, and that spores may adhere particularly abundantly to the 'beard'. Fruit matrix of tomato infected with tobacco mosaic virus clings abundantly to the hairy seeds, whereas the dry smooth surface of tobacco seeds is unfavourable for such contamination which is virtually unknown in this host.

An essential feature in relation to infection and transmission of pathogens is the relative size of the embryo within the body of the seed, i.e. the size of embryo in relation to endosperm. Differences in shape, and position of the embryo in the seed may also be relevant. Martin (1946) who studied the gross internal morphology of seeds of more than a thousand genera of angiosperms, has classified seeds according to these criteria. His designations of size and shape of embryo are given in Fig. 10.12.

Examples with reference to Martin's classification (Martin and Barkley, 1961):

In *Caryophyllaceae* the embryo is peripherally curved, except in *Dianthus*, and about half the size of the seed.

In *Chenopodiaceae* (achene), including beet, *Chenopodium*, and spinach, the embryo is arched, ring-formed or spirally coiled; endosperm, which is starchy, is present in a few genera.

In *Compositae* (cypsil) the embryo has well-developed cotyledons and a broad stalk; there is little endosperm.

In *Cruciferae*, including cabbage, mustard, radish, rape, etc., the embryo is dominant, bent or folded; there is no endosperm in significant amount.

In *Gramineae* (caryopsis) the embryo is small, up to a quarter, lateral or basal-lateral; the endosperm is starchy, but varies from floury to flinty or rarely fleshy.

In most *Leguminosae*, including Phaseolus bean, soybean, pea, cowpea, chickpea, etc., the embryo is dominant to total, the interior consists mainly of broad, thick cotyledons, and the seed coat is thin, smooth and tough.

In *Papaveraceae* the embryo is minute to medium small, basal.

In *Solanaceae*, including nightshade, tobacco and tomato, the embryo is quarter to half in size, coiled or bent, in fleshy endosperm.

In *Umbelliferae* (mericarp) the embryo is very small, basal, in fleshy, firm endosperm.

Infection of the embryo, in particular of the thick nutritious cotyledons of *Leguminosae* by fungi and bacteria is, in comparison with other families, remarkably common, and the prevalence of seed-borne diseases in the pathology of such hosts as bean, pea and soybean may largely be assigned to the dominance of the embryo in the seed; parasites have ample access to the cotyledon in maturing seeds, and transmission is highly favoured by localisation of pathogens in the cotyledons.

In *Gramineae*, on the other hand, the embryo is small and chances for embryo infection more limited. If the embryo does become infected, as it may at earlier stages in the development of the caryopsis, it is usually killed resulting in seed abortion and often in considerable yield losses (Section 2.5).

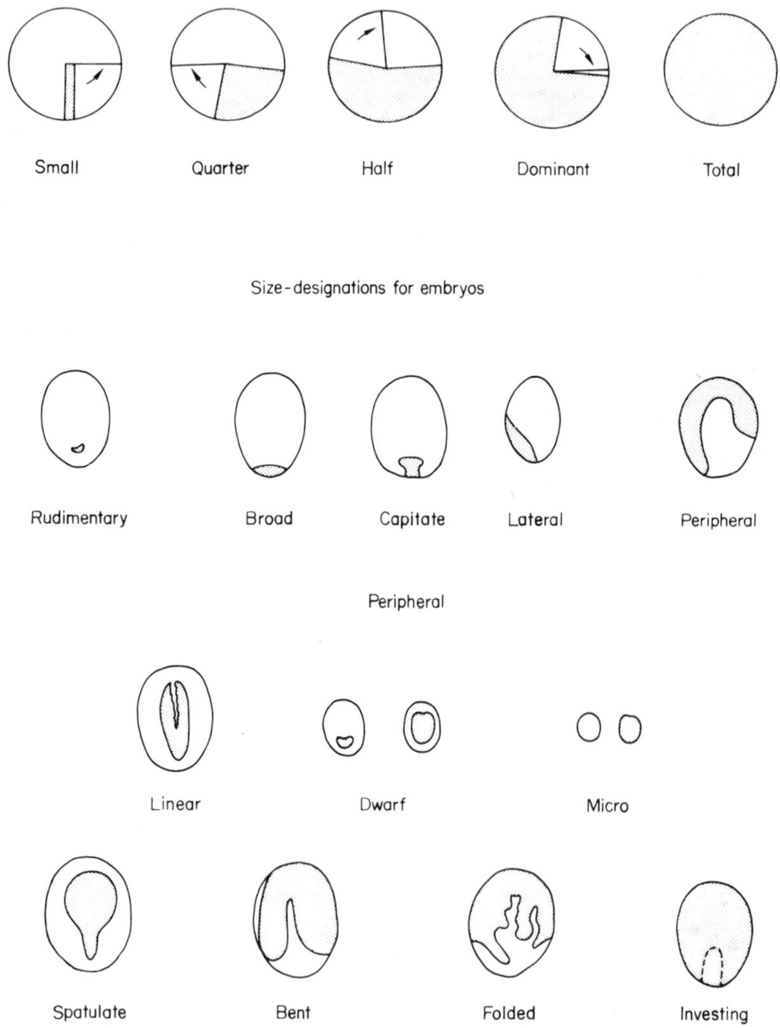

Fig. 10.12. Types of seed according to Martin (1946). The size designations refer to the embryo–endosperm size ratio and represent these volumetrically in quarter units. Two main groups, 'peripheral' seeds and 'axile' seeds are based on the size, shape and position of the embryo in seed. (Martin, 1946; from Bhatnager and Johri, 1972)

10.4 VASCULAR CONNECTIONS BETWEEN MOTHER PLANT AND SEED

In principle, the vascular anatomy of the flower is similar to that of any foliage shoot. Already in the early leaf primordia provascular strands are differentiated. In the flower each floral structure has a distinct vascular bundle or bundles (Fig. 10.13). Usually each petal and stamen is supplied by a single bundle, whereas the ovule-bearing organ, the carpel, is normally supplied by three major bundles — one abaxial and two adaxial

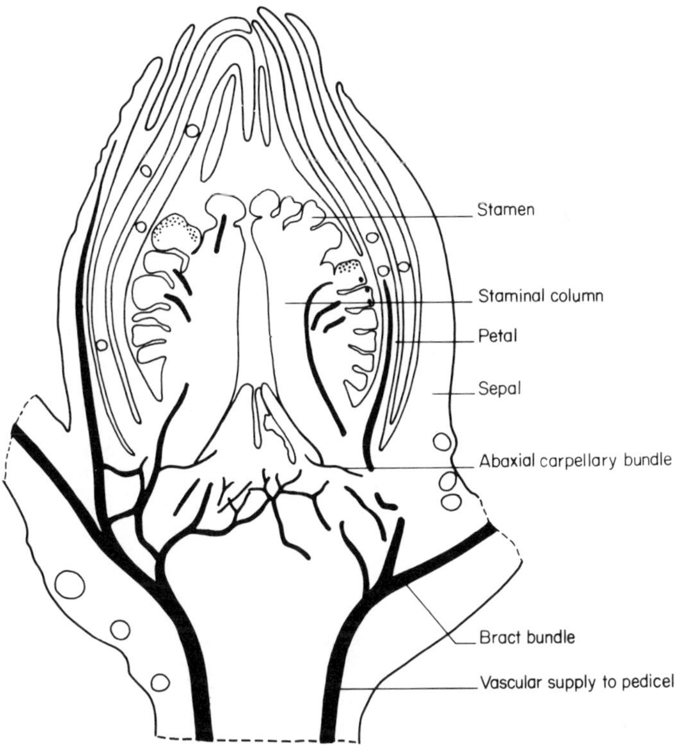

Fig. 10.13. Longitudinal section of flower bud of cotton showing vascular supply to various organs. (Gore, 1935)

(Fig. 10.14). The ovules are borne on the adaxial surface of the carpel lamina. From the carpel a vascular bundle continues through the funiculus and, in the anatropous ovule, further through the raphe, ending in the majority of cases in the undifferentiated tissue of the chalazal region (Fig. 10.15). In ovules of some families the vascular bundle may extend unbranched or after branching in the outer integument of the ovule, and rarely the inner integument. In some seeds the vascular strands are rather extensive, e.g. in cucurbits, chickpea, castor bean.

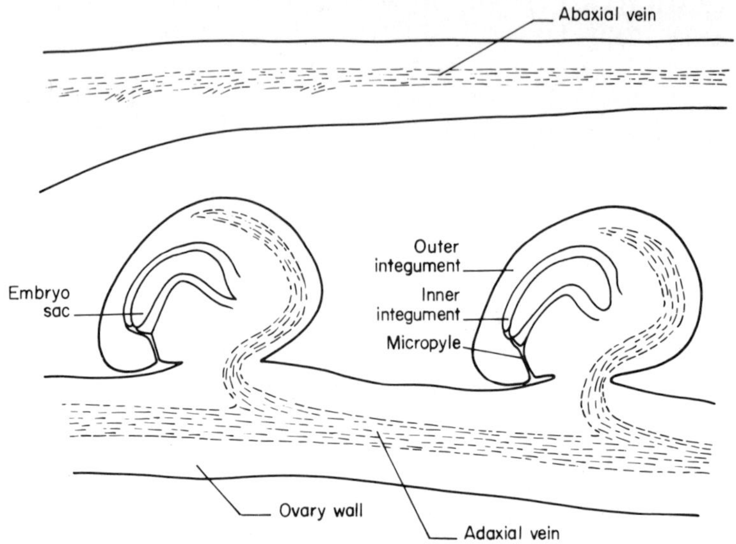

Fig. 10.14. Longitudinal section of portion of a carpel of pea showing vascular supply of developing ovules. (Hayward, 1938)

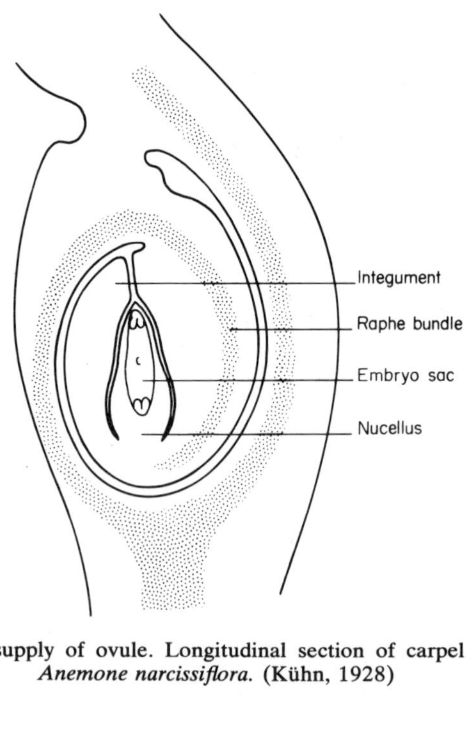

Fig. 10.15. Vascular supply of ovule. Longitudinal section of carpel base and ovule of *Anemone narcissiflora*. (Kühn, 1928)

The raphe is a continuation of the funiculus, often fused with the seed coat, and contains vascular strands thus providing additional chances for transmission of vascular-infecting fungi. Bean, cotton, cucurbits, flax, pea and tomato have seeds with raphe and are known to transmit vascular-infecting fungi such as *Fusarium oxysporum* and, in cotton, flax and tomato, *Verticillium alboatrum* and *V. dahliae*.

The vascularisation is more extensive in fruit coats than in seeds. This may be relevant to seed transmission of vascular pathogens in hosts having seeds in which the fruit coat acts like the seed coat in true seed, such as in the cypsil, caryopsis, etc. More information is needed on the precise location of vascular pathogens such as *Hymenula cerealis, Fusarium moniliforme, F. oxysporum, Verticillium alboatrum, V. dahliae*, and others, in cypsils, caryopses, etc. For example, the location of *Verticillium* in the achene of sunflower or of *F. moniliforme* in the caryopsis of maize.

10.5 STRUCTURE OF PROTECTING LAYERS IN SEED

Commercial seeds always possess a hard protective cover. In true seeds it is usually the modified integument(s), whereas in dry fruit such as achene, caryopsis, etc., it is the modified ovary wall, rarely also other accessory structures, in particular glumes or bracts. The integument(s) as well as the ovary wall undergo a series of changes before forming the natural protective cover. The true seed coat may be formed by all the layers of the integument(s) or by only one or a few of them, others being degenerated during development. In some seeds one or more of the integumental layers multiply, forming most of the seed coat as in cucurbits. In all these changes in the integuments or ovary wall, finally, a protective zone of one or more thick-walled and lignified cells is organised (Fig. 10.16) besides the differentiation of other layers. Distinct continuous cuticles which are sometimes very thick are laid down on different parts of seed, namely seed coat, endosperm and embryo. The thick-walled layers, e.g. palisade cells of legumes and crucifers, main sclerenchymatous layer of cucurbits, the fibrous layers of composites, and of grasses and cereals, and the long, thick-walled epidermis of tomato, *and* the thick cuticles, when present, are the main resistant layers in seed.

In Subsection 10.2.3 it is pointed out that depth of infection in seeds is largely related to time of infection and that the morphological barriers in seeds depend on this time factor in becoming effective.

The location of pathogens in seeds depends, naturally, also on the type of their parasitic nature. The deepness, into which a pathogen can penetrate is, however, not necessarily correlated with its virulence, and even

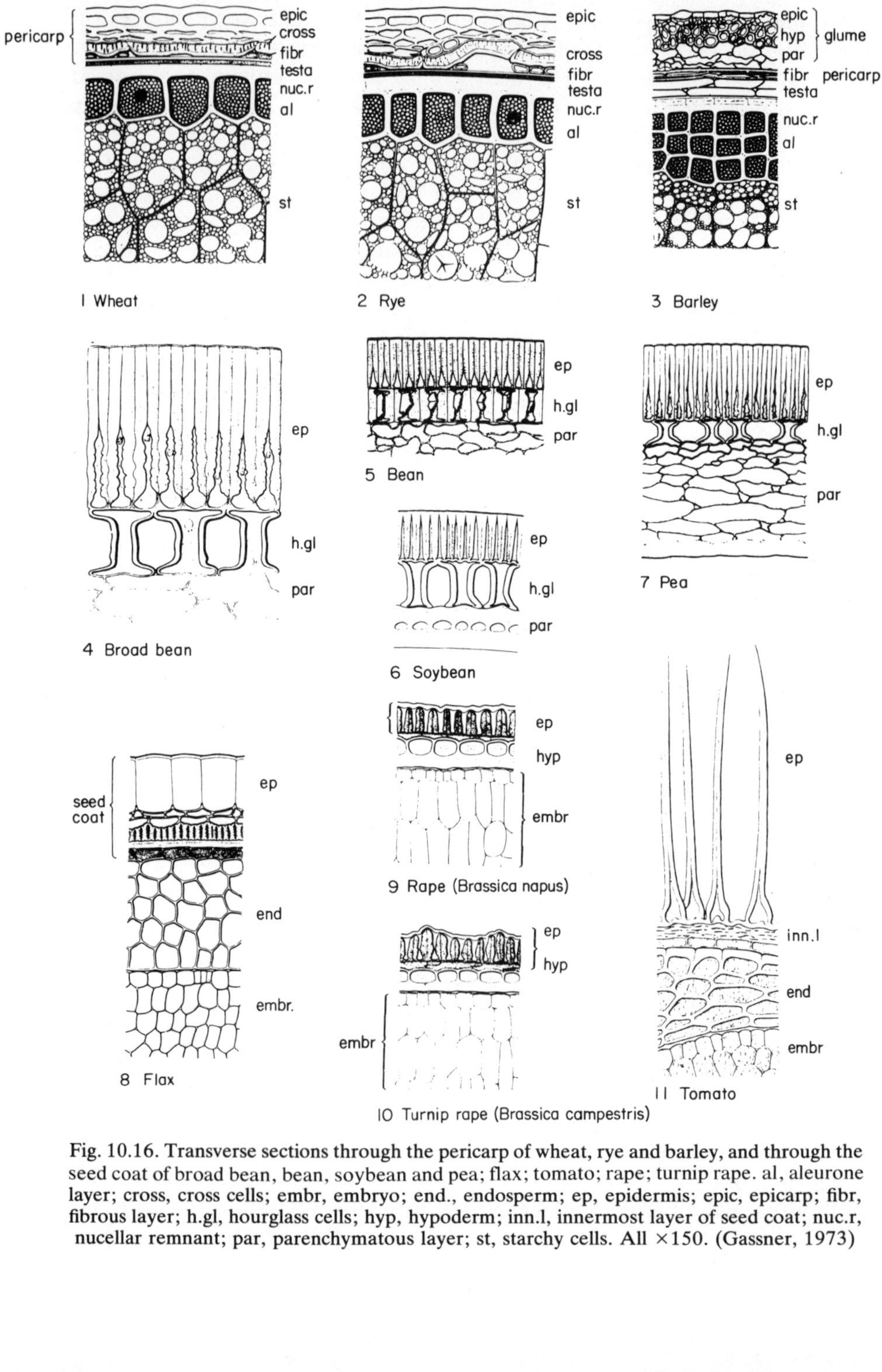

Fig. 10.16. Transverse sections through the pericarp of wheat, rye and barley, and through the seed coat of broad bean, bean, soybean and pea; flax; tomato; rape; turnip rape. al, aleurone layer; cross, cross cells; embr, embryo; end., endosperm; ep, epidermis; epic, epicarp; fibr, fibrous layer; h.gl, hourglass cells; hyp, hypoderm; inn.l, innermost layer of seed coat; nuc.r, nucellar remnant; par, parenchymatous layer; st, starchy cells. All ×150. (Gassner, 1973)

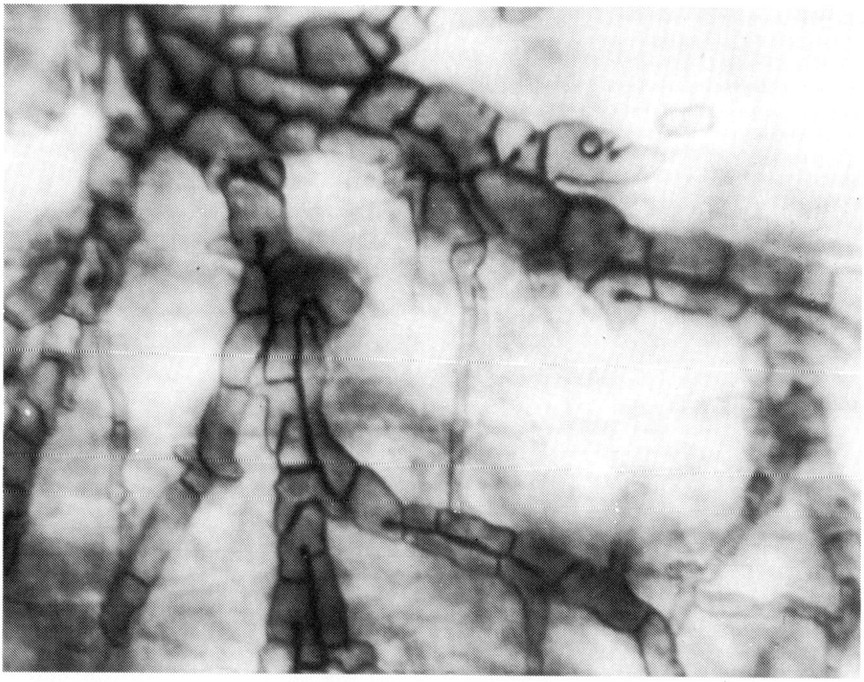

Fig. 10.17. Mycelium of *Alternaria tenuis* under the pericarp of wheat. (Courtesy of Clyde M. Christensen)

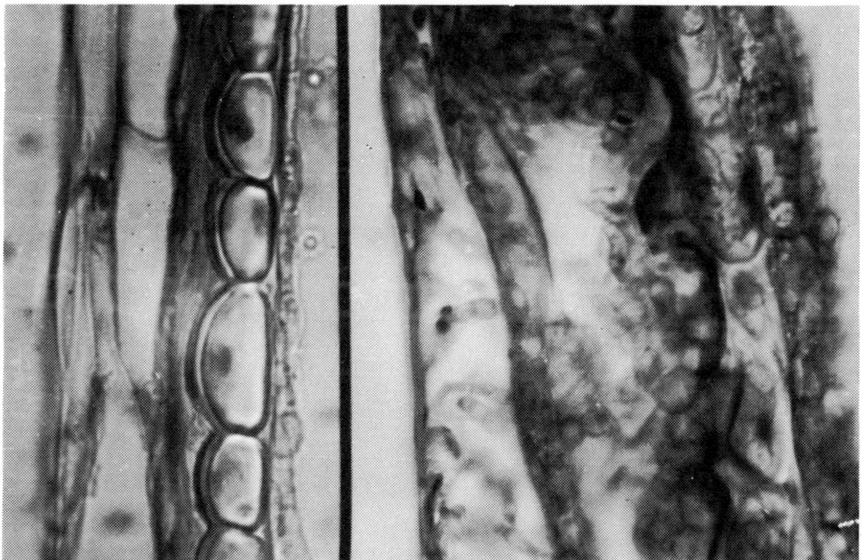

Fig. 10.18. Inner part of pericarp and testa of the cypsil of sunflower. Left, the thick cuticle of the endosperm; right, the cuticle partly digested by *Alternaria tenuis*. Longitudinal sections

organisms regarded as saprophytes may be harboured within the tissues of the seed. *Alternaria tenuis*, commonly regarded as a saprophyte, or a weak parasite in some hosts, is harboured internally in many kinds of seed, including seeds of cereals (see Fig. 10.17), crucifers and composites. According to Singh (unpublished) the fungus may in sunflower penetrate even into the embryo and it occurs abundantly in all inner layers of the pericarp and outside the thick cuticle of the endosperm which forms a very persistent barrier (Fig. 10.18). In some of the seeds the cuticle may be comparatively weak or uneven in thickness, and this weakening of the barrier appears likely to be a result of digestion of the cuticle by the fungus. In such cases the fungus was observed to have formed mycelial cushions in the inner pericarp, indicating its increased activity and power to penetrate.

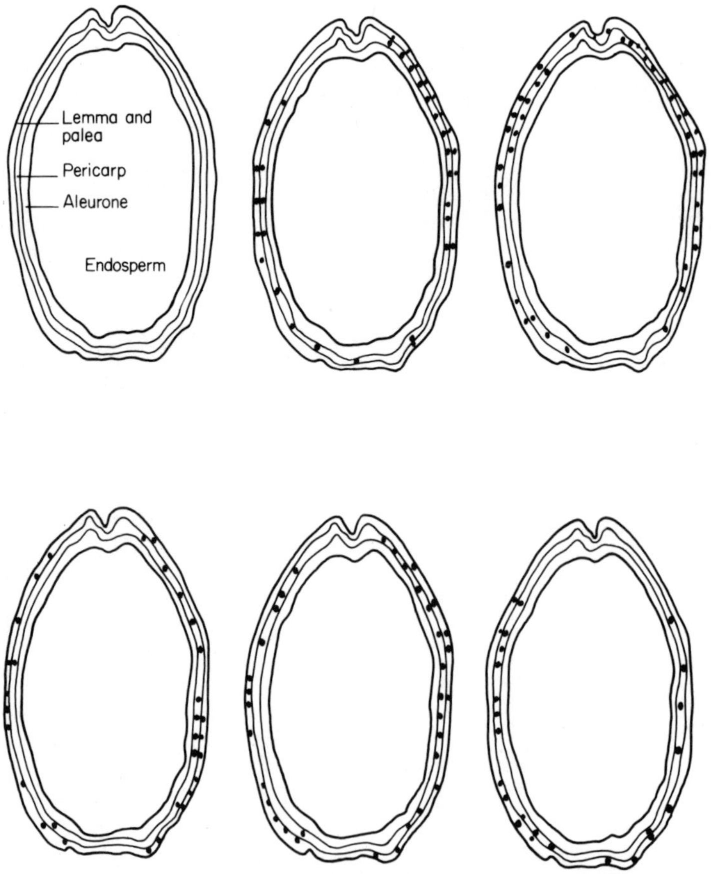

Fig. 10.19. Fungal mycelium in an individual apparently healthy barley grain, as observed in each of five different layers of a typical series of 10μ longitudinal sections. (Warnock and Preece, 1971)

Warnock and Preece (1971) studied by detailed cryostat sectioning the location of internal fungal mycelium in apparently healthy grains of barley (Fig. 10.19). The fungi isolated were predominantly 'storage fungi', including *Aspergillus*, 5 per cent; *Mucor*, 15 per cent and *Penicillium*, 99 per cent, all conventionally regarded as saprophytes. Hyphae were observed in the parenchyma layer of the lemma and palea which enclose the mature caryopsis, and within the pericarp layer of the caryopsis itself, but not in other parts of the grain.

The presence of 'saprophytes', or as they might be considered, apathogenic weak parasites, may be important in relation to establishment of virulent pathogens in seed, and to the transmission rate of these pathogens under field conditions.

PART II

Mechanisms of Seed Transmission

Introduction to Part II

In any observation of the mechanisms of seed transmission distinction must be made between two processes or conditions:

(1) Establishment of a pathogen in, on or with the seed — this implies that the pathogen is *seed-borne*.
(2) Transference of the seed-borne pathogen and establishment of infection in plant(s) from the seed — this implies that the pathogen is *seed-transmitted*.

A seed-borne pathogen may or may not be seed-transmitted. It is a reasonable assumption that seed transmission is established scientifically for a pathogen only when any other means of transmission is excluded, such as transmission by air, soil, plant residues, vectors, etc.

Seed-borne organisms or viruses may be carried with seeds in principally two different ways:

(1) The seed may be *contaminated* (infested), i.e. the pathogen may be carried adhering to the surface of the seed, in fungi usually as spores; or the pathogen may be mixed with the seed, for instance as sclerotia, galls, or as fruiting bodies inside or on the surface of plant debris.
(2) The seed may be *infected*, the pathogen having penetrated into the tissues of the seed, and often established in a resting stage — fungi for instance as dormant mycelium.

The first condition, seed contamination, is in most cases brought about by simple mechanical dissemination of the inoculum. This may take place when the seed crop is handled during weathering, during transportation after harvest, and often during the operation of threshing (many smut fungi), etc. Dissemination may also take place on the crop in the period of seed maturation, by wind, water or insects.

Establishment of seed infection is more complicated. Many factors, particularly physiological conditions pertaining to the pathogen and the host, and weather conditions, must work together to bring about infection. Few seed-borne pathogens have been thoroughly investigated with regard to the precise course of transmission, including establishment of infection and development of the disease in the subsequent crop; loose smut of wheat, *Ustilago tritici*, and barley, *U. nuda*, blind seed disease of ryegrass, *Gloeotinia temulenta*, common bacterial blight, *Xanthomonas phaseoli*, in bean are among the relatively few exceptions.

Nevertheless, a better knowledge of the mechanics of seed transmission may lead to better methods of controlling diseases, such as proper cultural practices affecting the microclimate or the edaphic factors, improvement of seed health testing procedures in field and laboratory, selection of seed disinfection methods, etc.

In classifying different forms of seed transmission, attention must be given to the following aspects and relationships:

(1) *The growth stages of the seed crop, particularly those pertaining to the seed primordium (ovulum)* from the beginning of the flowering and the moment of fertilisation until the maturing of the seed.

(2) *The weather conditions* which may be critical as to whether or how a given pathogen gets into contact with the seed primordium, spreads, and gains foothold within its tissues.

(3) *The time of infection,* as related to weather conditions in the period during which the coincidence of pathogen and host may happen to occur.

(4) *The paths of infection* through which a given pathogen may enter the seed.

(5) *The affected parts of the seed or fruit,* the localisation of the infection being dependent on the time of infection and the weather conditions during the development of infection.

(6) *The conditions and course of infection in the plant developing from the infected seed* — the last decisive link in the process of transmission.

The *paths* of infection cannot be discussed without referring to defined parts of the seed or fruits being affected, i.e. to the *seat* or the *site* of infection. Vice versa, in discussing the affected parts or the conditions pertaining to the process of infection, we will often have to refer to the paths of infection. Hence, repetitions cannot be avoided throughout the descriptions of conditions and processes.

Any discussion on the paths and site of infection of seed-borne pathogens must be related to the morphology and anatomy of the fruits and seeds. The outer layers of these structures are particularly important, whether it is the pericarp, the seed coat or both layers combined. 'Seeds' in the commercial sense include botanically different kinds of entities such as botanically true seeds of cabbage, pea, poppy and tomato, and the most varied forms of fruits such as the caryopsis of *Gramineae*, the achene of *Compositae*, the fruit clusters (multiple fruits or glomerules) of beet and the nuts of hazel and walnut.

How the pathogen invades and establishes itself is largely dependent on the development and final establishment of the ovary wall or pericarp, and the integuments or seed coat, and the structural relationship between these layers. In any attempt to understand and evaluate the nature and significance of a definite action of a seed-invading organism or virus in relation to a certain 'seed' structure, we must take into consideration such essential differences in the development and function of these protecting layers.

The invasion and establishment of pathogens must be studied in their relation to the successive phases, the ontogenesis of the seed or fruit. The integration of this last process in the life of a plant is envisaged expressively in the following definition of a seed, in which Macmillan (1902) brought into the foreground this ontogenetic aspect:

> A seed is an ultimate, trigenerational symbiotic
> unit in the plant life-history,
> integrated from tissues and structures
> belonging to two sporophytic generations
> and the intervening gametophytic phase.

Or, with a little digression from the academic approach of this book, another definition of a seed may be quoted (Georgie Starbuck Galbraith, *The New York Times*):

> This was the goal of the leaf and the root.
> For this did the blossom burn its hour.
> This little grain is the ultimate fruit.
> This is the awesome vessel of power.
>
> For this is the source of the root and the bud. . . .
> World unto world unto world remolded.
> This is the seed, compact of God,
> Wherein all mystery is enfolded.

11 Environment and Plant-Seed Transmission

INTRODUCTION

The predisposition of the young seed to a given pathogen may change from one day to another. From the beginning of flowering until the maturation of the seed, a series of changes rapidly occurs in the developing structure. The weather conditions during this period of the seed crop decide the extent of dissemination, process of inoculation and eventually the actual infection just at the point when the seed primordium or maturing seed is susceptible, and the nature of this infection, its extent and location in the seed. Dry weather and favourable winds during the flowering of barley and wheat disseminate loose smut spores but different conditions may be decisive for the establishment of infection, such as higher air humidity.

11.1 STANDARD TERMINOLOGY OF GROWTH STAGES

In recent years a standard scale of growth stages in cereals, based on that made by Feeke (1941) for wheat, has been extensively used in research work on various cereals. The scale has been improved particularly by the illustration of the stages made by Large (1954) from plants in the field. It is used for defining the course of growth when inoculations of pathogens or disease assessments are made. The scale furnishes precise references for descriptions of development of diseases and rationalises comparisons of observations and experimental results between authors.

Under the aegis of the FAO illustrated parallel scales of growth stages for the following major crop plants have been worked out: wheat, oats, barley and rye combined in one sheet; maize, rice, tobacco and cotton in separate sheets (Chiarappa, 1970).

Feeke's scale of growth stages and Large's illustration of these are given in Table 11.1 and Fig. 11.1.

Zadoks, Chang and Konzak (1974) proposed a revised code for the growth stages of cereals, based on the decimal system.

Table 11.1. Growth stages in cereals. Feeke's scale, cf. Fig. 11.1.

Stage		
1	One shoot (number of leaves can be added) = 'brairding'	Tillering
2	Beginning of tillering	
3	Tillers formed, leaves often twisted spirally. In some varieties of winter wheats, plants may be 'creeping' or prostrate	
4	Beginning of the erection of the pseudo-stem, leaf sheaths beginning to lengthen	
5	Pseudo-stem (formed by sheaths of leaves) strongly erected	
6	First node of stem visible at base of shoot	Stem extension
7	Second node of stem formed, next-to-last leaf just visible	
8	Last leaf visible, but still rolled up, ear beginning to swell	
9	Ligule of last leaf just visible	
10	Sheath of last leaf completely grown out, ear swollen but not yet visible	
10.1	First ears just visible (awns just showing in barley, ear escaping through split of sheath in wheat or oats)	Heading
10.2	Quarter of heading process completed	
10.3	Half of heading process completed	
10.4	Three-quarters of heading process completed	
10.5	All ears out of sheath	
10.5.1	Beginning of flowering (wheat)	Flowering (wheat)
10.5.2	Flowering complete to top of ear	
10.5.3	Flowering over at base of ear	
10.5.4	Flowering over, kernel watery ripe	
11.1	Milky ripe	Ripening
11.2	Mealy ripe, contents of kernel soft but dry	
11.3	Kernel hard (difficult to divide by thumb-nail)	
11.4	Ripe for cutting. Straw dead	

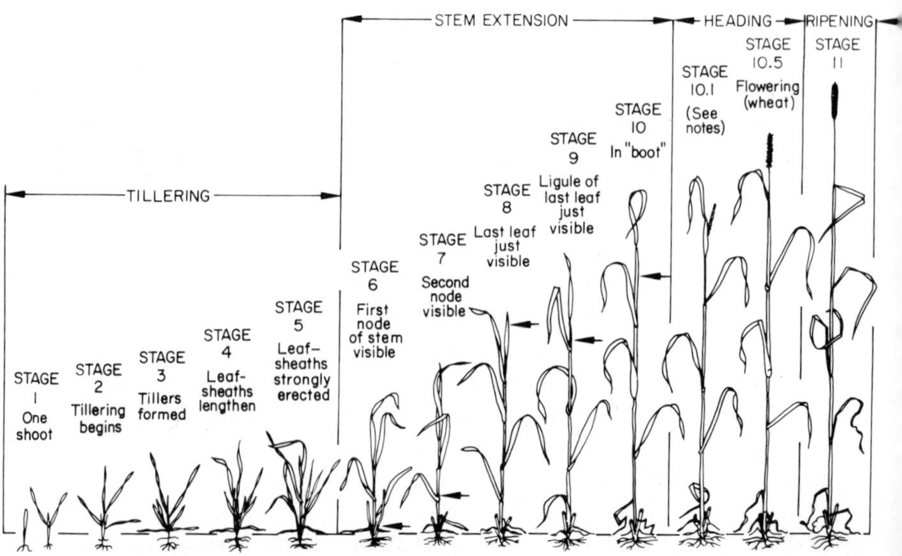

Fig. 11.1. Growth stages of wheat, according to Feeke's scale (1941), from Large (1954), cf. Table 11.1

11.2 TIME OF INFECTION AND GROWTH STAGE OF HOST

The degree of infection of seed may be related to the time of infection as defined by the growth stage of the host. Brönnimann (1968) working in Switzerland related a range of growth stages of wheat, defined according to Feeke's scale, to infection by *Septoria nodorum* (perf. state *Leptosphaeria nodorum*), glume blotch of wheat. He established a clear relationship between the growth stage of the host at the time of infection with the rate of seed infection in terms of per cent infected ears (Fig. 11.2), as well as in terms of thousand grain weight, number of grains per ear and average of ears (Fig. 11.3).

The dynamics of infection were essentially the same in spring wheat and winter wheat, although there was some difference in the rhythm of disease development. Only inoculations of stages 7–11 in winter wheat led

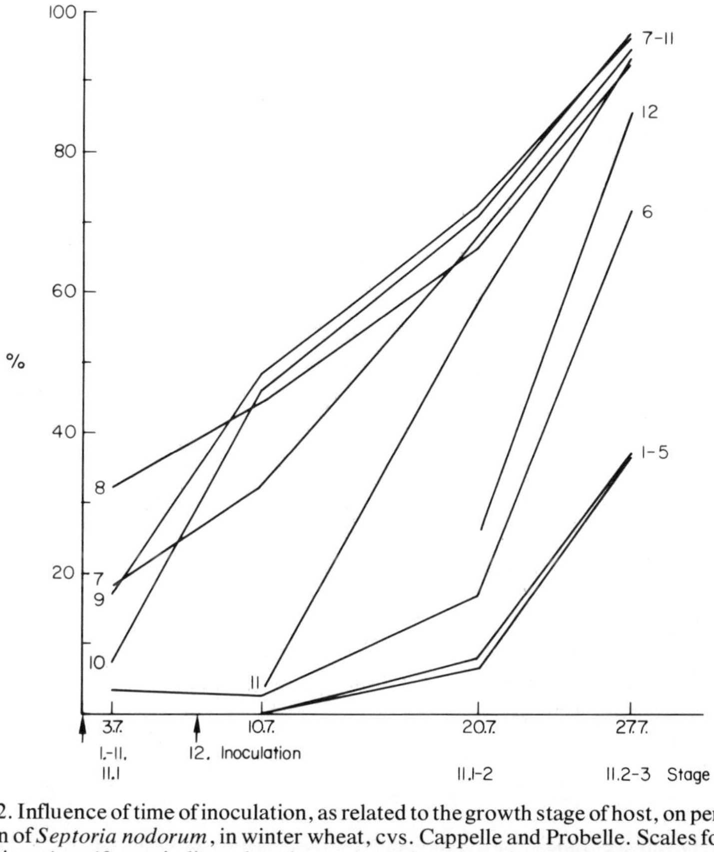

Fig. 11.2. Influence of time of inoculation, as related to the growth stage of host, on per cent ear infection of *Septoria nodorum*, in winter wheat, cvs. Cappelle and Probelle. Scales for time of inoculation, 1 to 12, are indicated at the curves (cf. text at abscissa in Fig. 11.3, top row). Abscissa: dates of record and code numbers of growth stage; the arrows indicate dates of inoculation. Ordinate: per cent ear infection. (Brönnimann, 1968)

to heavy infections of ears and to considerable effect on grain, earlier inoculations having little effect on disease development. Secondary infections, however, did occur at increasing maturation of the crop, where early inoculation was made (curves 1–5, Fig. 11.2). Direct inoculation of ears led to heavy infections, except during the milk stage, the last stage at which inoculation was made. The largest losses in winter wheat were observed when infection took place during the heading stage (from stage 10.1) yield reduction being up to 46 per cent and in spring wheat 40 per cent. Inoculation of spring wheat at any growth stage resulted in infection

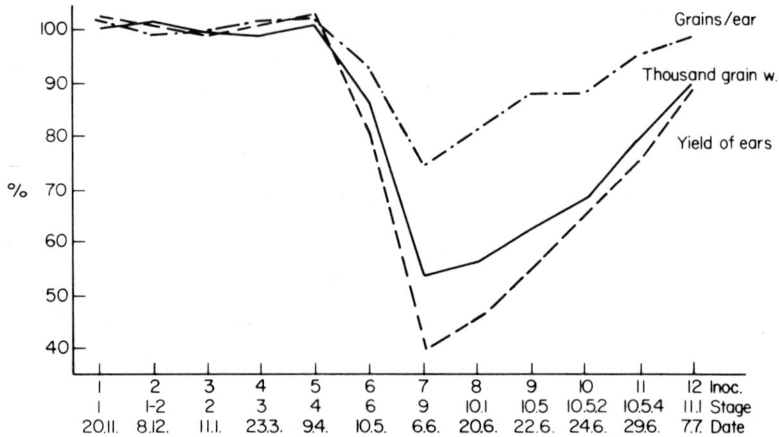

Fig. 11.3. Influence of time of inoculation of *Septoria nodorum* in winter wheat, cvs. Cappelle and Probelle, as related to the growth stage of host, on yield in terms of thousand grain weight, number of grains per ear, and average number of ears. Abscissa: time of inoculation indicated by the related scale numbers, by the corresponding code numbers for growth stage of host, and by the dates of performed inoculation. Ordinate: values related to untreated wheat. (Brönnimann, 1968)

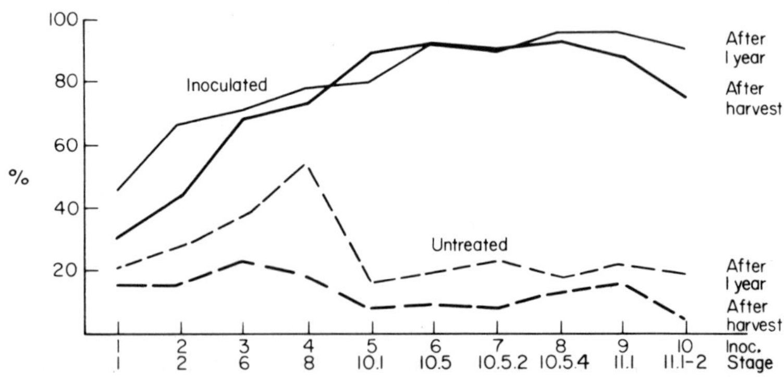

Fig. 11.4. Influence of time of inoculation of *Septoria nodorum* in spring wheat, cv. Hinal, as related to the growth stage of host, on per cent infection of the grain. Abscissa: time of inoculation indicated by the related scale numbers and by the corresponding code numbers for growth stage of host. Ordinate: per cent infection of the grain. (Brönnimann, 1968)

of the grain, increasing particularly from stage 6 to stage 10.1 (Fig. 11.4).

The effect of inoculation as related to growth stage, obtained by Brönnimann, was confirmed in England by Cooke and Jones (1970) and Jones and Odebunmi (1971). When inoculum of *S. nodorum* was applied to spring wheat cv. Opel at growth stage 10, the mean yield per head was reduced by 70 per cent.

In Germany Pirson (1960) also established a relationship between the time of inoculation of *S. nodorum* in winter wheat, and the amount of infection, expressed by the number of conidia produced in the spikes, and the reduction of grain weight. Late inoculation, 10 July, probably at stage 11, had little or almost no effect, while early inoculation, 10 June, which may have been at stage 9, led to a reduction of the grain weight of about 40 per cent, well in agreement with the results recorded in Switzerland (cf. Fig. 11.3).

11.3 TIME OF INFECTION AND WEATHER CONDITIONS

The weather conditions in the period at flowering and seed development are decisive for infection. The exact time of infection within this period again determines the extent of infection in individual seeds and number of seeds infected.

In Scotland Wilson, Noble and Gray (1945) inoculated flowering plants of *Lolium perenne* with *Gloeotinia temulenta*, blind seed disease. The plants were placed in a greenhouse or in cold frames under different conditions of temperature and humidity; the following results were recorded:

		Infection	*Germination*
Warm and dry	(18°C, 65% relative humidity)	poor	excellent
Warm and humid	(18°C, 81% relative humidity)	poor	excellent
Cool and dry	(12–13°C, 77% relative humidity)	severe	good
Cool and humid	(12–13°C, 100% relative humidity)	severe	poor

They found that cool weather conditions favoured the disease by (1) prolonging the duration of the formation of the apothecia, (2) prolonging the period of pollination and maturation of seed, and also by (3) increasing the extent of secondary conidial infection.

The degree of seed infection in cotton by *Colletotrichum gossypii (Glomerella gossypii)*, pink boll rot, at the time the boll opened was directly related to the amount of rainfall (Arndt, 1956). The same correlation was established for seed infection of lucerne, at the time of harvest, by *Phoma medicaginis* (syn. *Ascochyta imperfecta*), spring stem break (Mead and Cormack, 1961).

Also bacterial blights and anthracnoses of bean, soybean and pea are strongly influenced by rainfall and high air humidity in both their development on the seed crop and establishment in the seed. Consequently seed production has been shifted to semi-arid areas in the U.S.A.

Development of loose smut of barley, *Ustilago nuda*, is dependent not only on weather conditions but on the type of flower as well. By 1888 J. L. Jensen (1888) had observed that barley can be attacked by loose smut only if the glumes are open — if the flowers are open, infection will occur, if they are closed, infection will be prevented, hence open flowering cultivars are more severely attacked than closed flowering. If the weather is cold and rainy during the time of flowering, the flowers remain closed, and there will be little infection, also because the rain will wash off the spores. If the weather is warm and dry, infection will occur.

A similar relationship exists between weather conditions and the development of ergot in rye, but with an opposite effect. In cold and rainy weather the fertilisation of rye occurs late or does not occur at all. Consequently the flowers remain open for a somewhat longer period of time, and hence the unfertilised and susceptible ovaries are exposed to infection for a long time. In dry and warm weather the flowers are open up to about half an hour only and are rapidly and promptly fertilised.

According to observations by various workers (Greaney and Machacek, 1946; Greaney and Wallace, 1943), warm and humid weather favours development of headblight incited by *Drechslera sorokiniana* in wheat. Andersen (1952) found that temperatures from 25°C to 30°C were the most favourable for infection and headblight development (Fig. 11.5). The incubation period for development of visible symptoms was 1.5 days at these temperatures at the flowering stage, when the plants were exposed to continuous wetness, 2.5 days at 20°C, and 7 days at 15°C. If air temperatures of about 25°C prevailed, then a 12-hour period of moisture resulting from heavy dew, small showers, etc., would lead to heavy field infection, but the disease would not increase unless such conditions occurred frequently during the heading period. He established that as the plants near maturity they become more susceptible and increasingly infected but at these later stages of the crop, infection does not lead to killing of the heads or spikelets but to discoloration and shrivelling of seed and establishment of the fungus in the seed.

Hino and Furuta (1968) studied in Japan the influence of season on flower infection by *Fusarium moniliforme* in rice and subsequent seed transmission. In plants flowering in July (highest temperature), August, and September (lowest temperature) the average embryo infection percentages were 9, 8 and 0.03 respectively, and there was a high correlation between flowering date at inoculation and damage on seedlings after sowing; maximum infection was recorded from seeds of July-inoculations, minimum from September-inoculations.

The water content of the grain in cereals is of importance to establishment of infection. If maturing grains of rye contain 20 to 30 per cent water for a period of 4–5 days, *F. nivale*, snow mould, has time enough to establish itself in the pericarp (Appel and Fuchs, according to Ravn, 1922). The use of combine harvesters in countries with humid weather conditions during harvest greatly influences establishment of pathogens and storage fungi in so far as the seed may be exposed to favourable conditions for a protracted period.

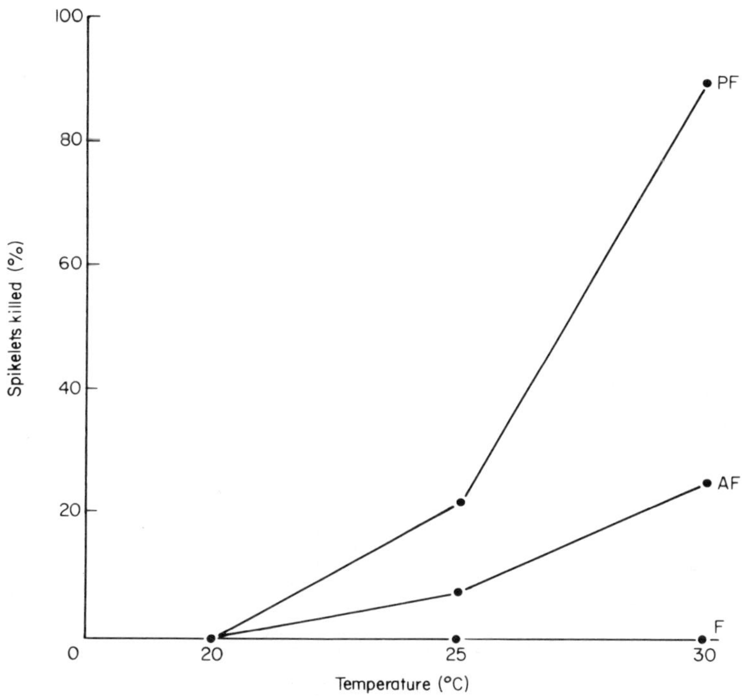

Fig. 11.5. The comparative susceptibility of wheat, cv. Newthatch, in different stages of development to infection by ***Drechslera sorokiniana.*** Plants were exposed to continued wetness for 36 h after inoculation. F flowering; AF after flowering; PF past-flowering stage. (Andersen, 1952)

In Sweden *Cercosporidium punctum* (syn. *Passalora kirchneri*), leaf and stem spot, may be disastrous to seed production of dill. The microclimate has considerable influence on seed infection and seed yield (Neergaard and Helmbring, unpublished). Plants in the outskirts of the field are much less affected than plants inside the crop because of differences in humidity. The plants on the outskirts are sooner dried after night dew or rain than those inside the crop, where ventilation is poor. A seed crop arranged as narrow plots alternating with low crops between them provides better aeration.

The occurrence of *Alternaria brassicicola*, black spot, and *Phoma lingam*, black leg, on different groups of cultivars of head cabbage in Denmark also has bearing on the problem of weather conditions and infection time. No genetical differences in resistance to these fungi could be established between the different types of cabbage, including savoy cabbage, spring cabbage, red cabbage and white cabbage. Nevertheless, a statistical record through a seven-year period showed that seed lots of the first two types are considerably less infected by *A. brassicicola* than the last two (Neergaard, 1945), and later observations confirmed this. Similar observations were made on black leg, *P. lingam*. Through a period of 14 years seed lots of white and red cabbage were more frequently infected and in higher percentages than those of savoy cabbage and spring cabbage (Neergaard, 1969). These differences are due to the fact that the seed of savoy cabbage and spring cabbage in Denmark is earlier harvested (August), than red and white cabbage which are harvested about a month later. The late seed crops are more exposed to infections as the morning dew remains longer on them than on the earlier crops. (The distinction between the four types of head cabbage in relation to time of harvest is becoming less sharp in modern times. New very early cultivars of the previously late types have been developed and consequently the differences as to degrees of escape are likely to become less.) The escape of infections in the early harvested types is an example of pseudoresistance or klendusity.

In the State of Minnesota earlier cultivars of wheat are rather more affected by fusarial headblight, mainly caused by *Fusarium graminearum*, than late cultivars, probably also due to escape in the latter (Hanson, Ausemus and Stakman, 1950). The severity of disease development is particularly influenced by conditions of humidity during the critical period in the development of the seed, i.e. from the flowering to the medium dough stage in the early part of the summer in Minnesota. There are, however, also true differences between wheat cultivars in resistance at this stage.

Pirson (1960) found that early cultivars of wheat in Germany are less damaged by glume blotch, *Septoria nodorum*, than late cultivars. This is probably also essentially due to escape.

Hyde (1950) working on grain storage investigated 278 wheat samples from 89 localities in different parts of the world in 1947 and 1948. She found mycelium of different fungi below the epidermis in all seed samples except five originated from dry districts. The extent of infection in each sample was clearly dependent on the weather conditions of the place of production. Localities with a high relative humidity gave seed samples with severe infection, localities with low relative humidity yielded weakly affected samples. There was positive correlation between the amount of mycelium in the seeds and the level of relative humidity during the last weeks before harvesting. The fungi most frequently isolated were the following (Hyde and Galleymore, 1951): *Alternaria tenuis* (64% of the samples), *Mycogone* sp. (8%), *Cladosporium herbarum* (6%), *Aureobasidium pullulans* (5%), *Fusarium* spp. (2%), *Botrytis cinerea* (2%) and *Stemphylium botryosum* (1%).

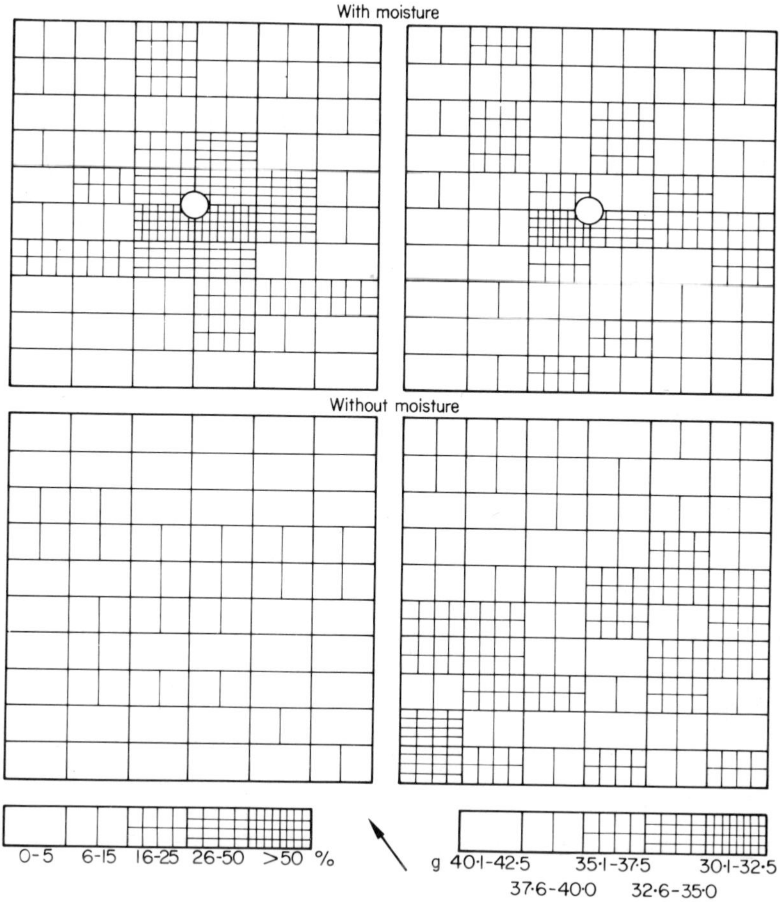

Fig. 11.6. Influence of wetting from a mist sprayer, placed in the centre of a 6 × 6 m plot, on infection by *Septoria nodorum* in winter wheat, cv. Francest. Inoculation from strewn infected harvested remnants. Left: per cent infection of ears in the milk and soft dough stage. Right: thousand grain weight (grammes). Arrow: prevailing direction of wind. (After Brönnimann, 1968)

The amount of seed infection may be correlated directly with the amount of summer rainfall. The number of wheat seed lots infected by *Septoria nodorum*, glume blotch of wheat, and the average percentage of seed infection and the highest infection recorded all increase in England as summer rainfall increases (Hewett, 1965). The influence of the amount of rainfall corresponded to regional distribution of summer rainfall. In the west and south of England, where rainfall is usually 50–100 per cent greater than in the eastern counties, infection was higher through the five-year period studied, 1959–1963. These observations are well in accordance with reports on wet summer weather in connection with the occurrence of glume blotch in various countries. Ogilvie (1943) found that the disease, as ear blackening, in England coincided with a July rainfall of 100 mm (4 in) or more, Hewett (1965) that infection occurs in England and Wales when the July rainfall is more than about 75 mm (3 in), Becker (1956) in Holland and Block (1959) in Germany that a July rainfall of 100–125 mm (4–5 in) is required.

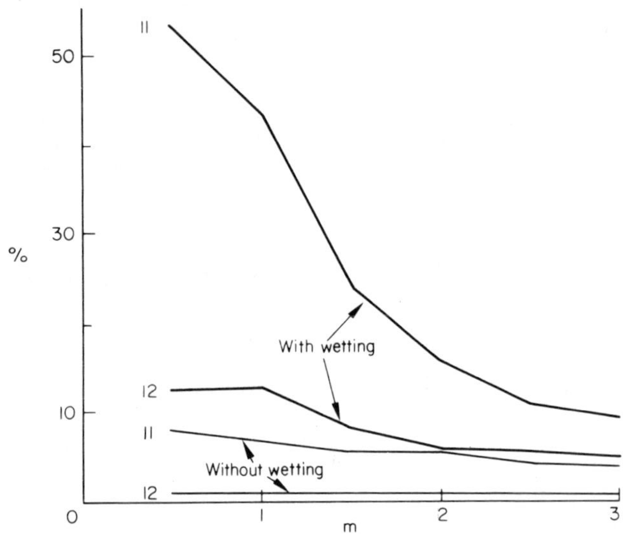

Fig. 11.7. Influence of wetting from a mist sprayer, placed in the centre of a 6 × 6 m plot, on infection by *Septoria nodorum* in winter wheat, cv. Francest. Abscissa: distance from centre. Ordinate: per cent infection of ears in the milk and soft dough stage. Curves 11: infected from strewn harvest remnants. 12: infected from spore suspensions. (After Brönnimann, 1968)

These records of the importance of moisture for development of glume blotch in wheat have been experimentally substantiated by Brönnimann (1968). In a field plot where infected empty spikes and other harvest remnants had been evenly spread in spring, a mist sprayer (aerosol), placed at the same time in the centre of the plot, provided moisture for a period of 16 hours a day making a daily total of 3 l. of water. The effect of the wetting procedure appeared after flowering, particularly at the milk and soft dough stages, and is presented in Fig. 11.6. The average amount of infection decreased markedly with increasing distance from the source of wetting (Figs 11.7 and 11.8). Close around the mist sprayer the amount of water delivered per month corresponds well to the amounts of summer rain indicated as favourable to development of heavy attacks of *S. nodorum*.

The rate of increase of inoculum in development of *S. nodorum* epidemics is directly related to the extent of the period from infection to sporulation of the parasite in the host. Shearer and Zadoks (1972, 1974) studied the effect of temperature and moisture on this period of the fungus in leaf tissue of a susceptible wheat cv. Felix in the three-leaf stage (growth stage 1.3 according to Feeke's scale). The shortest 'latent' period was 6 days at 23°C when the plants were placed in continuously water-saturated air, whereas maturity of pycnidia was delayed by 5–6 days when 12 h saturated air alternated with 12 h at 85–90 per cent relative humidity (Fig. 11.9). No sporulation occurred when the plants were kept continuously at 85–90 per cent relative humidity. In field, under fluctuating conditions, the duration of leaf wetness greatly influences the length of the period from infection to sporulation. The results are relevant to the possible rate of transmission from infected seeds to the young seedlings. Although indicative, the results obtained with seedlings are not directly applicable without further tests to the rate of increase in mature plants.

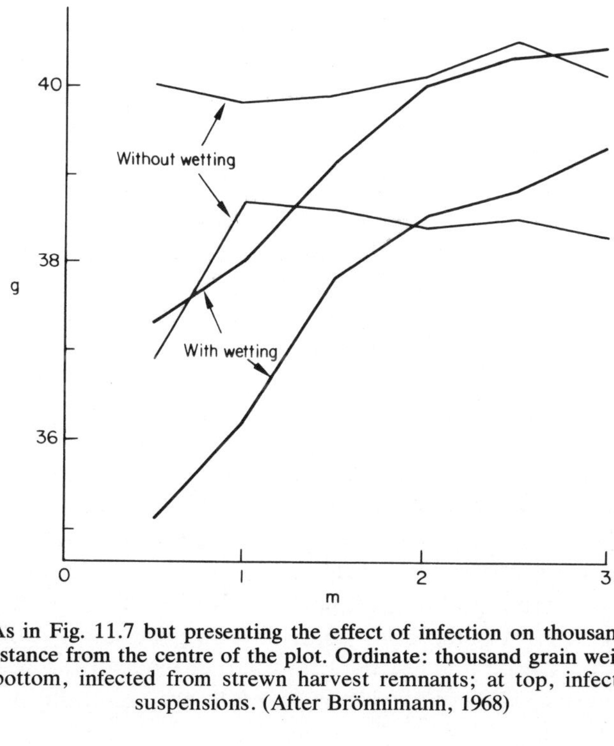

Fig. 11.8. As in Fig. 11.7 but presenting the effect of infection on thousand grain weight. Abscissa: distance from the centre of the plot. Ordinate: thousand grain weight (grammes). Curves at bottom, infected from strewn harvest remnants; at top, infected from spore suspensions. (After Brönnimann, 1968)

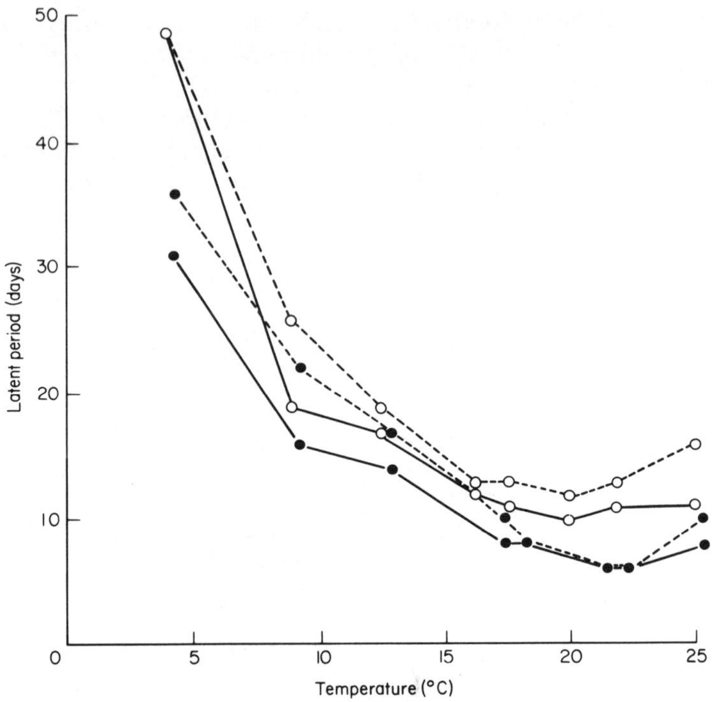

Fig. 11.9. The effect of temperature and moisure, and of two quantities of inoculum, on the period between inoculation and sporulation of *Septoria nodorum* in seedlings of wheat, cv. Felix, under constant conditions of temperature and humidity (see text). The higher quantity of inoculum (unbroken curve) was 5×10^5, the lower (broken curve) 5×10^4 per ml^{-1}. Black dots, wet; white dots, alternate humidity. (Shearer and Zadoks, 1972)

The amount of seed infection and seed transmission of *Phoma betae* in sugar beet is also highly dependent on the amount as well as the time of precipitation, i.e. whether rainfalls prevail in the early or in the late part of crop development. Lysle D. Leach (personal communication, 1973) compared the influence of rainfall in different years and different countries, distinguishing between three periods, 31–60 days before harvest, 1–30 days before harvest, and the harvest period which extended from cutting of the seed stalks to threshing of the dried seed and ranged from 10 to 30 days depending on the area and season. High precipitation during the harvest period leads to high infection percentage of both untreated seed, seed pretreated in sodium hypochloride and the seedlings in soil. High precipitation early in the growth season, even when restricted to the first 30-day period, leads to relatively high percentages in untreated seed; but these infections were found to be superficial as revealed by two facts: the infections were almost completely controlled by pretreatment of the seed in sodium hypochloride and the infection percentages of the seedlings in soil were very low. The conclusion is that late precipitation, during harvest, leads to high percentages of both seed infection and seed transmission, whereas early precipitation leads to fairly high percentages of seed infection but only to low rates of transmission to the seedlings.

The fluctuating intensity of yellow slime disease, or Rathay's disease (*Corynebacterium rathayi*) on cocksfoot in Denmark in the period 1912–1960 has been studied statistically by Skou (1965). He found a clear correlation between the yearly percentage of infected seed lots, as recorded by the Government Seed Testing Station in Copenhagen, and the amount of rainfall in the early part of the summer of the years prior to that of the seed harvest. The disease is always conspicuously developed in the seed fields in May–June, and in this period the disease is vigorously spread in wet weather. On average the rainfall was 34 per cent (54 mm) greater in May–June of the two years preceding those with maxima of the disease than corresponding with minima of the disease in terms of per cent infected seed lots; thus the amount of rainfall in the harvest year itself was found consistently to be less important than that in the two or more preceding years.

Temperature has a marked effect on development of seed infection of viruses. Crowley (1957) established the effect of temperature on seed transmission of bean common mosaic virus. Bean cultivar Canadian Wonder was inoculated at the primary leaf stage, and one week later the plants were placed under controlled temperature conditions which were varied as in Table 11.2.

Table 11.2. Relation of temperature to transmission of bean common mosaic virus (Crowley, 1957)

Temperature (°C)		*Seeds transmitting the virus*	*Total of harvested seeds*
before fertilisation	*after fertilisation*		
20	20	23	138
20	17–18	29	114
17–18	20	0	140
17–18	17–18	0	59

Only the temperature before fertilisation was important in determining the percentage of infected seeds, the higher of the two temperatures being favourable to development of seed infection, the lower having even a preventative effect.

Barley stripe mosaic virus in barley is favoured by relatively high air temperatures. Kankam and Arny (1960) established, by determining virus concentration with *Chenopodium amaranticolor* as indicator host, that virus concentration increased and symptoms appeared earlier and were more severe at 24°C than at 16°C. Singh, Arny and Pound (1960) demonstrated that air temperature had a marked influence on seed transmission of stripe mosaic in barley. They placed four resistant cultivars (a susceptible cultivar failed to produce seeds) at 16, 20, 24 and 28°C, and some of the infected plants were subsequently moved from these temperatures to 28, 24, 20 and 16°C, respectively. At 16°C only one of the cultivars gave seed transmission, at 20 and 24°C some transmission was obtained in all cultivars and, when the plants were moved, all cultivars gave seed transmission predominantly in accordance with the final temperature conditions. If plants were placed at 28°C for 12 days, the virus invaded the seed and was transmitted even when the plants were placed at 16°C for the remaining growth period.

Frosheiser (1974) found that transmission of alfalfa mosaic virus through both pollen and ovules was much less at 29°C than at 18 and 24°C.

11.4 TIME OF INFECTION AND FLOWERING PERIOD OF HOST

The moment of pollination, i.e. the moment when the pollen grain starts germinating on the stigma, as well as the moment of fertilisation, are both crucial points in the course of flower infection by many pathogens. Flower infection of loose smut of wheat, *Ustilago tritici*, may occur already on the first day of flowering, but at this time only 50 per cent of the seed primordia become infected. According to Piekenbrock (1927) the favourable period lies between two and five days after the beginning of flowering, just after fertilisation. Afterwards the infection percentages decrease rapidly and after a few days come down to zero. This is in agreement with the observations of Ohms and Bever (1956) who found that infection of loose smut in winter wheat occurred one or two days before and at anthesis (presumably stage 10.5.2, according to Feeke's scale, see Fig. 11.1). The rate and intensity of infection decreased rapidly in the five days following anthesis, presumably ending at stage 10.5.4 (Fig. 11.10). Results obtained by Gothwal (1972*a*) are quite in agreement. Freeman and Johnson (1909) found that infection hardly took place when the grain had attained two-thirds of its terminal size; this size is reached in about five days after flowering, or probably just before stage 11.1.

Freeman and Johnson (1909) found that infection by *Ustilago nuda*, loose smut of barley, also predominantly occurred immediately after fertilisation; Campbell and Tyner (1959) recorded maximum infection when the plants were inoculated two days after heading (stages 10.5.1–10.5.2); Shands (according to Dickson, 1956) established by introducing the spores into the flowers one day before pollination (10.5.1?), during pollination (10.5.2) and until six days after pollination (11.1?) the highest infection percentage at inoculation just before or during the pollination, pre-

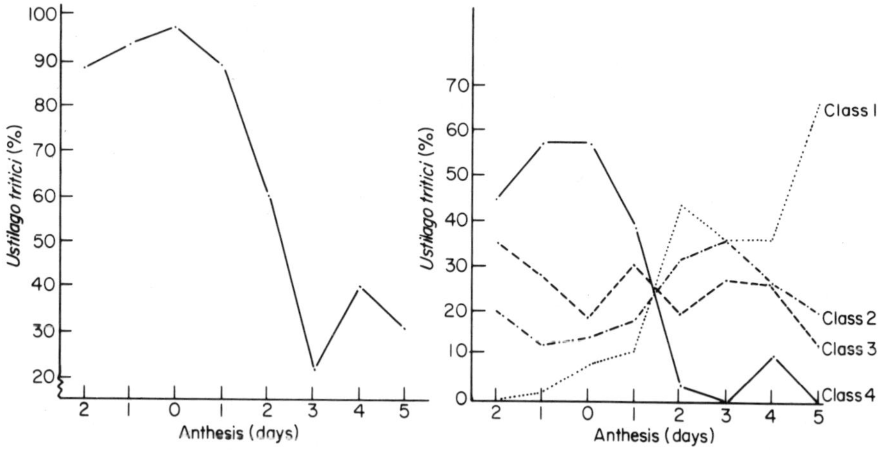

Fig. 11.10. Left: Time of infection in relation to infection percentage of *Ustilago tritici* in winter wheat. Right: Time of infection in relation to amount of hyphae present in the infected embryos. Classes for infection of the scutellum: 1, from trace to 1/4; 2, from 1/4 to 1/2; 3, from 1/2 to 3/4; and 4, from 3/4 to complete. (From data published by Ohms and Bever, 1956)

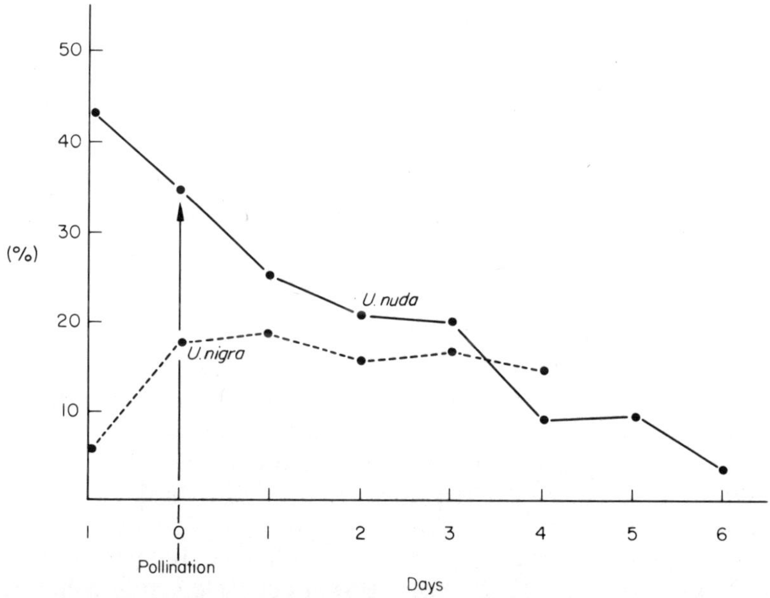

Fig. 11.11. Infection time in relation to infection percentage of *Ustilago nuda* (data from Shands, in Dickson, 1956), and *Ustilago nigra* (data from L. M. Josephson, in Dickson, 1956), both on *Hordeum vulgare*.

sumably at stage 10.5.2 (Fig. 11.11). This seems to fall well into line with Pedersen's observations (1956*b*) on loose smut in barley. He observed that straightforward infections occur in the form of hyphal growth directly towards the integument when the embryo is still young. In later stages of the seed, after a few days or about a week (i.e. at stage 11.1), the infective hyphae assume a more irregular growth, become ramified, bulbous, swollen, some cells having a dumbbell form. Apparently they lack the chemotactic stimulus presumed to emanate from the embryo and to act as growth hormones, according to the hypothesis suggested by Pedersen (1956*b*).

Pariĭ (1972) found that the most effective growth stage for infection of winter wheat by *Ustilago tritici* was at full flowering, and of barley by *U. nuda* at the beginning of flowering.

In contradistinction to the loose smuts of the embryo infecting type, ergot, *Claviceps purpurea*, is generally capable of infecting the ovule of rye only prior to fertilisation of the flower (Engelke, according to Ravn, 1922). Gäumann (1950) emphasised the significance of the opening period of the flowers. The opening period is in rye 20–35 min and in wheat 8–35 min. This implies that the stigmata are exposed approximately for an equal length of time to inoculation in these two cereals. However, rye is cross-fertilised, wheat self-fertilised. When flowers of wheat and barley open they are normally already fertilised, whereas rye flowers at opening expose the stigma, fully receptive to pollen grains and to the inoculum of the ergot fungus. This difference between rye on the one hand and barley and wheat on the other, accounts essentially for the more frequent occurrence of this parasite in rye than in wheat and barley.

Burton and Lefebvre (1948) established that *Claviceps paspali* infects flowers of Bahia grass, *Paspalum notatum*, also only before the pollination and fertilisation. They found that male sterile hybrids are much more severely affected than their parents. Apparently pollination and fertilisation initiate a process that makes the susceptible flowers resistant to the fungus. By inoculation 30 min after pollination no sclerotia were developed.

Ramakrishnan (1948) and Futrell and Webster (1965) found that sorghum likewise was susceptible to ergot, asali or sugary disease, caused by *Sphacelia sorghi*, only between the time the flowers open and the time of fertilisation. Similarly Fuentes *et al.* (1964) were unable to infect maize with *Claviceps gigantea* after fertilisation had occurred.

In recent years the new high yielding male sterile hybrids of wheat have shown the same type of susceptibility with the consequence of a substantial increase in frequency of ergot in these cultivars (Johnson and Smith, 1968; Stoskopf and Rai, 1972). Many flowers of male sterile cultivars are open for one day or more before fertilisation occurs.

In Denmark where different fodder grasses are grown extensively for seed, *Poa pratensis* are attacked much more frequently and severely by ergot than grasses of other genera. This predominance may be related to the apomictic nature of *Poa*; the ovule is likely to be susceptible for a protracted period of time (Neergaard, unpublished). Also Dallis grass, *Paspalum dilatatum,* reproduces apomictically and is very susceptible to *Claviceps paspali* (Fukushiro and Furuta, 1970).

However, there may possibly be exceptions from the rule that fertilised ovaries are resistant to infection. Puranik and Mathre (1971) recorded infections up to nine days after pollination in male sterile barley, the infection percentages falling abruptly five days after pollination. Their barley was highly susceptible as compared with wheat (10 conidia/ml. per floret being inoculum sufficient for infection on barley, while 1,000 conidia/ml. were needed on wheat). The infections obtained after pollination may be due partly to this high degree of susceptibility but may also partly be explained by the possibility of late fertilisation due to a protracted period of time required between the moment of pollination and the moment of completed fertilisation in the barley cultivar studied, as the period of fertilisation may vary considerably in cereals. Finally, it is possible that cultivars may vary as to how promptly the resistance mechanisms in the fertilised ovary become effective.

Drechslera graminea, barley leaf stripe disease, infects barley most severely at the time immediately prior to flowering, the infection percentage becoming lower during flowering and rapidly decreasing during the days immediately after flowering (Genau, 1928). Fazli and Schroeder (1966) made corresponding observations on *D. oryzae* in rice. Inoculations at the flowering stage, milk stage, soft dough stage and mature stage led in this order to decreasing severity of disease and decreasing losses in terms of number of kernels developed, percentage of discoloured kernels and total weight of kernels. In contrast, Andersen (1952) showed experimentally that ears of wheat were practically not infected by *D. sorokiniana* prior to flowering, they were affected weakly during flowering and increasingly after flowering ('after-flowering' and 'past-flowering' stages), cf. Fig. 11.5.

Fusarium graminearum (Gibberella zeae), scab or headblight, was found by Andersen (1948) to be capable of infecting wheat spikes only weakly when inoculated before flowering, 6 per cent spikelets being infected, severely (79 per cent) during flowering, still more severely (98 per cent) in the last part of flowering, the 'after-flowering stage', and most severely (100 per cent) by inoculation at the end of flowering. However, Schroeder and Christensen (1963) demonstrated that the relation between susceptibility to scab and maturity of the grain is highly dependent on cultivar. One cultivar may be most susceptible at the flowering stage, another at the milk stage, and still another at the soft dough stage; moreover, cultivars most resistant at the same stage of maturity may vary in degree of resistance. Seto (1937) found for rice that the most favourable stage of development for seed infection of *F. graminearum* is the flowering stage (74 per cent). During three weeks after this stage he obtained infection at a reduced level. Also for rice possible differences between cultivars should be considered.

As far as the coincidence between the stage of flowering and inoculation is concerned, it should be taken into consideration that there are considerable differences within the plant kingdom as to the length of the period between the moment of pollination and the moment of fertilisation. This may be illustrated by a few examples:

Barley, 0.5–2 hours (Pope, 1943)
Rye, 7 hours (Sörlin, 1939)
Bean, 8–9 hours (Sörlin, 1939)
Rice, 12–14 hours (Juliano and Aldama, 1937)
Wheat, 2 days (Sörlin, 1939)
Apple, 2–4.5 days or more (Sörlin, 1939)
Cabbage, 5 days (Thompson, 1933)
Beech, 3 weeks (Benson, 1894)
Birch, 1 month (Benson, 1894)
Alders, 3 months (Benson, 1894)
Hazel, 4 months (Benson, 1894)
Certain species of orchids, about 6 months (Duncan and Curtis, 1942)
Certain species of oaks, 11–14 months (Benson, 1894; Bağda, 1948)

In *Ginkgo biloba* fertilisation and embryo growth is sometimes delayed until after ovules are shed (Eames, 1961)!

For most of the cultivated plants, for instance cereals and the ordinary vegetables, a period of one to a few days between pollination and fertilisation may be anticipated. These two crucial moments in the development of the ovule are undoubtedly of great importance in the infection biology of many seed-borne pathogens, not least in viruses. Experimental investigations with special regard to this relationship are needed.

11.5 TIME OF INFECTION AND LOCALISATION IN SEED

Extensive studies on Fusarium blight in rye and blind seed disease in ryegrass have been carried out with special regard to the relationship between the time of inoculation and the topographic development of infection.

According to Baltzer (1930) *Fusarium culmorum*, scab or Fusarium blight, produces the heaviest infection on rye during the flowering. At this time the hyphae penetrate into the hilum and the primordia of the coleoptile. The hilum is blocked up and transportation of nutrition is hampered leading to incomplete development of the grain. If the infection takes place as late as during the maturing of the grain, between yellow maturing and full maturing, the grain is less affected. At this time the germ tubes of the conidia are capable of penetrating only into the pericarp, otherwise the development of the kernel is normal. At any point between these two phases in the development of the crop, the time of flowering and the time of maturing of the grain, infection leads to damage of gradually decreasing effect on grain weight, percentage of full grown grains and germination capacity. Djerbi (1971) established for wheat the same relationship between the time of infection and the extent to which *Fusarium culmorum* penetrates into the grain, cf. Section 13.5, Pericarp infection.

Wilson, Noble and Gray (1945) studied a range of stages in seed primordia of *Lolium perenne* infected by blind seed disease, *Gloeotinia temulenta*: (1) infection immediately after fertilisation of the ovary; (2) infection about one week after fertilisation; (3) infection after development of the endosperm and differentiation of the embryo into scutellum, plumule and radicle, but before the pericarp had dried; and finally (4) infection when the caryopsis was almost mature.

If infection occurs at one of the first three stages, the seed primordium is killed, the hyphae spreading in the embryo. At the first stage, immediately after fertilisation, the infection occurs presumably just below the stigma, and conidia may be developed in the embryo. At the third stage the hyphae spread in the tissues of the pericarp and seed coat and penetrate from these tissues into the embryo, and conidia are developed on the surface. In the first two types of infections the seed is shrivelled. In the third type it is as large as uninfected seed but being functionally a sclerotium fails to germinate, is moribund — a 'blind' seed. If infection occurs at the fourth stage, the mycelium spreads in the tissue of the pericarp and the conidia develop on the surface of the seed, whereas the germination capacity of the seed will not be affected.

Two seed-borne flax fungi may be selected as other examples where the amount of infection in different tissues of the seed is dependent on the time of infection. The mycelium of *Aureobasidium lini* extends more or less deeply into the outer layers of the seed, dependent on the time of infection (Johansen, 1943). If the seed is infected early the mycelium may grow into the embryo and kill it. If the seed coat has developed further, the mycelium is predominantly confined to this tissue but may penetrate into the deeper fibrous layers. From the seed coat the fungus is transmitted to the cotyledons during the germination.

Another flax pathogen, *Septoria linicola (Mycosphaerella linorum)*, pasmo, has a very similar development. When the ovary is infected during the flowering no development of the fruit occurs. When the infection takes place after fruit formation the fungus grows from outside through the capsule and the funiculus into the seed coat. According to the time of infection between these two stages in the seed development, the fungus causes more or less damage to the seed (Johansen, 1943).

Davis (1952) established that the mycelium of *Acroconidiella eschscholtziae* penetrates into the capsule of *Eschscholtzia californica* at the point of attachment, sometimes without producing any visible symptoms on the capsule. Within the fruit the parasite kills the young seeds or penetrates into the seed coat of maturing seeds. She demonstrated that the mycelium penetrates through the outer parenchyma layer and the deeper located crystal layers (i.e. the inner epidermis of the outer integument) into the embryo. However, when the seed has attained a certain degree of maturity, the fungus is no longer capable of penetrating the crystal tissue. Consequently, in mature seeds the infection is confined to the outer epidermis and the parenchyma tissues below.

11.6 TIME OF INFECTION AND SEED-BORNE FUNGUS FLORA

The time of infection in relation to the stage of the developing seed is decisive not only with regard to the extent of seed infection, the point of entry infection and the parts of seed being involved, but may also determine the composition of the seed-borne fungus flora. Kilpatrick (1952) examined the seed-borne fungi of ten cultivars and strains of soybean in Mississippi, representing five maturation groups, harvested at intervals of one week. He reported that *Cercospora kikuchii*, purple blotch, was isolated more frequently from the late maturing cultivars than from the early maturing. The reverse was encountered in the case of *Phomopsis sojae*, pod and stem blight. From green immature seeds of soybean only very few fungi were isolated, while 75–100 per cent of mature seeds with a low water content were infected. Wilcox, Laviolette and Athow (1974) established that delayed harvest in Indiana, U.S.A., results in significant increase of seed infected by *P. sojae* but not of seed infected by *C. kikuchii*.

11.7 TIME OF INFECTION AND SEED TRANSMISSION OF VIRUSES

For several virus diseases there is a definite relationship between the time of infection of the crop and the amount of seed infection and transmission. Athow and Bancroft (1959) related the amount of seed transmission of tobacco ringspot virus in soybean to the time when the first symptoms were observed on the plants in the field (Table 11.3). They found that early infection, according to the symptoms observed, resulted in a high percentage of plants transmitting the virus by seed and a high percentage of seed transmission by these plants. Late infection, according to symptoms recorded close to or during the flowering period, gave proportionately low percentages of transmission. This marked effect of age of plants at inoculation and infection on the rate of seed transmission was confirmed by Owusu, Crowley and Francki (1968).

Table 11.3. Effect of time of infection the mother plant on seed transmission of tobacco ringspot virus in soybean (Athow and Bancroft, 1959)

Age of plants when symptoms first observed (days)	*Number of days before flowering*	*Plants showing seed transmission (%)*	*Seed transmission by these plants (%)*	*Average seed transmission (%)*
38	17	78	91	56
46	9	15	54	10
52	3	16	11	4

Already Fajardo (1930) found that bean common mosaic virus is seed-transmitted only when the mother plants are infected before flowering but Nelson (1932) observed that in rare instances the very susceptible bean cultivar Refugee may transmit the virus through seeds even when the mother plants show no symptoms up to the blossom time.

Cowpea (aphid-borne) mosaic virus in cowpea is another example of decreasing rate of seed transmission with the age of the mother plant (Table 11.4); plants inoculated so late as 10 days prior to flowering or after flowering yielded no seed transmission (Khatri and Chohan, 1972).

Table 11.4. Rate of seed transmission of cowpea mosaic virus in relation to time of inoculation of parent plant (Khatri and Chohan, 1972)

Age of the plant when inoculated (days)	*Seed transmission (%)*	
	I	II
10	35	29
20	12	16
30	5	7
40	0	0
50	0	0
Uninoculated	0	0

Also, for barley stripe mosaic virus in wheat and barley Hagborg (1954) found that early inoculation results in seed transmission while inoculation at the time of heading does not, and Singh, Arny and Pound (1960) compared plants inoculated 20, 40, 50 and 60 days after planting (corresponding to 30 days and 10 days before heading, at the heading stage, and 10 days after this stage, or inversely 40, 20 and 10 days before the soft dough stage and at the soft dough stage). In a susceptible cultivar, Oderbrucker, inoculation 10 days before heading led to the highest percentage of seed transmission, the percentage falling abruptly at later inoculations (Table 11.5) and in four resistant cultivars inoculations 10–20 days before heading were most favourable for seed transmission. Crowley (1959) obtained comparable results (Table 11.5), and Timian (1967) recorded in field and glasshouse experiments no seed transmission of barley stripe mosaic virus in barley inoculated 7 days before anthesis or later.

Table 11.5. Effect of age of plant at time of inoculation on seed transmission of barley stripe mosaic virus in three susceptible cultivars of barley (Crowley, 1959; Singh, Arny and Pound, 1960)

Number of days before and after heading stage (flowering)		*Average seed transmission (%)* Mars (Crowley, 1959)	Campana (Crowley, 1959)	Oderbrucker (Singh *et al.*, 1960)
Before	30	—	—	47
	20	44	17	—
	12	24	2	—
	10	—	—	61
	3	15	0.7	—
	0	—	—	24
After	3	16	0.7	—
	10	0.3	0.7	6
	17	0.6	0.3	—

Lettuce mosaic has given similar results (Table 11.6): decreasing amounts of seed transmission from plants inoculated by aphids when they were 4, 9, 13 and 21 weeks old (Couch, 1955). The earliest infected plants developed the highest rate of transmission, while seed from plants inoculated after flowering were virus-free.

Table 11.6. Effect of age of plant at time of inoculation on seed transmission of lettuce mosaic virus (Couch, 1955)

Age of plants when inoculated (weeks)	*Seed transmitting lettuce mosaic virus* (%)
0 (seed-transmitted)	7.4
4	7.9
9	4.9
13	4.6
21	0.0

11.8 INFECTION OR CONTAMINATION DURING HARVEST

After maturation of the seed there are further possibilities of infection. This may occur during the curing of the seed crop, while the harvested crop is placed in shocks, piles or stacks, and during threshing and subsequent processing such as artificial drying, and during storage. This entire process may provide conditions favourable for spreading of pathogens and saprophytes and for establishment of contamination or infection. The conditions of humidity as well as the mechanical treatment during threshing are particularly important. Processing machinery may convey transmission of contamination. *Drechslera graminea* in barley and bunt in wheat can be transferred from an infected seed lot to a healthy one during threshing or mechanical grading (cf. Section 19.9).

12 Entry Points of Seed Infection

INTRODUCTION

The seed primordium or the maturing seed may be infected either (1) directly from the mother plant, or (2) by transmission from outside.

Infection directly from the mother plant may be introduced through the flower or fruit stalk (pedicel, peduncle) and the seed stalk (funiculus), or directly from the seed surface (integuments).

Infection from outside may be introduced through the stigma, the ovary wall or pericarp, and the flower or fruit stalk, and later through the seed coat. A pathogen may penetrate several of these parts of the seed, and in turn infect them.

12.1 INFECTION DIRECTLY FROM THE MOTHER PLANT

Most if not all embryo-infecting viruses, as well as some vascular infecting bacteria, invade the ovule or young seed through the funiculus.

Invasion of bean common mosaic virus into the ovule of bean has been demonstrated by Schippers (1963). All embryo-borne viruses can be assumed to invade the ovule or seed primordium through the funiculus in case of infection from the mother plant.

Burkholder (1921) and Zaumeyer (1930) demonstrated that *Xanthomonas phaseoli*, common bacterial blight, penetrates the ovule through the funiculus (Fig. 12.1). The bacteria enter the sutures of the pod from the vascular system of the pedicel and pass into the raphe leading into the seed coat; they also enter through the micropyle. Apparently healthy pods may at the time of harvest contain infected seeds (Burkholder, 1921). The seeds have typical bacterial spots and there is a yellow discoloration at the hilum where the pathogen penetrated into the pod through the vascular system. *Corynebacterium flaccumfaciens,* bacterial wilt, also penetrates into the ovule of bean through the vascular bundles of the flower stalk, or the bacteria follow the sutures on the pods, penetrate the hilum and form yellow masses under the seed coat and may be visible through this in white-seeded cultivars as a bright-yellow discoloration (Zaumeyer and Thomas, 1957).

Skoric (1927) showed that *Pseudomonas pisi* in pea penetrates through the micropyle into the seed coat (Figs 12.2 and 12.3), Naumann (1963) found that *P. lachrymans*, angular leaf spot of cucumber, penetrates through the funiculus and micropyle into the seed coat, endosperm and embryo of cucumber (Fig. 12.4), and Cook, Larson and Walker (1952) found evidence for invasion of the black rot organism of cabbage, *Xanthomonas campestris*, through the funiculus into the seed coat (Fig. 12.5).

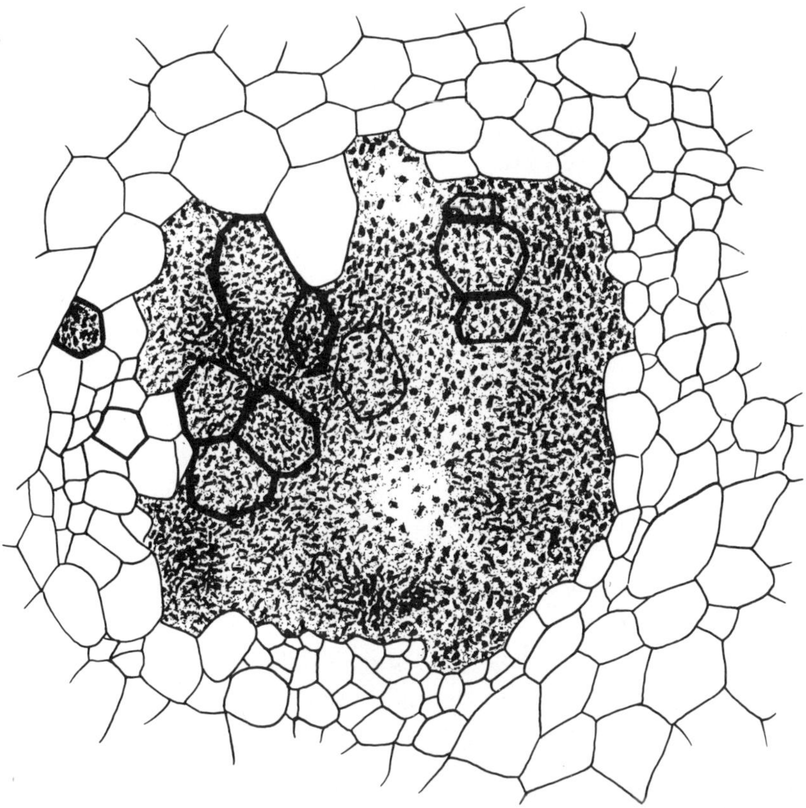

Fig. 12.1. Cross-section of a Phaseolus bean funiculus, showing disintegration of tissue caused by *Xanthomonas phaseoli.* The bacteria have invaded the xylem elements of the funiculus and disintegrated the tissue with the formation of a large lysigenous cavity, in the lower part of which portions of the disintegrated walls may be seen. ×1,650. (Zaumeyer, 1930)

Vascular infecting species of *Fusarium* provide examples of serious seed-borne fungi that penetrate from the vascular system through the funiculus into the seed. Rudolph and Harrison (1945) investigated different seed-invading species in cotton. They isolated *Fusarium moniliforme*, *F. oxysporum, F. equiseti* and other unidentified species from vascular bundles in all parts of cotton, including the internal tissues of the seed.

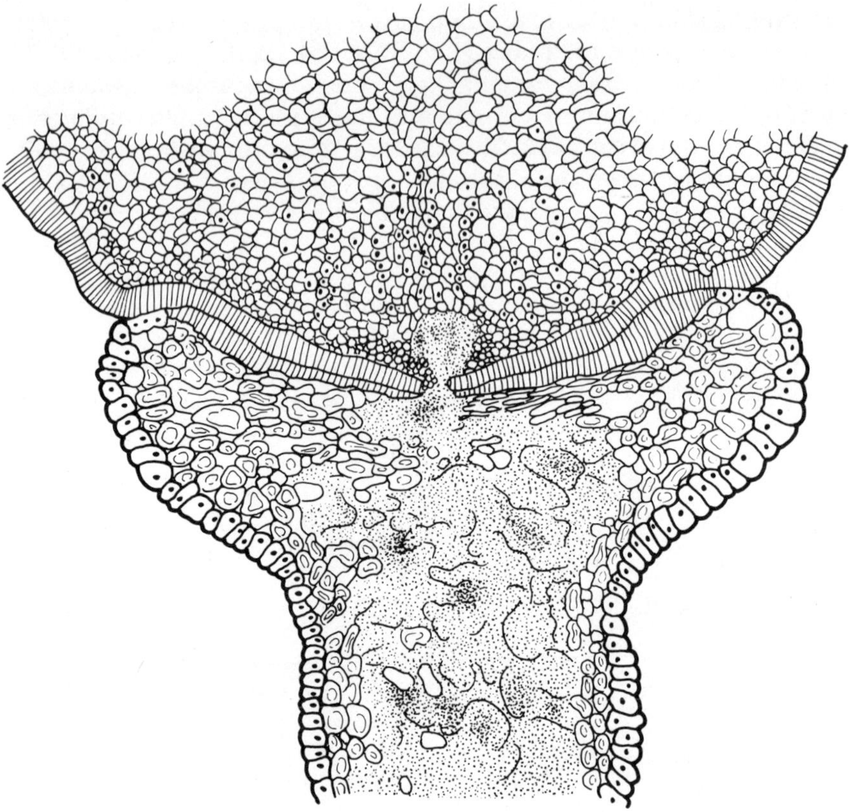

Fig. 12.2. Longitudinal section of the funiculus and lower part of a pea seed showing penetration of *Pseudomonas pisi* through the micropyle into the seed coat. (Skoric, 1927)

However, there was no indication that the infected plants were diseased, or that the seed was damaged by the presence of the fungus — only *F. equiseti* caused in some cases a brown decay of the maturing seeds. The saprophytic or apathogenically parasitic infections were amazingly frequent, 5–25 per cent of the seed being infected.

F. oxysporum f.sp. *lycopersici*, Fusarium wilt of tomato, produces vascular mycosis and penetrates through the fruit stalk and funiculus into the seed. Fusarium wilt of stocks, *F. oxysporum* f.sp. *matthiolae*, furnishes a similar example (Baker, 1952). Van der Spek (1972) established that seeds of beet and spinach are colonised by *Verticillium dahliae* invading from the vascular system of the mother plant.

Other systemic infections without external symptoms may be produced by *Epichloë typhina*, choke or cat's tail disease of grasses, which is seed-transmitted in *Festuca rubra* (Sampson and Western, 1954), and by *Chaetomium globosum* in *Lolium temulentum*, where the fungus grows into the ovule and later develops from the seed further into the seedling

(Günnewig, 1933). Katz (1949) confirmed this observation and also demonstrated symptomless, systemic invasion of *Alternaria lolii-temulenta* in *Lolium temulentum*. Another seed-borne endophyte, almost universally present in aerial parts of plants, is *Aureobasidium pullulans*. It is carried in the embryo, testa and pericarp of sycamore, *Acer pseudoplatanus* (Pugh and Buckley, 1971).

Ramularia cyclaminicola has occurred seriously in cultures of cyclamen in North America as the causal agent of a typical vascular mycosis, Ramularia leaf spot or stunt. The disease may develop as a stunt, as a wilt or as a leaf spot disease in which the fungus sporulates in the deteriorated tissues of the spots (Baker, Dimock and Davis, 1950). The hyphae penetrate through the vascular system into the young seeds where they remain as dormant mycelium in or below the seed coat (Jacques, 1942).

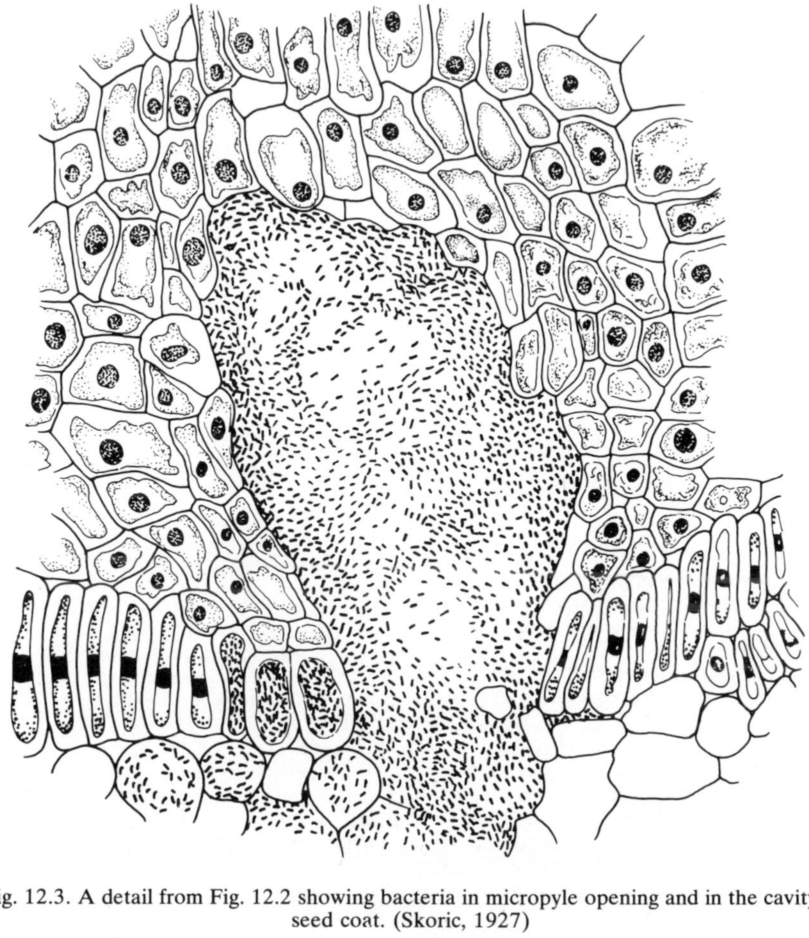

Fig. 12.3. A detail from Fig. 12.2 showing bacteria in micropyle opening and in the cavity in seed coat. (Skoric, 1927)

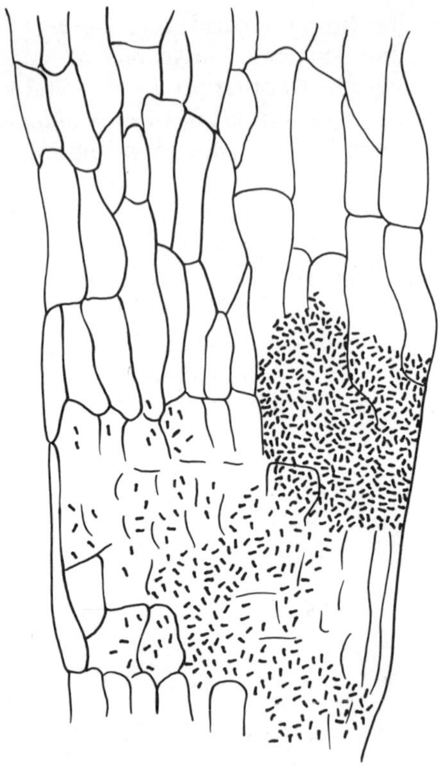

Fig. 12.4. Bacteria of *Pseudomonas lachrymans* advancing in and between the parenchyma cells of the funiculus of cucumber. (Wiles and Walker, 1951)

→

Fig. 12.5. Method of invasion of *Xanthomonas campestris* into seed of cabbage. (1) Longitudinal section through a portion of a healthy cabbage seed pod illustrating the relation of the funiculus to the maturing seed and pod walls: C, cotyledons; F, funiculus; FM, false membrane; H, hypocotyl; P, pod; SC, seed coat; X, xylem in the funiculus. Approx. ×94. (2) Vascular bundle in the wall of the seed pod in which the bacteria have completely disrupted all the xylem. The darker areas at the margins of the invaded region are an accumulation of dead bacteria, cell remains, etc. Approx. ×185. (3) Mesophyll tissue at the base of the funiculus extensively invaded by the bacteria. Some remains of spiral thickenings can be seen in the upper right corner. Approx. ×400. (4) Section through the bend of the funiculus in which the bacteria have caused considerable destruction of the xylem and have invaded the surrounding mesophyll tissue intercellularly. Approx. ×400. (5) Section of the funiculus at its junction with the seed, showing the destruction incited by the bacteria in invaded xylem elements. Approx. ×400. (6) Portion of an immature seed from a healthy pod, demonstrating the extension of some xylem elements into the immature seed coat from the funiculus. Approx. ×185. (Cook, Larson and Walker, 1952)

Goodey (1945*b*) traced invasion of stem eelworm, *Ditylenchus dipsaci*, from weakly affected mother plants of onion through the pedicel and placenta into the ovary wall and later through the short funiculus from which the nematodes may become attached to the seed in the vicinity of the hilum.

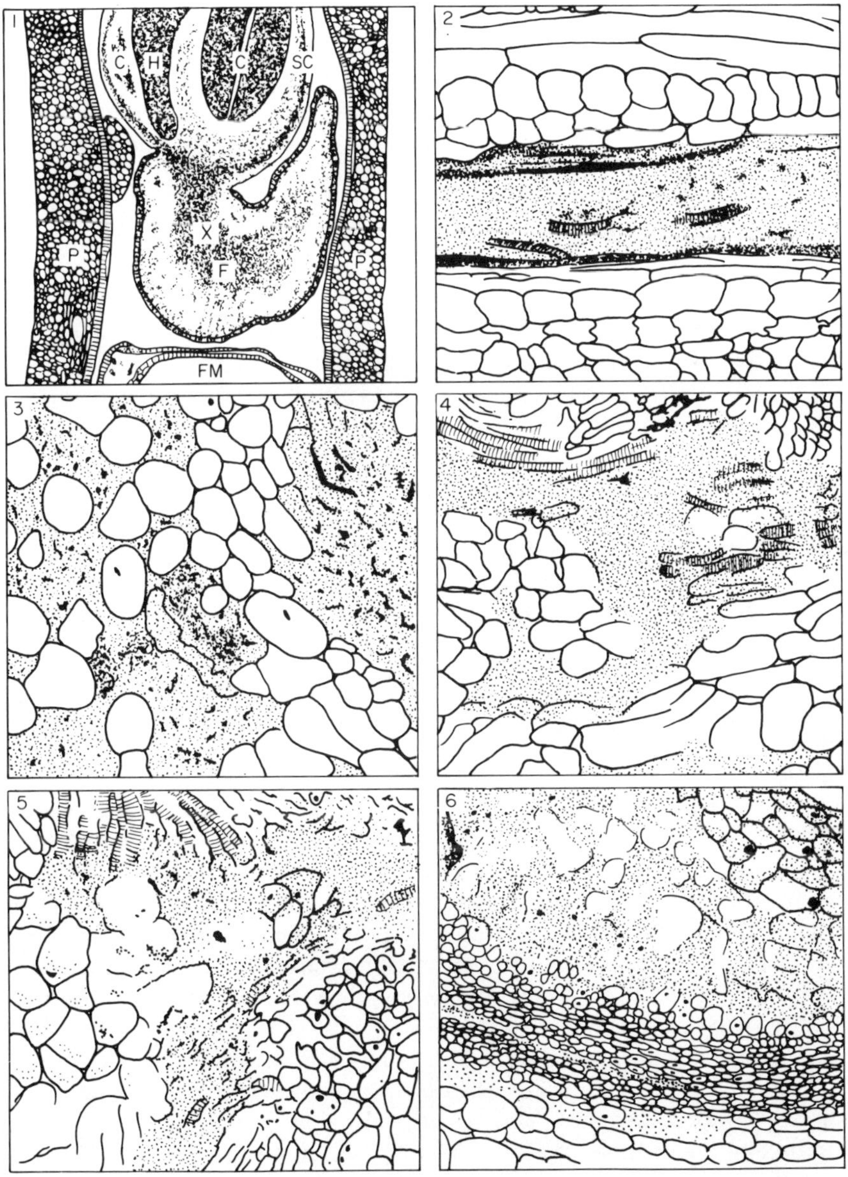

12.2 INFECTION FROM OUTSIDE

Inoculum may be transmitted from outside to the undifferentiated flowers, fruits or seeds and to the maturing fruit or seed, either as a result of dissemination from other plants or from local infections on the mother plant. Dispersal of the inoculum may occur by wind, rain drops, insects or, less frequently in cases of true seed infection, during the harvest and processing of the seed crop.

12.2.1 Stigma as path of infection

Distinction should be made between two kinds of blossom infection. The first, developing through the stigma and the style, leads to infection of the embryo. The second may infect the style and reach ovary and ovule but does not lead to infection of the embryo, infection being restricted to other parts of the seeds, for instance the seed coat from which locus the seedling later may be infected.

12.2.1.1 INTRAEMBRYAL INFECTION THROUGH STIGMA

There are several references in the literature to transmission through pollen, and here we may expect the possibility of the embryo type of blossom infection. Evidently this is true as far as pollen in relation to transmission of viruses is concerned. Blakeslee (1921) established pollen transmission of a seed-borne virus in *Datura stramonium*. Ikeno (1930) demonstrated that a mosaic of *Capsicum annuum* is pollen-transmitted, Reddick (1931) and Nelson and Down (1933) that bean common mosaic virus in bean may be transferred in this way.

Generally there seems to be a parallelism between pollen transmission and seed transmission of viruses. Bean yellow mosaic virus is transmitted by pollen and seed of *Melilotus alba* but neither by pollen nor by seed of *Phaseolus vulgaris* (Frandsen, 1952), whereas in this host bean common mosaic virus (*Phaseolus Virus* 1) is transmitted by both pollen and seed (Nelson and Down, 1933; Medina and Grogan, 1961). Bennett (1959) established that Lychnis ringspot virus is pollen- and seed-transmitted in *Lychnis divaricata* and *Silene noctiflora*. Gold *et al.* (1954) found rod-shaped particles of barley stripe mosaic virus in pollen of barley and in seed produced from healthy pistils pollinated by pollen from diseased plants. Other viruses that have been shown to be transmitted through pollen and seed are elm mosaic virus in American elm, *Ulmus americana* (Callahan, 1957), apple chlorotic leaf spot virus in raspberry, lettuce mosaic virus in lettuce (Ryder, 1964), raspberry ringspot virus in *Fragaria virginiana* and *Rubus* sp., tomato black ring virus in *Rubus* sp. and soybean, and arabis mosaic virus in *Rubus* sp. (Lister and Murant, 1967), cucumber mosaic virus in chickweed, *Stellaria media* (Tomlinson and Carter, 1970*b*), potato spindle tuber virus in potato and tomato (Fernow, Peterson and Plaisted, 1970; Singh, 1970), yellow dwarf virus in onion (Härdtl, 1972*a*), and alfalfa mosaic virus in lucerne (Frosheiser, 1974). Several stone fruit viruses are likewise pollen- and seed-transmitted (cf. discussion Section 13.1). Tobacco rattle virus may occur in pollen of systemically infected Petunia plants (Cadman, 1962) but it is uncertain whether seed transmission occurs.

On the other hand, sowbane mosaic virus infects pollen of *Atriplex coulteri* but is not seed-transmitted in this species but in other species of *Atriplex* (Bennett and Costa, 1961). Two viruses of grapevine are pollen-borne but not seed-borne in this host but they are pollen- and seed-borne in *Chenopodium amaranticolor* (Brückbauer and Rüdel, 1961; Dias, 1963). Apple chlorotic leaf spot virus is pollen-borne in *Chenopodium quinoa* and *C. amaranticolor* but not seed-borne in these hosts (Cadman, 1965). Phatak (personal communication, 1971) established pollen infection of potato virus X in *Petunia hybrida* but seed transmission has not been established.

C. O. Smith (1922) and Ark (1944*a*) have shown that pollen on *Juglans regia* may be contaminated by *Xanthomonas juglandis*, walnut blight, and Ark found that most of the flowers that were experimentally pollinated by contaminated pollen developed infected nuts. Ivanoff (1933) demonstrated that pollen of maize may be infected or contaminated by *X. stewartii*, bacterial wilt of maize. Hence, pollen transmission of this organism may probably also occur.

Loose smuts of barley and wheat, *Ustilago nuda* and *U. tritici*, are the classical examples of seed-borne fungi of the embryo-infecting type of blossom infection (Maddox, 1895, 1896; Sato and Yamada, 1896; Nakagawa, 1898). It has been claimed (Brefeld, 1903; McAlpin, 1910; Lang, 1910, 1917) that the infective diploid hyphae, produced after conjugation between germinating hyphae of the teliospores, penetrate between the papilla of the stigma and grow intercellularly through the style. However, Lang (1910) did observe that the germ tubes of *U. tritici* could not at first penetrate the sound stylar tissues of wheat flowers; only when the papillae of the stigma had begun to wither and collapse did the germ tubes penetrate between the cells. Reinvestigations of the infection procedure (Ruttle, 1934; Vanderwalle, 1942; Batts, 1955*a*; Pedersen, 1956*b*) did not confirm that the stigma is one of the paths of infection. Kozera (1968) found that although teliospores of *U. nuda* germinated well on the stigmas of both resistant and susceptible barley cultivars, no hyphae were seen penetrating through the style into the ovary. Thus, all the evidence points to penetration through the ovary wall as the only method of invasion. This type of infection will be further discussed below.

Also *U. violacea*, anther smut of carnation and other *Caryophyllaceae* (*Melandryum album*, *Dianthus*, *Saponaria*, *Stellaria* and others), has been claimed to be seed-transmitted and to infect through the stigma

(Brefeld and Falck, 1905) but this is evidently not so. The infection may occur through flowers, resulting in systemic invasion of the mother plant (H. G. Baker, 1947). The fungus also infects through seedlings, axillary buds and underground shoots (Hecke, 1907, 1926; Werth, 1911; Zillig, 1921; Liro, 1924; Blumer, 1941). When female plants of *Silene dioica* and *S. alba* are infected, the flowers produce anthers which in healthy female plants are only rudimentary; the anthers become filled with teliospores as in those of infected male plants (Werth, 1911). The mycelium is perennial and overwinters in the root stocks (Baar, 1903). Blumer (1941), who investigated *U. violacea* on *Melandryum,* found that normal and infected anthers may occur side by side in the same flower, while Werth (1911) recorded pollen grains and smut spores occurring side by side in the same anther. However, transmission through pollen and stigma with establishment of seed infection and transmission as claimed by Brefeld (according to Blumer) has not been confirmed.

Diehl (1950) demonstrated by infection experiments that *Balansia obtecta (Hypocreales)* infects the seed primordium through the stigma and style, establishing itself as dormant mycelium in the embryo and between the embryo and the endosperm of *Cenchrus echinatus.* These results need confirmation.

12.2.1.2 EXTRAEMBRYAL INFECTION THROUGH STIGMA

Examples of blossom infection, not leading to infection of the embryo but to infection of other parts of the seed, are loose smut of oats, *Ustilago avenae* (Falck, 1908; Zade and his co-workers, 1922, 1924; Gage, 1927) and barley leaf stripe disease, *Drechslera graminea* (Vogt, 1923). Both are examples of 'blossom and seedling infections' (Zade, 1924). They are transmitted to the stigma on which the spores are readily caught on the sticky surface. The hyphae grow along the style, do not grow into the embryo and endosperm, but spread into the parenchyma of the glumes and the epidermis of the caryopsis, where they establish themselves as a very tenacious dormant mycelium. Paxton (1922) demonstrated that dormant mycelium of *D. graminea* in 16-year-old vegetative herbarium material was able to develop conidia.

Botrytis anthophila, anther mould of red clover, belongs to the same type of blossom infection according to Silow (1933) and Bondarzew (1914), who found that the invasion of the germ tubes follows closely the path of the pollen germ tubes. Honeybees place the conidia along with the pollen on the stigmatic surfaces of healthy flowers. The conidia germinate, the hyphae penetrate the style, and the fungus establishes itself as 'intraseminal' mycelium in the seed primordium; the ovary develops normally, while the mycelium persists within the seed coat (Silow, 1933). Jost, Volken and Kern (1964) consider infection of the ovary through the stigma and style unlikely as the female organ contains one or more substances inhibiting growth of the fungus.

12.2.2 Nectary as path of infection

There is a possibility of the nectary being an entrance for pathogens, well known from *Erwinia amylovora*, fire blight of fruit trees in North America and Northern Europe. This pathogen infects ovaries and young twigs through the nectaries but is not known to be seed-transmitted. In cotton, however, entrance of boll rot fungi into the boll occurs largely via the inner and outer involucreal nectaries, and removal of nectaries is considered in breeding for resistance against boll rot (Cauquil and Ranney, 1967).

12.2.3 Ovary wall, pericarp and integuments of seed coat as paths of infection

Loose smuts of barley and wheat, *Ustilago nuda* and *U. tritici*, exemplify this type of seed invasion, as established by various workers (Ruttle, 1934; Vanderwalle, 1942; Simmonds, 1946; Batts, 1955*a*; Pedersen, 1956*b*; Loiselle and Shands, 1960; Kozera, 1968; Shinohara, 1972). The teliospores are disseminated by wind to young florets of healthy heads. They lodge on the feathery stigma, the style and on the ovary wall, and within one day the spores germinate as basidia or promycelia. The infection occurs in the lower part of the ovary or on the brush end of the kernels; the infective hyphae penetrate directly through the wall. The penetration may occur through any part of the epidermis, sometimes between the vertical walls of the epidermal cells, often through the centre of the cell wall. At the point of entry an appressorium is formed (Batts, 1955*a*; Shinohara, 1972), from which vigorous pointed hyphae enter the cells immediately below. Within two days some of the hyphae penetrate the epidermal cell layer and a few of the subepidermal ones. The hyphae continue transversing the parenchyma of the pericarp and grow directly towards the integuments, very likely due to chemotactic action originating from the young embryo as suggested by Pedersen.

The hyphae now follow the collapsed integuments towards the embryo (Pedersen, 1956*b*), they cross the integument into nucellular tissue (Shinohara, 1972), and after about 11 days the mycelium has entered the nucellus region, which it permeates, and after about 18 days it passes towards the base of the grain (Batts, 1955*a*). Three to four weeks after the germination of the teliospores the hyphae reach the embryo itself. On the dorsal side of the grain the fungus grows from the testa, nucellus and aleurone to the scutellum and passes to the meristem of the embryo (Fig. 12.6). The mycelium has not been found in the radicle (Batts, 1955*a*). In

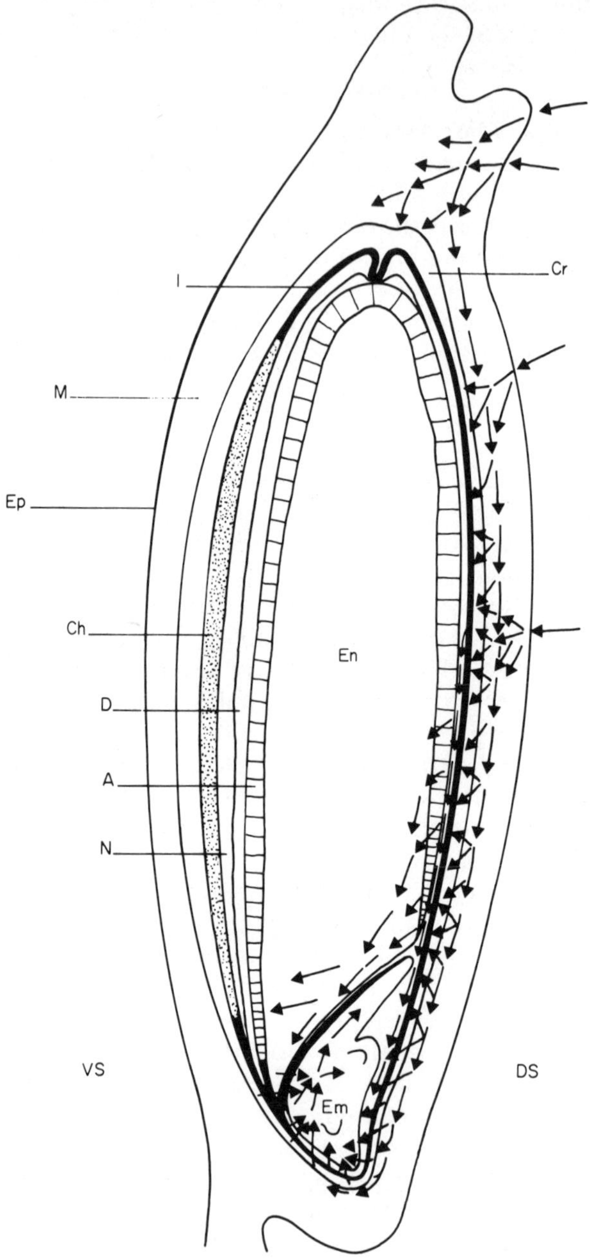

Fig. 12.6. The path of embryo infection of *Ustilago nuda* in the developing caryopsis of barley: A, aleurone layer; Ch, chalaza; Cr, cross-layer; D, disorganised tissue; Em, embryo; En, endosperm; Ep, epidermis; I, integument; M, middle layer; N, nucellar layer; DS, dorsal side; VS, ventral side. (Shinohara, 1972)

the later stage of infection, just before ripening, hyphae may be found in the endosperm, especially at the base (Pedersen, 1956*b*), cf. Fig. 12.6. Batts (1955) and Shinohara (1972) found the mycelium to be intracellular in the pericarp and testa and intercellular in the aleurone, scutellum and other tissues, whereas Lang (1917) and Vanderwalle (1942) found all mycelium to be intercellular. During the maturation of the caryopsis the mycelium is changed to a thick-walled, swollen, dormant mycelium.

The bean anthracnose fungus, *Colletotrichum lindemuthianum*, and the pea blight fungus, *Ascochyta pisi*, may penetrate the pod at different times during its development, and directly invade the maturing seeds; likewise *Phoma medicaginis (Ascochyta imperfecta)*, spring black stem, in lucerne (Mead, 1964). Bean pathogenic bacteria such as *Xanthomonas phaseoli*, common bacterial blight, may also penetrate through the tissues of the pods directly into the seeds (Burkholder, 1921). A similar development occurs in seed crops of crucifers infected by *Alternaria brassicae, A. brassicicola, A. cheiranthi* and *A. raphani*.

The flax parasites, *Colletotrichum lini*, anthracnose, and *Aureobasidium lini (Polyspora lini)*, browning or stem break, may invade the dying petals and maturing capsules of the host, penetrating through the wall of the capsule and through the funiculus into the seed coat (Lafferty, 1921; Johansen, 1943; Muskett, 1950). Muskett points out that these stages of infection are not always easy to detect in the field, and that field inspection of the seed crop in these cases does not provide complete information about infection of the seed.

Rhizoctonia solani penetrates from the pericarp of *Capsicum frutescens* into the placenta and further through the funiculus into the embryo and endosperm (H. G. Baker, 1947). The fungus cannot penetrate through the funiculus if cutinisation of the epidermis of the endosperm has occurred, unless fissures have been developed in the cuticle through growth pressures leaving a path of infection. Such fissures may occur in the thin layer of endosperm tissue around the radicle of the embryo.

Dying petals of crucifers may provide a foothold for mycelial invasion of *Sclerotinia sclerotiorum*, originating from wind-borne ascospores, into the primordium of siliquas (McLean, 1958).

12.2.4 Flower and fruit stalk as paths of infection

Like *Colletotrichum lini* and *Aureobasidium lini* on flax, mentioned above, *Acroconidiella eschscholtziae* on *Eschscholtzia californica* may also penetrate into the capsule from infected petals through the point of attachment without producing visible symptoms on the capsule (Davis, 1952). The mycelium may also penetrate in turn the fruit stalk, the capsule and the seed coat.

Septoria linicola (Mycosphaerella linorum), pasmo, penetrates also through the fruit stalk into the capsule and from the placenta through the funiculus into the seed coat without producing any external symptoms on the capsule (Loughnane, McKay and Lafferty, 1946). Sometimes pycnidia may develop profusely on the seed coat around the hilum.

Also vascular infecting fungi invade through the funiculus, as already indicated.

13 Infected or Contaminated Parts of the Seed

INTRODUCTION

The course of infection, including the selection of the point of entry and the colonisation of the seed, varies greatly from one pathogen to another. For instance, three major seed-borne parasites leading to fatal systemic infection of barley differ with regard to the development and the site of infection. Loose smut of barley, *Ustilago nuda*, has blossom infection leading to establishment of dormant mycelium in the embryo; covered smut, *U. hordei*, is predominantly confined to purely mechanical contamination of the surface of the pericarp; and, finally, *Drechslera graminea*, barley leaf stripe, establishes itself as dormant mycelium within the pericarp.

On the other hand, the anatomy of seed varies from one host plant genus or species to another. Seed with a large embryo, such as seeds of crucifers and legumes, may be expected often to carry the infection of fungi in the embryo. Seeds in which the embryo is a small part of the seed body, for instance in *Umbelliferae* and grasses, carry such infections much less frequently in the embryo. A small embryo is likely to be killed more readily than a large one, cf. Section 10.3.

Finally, the infection conditions, particularly weather conditions during the development and maturation of the seed, are factors decisive to the temporal development of infection and the organs being infected.

In assessing the capacity of an organism to infect different parts of a seed, these modifying factors should be taken into consideration, and references to seed parts being infected by a certain parasite taken with some reservation as they may include one of several possibilities. A general picture of the process of seed infection may best be achieved by following the development of the infection as started at a range of different stages in the development of the ovule and maturing seed.

The following parts of seed, this term taken in its botanical sense, may be infected or contaminated (Table 13.1).

For certain kinds of commercial 'seeds' such as those being caryopses or achenes, in which the pericarp acts as a seed coat, the delimitation between 'seed infection' and 'seed contamination' naturally may be modified accordingly.

Finally, a seed lot may be contaminated in the form of loose infected plant debris or parts of pathogenic organisms, e.g. sclerotia, mixed with the seeds (concomitant contamination).

Table 13.1. Location of infection and contamination of seeds

<table>
<tr><td rowspan="2">Seed infection</td><td>Embryo</td><td>Embryal infection</td></tr>
<tr><td>Endosperm
Seed coat/pericarp
(a) infection of seed coat/pericarp</td><td>Extraembryal infection</td></tr>
<tr><td rowspan="2">Seed contamination</td><td>(b) contamination of seed coat/pericarp</td><td rowspan="2">Contamination</td></tr>
<tr><td>Bract, infection or contamination</td></tr>
</table>

13.1 OVULE INFECTION

Demonstration of ovule infection is of interest principally as an establishment of the important fact that infection occurs in an early stage of the seed primordium. Ovule as a technical term is sometimes used rather loosely in literature; it is sometimes assumed to include also the immature seed. Here the word is used in its restricted sense, the nucellus enclosed in its one or two integument, i.e. the structure which after fertilisation develops into a seed; the young ovule, before differentiation of the embryo has taken place (Section 10.2).

Some of the seed-borne viruses that invade the ovule are not capable of surviving the development of the seed and consequently they are not transmitted. Bean common mosaic virus has been shown to occur in the ovule and persist in the embryo. Crowley (1957) established, by using dissected bean seeds as material for inoculation, that the virus was located in the embryo. He showed through the effect of prior- and post-fertilisation temperatures on seed transmission that the virus is capable of infecting the megaspore or embryo sac before fertilisation, not later (Table 11.2). Schippers (1963) studied the time of invasion of the virus into the ovule of the bean cultivar Beka. He divided flower buds into three groups (phases) according to their reaction to infection:

(1) Those which are infected during an early stage and may produce infected seeds (Phase 1).

(2) Those which become infected during a later critical stage and are aborted or are lost by acro-necrosis of main stem and shoots (Phase 2).

(3) Those which have passed the critical stage at the time of their infection and which later, in spite of infection of the plant, never produce infected seeds (Phase 3).

Anatomical studies revealed that infection of the ovule was established during the period at which the megaspore was still mono-nucleate (Phase 1). Virus particles could be detected in ovaries when they had differentiated into ovary walls and ovule-primordia, not before, possibly owing to their meristematic nature. When the embryo sac developed from the mono-nucleate to the eight-nucleate stage (Phase 2), cf. Fig. 10.3(4), the flower bud became necrotic and was dropped as a reaction to infection. Flower buds which had passed the critical period of necrosis (Phase 3) at the time they became infected produced healthy seeds. The limited period for establishment of ovule and embryo infections accounts for the restriction of transmission of the virus to about 15 per cent of seed from diseased plants, in this cultivar.

With tomato aspermy disease seed transmission is made impossible by virus-induced abortion (Caldwell, 1952). The virus interferes with the normal meiotic processes of both mega- and microsporogenesis.

Some examples of pollen transmission of viruses have already been mentioned (Subsection 12.2.1.1). Bean common mosaic virus may be transmitted in this way and infect the ovule (Reddick, 1931; Nelson and Down, 1933; Medina and Grogan, 1961). Transmission of Prunus necrotic ringspot virus (peach ringspot virus) and prune dwarf virus (cherry sour yellow virus) in cherry through pollen, either alone or in combination, has been demonstrated (Way and Gilmer, 1958; Gilmer and Way, 1961) — 20–25 per cent of the seed was shown to produce infected seedlings. Pollen transmission and ovule infection has been established for other viruses that infect stone fruit trees (for references, see Gilmer and Way, 1961).

As well as inducing abortion by their lethal effect on the megaspore, as in tomato aspermy, viruses may disturb development and function of the microspore or kill it thus producing pollen sterility. This is known to occur at different rates with tobacco streak virus in *Datura stramonium* (Blakeslee, 1921), tobacco ringspot virus in tobacco (Valleau, 1932*b*), tomato aspermy virus in tomato (Caldwell, 1952) and barley stripe mosaic virus in barley (Inouye, 1962) and wheat (Inouye, 1962).

It has been shown that some viruses may escape from ovules infected by pollen transmission and may invade the mother plant. Gilmer and Way (1963) found that pollen infected with prune dwarf virus infected two sweet cherry trees and three of sour cherry out of a total of 28 attempts of transmission but symptoms were recorded only in the third growing season after pollination. Also George and Davidson (1963) reported that Prunus necrotic ringspot virus as well as prune dwarf virus may be transmitted in cherry by pollen from tree to tree, and Cameron, Milbrath and Tate (1973) that Prunus necrotic ringspot virus was spread between and within sour cherry, Italian prune, and 'Brooks' prune by pollen transmission; none of deflowered trees acquired the virus. Cadman (1963) showed that raspberry cv. Lloyd George could be infected by pollen from raspberry plants infected by apple chlorotic leaf spot virus. Prunus necrotic ringspot is pollen transmissible also in Buttercup squash, *Cucurbita maxima*, which thus may account for natural spread in cherry orchards (Das and Milbrath, 1961).

Most viruses, however, are unable to survive in the microspore or megaspore. For further references on pollen transmission of virus, see Subsection 12.2.1.1.

Many parasitic fungi that invade the ovule cause deterioration of its tissues. This has already been mentioned for *Gloeotinia temulenta* in blind seed disease of ryegrass and holds true also for *Fusarium culmorum* invading rye and wheat flowers (Baltzer, 1930; Djerbi, 1971).

Epichloë typhina, choke disease of grasses, has been shown to be seed-transmitted in red fescue (Sampson, 1933; Wernham, 1942; Sampson and Western, 1954). In pedigree material of red fescue, investigated at Aberystwyth in Wales, 99 per cent infected plants were produced. In these plants the fungus was traced to all parts of the flower, the palea, lodicules, stamen and young ovaries. It was established that the embryo sac became infected when it encroached upon the nucellus, and that the mycelium finally formed a layer around the aleurone cells and occasionally penetrated the embryo; it may also penetrate the endosperm. Also pollen may be contaminated, but transmission by this means has so far not been observed. In most grasses, however, *E. typhina* kills the ovule and causes sterility. In cocksfoot, unlike red fescue, this fungus seems not to be seed-transmitted. In cocksfoot the stromatic sheath is so complete that total suppression of the panicle occurs by far in most cases, while in red fescue the inflorescence is not completely suppressed; hence viable and often infected seeds are produced on heavily affected stems (Western and Cavett, 1959).

Some parasites may penetrate into the ovule without injuring it but may not persist there. Cook (1930) found mycelium of *Peronospora destructor*, downy mildew of onion, in the ovule of onion, but it was not established whether the fungus was able to keep alive in the mature seed. Later investigations on seed transmission of this fungus (Stuart and Newhall, 1935; Yarwood, 1943) led to conflicting evidence and definite proof is still wanted.

13.2 EMBRYO INFECTION

The location of the pathogen within the embryo is dependent on the species, even the race or strain of the pathogen, and the species or even the cultivar of the host. The following parts may be infected: (1) the cotyledons, in *Gramineae* the scutellum, (2) the plumule, (3) the hypocotyl, and (4) the radicle. See Fig. 13.1.

Bean common mosaic virus has been established in 83 per cent of excised embryos of bean seeds (Crowley, 1957), and is carried in both cotyledons and plumule (Ekpo and Saettler, 1974). Athow and Bancroft

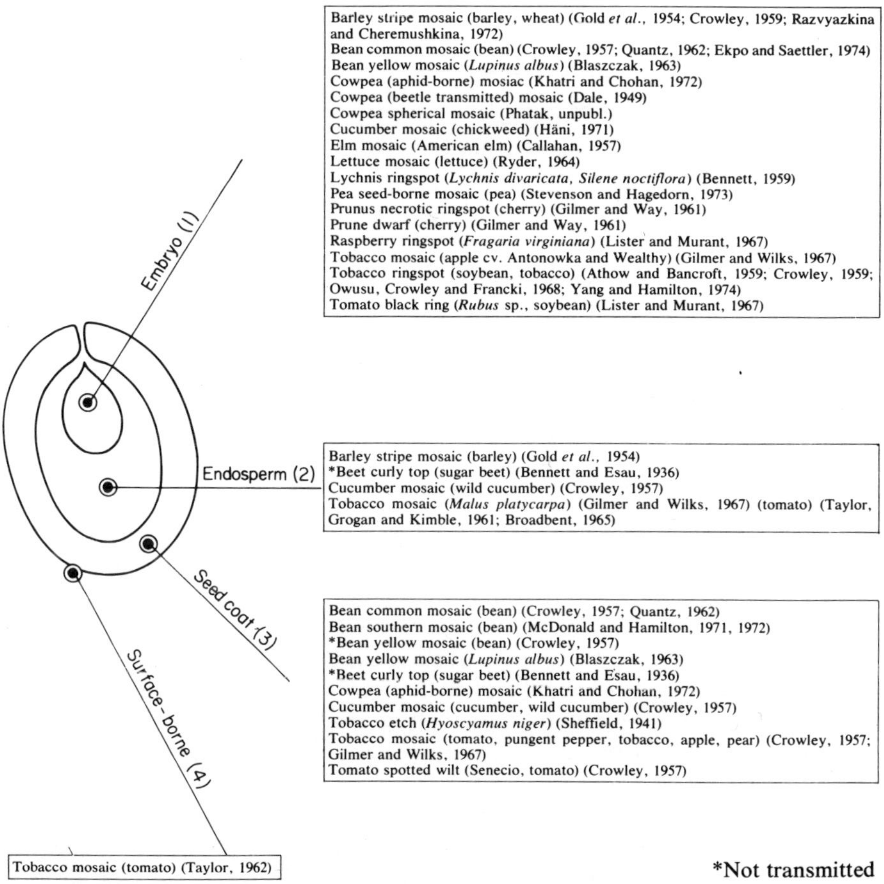

Fig. 13.1. Location of seed-borne viruses

(1959) demonstrated tobacco ringspot virus in the embryo tissues of soybean. The rod-shaped particles of barley stripe mosaic virus were found by Gold *et al.* (1954) in the embryo of barley and by Razvyazkina and Cheremushkina (1972) in the cytoplasm of the embryonic shield and the plumule.

Most seed-transmitted viruses are able to invade the host systemically and establish a compatible relationship with the tissues of the embryo allowing both virus and host to perpetuate themselves. However, there are viruses that do invade the embryo, but they do not survive as the seed matures. Cheo (1955) reported to have established bean southern mosaic virus in immature seeds of Logan bean, but unfortunately, in his study he did not take the precaution to decontaminate the dissected seed parts, a procedure essential for reliable results. McDonald and Hamilton (1972) also recovered the virus from immature embryos of Logan bean but when these were treated to remove surface contamination, infectivity was eliminated.

Stelzner (1942) reported that embryos newly excised from fresh seed of *Datura* and potato plants infected with potato virus X were in turn infected to an extent of 50–100 per cent. When the seed was stored for a short time the virus became avirulent. Sap pressed from immature seed of tobacco and tomato plants infected by potato virus Y was shown to be infective, whereas sap from seed stored for 6 days was not. Kausche (1940) found a similar process of inactivation for tobacco mosaic virus in seed of tobacco cv. Samsun.

A fairly restricted number of fungi are embryo-borne.

The embryo type of infection of loose smut was established by Maddox (1896) in Australia for *Ustilago tritici* and later confirmed by Brefeld (1903) in Germany. Investigations during the last 20 years have thrown light upon the significance of the location of the loose smut fungi, *U. nuda* and *U. tritici*, within the embryo of barley and wheat, in relation to predicting the amount of disease in the crop, and as an expression of susceptibility or resistance to loose smut. Loiselle and Shands (1960) compared the location of *U. nuda* in kernels of resistant and susceptible barley cultivars and found that in resistant kernels mycelium was present in the chalaza and the parenchyma associated with the vascular bundle while in a susceptible cultivar it occurred in the integument, aleurone, endosperm and embryo. Batts and Jeater (1958*a*) and Popp (1959) demonstrated by artificially infected susceptible and resistant wheat cultivars that there were three types of reaction to races of the wheat loose smut fungus. Some show resistance of the entire embryo — they are classified as immune — others show susceptibility of all tissues of the embryo except the plumule bud, and a third group of cultivars react by susceptibility of all the tissues including the plumule bud. Only infection within the plumule bud is directly correlated with smut in plants grown from this seed, and it was shown that smut estimates based on this criterion are highly accurate. However, this may not always be so. In certain cases distinction should be made between embryo resistance and adult plant resistance. Keystone barley in Canada is extremely embryo susceptible but mature plant resistant and the embryo count method is therefore no longer used for the determination of *U. nuda* in this cultivar (Wallen, 1964).

In late infection of *Gloeotinia temulenta* in ryegrass, i.e. more than one week after fertilisation, the mycelium spreads into the different parts of the embryo and endosperm, into the scutellum, plumule and radicle. The seed is least viable when the hyphae occur in the scutellum, where they are found in most of the ungerminating seeds (Wilson, Noble and Gray, 1945). *Diplodia zeae,* stalk rot, also occurs as mycelium in the embryo (Heald, Wilcox and Pool, 1909; J. H. Miller, 1952). Miller found the mycelium in apparently healthy maize between the scutellum and the interior parts of the embryo, and between the coleoptile and the plumule and immediately below the coleorhiza. In cotton *D. gossypina*, different *Fusarium* spp., and *Colletotrichum gossypii* are frequently encountered in the embryo (Roncadori, McCarter and Crawford, 1971).

Jacobsen and Williams (1971) frequently found *Phoma lingam* in the embryo of cabbage where the hyphae usually were limited to the margins or peripheral layers of the cotyledons but occasionally also the radicle was infected.

Vidhyasekaran, Subramanian and Govindaswamy (1970) detected *Fusarium moniliforme* in the embryo in 80 per cent of stored rice; as an ectophyte the fungus survived only for a short period. *Curvularia pallescens* and *C. lunata* had invaded the embryo in 38 and 25 per cent of the seeds, respectively. Djerbi (1971) found intercellular mycelium of *F. culmorum* in the embryo of wheat, particularly in the scutellum. Singh, Mathur and Neergaard (1974*a*) recorded 13 per cent embryo infection in *Sesamum indicum*, showing a total of almost 90 per cent infection of *Alternaria sesami*.

Arya and Sharma (1962) reported that mycelium of *Sclerospora graminicola*, the causal organism of green ear disease of bajra, *Pennisetum glaucum*, occurs in the embryo of seeds from infected portions of the ears.

The pathogen of crazy top of maize, *Sclerophthora macrospora (Sclerospora macrospora)*, also occurs within the embryo, as well as in the endosperm; plants from infected seeds developed the disease (Ullstrup, 1952; Ullstrup and Sun, 1969).

Sclerospora sacchari and *Sclerophthora rayssiae* var. *zeae,* downy mildew of maize, are borne in the embryo of maize, the aseptate, swollen, irregular hyphae being present mostly in the upper part of the plumule and adjoining coleoptile. Singh, Joshi and Chaube (1968) showed that crushed infected embryos could reproduce the diseases when placed on young unfolded leaves of 20-day-old plants, and that seeds of diseased plants, raised in autoclaved soil and kept in a saturated atmosphere, developed plants attacked by the respective downy mildews.

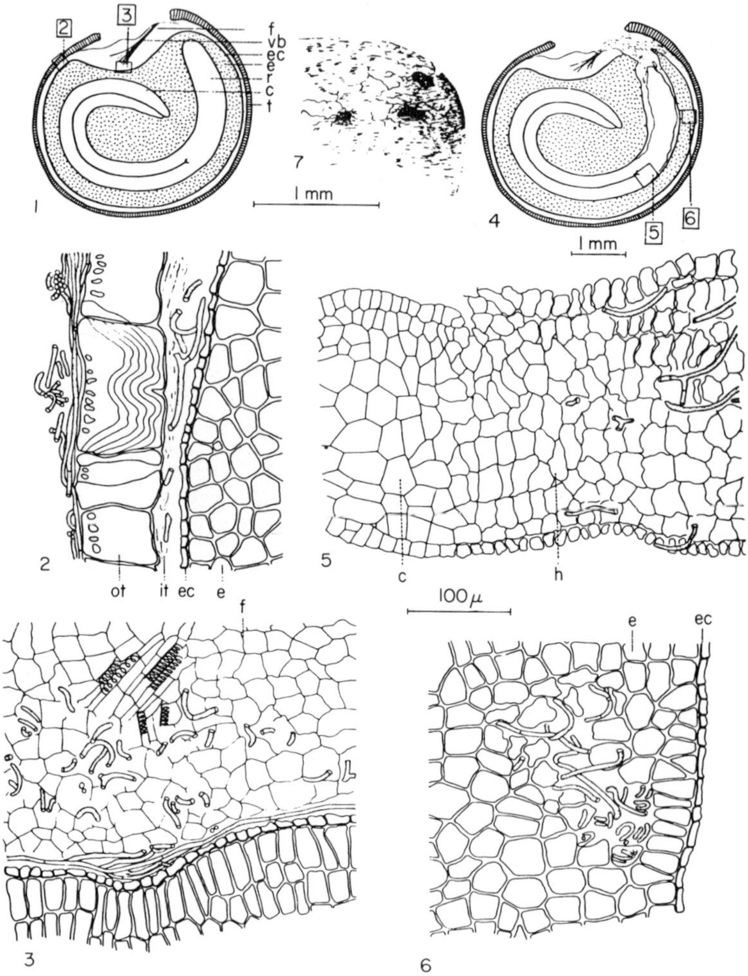

Fig. 13.2. Location of infection by *Rhizoctonia solani* in seed of *Capsicum frutescens*. (1)–(3) Seed with infection of funiculus and testa. (2) Mycelium on exterior and in inner layer of testa. (3) Mycelium in remnant of funiculus. (4)–(6) Seed with infection of embryo and endosperm. (5) Mycelium in hypocotyl. (6) Mycelium in endosperm. (7) Surface of seed coat, showing mycelium and small sclerotia. Framed indications 2, 3, 5 and 6 refer to enlargements in the subsequent figures. c, cotyledon; e, endosperm; ec, endosperm cuticle; f, remnant of funiculus; h, hypocotyl; it, inner layer of testa; ot, outer layer of testa; r, radicle; t, testa; vb, vascular bundle of funiculus. Camera lucida drawing by K. C. Baker. (K. F. Baker, 1947)

In California *Phytophthora cinnamomi* may affect mature fruits of avocado pear, *Persea americana*, when these are lying on the ground; the seed becomes infected, the fungus penetrates into the embryo, particularly into the cotyledons, and the cotyledons and plumule bud may be killed. The disease may be established in the nursery from infected seed (Zentmyer, personal communication, 1968).

In India *Phytophthora parasitica* var. *sesami*, sesamum blight, has been demonstrated as thickened mycelium in the embryo of *Sesamum orientale*, from which it could be isolated until four months after harvest. Seed transmission was not definitely proved but the disease was observed in many cases in fields which were never put under plough before sowing of sesamum (Sehgal and Prasad, 1966).

In sunflower seed *Plasmopara halstedii* may rarely be present in the embryo (Cohen and Sackston, 1973*a*).

For many pathogens the infection has been found primarily or exclusively to be confined to the cotyledons. Baker and Davis (1950*b*) found dormant mycelium of *Acroconidiella tropaeoli* (syn. *Heterosporium tropaeoli*), leaf spot and stem canker of nasturtium, in the cotyledons of *Tropaeolum majus* (Fig. 13.12). Atkinson (1950) established embryo infection of *Alternaria raphani* in radish, whereas *A. brassicicola* in *Brassica oleracea* has been found consistently confined to the seed coat (Domsch, 1957), and never encountered in the embryo below the infected seed coat (Boek, 1952). K. F. Baker (1947) established the location of inter- and intracellular mycelium of *Rhizoctonia solani* in the embryo and endosperm of *Capsicum frutescens* (Fig. 13.2).

The thick nutritious cotyledons of leguminous seeds often become more or less deeply infected and may be regarded as the main site for a number of fungi infecting the embryo of legumes, for instance *Ascochyta pisi*, leaf and pod spot, in pea, *Colletotrichum lindemuthianum*, anthracnose, in bean (Zaumeyer and Thomas, 1957), and *C. dematium (C. truncatum)*, anthracnose, in soybean (Tiffani, 1951). Dekker (1957) found that in 75 per cent of pea seeds infected by *Ascochyta pisi* the fungus had penetrated deeper than the seed coat and that 40 per cent of the epicotyls were infected.

Some parasites replace the whole seed or fruit by a sclerotium which may accompany the healthy seeds as impurities. Examples are *Ciboria alni* on *Alnus* and *C. betulae* on *Betula*, both members of the subgenus *Stromatinia*, characterised of being capable to transform seed primordia into sclerotia; different species of *Claviceps* and *Ustilaginoidea*, which transform the ovary of grasses into a sclerotium, e.g. *C. purpurea* on many grasses, *U. virens* on rice and other hosts.

Gupta (1962) showed that severe infection of *Protomyces macrosporus* in *Coriandrum sativum* leads to hypertrophied fruits in which the embryo is completely lacking. In less hypertrophied fruits the embryos are poorly differentiated and small. Only in fruits where the chlamydospores of the parasite are confined to the pericarp is a normal embryo present, and the seed germinates normally.

Embryo infection has been reported for some bacterioses, without precise indication of the location of the infection, e.g. *Corynebacterium michiganense*, bacterial canker in tomato (Bryan, 1930). *Xanthomonas malvacearum,* bacterial blight or black arm disease of cotton, invades mature seeds of *Gossypium* through the basal end at the chalaza and overwinters there (Tennyson, 1936; Brinkerhoff and Hunter, 1963), and *X. phaseoli* may pass from the seed coat into the region of the cotyledons which are subsequently invaded when the seed germinates (Zaumeyer and Thomas, 1957). In histological preparations of cucumber seed Naumann (1963) found *Pseudomonas lachrymans*, the causal agent of angular leaf spot, in the embryo where the bacteria were confined to the outermost layers of the radicle.

Diercks and Klewitz (1962) and Schreiber (1974, in litt.) established *Ditylenchus dipsaci*, stem eelworm, beneath the testa in lesions on the cotyledons near the hypocotyl of *Vicia faba*, see Fig. 6.4. Artificial inoculation on cotyledons reproduced dark brown necrotic patches of the type observed on naturally infected seeds.

13.3 ENDOSPERM INFECTION

Gold *et al.* (1954) found particles of barley stripe mosaic virus also in the endosperm but Inouye (1962), who established that the frequency of endosperm infection was almost equal to embryo infection, found evidence for the assumption that the virus in the endosperm is not responsible for seed transmission. Crowley (1957) found cucumber mosaic virus to be present in the endosperm of immature seed of wild cucumber, *Echinocystis lobata* — there is no endosperm in mature seed as this tissue is absorbed by the developing embryo. For further references see Fig. 13.1.

Weston (1920) found mycelium of *Sclerospora philippinensis*, Philippine downy mildew, in the endosperm and seed coat of apparently healthy maize grains, whereas the fungus was not present in the embryo.

→

Fig. 13.3. Location of Stewart's wilt bacteria in maize seed. Above: Bacterial cavities of various sizes, the largest shown at (a), situated in the chalazal region of a kernel. The endosperm is shown to the right of these cavities at (b). Some of the vascular tissue of the kernel is shown at (c), adjoining a bacterial cavity. ×90. Below: Enlarged view of a bacterial cavity situated in the chalazal region of a maize kernel. Remnants of spiral vessels in the cavity are shown at (a). The cells near the border line of the cavity at (b) appear to have collapsed. ×300. (Ivanoff, 1933)

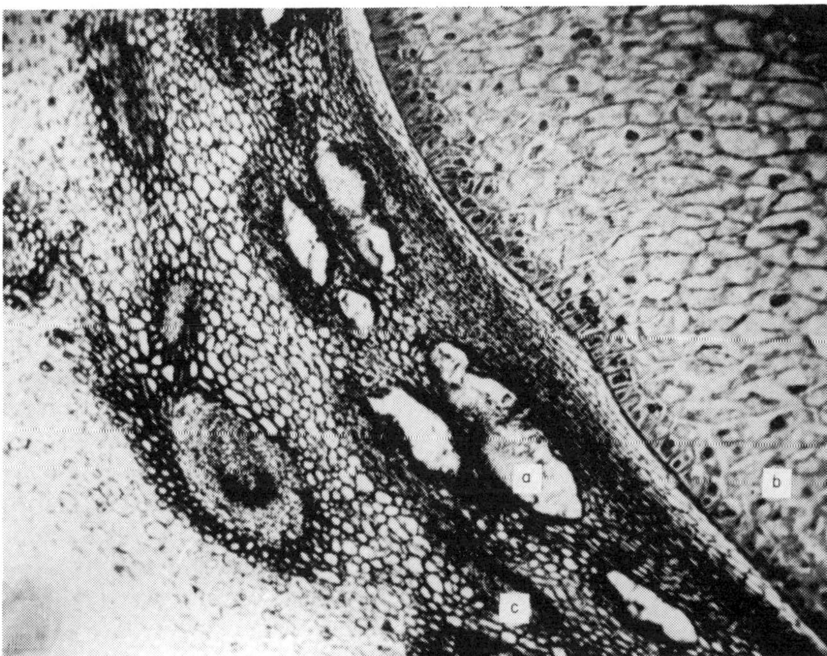
a
b
c

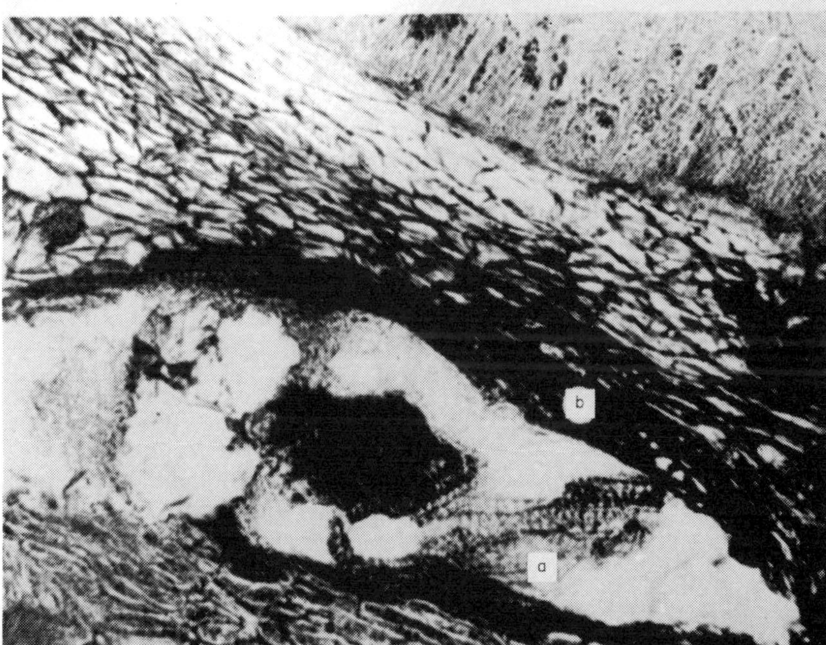
b
a

Heald, Wilcox and Pool (1909) established that mycelium of *Diplodia zeae* is present in both endosperm and embryo of maize, and J. H. Miller (1952) that the fungus first invades the tissues of the embryo and thereafter penetrates into the starchy and horny endosperm and the pericarp. Meffert (1950) reported that *Drechslera papaveris (Pleospora papaveracea,* syn *P. calvescens)* is transmitted in the endosperm of *Papaver somniferum* and Fazli and Schroeder (1966) that *D. oryzae* invades the endosperm of rice. K. F. Baker (1947) observed inter- and intracellular mycelium of *Rhizoctonia solani* in the endosperm as well as in the embryo of *Capsicum frutescens*. *Fusarium moniliforme* is often present in the endosperm of sorghum (Mathur, Mathur and Neergaard, 1974).

Erwinia stewartii, bacterial wilt of maize, is transmitted to the endosperm of the host. Rand and Cash (1921) isolated the bacterium from the tissues of the endosperm, and Ivanoff (1933) demonstrated that it occurs intracellularly in the chalazal tissues (Fig. 13.3) and between the innermost row of cells of the chalazal region and the marginal layer of the endosperm and in the endosperm. Cormack and Moffart (1956) found that *Corynebacterium insidiosum*, bacterial wilt of lucerne, was confined mainly to the aleurone layer of the endosperm, Naumann (1963) that *Pseudomonas lachrymans*, angular leaf spot in cucumber, infected the endosperm in the outermost layers near the micropyle.

13.4 SEED COAT INFECTION

Some highly infective seed-borne viruses are carried in the seed coat. Crowley (1957) demonstrated the viruses of bean common mosaic, cucumber mosaic, tomato spotted wilt and tobacco mosaic in the testa of their hosts (see Fig. 13.1). Tobacco mosaic virus was found by the indexing procedure to be present in the seed coat of freshly harvested seeds of tobacco and tomato, but in very low concentrations in tobacco seeds stored for 9 months, and it was not present in tomato seeds stored for 4 months (Taylor, Grogan and Kimble, 1961; Taylor, 1962). Some, perhaps most viruses, carried in the seed coat, are not transmitted. Sheffield (1941) found intranuclear inclusion bodies of tobacco etch virus in the seed coat of tobacco but not in the embryo and she established that the virus is not seed-transmitted. Curly top virus and bean yellow mosaic virus are present in the seed coat of sugarbeet (Bennett and Esau, 1936) and bean (Crowley, 1957), respectively, but not in the embryo, and these viruses are also not seed-transmitted.

The seed coat is the common site of infection of most seed-transmitted Fungi Imperfecti. Such infections, established as dormant mycelium, have been demonstrated for *Alternaria brassicicola* in *Brassica* (Boek, 1952; Domsch, 1957), *A. brassicae* (Chupp, 1935), *A. longipes* in tobacco (Heursel, 1961), *A. raphani* in radish (Atkinson, 1950), *A. sesami* in *Sesamum indicum* (Singh, Mathur and Neergaard, 1974*a*), *Cercospora kikuchii* in soybean (Ellis, Ilyas and Sinclair, 1975), *Colletotrichum lindemuthianum* in bean (Grümmer and Mach, 1955), *C. dematium* in soybean (Schneider *et al.*, 1974), *Drechslera oryzae* in rice (Nisikado and Nakayama, 1943; Fazli and Schroeder, 1966), *D. papaveris* in poppy, mainly between the layer of cross cells and that of reticular cells (Schmiedenknecht, 1958), *D. sorokiniana* in wheat (Weniger, 1923), *Acroconidiella eschscholtziae* in *Eschscholtzia californica* (Davis, 1952), *Cercospora medicaginis (C. zebrina)* in *Medicago hispida* (and/or *M. maculata*?) (Hopkins, 1921), *C. kikuchii* in soybean (Kilpatrick, 1952; Deutschmann, 1953), *Colletotrichum lini, Aureobasidium lini* and *Septoria linicola* in flax (Lafferty, 1921; Johansen, 1943; Loughnane, McKay and Lafferty, 1946; Muskett, 1950), and *Botrytis cinerea* in flax (Van der Spek, 1965), *Fusarium oxysporum* f.sp. *matthiolae* in stocks (Baker, 1952). In red clover Bennum (1972) established the presence of *Botrytis anthophila* as intercellular mycelium in the seed coat, predominantly in the stone cell layer. In cabbage *Phoma lingam* is most commonly found in the outer epidermis and subepidermal parenchyma, occasionally in the inner epidermis of the outer integument and in the inner integument (Jacobsen and Williams, 1971). K. F. Baker (1947) found mycelium of *Rhizoctonia solani* in the seed coat of tomato and pungent pepper, *Capsicum pubescens* (Fig. 13.2).

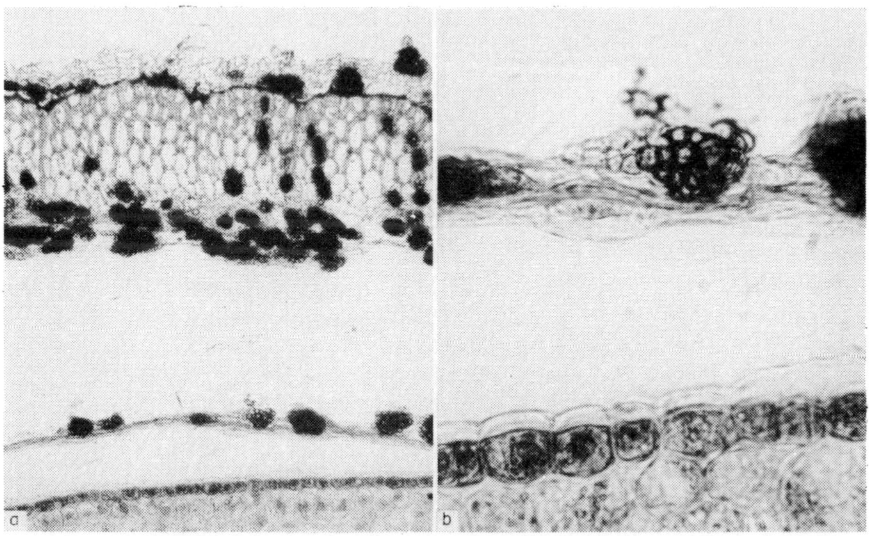

Fig. 13.4. Microsclerotia of *Verticillium alboatrum* in hull and testa of sunflower, longitudinal section. Left, ×200; right ×800. (Courtesy of W. E. Sackston)

Domsch (1957) established that *Alternaria brassicicola* penetrates the epidermis of the seed coat of maturing seeds of *Brassica oleracea*. The large epidermal cells may be densely colonised by heavily sporulating mycelium. The fungus is not capable of penetrating into the deeper layers of mature seed, apparently because a layer of thick-walled palisade cells (stone cells) in the seed coat immediately beneath the epidermis forms a mechanical barrier.

Seed coat infection of *Ascochyta pisi*, leaf and pod spot, in pea is just as important as infection of the embryo. Dekker (1957) obtained with seeds from which the seed coat had been removed 5 per cent diseased plants, whereas the same number of non-peeled seeds yielded 18 per cent.

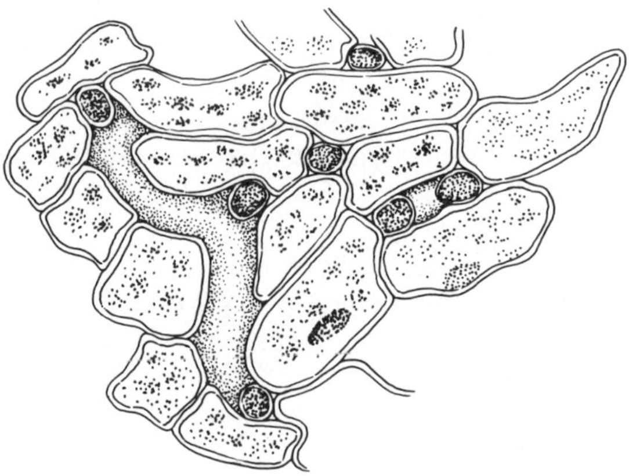

Fig. 13.5. Section of the seed coat of *Pisum sativum* showing mycelium of *Peronospora viciae* in the intercellular spaces. (Melhus, 1931)

In some cases fruiting bodies are developed during the maturation of the seed, as for instance macroconidial sporodochia of *Gloeotinia temulenta* in blind seeds of *Lolium perenne* (Wilson, Noble and Gray, 1945) and pycnidia of *Septoria* spp. on seeds of celery and parsley. Microsclerotia of *Verticillium alboatrum* have been shown to be confined to the hull and testa of sunflower (Sackston and Martens, 1959), see Fig. 13.4; Klisiewicz (1963) established that *Fusarium oxysporum* f.sp. *carthami* is transmitted in the achene of safflower, *Carthamus tinctorius*, and traced hyphae in relatively deep infected achenes in the seed coat, where they were found in the sclerenchyma and the spongy parenchyma.

Some of the few cases of established seed transmission within *Phycomycetes* provide examples of seed coat infection: *Sclerospora philippinensis* in maize (Weston, 1920), *Peronospora farinosa (P. schachtii)* in beet (Leach, 1931), *P. viciae* (syn. *P. pisi*) in pea (Melhus, 1931, see Fig. 13.5; Campbell, 1935).

Seed-borne bacteria are often referred to as 'deep-seated' but are usually confined to the seed coat. *Xanthomonas phaseoli,* common bacterial blight of bean, exemplifies this relationship, penetrating through the funiculus and raphe into the seed coat (Burkholder, 1921; Zaumeyer, 1930), see Fig. 13.6, and *Pseudomonas pisi* is also carried in the seed coat (cf. Fig. 13.7, Skoric, 1927). *X. campestris,* black rot of crucifers, invades to some extent the seed coat through the vascular system in the funiculus (Walker, 1950) and this holds true also for the bacterial blight organism of

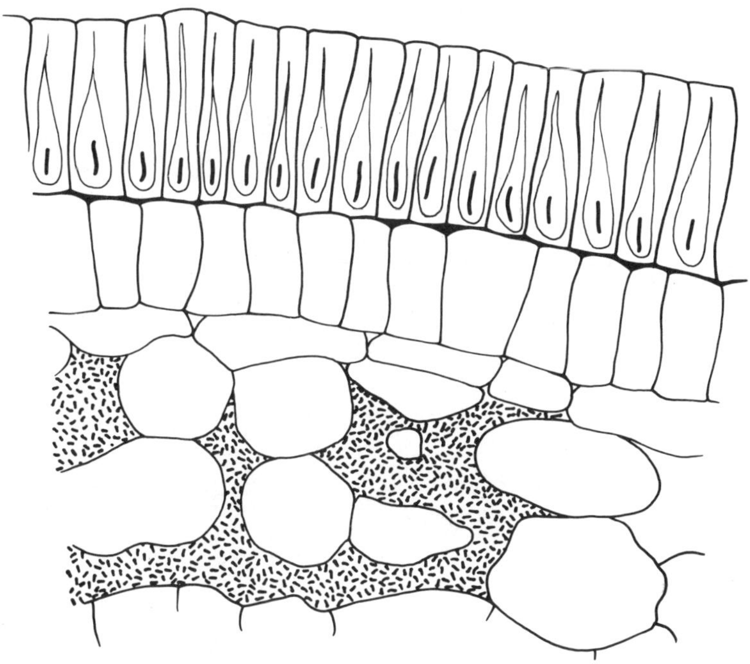

Fig. 13.6. Bacteria of *Xanthomonas phaseoli* in the larger intercellular spaces of the third layer of the seed coat of Phaseolus bean. ×1,250. (Zaumeyer, 1930)

stocks, *X. incanae* (Kendrick and Baker, 1942; Baker, 1952). Dunleavy (1962) reported that *Corynebacterium* sp. which causes a severe stunt in soybean, was transmitted in vascular elements of dormant seed obtained from eight soybean-producing states in the U.S.A. Brinkerhoff and Hunter (1963) found *X. malvacearum*, black arm, in both the chalazal and micropylar halves of the seed coat in *Gossypium hirsutum* and *G. barbadense*, and *Corynebacterium michiganense* invades through the chalaza and continues up to the innermost cells of the seed coat but not into the endosperm of embryo (Patiño-Méndez, 1967).

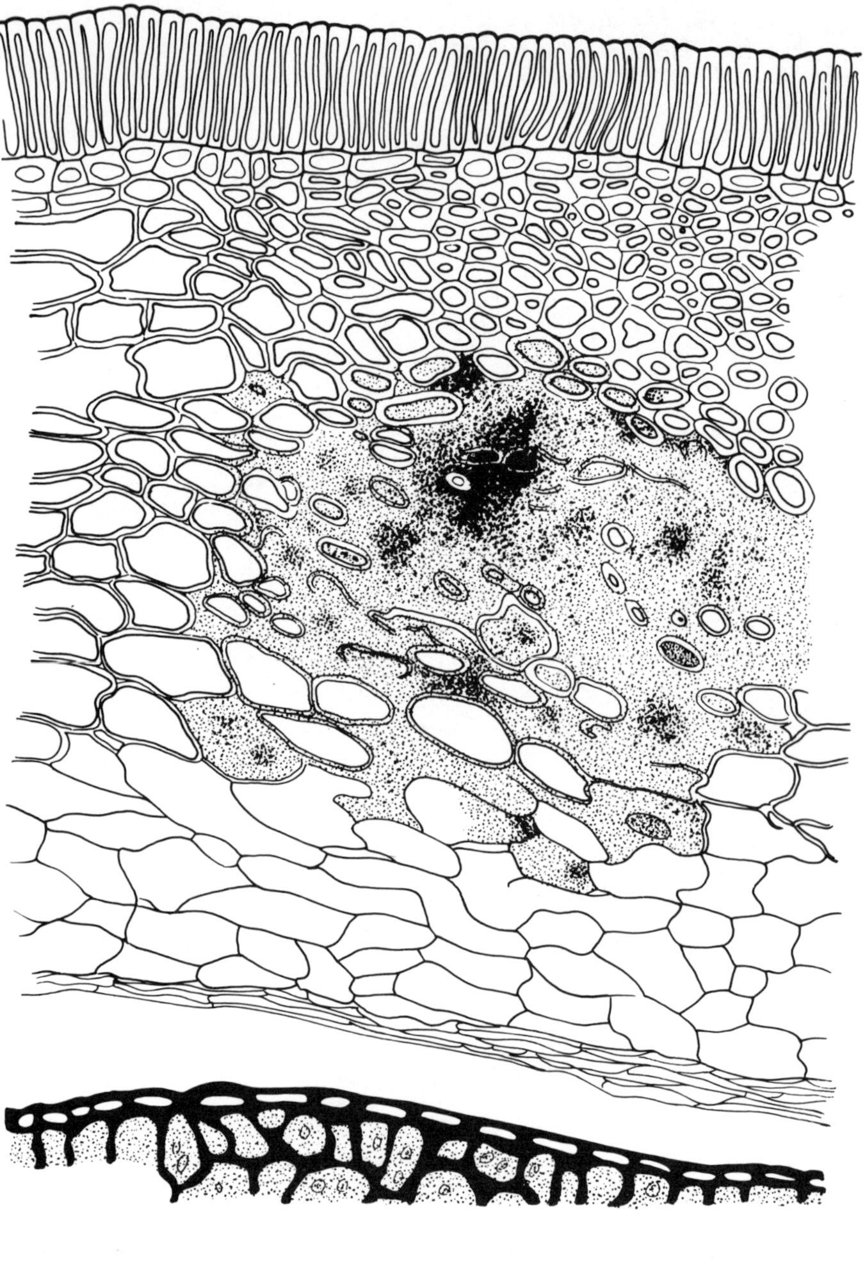

Fig. 13.7. Cross-section of seed coat of a ripe pea seed showing the location of *Pseudomonas pisi* bacteria. (Skoric, 1927)

13.5 PERICARP INFECTION

From a practical point of view only those cases of pericarp (or fruit coat) infection are of interest in which the pericarp or parts of it accompany the seed for sowing. In many hosts infection of the fruit may suppress development of seeds as in fruits of apple, pear, plum, etc., mummified by *Monilia*. In other cases the infected pericarp accompanies the seed which may remain intact. In hazel nuts, infected by *Monilia fructigena* (Buchwald, 1943), the kernel may be intact or it may be infected, or at a later time transmission of the fungus to the seedling may be expected to occur, because the infected nut shell is present with the seed.

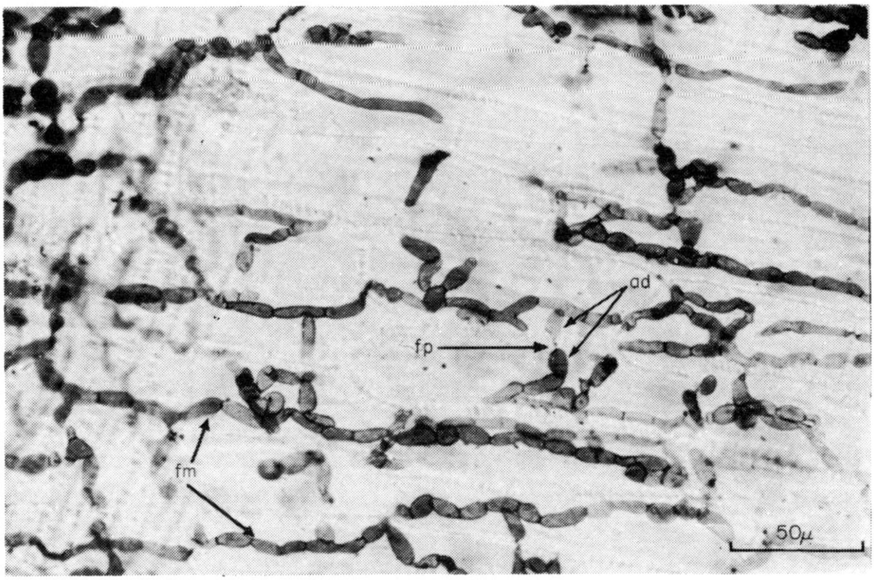

Fig. 13.8. Intracellular necklace-like hyphae of *Fusarium culmorum* in the pericarp of wheat. The hyphae are vigorously pointed when they penetrate through the walls between neighbouring cells, as shown at fp. ×400. (Courtesy of M. Djerbi)

In the caryopsis of *Gramineae* and achene of *Compositae* the pericarp acts like seed coat. Many species of *Drechslera* parasitic in cereals and grasses establish themselves as dormant mycelium in the pericarp as they do in the seed coat, *D. sorokiniana* in wheat (Weniger, 1923) and barley (Mead, 1942*a*), *D. oryzae* in rice (Nisikado and Nakayama, 1943; Fazli and Schroeder, 1966), *D. graminea* in barley, particularly in the apical part of the grain (Vogt, 1923; Genau, 1928; Platenkamp, 1975).

Djerbi (1971) carried out histological and ecological investigations on the mechanism of infection of *Fusarium culmorum* in the caryopsis of wheat. The pericarp is more or less deeply invaded. Infection may be found (1) only in the outer layers, confined to the epicarp (Figs 13.8 and 13.9); (2) any layers of the pericarp and the integuments may be invaded; (3) all parts of the caryopsis may be invaded, including the hyaline band and the aleurone layer (Fig. 13.10), where mycelial cushions may develop, and the embryo may be partially or entirely covered by a mycelial network (Fig. 13.11). The extent of invasion reflects the actual time of inoculation and infection. When this takes place near maturity of the caryopsis, only the outer layers will be invaded and the hyphae will not cross the 'first morphological barrier' provided by the endocarp; when infection takes place at an earlier date, colonisation may reach down to the layers of the integuments, which constitute the 'second morphological barrier'; when the caryopsis is invaded still earlier the hyphae are not arrested by this barrier — the endosperm and often the region surrounding the embryo are invaded, particularly at the ventral side of the caryopsis.

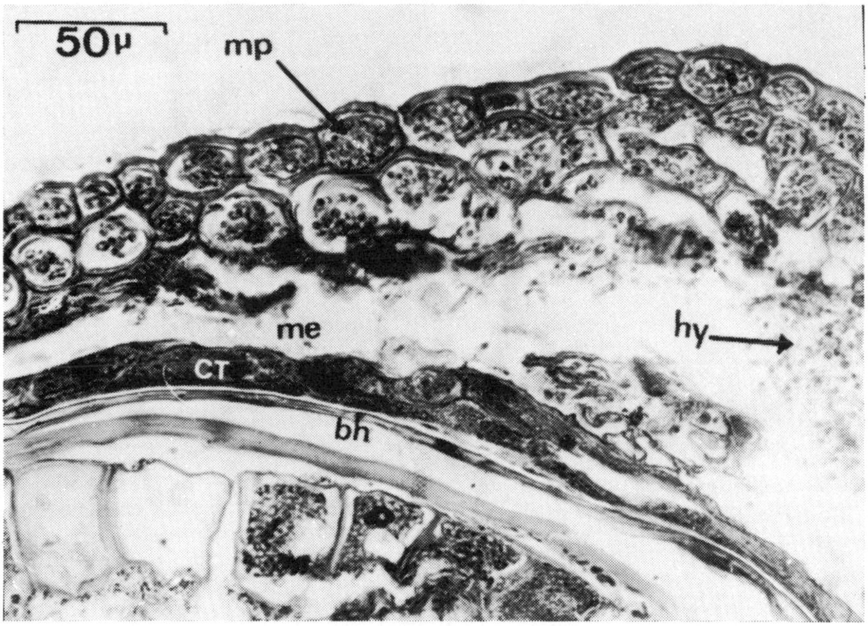

Fig. 13.9. Transversal section of the grain of wheat infected by *Fusarium culmorum*. The cells of the pericarp are partially or completely invaded by the mycelium (mp), and also the mesocarp (me) is invaded (hy). A barrier against further penetration is provided by the layer of transversal cells of the endocarp (CT). ×400. (Courtesy of M. Djerbi)

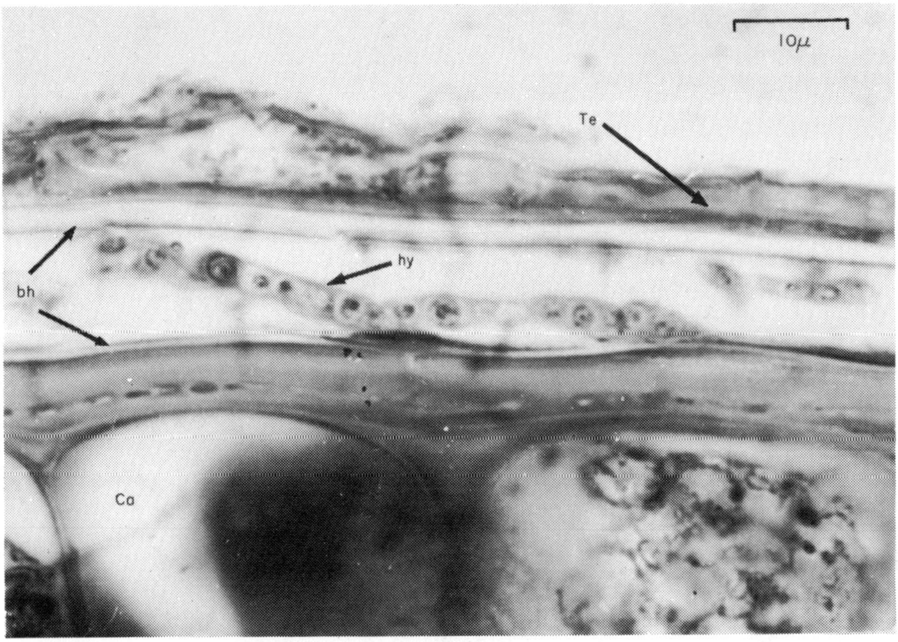

Fig. 13.10. An infective hypha in the 'hyaline band' between the seed coat (Te) and the aleurone layer (Ca). ×1,500. (Courtesy of M. Djerbi)

Oospores of *Sclerophthora macrospora*, crazy top, may be carried in the pericarp of maize and ragi, *Eleusine coracana*, oospores of *Sclerospora sorghi*, downy mildew, in the pericarp of *Sorghum vulgare* (Safeeulla, 1971) and as mycelium in maize (Jones, Leeper and Frederiksen, 1972), *Sclerospora philippinensis* is carried as mycelium in the pericarp of maize (Advincula, 1975) and *Plasmopara halstedii*, downy mildew, as dormant mycelium in the inner parenchymatic layers of the pericarp of sunflower (Delanoe, 1972), or as oospores in the pericarp (Cohen and Sackston, 1973*a*); *Acroconidiella tropaeoli* is carried as dormant mycelium in the pericarp of *Tropaeolum majus* (Baker and Davis, 1950*b*), see Fig. 13.12, *Sclerotinia sclerotiorum* as mycelial strands in the thick-walled fibrous cells of the pericarp and in the thin-walled parenchyma cells of the testa and inner layers of the pericarp of sunflower (Tollenaar and Bleiholder, 1971), while chlamydospores of *Protomyces macrosporus* are carried in the pericarp of weakly infected fruits of *Coriandrum sativum* (Gupta, 1962).

Fusarium oxysporum f.sp. *carthami* spreads intra- and intercellularly in tissues of the pericarp and seed coat of safflower. Intercellular growth was found limited to the sclerenchymatous hypoderm tissue and parenchyma (Klisiewicz, 1963).

Fig. 13.11. Wheat embryo partially covered by mycelial masses of *Fusarium culmorum*. (Courtesy of M. Djerbi)

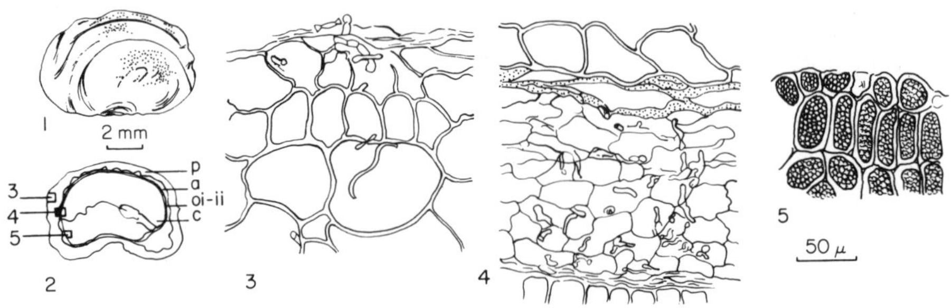

Fig. 13.12. Infection of *Acroconidiella tropaeoli* in the pericarp of *Tropaeolum majus*. (1) External view of fruit showing dark mycelial masses in the pericarp. (2) Section of fruit indicating location, 3, 4 and 5, of enlargement shown in (3), (4) and (5). a, air spaces; c, cotelydon; ii, inner integument; oi, outer integument; p, pericarp. (3) Section of pericarp showing mycelium. (4) Section showing infected tissues of the integuments between inner pericarp and cotyledon. (5) Section of infected cotyledon. Camera lucida drawing at two magnifications: (1)–(2) comprise one group, (3)–(5) another. (Baker and Davis, 1950)

Septoria apiicola has been demonstrated as mycelium in the pericarp and testa of celery (Sheridan, 1966).

In the pericarp and seed coat of *Beta vulgaris* Leach (1931) found mycelium as well as oospores of *Peronospora farinosa (P. schachtii)*; 27 seed samples from infected mother plants and 18 commercial seed samples yielded an average of 0.2 per cent seedlings showing primary infection of the pathogen.

13.6 CONTAMINATION OF SEED COAT AND PERICARP

Many pathogens are carried on the surface of the seed. This holds true also for certain viruses, tobacco mosaic virus providing an interesting example. Taylor, Grogan and Kimble (1961) showed that the virus is held externally on the testa of tomato, and from the testa the seedling is contaminated and infected during transplanting. With tobacco, on the other hand, the virus could not be demonstrated externally on the testa, perhaps because it may be so easily dislodged that it would not survive long enough to contaminate the seedling. This difference between the conditions of survival of the virus on the seed coat of the two hosts could be related to the different types of seed of tobacco and tomato. Freshly extracted tomato seeds are surrounded by a gelatinous matrix which when it dries thus seals the virus on to the hairy seed coat, whilst the testa of a tobacco seed is brittle and sheds freely and cleanly from the pod when it is mature (Taylor, 1962).

Spores and other reproductive organs are often carried on the surface of the seeds. In this way smut fungi may be transmitted as well as most of the seed-borne species of *Alternaria*, e.g. *A. brassicicola* on crucifers, *A. linicola* on flax, *A. tenuis* in numerous hosts, *A. longipes* on tobacco (Heursel, 1961). *Fusarium*, e.g. *F. avenaceum*, *F. culmorum*, *Drechslera*, e.g. *D. sorokiniana*, *Stemphylium*, e.g. *S. botryosum*, *Ulocladium*, e.g. *U. consortiale*, *Aureobasidium lini* (Lafferty, 1921), and many other seed-borne fungi. Seeds of soybean may be encrusted by the oospores of *Peronospora manshurica*, downy mildew (Hildebrand and Koch, 1951; Johnson, Chamberlain and Lehman, 1954; Dunleavy, 1971), and Campbell (1935) observed oospores of *P. viciae (P. pisi)* adhering to seeds of pea. Uredospores of *Puccinia antirrhini* have been found to adhere to superficial irregularities of the seed of *Antirrhinum majus* and shown to act as source of infection (Simonyan, 1961).

Many bacteria are carried on the surface of the seed, e.g. *Xanthomonas vesicatoria* on tomato, *Corynebacterium flaccumfaciens*, bacterial wilt, in bean, *Pseudomonas lachrymans*, angular leaf spot, on cucumber, *X. incanae* on *Matthiola incana*, *X. campestris* on cabbage (Fig. 13.13).

Sclerotia of *Botrytis cinerea* and *Sclerotinia sclerotiorum* may be attached to the seeds.

Goodey (1945*a*) found *Ditylenchus dipsaci* attached to the seed of onion in the region of the hilum.

Contamination of the pericarp is of practical importance in species where pericarp and seed coat are fused. Examples are *Cichorium* with teliospores of *Puccinia cichorii*, carrot with conidia of *Alternaria radicina, Tropaeolum* with conidia of *Acroconidiella tropaeoli* (syn. *Heterosporium tropaeoli*), rice carrying *Drechslera oryzae*, and numerous seed-borne smut fungi of *Gramineae*. Mycelium of *Drechslera graminea* and *D. teres* may be found on the pericarp of barley, within the

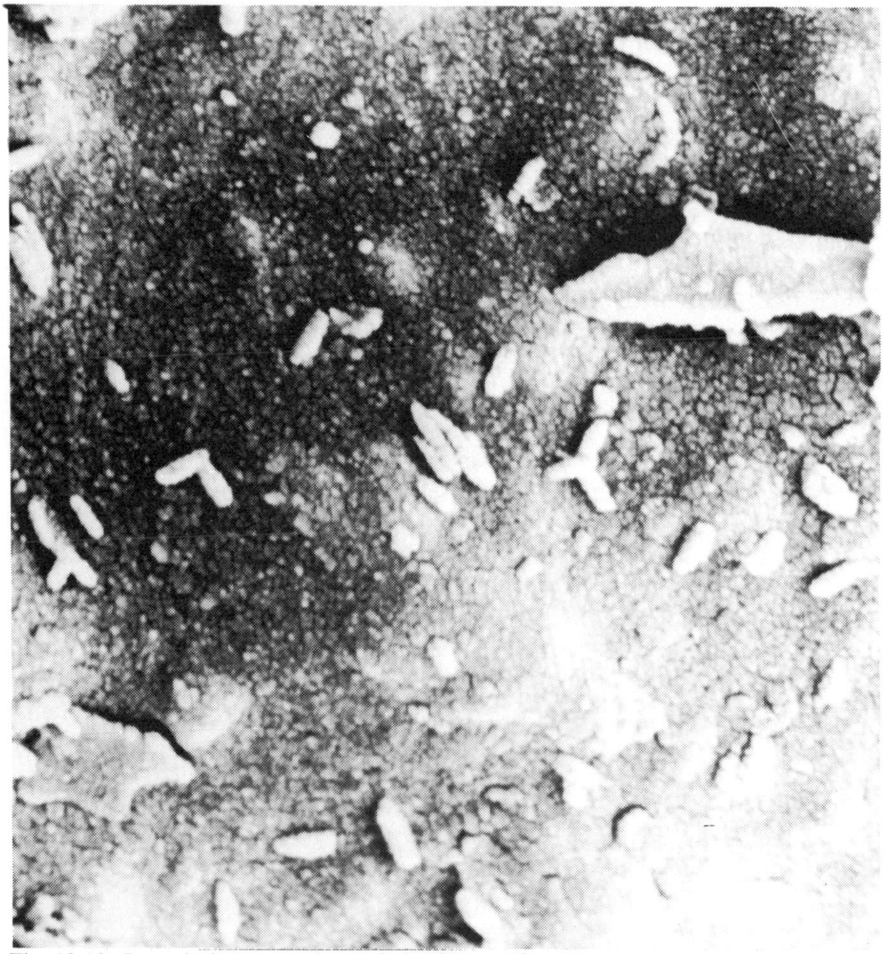

Fig. 13.13. Bacteria borne on the surface of a cauliflower seed, from a seed lot carrying inoculum of *Xanthomonas campestris* as found by testing the seed. The rod-shaped bacteria shown are likely to be of this pathogen. Scanning electron micrograph. Magnification ×3,000. (Courtesy of J. W. Guthrie)

glumes (Ravn, 1900). Also *D. avenae* is carried as resting mycelium between the glumes and the pericarp (Rathschlag, 1930). Oospores of *Sclerospora sorghi* may stick to the surface of the seed of sorghum (Ramakrishnan, 1963), and *Protomyces macrosporus* may be carried abundantly on the cremocarps (fruits) of coriander (Gupta and Neergaard, 1970).

13.7 BRACT INFECTION

In grasses the glumes may be the site of pathogens thus carried with the seed. Ravn (1900) demonstrated *Drechslera teres (Pyrenophora teres),* net blotch, in the inner glumes of barley, particularly in the lower. Resting mycelium is carried in the inner glumes, as exemplified by *D. oryzae (Cochliobolus miyabeanus),* brown spot, in rice (Ocfemia, 1924; Nisikado and Nakayama, 1943), and *D. graminea*, leaf stripe disease in barley, located in the inner epidermis of the glumes (Vogt, 1923; Genau, 1928), see Fig. 13.14. *Ustilago avenae* is transmitted in the parenchyma of the inner wall of the lemma of oats (Zade, 1922), and *U. nigra* usually as spores enclosed within the lemma and palea of barley (Dickson, 1956). *Sclerospora sorghi* is borne abundantly as oospores in the persistent glumes of sorghum, and such seeds grown in sterilised soil develop seedlings with the downy mildew (Safeeulla, 1971). Rao (1972*b*) showed by dissection of rice seeds that white tip nematodes, *Aphelenchoides besseyi*, remained coiled up inside the palea and on the surface of the lodicules which appeared dried up and papery as the seed matured.

McKay and Pool (1918) found that *Cercospora beticola*, leaf spot, may be transmitted on bracts of *Beta vulgaris*, and *Puccinia malvacearum*, hollyhock rust, is carried in the same way in *Malva rotundifolia* and *Althaea rosea* (Taubenhaus, 1911).

Xanthomonas oryzae is, in Japan, normally found in the tissues of the husk, whereas it has not been found in unhulled rice (Tagami *et al.*, 1963; Wakimoto, 1955). Also, in China this pathogen has been located in the glumes (Fang, Lin and Chu, 1956).

13.8 CONCOMITANT SEED CONTAMINATION

Small pieces of infected plant debris, loose reproductive organs of pathogens and other impurities, including dead infected seeds, may occur mixed with the living seeds and be present when the seed is sown. This 'false' seed transmission may be very important, often being responsible for spread of diseases that would not be suspected of being seed-borne. Thorough cleaning of the seed reduces substantially the amount of this kind of seed-borne inoculum but cannot be expected to prevent contamination.

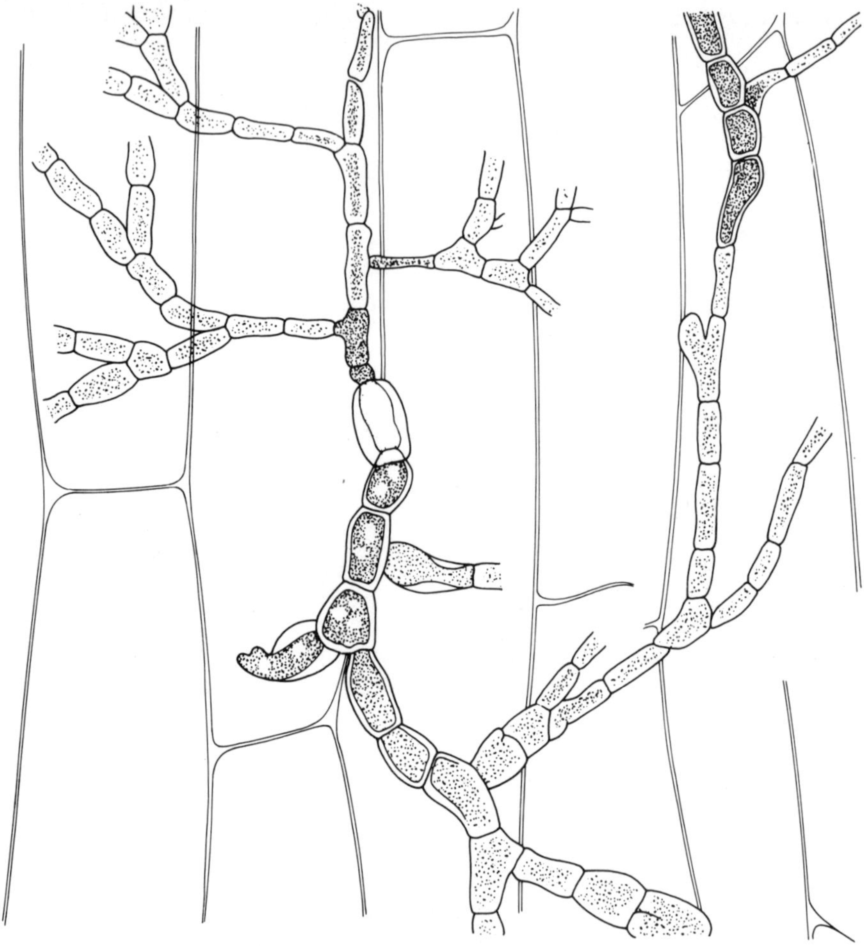

Fig. 13.14. Resting mycelium of *Drechslera graminea,* germinating on the epidermis of the pericarp of barley. (Vogt, 1923)

Perithecia of *Gaeumannomyces graminis* (syn. *Ophiobolus graminis*), take-all disease of cereals, may occur on small broken pieces of the culm along with the seed of wheat (Brittlebank, 1921). *Melampsora lini,* flax rust, may also occur on small pieces of plant debris mixed with flax seeds, and in some cases contaminated plant parts may be attached to the seed coat (Muskett, 1950). This also applies to *Septoria linicola*, the pasmo organism (Christensen, Henderson and Aragaki, 1953): 'In Minnesota, most seed lots of flax contain infected plant debris, often with numerous pycnidia, sometimes from 50 to 75 per mm^2, and each pycnidium may contain from 1,000 to 20,000 spores'. *Itersonilia pastinacae,* parsnip canker, is another fungus chiefly carried in trash fragments contaminating the seed (P. R. Smith, 1966).

Nash and Snyder (1964) found that infested soil accompanying commercial bean seed lots contained chlamydospores of *Fusarium solani* f.sp. *phaseoli*, Fusarium bean root rot; and the wilt organism of China aster, *Fusarium oxysporum* f.sp. *callistephi*, may accompany the seed as small bits of spore crusts (Baker, 1952).

Phoma medicaginis var. *pinodella* can be carried in dust with pea seed as chlamydospores capable of producing foot rot in field plantings (Baker, 1972) and in India Sattar (1933) invariably found broken pieces of plant tissue bearing mature pycnidia of *Ascochyta rabiei* in seed samples from districts where the disease was prevalent. He pointed out that by the crude method of winnowing it is difficult to clean the seed from such contaminations.

Snapdragon rust, *Puccinia antirrhini*, is strongly suspected by circumstantial evidence to be not so much seed-borne as 'seed packet borne'. Baker (1952) states that spores can be readily detected in the bottom of almost any seed packet. J. Walker (1954) recorded 1–2 million uredospores per ounce of seed in most of 139 contaminated seed samples. The seedlings are resistant but the later adult plants are not. In Rhodesia, where the disease appeared in plants grown from introduced seed (Hopkins, 1940), the packet very often was stuck on the tally at the end of the row, and from this source older plants nearby were infected (Hopkins, personal communication, 1957).

Telial spores of hollyhock rust, *Puccinia malvacearum*, may be carried on fractions of capsules of *Malva rotundifolia* (Taubenhaus, 1911) or on bits of the calyx and stems not wholly removed in cleaning hollyhock seed (Baker, 1952).

Sclerophthora macrospora, downy mildew, on grasses and oats may occur as oospores on pieces of diseased tissue (Summers, 1952), and oospores of another downy mildew fungus, *Sclerospora graminicola*, may be found in debris and siftings from seed of *Pennisetum typhoides* (Safeeulla, personal communication, quoted by Shaw, 1970).

The soybean smut fungus, *Melanopsichium missouriense*, may be carried as a concomitant in the form of hard charcoal-like galls with soybean seed (Whitehead and Thirumalachar, 1960; Gerdemann and Chamberlain, 1969).

Some fungi develop sclerotia that accompany the seed, particularly *Sclerotinia*, e.g. *S. sclerotiorum* in many crop seeds, *S. trifoliorum* and *S. spermophila* in clover, *Claviceps purpurea*, e.g. in rye, and others, and *Typhula*, e.g. *T. trifolii*, on clover (Røed, 1956). Often such sclerotia are produced in stems and mixed with seeds during threshing, a fact pointed out for *Typhula* 75 years ago by Rostrup (1902).

The bacterial wilt organism of lucerne, *Corynebacterium insidiosum*, provides an important example of bacteria that may spread on plant debris mixed with seed. For this reason the debris is taken into account in plant quarantine regulations of several countries (Neergaard, 1970*c*).

Grogan and Kimble (1967) pointed out the role of seed surface contamination in the transmission of *Pseudomonas phaseolicola* in *Phaseolus vulgaris*, and they as well as Guthrie (1970) showed experimentally that halo blight free bean seed contaminated with dry infected plant refuse may produce infected plants — provided the refuse is applied within a certain limit of dilution (Guthrie, 1970).

Stem eelworm, *Ditylenchus dipsaci*, may be carried in small parts of plants with red clover, lucerne and onion seeds (R. H. Smith, 1922; Goodey, 1950; Nielsen, 1964).

Ditylenchus angustus, Ufra disease of rice, Dak pora, is carried between the glumes and the kernel in the form of 'nematode-wool', i.e. small wool-like balls consisting of larvae of the fourth stage. These nematode balls are very resistant to extreme conditions of temperature and humidity and convey the pathogen to the subsequent crop (Swarup, personal communication, 1967).

Cysts of soybean cyst nematode, *Heterodera glycines*, may be carried in seed bags, either loose or in so-called peds, i.e. small irregular accumulations of soil which are formed during the harvesting of seed (Epps, 1968, 1969). Beet cyst nematode, *H. schachtii,* may also be carried as a concomitant contaminant with sugarbeet seed (Sethi, 1971, according to Lambat, personal communication, 1972).

Also galls of nematodes may be present mixed with the seed, for instance galls of wheat nematode, or ear cockle, *Anguina tritici*, in wheat and, less frequently, in rye, *A. agrostis*, bent grass nematode, in *Agrostis*, *Festuca*, *Phleum* and *Dactylis glomerata*.

Pathogens may be borne secondarily, upon other seed-borne pathogens. *Dilophospora alopecuri* is transmitted in rye mainly on galls of *Anguina tritici*, which, however, may be removed by cleaning. In seeds of the cultivar Chewings fescue, *Festuca rubra* var. *fallax*, imported from New Zealand to Scotland, a *Corynebacterium* was borne on galls of *A. agrostis* (Noble, 1957). In India, China and Egypt it has been established that *Corynebacterium tritici* is transmitted on the galls of *A. tritici* and that the bacterium is always associated with the nematode where attacks of tundu disease on wheat are encountered. Without the nematode, the bacterium is harmless (Noble, 1957; Swarup, personal communication, 1967).

14 Seed-Plant Transmission, Establishment of Infection and Course of Disease

INTRODUCTION

The establishment and development of infection within the seedling or subsequent adult plant is the last decisive link in the process of seed transmission. Seed transmission has been established only if this completion of the infection course has been positively demonstrated to the exclusion of other means of transmission. An organism or a virus may be seed-borne and yet not seed-transmitted.

Different seed-borne stages of an organism may relate differently to seed transmission.

Neill and Hyde (1942) found that *Gloeotinia temulenta*, blind seed disease, may be transmitted as mycelium in seed of *Lolium perenne* where it keeps alive for up to two years, whereas the macroconidia may develop on and in the seed but keep alive only for a very short period (about four months), thus having no significance for transmission of the infection (Neill and Hyde, 1939), although the infected, 'blind' seed produces apothecia.

Uredospores of different rust fungi, for instance safflower rust, *Puccinia carthami*, and beet rust, *Uromyces betae*, may occur on seed and yet not persist to secure transmission. But teliospores may persist, as in both cases of safflower rust (Calvert and Thomas, 1954) and beet rust.

It has been discussed whether black stem rust, *P. graminis*, may be successfully transmitted by seed. The fungus may occur profusely as sori of uredo- and teliospores on wheat grains, but Hungerford (1920) established in his extensive investigations that transmission does not occur. At the Institute of Seed Pathology for Developing Countries, Copenhagen, germination of seed-borne teliospores, originating from wheat seed imported to India from Kenya in 1967, has been demonstrated. Further investigations are needed to establish whether seed-borne teliospores may germinate and develop basidiospores capable of infecting *Berberis*, or whether uredospores may germinate and secure transmission. The odds would be extremely small.

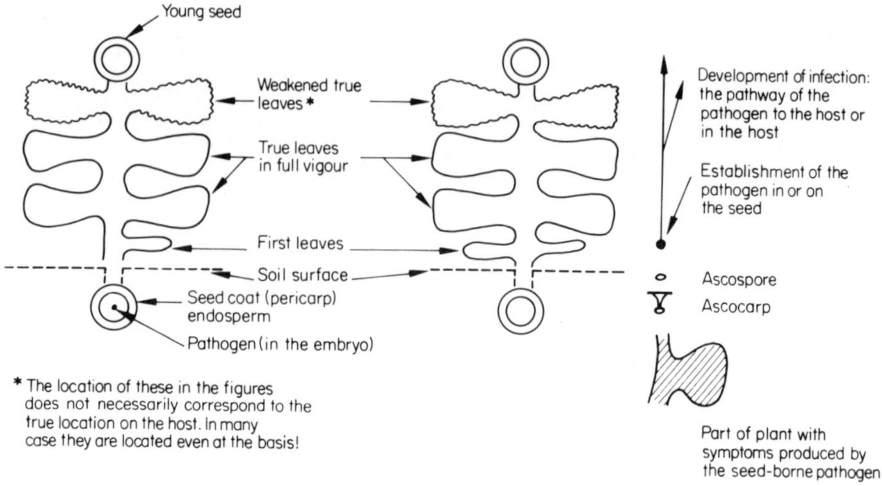

Fig. 14.1a. Types of development of seed transmission (see text and Fig. 14.1b)

Eight principal types of disease cycle and infection course will now be discussed. These are classified partly according to the way in which the pathogen is carried with the seed, partly according to the subsequent development of the pathogen during the growth of the plant grown from the infected or contaminated seed (see Table 14.1).

(1) *Intraembryal infection followed by systemic infection.* The embryo is infected and the pathogen becomes active by germination of the seed and penetrates through the stem of the plant or follows closely behind the growing points of the plant. Examples: *Ustilago nuda* [Fig. 14.1b(1)] and *U. tritici*, *Xanthomonas phaseoli* [Fig. 14.1b(2) and (3)], *Colletotrichum dematium* in soybean [Fig. 14.1b(4) and (5)], and bean common mosaic [Fig. 14.1b(6)].

(2) *Intraembryal infection followed by local infections.* The embryo is infected, as in *Ascochyta pisi* [Fig. 14.1b(7)] and anthracnose fungi of legumes. The conidia are spread from primary infections to leaves, petioles and stems, where they germinate, the hyphae penetrating the host producing local lesions.

(3) *Extraembryal infection followed by systemic infection.* The seed is infected outside the embryo (endosperm, seed coat, pericarp), but the pathogen grows into the young plant during the germination of the seed and penetrates further up during the growth of the plant. Examples: *Botrytis anthophila* in red clover [Fig. 14.1b(8)], *Drechslera graminea* in barley [Fig. 14.1b(9)], cucumber mosaic in cucumber and particularly in wild cucumber.

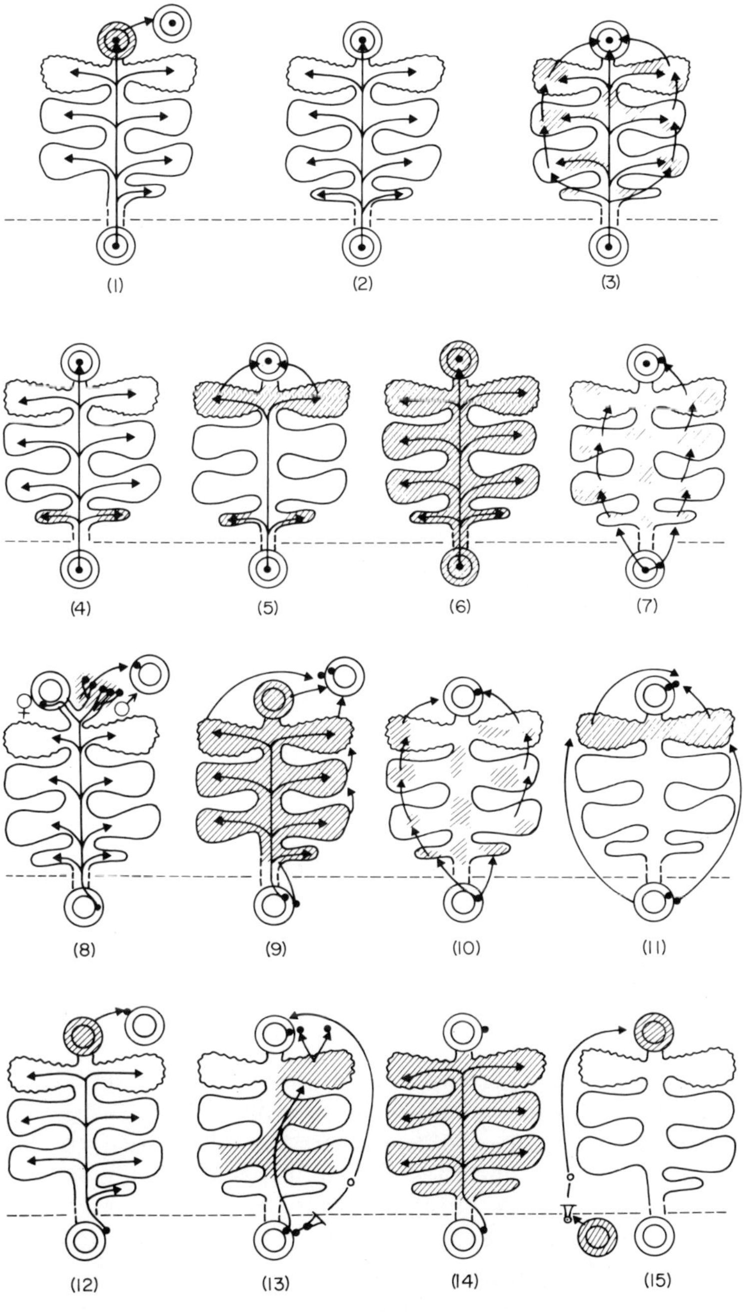

Fig. 14.1b. Fifteen cycles of seed-transmitted diseases (see text and Fig. 14.1a)

(4) *Extraembryal infection followed by local infections.* The seed is infected outside the embryo. During the germination the pathogen is carried either passively on the cotyledons, or in the seed coat, or it is transmitted to the young plant by air currents, rain splashes, insects, etc., and sooner or later penetrates into the host. Examples: *Alternaria brassicicola* [Fig. 14.1b(11)], *Alternaria radicina, Septoria apiicola* [Fig. 14.1b(10)], *Phoma lingam, Xanthomonas phaseoli.*

(5) *Seed contamination followed by systemic infection.* The seed is contaminated, and the pathogen penetrates during the germination of the seed into the young plant and further upwards through this. Examples: *Ustilago hordei* [Fig. 14.1b(12)] and *Tilletia foetida, T. caries, Urocystis occulta*, tobacco mosaic in tomato.

(6) *Seed contamination followed by extramatrical saprophytism or a dormant stage and thereafter by local infections.* The seed is contaminated and the pathogen lives for a period of time saprophytically or persists in a resting stage in the soil or in plant residues and infects the host at a later time. Examples: *Sclerotinia sclerotiorum* [Fig. 14.1b(13)], and many seed-borne pathogenic soil invaders.

(7) *Seed contamination followed by extramatrical saprophytism and then by systemic infection.* The seed is contaminated and the pathogen lives for a shorter or longer period of time saprophytically, thereafter invading the host. Example: *Fusarium oxysporum* f.sp. *callistephi* [Fig. 14.1b(14)].

(8) *Seed contamination by structures from organspecific seed infection followed by an extramatrical non-parasitic phase and later by direct organspecific seed infection.* The ovary is transformed into a sclerotium or a gall, and the pathogen lives for a period of time saprophytically, respectively saprozoically or as a resting stage in the soil. Later on the pathogen infects the host. Examples: *Claviceps purpurea* [Fig. 14.1b(15)], *Gloeotinia temulenta*, *Ustilaginoidea virens*, *Ciboria betulae*, and, with certain reservations, *Anguina tritici*.

These types of disease cycle are not sharply separated. One organism or virus may belong to more than one infection type. Additional types may exist.

The mechanics of seed germination may have a bearing upon the mode of transmission of inoculum from seed to seedling. In this respect there are two types of host, *epigeal* in which the cotyledons are carried above ground, and *hypogeal* in which the cotyledons, still being covered by the seed coat, remain in the soil. Epigeal cotyledons become green and may function like true leaves, hypogeal cotyledons remain pale and serve as storage and absorption organs. Examples of epigeal hosts, which constitute the great majority of dicotyledonous plants, are composites, cucurbits, crucifers and the nightshade family, while hypogeal hosts include legumes such as pea, vetch and *Phaseolus multiflorus*, and also *Tropaeolum majus*, oak, chestnut and walnut. Bean and soybean are intermediate forms, the cotyledons are elevated above ground, remain fleshy and do not grow. Grasses and cereals are hypogeal.

In hypogeal hosts the fleshy cotyledons or the starchy grain body of cereals and grasses act as a starting point, often as the food base, for invasion into roots and stem of the seedling, as *Acroconidiella tropaeoli* in *Tropaeolum majus* (Baker and Davis, 1950*b*), *Ascochyta pinodella* and *Mycosphaerella pinodes* in pea, or *Xanthomonas translucens* in wheat (Wallin, 1946*b*). Often these exemplify intraembryal infection followed by either local or systemic infections. Inoculum may also be carried as contamination, e.g. *Tilletia caries*, that brings about systemic infection by infective hyphae invading the coleoptile of wheat.

Table 14.1. Types of seed transmission

Occurrence at the seed		*Type of cycle*	*Pathogen*	*Infection in the crop*	
Infection of the seed	in the embryo	(1)	*Bean common mosaic* [Fig. 14.1b(6)]		systemic infection
			Ustilago nuda [Fig. 14.1b(1)]		
			Xanthomonas phaseoli [Fig. 14.1b(2)]	local infection	
			Colletotrichum dematium [Fig. 14.1b(4)]		
		(2)	*Ascochyta pisi* [Fig. 14.1b(7)]		
	outside the embryo	(3)	*Botrytis anthophila* [Fig. 14.1b (8)]		
			Drechslera graminea [Fig. 14.1b(9)]		
			Colletotrichum dematium [Fig. 14.1b(5)]	local infection	
		(4)	*Xanthomonas phaseoli* [Fig. 14.1b(3)]		
			Septoria apiicola [Fig. 14.1b (10)]		
			Alternaria brassicicola [Fig. 14.1b(11)]		
Contamination (infestation) of the seed		(5)	*Ustilago hordei* [Fig. 14.1b(12)]		systemic infection
		(7)	*Fusarium oxysporum* f. sp. *callistephi* [Fig. 14.1b(14)]		
		(6)	*Sclerotinia sclerotiorum* [Fig. 14.1b (13)]	local infection	
		(8)	*Claviceps purpurea* [Fig. 14.1b 15)]		

In epigeal hosts there appears to be three principal possibilities: (1) direct invasion into the seedling, such as systemic xylem invasion by *Fusarium oxysporum* f.sp. *callistephi, F. oxysporum* f.sp. *matthiolae* and *Verticillium alboatrum* (Sackston and Martens, 1959; Snyder and Wilhelm, 1962), which provide examples of seed infection followed by systemic infection, or (2) invasion into the soil for a longer or shorter time, and then followed by local or systemic invasion of the host such as exemplified by the soil inhabitants *Rhizoctonia solani* and *Verticillium alboatrum* and the soil invader *Alternaria zinniae*, or finally (3) subsistence for a limited period of time in host refuse, including decaying seeds, as in *Phoma lingam, Xanthomonas campestris* and *Pseudomonas phaseolicola*; inoculum of these three parasites is carried as contamination or infection and leads to local or systemic infections. Epigeal germination may be expected to favour seed transmission of pathogens that specifically infect aerial part such as *Pseudomonas phaseolicola* and *Xanthomonas carotae*.

Before discussing the above types of disease cycle attention should be drawn to the important fact that other means of transmission than seed may play a major role. The significance of seed transmission must be evaluated with due consideration of other factors of transmission and dissemination. Figure 14.2 outlines a conspectus of interacting courses of transmission of seed-borne pathogens.

→

Fig. 14.2. Courses of transmission of seed-borne pathogens. Definitions of terms used in the diagram:

Heterophytic: transmission to plants other than the one developed from the seed carrying the pathogen.

Heterospecific: transmission to other species of host plants than the species of the seed carrying the pathogen.

Homophytic: transmission from the seed carrying the pathogen to the plant developing from this seed.

Homospecific: transmission to the same host plant species as that of the seed carrying the pathogen.

Direct-systemic infection: systemic infection from the seed into the plant developing from that seed.

Indirect-systemic infection: systemic infection established after an (often short) intervening period of passage through soil, after which the pathogen invades the seedling developed from the seed carrying the pathogen.

Soil inhabitant: a pathogen which may live for an indefinite period of time being independent of presence of hosts.

Soil invader: a pathogen which in case its host is not present can live in soil for a limited period of time only.

Soil transient: a pathogen which passes through soil for a definite (usually very short) period of time, from its source towards the host to invade this host, as in germ tubes of a smut fungus growing from the seed to invasion into the seedling.

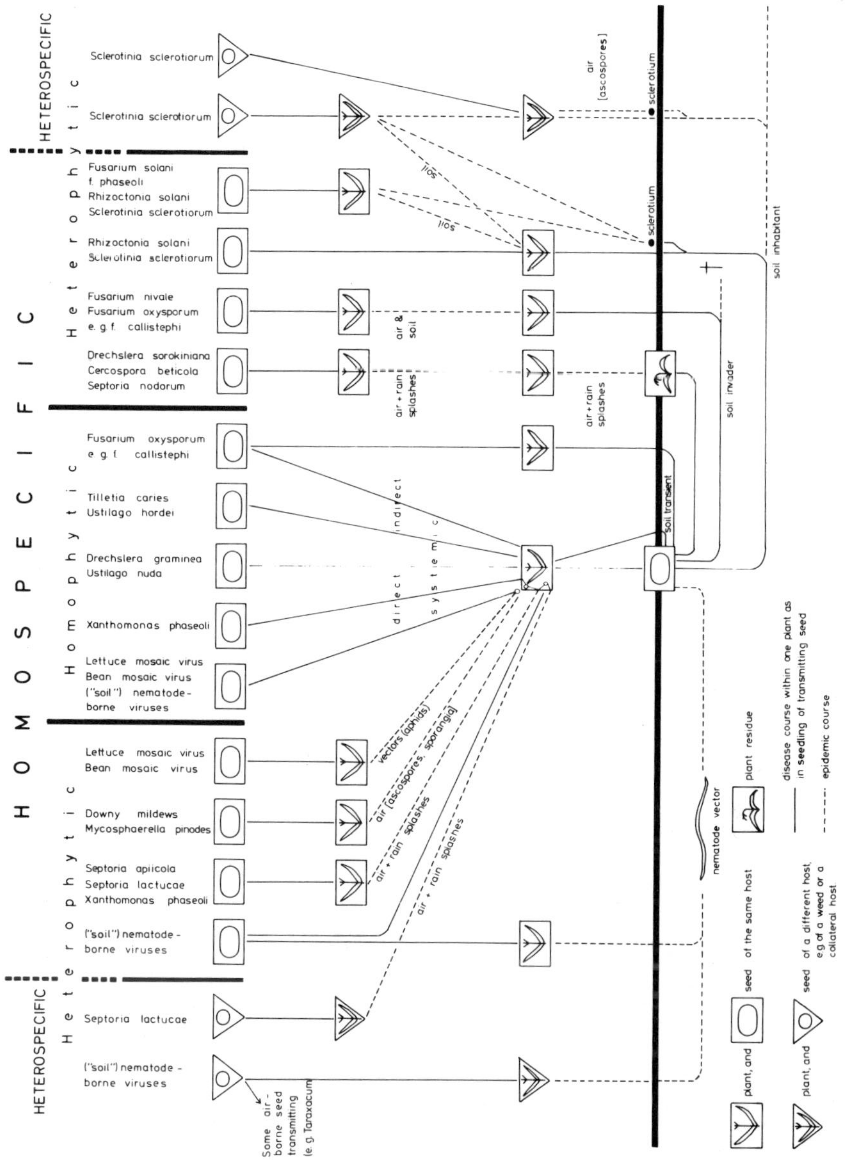
HOMOSPECIFIC
HETEROSPECIFIC
Heterophytic
Homophytic
Sclerotinia sclerotiorum
Fusarium solani f. phaseoli
Rhizoctonia solani
Sclerotinia sclerotiorum
Rhizoctonia solani
Sclerotinia sclerotiorum
Fusarium nivale
Fusarium oxysporum e. g. f. callistephi
Drechslera sorokiniana
Cercospora beticola
Septoria nodorum
Fusarium oxysporum e. g. f. callistephi
Tilletia caries
Ustilago hordei
Drechslera graminea
Ustilago nuda
Xanthomonas phaseoli
Lettuce mosaic virus
Bean mosaic virus
("soil") nematode-borne viruses
Lettuce mosaic virus
Bean mosaic virus
Downy mildews
Mycosphaerella pinodes
Septoria apiicola
Septoria lactucae
Xanthomonas phaseoli
("soil") nematode-borne viruses
Septoria lactucae
("soil") nematode-borne viruses
Some air-borne seed transmitting (e.g. Taraxacum)
air [ascospores]
sclerotium
soil
air & soil
air + rain splashes
direct
indirect
systemic
vectors (aphids)
air [ascospores, sporangia]
air + rain splashes
soil inhabitant
soil invader
soil transient
nematode vector
plant, and seed of the same host
plant, and seed of a different host, e.g. of a weed or a collateral host
plant residue
disease course within one plant as in seedling of transmitting seed
epidemic course

14.1 INTRAEMBRYAL INFECTION FOLLOWED BY SYSTEMIC INFECTION

As a first example [Fig. 14.1b(1)] loose smut of barley, *Ustilago nuda*, is selected (Fig. 14.3). The dormant mycelium is carried in any part of the embryo except the radicle. As the seed germinates, the hyphae grow intercellularly behind the growing point, upwards through the stem into the leaves [Fig. 14.3(2)] and also into the primordia of the ears [Fig. 14.3(3)] which are laid down soon after germination of the seed, and downward into the roots. In this phase of the development the host hardly shows any macroscopically visible symptoms but nevertheless the presence of the parasite in the tissues may produce various diffuse symptoms.

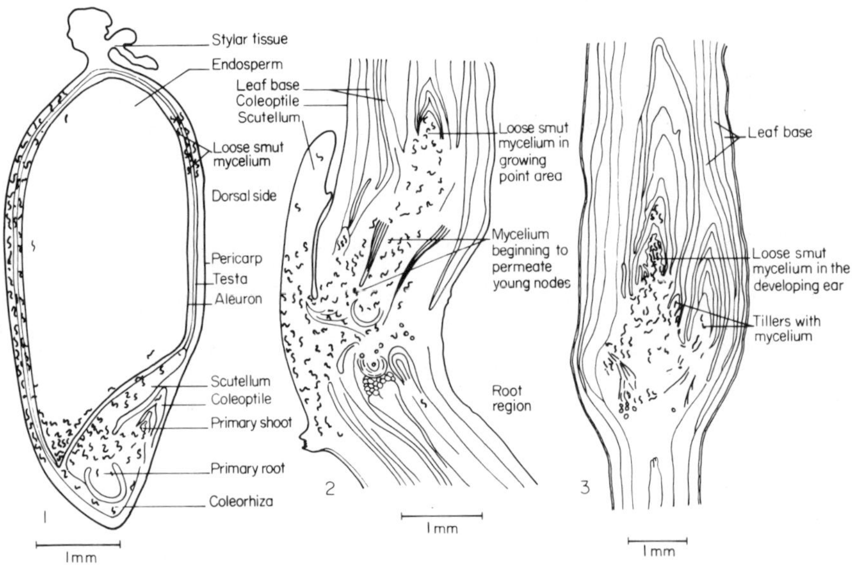

Fig. 14.3. Diagrammatic representation of the position of mycelium of *Ustilago nuda* in barley—longitudinal sections. (1) A grain, 33 days after inoculation. (2) A three-week-old seedling. (3) Crown node of a five-week-old seedling. (Malik and Batts, 1960*a*, 1960*b*)

The intercellular spaces, the vascular bundles and the internodia are more developed than in healthy plants. The stomata are increased in number and size, and consequently respiration and transpiration are abnormally increased (Kourssanov, 1926; Kluchnicova, 1928). Sometimes external symptoms appear before heading. The leaves may become yellowish-green to yellow on winter barley, ochre to brownish on summer barley. In the spike the mycelium develops sporiferous branches, giving rise to teliospores which begin to be formed some weeks before the ears emerge.

The smutted heads extrude earlier than the healthy heads and continue to appear throughout the flowering period of the crop, being carried higher out of the leaf sheath than the heads of healthy plants (cf. covered smut). Eventually the teliospores, short-living under prevailing natural conditions, are dispersed by wind to healthy flowers, where they germinate, conjugate and produce infective hyphae. These penetrate through the young ovary wall and eventually establish themselves as dormant mycelium in the embryo before maturation of the seed.

Common bacterial blight of bean, *Xanthomonas phaseoli*, is another example of this type of disease cycle. The life-cycle of the parasite in bean has been studied by Burkholder (1921) and Zaumeyer (1930, 1932). The bacteria occur in the embryo, particularly around the cotyledons and, being disregarded in this discussion, in and on the seed coat. The presence of the pathogen in the seed is detectable by weak yellow spots at the hilum. As the seed germinates, the bacteria invade through rifts in the cuticle of the cotyledons and penetrate intercellularly until the vascular system is reached. From these tissues they may move into the leaves, finally producing lesions and sometimes stem cankers. However, the disease may also occur without any external symptoms. No lesions, neither on leaves, stems nor pods, reveal that the bacteria have penetrated through the xylem tubes, into the pod and through the hilum or micropyle into the seed, infecting the embryo by proceeding into the cotyledon region [Fig. 14.1b(2)]. Secondary inoculum developed in primary lesions may be spread, carried by wind-borne rain, dust, implements, animals and man, and may reach the pods [Fig. 14.1b(3)]. Here the bacteria may penetrate through the wall of the pod contaminating the seeds or infecting them through the seed coat. The bacteria may also enter the seed through the vascular elements by way of the funiculus and raphe or through the micropyle.

Also other vascular infecting bacteria may complete their life-cycle without producing any external symptoms. *Xanthomonas phaseoli* var. *fuscans*, fuscous blight of bean, *Corynebacterium flaccumfaciens*, bacterial wilt of bean, and others may occur symptomless in Pinto beans (Thomas and Graham, 1952). This also holds true for *X. campestris*, black rot of cabbage. Seed infection by this pathogen is developed to the greatest extent in weakly infected seed-bearing plants, on which symptoms do not appear before the seed has attained the dough stage (Walker, 1950). Bacteria may spread widely in even very resistant plants without showing symptoms. This important fact has been demonstrated by using the radioisotope technique in studies on bacterial wilt of maize, *Erwinia stewartii* (Warren, 1951). Seed from symptomless plants of pepper, *Capsicum annuum*, affected by the bacterial-spot pathogen, *X. vesicatoria*, yielded the bacterium (Crossan and Morehart, 1964). Leben (1963) showed that the bacterium is capable of multiplying on non-diseased seedlings of tomato derived from infected seed and protected from the microflora of the air. Such a resident phase of a pathogen may influence development of epidemics.

Colletotrichum dematium (C. truncatum), the agent of stem anthracnose of lima bean and soybean, is transmitted in the cotyledons of soybean in the same way as different other anthracnose organisms of legumes, for instance *C. lindemuthianum*. When the seeds germinate, some of the seedlings are killed before they emerge (pre-emergence killing); others may emerge and develop normally for some time, but eventually lesions are produced in the young leaves, the stem may be girdled and the plant succumbs (seedling blight). However, the mycelium may also penetrate the host without developing visible symptoms [Fig. 14.1b(4)].

Another typical development of the infection course has been exhaustively studied by Tiffani (1951) and exemplifies in a remarkable way a certain type of seed transmission. From the cotyledons the hyphae grow into the stem and remain here intercellularly between the cortical cells and the epidermis tissue. In the young plant only the tissues near the place where the cotyledons were attached are infected. After 1.5 months the hyphae have only reached 2 mm from the original locus of infection in the stem where the cotyledon was attached. Not until one week before flowering does the mycelium resume growing in the cortical cells, advances further upwards into the stem, through the petioles and continues into the blade of the lower leaves, in which finally the hyphae occur in the epidermis cells. In the immature pods the mycelium invades the pericarp and cavities and further into the primordia of the cotyledons. As the pods mature, acervuli may appear on some, while other infected pods may remain symptomless. Special circumstances are required to induce development of conidia and infection of pods and seeds by conidia. Such conditions may be present when the plants or parts of the plant are weakened or submitted to extreme humid weather conditions [Fig. 14.1b(5)].

Symptomless infection courses are far more frequent and of greater significance in the life-cycle of seed-borne organisms than generally assumed. Other categories of diseases may be found to follow the same pattern of development. Further investigations on the occurrence of this type of infection course are needed.

Seed-borne viruses contribute other examples of this type of disease cycle. In bean common mosaic the embryo is infected. As the seed germinates, the virus spreads systemically in the young plant, mosaic symptoms soon develop, and the whole plant assumes the characteristic appearance of the disease, including mottling, curled puckered, downward-cupped leaves, yellowing, distortions and stunting [Fig. 14.1b(6)]. At the beginning of flowering, before fertilisation, the ovule is invaded, and the virus persists in the embryo but is inactivated in pod-walls and seed coats during the maturation of the seed and drying (Schippers, 1963). Schippers established the virus in 80 per cent of the ovaries in plants 20 days after inoculation, but only some of the ovules were invaded. Seed transmission is restricted to about 15 per cent from diseased plants in the cultivar Beka studied by him; in other cultivars up to about 80 per cent seeds may transmit the virus.

14.2 INTRAEMBRYAL INFECTION FOLLOWED BY LOCAL INFECTIONS

Ascochyta pisi, Ascochyta blight or leaf and pod spot of pea, is carried as mycelium in the cotyledons and plumule as well as in the seed coat (Dekker, 1957), see Fig.14.1b(7). The primary lesions appear on the first leaves, and particularly on the stem near the unfolding first true leaf, the infection taking place as the seed germinates. The lesions are circular to oval, definite, somewhat sunken, whitish to grey-brown or tan-coloured with dark-brown, often prominent margins, sometimes papery. Individual leaves are killed. On the stems and petioles the lesions are somewhat elongated; in stem lesions the fungus occurs exclusively in the epidermis and the outer layers of the cortex (Dekker, 1957). Under moist conditions pycnidia develop in the spots, sometimes sparsely, sometimes copiously but may be lacking in a dry environment. In moist weather gelatinous tendrils of conidia are extruded from the pycnidia, the spores are discharged and disseminated by free water in the form of rain, dew, wetting fog, etc. Being dependent on free water the conidia are not disseminated for long distances, usually not exceeding 3–4 m within several weeks (Dekker, 1957). On mature plants there may be numerous leaf spots that may coalesce forming a general blight. On pods the lesions are more or less circular, sunken, tan-coloured with dark margins. The pod tissue is killed forming cankers extending through to the seeds; these may be invaded by mycelium from the pod wall or by aerial mycelium spreading in the cavity within the pod. The seed coat as well as the embryo, including the cotyledons and plumule, become infected. Dekker further showed that *Ascochyta pisi* may occur in parts of the host which are free of lesions. He isolated the fungus from plants which did not show symptoms although they had developed from heavily infected seed. His observations 'suggest that the fungus grows upward with the meristem of the embryo, thus leaving behind in the stem a chain of mycelium which may or may not be interrupted'. From symptomless plants apparently healthy pods were harvested, and seed from these were disinfected externally with corrosive sublimate and incubated; 2 out of 500 such seeds yielded the fungus thus demonstrating internal infection.

The life-cycle of the two other species of the 'Ascochyta complex' of pea, *Phoma medicaginis* var. *pinodella (Ascochyta pinodella)* and *Mycosphaerella pinodes* follows essentially the same pattern but with production of different symptoms, including severe foot rots, and with the important exception that in *M. pinodes* perithecia are formed. Ascospores are discharged during the growing season, in northern U.S.A. from early in June, and carried by air currents for one kilometre or more, hence bringing into effect a much wider dissemination than effected by pycnospores.

Other examples of this type of life-cycle include anthracnose fungi of legumes such as *Colletotrichum lindemuthianum* in common bean and *C. dematium* in soybean.

14.3 EXTRAEMBRYAL INFECTION FOLLOWED BY SYSTEMIC INFECTION

Botrytis anthophila, blossom blight of red clover, exemplifies seed infection located outside the embryo [Fig. 14.1b(8)]. The fungus is established as dormant mycelium in the seed coat (Silow, 1933; Minyaeva, 1938; Bennum, 1972). As the seed germinates, the hyphae grow from the seed coat into the root, then through the hypocotyl, predominantly intercellularly, into all parts of the host including the meristem but excluding the vascular system (Bennum, personal communication, 1974). Russian investigators (Bondarzew, 1914; Jaczevski, 1916) claim that the presence of the fungus in the seed stimulates germination, suggesting a symbiotic interrelationship between the two organisms at this point. However, Minyaeva (1938) found that infected plants, although of the same height as healthy ones, are lighter in weight and with fewer stems and heads. As the plant grows on, the fungus may be isolated from almost any part of the host (Minyaeva, 1938; Bennum, 1972) and the mycelium may overwinter in the plant (Hey, 1945). Symptoms become visible only during flowering. The mycelium invades young flowers and penetrates the anthers, where the pollen grains are destroyed and conidia formed abundantly so that the anthers become grey. Conidia may be found abundantly in red clover pollen. Stapel (1969) found that at least 40 per cent of 40,000 Danish pollen samples collected by bees were contaminated, and Bennum (1972) found the conidia in more than 70 per cent of 659 pollen samples. The parasite may substantially reduce fertility of red clover, in one case in Sweden from normally 80–90 per cent to 6 per cent in diseased plants (Bergström, 1940). The spores are spread by wind or by bees (Silow, 1933) to stigmas of healthy flowers, where they germinate. It has been claimed (Bondarzew, 1914; Silow, 1933) that the hyphae penetrate the style (Fig. 14.4), eventually reaching the ovary, thus producing infection of the seed, but Jost, Volken and Kern (1964) consider infection throughout the style unlikely because substances produced in the female organs inhibit growth of the fungus; Bennum (1972) never observed hyphae in the stigma or germinating conidia on the stigma in spite of the presence of numerous conidia from surrounding infected anthers. She found evidence for infection of the seed coat from the mother plant through funiculus and hilum [Fig. 14.1b(8)].

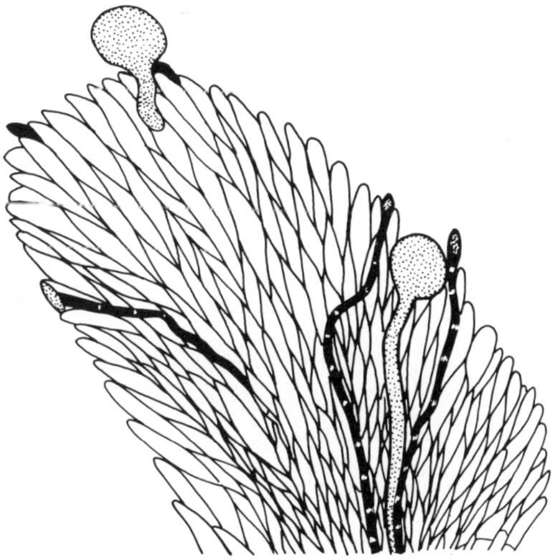

Fig. 14.4. Stigma of red clover showing germinating pollen grains and conidia of *Botrytis anthophila*. ×250. (Silow, 1933)

Drechslera graminea, barley leaf stripe, is different from *Botrytis anthophila* in one important respect: symptoms appear during the growth of the parasite in the host. The starting point of the infection is the same, the seed infected outside the embryo. The resting mycelium persists mainly in the epidermis of the pericarp. As the seed sprouts, the hyphae penetrate through the coleoptile, coleorhiza or root and grow upwards through the seedling, according to Ravn (1900) permeating the plant extensively, hyphae being present in meristematic tissues and penetrating intercellularly into the newly developed leaves and shoots. Symptoms may develop in the first leaf, but usually do not appear until the 4–5 leaf stage. Pale, then greyish-yellow streaks and brown lines appear, later the withering leaf blade splits longitudinally, and finally affected leaves may remain hanging as brown, brown–blackish-grey shreds or ribbons. Growth is checked, the plant often being only half of the normal size. According to Smith (1929), however, the path of infection within the seedling takes a different course which is not typically systemic. The hyphae penetrate directly from the coleoptile into the sheathing base of the first leaf, this base being in close contact with the inner face of the coleoptile. The hyphae proceed across the sheathing base of the first leaf into the second leaf in contact with it, and this process of penetration from

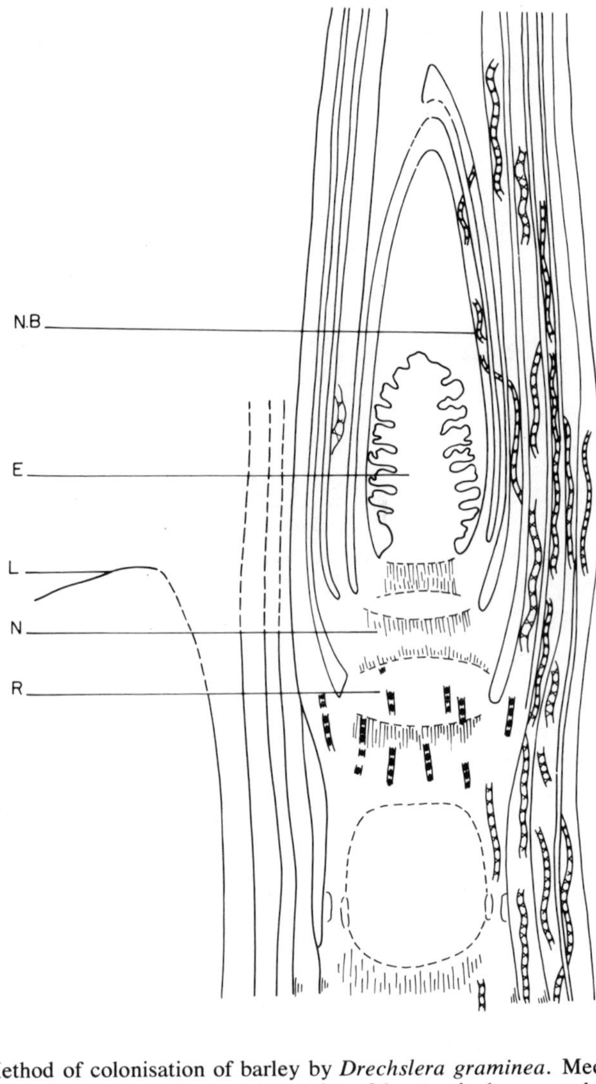

Fig. 14.5. Method of colonisation of barley by *Drechslera graminea*. Median longitudinal section of a young barley plant. The growing region of the stem is shown, enclosed in the sheath parts of leaves. The blade part of only one leaf (L) comes into the figure. The nodal regions of the stem are indicated by light shading, as at N. E, rudiments of the ear. The hyphae of the fungus are indicated by snake lines, inserted for the most part only in the right-hand side of the figure. R, mycelium in the stem. At NB mycelium ready to proceed into the young ear from the uppermost leaf. (Smith, 1929)

outer leaf to inner leaf is repeated for all the leaves subsequently throughout the growth of the plant.

Finally, the young ear, in its turn, is infected by hyphae from the uppermost leaf (Fig. 14.5); they penetrate across the glumes reaching the surface of the ovary, where they penetrate the pericarp until the level of the testa (Smith, 1924, 1929; Smith and Rattray, 1930). As it appears, the fungus does not behave as many smut fungi do, the mycelium spreading from the apical growing-point of the stem to the leaves and other parts. On the contrary, the leaves are first infected, and the stem-apex only in the final phase of the disease. Often diseased shoots do not reach the stage of earing, but are killed, or if the ear gets through, the grains are present but not filled, and the awns and glumes are limp and darker than normal. Some ears may be arrested in development because the top internode of the axis has ceased growing before the ear is completely out of the sheath, the awns are twisted, the grains small and soft, not properly matured (Verhoeven, 1921; Millard and Burgess, 1927). On the killed tissues conidia are developed which spread to healthy flowers, where they germinate, generally near the awn end of the glume, giving rise to mycelial growth between glume and pericarp and in the pericarp itself [Fig. 13.14; Fig. 14.1b(9)]. According to Genau (1928) infection takes place particularly in the period immediately prior to flowering or on the first day of flowering.

The spore dissemination of *Ustilago avenae*, black loose smut of oats, starts already during the earing, continues as the kernels mature, and still occurs during the process of threshing. As a consequence of this protracted spreading period of the inoculum, the parasite may be seed-transmitted in two ways. The kernels may be actually infected, the mycelium persisting in a dormant stage on the pericarp, or the seed may be passively contaminated by the teliospores which are carried beneath the lemma and palea. The first case is parallel to the other two examples given in this section, the infection following almost the same pattern. The parasite invades the young seedling, the host is systemically penetrated by the mycelium, mainly intracellularly in the coleoptile and first internodes, and generally intercellularly in the first leaf and deeper tissues. The hyphae grow from the base of the third or fourth leaf into the region of the growing point where they can be found two to three weeks after sowing, and before elongation of the internodes. Later the mycelium is carried upward by the growth of the internodes and finally enters the seed primordia, where teliospores are formed by segmentation of sporiferous mycelium (Falck, 1908; Zade, 1922, 1924; Mills, 1966). In its main feature this life-cycle parallels the one of *Botrytis anthophila* in red clover, with the difference that in red clover the male reproductive organs are changed into the spore masses — in oats the female reproductive organs undergo this transformation.

The infection course of *Peronospora manshurica*, downy mildew of soybean, may also be classified as belonging to this type (Jones and Torrie, 1946; Hildebrand and Koch, 1951).

14.4 EXTRAEMBRYAL INFECTION FOLLOWED BY LOCAL INFECTIONS

The celery leaf spot fungus, or late blight of celery, *Septoria apiicola*, infects seed coat and pericarp, but not the embryo [Fig. 14.1b(10)]. During the maturation of the seed, pycnidia are formed, embedded in the pericarp. When the seed germinates, the pycnidia absorb water, conidia are extruded through the ostiolus and disseminated by spattering and wind-blown water on to the cotyledons and young leaves. The first greyish to brown spots may appear on the seedlings from 9 to 12 days after germination, pycnidia may develop but usually do not mature while the plant is in its seedling stage but later, after transplanting. In dry periods newly developed leaves remain healthy, while under humid weather conditions leaf after leaf is readily infected. Such conditions are established by a mean relative humidity of 90 per cent at 14°C for a minimum of two days (Sheridan, 1968). In second-year plants the disease may intensify and culminate. In countries where humid weather conditions prevail (night dew), as in Northern Europe, dissemination to all overground parts of the plant may occur in the period of flowering and seed maturation, and seed primordia and maturing seeds may largely be infected [Fig. 14.1b(10)].

The life-cycle of *Alternaria brassicicola* on *Cruciferae* follows a similar pattern [Fig. 14.1b(11)]. The parasite is harboured as dormant mycelium in the seed coat, or as conidia contaminating the surface of the seed. The fungus is parasitic on weakened tissues. As long as the plant is young but beyond the seedling stage, it may be found only in weak, scarcely visible necroses, often on the stem at the level of the ground or in leaf spots in weakened leaves. Not until second-year plants reach the flowering stage, and particularly in the subsequent period of maturation, does the fungus resume activity under favourable humid weather conditions, because the vitality of the host now is gradually decreasing. The conidia are spread to siliquae, the germ tubes penetrate through the pericarp into the seed coat of the maturing seed, and the hyphae establish themselves as dormant mycelium [Fig. 14.1b(11)].

Alternaria radicina in carrot and other *Umbelliferae* and presumably most of the seed-borne species of *Alternaria* and *Stemphylium* have a similar life-cycle (Neergaard, 1945).

The life-cycle of the blight inciting *Xanthomonas* in bean may develop not only as systemic infection in the way described above, but also as local infections.

14.5 SEED CONTAMINATION FOLLOWED BY SYSTEMIC INFECTION

Smut fungi with seedling infection are prominent examples of this type of infection course [Fig. 14.1b(12)]. The disease cycle of covered barley smut, *Ustilago hordei*, has been investigated by Rump (1926) and Tapke (1940). The teliospores, carried beneath the lemma and palea (the hull), germinate by basidia on which sporidia are formed. These in turn germinate, and shortly after the hyphae penetrate into the coleoptile, which is susceptible only as long as it remains unsplit — after protrusion of the first green leaf, the chances for penetration are practically over. The spores residing on the pericarp beneath the hull may germinate and start infection even during storage if the grain is kept at a relative high atmospheric humidity. Spores that have reached the seed during threshing may send infection hyphae beneath the hull under certain conditions of moisture. Tapke found that subhull inoculum is more effective than superficial inoculum, the success of which is dependent on a protracted moderate humid weather period following emergence. He showed that when seed is sown and the inoculum is beneath the hull, the pathogen is frequently able to establish deep-seated infection of the seedling before it emerges. When established in the coleoptile, the hyphae grow between the cells in the direction of the longitudinal axis. From the coleoptile they penetrate into the parenchymatic tissue of the plumule, where they may be found two weeks after the germ tubes of the sporidia penetrated into the seedling. The mycelium advances, still intercellularly, through the parenchyma, until it penetrates into the flower primordia. 34 days after the seedling had been infected, Rump (1926) found abundant mycelium at the basis of the primordia of the heads. Diseased plants are not so tall as healthy (cf. loose smut). The affected ears emerge about the same time as healthy ears but remain shorter. The teliospores are usually retained within the persistent membranes of the grain and the basal parts of the glumes for some time, but dissemination of the spores and inoculation of the seed continue, according to Tapke (1940) throughout the development, maturation, curing and drying of the cut standing and grain in sheaves, and culminate with the completion of threshing [Fig. 14.1b(12)].

In covered smut of oats, *Ustilago hordei (U. kolleri, U. levis)*, the promycelium often functions as infective hyphae (Brandwein, 1944), and in bunt or stinking smut of wheat, *Tilletia caries* and *T. foetida*, the infective hyphae are developed from primary as well as from secondary sporidia, see Fig. 14.6. Stem smut of rye, *Urocystis occulta*, has a similar cycle.

Tilletia caries is a soil transient, being capable of surviving without the host for a few weeks only. In damp soil the teliospores have a longevity of four to eight weeks, according to Riehm (1913), Bonne (1931) and Martin

(1938), confirmed by Weltzien (1957*a*). Whole spore balls have a maximum longevity of only eight to ten weeks. Swinburne (1963) found that *T. caries* infected the pericarps of wheat four days after sowing, while the coleoptiles remained healthy until the seventh day, at 15–18°C. Gassner and Niemann (1954) who used pregerminated sporidia as inoculum, established susceptibility of the pericarp and young coleoptile from the stage of the ungerminated grain until the coleoptile was about 20 mm long (Fig. 14.7). Soon after this stage the coleoptile splits, as the primary leaf protrudes, and it becomes practically immune. Swinburne (1963) observed that the hyphae grow inter- and intracellularly through the coleoptile, first and second leaves, pass through successive leaf bases reaching eventually the tissues immediately beneath the growing point. When the fifth leaf is

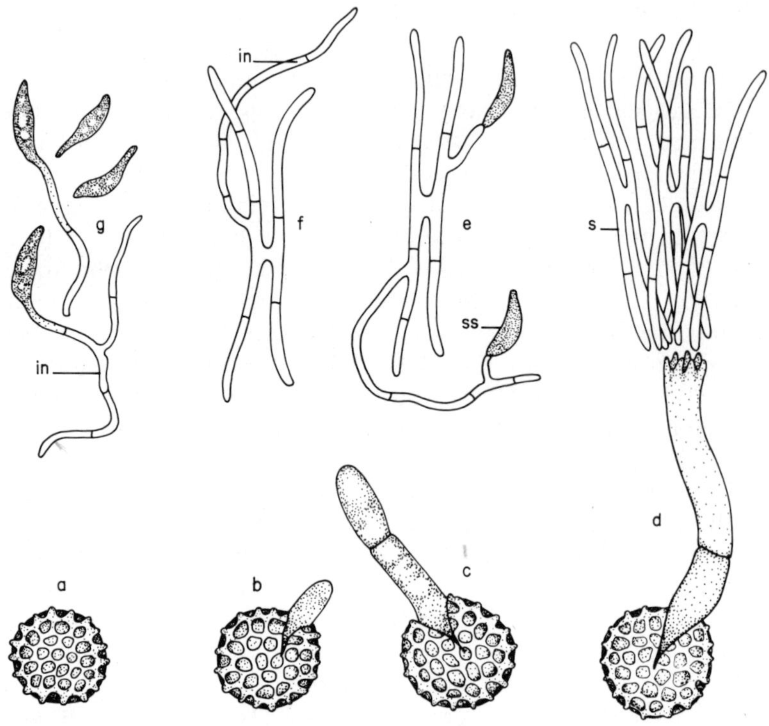

Fig. 14.6. Stages in the germination of teliospores of stinking smut, *Tilletia caries.* a, spore surface showing characteristic reticulate ridges; b, spore in early stage of germination with young mycelium protruding from the ruptured spore wall; c, a later stage in the formation of the promycelium; d, mature promycelium with a tuft of 8 uninucleate primary sporidia; s, basidiophores which have copulated in couples, united by a copulation tube and forming H-pieces, borne on the tip of the promycelium; e, a separated couple of conjugated sporidia which has produced binucleate conidia or secondary sporidia, ss, borne on a lateral secondary stigma; f, a separated couple of sporidia which has given rise directly to an infection hypha, in; g, several secondary sporidia which have started to germinate or have produced infection threads. (Heald and Woolman, 1915)

about to emerge and the basal internodes begin to elongate the hyphae penetrate into the growing point. At the sixth leaf stage some of the flower primordia may be penetrated intercellularly by a few hyphae in each flower. Trione (1972) tried to isolate the fungus from leaves, ovarian tissues and nodes but obtained cultures only from ovaries and terminal internode tissue. He hypothesises that the dikaryon develops only in localised areas rich in nutrients. When the ears emerge the hyphae are producing teliospores and the spore masses are gradually replacing the ovuliferous tissue. When infected ears emerge they look normal but symptoms become apparent when the ears get older. Infected ears have a loose, open appearance, bunt balls being fatter than healthy grains; the axis of the spike is elongated thereby widening the spaces between the individual spikelets. The haulms of affected plants are shortened.

Another example has been established by Hildebrand and Koch (1951) who found evidence of systemic infection of downy mildew of soybean, *Peronospora manshurica*, by planting oospore-encrusted soybean seeds.

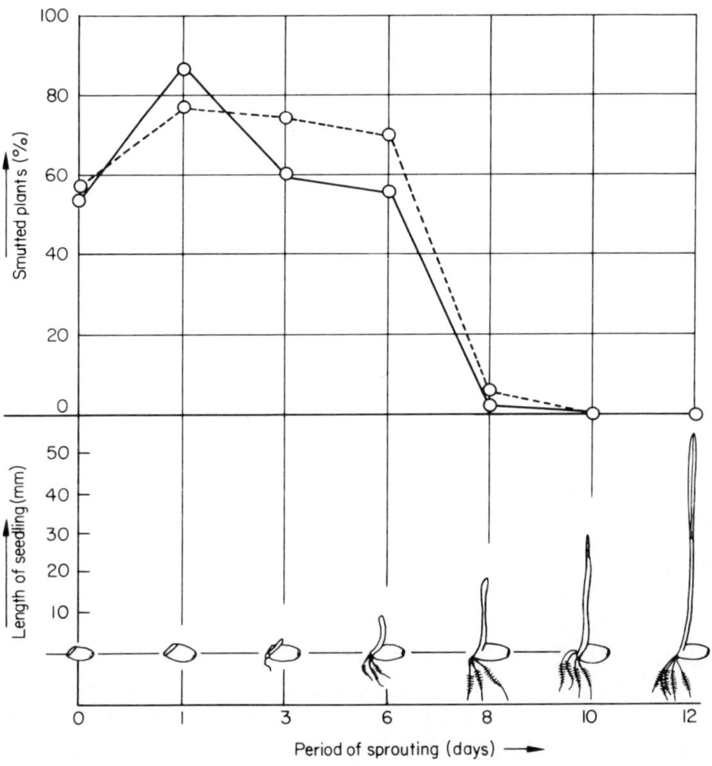

Fig. 14.7. Period of susceptibility of wheat seedlings to *Tilletia caries*. Unbroken curve, seedlings grown at 12°C in light; broken curve, seedlings grown at 12°C in darkness. (Gassner and Niemann, 1954)

14.6 SEED CONTAMINATION FOLLOWED BY EXTRAMATRICAL SAPROPHYTISM OR A DORMANT STAGE AND SUBSEQUENTLY BY LOCAL INFECTIONS

Within a considerable range of hosts *Sclerotinia sclerotiorum* is largely disseminated as sclerotia mixed with the seed. The sclerotia may adhere to the seed or be loosely mixed with it (concomitant contamination). In the soil mycelium arising from sclerotia may infect the host, or the sclerotia may remain dormant until favourable conditions enable development of apothecia. From these apothecia ascospores are discharged that may produce infection of the host, and this may be in the form of infection of flowers leading to subsequent infection or contamination of the seed, as in sunflower and crucifers. In growing plants the hyphae penetrate mainly through wounds and invade weakened parts of the plant, the tissues of which are killed. The hyphae may also invade and advance through the vascular bundles and develop sclerotia either in cavities of the plant, for instance in stems and in fruits such as pods and siliquae, or they may under humid weather conditions develop on the surface of the host. During threshing the sclerotia become mixed with the seed [Fig. 14.1b(13)].

14.7 SEED CONTAMINATION FOLLOWED BY EXTRAMATRICAL SAPROPHYTISM AND SUBSEQUENTLY BY SYSTEMIC INFECTION

Under moist conditions *Fusarium oxysporum* f.sp. *callistephi*, the aster wilt organism, produces crusts of conidia on the bases of the mother plant and these are further spread by water. In threshing bits of spore crust are broken free and lodge on the seed, or during early morning threshing the spores are sticky from dew and may be smeared on to seeds. From the contaminated seed the fungus grows into the soil and invades the seedling through the roots (Baker, 1952). Once introduced into the soil with the seed, the pathogen may persist there for many years [Fig. 14.1b(14)].

14.8 SEED CONTAMINATION BY STRUCTURES FROM ORGANSPECIFIC SEED INFECTION FOLLOWED BY AN EXTRAMATRICAL NON-PARASITIC PHASE AND LATER BY DIRECT ORGANSPECIFIC SEED INFECTION

Typical of this disease cycle is *Claviceps purpurea*, ergot [Fig. 14.1b(15)], the sclerotia [Fig. 14.8(1)] of which fall to the ground during the harvesting operations or are disseminated loosely mixed with the seed (concomitant contamination). In the spring, one to about fifty drumstick-shaped perithecial stromata emerge on each sclerotium [Fig. 14.8(6)], which thus may produce more than a million ascospores (Rostrup, 1902). Sclerotia stored under dry conditions and sown in spring cannot germinate until the following spring (Aderholt, according to Ravn, 1922). The ascospores [Fig. 14.8(4)] are wind-disseminated from the fruiting bodies during the flowering of the hosts, the flower tissues being susceptible for a short period prior to fertilisation (cf. Section 11.4) and fresh infections may continue for about two weeks after the appearance of the first flowers. The spores germinate on the meristematic tissues of the base of the

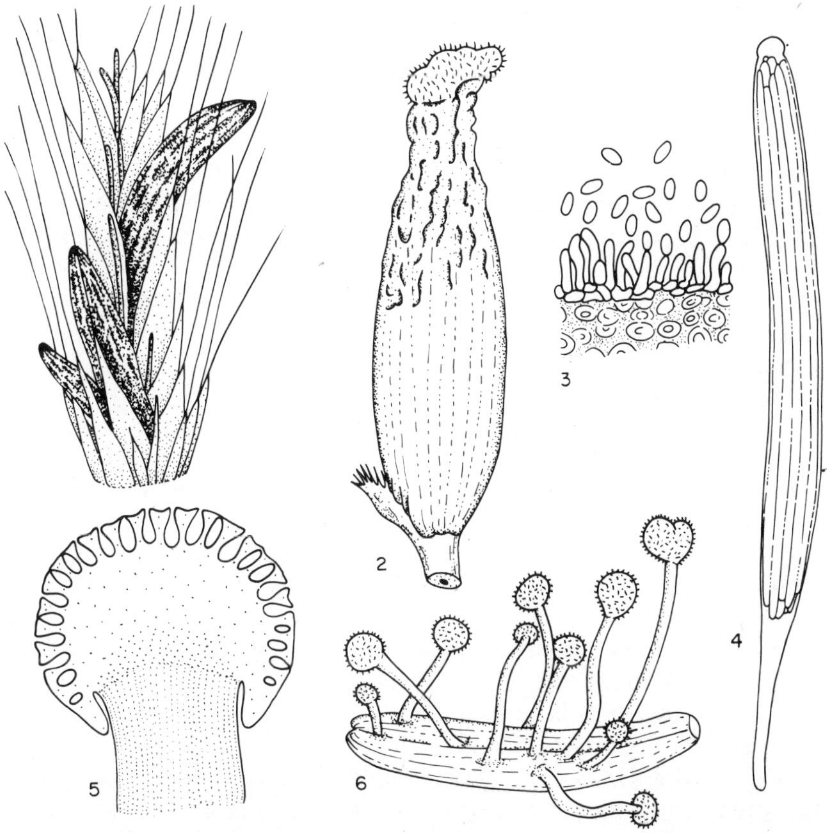

Fig. 14.8. Stages in the development of ergot, *Claviceps purpurea,* on rye. (1) Ergots (sclerotia) appearing as dark spurs and replacing the rye grains. (2) Young infected rye ovary on whose surface conidia are formed. (3) Conidiophores and conidia developed on the surface of young infected ovary. (4) Mature ascus of *C. purpurea*, bearing eight thread-like ascospores. (5) Diagram of apex of perithecial stroma, showing perithecia arranged near the periphery. (6) Sclerotium in spring, bearing several club-shaped perithecial stromata. (Wolf and Wolf, 1947)

ovary, and the hyphae invade the host tissues. Campbell (1958) established that the parasite penetrates the ovary of barley near the base of the ovule. The hyphae spread intercellularly in the ovary wall, intracellularly in the ovule, and simultaneously develop a hyphal mass on the ovary surface and on this hymenium. The honeydew conidial stage [Fig. 14.8(3)] promptly develops as a sticky pale yellow mass. The honeydew is attractive to insects which disseminate the inoculum. The conidia give rise to secondary infections on the same host species or on species which flower slightly later than the one currently infected. The conidial infection follows the same course as the ascospore infection. The secondary sclerotia from conidial infection are smaller than the primary sclerotia originating from ascospore infections. As only the ovary is affected, the infection is organspecific.

Some species of *Claviceps* affect organs of the inflorescence, outside the ovary (Diehl, 1950; Langdon, 1954); they are still organspecific. *C. phalaridis,* parasitising *Phalaris tuberosa* and other grasses in Australia is, however, systemic in its hosts (Walker, 1970). Also the related genera *Balansia*, *Balansiopsis* and *Atkinsonella* are systemic endophytes and attack floral and inflorescence organs in addition to the ovary (Diehl, 1950). *C. phalaridis* is closely related to these genera but cannot be included in them, neither in *Claviceps* as provisionally done (Walker, 1970). Although affecting only floral and inflorescence organs, the life-cycles of these fungi are not strictly organspecific because of their systemic nature.

Ustilaginoidea differs from *Claviceps* chiefly in the imperfect state and undergoes an analogous life-cycle in *Gramineae* of warmer regions. In *Ustilaginoidea virens* (syn. *U. oryzae*), false smut of rice, and *U. setariae* on *Setaria* the ovary is replaced by a sclerotium or pseudomorph, on the surface of which spherical smut-spore-like, olive green to blackish-brown chlamydospores are formed. After a resting period the sclerotia form stipitate spherical stromata in which the perithecia are produced (Hashioka, Yoshino and Yamamoto, 1951). Ascospores are believed to initiate primary infection during a short period just before heading of rice (Yoshino and Yamamoto, 1952; Ikegami, 1960). Raychaudhury (1946) established that infection may take place at a rudimentary stage in the development of the inflorescence; inside the spore masses he found style, stigma and anthers intact and pollen grains preserved, indicating failure of fertilisation. This observation was confirmed by Hashioka, Yoshino and Yamamoto (1951), who recorded that practically all of 6,000–7,000 smut balls contained intact anthers.

The cycle of blind seed disease of ryegrass, *Gloeotinia temulenta*, has been described by Wilson, Noble and Gray (1945) in Scotland. Infected seeds display no external symptoms, but on removing glumes the kernel is seen to be shrunken, rusty-brown, carrying masses of macroconidia, and penetrated by mycelium. No seedling develops from such a 'blind seed' but in the following summer when the grass comes into flower, an apothecium arises from the dead seed; shortly before this, microconidial sporodochia also appear on the seed. Griffiths (1958) showed that the fungus is obligately heterothallic, and that apothecia are produced from ascogonial stromata fertilised by microconidia of a compatible strain; one infected seed often harbours several strains. The production of apothecia is well synchronised with the development of the host. It normally begins at the time when the earliest ryegrass flowers open, and ceases just after flowering is over. Cold wet weather conditions favour the disease by prolonging duration of apothecial formation as well as the period of pollination and ripening. An average apothecium continues to discharge spores for almost 10 days, and an average spike completes flowering in about 10–12 days under normal Scottish weather conditions. The ascospores from the apothecia are spread by air currents, they lodge on the ovary between the open glumes, germinate immediately, and in about a week after the ascospore infection the life-cycle is completed by development of profuse masses of macroconidia, the so-called slime stage of the fungus. These rather shortliving spores serve exclusively for further epidemic spread of infection during the period of flowering. Their longevity after maturation of the seed is only a few months. Wilson, Noble and Gray observed spikes in which the top spikelet contained young seeds from which slime spores were exuding, while in the lowest spikelet flowers were still open.

Anguina tritici parasitises wheat causing gall formation of the kernels, or ear cockles. Each of these galls, loosely mixed with the seed, may contain thousands of second-stage juveniles. As water penetrates the gall, the eelworms resume activity in the soil, then penetrate the shoots of emerging wheat seedlings, such as those of winter wheat, where in temperate climates they hibernate. As the seedlings resume growing in spring the eelworms migrate upwards between the leaf-sheaths, and here they live ectoparasitically between the young leaves, sometimes causing characteristic twisting so that these as they grow do not easily separate from the ensheathing older leaves. The larvae around the growing points are mechanically carried to the developing ears along with the lengthening culms. When the ears are developed the larvae enter the endoparasitic phase, penetrating into the flowers, in the ovary of which they become mature after three to five days; their presence incites swelling of the ovary, the adults pair, and in further six to ten days oviposition occurs. During the following two to three weeks the young eelworms are hatched, they develop to the second juvenile stage while the gall dries up concurrently with the maturing of the healthy grain (Rostrup, Bovien and Thomsen, 1940; Goodey, 1965; Swarup and Gupta, 1971). The presence of the eelworm on wheat does not usually result in pronounced symptoms on the host — only faint signs are developed, the eelworms being present only in small numbers. The principal symptom is the gall formation, but for the reason that other parts of the host may also react by symptoms and because of the ectoparasitic phase, we are not dealing with truly organ-specific infection.

15 Environmental Physicochemical Factors Affecting Establishment and Course of Disease

INTRODUCTION

A discussion of the conditions leading to inception and establishment of diseases through transmitted seed falls naturally into the following two items:

(1) The external factors of the environment exerting an influence on the host, on the pathogen, and on the association of host and pathogen.
(2) The internal factors, i.e. the gene-controlled pathogenicity of the parasite, and the gene-controlled susceptibility of the host.

Temperature, humidity and other aspects of the physicochemical conditions of the soil, the adaphic factors, particularly soil reaction, soil type and soil fertility, are important environmental factors. 'Air fertility' may be added for greenhouse crops, as a supply of CO_2 may influence development of diseases positively or negatively; little is known about this. Conditions of light with regard to intensity, quality and periodism are important, particularly to the host; but the influence of light on fungi has been underestimated.

In the period prior to infection these factors exert influence on the predisposition of the host and the readiness of the pathogen. During the process of the establishment of infection and further course of the disease the influence of environment is directed upon the integrated totality of pathogen and host.

15.1 INFLUENCE OF PHYSICOCHEMICAL FACTORS ON THE PREDISPOSITION OF THE HOST

Foster and Walker (1947) investigated the influence of soil conditions on the predisposition of tomato plants to Fusarium wilt, *Fusarium oxysporum* f.sp. *lycopersici*, and Kendrick and Walker (1948) elaborated on the influence of these factors on predisposition to bacterial canker of tomato, *Corynebacterium michiganense*. Both are seed-borne diseases. Walker (1950) summarises the results as follows.

Tomato plants were grown under a variety of conditions. The roots were inoculated and the plants grown under conditions favourable to disease development. The plants were found to be predisposed to Fusarium wilt by (1) soil and air temperatures near the optimum for host plant growth; (2) low soil moisture; (3) short day length; (4) low light intensity; (5) nutrition low in nitrogen, low in phosphorus, high in potassium and low in pH. Susceptibility of the plants was decreased and resistance correspondingly increased by (1) soil or air temperatures above or below the optimum for plant growth; (2) very wet soil; (3) long daylight periods; (4) high light intensity; (5) nutrition high in phosphorus, low in potassium, high in nitrogen and high in pH. Thus it is evident that the degree to which tomato plants are affected by wilt after transplantation into infested soil may be determined by the conditions under which transplants were grown as well as by the conditions which prevail after infection. With relation to bacterial canker of tomato, temperature and low soil moisture had predispositional effects opposite to those in Fusarium wilt, while the effects of high and low light intensity were the same, respectively, for each disease.

Bawden and Roberts (1947, 1948) found that susceptibility to viruses was increased by dark treatment, e.g. by reducing to one-third the light intensity for 24 hours. Thus tomato and *Nicotiana glutinosa* became more susceptible to the viruses of bushy stunt and tobacco mosaic, and tobacco to the viruses of tobacco mosaic and tobacco necrosis. The virus content in the shaded plants was increased. Post-inoculation treatment had little effect and most often decreased susceptibility.

15.2 INFLUENCE OF PHYSICOCHEMICAL FACTORS ON THE COURSE OF DISEASE

The influence of physicochemical factors on the course of disease is dual in the sense that they operate on (1) the establishment of infection, and (2) the performance of the infected host.

Establishment of a systemic infection is a one-time phenomenon; there is a definite crucial point in the effect of the environment. Establishment of multiple local infections, in sequence and in accordance with the multiplication and dissemination rate of the pathogen, implies for development of an epidemic that suitable conditions occur repeatedly or continually over a protracted period. This differentiation is essential for understanding the specific effects of environment on development of disease.

The reaction of the infected host to environment is, of course, dependent on the type of host–parasite interaction, and the influence of environment on this specific interrelationship. In some systemic infections, such as loose smuts of wheat and barley, and probably bunt of wheat, the development of pathogen and host is so well co-ordinated that any change in growth rhythm applies equally to both organisms. Independent of any variation in growth conditions, the smutted spikes appear perfectly synchronised with the flowering of the host crop. In other systemic diseases growth conditions may increase or suppress the effects of the pathogen: according to circumstances the competition between pathogen and host may increase or reduce the rate of infection that leads to fatality, as in certain smuts, e.g. kernel smut of sorghum, where the ratio between smutted and healthy spikes may vary in infected plants; or the competition may result in general increase or reduction of the harmful effects of the pathogen as in virus diseases under the influence of temperature.

In the investigations on the effect of environment on development of the seed-borne diseases reviewed below, often distinction is not made between establishment and course of disease. More conclusive information in this respect is needed for some of these diseases.

15.2.1 Moisture

Seed-borne pathogens belonging to the seedling infecting type, such as many smuts, are greatly influenced by soil moisture. Other types of pathogens that infect above ground parts of the host at a later stage of its development, are predominantly dependent on air humidity for their establishment and course.

*Soil moisture**. The conditions of soil moisture affect the germination capacity of smut spores, and the vitality and hence the predisposition of the host. Spore germination of *Ustilago avenae*, loose smut of oats, and severity of the disease are at their maximum when the water content is 30 per cent of the water-holding capacity of the soil, severity being lower at 60 per cent, and lowest at 80 per cent (Bartholomew and Jones, 1923; Jones, 1923). Excessive soil moisture may lead to lack of oxygen, thus hampering spore germination (Jones, 1923). Also the predisposition of the seedling is influenced. The growth beyond the susceptible seedling stage is substantially retarded under extreme, dry soil conditions.

*There is considerable confusion in the terminology applied in plant pathology for water content of soil. Griffin (1966) surveyed some of the literature and objected to usage of such terms as 'per cent moisture holding capacity', while he finds 'per cent saturation' logical and satisfactory. He recommends terminology such as standardised in soil science. In the discussions reported here, however, any references to degrees of soil moisture may well be accepted as relative values.

In soil held at 20°C and 20 per cent of the water-holding capacity, 94 per cent of germinating oats were infected by covered smut of oats, *Ustilago hordei (U. kolleri)*, while in soil at 60 per cent of the water-holding capacity only 48 per cent were infected (Reed and Faris, 1924). The soil mois - ure optimum for infection was changed when other temperatures were applied. At 15°C maximum infection developed at a soil moisture of 15 per cent of the water-holding capacity, and at 25°C the most severe rate of infection was at a soil moisture of 35 per cent.

Similar results have been obtained for covered kernel smut of sorghum, *Sphacelotheca sorghi*. Low soil temperature and moisture and slightly acid soil during the pre-emergence period are favourable for infection (Kulkarni, 1922; Melchers and Hansing, 1938). Reed and Faris (1924) found that severity of disease decreased generally with increasing soil moisture, from 10 per cent of the water-holding capacity and upwards at a constant temperature, 19–20°C, and a constant soil reaction, pH 7.2. However, also in this disease the infection degree is highly dependent on the soil temperature. Hsi (1958) demonstrated the importance of the post-emergence environment, particularly where irrigation is practised; under drought conditions in eastern New Mexico post-emergence irrigation substantially affected the amount of infection. Early-irrigated plants produced considerably more disease than late-irrigated plants; in a late-irrigated field heads of the main stalk escaped smut, while on the same plant those of the young branches that had developed after irrigation did not. This effect on the same individual plant may be due to differences between the growth rate of the host tissue and that of the fungus.

Also head smut of sorghum, *S. reiliana*, is favoured by low soil moisture, 15 per cent water content (on oven dry basis) being more favourable to infection than 25 per cent (Christensen, 1926).

In contrast to these experiences, covered smut of barley, *Ustilago hordei*, seems to be more favoured by higher soil moisture than for instance covered smut of oats, *U. hordei (U. kolleri)*, mentioned above. Faris (1924*a*) recorded more severe covered smut of barley at 50 per cent than at 40 per cent of the water-holding capacity, and Rump (1926) established the optimum soil moisture as high as at 66 per cent.

Ling and Moore (1937) and Ling (1941) found severe disease development of stalk smut of rye, *Urocystis occulta*, as for most cereal smuts, in relatively dry soil, at 25 per cent of the water-holding capacity (air-dry basis), and recorded decreasing degrees of severity at 50 per cent, 65 per cent and 75 per cent. Correspondingly, Hafiz (1948) found that *U. agropyri*, flag smut of wheat, does not develop on wheat in moist soil, at 70 per cent of the water-holding capacity. Ling (1941) established, by comparing soil moisture after seedling emergence, that the development of stalk smut after the initial infection of the pre-emergence seedling is still favoured by a relatively dry soil, as appears from Table 15.1.

Table 15.1. Relation of soil moisture to development of stalk smut of rye (Ling, 1941)

Soil moisture WHC *pre-emergence*	*post-emergence*	*Infection per cent* (2 tests)
25%	25%	35/26 : 31
25%	50%	29/24 : 27
25%	75%	14/19 : 17

On the other hand, Purdy (1964), comparing the effect of six post-emergence soil moisture levels, obtained no material influence on the amount of flag smut of wheat in terms of percentage infection.

Finally Rabien (1927) found that bunt or stinking smut of wheat, *Tilletia caries*, may be established over a great range of soil moisture during the germination of the grain. Soil kept at 40 per cent of the water-holding capacity which is normally favourable to germination of wheat, gave a bunt percentage of 55, while in dry soil, at 20 per cent of the water-holding capacity, 22 per cent bunt was recorded, and at the other extreme, in very moist soil, at 80 per cent of the water-holding capacity, only 11 per cent.

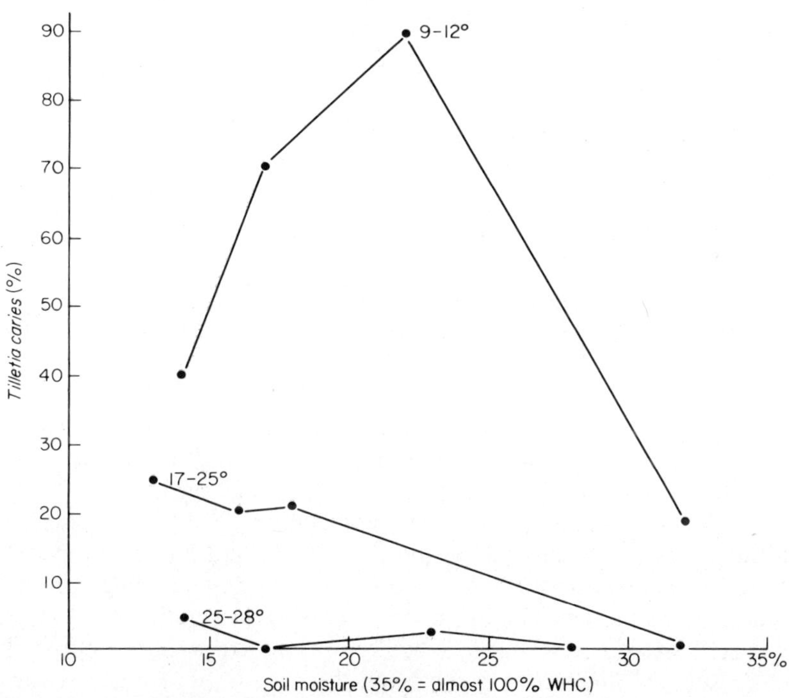

Fig. 15.1. The relation between the water-holding capacity and the degree of infection by *Tilletia caries* in wheat at different soil temperatures. (According to data from Hungerford, 1922)

Similar results have been obtained by other workers (Appl, 1915; Hungerford, 1922, cf. Fig. 15.1; Woolman and Humphrey, 1924; Caspar, 1926; Leukel, 1937). Water-saturated soil appears to have an inhibitory effect on spore germination (Woolman and Humphrey, 1924; Leukel, 1937). Nieves (1930) found under conditions prevailing in Argentina that the highest infection of *T. foetida* was obtained when the soil became progressively drier; regular, even slight rainfall during seed germination prevented high infection. Purdy (1963*a*) studied the influence of soil moisture on the post-infection development of wheat bunt. He found that variable soil moisture during this period of the disease course did not affect the development of infection in two susceptible cultivars, Red Bobs spring wheat and Federation. Once infection of wheat bunt is established, the amount of post-infection soil moisture is immaterial, as Purdy (1964) found for flag smut of wheat.

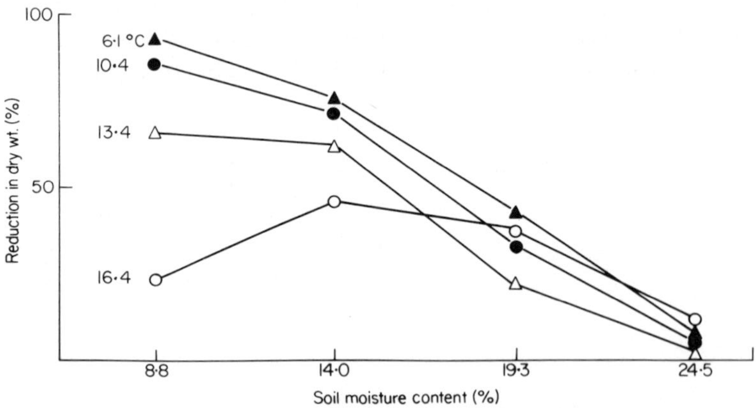

Fig. 15.2. The relation between soil moisture content at four temperatures and disease intensity of *Fusarium nivale* in terms of dry weight in seedlings of wheat, cv. Atle. (Millar and Colhoun, 1969)

Shen (1940) recorded severe development of *Fusarium culmorum* in wheat seedlings in dry soil with 30 per cent water-holding capacity, and less severe disease at 50 and 80 per cent water-holding capacity. Colhoun, Taylor and Tomlinson (1968) obtained similar results. The disease on seedlings was favoured by lower soil moisture and higher temperatures; wet cool soils gave no market symptoms. In barley seedlings the disease intensity was somewhat higher under comparable conditions. At high soil moisture content no typical foot rot symptoms were found, but in dry or medium wet soil and with high air humidity severe foot rot was recorded.

Disease incidence of *Fusarium nivale* in wheat was reduced at four different temperatures with increase in soil moisture, as shown in Fig. 15.2. (Millar and Colhoun, 1969), and likewise the incidence of disease caused by *F. avenaceum* in wheat was reduced with increase in soil moisture at two different temperatures (Colhoun, Taylor and Tomlinson, 1968). Also infection of wheat coleoptiles by seed-transmitted *Septoria nodorum* decreases with increased soil moisture, cf. Fig. 15.3 (Holmes and Colhoun, 1971).

During germination of barley seed, carrying *Drechslera graminea*, relatively dry soil, less than 40 per cent saturated, is more conducive to leaf stripe disease than very wet soil at a level of about 90 per cent saturation (Leukel, 1933). This effect indicates that the pathogen may have been present as contamination rather than infection.

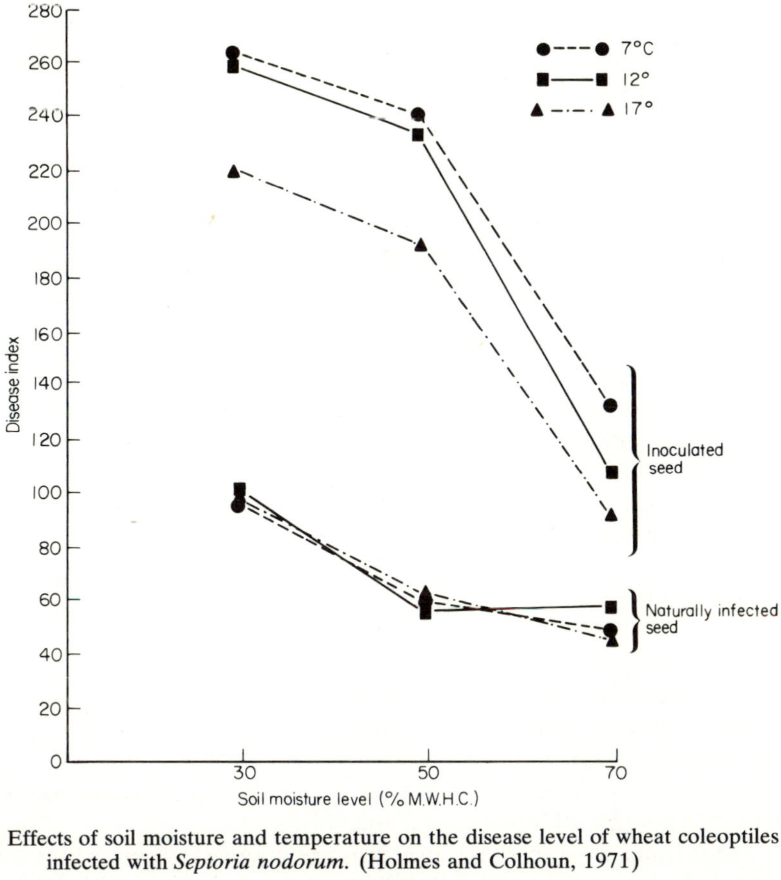

Fig. 15.3. Effects of soil moisture and temperature on the disease level of wheat coleoptiles infected with *Septoria nodorum*. (Holmes and Colhoun, 1971)

Air moisture — precipitation. Infection on above-ground parts of plants is usually highly dependent on the relative humidity of the air. Most fungi require practically 100 per cent relative humidity for spore germination. Bacterial pathogens require water droplets for spread. Seed-borne diseases such as many anthracnoses, fusarioses, alternarioses and bacterioses in cereals, pea, bean, tomato, cotton, flax, carrot and various ornamentals occur as weak, scarcely visible infections during a major part of the life of the host. Only under favourable weather conditions, particularly continuous moist weather as in dew, fog or rain, is sporulation stimulated, inoculum increased, dissemination, germination of spores and establishment of infection favoured. Precise information is scarce on the influence of relative humidity and temperature on the development of seed-borne bacteria and fungi as related to different stages in the development of the crop. More details are needed of the microclimate prevailing under different conditions due to the density of the crop, as affected by the distance between plants in the field (cf. Subsection 19.6.6), development of foliage, presence of weeds, extent and location of individual fields (cf. Subsections 19.1.2 and 19.6.6) and possibilities of ventilation in the crop, to different techniques of irrigation (cf. Subsection 19.5.2), etc.

Also, extremely dry conditions may have a decisive effect on disease development, especially for instance infection by wilt organisms where such conditions may greatly aggravate damage to infected plants.

Some macroclimate conditions particularly pertaining to precipitation are discussed in Section 11.3.

Bacterial epidemics may be promoted substantially by hail, a well-known example of this relationship being fireblight, *Erwinia amylovora*. In Idaho, U.S.A., Guthrie (1970) observed secondary spread of halo blight, *Pseudomonas phaseolicola*, in 1966 after hail storms but not after normal rainfall in subsequent years. On the other hand, Daft and Leben (1972) in Ohio, U.S.A., have shown that a succession of wind and rain storms result in outbreak of bacterial blight of soybean, *P. glycinea*, in epidemic proportions if inoculum is present early in the season, as it may be in the cotyledons. Wind of such strength mechanically causes lesions which predispose the leaf tissues to infection.

15.2.2 Temperature

Soil temperatures are generally more constant than air temperatures since they are linked with other factors such as aeration and water-holding capacity. The soil conditions of significance for establishment of infection by fungi are ordinarily those that favour fungal growth thus enabling penetration and infection (see Table 15.2).

Smut fungi with seedling infection are greatly influenced by soil temperature in the period before seedling emergence. Heuser (1922) found the minimum, optimum and maximum temperatures for spore germination of *Tilletia caries* to be 5, 16–18 and 25°C, respectively, but Meiners and Waldher (1959) recorded that the optimum temperature for spore germination of the fungus from *Agropyron trachycaulum* was 5°C rather than 10 or 15°C. These authors also studied the influence of temperature on spore

Table 15.2. Relation of edaphic factors on seed-transmitted infections — Summary. WHC = water-holding capacity; MC = moisture content

			Soil moisture (pre-emergence)			
tempera-ture (°C)	*WHC (%)*	*MC (%)*	*optimum conditions*	*response*	*host*	*pathogen*
16		15 25	low	16% infection 0% infection	sorghum	Sphacelotheca reiliana
28		15 25		47% infection 23% infection		
36		15 25		11% infection 1% infection		
19 20	10 40		low (pH 7.2)	} decreasing severity	sorghum	Sphacelotheca sorghi
9 – 12 9 – 12 9 – 12		14 23 32	medium (low temp.) low (high temp.)	40% infection 90% infection 20% infection (cf. Fig. 15.1)	wheat	Tilletia caries
	20 40 80		medium	22% infection 55% infection 11% infection	wheat	
	25 50 65 75		low	most severe infection } decreasing severity	rye	Urocystis occulta
	70		low	no disease	wheat	Urocystis agropyri
	30 60 80		low	maximum } decreasing } spore germination and disease	oats	Ustilago avenae
	40 50 60		high	severe infection } increasing severity	barley	Ustilago hordei
20 20	20 60		low	90 – 100% infection no infection	oats	Ustilago hordei
15 25	15 25		low	most severe infection	oats	
	40 90		low	severest infection decreasing	barley	Drechslera graminea
	30 50 70		low	severe infection } decreasing severity	wheat	Fusarium avenaceum
	30 50 80		low	severe infection } decreasing severity	wheat	Fusarium culmorum
		9 14 19 25	low	severe infection } decreasing severity (cf. Fig. 15.2)	wheat	Fusarium nivale
pre-emergence	WHC % 25 25 25	WHC % 25 50 75	low post-emergence	31% infection 27% infection 17% infection	rye	Urocystis occulta

Table 15.2 *continued*

Soil temperature (pre-emergence)

temp.(°C)	*response*	*host*	*pathogen*
5 16 – 18 25	minimum optimum for spore germination maximum	wheat	Tilletia caries
5	optimum for spore germination	wheat	Tilletia caries (from Agropyron trachycaulum)
5 10	optimum for infection	wheat cv.Dawson cv.Marquis	Tilletia caries/T.foetida
4 9 18	less severe infection most severe infection least severe infection	wheat	Tilletia caries/T.foetida
18 – 20	optimum for infection	wheat	Urocystis agropyri
5 10 – 25 30	optimum for infection	barley	Ustilago hordei
5 15 – 20 >30	optimum for infection maximum for infection	barley	Ustilago nigra
diff.	no reaction	barley	Ustilago nuda
6 >16	optimum for infection maximum for infection	wheat cv.Atle	Fusarium nivale

Soil reaction

pH	*response*	*host*	*pathogen*
5.8 7.9	6% infection 42% infection	wheat	Tilletia caries
4.6 7.4 8.6	12% infection 92% infection 18% infection	oats	Ustilago hordei
acid alkaline	considerable infection double infection	barley	Ustilago hordei
5 (20°) 6 (15°)	83% infection 71% infection (most severe)	barley	Ustilago hordei
acid	favourable for infection	wheat	Fusarium nivale
acid	favourable for infection	tomato pea cotton	Fusarium oxysporum f.sp.lycopersici f.sp.pisi f.sp.vasinfectum
alkaline	favourable for infection	tomato cotton	Verticillium alboatrum

germination capacity and speed as encountered within a considerable range of accessions of three other *Tilletia* species, *T. bromi-tectorum* (37 accessions), *T. guyotiana* (14 accessions) and *T. fusca* (11 accessions), grown on soil-extract agar. They noted a great variation in the time required for germination of a given species at a given temperature. For instance, the spores of one accession of *T. bromi-tectorum* showed a high percentage of germination within 10–14 days, while others needed 34 up to 44 days to attain a corresponding amount of germination. Out of the tested temperatures, 2, 5 and 10°C, the optimum was clearly 5°C, as it also was for six other species of *Tilletia*, among which was *T. contraversa*, dwarf bunt of wheat. The reduction in germination energy encountered may, in part, be due to ageing of the spores, but this appeared not consistently to be the cause.

Infection of bunt may take place at temperatures from about 5°C to about 20°C. Faris (1924*b*) found the optimum temperature for infection by *T. caries* and *T. foetida* on wheat to be about 5–10°C in two cultivars; Leukel (1937) recorded 10°C for four cultivars. Gibs (1924) obtained similar results, recording highest infection at 9°C soil temperature, less at 4°C and the least at 17.7°C, while Feucht (1932) found that 5–6°C was the optimum temperature for infection of the two fungi on cv. Strubes Dickkopf. Temperatures below 5°C and above 20°C are highly unfavourable (Leukel, 1937). In more recent investigations Kendrick and Purdy (1959) confirmed soil temperatures of 5 and 10°C to be optimal for infection of wheat bunt from seed-borne spores; this applies for any of the 28 known races of bunt fungi (Kendrick and Purdy, 1962). At higher temperatures, 15 and 20°C, less bunt is produced, but the incubation period required for maximal infection is reduced from 25 days at 5°C to 5 days at 20°C. In general, temperatures from 5 to 15°C favour infection in wheat seedlings by both wheat bunt fungi, with 6–10°C being optimum. Purdy and Kendrick (1963) found, however, that percentages of infection can be increased at 20°C if spores have germinated immediately before the wheat has been sown. The optimum temperatures for infection are considerably lower for infection of the two bunt fungi than the optimum temperature for growth of wheat. In Switzerland the wheat cultivar Plantahof attained a length of 5 cm in 5 days at 27°C; 27 days are needed to attain this length at 5°C (Gäumann, 1946), while the mentioned optimum temperatures for infection, particularly 5°C, are only slightly above the minimum temperature for the growth of the fungus. A 9–10°C germination of the spores is rapid compared with that of the seed thus accounting for high infections (Gibs, 1924). At higher temperatures, such as 15 and 20°C, the seedlings pass out of the susceptible stage more rapidly.

The pre-emergence soil temperature requirement of another wheat smut, flag smut of wheat, *Urocystis agropyri* (*U. tritici*) is different from that of wheat bunt. The optimal temperature range for infection of flag smut is as high as 18–23°C (Griffiths, 1924; Noble, 1924; Verwoerd, 1929).

Also, covered smut of barley, *Ustilago hordei*, is greatly influenced by soil temperatures in the pre-emergence stage. The parasite invades the host only during the early seedling stage (Appel and Gassner, 1907*b*; Schellenberg, 1911; Rump, 1926; Tapke, 1940), the most favourable temperature range for development of the smut being 10–25°C, and fluctuating temperatures within the optimum further increase infections (Faris, 1924*a*). Schafer, Dickson and Shands (1962*b*) confirmed the level of the optimum soil temperature range by testing the influence of soil and air temperature on a highly susceptible cultivar Odessa and a moderately resistant cultivar Persicum. More than 95 per cent covered smut developed in Odessa at soil/air temperatures 12/16, 20/20, 20/24 and diurnal alternations between 12/16 and 24/24. The resistant cultivar Persicum developed no smut at 12/16 but maximum infection at the much lower level of 24 per cent when temperatures were alternating between 24/24 and 12/16. It is obvious that covered smut of barley requires higher soil temperatures than wheat bunt. Moreover, particularly at the higher temperatures, *U. hordei* developed leaf smut, the resistant cultivar particularly at the alternating temperature between 24/24 and 12/16°C.

For infection by *U. nigra,* black smut or semi-loose smut of barley, minimum has been found to be below 5°C, optimum 15–20°C (Leukel, 1936) and fairly high percentages of infection occurred at soil temperatures of 10–25°C. In contrast *U. nuda*, loose smut of barley, showed little if any significant reaction to soil temperature.

Sphacelotheca reiliana is greatly influenced by soil temperature. At 25 per cent moisture content the optimum is 28°C and the minimum 20°C; at 15 per cent moisture content the optimum is 28°C but the minimum 12–16°C; the maximum is 36°C at 25 per cent and above 36°C at 15 per cent moisture content (Christensen, 1926).

The influence of soil temperature on development of head blight or scab, *Fusarium graminearum (Gibberella zeae)*, on maize, wheat and other cereals has been extensively investigated (Dickson, 1923, 1925; Dickson, Eckerson and Link, 1923; Jones, 1924; Dickson and Holbert, 1926). In North America *F. graminearum* seedling blight is escaped in the *southern* part of the maize belt and in the *northern* part of the wheat belt. The temperature curve for the growth rate of the fungus on agar (optimum 26°C) is quite different from the temperature curve for the disease development on maize (optimum about 8°C), but much more like the temperature curve for the disease on wheat (optimum about 28°C) (Fig. 15.4). In wheat the disease is most severe at the optimum temperatures of seedling growth (24–28°C) at which, however, the seedling grows too fast to become well established, whereas at lower temperatures (8–16°C) it becomes more robust and resistant. On the other hand, the optimum temperatures for development of maize plants are 24–28°C; at this temperature the plant becomes vigorous, more resistant and hence only slightly affected, in spite of the fact that this temperature range is optimum also for the pathogen, favouring mycelial growth and production and germination of conidia. It should be noted, however, that the temperature curves for disease development are based on a definite constant soil moisture. The curve may be quite different if soil moisture is changed.

Data published by McKeen (1951) provide the same inversion of the temperature curve, as shown for maize blight in Fig. 15.4, for three seed-borne pathogens of maize, including *F. graminearum*, discussed above.

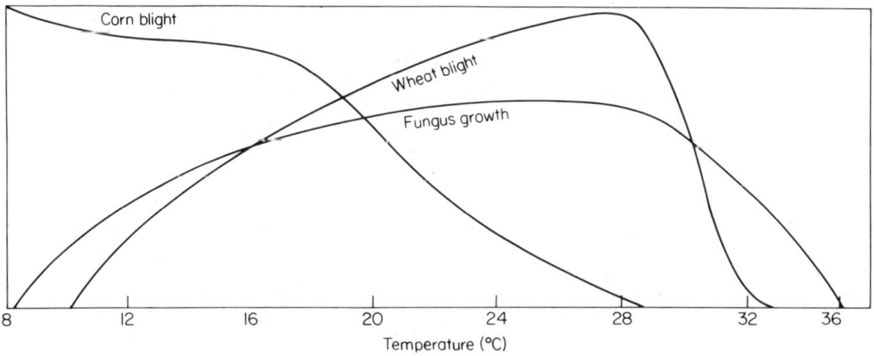

Fig. 15.4. Influence of soil temperature on percentage of infection by *Fusarium graminearum* on seedlings of maize and wheat, compared with the growth rate of the fungus on agar medium. Schematic. (Jones, 1924)

Temperature requirements for their growth on potato dextrose agar are distinctly different from their rates of post-emergence blight (Fig. 15.5).

Development of Fusarium wilt of tomato, *F. oxysporum* f.sp. *lycopersici*, is highly dependent on soil temperature, 27°C being optimum (Clayton, 1923). For Fusarium wilt of cotton, *F. oxysporum* f.sp. *vasinfectum*, the optimum temperature region is 28–32°C (Chester, 1950*a*), for Fusarium wilt of flax, *F. oxysporum* f.sp. *lini* 24–28°C (Muskett and Colhoun, 1947). On the other hand Verticillium wilt of for example tomato, *Verticillium alboatrum*, is favoured by lower temperatures, the optimum in soil being 20°C (Bewley, 1922; cf. Ikramov and Mirdadaev, 1974), although the optimum may be somewhat modified by air temperature and may vary with the strain of the fungus (Edgington and Walker, 1957).

Calvert and Thomas (1954) have shown that the extent of infection by safflower rust, *Puccinia carthami* on *Carthamus tinctorius*, from seed carrying teliospores is greatly influenced by prevailing temperatures during the first days after sowing, lower temperatures (5–10°C) being more favourable than higher (20°C), and Schuster (1956) found that at soil temperatures of 30°C and 35°C infection of the seedling is prevented. Lehman (1953) established that at 13°C seedlings from soybean seed encrusted in oospores of *Peronospora manshurica*, downy mildew, were systemically infected at a rate of 40 per cent, whereas no infection occurred at 18°C or above, and Pederson (1961) found that a higher percentage of systemic infection resulted when inoculated seed was germinated at 15°C and 20°C than at 25°C and 30°C.

In many seedling diseases resistance is decreased with temperatures favourable to rapid growth of the seedling. Many seedling pathogens are able to decompose hemicelluloses and protopectins, but not cellulose or lignin. Temperature has a great influence on the chemical composition and the ultramicroscopic structure of the cell walls. Gäumann (1946) states that cell walls in wheat plants grown at 33°C are decomposed about $1^1/_2$ times faster by the enzymes of *Fusarium avenaceum* than cell walls in wheat plants grown at 3°C, and in rye seedlings decomposition at the low temperature is twice as long as at the high temperature. He found considerably more agreement between the temperature curves for infection percentages for seedling fusarioses in rye (*F. nivale*) and wheat (*F. graminearum*; *F. culmorum*) on one hand, and the temperature curve for decomposition of the cell walls on wheat seedlings on the other, than between the temperature curves for infection percentages and, for instance, the temperature curves of the growth intensity of the related *Fusarium* species (based on dry weight of mycelium).

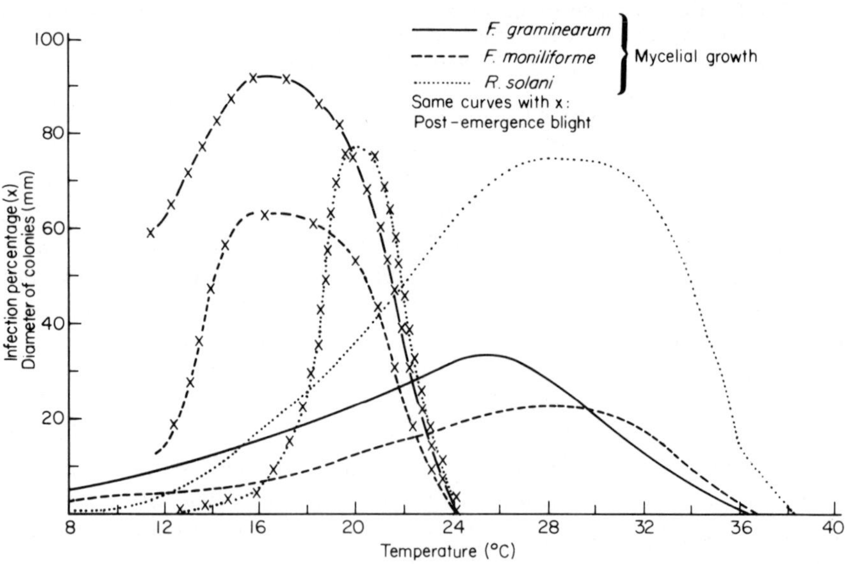

Fig. 15.5. Influence of soil temperature on percentage of infection by *Fusarium graminearum*, *Fusarium moniliforme* and *Rhizoctonia solani* (post-emergence blight) on maize, compared with the growth rate of the fungi on potato dextrose agar. (Data published by McKeen, 1951). Compare Fig. 15.4

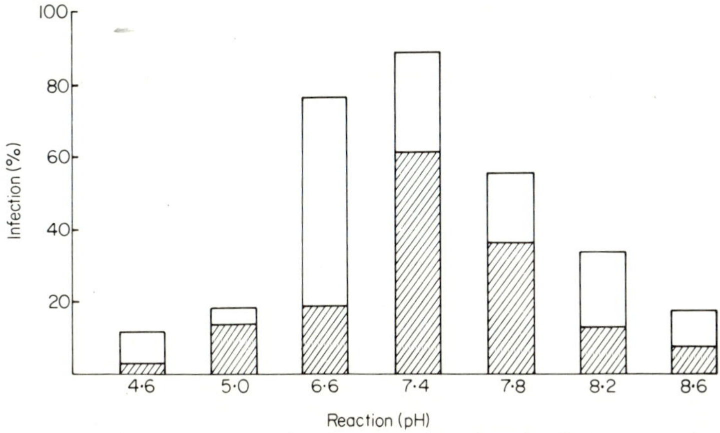

Fig. 15.6. Influence of soil reaction on development of *Ustilago hordei* in two species of oats. Hatched plus unhatched: *Avena sativa* cv. Viktor. Hatched: *Avena nuda* var. *inermis*. (Reed and Faris, 1924)

Host and pathogen are not merely two units in their reaction to such factors as temperature. They are two groups of units, one consisting of a range of races or biotypes, the other of genetically different individuals or cultivars. Varietal differences in wheat in response to the temperature optimal for infection of stinking smut have already been mentioned.

Another example is the influence of temperatures on establishment of infection by *Drechslera graminea*, barley leaf stripe disease. This influence varies according to biotype or race of the pathogen and to cultivar of the host (Isenbeck, 1938). In infection experiments, including three biotypes of the fungus and four cultivars of barley, incubated at 5, 10, 15, 20 and 25°C, infection percentages varied considerably according to biotype and cultivar, the highest infection percentages being encountered at 5 or 10°C. It was the host–parasite combination that mattered in the reaction to temperatures. Leukel (1933) found that temperatures of 15°C or below during emergence favour development of the disease from naturally inoculated seed while temperatures above 20°C tend to inhibit the disease.

15.2.3 Soil reaction

Data reported on the effect of pH of soils on disease development must be considered in combination with other soil factors, such as water retention and nutrient supply, particularly where observations are conflicting.

In infection experiments with *Avena nuda* var. *inermis* and *A. sativa* cv. Viktor grown in sand at 19–22°C and 20 per cent humidity, Reed and Faris (1924) recorded the following infection percentages of covered smut of oats, *Ustilago hordei (U. kolleri)* for the two hosts: pH 4.6: 4 and 12, pH 7.4: 64 and 92, and 8.6: 8 and 18 per cent, cf. Fig. 15.6. Neutral soil reaction is favourable for development of the disease.

Faris (1924*a*) found marked differences in development of covered smut of barley, *U. hordei*, at four different soil reactions. In acid soil the smut percentages were approximately double those in alkaline soil. Soil reaction may also affect the optimum level of moisture and temperature for infection.

Leukel (1937) obtained in infection experiments with *Tilletia caries* 6 per cent diseased plants at a soil reaction of pH 5.6 and 42 per cent at pH 7.9, whereas soil reaction had no consistent effect of development of black- or semi-loose smut of barley, *U. nigra* (Leukel, 1936).

On synthetic media fungi usually develop most vigorously at neutral reaction, but no conclusion can be drawn from this with regard to the requirements of fungi in soil. Among seed-borne fungi acid soil reaction is preferred by *Fusarium culmorum* (Blair, 1937; Shen, 1940), neutral soil reaction by *Rhizoctonia solani* (Blair, 1943) and *Urocystis occulta* (Ling, 1941), alkaline soil by *Gaeumannomyces graminis* (syn. *Ophiobolus graminis*) (Garrett, 1936).

Two vascular infecting fungus species are favoured by quite different soil reactions, *Fusarium oxysporum* f.sp. *lycopersici*, Fusarium wilt of tomato, and *F. oxysporum* f.sp. *vasinfectum*, Fusarium wilt of cotton, by acid soil, *Verticillium alboatrum*, causing wilts also in both tomato and cotton, by alkaline soil. The two diseases on cotton are commonly referred to as Fusarium wilt of acid soils, and Verticillium wilt of alkaline soils (cf. Noble and Richardson, 1968). Also *F. oxysporum* f.sp. *pisi*, pea wilt (race 1), and pea near wilt (race 2), is favoured by acid soil. Development of near wilt has been shown to be reduced by alkaline soil to such an extent that the pea cultivars New Era, New Season and New Wales, known as resistant against race 2 of the fungus, proved to be so only in alkaline soil. When the pH was lowered to 4.8 and 5.6 they were susceptible. However, other soil conditions such as mineral conditions may modify the influence of pH. Another pea cultivar, Koroza, remained resistant also in acid soil (Hubbeling and Mathur, unpublished). In testing for resistance such predispositional factors must be taken into consideration.

For *Fusarium nivale* in wheat Millar and Colhoun (1969) found that within the range of soil pH from 4.8 to 8.1 disease severity increased with a decrease in pH value.

15.2.4 Light intensity

Light may exert a considerable influence on the predisposition of the host. At low light intensity tomato becomes more susceptible to *Fulvia fulva* (syn. *Cladosporium fulvum*) and to tobacco mosaic virus (cf. Section 15.1), while barley becomes more resistant to *Drechslera graminea*. In infection experiments with three biotypes of *D. graminea* on two cultivars of barley, Isenbeck (1938) found increasing infection at increasing light intensity. However, different combinations of host cultivar and parasite biotype reacted differently to the same series of light doses in the same way as they did to different temperatures, mentioned above.

Fungi are heterotrophic, most of them grow independently of light, and spores usually germinate equally well in light and darkness. Stinking smut of wheat, *Tilletia caries*, and certain other fungi are exceptions from this rule. Hahne (1925) found that germination of stinking smut spores was strongly stimulated by daylight; increasing rate of daylight gave increasing percentage of spore germination. This was true for spores taken from isolates but in nature spores of this fungus must germinate in darkness, below the soil surface. It has been suggested that the light sensitivity is merely a manifestation indicating that the effect of light in nature has been replaced by some other factors. It has been established that light is essential to spore germination of *T. contraversa*, dwarf bunt of wheat (Böning, Wagner and Minckwitz, 1953; Gassner-Hoechst, 1953; Pichler, 1953; Baylis, 1958; Meiners and Waldher, 1959), and light stimulates spore germination of *T. fusca*, *T. holci* and *T. caries*, placed on soil-extract agar, but has little or no effect on *T. asperifolia*, *T. bromi-tectorum* and *T. elymi*, and suppresses germination of *T. scrobiculata* on this medium (Meiners and Waldher, 1959). Krajný (1974) found that light inhibited teliospore germination in *Ustilago tritici*.

The fact established by Brandt (1964) and Gafoor and Heale (1971) that formation of microsclerotia of *Verticillium* is delayed or suppressed by light (near ultraviolet radiation) has bearing on seed transmission. The light conditions for development of seed-borne microsclerotia may be different according to the type of crop, i.e. whether the seed is exposed to light during maturation or not. Hence, microsclerotia may be expected to occur on seed of sunflower as in fact they do (Sackston and Martens, 1959; Bruni, 1970), see Fig. 13.4, and in cotton microsclerotia formed within stem contaminate seeds (Evans, Wilhelm and Snyder, 1966).

15.2.5 Photoperiodism

15.2.5.1 RELATION OF PHOTOPERIODISM TO SUSCEPTIBILITY OF HOSTS

Rodenhiser and Taylor (1940*a*, 1940*b*) found increasing disease severity of stinking smut of wheat in two cultivars of wheat, Hope and Marquis, when day length was increased (Table 15.3). In further studies (Rodenhiser and Taylor, 1943) the wheat cultivar Canus was found resistant to each of three races of *Tilletia foetida* and five races of *T. caries* under natural daylight conditions, while resistance was broken down under continuous light.

By increasing day length for the winter rye cultivar Dakold infected by

Urocystis occulta, stalk smut, Ling (1941) also found increasing disease severity, whereas increased day length did not increase disease severity in the spring cultivar Prolific.

Day length has been found to have a marked influence on development of seed-transmitted downy mildew, *Peronospora parasitica*, in cabbage. With more than 16 hours of daylight no infection developed under various conditions of temperature and relative humidity, whereas strong systemic infection occurred at the cotyledon stage with less than 16 hours of daylight, other factors remaining the same (Polyakov and Vladimirskaya, 1964).

Sackston and Shepard (1973) showed that symptoms of *Verticillium dahliae* wilt in sunflower cv. Sunrise were more severe and appeared sooner under long days, 16 h, or short days interrupted by a dark period, than under short days, 10 h, and that this effect was independent of the initiation of flowering in this host.

Table 15.3. Relation of day length to amount of stinking smut of two cultivars of wheat

	Infection (%)	
Day length (h)	*cv. Hope*	*cv. Marquis*
8	1	2
10–11	18	18
24	64	33

15.2.5.2 RELATION OF PHOTOPERIODISM TO SPORULATION OF PARASITE

Direct influence of day length on sporulation has been established for *Leptosphaeria avenaria* f.sp. *triticea* (Hogenson and Hosford, 1971). Less than 12 h of daily light of wavelengths greater than 560 nm stimulates production of immature perithecia but not ascospores. These were abundantly formed by exposure to light within the range of 12–24 h per day for approximately 50 days. These long day requirements agree with the natural conditions in North Dakota, U.S.A., where mature perithecia first appear on maturing wheat in late summer and/or autumn.

15.2.6 Diurnal light cycles

Light, very often diurnal periodicity of light, has a marked influence on sporulation of fungi, as shown by Charles M. Leach and other authors (see Subsection 29.5.4.6). In Fungi Imperfecti diurnal sporulators (Leach, 1967*a*, 1967*b*) sporulate when daily exposed to light for a period resulting in formation of conidiophores, the 'inductive phase', followed by a period of darkness which results in formation of conidia, the 'terminal phase' (Leach, 1967*a*, 1967*b*). Some fungi, on the other hand, sporulate under continuous light, some even under continuous darkness. For many fungi also a certain fluctuation of temperature is required for induction of sporulation.

Many pathogenic fungi show a peak of spore dissemination in the morning hours, after a night of spore production. The diurnal pattern of spore dissemination is not only dependent on fluctuation of light and temperature but on other factors as well, such as wind velocity, relative humidity and frequency of dew.

Alternaria porri, A. porri f.sp. *solani* and *A. tenuis* have a diurnal periodicity with a pronounced peak for spore dissemination in the morning hours (Rotem, 1964; Meredith, 1966), as shown in Fig. 15.7 in which also weather factors are considered, and the same holds true for *Trichoconis padwickii*, stackburn of rice (Sreeramulu and Vittal, 1966).

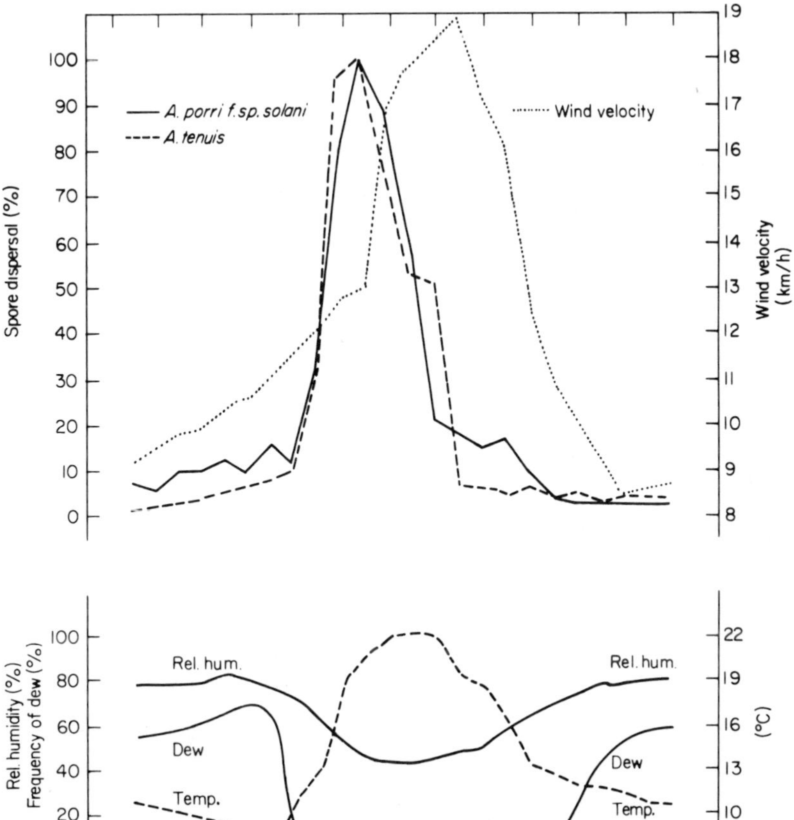

Fig. 15.7. Diurnal periodicity of *Alternaria porri* f.sp. *solani* and *A. tenuis* spore dispersal, compared with diurnal periodicity of wind velocity, temperature, relative humidity and frequency of dew. (Adapted from Rotem, 1964)

Diurnal cycle of sporulation is well known also for downy mildews, including tobacco blue mould, *Peronospora tabacina* (Yarwood, 1937*a*; Cruickshank, 1963*c*). Hirst (1953) found a similar very regular diurnal periodicity curve for *Phytophthora infestans* and *Polythrincium trifolii* indicating a substantially increased spore production of these fungi during night, and Leach (1975) for *Drechslera turcica*.

Sowing time, fertilisation and other conditions of importance to cultural practice are discussed in Chapter 19.

16 Environmental Biotic Factors Affecting Establishment and Course of Disease

INTRODUCTION

Interaction between organisms operates everywhere in nature. The presence of other organisms may considerably influence the course of seed transmission as well as establishment of secondary infection in spread of disease. Some organisms may stimulate or impede the pathogenicity of other organisms. Others may act as vectors, insects providing most of the examples of this category (Fig. 16.1); some of these are obligatory vectors.

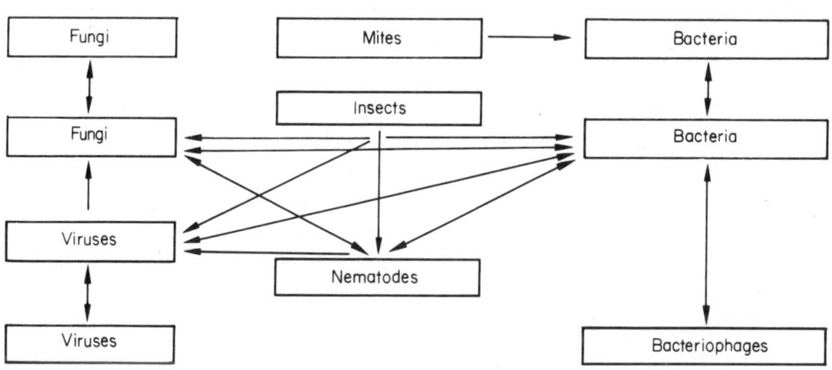

Fig. 16.1. Biotic factors in relation to seed-borne pathogens.

16.1 RELATIONSHIPS BETWEEN INSECTS AND BACTERIA

Erwinia stewartii, the incitant of the disastrous bacterial wilt of maize in North America, is disseminated from contaminated seeds and enters through wounds produced by various insects. These are larvae of flea beetles, particularly brassy flea beetle, *Chaetocnema pulicaria* Melsh (Rand, 1923; Poos and Elliott, 1936), and toothed flea beetle, *C. denticulata* Ill. (Rand and Cash, 1921; Poos and Elliott, 1936), also larvae of southern corn root worm, *Diabrotica duodecimpunctata* Oliv. (Rand and Cash, 1924, 1933; Poos and Elliott, 1936), *D. longicornis* Say, *Phyllophaga* sp. (Ivanoff, 1933), seed corn maggot, *Hylemya cilicrura* Rond., and common wheat wireworm, *Agriotes mancus* Say (Frutchey, 1936). Reviews of the relevant investigations have been given by Leach (1940) and Pepper (1967). It has been established experimentally that infection follows injury of the roots of seedlings by larvae, and that flea beetles are responsible not only for this primary infection but also for secondary spread of wilt in midsummer. The brassy flea beetle, *Chaetocnema pulicaria*, is the primary vector and the only one of importance as over-wintering carrier of the bacteria (Elliott and Poos, 1934, 1940; Poos, 1939).

16.2 RELATIONSHIPS BETWEEN INSECTS AND FUNGI

The importance of insect transmission of plant pathogens including parasitic fungi, is well known (for references, cf. Leach, 1940; Carter, 1966).

Examples of insect-transmitted seed-borne fungi are *Diplodia zeae*, *Fusarium moniliforme* and *F. graminearum*, for which the corn earworm (*Heliothis zeae* Boddie) and the European corn borer (*Pyrausta nubilalis* Hbn.) provide entries for infection and aid in the rapid spread of the fungi within maize plants (Taylor, 1952; Christensen and Wilcoxson, 1966).

The presence of insects may, under certain critical circumstances, be quite essential for establishment of infection. In southern Delaware most of the late cultivars of soybean are infected in low percentages by *Diaporthe phaseolorum* var. *zeae*. Normally the parasite gains access to the seed through the adaxial vein and the funiculus, and colonises the endocarp while the exocarp is seldom penetrated. However, Crittenden (1968) reported that the corn earworm, in attacking the late cultivars Hill and Dorman, makes holes through the exocarp, mesocarp and endocarp thus exposing the seed to the fungus. He found that this may lead to a many-fold increase of pod infection. In the absence of the insect the two cultivars remain almost free of infection.

A relationship has been claimed to exist between a yeast fungus *Nematospora* and various insects feeding on cotton and *Leguminosae*. Wingard (1925) observed dissemination of the seed-borne *Nematospora coryli* by southern green stinkbug, *Nezara viridula* (L.), that feeds upon the pods of lima beans thus causing inoculation. Sunken, necrotic lesions are produced on the seeds (stigmatomycosis). This association has been suggested by Johnson, Chamberlain and Lehman (1954) also for *Nematospora coryli*, yeast spot, and southern green stinkbug on soybean. However, the insects themselves also act as independent incitants of such symptoms. Other fungi of stigmatomycosis are encountered on cotton, where at least four different species have been identified (Ashby and Nowell, 1926; Pearson, 1947). Two of these, *N. gossypii* and *N. coryli*, were most commonly associated with the disease on cotton and were found to be disseminated and introduced into the cotton bolls by different species of plant bug, the cotton stainers, *Dysdercus* spp. This relationship has been observed by various investigators in North America, in Uganda, South Africa, Belgian Congo and in Transvaal (review by Leach, 1940).

The role of the soil microfauna in development of seed-transmitted root rot diseases is a neglected field. In north-central U.S.A. *Drechslera sorokiniana* and *Fusarium* spp. are commonly associated with infestation of the bluegrass billbug, *Calandra parvula*, which is common and serious in cereals and grasses. In these hosts the insect enhances development of basal stem and root rot, produced by the above soil- and seed-borne pathogens (Hanson, Milliron and Christensen, 1950).

16.3 RELATIONSHIPS BETWEEN INSECTS AND VIRUSES

The paramount importance of insects as vectors responsible for secondary spread of seed-transmitted viruses is well known, cf. Section 3.2. Their possible role in bringing about primary infection of seed-transmitted virus, for instance by inoculating virus contained in the seed coat adhering to young seedlings such as tobacco mosaic virus in tomato, has not been established.

16.4 RELATIONSHIPS BETWEEN INSECTS AND NEMATODES

Rao and Yadava (1972) found that rice gall midge, *Pachydiplosis oryzae*, increases incidence of white tip nematode, *Aphelenchoides besseyi*, in rice.

16.5 RELATIONSHIPS BETWEEN MITES AND BACTERIA

Lapis (1972) found that soybean plants infected with mites (*Tetranychus telarius*) showed resistance or immunity to bacterial blight, *Pseudomonas glycinea*. This plant reaction increased with increases in number of mites and in time of exposure of plants to them prior to infection by the bacteria.

16.6 RELATIONSHIPS BETWEEN NEMATODES AND BACTERIA

Corynebacterium tritici, yellow ear rot, and *Anguina tritici*, ear cockle, are associated in producing a disease of wheat. This association between the organisms is an obligatory condition of pathogenicity. Alone neither one produce symptoms (Cheo, 1946; Vasudeva and Hingorani, 1952; Gupta and Swarup, 1972). Bacteria carried within the galls can remain viable for years (Subsection 20.2.3).

Nematodes may act as facultative vectors as well as synergists of pathogenic bacteria. Hawn (1963) reported transmission of *Corynebacterium insidiosum*, bacterial wilt of lucerne, by stem eelworm, *Ditylenchus dipsaci*, which seriously increased percentages of the wilt, and Hunt *et al.* (1971) established that *Meloidogyne hapla* either aided entry of bacteria into roots of lucerne or hastened the colonisation of *C. insidiosum* thus increasing substantially incidence of bacterial wilt. Pitcher (1963) found a similar relationship between *C. fascians* and *Aphelenchoides ritzemabosi*.

16.7 RELATIONSHIPS BETWEEN NEMATODES AND FUNGI

An important and striking interaction exists between certain pathogenic nematodes and certain pathogenic fungi. Their combined pathological potential may be far greater than the sum of their individual effects. Especially important is the relationship between root knot nematodes and the seed-borne cotton pathogen *Fusarium oxysporum* f.sp. *vasinfectum*. Hollis (1958) found a clear correlation between the extent of vascular browning in cotton plants, infected by this *Fusarium*, and the extent of infection of root knot nematodes. A. L. Smith (1953*a*, 1954) and co-workers, and Martin, Newsom and Jones (1956) established a positive correlation between yield and resistance against Fusarium wilt and root knot, *Meloidogyne incognita* and *M. incognita* var. *acrita*, as well as between root knot severity index and wilt percentage.

It has been demonstrated that not only root knot nematodes but also reniform nematode, *Rotylenchus reniformis*, increases wilt severity in wilt susceptible cultivars, and induces wilt in those which are resistant (Neal, 1954; Jones, Newsom and Finley, 1959; Khadr, Salem and Oteifa, 1972) and that the sting nematode, *Belonolaimus longicaudatus*, reduces Fusarium wilt resistance in cotton (Holderman and Graham, 1953; Smith and Dick, 1960; Smith, 1961). In comparative experiments Cooper and Brodie (1963) studied the predispositional effect of the root knot nematodes, *Meloidogyne incognita*, and the sting nematode, *B. longicaudatus*. The two nematodes are basically different in their host–parasite relationship, the one a gall-producing sedentary endoparasite, the other a root-stunting migratory ectoparasite — nevertheless, both predisposed resistant as well as susceptible cotton cultivars to Fusarium wilt. The precise mechanism of this effect is not yet fully understood. The explanation may be that the nematodes provide entry for the fungus through wounds, as in tomato seedlings inoculated with *M. incognita* and *Fusarium oxysporum* f.sp. *lycopersici*. Soon after inoculation nematode larvae penetrate the meristematic tissues of the roots causing injury, and subsequently the fungus invades the apical meristem and vascular tissue of the root causing wilt (Kawamura and Hirano, 1968). On the other hand, perhaps nematodes cause a particular physiological host response which lowers resistance, as suggested for the relationship between nematodes and Fusarium wilt of tomato (Jenkins and Coursen, 1957). Well in accordance with this view are the findings by Perry (1963) and Powell (1963) who demonstrated that a material increase of Fusarium wilt of cotton, *F. oxysporum* f.sp. *vasinfectum*, is obtained only when infection of the fungus is established some time after the infection of the nematode *M. incognita* has been established in the host. Combined inoculation with both pathogens does not lead to increased development of wilt.

The association between rot knot nematodes (referred to *Heterodera marioni*) and Fusarium wilt of tomato, *F. oxysporum* f.sp. *lycopersici*, is of a similar nature to that of the cotton disease complex (Young, 1939). A Fusarium wilt-resistant tomato cultivar Chesapeake was affected by this fungus only in the presence of the root knot nematodes, *M. incognita*, resulting 100 per cent, and *M. hapla* giving 60 per cent wilting (Jenkins and Coursen, 1957). However, Binder and Hutchinson (1959) were not able to confirm this effect, perhaps because they used different seed lots, nematodes and numbers of nematodes in their experiments.

A third form of this wilt-producing *Fusarium*, *F. oxysporum* f.sp. *tracheiphilum*, is seed-borne on cowpea, spores being carried on the surface of the seed (Kendrick, 1931). In soil with which the fungus was mixed, the presence of *Meloidogyne javanica* was shown to increase wilt severity, as compared with soil not infested with the nematode. Also in soybean, attacks of *F. oxysporum* are increased by *M. incognita* and even more so by *Heterodera glycines*. Simultaneous attacks of *F. oxysporum* and *H. glycines* lead to higher populations of the nematode than when the fungus is absent (Ross, 1965). Root knot nematodes, including *M. incognita*, likewise increase the susceptibility of resistant watermelon cultivars to wilt caused by *F. oxysporum* f.sp. *niveum* (Sumner and Johnson, 1973).

Finally, synergistic interaction between *F. oxysporum* f.sp. *pisi* race 1 and *Pratylenchus penetrans* causes breakdown of wilt resistance in pea cv. Wisconsin Perfection (Oyekan and Mitchell, 1971). Also near wilt caused by race 2 is enhanced by the presence of this nematode (Seinhorst and Kuniyasu, 1971).

The aggravating effect of nematodes when associated with *Fusarium* may, as already mentioned, be due in part to the establishment of infection by root wounding providing avenues of entrance (Armstrong and Armstrong, 1948). However, the effect of nematodes on host susceptibility in later stages of plant development may be of no less importance (A. L. Smith, 1954; Jenkins and Coursen, 1957). Other factors which still need critical studies have been pointed out: in feeding, nematodes secrete various metabolic products into the host cells. Some effects of these substances on the host are increase in respiration, activation of oxidising enzymes and degradation of proteins (albumins) by proteolytic enzymes. Due to the impairment of the host plant, caused by the altered metabolism, the plant readily succumbs to wilt (Myuge, 1956). Also, the influence of root knot nematodes on nutrition of the host has been shown to be important. Infection of *Meloidogyne incognita* has been reported to decrease the potassium content in plants of lima bean, *Phaseolus lunatus*, nematode damage being correlated with the amount of potassium available to the host, because the nematodes interfere in some manner with the potassium nutrition of the plant (Oteifa, 1952). It has been demonstrated that low potassium levels have increased the severity of Fusarium wilt in cotton (Young, 1938) and tomato (Walker and Foster, 1946).

Synergism between nematodes and another wilt organism, *Verticillium dahliae*, has been reported. The fungus increased the rate of reproduction of *Pratylenchus penetrans* in the roots of eggplant and tomato (Mountain and McKeen, 1962*a*), and the nematode increases the incidence of Verticillium wilt in these plants (McKeen and Mountain, 1960; Mountain and McKeen, 1962*b*, 1965). In tomato *Tylenchorhynchus capitatus* and *Meloidogyne incognita* increase the incidence and severity of Verticillium wilt, *Verticillium alboatrum* (Overman and Jones, 1970), and susceptibility of cotton to this wilt is increased by *Meloidogyne incognita* var. *acrita* (Bazan de Segura and Aguilar, 1955; Miller, Nigh and Alcorn, 1967; Khoury and Alcorn, 1973).

16.8 RELATIONSHIPS BETWEEN NEMATODES AND VIRUSES

Nematodes act as obligate vectors for several seed-borne viruses. The vectors of the ubiquitous tobacco rattle virus, a rod-shaped virus, are all restricted to members of the genus *Trichodorus*, namely *T. pachydermus* (Sol and Seinhorst, 1961; Van Hoof, 1962); *T. primitivus* (Sänger, 1961; Harrison, 1961); *T. teres* (Van Hoof, 1964); *T. viruliferus* (Van Hoof, 1964); *T. similis* (Cremer and Kooistra, 1964); *T. allius*, *T. christiei* and *T. porosus* (Ayala and Allen, 1966). Van Hoof (1970) demonstrated that even after storage in soil for three years without access to plant roots, *Trichodorus* can transmit tobacco rattle virus to tobacco.

Four other seed-borne viruses, all of them polyhedric, are transmitted by two genera of nematodes as obligate vectors: *Longidorus* and *Xiphinema*. Arabis mosaic virus is transmitted by *Xiphinema coxi* (Fritzsche and Schmidt, 1963) and *X. diversicaudatum* (Murant and Lister, 1967); tobacco ringspot virus has *X. americanum* (Fulton, 1962; McGuire, 1973) and *X. coxi* (Van Hoof, 1971) as vectors. On the other hand, tomato black ring virus is transmitted by *Longidorus attenuatus* (Harrison, 1964) and *L. elongatus* (Harrison, Mowat and Taylor, 1961; Murant and Lister, 1967), raspberry ringspot virus by *L. macrosoma* (Harrison, 1962, 1964) and *L. elongatus* (Murant and Lister, 1967). Harrison and Winslow (1961) and Taylor and Raski (1964) showed that arabis mosaic virus and grapevine fanleaf virus persist in their vectors, *Xiphinema* spp., for at least 8 months, or probably for the lifetime of the nematode, and McGuire (1973) that tobacco ringspot virus persists for 9 months in *X. americanum*.

Bergeson and Athow (1963) found that the percentage of transmission of tobacco ringspot virus in cucumber by *X. americanum* was proportional to the population of this vector in the soil. However, insect transmission was more important.

16.9 RELATIONSHIPS BETWEEN FUNGI

The composition of the soil microflora may to a considerable extent determine the development of pathogenic fungi introduced into the soil with seed. Competition occurs between the organisms on access to nutritional material, and the relationship between different organisms may be positively dependent — stimulating, as in synergism, or hampering, as in antagonism.

It is well known that contaminated seed sown in disinfested soil may lead to more severe development of diseases than contaminated seed sown in untreated soil, where antagonistic soil organisms have not been eliminated. *Drechslera sorokiniana,* which is very sensitive to the effect of other fungi, developed vigorously in sterilised soil but could be almost completely suppressed when a little unsterilised soil was added (Henry, 1931, cf. Campbell, 1956).

Trichoderma viride has been reported by Weindling and co-operators, and other authors, as being strongly antagonistic to many fungi including seed-borne parasites, and to be able to produce specific antibiotics (Weindling, 1934; Weindling and Emerson, 1936; Weindling and Fawcett, 1936). More recently Webster and Lomas (1964) who re-examined Weindling's strain identified the fungus as *Gliocladium virens*. They found, by testing more than 40 isolates of *T. viride* (perf. state *Hypocrea rufa*) from various sources, no evidence of antibiotic production.

However, Dennis and Webster (1971*a*, 1971*b*) found in their extensive laboratory studies on antagonistic properties of *Trichoderma* that nine species including *T. viride* produced non-volatile antibiotics active against a range of fungi. Furthermore, Dennis and Webster found that nine partly different species of *Trichoderma*, including *T. viride*, produced volatile components inhibiting to the growth of other fungi, including *Rhizoctonia solani*.

Weindling and Fawcett (1936) obtained effective control of damping-off in Citrus seedlings, caused by *Rhizoctonia solani*, by use of *Gliocladium virens* in soil at pH values below 4.3. Hashioka and Fukita (1969) demonstrated strong antagonism of *G. deliquescens* (perf. state *Hypocrea lutea*) to three rice pathogens, including *Drechslera oryzae* and *Pyricularia oryzae*. They studied the mode of mycoparasitism of this species on the rice pathogens as well as the mycoparasitism of *Trichoderma longibrachiatum* (perf. state *Hypocrea schweinitzii*), *T. polysporum* (perf. state *Hypocrea* sp.), and *Acremonium* sp. (perf. state *Hypocrea austrograndis*) on the mentioned rice pathogens. When the rice pathogens were in close proximity with *Gliocladium* or *Acremonium*, severe malformation or disorganisation of host hyphal cells occurred but no direct invasion by infection hyphae. With *Trichoderma*, on the other hand, the parasite invaded the host hyphae and extended within them, thus establishing a direct parasitic relationship in contrast to the indirect mycoparasitism of *Gliocladium* and *Acremonium*. Norton (1954) reported strong antagonism in soil by *T. viride* (as *T. lignorum)* and *T. terricola* against *Macrophomina phaseolina*.

Chaetomium, a genus of common saprophytic soil inhabitants, is frequently seed-borne in many kinds of seed. The presence of species of this genus with seed may possibly influence seed transmission of pathogens.

In investigations at the University of Minnesota, the severe oat blight, produced by the seed-borne pathogen *Drechslera victoriae*, was prevented when certain Brazilian isolates of *Chaetomium cochlioides* were introduced into unsterile soil (Tveit and Moore, 1954). Tveit (1953) and Tveit and Wood (1955) obtained complete control of seed-borne *Fusarium nivale* on oats in sterile soil by a strain of *C. cochlioides*, but only slight or no effect by two other strains of the antagonist.

Higher plants exert a pronounced influence on the soil microflora around their roots, in the so-called rhizosphere where microorganisms are more abundant than in soil free from the influence of the roots (Hiltner, 1904). The effect is most intense at the root surface, the rhizoplane, and falls off sharply with increasing distance. Antagonistic as well as symbiotic interactions between microorganisms are substantially intensified in the rhizosphere. There is evidence of a selective action by roots, varying with species according to host or non-host relationship to pathogenic fungi, and varying with cultivars susceptible as against resistant, with age of plant or part of plant, and finally with the conditions of the soil (Katznelson and Richardson, 1943).

Reyes and Mitchell (1962) found in their studies on population dynamics in unplanted soil and plant rhizospheres that populations of *Fusarium oxysporum* f.sp. *lycopersici* and *F. solani* f.sp. *pisi* are stimulated in the rhizosphere of susceptible plants but not in that of non-host plants of various families. They established that *F. solani* f.sp. *pisi* appears within a zone of 2 mm of the root susceptible pea plants, particularly around young portions. The rhizosphere effect was evident four days after planting, attained its maximum the eighth day after planting and was still apparent 64 days after planting. The stimulation occurred at 20°C, but not at 16°C and 24°C and occurred in loam, in mulch and in compost, but it did not occur when soil moisture was as low as 4 per cent. At moisture levels suitable to plant growth the number of rhizosphere microorganisms are greater at the lower levels than at the higher levels — outside the rhizosphere the reverse is the case. It has been shown by Katznelson, Rouatt and Payne (1955) that amino acids from roots were liberated by lowering the soil moisture level, thus suggesting that this may in part account for the increase in numbers of rhizosphere organisms.

An important approach in the study of the physicochemical conditions in relation to soil microorganisms has been made in the U.S.A. Carbonaceous organic matter highly influences the soil microflora. For instance, bean rhizosphere microorganisms antagonistic to *Fusarium oxysporum* f.sp. *phaseoli*, *Rhizoctonia solani* and *Verticillium alboatrum* have been increased by adding to the soil mature oat straw and oat straw combined with NH_4NO_3 to produce an N:C ratio of 30 or 50. When high N-applications alone or with oats were introduced, the number of antagonists substantially decreased. The importance of organic matters in the crop soil in relation to the soil microflora has been studied extensively by Kaufman and Williams (1963). They established that fungi were significantly affected by the carbon:nitrogen (C:N) ratio of the soil. A range of C:N ratios from 1:1 to 40:1 were investigated as well as the effect of two rates of application of wheat straw as the added carbon source, namely 2 and 10 tons/acre.

Addition of carbonaceous residues to soil results in an increase in total numbers of soil fungi; this increase is further enhanced by addition of a readily available nitrogen source (Waksman and Starkey, 1924; Jensen, 1931; Waksman, 1944). Increase in population involves as a prerequisite utilisation of nitrogen, which is therefore a limiting factor. The demand is dependent upon the carbon supply (Cochrane, 1958). Kaufman and Williams demonstrated that a narrow C:N ratio resulted in substantial increase of the population of *Chaetomium*, as already mentioned, a genus containing species very antagonistic to certain plant pathogens. However, general conclusions cannot be drawn easily. Some investigators found a wide C:N ratio to be most desirable for reduced disease incidence as far as certain soil-borne pathogens are concerned (Garrett, 1940; Butler, 1953*a*). This was due to the fact that their survival was limited by the available nitrogen supply, just as is the survival of their antagonists. As to the effect of the heavy rate of wheat straw application, adopted by Kaufman and Williams, the total numbers of fungi were increased but the total of types of fungi, the number of species encountered, decreased — the microflora was quantitatively increased but qualitatively reduced.

Practical results in biological control may be obtained mainly by establishing conditions favourable to the natural population of antagonistic microorganisms in the soil rather than by introducing new ones into it. Studies on the effect of green manures, amendments, fertilisers, various crops, details of the physical conditions such as pH and soil type are more likely to lead to results of benefit to agricultural practices.

Interaction between fungi may be encountered everywhere under suitable growth conditions, also in aerial plant parts and within host plants. Among the common seed-borne saprophytes, *Cladosporium herbarum* has been shown to have an antagonistic effect on development of *Botrytis cinerea* in young fruits of tomato by colonising the dead petals in advance of the pathogen. *Cladosporium* is able to grow under drier conditions than many other fungi including *Botrytis* and therefore it colonises dead tissues more readily. When growth substances were used to avoid fall of tomato fruits, colonisation by the saprophyte was favoured, and infection by *Botrytis* was reduced from about 50 per cent in untreated to about 2 per cent in treated plants (Newhook, 1957).

An important aspect of antagonism or synergism is connected with the effect of infection by one pathogen prior to infection by another pathogen.

Tilletia in wheat may predispose to infection by *Fusarium*, *Drechslera* and *Uromyces* (Fischer and Holton, 1957), but may reduce infection of powdery mildew (Sempio, 1938). Bamberg *et al.* (1947) demonstrated the interaction between *Tilletia foetida* and *T. caries*, both of which infect singly, but eliminate the other. Seeds inoculated with *T. foetida* and sown in soil infested with *T. caries* developed seedlings much less infected by *T. caries* than did non-inoculated seedlings. However, Berend (1973) found no evidence of antagonism when seed was inoculated with a mixture of both species but in some cases the hybrid *T. intermedia* appeared.

Morton and Peterson (1960) reported from field observations that early infection of *Drechslera sorokiniana*, spot blotch, in leaf blades of barley inhibited later development of another seed-borne fungus, *Septoria passerinii*, Septoria leaf blotch, in leaf sheaths. In further studies, based on inoculation of the two fungi on excised leaves and impregnation of the leaves with cell-free extracts of either fungus, Wibe and Morton (1962) confirmed the results of the observations and found that some substance(s) produced by *D. sorokiniana* may cause some of the inhibition.

Conversely, wheat is predisposed to *Septoria nodorum*, glume blotch, by attack of *Puccinia recondita* f.sp. *tritici*, brown rust. The interaction was experimentally studied in four cultivars (Van Der Wal, Shearer and Zadoks, 1970).

Cephalosporium inoculated to hosts of *Fusarium oxysporum* prior to inoculation of this pathogen has been shown to have an inhibitory effect on development of the vascular *Fusarium*. *Cephalosporium* prevented symptoms of Fusarium wilt in tomato (Allison and Phillips, 1963; Roy and Patel, 1963) and in cotton (Allison and Phillips, 1963). In cowpea Fusarium wilt was decreased (Long, 1963).

The problem of symptomless carriers has bearing on interaction of pathogens in systemic disease. Armstrong and Armstrong (1948) established experimentally that many wilt-producing *Fusarium* species may produce symptomless, systemic infections in a wide range of hosts, for instance in tobacco, soybean, cotton, tomato and *Antirrhinum*. These plants were inoculated with species of *Fusarium* against which they were resistant, and it was demonstrated that the inoculated plants may act as carriers.

Table 16.1. Cross-protection against *Fusarium oxysporum*. (Data from Davis, 1964)

Inoculation	*Inoculum* *Forma specialis of Fusarium oxysporum*	*Results* *after preinoculation*	*without preinoculation*
Preinoculation to non-susceptible hosts, see below	batatas conglutinans dianthi lini lycopersici niveum tracheiphilum vasinfectum	No symptoms	No symptoms
Inoculation to susceptible hosts (indicated in parenthesis), 7 days later	conglutinans (cabbage) dianthi (carnation) lini (flax) lycopersici (tomato) niveum (watermelon)	Symptoms delayed or failed within 4 weeks	Severely affected within 10 days

Davis (1964) obtained cross-protection of Fusarium wilt with eight formae speciales of *Fusarium oxysporum*. He inoculated seedlings of wilt suscepts in test tubes with forms non-pathogenic to the host and one week later with the one pathogenic to the host species concerned according to the scheme presented in Table 16.1. Davis found that the non-pathogenic forms readily invaded the root a few millimetres from the point of inoculation.

Katan (1971) showed that apathogenic infection of hosts acting as symptomless carriers does occur under natural conditions. The roots and stems of various weeds in soil, naturally infested with *Fusarium oxysporum* f.sp. *lycopersici*, were colonised by the pathogen. The fungus was isolated from naturally infected symptomless carriers, including *Amaranthus graecizans, A. retroflexus, Digitaria sanguinalis, Malva sylvestris* and *Oryzopsis miliacea*. Inoculation on these and other weeds as well as some cultivated plant species confirmed the relationship. All tested species were susceptible as symptomless carriers.

The presence of saprophytes, apathogenic parasites or symbionts as systemic infections in seed-bearing plants and in seeds as demonstrated for certain species of *Chaetomium* and *Alternaria* (Günnewig, 1933; Katz, 1949), *Fusarium*, or unnamed endophytic fungi (Lloyd, 1959*b*), is of considerable interest with respect to interaction between apathogenic and pathogenic organisms, and the influence of apathogenic organisms on the predisposition of a host to a given parasite. Combined infection experiments with apathogenic and pathogenic organisms may give valuable information on the role of the former in the development of parasitic diseases in plants and seeds, including the effect on seed transmission.

16.10 RELATIONSHIPS BETWEEN BACTERIA AND FUNGI

Xanthomonas malvacearum, bacterial blight or black arm, intensifies development of *Colletotrichum gossypii*, anthracnose or boll rot in cotton. When the fungus is associated with the bacterium, the number of conspicuous lesions is substantially increased as compared with disease development without this association. Apparently the bacterium clears the way in the bolls for infection of the fungus (Weindling and Miller, 1941).

In Canada synergism has been demonstrated between *Pseudomonas glycinea*, bacterial blight, and *Septoria glycines*, brown spot, in soybean. Brown spot is markedly increased in the presence of the blight (Benedict, 1964).

In Venezuela, a synergistic effect between *Macrophomina phaseolina* and *Xanthomonas phaseoli* in bean has been recorded. Damage by combined infection was twice that caused by the combined sum of each (Diaz Polanco, 1972).

On the other hand, many species of *Bacillus* have been shown to have marked antagonistic effects against pathogenic fungi (Novogrudskij, 1937; Ledingham, Sallans and Simmonds, 1949; Vasudeva and Roy, 1950; Newhook, 1951; Vasudeva, Jain and Nema, 1952). Vasudeva and Roy (1950) found *B. subtilis* to be antagonistic to *Fusarium udum*, wilt of pigeon pea, and Rangaswami (1963) that *B. mycoides* is antagonistic to *Drechslera oryzae*.

Russian workers have reported antagonistic effect by inoculation of flax seed with soil-inhabiting bacteria against seed-borne *Fusarium oxysporum* f.sp. *lini* and *Colletotrichum lini* (Novogrudskij, 1936; Novogrudskij *et al*., 1937; Beresova, 1939) and *Fusarium* on seeds of *Pinus sylvestris*.

Elimination of saprophytes on seed may reveal their antagonistic effect. Ledingham, Sallans and Simmonds (1949) found that wheat from seed treated with formaldehyde or lactic acid is more severely attacked by *Drechslera sorokiniana* than from untreated seed, particularly in sterile soil, because of the reduction of the seed-borne antibiotic microflora, especially the bacteria. Simmonds (1947) observed that apparently protective bacteria, lysogenic to *D. sorokiniana*, originate from the natural surface flora of wheat seed. He established that 24 hours of incubation of wheat seeds in a moist chamber prior to planting reduced the percentage of *D. sorokiniana* on the seedlings, undoubtedly because the moisture favoured development of antagonistic bacterial saprophytes from the pericarp.

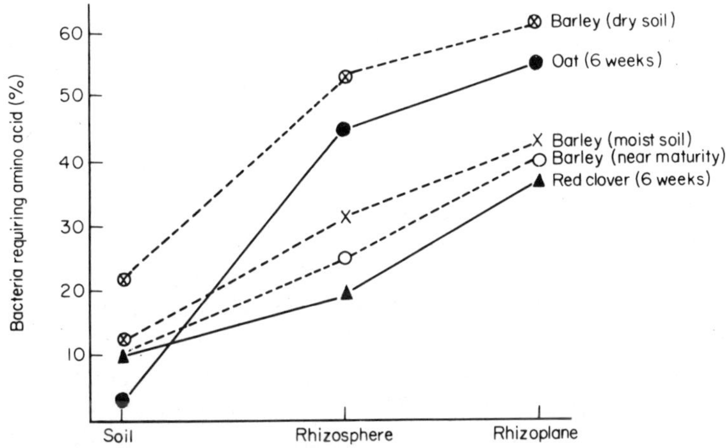

Fig. 16.2. Population gradients of bacteria requiring free amino acids, as related to distance from plant root surface. (According to data by Rouatt, from Lochhead, 1959)

Bacteria in soil are important as antagonists against plant pathogenic fungi. The bacterial flora is greatly influenced by the conditions of the rhizosphere (Fig. 16.2).

Antagonism of actinomycetes from various soils was found to occur *in vitro* in 18–31 per cent of the isolates investigated against one of three specific pathogens, *Rhizoctonia solani* and two species of *Pythium* (Cooper and Chilton, 1949). Lachance and Perrault (1953) established that

19 soil-borne bacteria and 11 soil-borne actinomycetes exerted a more or less pronounced antagonistic effect on *Fusarium oxysporum* f.sp. *lini*. The organisms delayed the onset of wilt symptoms, but failed to protect the plants through their entire growth. Kublanovskaya (1952) suppressed Fusarium wilt of cotton, *F. oxysporum* f.sp. *vasinfectum* by adding actinomycetes to infested soil. From *in vitro* investigations (Christensen and Davies, 1940) a mutagenic effect exerted by metabolics of bacteria has been observed. Bacteria-staled media caused very abundant mutations in monosporous lines of *Drechslera sorokiniana*, some showing greater pathogenicity, some less.

From other investigations Christensen and Allison (1958) report the effect of a specific antibiotic, filipin, on *D. sorokiniana* in agar media. 0.5 ppm in the medium brought about a canker-like growth of spores and an increase in development of mutants which arose to be 10–100 times as frequent in the medium containing filipin as in the medium without the antibiotic. Particularly significant is that some of the mutants induced by certain antibiotics were more virulent on wheat and barley than the original strains of the pathogen.

Soil reaction has a great influence on the qualitative population of bacteria. Papavizas (1963) found that at pH 4.7 true bacteria were predominant as antagonists, while at 7.1 streptomycetes were predominant. Also the soil type is of importance as a factor modifying antibiotic effects. Streptomycin is strongly absorbed by clay and soil organic matter and loses activity under such conditions (Siminoff and Gottlieb, 1951).

16.11 RELATIONSHIPS BETWEEN BACTERIA

Adam and Pugsley (1935) found that a yellow, non-pathogenic bacterium had an antagonistic effect on the halo blight bacterium, *Pseudomonas phaseolicola*, and Scherff (1973) that *Bdellovibrio bacteriovorus*, a small comma-shaped bacterium parasitic on other gram-negative bacteria, inhibits development of bacterial blight, caused by *P. glycinea*, in soybean. *Streptomyces* is antagonistic to many bacteria. Mitchell, Zaumeyer and Andersen (1952), Mitchell, Zaumeyer and Preston (1953), Mitchell and Zaumeyer (1954), and Gray (1955, 1956) demonstrated a strong effect of streptomycin against *P. phaseolicola* and *Xanthomonas phaseoli*, and this antibiotic is also effective against bacterial spot of tomato, *X. vesicatoria*. Other antibiotics from bacterial metabolites, streptothricin and pleocidin, can be used against bacterial blight of bean (Gray, 1958). In contrast, *Achromobacter* sp. has a strong synergistic effect on *P. phaseolicola*; the number of lesions produced by this pathogen in Phaseolus bean increased up to four-fold when the contaminant was present (Maino, Schroth and Vitanza, 1974).

16.12 RELATIONSHIPS BETWEEN BACTERIOPHAGES AND BACTERIA

Some bacteriophages have been tested in their effect against seed-borne bacterial pathogens. In extensive field experiments maize seed treated with a phage and subsequently inoculated with *Xanthomonas stewartii* developed 1.4 per cent infected plants against 18 per cent infected plants from untreated seeds (Thomas, 1935, 1940). The impact of seed-borne bacteriophages on seed-borne bacteria is discussed in Section 3.5.

16.13 RELATIONSHIPS BETWEEN BACTERIA AND VIRUSES

Hedges (1944) established in her extensive investigations that symptoms of common blight of beans, *Xanthomonas phaseoli*, is enhanced synergistically when the bacterium is associated with bean common mosaic virus.

The bacterium seems to be able to act not only as a passive carrier of the virus, but as a true vector. Two isolates of *X. phaseoli*, one typical and one yellow variant which occurred in a masked form in the trifoliate leaves of bean, produced 75 per cent infection of bean common mosaic. The typical form of *X. phaseoli*, kept as a stock culture on steamed potato cylinders, retained its capacity to produce mosaic for at least six weeks. Expressed juice of the virus has a longevity of only two days at room temperature, but the stock culture of the bacterium was kept for about two weeks at this temperature. The yellow variant of the bacterium, after having lost its initial ability to transmit the virus, was able to cause bacterial blight of a hitherto immune bean cultivar, thus indicating a mutual benefit between the two pathogens as to pathogenic effects. Continuous serial passages containing both pathogens increased disease severity, but revealed the dominance of the virus.

Panzer and Nickeson (1959) inoculated 36 accessions and 12 commercial cultivars of bean with the two seed-borne pathogens separately and combined. The severity of both diseases in terms of symptom expression was increased when the pathogens were combined. Particularly striking was the observation that while 22 accessions were resistant against the bacterial blight and 17 were resistant to the mosaic, only 5 accessions were resistant to both pathogens combined.

Lovrekovich, Lovrekovich and Stahmann (1968*b*) reported that tobacco leaves infected with tobacco mosaic virus developed definite resistance to wildfire, *Pseudomonas tabaci*. They found that virus-infected leaves characterised by resistance to *P. tabaci* showed an increase in peroxidase activity of up to 200 per cent over the activity of healthy, susceptible leaves. It has been suggested that this enzyme plays a role in resistance of tobacco to *P. tabaci* (Lovrekovich, Lovrekovich and Stahmann, 1968*a*).

16.14 RELATIONSHIPS BETWEEN VIRUSES AND FUNGI

In New Zealand cauliflower mosaic virus has been found to break down resistance against dry-rot of swede (*Phoma lingam*) in so far as mosaic susceptible cultivars are concerned (Smith, 1960). For instance, the swede cultivar Gartons Parkside, a cauliflower mosaic susceptible cultivar, is highly resistant in the absence of the virus but highly susceptible when the plants are severely affected by the virus.

According to field observations in various countries Cercospora leaf spot of sugarbeet is known often to be prevalent on plants showing symptoms of virus yellows. Crane and Calpouzos (1969) confirmed by combined inoculations that beet yellows virus increases the susceptibility of sugarbeet to *Cercospora beticola*. Within four weeks 14 per cent of the leaves were killed by the fungus alone, whereas 50 per cent of the leaves were killed when they had been pre-inoculated with the virus 24 hours before they were inoculated with the fungus. Also the number of Cercospora lesions was substantially increased by the virus preinoculation.

16.15 RELATIONSHIPS BETWEEN VIRUSES

Ross (1969*b*), Schmitthenner and Gordon (1969) and Quiniones, Dunleavy and Fisher (1971) studied interaction of the seed-borne soybean mosaic virus and bean pod mottle virus, which is not seed-transmitted, in different cultivars of soybean. Plants inoculated with both viruses were more severely stunted than those inoculated with either one. Yield reduction was synergistically increased by both viruses combined as compared with infection by either of the two viruses alone. Seed transmission of soybean mosaic virus was reduced in some cultivars, and increased in others by the presence of bean pod mottle virus (Table 16.2). Kuhn and Dawson (1973) found that the rate of seed transmission of soybean mosaic virus in cowpea was increased by another virus which is not seed-transmitted, cowpea chlorotic mottle virus, from 12 to 20 per cent.

McKinney (1956) reported striking synergistic reaction in dual infections with the viruses of barley stripe mosaic which is seed-borne and brome mosaic which is not. In several cultivars of spring and winter wheat, as well as in the barley cultivar Atsel, an accentuated stunting was produced. Both avirulent and virulent strains of barley stripe mosaic virus react by association with brome mosaic virus. When virulent strains are involved, the plants develop stunting and are frequently killed.

Table 16.2. Yield reduction in soybean caused by soybean mosaic virus (SMV), bean pod mottle virus (BPMV), and these viruses combined, and seed transmission of SMV without and in combination with BPMV. (Data from Ross, 1969*b*, Schmitthenner and Gordon, 1969 and Quiniones, Dunleavy and Fisher, 1971)

	SMV				*BPMV**				*SMV + BPMV*			
	Ross (1968)		S. & G. (1969)	Q. *et al.* (1971)	Ross (1968)		S. & G. (1969)	Q. *et al.* (1971)	Ross (1968)		S. & G. (1969)	Q. *et al.* (1971)
Percentage	cv. 1	cv. 2	one cv.	three cvs.	cv. 1	cv. 2	one cv.	three cvs.	cv. 1	cv. 2	one cv.	three cvs.
Yield reduction	18	43	20	18	26	14	40	10	73	81	90	66
Seed transmission of SMV	11		—	27	—	—	—	—	6		—	39

* Not seed-transmitted.

Intermezzo: Integration of Disease Disposing Factors

This is an 'Intermezzo' between Chapters 15 and 16 which deal with the 'external' factors, and Chapters 17 and 18 which discuss the 'internal' factors affecting establishment and development of seed-borne diseases. It emphasises the integral nature of these relationships to disease development.

The diagram (Fig. I.1) shows the abounding complexity of these effects but is nevertheless an oversimplification.

The effect of each of the environmental factors is not, as suggested by the figure, independent. Each factor is under the influence of a wealth of other factors. The effect of a certain level of soil moisture on development of smuts or Fusarium diseases in cereals varies with temperature (Subsection 15.2.1), and with soil reaction (Subsection 15.2.3). Also light exercises a co-influence such as day length on severity of bunt in wheat (Subsection 15.2.5).

Biotic factors also are interdependent, such as those acting between different fungi, between fungi and bacteria, between pathogenic fungi and nematodes, and in other such interrelationships, discussed in Chapter 16; and all organisms are dependent on physicochemical environmental conditions, e.g. vector populations (aphids, nematodes), and soil microflora the composition of which fluctuates according to the infinite range of edaphic conditions thus determining antagonistic effects (Section 16.9).

Likewise the 'internal' factors of pathogen and host form a coherent system of interdependent variables, determining the reaction of pathogen, of host, and of pathogen and host combined, to environment, as well as the interaction between host and pathogen.

The reaction of different host cultivars to a pathogen may depend on a range of external factors, such as soil reaction or presence of certain nematodes to resistance of pea to *Fusarium oxysporum* f.sp. *pisi* (Subsections 15.2.3 and 16.7). The effect of *F. moniliforme* in many hosts, in particular in rice and other cereals, depends largely on the effect of growth substances, the different gibberellins. They may cause increase of the height of host (Subsection 17.6.2); the pathogen also causes other symptoms, with or without the effect of growth substances, such as stunt and rot. The host reacts by such different symptoms according to its inherent characters, as well as it may react according to biotype of the pathogen, under influence of external factors such as temperature and soil mixture. The effect of vivotoxins depends on the inherent characters of the host as well as those of the pathogen. Victorin produced by *Drechslera victoriae* is effective on susceptible oats (Section 17.3), and a certain abnormal condition of the pathogen, brought about by infection of a supposed mycovirus, may increase the toxin-induced pathogenicity (Section 17.4).

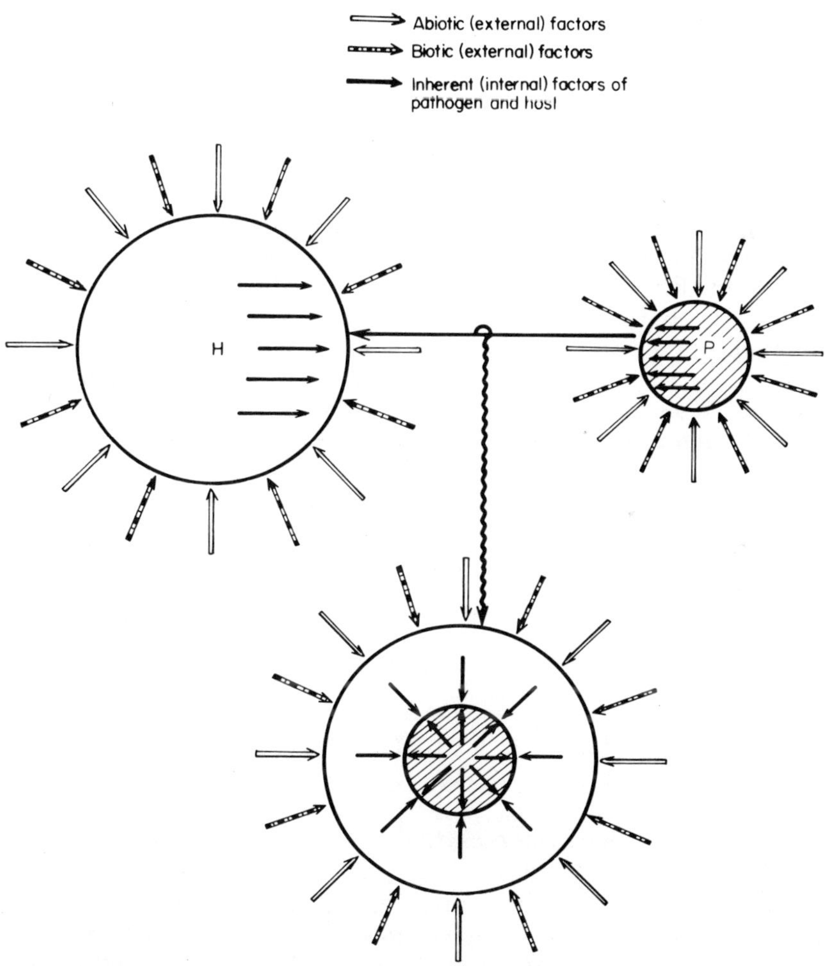

Fig. I.1. The principal effects of 'external' factors of environment, and 'internal' factors of pathogen and host, on establishment and course of disease. H, host; P, pathogen

17 Factors Inherent in the Pathogen Affecting Establishment and Course of Disease

INTRODUCTION

Variation in the infection capacity of a pathogen may be due to different relationships: the actual stage in the life-cycle of the pathogen, the variability in terms of biotype or race and the virulence of the pathogen in terms of the amount of inoculum needed to establish infection. These factors will be discussed in the following sections: (1) pathogenicity as related to the life-cycle or ontogenesis of the pathogen, (2) pathogenicity as related to physiologic or pathogenic races or strains of the pathogen (pathotypes), (3) pathogenicity as related to infection potential of the pathogen, (4) pathogenicity as related to the actual condition of the pathogen. These aspects are all important with regard to seed transmission of the pathogen and to their bearing on further secondary spread of disease.

In a general assessment of seed-borne diseases two further aspects, discussed below, are of fundamental importance, the genetical and the physiological implications of pathogenicity.

17.1 PATHOGENICITY AS RELATED TO ONTOGENESIS OF THE PATHOGEN

The various stages throughout the life-cycle of a parasite may be different in their capacity to bring about infection. In smuts the dikaryophase is pathogenic, the haplophase non-pathogenic. Teliospores of many smut fungi, for instance *Ustilago hordei*, covered barley smut (Rump, 1926), are not able directly to infect the host. From the diploid teliospore a haploid promycelium is developed, on which primary sporidia are formed which again under favourable conditions may develop secondary sporidia by budding, etc. This constitutes the saprophytic phase of the fungus. The parasitic phase is introduced after intersporidial or intermycelial anastomosis and subsequent germination in the form of dikaryotic, infective so-called fusion hyphae that are capable of penetrating into the host. On the other hand, in the closely related covered smut of oats, *U. hordei (U. kolleri)*, already the promycelium is capable of penetrating into the coleoptile of the host (Brandwein, 1944). Obviously, in the short period of sporidial or promycelial germination, during which the seedlings are predisposed, the conditions in the soil are of paramount importance to the extent to which establishment of infection may succeed.

In the embryo-infecting smut fungi, loose smuts of barley and wheat, *U. nuda* and *U. tritici*, the teliospores germinate as a promycelium or probasidium. This promycelium produces, in the case of the barley loose smut fungus, hyphae from pairwise conjugated cells, while in the case of the wheat loose smut fungus, hyphae grow out from the individual cells of the promycelium, and the cells of these branches later conjugate in pairs to initiate the dikaryotic phase and the infective hyphae.

Another example of infection capacity alternating during the life-cycle may be taken from the group of heteroecious rust fungi. Teliospores of *Puccinia graminis*, black stem rust of small grains and grasses, may be carried as sori on the grain, but they are not capable of infecting the young wheat plants; neither would the basidial spores, if they could develop from the teliospores, be capable of infecting the young wheat plant but only *Berberis*. The theoretical possibility of seed-borne teliospores as the starting point is, however, extremely unlikely. Uredospores as another possibility of seed transmission is also unlikely, although one instance is known where *Puccinia graminis* was introduced from India to the U.S.A. (Washington State) on *Agropyron semicostatum*, covered with pustules on glumes and rachis. The teliospores did not germinate, but germination of one uredospore was observed (Leppik, 1962*a*). In autoecious rust fungi the situation is, of course, different. *Puccinia carthami,* safflower rust, is readily seed-transmitted and apparently only by teliospores (Calvert and Thomas, 1954).

In many seed-borne fungi distinction may be made between active parasitism in 'earlier parts' of the life-cycle as against 'later stages' of parasitic inactivism with development of the perfect state after the tissues of the host have died. Rather seldom do we find the perfect state of fungi in or on the seed: among the few incidences are cleistothecia of *Erysiphe*; immature perithecia of *Pleospora herbarum*; perithecia of *Gibberella fujikuroi* in rice seed; perithecia of *Didymella bryoniae* in seeds of Cucurbitaceae, including cucumber and pumpkin; oospores of *Sclerophthora macrospora* on pieces of diseased or rather dead tissue of grasses or oats mixed with the seed. In *Ascomycetes* the imperfect state is practically always responsible for establishment of primary infection.

17.2 DIFFERENTIATION OF THE PATHOGEN INTO PATHOGENIC RACES (PATHOTYPES)

The existence of a great number of physiologic forms or races, including many pathogenic races or pathotypes, has been established for very many seed-borne fungi. The problems involved are many, but the implications pertaining to seed health testing have largely been neglected. Some theoretical aspects of practical importance will be discussed.

Barley leaf stripe fungus, *Drechslera graminea*, is extremely variable. Christensen and Graham (1934) in America classified a total of 1,200 single spore isolates with regard to cultural characters, morphology of conidia and — identifying 20 races — tested them for aggressiveness on various cultivars of barley. Isenbeck (1938), on the other hand, referred 24 German single spore isolates, tested for pathogenicity on five barley cultivars, to 6 physiologic 'race groups'.

Distinction should be made between (1) pathogenic races that are different in degrees of infection capacity or virulence, 'quantitatively' different ('aggressive races', Van Der Plank, 1968), and (2) pathogenic races that are more 'qualitatively' different, i.e. races possessing different host ranges ('virulent races', Van Der Plank, 1968).

Pathogenic races of the first category may probably be encountered within most parasitic fungi. From various points of view (quarantine, breeding for resistance) the latter category is of great importance.

The terminology and nomenclature of pathogenic races are rather vague or inconsistent. *Formae speciales* are identified by means of differential host genera or species, pathogenic races (or pathotypes s.str.) by means of differential host varieties (cultivars). Sometimes, however, 'race group' has been used rather than 'f.sp.' as in 'maize-group' of *Sphacelotheca reilliana*, while for other pathogens 'race group' has different meanings. Designations of pathogenic races are inconsistent, ranging from Arabic numerals, 1, 2, 3 (rust fungi), Roman numerals, I, II, III (*Ascochyta pisi*), to Latin letters, A, B, C (*Ascochyta pisi,* also for *Phytophthora infestans**) or Greek letters, alpha, beta, gamma (*Colletotrichum lindemuthianum*) or combinations of letter and numeral P_1, P_2, P_3 (*Claviceps purpurea*). There are conflicting conceptions on many smut fungi such as the closely related oat smut fungi *Ustilago avenae (U. nigra)* and *Ustilago hordei (U. kolleri)* which hybridise very easily, and these two species and another closely related form on barley, *Ustilago hordei (U. levis)* (Rodenhiser, 1928; Fischer, 1943).

In pathogenic races of many species of *Ustilaginales* the starting point of infection is a population of dikaryotic-diploid chlamydospores or

*Now numerals in brackets, indicating resistance genes in the host, are used to characterise races: (1, 3), (1, 2, 3), etc. (Black, 1952).

teliospores. As already discussed, these spores 'germinate' usually developing monokaryotic promycelia. In order to be able to develop infective hyphae a gametic fusion between the haplonts must occur, the monokaryotic sporidia arisen from the promycelia must cross with other compatible sporidia in order to produce infective hyphae. Consequently a new generation of a smut fungus of this biological type consists of a group of related biotypes, and, in common usage of terms, such a pathogenic race is a group of teliospores possessing the same virulence to a certain range of differential hosts. In the smuts only the haplophase can be propagated as a clone and maintained in that condition. Briefly, pathogenic races of smut fungi, having dikaryotic teliospores as the original starting point of infection, are biologically different from pathogenic races of *Uredinales*, which have dikaryotic uredospores, that can be propagated clonally, or from those of *Ascomycetes* and Fungi Imperfecti, for instance *Colletotrichum lindemuthianum*, which have conidia as starting point and are clones.

The variability in clones of fungi should be well understood as a very important factor in the development of new races. Buxton (1959) who elaborated on the mechanism of variation in *Fusarium oxysporum* f.sp. *pisi* points out three major ways in which this fungus appears to produce variants: (1) by heterokaryosis; different associations of nuclei or changes in the nuclear population of the multinucleate thallus are brought about by anastomosis between hyphae, resulting in intermingling of haploid nuclei, and appearance of a variety of genotypes; (2) by recombination through a parasexual mechanism brought about by fusion of two unlike haploid nuclei followed by multiplication of the diploid nuclei so formed, mitotic crossing-over in the homokaryotic diploid strains resulting in the appearance of genetic recombinants; and (3) by adaption in response to changes in environment; adaptive pathogenicity has been obtained by retaining conidia from a genetically uniform pure (uninucleate) single-spore isolate of race 1 in root exudate from a pea cultivar resistant to this race. The last observation needs confirmation.

Heterokaryosis may be important. *Drechslera oryzae*, for instance, is heterokaryotic, mycelial cells having 1–9 nuclei per cell but most commonly 2. Single conidial and single hyphal tip cell isolates have shown pathogenic variation on the same set of rice cultivars, a fact that highly affects race identification based on single spore isolates (Vorraurai and Giatgong, 1971).

Leaving these genetic aspects of variability in fungi, we may look at some of the implications of international spread of pathogenic races by seeds. Introduction of pathogenic races or pathotypes to regions, where local cultivars are resistant to endemic races of the same species, may have serious consequences. Certain North American wheat cultivars are extremely susceptible to certain European pathogenic races of *Tilletia caries*. If these races happened to be introduced to North America, it would probably be necessary to breed new resistant cultivars, a time-consuming operation. In *Ustilago hordei*, covered smut of barley, which has a world-wide distribution, only a few races have a wide geographic distribution.

Consistent use of very resistant cultivars favours development of highly virulent races. A race new to a given locality may become epiphytotic in an environment different from that of the place of its origin.

Information on local pathogenic races of seed-borne fungi and their infection capacity to cultivars grown in other countries is of great practical importance. In laying down quarantine provisions due attention should be paid to differences between countries or world parts in the occurrence of pathogenic races; this, unfortunately, has been largely neglected. Blast of rice, *Pyricularia oryzae*, is widespread in rice-growing areas and often destructive. Many pathogenic races have been encountered, some of which are only locally spread, being a potential danger to other rice-growing countries. Only very few rice-growing countries have specified quarantine provisions against this fungus. A race of *Plasmopara halstedii* on sunflower, a crop plant of increasing importance in some countries, has been introduced to the U.S.A. and encountered at Ames Regional Plant Introduction Station, Iowa (Leppik, 1962*b*), on seed imported from Turkey and Eastern Europe. From Ethiopia a race of *Ascochyta pisi*, evidently new, was detected at the same station on seeds of *Lathyrus sativus* and *L. cicera*, on which the pathogen is very destructive. Seeds imported for breeding purposes are known to be vehicles for pathogenic races new to the area into which the seed is imported, often in very small quantities.

Centres of origin of cultivated plants, so important as sources of material for the plant breeder, are often also, as pointed out by Branquinho de Oliveira (1963) and by Leppik (1970), centres of origin of the related pathogens. This pertains to the international spread of pathogenic races. Gene centres of cultivated plants and their progenitors serve as main source of genotypic resistance to diseases, and usually gene centres of plants contain the highest number of pathogenic races (Leppik, 1970). This also applies to seed-borne pathogens, and there is a great need for studies on distribution of races in gene centres of cultivated plants. There is a real danger that seeds, collected from a gene centre, may contain pathogens or virulent races new to those regions into which plant material for breeding, etc., is imported.

An incomplete survey of seed-borne fungi containing pathogenic races is presented below, merely for general orientation. Not all of the original references have been checked, and consequently the list may include some species that contain merely 'quantitatively' different pathogenic races. For instance, *Sclerotinia trifoliorum* has been suspected of constituting qualitatively different pathogenic races but Frandsen (1946) found in his extensive infection experiments only quantitatively different pathogenic races, and therefore this parasite has not been included in the list. Frandsen (1946), Kreitlow (1949) and others have pointed out that a sclerotium may be composed of the mycelium of one or more biotypes of the fungus. It should be emphasised, however, that also appearance of a 'new' highly virulent race of a pathogen, although restricted to a host range hitherto known for the pathogen, nevertheless may have comparable disastrous effects in intensifying epidemics as well as endemic occurrence.

As the number of established pathogenic races is directly dependent on the number of differential host cultivars being used, the numbers of races given by various authors are not always comparable. Much effort is being made to attain international standardisation of the testing technique; this has been attained for certain pathogens of international importance, but is still wanted for others. The separation of Eastern European countries still prevailing has regrettably delayed international co-operation in the effort to meet the danger from such seed-borne races. These show no respect for political boundaries.

Several pathogenic races may be encountered in one locality and may, consequently, occur also on one seed lot. Physiologic races are different not only as to host range but also in their reaction to physical conditions prevailing in the period of the establishment of infection, as in the reaction of *Drechslera graminea* to conditions of temperature, light and other factors, discussed in previous sections. Thus the ratio between pathogenic races of, for instance, this fungus in a given seed lot of barley may, according to conditions of infection, change in the following crop and in the harvested seed lot.

17.2.1 Pathogenic races (pathotypes) of seed-borne organisms

FUNGI

(1 Myxomycetes (Plasmodiophoromycetes))

Plasmodiophora brassicae — Johnston (1968): 3 races; Johnston (1970): 4 races; Paterson (1972): 5 races; Strandberg and Williams (1967): 7 races; Jönsson (1972): *16 races*, see also Buczacki and Humphrey (1973): 2 further races.

(2 Oomycetes)

Bremia lactucae on lettuce — Ling and Tai (1945) proposed five *formae speciales* with hosts from seven genera of Compositae, two of these occurring on species of *Lactuca*. One infects *L. sativa* and *L. indica*, another, *B. lactucae* f.sp. *chinensis*, infects *L. chinensis*. Jagger and

Whitaker (1940) and Sleeth and Leeper (1966) report a total of at least *6 races* on *L. sativa*.

Peronospora manshurica (P. sojae) — Geeseman (1950): 3 races; Lehman (1953): 4 races; Dunleavy (1959): 23 races; Buitrago, Montes, and Orozco (1970): 1 additional race; Dunleavy (1971): 15 more races, a total of *39 races*.

Peronospora tabacina (P. hyoscyami) on tobacco — Hill (1963), Ibrahim-Nesbat, Zalpoor and Keivani (1969): 2 races; Shepherd (1970): *3 races*, or rather *formae speciales*.

Phytophthora megasperma var. *sojae* — Hartwig, Kelling and Edwards (1968): 2 races; Schmitthenner (1972): 3 races; Schwenk and Sim (1974): *4 races*.

Plasmospora halstedii — several (at least 4) races according to Savulescu (1941), Savulescu and Vanky (1955–1956) and Leppik (1962*a*, 1962*b*).

(3 Ascomycetes)

including Fungi Imperfecti

Alternaria longipes — 4 'homokaryonised' races (Suzuki, 1971).

Alternaria porri f.sp. *solani* (syn. *A. solani*) — Neergaard (1945): 2 *subformae speciales*.

Alternaria triticina — Garg, Arya and Kant (1972): 6 races.

Ascochyta pisi — Wallen (1957): 4 races; Hubbeling and Huijberts (1969): 5 races.

Ascochyta rabiei — Vir and Grewal (1975): 2 races.

Cercospora beticola — Solel and Wahl (1971), found evidence of pathogenic races. Whitney and Lewellen (1975): 2 races.

Cercospora oryzae — Ryker and Cowart (1948): 10 races.

Cercospora sojina — Ross (1968): 4 races.

Claviceps purpurea — Stäger (1903, 1923), Barger (1931): 3 races (P_1, P_2, P_3); Mastenbroek and Oort (1941): 1 additional race (P_4); Baldacci and Forlani (1948): 1 additional race (P_5); 2 races, previously referred to *C. microcephala* are now regarded as 2 additional races (P_6 and P_7), see Sprague (1950), a total of *7 races*. These races, however, being defined by host genera, are rather *formae speciales*.

Cochliobolus carbonus — Ullstrup (1944) and Robles (1949): 2 races.

Cochliobolus heterostrophus (con. state *Drechslera maydis*) — Smith, Hooker and Lim (1970): 2 races; Nelson *et al.* (1973): *3 races*.

Cochliobolus sativus (con. state *Drechslera sorokiniana*) — Christensen (1922): doubtful whether races are qualitatively different; Robles (1949): *8 races* on *Zea*.

Colletotrichum lindemuthianum — Barrus (1918): 2 races; Burkholder (1923): 1 additional race; Leach (1923): 8 races; Budde (1928), Peuser

(1932), Schreiber (1932): *34 races*; Yerkes and Ortiz (1956): *10 new races*. Hoffmann, Schnock and Krüger (1974): *1 additional race*.

Colletotrichum lini — Schwinghamer (1954, 1956): 5 races; Rataj (1960): *16 races*.

Colletotrichum orbiculare — McLean (1967): 2 races.

Drechslera graminea — see *Pyrenophora graminea*.

Drechslera maydis — see *Cochliobolus heterostrophus*.

Drechslera sorokiniana — see *Cochliobolus sativus*.

Drechslera teres — see *Pyrenophora teres*.

Drechslera turcica — see *Trichometasphaeria turcica*.

Fulvia fulva (syn. *Cladosporium fulvum*) — Patrick, Kerr and Bailey (1971): 12 races, classified in three groups (Hubbeling, 1971).

Fusarium oxysporum — Armstrong and Armstrong (1968) listed 69 named *formae speciales* (seven of which, however, with the reservation that they have not been adequately established). Many of the listed *formae speciales* are seed-borne (cf. Noble and Richardson, 1968), and several of these comprise more than one pathogenic race, namely the following (for references, see Armstrong and Armstrong (1968), if not otherwise indicated):

Fusarium oxysporum f.sp. *callistephi* — 4 races (Armstrong and Armstrong, 1971*a*).

Fusarium oxysporum f.sp. *carthami* — 3 races (according to Klisiewicz and Thomas, 1970*a*, 1970*b*); *4 races* (Klisiewicz, 1975).

Fusarium oxysporum f.sp. *lini* — Millikan (1949) indicates at least 3 definite races but Armstrong and Armstrong (1968) indicate only one. However, in India *10 races* have been distinguished based on 8 differential cultivars (Thakur and Husain, 1972).

Fusarium oxysporum f.sp. *lupini* – 4 races (Banihashemi and Dezeeuw, 1975).

Fusarium oxysporum f.sp. *lycopersici*—3 races (Yamamoto *et al.*,1974).

Fusarium oxysporum f.sp. *melonis* — 3 races.

Fusarium oxysporum f.sp. *niveum* — Crall (1963): 2 races.

Fusarium oxysporum f.sp. *pisi* — Haglund and Kraft (1970): 5 races; Armstrong and Armstrong (1971*b*): 6 races; Armstrong and Armstrong (1974): *11 races*.

Fusarium oxysporum f.sp. *tracheiphilum* — 3 races.

Fusarium oxysporum f.sp. *vasinfectum* — 5 races; Armstrong and Armstrong (1974): 6 races.

Kabatiella caulivora — Hanson (1950): at least 2 races.

Pyrenophora graminea (con. state *Drechslera graminea*) — Christensen and Graham (1934): 20 races with differences of aggressiveness. Arny (1945) demonstrated 2 qualitatively different races; Singh and Saksena (1971) 13 races in India.

Pyricularia oryzae — Latterell *et al.* (1954): 2 races; Latterell, Tullis and Collier (1960): 15 'race groups'; Atkins (1962): 1 additional 'race group'; a total of 16 'race groups'; Chiu, Chien and Lin (1965): 19 races, based on 16 differential cultivars; Bandong and Ou (1966): 26 'race groups'; Chien (1967): 27 races; and in the Philippines the International Rice Research Institute (IRRI, 1964–1969) has identified 116 races. Matsumoto, Kozaka and Yamada (1969) established 34 races, based on 327 isolates from 13 countries, and the number of races is steadily growing. Efforts have been made to standardise differentials for testing. Eight cultivars have been selected, and *32 race groups* characterised (Ou, 1972, literature review of pathogenic races). According to Suzuki (1967, personal communication) identification of the pathogenic races will have to be abandoned because of their pronounced variation in pathogenicity. He believes that the pathogenic races cannot be identified because persistent heterokaryosis makes it impossible to obtain type cultures.

Rhynchosporium secalis — Caldwell (1937): 6 races; Schein (1958): 7 races, in Britain 8 'race groups' (Owen, 1963), 2 races (Williams and Owen, 1973); Jackson and Webster (1976) differentiated *75 races*.

Trichometasphaeria turcica (con. state *Drechslera turcica*) — Lefebvre and Sherwin (1945): *4 races*, differing only in their reaction on Sudan grass and sorghum. Robert (1960) found physiologic specialisation but specificity was not stable enough for use as a criterion in establishing races.

(4 Basidiomycetes)

(4.1 Ustilaginales)

Sphacelotheca cruenta — Melchers (1930): 2 races; Rodenhiser (1937): *3 races*.

Sphacelotheca reilliana — Al-Sohaily and Mankin (1960), Al-Sohaily, Mankin and Semeniuk (1963): sorghum-group with 4 races, maize-group with 1 race.

Sphacelotheca sorghi — Tisdale, Melchers and Clemmer (1927): 2, probably 3, races; Vaheeduddin (1938): 1 race produced by crossing be tween 2 single spore isolates; Vaheeduddin (1951): 2 more races, a total of *6 races*.

Tilletia caries — Flor (1933): 7 races; Rodenhiser and Holton (1937): 11 races; Holton and Rodenhiser (1942): 3 more races; Rodenhiser and Holton (1945): a total of *16 races*. Holton and Heald (1941) in reviewing the literature on the subject registered a total of *73 races* in various countries, but probably repetitions are included. Kendrick and Holton (1961) discussed racial population dynamics in both *T. caries* and *T. foetida*, and Kendrick (1964) established a total of *30 races of these two species together*, classified in 6 groups based on resistance types. Schuhmann (1966) found in 16 German accessions 3 race groups, based on 12 wheat cultivars; cf. also Dodov and Todorova (1974) who established 4 race groups of this species and *T. foetida* in Bulgaria, some races being different from those found in the U.S.A. In Turkey Özkan and Finci (1973) identified 8 races.

Tilletia contraversa — Hoffmann, Kendrick and Meiners (1962): 8 races, 4 of which have counterparts among the common bunt races. Hoffmann, Kendrick and Metzger (1967) revised the classification of the races and established 2 new races, making of total of *10 races*. Schuhmann (1966) established 8 race groups among 60 accessions of German origin.

Tilletia foetida — Rodenhiser and Holton (1937): 8 races; Holton and Rodenhiser (1942): 2 more races; Rodenhiser and Holton (1945): a total of 15 races. Holton and Heald (1941) in reviewing the literature registered a total of *66 races* in various countries (compare *T. caries*), among which are probably several repetitions. Kendrick (1961): 10 races. In Turkey Özkan and Finci (1973) identified *29 races*.

Urocystis agropyri (U. tritici) — Holton and Johnson (1943): 2 races; Hafiz (1951): 4 race groups, of following geographic origins: (1) collected in the U.S.A., (2) collected in the U.S.A., Italy and Australia, (3) collected in China and Cyprus, and (4) collected in Pakistan. Yu, Wang and Tsiang (1936): 5 races; Yu, Wang and Fang (1945): a total of *12 races*; Johnson (1959) and Halisky (1965) identified 4 races. For review, see Purdy (1965).

Ustilago avenae — Fleroff (1923), Reed (1924, 1927, 1929): 11 races; Reed and Stanton (1932): 9 races; Sampson (1929), Western (1936), Tervet (1940): 3 races; Reed (1940): 29 races; Reed and Stanton (1942): 30 races; Reed, Stanton and Wilds (1947): 31 races; Holton and Rodenhiser (1948): 15 races; Cherewick (1951): 23 races; Hansing (1954): 4 additional races; Luke, Pfahler and Hadden (1961): 2 additional races, making a total of *29 races*.

Ustilago bullata — Fischer (1940): 8 races; Meiners and Fischer (1953): 4 additional races and Meiners (1952): 1 additional race, making a total of *13 races* (or rather *formae speciales*, being based on host genera and species).

Ustilago crameri — Wang (1944): 6 races.

Ustilago hordei (U. kolleri, U. levis) on barley — Tapke (1945): 13 races; Yu and Fang (1945): 9 races (in China); Cherewick (1951): *16 races*. 13 races recently studied in Canada (Emara, 1972). On oats — Reed (1929): 5 races; Reed (1940): 14 races; Holton and Rodenhiser (1948): 7 races; Reed and Stanton (1942): *14 races*; Cherewick (1951): 10 races; Hansing (1954): *1 additional race*.

Ustilago nigra — Josephson (1942): 5 races plus 3 doubtful; Tapke (1945): 7 races; Cherewick (1951): *14 races*; Tapke (1951): *4 more races*.

Ustilago nuda — Cherewick (1953): 10 races; Tapke (1953): 5 races; Tapke (1955): 4 races. Cloninger and Poehlman (1954): 13 race groups; Halisky (1965): 20 races.

Ustilago striiformis — Davis (1930): 2 races; Fischer (1940), Bever (1947): 11 races; Moore (1948): *11 races*; Govindhu and Fischer (1962): 6 physiological varieties within which probably pathogenic races exist.

Ustilago tritici — Grevel (1930): 4 races; Moore (1942): 5 races; Cherewick (1951): 10 races; Bever (1947, 1953): 19 races; Gaskin *et al.* (1957): *20 races*; Mitov and Savov (1971): 5 races in Bulgaria. Tusa and Răudulescu (1973): 22 race groups in Roumania.

(4.2 Uredinales)

Melampsora lini — Flor (1935): 14 races; Straib (1939, 1941, 1942): 24 races; Flor (1946): 105 races; Arif (1954): 9 additional races; Waterhouse and Watson (1944): possibly 5 additional races; Cruickshank (1952): 12 New Zealand races (possibly partly or totally different from the others); Flor (1954, 1956*a*): 239 races; Zimmer and Hoes (1974): *370 races*.

Puccinia antirrhini — Yarwood (1937*b*): 2 races.

Puccinia carthami — Thomas (1955, 1958): 4 races; Ashri (1961): one or more additional races.

(4.3 Aphyllophorales)

Thanatephorus cucumeris, syn. *Corticum solani* (imp. state *Rhizoctonia solani*) — Schultz, 1936; Sanford, 1938; Storey (1941); Person (1944); Flentje and Saksena (1957), Papavizas and Davey (1962); Flentje and Stretton (1964); El Zarka (1965): many unnamed biotypes, some with a wide host range, others exhibiting a more selective parasitism. The biotypes have been classified into six anastomosis groups (AG 1, 2, 3, 4, 5 and 6) which show differences in host range, cultural characteristics and heterokaryon compatibility (Schultz, 1936; Richter and Schneider, 1953; Anderson *et al.*, 1972; Ogoshi, 1973).

BACTERIA

Pseudomonas glycinea — Cross *et al.* (1966) reported 7 races.

Pseudomonas phaseolicola — Walker and Patel (1964*b*) and Patel and Walker (1966) established 2 races.

Pseudomonas pisi — Taylor (1972): 2 races.

Xanthomonas malvacearum — Verma and Singh (1970) recorded 2 new races, Nos. 16 and 17, making a total of *17 races* defined.

Xanthomonas translucens — Wallin (1946*a*): 6 races.

Xanthomonas vesicatoria — Cook and Stall (1969): 3 races; Kimura *et al.* (1973): *4 races*.

VIRUSES

Probably all seed-borne viruses include strains pathogenic to specific host ranges of cultivars or species. For bean common mosaic virus an international standard list of 11 cultivar differentials has been established and 11 strains have been identified (Silbernagel, 1969; Hubbeling, 1972). A severely virulent strain of lettuce mosaic virus, isolated by Zink, Duffus and Kimble (1973), caused systemic infection in cv. Gallega which is resistant to common lettuce mosaic virus. The severe variant was, however, not seed-transmissible.

NEMATODES

Ditylenchus dipsaci — Some forms are extremely polyphagous while in contrast others, such as those attacking lucerne and onion, have a restricted host range and may be regarded as 'pathogenic races' or *formae speciales*. By more critical evaluation of minute morphological characters these and other 'races' may possibly be referred to specific rank (Thorne, 1961; see also Webster, 1967).

17.3 INOCULUM POTENTIAL

Some fungi are capable of infecting and producing disease if one single spore under favourable conditions germinates on the host; other fungi may gain a foothold only if a certain amount of spores germinate simultaneously. The amount of inoculum necessary for establishment of infection, expressed as inoculum potential, varies according to (1) virulence of the pathogen, variable also according to pathogenic races, and (2) susceptibility of the host, e.g. according to the resistance of one cultivar as against others.

Inoculum may be any infective material: propagules such as conidia, chlamydospores, mycelium, sclerotia, bacteria, virus particles, nematodes. The amount of inoculum may be in terms of spore load or load of bacteria, etc., or in terms of any quantity of pathogenic thallus, or in number of virus particles, usually given as the concentration of the virus. The minimum amount of inoculum needed is determined by the numerical threshold of infection.

Originally 'inoculum potential' was defined as the number of infective particles present in the environment of the uninfected host (Horsfall, 1932*a*), but particularly in studies pertaining to epidemiological problems (Van Der Plank, 1960), this definition has been modified also to include, for instance, individual infected plants. In seed pathology, accordingly we must include infected seeds, both the infected seeds as units and the variation of inoculum within as well as on individual seeds, for instance, the amount and virulence of mycelium within the seed coat of a seed.

In discussing inoculum potential, distinction should be made between (1) the intensity factor, and (2) the capacity factor (Dimond and Horsfall, 1960). The intensity (or onsetting) factor in our subject particularly pertains to the establishment of primary seed-transmitted infection, while the capacity (or perpetual) factor predominantly pertains to the development of secondary infections, specifically to the increasing rate of disease.

Dimond and Horsfall (1960) point out that the 'inoculum potential is the resultant of the action of the environment, the vigor of the pathogen to establish an infection, the susceptibility of the host, and the amount of the inoculum present'. They add: 'this is precisely the index needed in describing the rise and decline of epidemics and in forecasting epidemics'. On our subject it may be added: this is precisely the index needed in describing the development of the seed-transmitted disease in a crop and in forecasting the course of development.

The procedure of measuring the inoculum potential effect has been named infectivity titration. The so-called dilution end-point is that dilution of infective propagules below which no host individual manifests symptoms as a result of inoculation. Such dilutions occur everywhere in nature. The results of such infectivity titrations have been interpreted, as pointed out by Garrett (1960), in two ways: (1) only one out of a number of infective propagules will be capable of causing infection and of having the opportunity of doing so, or (2) no single propagule is capable of infecting but only contributes as a unit in an action, in which a minimal number of propagules is required to infect the host co-operatively.

There are many examples of the first possibility, but there seems to be no proven example of the second (cf. Van Der Plank, 1975), except viruses with genetically complementary particles; see below, last paragraph of this subsection. The inoculum potentiality of individual propagules is often greater in multicellular than for instance in one-celled propagules. Spores of *Drechslera* and *Alternaria* so often seed-borne may readily cause infection individually — they are rather resistant to adverse conditions. Also the vigour of a sclerotium or that of an individual nematode favours them in their capacity of independent inoculum potentials. The vigour of nematodes in this relationship accounts partly for their enhancing effect in bringing about infection of *Fusarium oxysporum* in various hosts, including invasion of otherwise *Fusarium*-resistant hosts — the germ tubes from the cells of the Fusarium spore fail to accomplish the penetration of the host.

As to the mass effect of inoculum a few examples may be given.

Heald (1921) found in his classical studies that a minimum of 100 spores of *Tilletia caries* per grain of the wheat cultivar Jenkins Club was necessary to produce a smutted spike, whereas 500–5,000 spores per grain were needed to produce a corresponding attack in the smut resistant cultivar Marquis. He found that up to a certain point (0.5 g per 100 g grain) increasing amounts of inoculum produce increasing degrees of infection. Beck (1920), Riehm (1921), Mackie and Briggs (1923), Lobik (1926), Pichler (1954) and Glaeser (1961) also established the validity of this correlation, although some authors found other threshold values.

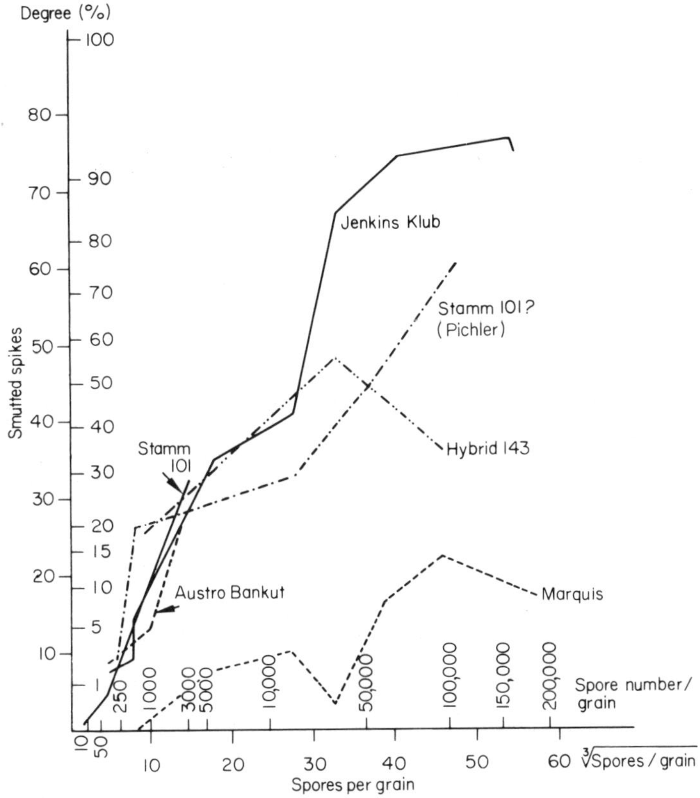

Fig. 17.1. Relation between spore number of *Tilletia caries* on the grain and the subsequent percentage of diseased spikes in the field, in different wheat cultivars. (Glaeser, 1961)

In Table 17.1 and Fig. 17.1 (Glaeser, 1961) the observations made by various authors on the relation between the amount of seed-borne inoculum and the level of disease in the field are compared. In Fig. 17.2 the linear increase of the disease in the field as related to the amount of inoculum carried on the seed is shown in a double logarithmic system of co-ordinates (Glaeser, 1961). This linear response curve can also be traced — within the limits of the upper threshold values — by smoothing out the curves in Fig. 17.1, where discrepancies from the expected regular response are likely due to experimental errors. The slope of the inoculum response curve is determined primarily by the degree of susceptibility of the host.

Table 17.1. Relation between the amount of seed-borne inoculum of *Tilletia caries* and *T. foetida* in wheat and the severity of the disease in the field, according to various authors

Cultivar	*Weight proportions*	*Spores per grain*	*Smutted spikes (plants)* (%)	*Remarks*	*References*
Veredeltem Rimpau Bastard Genealogischen	0.03% smutted grain	1,200	5.5	Naturally conta-minated seed	
Russischen Weizen	0.02% ,,	800	13.3		Beck (1920)
Wischenauer	0.01% ,,	400	2.3	Inocula-ted seed	
Kolbenweizen	1.0% ,,	2,000	5.5		
Jenkins Club (very susceptible)		100	1.8	Inocula-ted seed	
		450	2.5		
		530	6.0		
		5,330	32.8		Heald (1921)
Marquis (very resistant)		5,040	1.8	Inocula-ted seed	
Rimpaus	0.4% ,,	22,500	9		
	0.8% ,,	45,000	17		Riehm (1921)
Sommerweizen	1.6% ,,	90,000	39		
	3.2% ,,	180,000	31		
Marquis		8 (20)†	0.6*		Heald and Boyle (1923)
Bluestem		31 (78)†	0.6*		
	0.3 – 0.4 g/kg grain	7,500 – 11,000	10		Mackie and Briggs (1923)
	0.7 – 1 g/kg ,,	15,000 – 22,000	23 – 24		
	1.3 – 4 g/kg ,,	30,000 – 90,000	41†		
	33.3 g/kg ,,	750,000	59†		
		500 (2,000)†	3		Lobik (1926)
		2,500 (10,000)†	10		
		100 (400)†	0.5		Lehn and Vasters (Sorauer, 1932)
	½ smutted grain/ 80 g grain	1,700	1.3	Sowing in late spring reducing severity	Gassner (1938c)
	1 smutted grain/ 80 g grain	3,500	1.7		
Very susceptible cultivars	0.002 g/kg grains	50	2.5		Pichler (1954)
	0.01 g/kg ,,	225	2.8		
	0.02 g/kg ,,	500	18.9		
	1 g/kg ,,	22,500	28.5		
	2.5 g/kg ,,	49,500	47.5		
	5 g/kg ,,	112,500	74.8		
Stamm 101 (susceptible)		0	0		
		50	0.4		
		250	2.1		
		1,000	9.8		
		3,230	29.2		Glaeser (1961)
Austro Bankut (less susceptible)		0	0.02		
		50	0.3		
		290	3.0		
		1,000	5.6		
		3,500	24.6		

*Per cent smutted plants. †Corrections according to Bredemann (1929).

Munerati (1922) demonstrated that the location of inoculum of *Tilletia caries* on the grain is of importance. The nearer the spores are to the embryo, the greater the prospects of infection.

Leblond (1948) found clear positive correlation between the number of spores of *Ustilago avenae* and *U. hordei (U. kolleri)* combined and the infection percentage on oats in the field. In using his data by plotting the square root of the number of spores per seed against the infection per cent in the field, an approximation to a straight-line response is obtained.

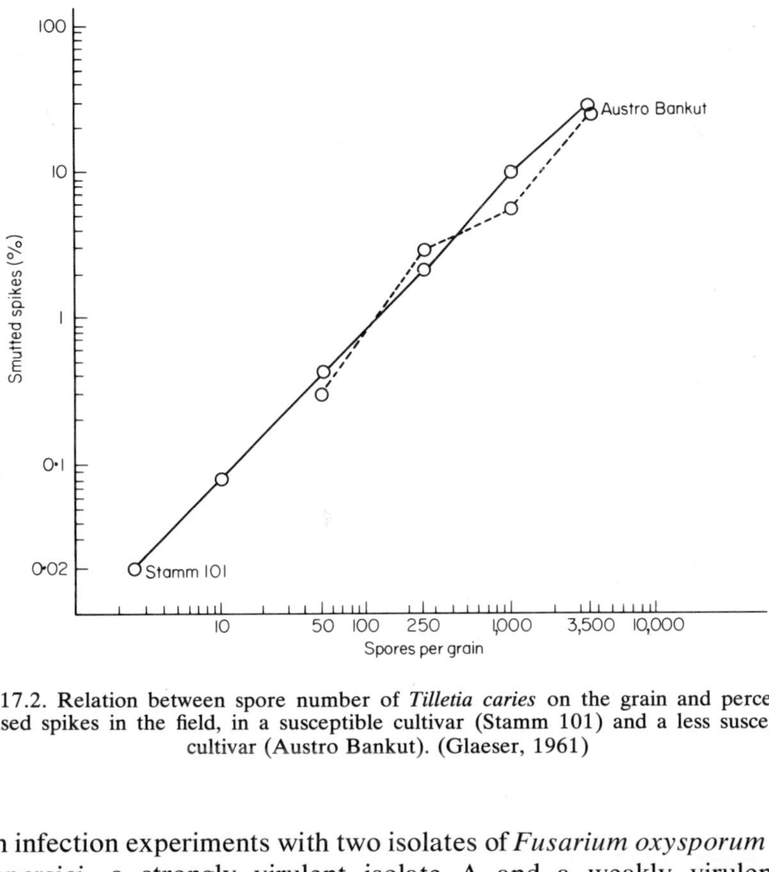

Fig. 17.2. Relation between spore number of *Tilletia caries* on the grain and percentage diseased spikes in the field, in a susceptible cultivar (Stamm 101) and a less susceptible cultivar (Austro Bankut). (Glaeser, 1961)

In infection experiments with two isolates of *Fusarium oxysporum* f.sp. *lycopersici*, a strongly virulent isolate A and a weakly virulent B, Haymaker (1928) applied a range of increasing numbers of conidia in suspensions. The more virulent isolate A gave increasing infection percentages in tomato, whereas the less virulent isolate B gave no increasing severity in disease by increasing the amount of inoculum.

Millar and Colhoun (1969) studied the influence of spore load of *F. nivale* in relation to development of disease in barley, oats and wheat. They found on the wheat cultivar Atle increasing disease incidence by

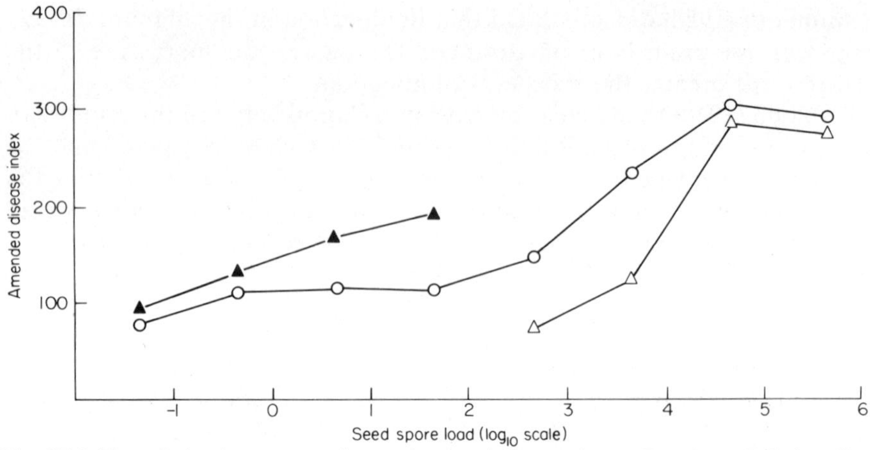

Fig. 17.3. The relation between seed spore load and disease due to *Fusarium nivale* in soil at 10.6% (——▲——), 17.3% (——○——) and 23.2% (——△——) soil moisture content. (Millar and Colhoun, 1969)

using increasing spore loads of *F. nivale* up to a load of about 500,000 spores per seed (Fig. 17.3).

Also, artificial and natural infection of wheat seed with *F. nivale* was compared (Colhoun and co-workers). When 30 per cent of the seed was inoculated with 10^3 spores per seed, the seed yield was reduced by about 8 per cent as compared with untreated. Another seed sample, infected naturally to an extent of 30 per cent of the seed as well, gave a yield of about 80 per cent of that obtained by another, uncontaminated sample of the same cultivar. The reduction of yield caused by this level of natural infection may thus be assumed to be approximately 20 per cent. For *F. culmorum* in barley a spore load of 10^4 spores per seed was found to be essential for getting severe disease.

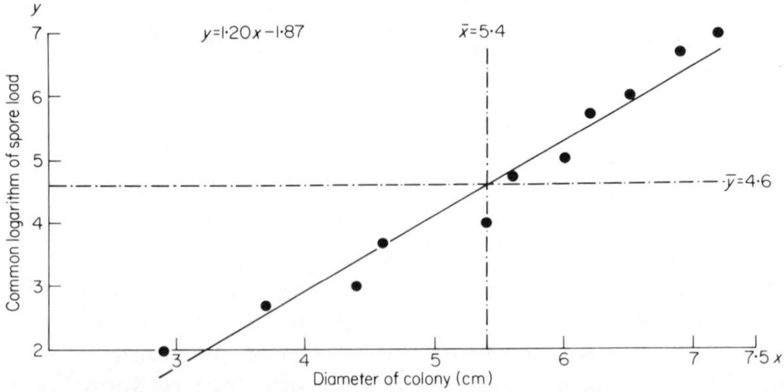

Fig. 17.4. Relation between the common logarithm of the spore load of *Fusarium culmorum* per 10 g of wheat seed and the colony diameter on modified Czapek-dox-agar, after incubation for 5 days at 25°C in the dark. (Malalasekera and Colhoun, 1969)

Malalasekera and Colhoun (1969) correlated the amount of inoculum of *F. culmorum* in terms of spore load per seed of wheat with the growth rate of mycelium developed from the spore-inoculated seed on an agar medium. The correlation between common logarithm spore load and colony size was highly significant (Fig. 17.4), as was the correlation between log spore load and disease recorded in terms of reduced dry weight of infected seedlings, and mean percentage of pre-emergence death of seedlings, grown from seed, which also was inoculated with logarithmically increasing spore loads. The relation of spore load to colony diameter on agar medium has direct bearing on techniques of measuring seed-borne inoculum by means of agar plate procedures.

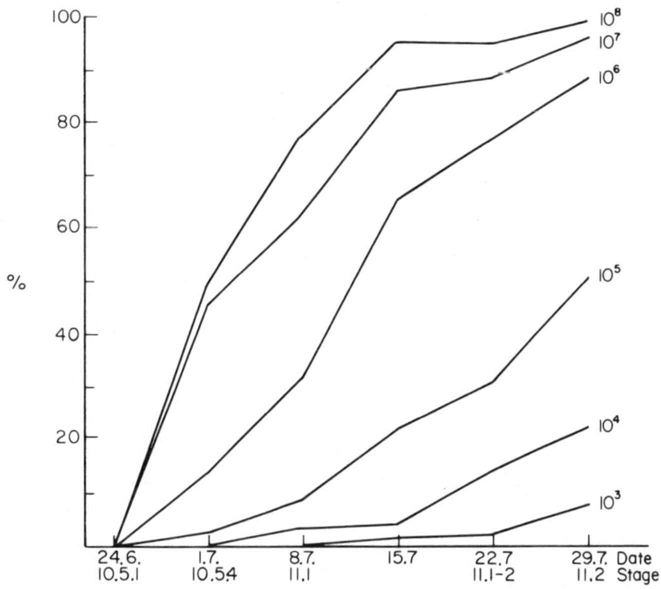

Fig. 17.5. Influence of amount of inoculum on infection of spring wheat, cv. Hinal, inoculated by *Septoria nodorum* in stage 10.5.2 (flowering). Abscissa: dates of record, growth stages according to Feeke's scale. Ordinate: infection of ears. Curves: concentration of conidia per ml. (Brönnimann, 1968)

Brönnimann (1968) inoculated spring wheat cv. Hinal in stage 10.5.2 (Feeke's scale), 'flowering complete to top of ear', using a range of spore concentrations of *Septoria nodorum* from 10^0 to 10^8 per ml. Concentrations from 10^0 to 10^3 per ml had no effect on development of attack and reduction of yield (Fig. 17.6). There was a marked influence of increasing inoculum potential on the increasing rate of the pathogen as expressed in the slope of the spore dosage response curve (Fig. 17.5). The range of increasing severity of attack, provided by spore concentrations from 10^4 to 10^8 per ml, is reflected in increasing yield losses in terms of thousand

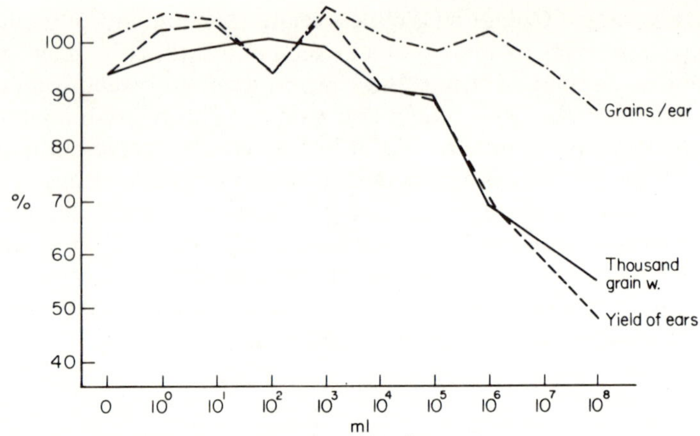

Fig. 17.6. Influence of amount of inoculum on infection of spring wheat, cv. Hinal, inoculated by *Septoria nodorum* in stage 10.5.2 (flowering). Abscissa: concentration of conidia per ml. Ordinate: yield related to untreated wheat. (Brönnimann, 1968)

grain weight (Fig. 17.6); infection percentage of the seed recorded in agar plate tests also increased gradually with inoculum potential from 10^4 to 10^8 spores per ml (Fig. 17.7). Brönnimann's results were confirmed in England by Cooke and Jones (1970).

Sheridan (1968) found that a concentration of 10,000 conidia per ml of *Septoria apiicola* was required for heavy infection at high humidity and 15–17°C. 2,500 conidia per ml produced lesions, but below this concentration production of lesions was uncertain. Percentage successful infection decreased with a decrease in number of spores applied. As few as 10 spores placed in a small drop of water on the upper surface of a leaf produced a lesion but one or two germinating spores failed to do so (10 inoculations).

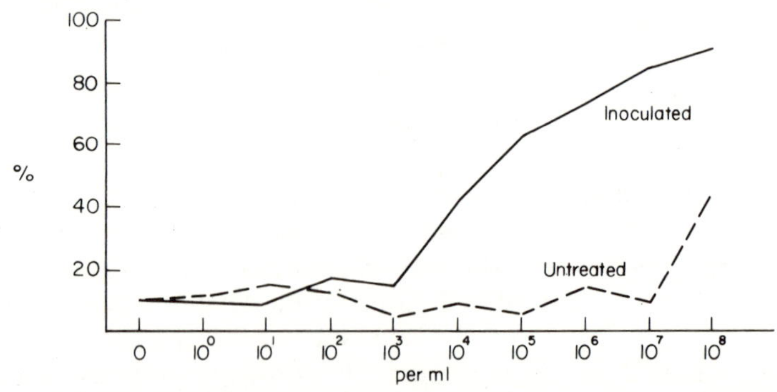

Fig. 17.7. Influence of amount of inoculum on infection of spring wheat, cv. Hinal, inoculated by *Septoria nodorum* in stage 10.5.2 (flowering). Abscissa: concentration of conidia per ml. Ordinate: infection of grains. (Brönnimann, 1968)

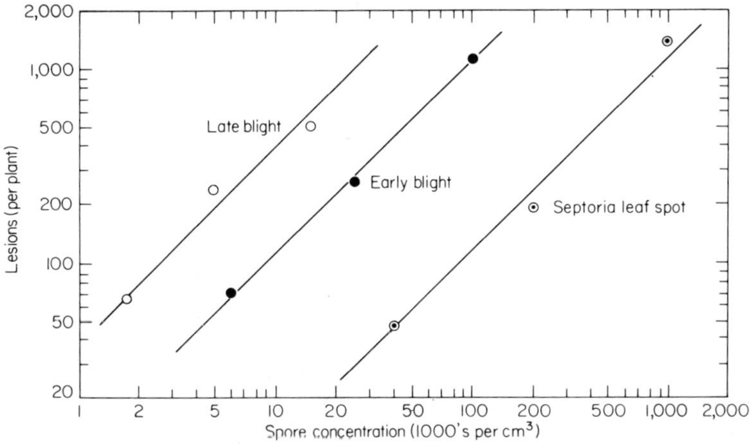

Fig. 17.8. Relation of spore concentration to number of lesions of three tomato diseases. Each point mean of 3 plants with 3 leaves each. (McCallan and Wellman, 1943)

McCallan and Wellman (1943) compared three major pathogens, *Phytophthora infestans*, late blight, *Alternaria porri* f.sp. *solani*, early blight and *Septoria lycopersici*, Septoria leaf blight, as to the number of spores 'necessary' to produce a lesion on tomato. They obtained one late-blight lesion per 15 spores, one early-blight lesion per 57 spores, and one Septorial-leaf-blight lesion per 525 spores (cf. Fig. 17.8).

The amount of inoculum needed depends also on susceptibility/resistance of the host. Kockman and Kubicka (1974) found that one conidium of *Phomopsis leptostromiformis* in the 'inoculum drop' was enough to infect yellow lupin, *Lupinus luteus*, 5 conidia were needed to infect white lupin, *L. albus* and *L. angustifolius*, and 25 conidia to infect *L. polyphyllus*.

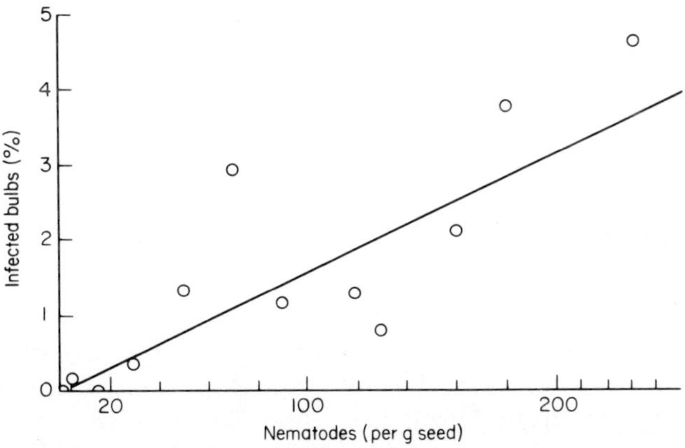

Fig. 17.9. Correlation of number of onion nematodes, *Ditylenchus dipsaci,* in seed, and percentage of infected bulbs in the subsequent crop. (Seinhorst and Koert, 1971)

Also in nematodes the number of individuals may be decisive in a course of infection. In studies on the complex disease caused by root knot nematode, *Meloidogyne incognita*, and Fusarium wilt fungus, *Fusarium oxysporum* f.sp. *lycopersici*, Kawamura and Hirano (1968) observed that the fungus invaded the apical meristem in roots of tomato when numerous larvae of the nematode had penetrated simultaneously but not when injury involved only one larva.

Seinhorst and Koert (1971) found a correlation between number of stem nematodes (*Ditylenchus dipsaci*) in onion seed and percentage of bulbs infected in the field (Fig. 17.9).

The inoculum level needed for infection by *Corynebacterium michiganense*, bacterial canker of tomato, has been studied by Thyr (1968). He established that tomato seedlings can become infected with as few as five cells, and when introduced directly into xylem tissue, one cell is enough to produce infection.

In viruses the inoculum may be expressed in terms of concentration (dilution) of host sap, in terms of number of virus particles or in weight units. Tobacco mosaic virus, which may be transmitted mechanically from contaminations on the seed coat adhering to young tomato seedlings, has a very high inoculum potential thus accounting for such incidences. Schramm and Engler (1958) found that as little as 5×10^{-16} g of the virus can be detected in tobacco by transmission tests. The weight of a tobacco mosaic virus particle is 8.2×10^{-17} g (Williams, Backus and Steere, 1951), hence the dose used is equal to 10 virus particles.

In some viruses there are two or more qualitatively different types of particle which must interact before infection can occur. Such two components have been demonstrated in cowpea mosaic virus (Van Kammen, 1968), and five nucleoprotein components have been found in alfalfa mosaic virus, three of which are necessary in combination to cause infection (Van Vloten-Doting, Dingjan-Versteegh and Jaspars, 1970). Also tobacco rattle virus possesses qualitatively different particles between which interaction is bound to bring about infection (Harrison and Woods, 1966; Sänger, 1968*a*, 1968*b*; Huttinga, 1972).

17.4 CONDITION OF INOCULUM

The actual condition of inoculum is pertinent to its potentiality. A pathogen may still be alive in seeds stored for some time, and yet it may have lost its virulence. Similarly, as saprophytic growth of pathogenic fungi may lead to loss of virulence, the effect of crop rotation on soil invaders may be due not exclusively to loss of growth vitality or gradual dying out of the pathogen as generally assumed — the effect may in some fungi be due merely to loss of virulence, the pathogen being capable of subsisting saprophytically. In contrast, a seed-borne pathogen which passes on from one generation of the host to another, possibly through many generations, in a resistant cultivar which acts as a symptomless carrier, or in a susceptible cultivar where it produces disease, is likely to be in a more virulent condition than if carried over in soil or dead plant residues for some time.

Lindberg (1959, 1971*a*, 1971*b*, 1973) established that transmissible agents of disease were present in isolates of three seed-borne fungi, *Drechslera victoriae*, Victoria blight of oats, *D. oryzae*, brown spot of rice, and *D. maydis*, southern maize leaf blight. The host fungi showed poor mycelial growth but increased toxin-induced pathogenicity to their respective hosts, as compared with healthy isolates.

17.5 GENETIC IMPLICATIONS OF PATHOGENICITY

The capacity of fungi to produce disease in plants is commonly controlled by genes. Our knowledge on inheritance of pathogenicity in fungi is largely based on studies of pathogenic races. The mere existence of differential virulence on the same host cultivar is indicative of specific factors governing pathogenicity. Primarily two groups of fungi have lent themselves for such investigations — the rust fungi and the smut fungi.

Following Craigie's (1927) discovery of heterothallism and the function of the pycnospores in *Puccinia graminis*, Newton, Johnson and Brown (1930) developed techniques for hybridising and selfing this fungus and demonstrated Mendelian segregation of its pathogenic characters (Johnson, Newton and Brown, 1934; Johnson and Newton, 1940). This revealed the gloomy prospect of the new pathogenic races unceasingly produced in nature by the process of crossing and segregation.

Extensive genetic studies have been carried out on flax rust, *Melampsora lini*, which is an autoecious long-cycled rust. Flor (1946, 1955, 1959, 1964), has shown that pathogenicity in this parasite is recessive to non-pathogenicity, and he propounded a hypothesis, according to which there is a gene-for-gene relationship between rust reaction in the host and pathogenicity in the fungus. Each race contains a number of different gene pairs, some in homozygous, some in heterozygous condition, and as there are more than twenty of these pairs there are innumerable possibilities for combination of these as to homozygoty and heterozygoty. Pathogenicity is expressed only when both recessive genes are combined in a gene pair. A race with one such pair is pathogenic on one host genotype; another race with two such pairs is pathogenic on two distinct host genotypes, thus being distinct from the first race; a third distinct race may be characterised by the first of the two gene pairs in heterozygous condition, thus being virulent only to the second host genotype, and so on, taking into account the total range of gene pairs. There is a complementary interaction of host genes and fungus genes. Resistance is produced when both host and fungus contain dominant genes, whereas susceptibility results when the fungus carries its recessive gene in homozygous condition. An exception is known: on flax cultivar Williston Brown virulence of the pathogen is dominant (Flor, 1956*a*).

There is a comprehensive literature dealing with inheritance of pathogenicity in smut fungi. In oat smuts, *Ustilago avenae*, loose smut, and *U. hordei (U. kolleri, U. levis)*, covered smut, Nicolaisen (1934) established that the specific factors of pathogenicity may be dominant, recessive or mixed dominant and recessive, depending on host cultivar. Transgressive segregation may produce races more virulent than the parent races (Nicolaisen, 1934; Holton, 1936, 1941). New races may appear by hybridisation between the two species (Holton, 1936; Utter, 1943; Bever, 1945), cf. Fig. 17.10. Utter obtained as many as 22 new races from a single cross between two races!

Inheritance of pathogenicity in *Tilletia caries* and *T. foetida*, bunt of wheat, is more complex than in most other smut fungi (Fischer and Holton, 1957). Pathogenicity is controlled by specific factors but the precise manner of inheritance is obscure. Holton (1959), who extensively studied the subject, concludes that multiple factors are involved and that transgressive inheritance may occur resulting in stronger as well as weaker pathogenicity.

Mutation in pathogens may lead to new races. Many pathogenic mutants have been recorded. Examples are recorded by Flor (1956*b*, 1958) in *Melampsora lini*, induced by UV and X-ray treatment of uredospores, by Christensen and Schneider (Stakman and Christensen, 1953) in *Drechslera sorokiniana* which mutates rather freely on artificial nutrient media, in *Ustilago maydis* (Stakman, Christensen and Hanna, 1929; Stakman, Tyler and Hafstad, 1933), in *U. hordei* (Holton, 1936), and in *Fusarium oxysporum* f.sp. *lycopersici*, induced by UV radiation (McDonnell, 1962).

Lincoln and Gowen (1942) induced mutants with increased pathogenicity by X-irradiation in *Erwinia stewartii*, and Ark (1946, 1951) obtained mutants of *Corynebacterium michiganense* with higher and lower virulence.

→

Fig. 17.10. Interspecific crossing between the loose smut of oats and covered smut fungi of oats and the appearance of a mutant, the buff smut fungus. Diagram indicating origin and subsequent appearance of the buff smut fungus which has hyaline teliospores. The shading indicates the cells in which the factors for colour are carried, and the absence of shading indicates those from which such factor or factors are absent. The nuclei are indicated by dots and circles. The plus and minus signs indicate sporidia of opposite sex. Buff smut is indicated in F_2 as two smooth hyaline teliospores. This smut type has been suggested to be a mutant, possibly due to loss of the factor or factors for colour in the present hybrid. (Holton, 1936)

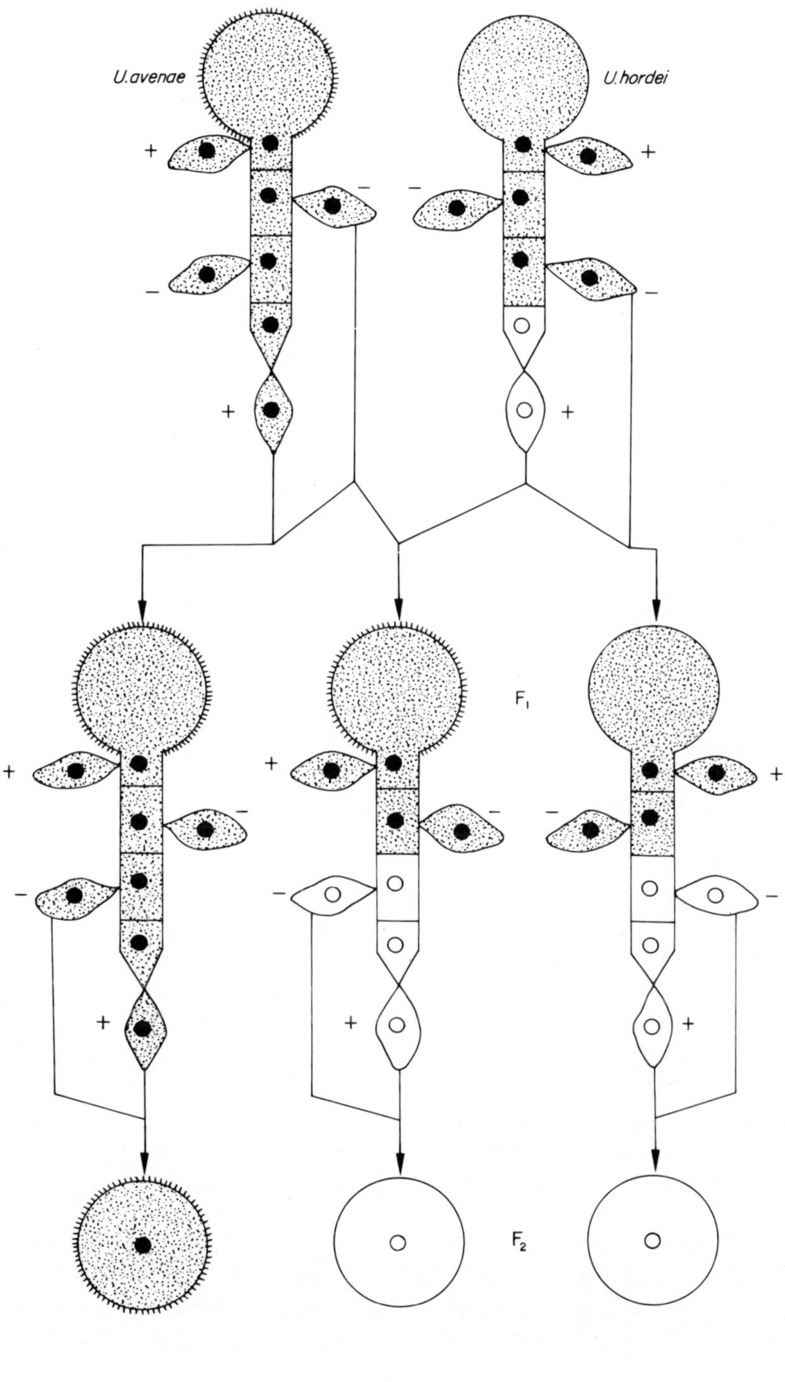
U.avenae
U.hordei
+
−
F₁
F₂

For viruses mutation is frequent. The variants differ from the parent strain in characters such as infectivity, symptoms produced in hosts, serological and immunological reactions and physicochemical characters (Kunkel, 1947). For tobacco mosaic virus Kunkel (1940) estimated a mutation rate of 0.5 per cent.

17.6 PHYSIOLOGIC IMPLICATIONS OF PATHOGENICITY

The biochemical actions of microorganisms are of basic importance in infectious plant diseases. They constitute a major part of the offensive mechanisms operating in host–parasite interactions, being the pathogen's counterpart to the corresponding defensive mechanisms of the host.

There are three activities in the metabolism of microorganisms that are important in pathogenesis of plants: secretion of enzymes, production of growth-regulating substances and production of toxins. These metabolites may act independently or together in the development of individual diseases.

17.6.1 Enzymes

The classical works on enzyme action in physiology of plant parasitism are those of De Bary (1886) on *Sclerotinia* and Ward (1889) on *Botrytis cinerea*, followed by the studies of Jones (1905) on tissue-macerating bacteria and Brown (1915, 1917, 1934, 1936, 1955) on *B. cinerea* and other pathogens. The chief enzymes involved in the macerating effect on plant tissues are the pectolytic enzymes. The following three groups of these are recognised: pectin(methyl)esterases, polygalacturonases and pectate and pectin transeliminases.

The macerating effect of these enzymes is through disintegration of the cell walls of plant tissues, and especially dissolution of the middle lamella; this layer has been assumed to consist mainly of pectates. However, the process has been little studied biochemically but it is accepted by a number of workers that essentially maceration does involve solubilisation of the middle lamella and the primary cell wall by polygalacturonases, possibly with the co-action of pectinesterases (Wood, 1967). Bateman (1964*a*) presented evidence indicating that maceration produced by *Rhizoctonia solani* in bean hypocotyls is brought about by the activity of polygalacturonase from the fungus together with pectinmethylesterase of the host.

In addition to pectic enzymes, cellulolytic enzymes may also be involved in the disintegration of cell wall material. Cellulase activity has been reported by Bateman (1963, 1964*a*, 1964*b*) in bean hypocotyl tissue infected by *R. solani*, in the later phases of the infection. Cellulase has been recorded as important in wilts — both *Fusarium oxysporum* f.sp. *lycopersici* and *Verticillium alboatrum* produce cellulase that may cause deterioration of cellulose, resulting in blockage of the vascular elements (Talboys, 1958; Husain and Dimond, 1958, 1960; Deese and Stahmann, 1962).

Wood (1967) has listed selected parasites in which cellulase, polygalacturonase and pectinesterase are produced *in vitro* and (the two first groups) *in vivo*. He concludes that probably the great majority of bacteria and fungi which cause degradation of tissue can secrete one or other of the last two categories of enzymes in culture.

17.6.2 Growth substances

Growth-regulating substances, originally discovered in tissue extracts from higher plants, have been shown to be produced in culture by many fungi and some bacteria. Outstanding examples of seed-borne fungi that produce such substances and also distort growth of plants are *Puccinia carthami*, *Gibberella* spp., and *Ustilago* spp.

In safflower, *Carthamus tinctorius*, *P. carthami* produces excess auxin leading to a striking elongation of the hypocotyl; infected hypocotyls weigh about 50 per cent more and contain several times more indolacetic acid than healthy hypocotyls (Daly and Inman, 1958). Tumorous tissues of maïze, infected by smut, *Ustilago maydis,* contain a much higher level of indolacetic acid than normal tissues (Turian, 1958, 1961; Turian and Hamilton, 1960).

Gibberellins are produced by *Fusarium moniliforme (Gibberella fujikuroi)* in rice, and cause a spectacular increase on the height of the plant, described as the bakanae effect. Twenty-nine chemically different substances have been differentiated in this group of growth-regulators (Yabuta, 1935; Yabuta *et al.*, 1941*a*, 1941*b*; Curtis and Cross, 1954; Stodola, 1958; Lang, 1970). In many plants, e.g. pea and bean, gibberellins produce elongation of internodes by increasing the length of individual cells.

The reverse effect, shortening of internodes, occurs in dwarf bunt, *Tilletia contraversa*. Other similar examples of change of growth are elongation in barley induced by loose smut, *Ustilago nuda*, and shortening in barley induced by covered smut, *U. hordei*.

17.6.3 Toxins

In recent years there has been an extensive debate on the terminology of such toxins that are active in the pathogenesis of plant diseases. The terms commonly accepted will here be briefly discussed.

Toxins, as used in plant pathology, are substances that damage or kill protoplasts of plants. Many plant pathogens are capable of producing metabolites that cause symptoms similar to those produced by natural infection.

The basic term *vivotoxin,* proposed by Dimond and Waggoner (1953*b*), has been generally accepted in recent years. Adopting the definition, with later modifications suggested by Dimond (1955) and particularly by Wood (1967), a vivotoxin is a substance, produced by a parasite, or a host plant in response to the parasite, or both, that operates in the causation of a disease, not being itself the primary agent. Since vivotoxin is a pathogenic agent it must satisfy rules similar to Koch's postulates for establishing pathogenicity. It must be purified, and it must reproduce at least part of the syndrome when introduced to healthy plants, and it must be present in diseased plants and not in healthy plants, and, finally, it must be present in concentrations sufficient to cause the observed symptoms.

Another term, *pathotoxin*, has been used with a slightly different meaning and may be used in the restricted sense of host-specific vivotoxins (Wheeler and Luke, 1963). Such a toxin is a metabolic product of a pathogen, which is toxic only to the host of that pathogen. It must be capable of causing all disease symptoms.

Phytotoxin is used in a broader sense, covering 'any microbial metabolites that interfere with or alter the metabolism of plant cells and so injure the plant symplast. Phytotoxins are produced either in the host, on its surface or even outside the host for example in soil' (Graniti, 1972). It is questionable whether enzymes should be included. Graniti (1972) would include enzymes if they are proved to possess inherent primary toxicity.

Still another term should be mentioned, *mycotoxin*. This term refers to toxins that originate from fungi and affect man and animals. The term has been internationally established in this restricted sense and is commonly used in medicine and pharmacy.

Phytotoxins, produced by microorganisms outside the plant, may damage plants directly or predispose them to pathogens. These toxins are commonly produced by soil saprophytes or weak parasites that are common as soil inhabitants. Decomposing plant residues may be toxic, for instance to young seedlings, due to activity of microorganisms. Patrick and Koch (1958, 1963) and Patrick, Toussoun and Snyder (1963) have demonstrated injury to lettuce, spinach and tobacco seedlings in field soil with high proportions of decomposing plant fragments, and the organisms isolated from the injured parts of affected plants were for the most part non-pathogenic. Extracts from decomposing plant debris were also shown to be active in reducing germination of seed and growth of seedlings, and in producing necroses. Such damage may predispose to invasion of parasites such as *Fusarium solani* f.sp. *phaseoli* and *Rhizoctonia solani*.

Few plant diseases have been thoroughly investigated as to vivotoxins involved. These substances probably operate in most diseases caused by parasites, and it may be assumed that in most parasitic diseases several vivotoxins are involved. In the following some well-known vivotoxins produced by seed-borne pathogens will be briefly discussed.

Victorin is produced by *Drechslera victoriae*. It is host-specific, i.e. a pathotoxin, highly toxic to oats susceptible to the parasite. Indicative of its high toxicity is that filtrates in a concentration of about 1:1,000,000 from cultures of the fungus reduce root growth of susceptible cultivars by 50 per cent, whereas a concentration of 1:5 fails to induce any effect on resistant cultivars (Wheeler and Luke, 1954). The action of the toxin is primarily to increase the permeability of susceptible tissues (Wheeler and Black, 1963); it also increases the respiratory activity of the host (Romanko, 1959), whereas transpiration is decreased because the toxin causes stomata to close (Wheeler, 1969; Turner, 1972). Romanko (1959) found indications that oat cultivars inactivate or bind the toxin in some way but this claim has not been confirmed; on experimental evidence Scheffer and Pringle (1964) explain the reaction of susceptible tissue as being adsorption of the toxin to a receptor on the surface of the cells; the receptor is lacking in resistant tissue. Nishimura and Scheffer (1965) found high concentrations of the toxin in spores of virulent races but not in those of avirulent races. Nelson, Scheffer and Pringle (1963) established that toxin production in *D. victoriae* is genetically controlled and that the toxin mechanism is conditioned by more than one gene. They found that susceptible cultivars of oats are more rapidly and more severely blighted when inoculated with high toxin producing strains. The inheritance of host reaction is monogenic (Litzenberger, 1949). Resistance to *Puccinia coronata*, crown rust, in oats of the Victoria type is always associated with susceptibility to *D. victoriae*.

The *PC-toxins* are other host-specific toxins. They are produced by *Periconia circinata*, the causal agent of the milo disease of sorghum. Two or more closely related compounds are produced in culture, each with a high degree of host-specificity (Scheffer and Pringle, 1961; Pringle and Scheffer, 1964). The toxins cause symptoms in susceptible cultivars similar to those in plants inoculated with the pathogen. The toxins inhibit growth of susceptible sorghum roots at a concentration of 1:10,000,000, while resistant seedlings tolerate more than 26,000 times higher concentrations (Mansour, 1968).

Ophiobolin (cochliobolin) is a metabolite of *Drechslera oryzae* (perf. st. *Cochliobolus miyabeanus*). It is a non-specific vivotoxin, which is able to produce some of the symptoms characteristic of the parasite (Orsenigo, 1957). Chemically the toxin has been differentiated into ophiobolin A, ophiobolin B and ophiobolin C (Canonica, 1972, referring to his earlier publications).

Piricularin is produced by the rice blast fungus *Pyricularia oryzae* and is also extremely active, dilutions such as 1:5,000,000 being able to increase respiration of rice plants, both resistant and susceptible cultivars (Tamari and Kaji, 1954). Piricularin is toxic not only to rice but to other higher plants and many microorganisms. An additional toxin, *α-picolinic acid*, is produced by the fungus (Tamari, Ogasawara and Kaji, 1966).

Colletotin is a non-specific toxin, produced by *Colletotrichum fuscum*, an important pathogen on *Digitalis lanata* and *D. purpurea*. Pathogen and toxin cause similar symptoms of leaf mottle and spotting; moreover, *D. purpurea*, which is more resistant to the pathogen, is also more resistant to the toxin (Goodman, 1960; Lousberg and Salemink, 1972). In leaves of *Digitalis* and tomato the toxin causes disintegration of the tissue, which suggests that it may activate enzymes degrading pectin substances (Lewis and Goodman, 1962).

The role of toxins in the pathogenesis of Fusarium wilt in tomato, caused by *Fusarium oxysporum* f.sp. *lycopersici*, has been intensively studied in the last decades. Several vivotoxins have been isolated, including *fusaric acid* and *lycomarasmin* (Gäumann and Jaag, 1947). They produce various symptoms in the host which may indicate that each toxin possibly contributes to the syndrome of Fusarium wilt. Fusaric acid causes leaf necrosis, while another toxin, *phytonivein*, which is known from *F. oxysporum* f.sp. *niveum*, wilt of watermelon, causes wilting (Nishimura, 1962), and lycomarasmin modifies the permeability of the cells to water and increases transpiration of the host (Gäumann and Jaag, 1947). However, Dimond and Waggoner (1953*a*) established that lycomarasmin is not available in sufficient quantities in the host to cause wilting. The toxin seems to be a product of lysis of mycelium, a fact which is in contrast to the early appearance of disease symptoms after inoculation of plants. As to fusaric acid the physiological effects of infection by the parasite is different from the effects of treatment by fusaric acid. Symptoms of fusaric acid injury are unlike those incited by the pathogen, with increased respiration as well as synthesis and activities of mitochondrial enzymes. Conversely, fusaric acid decreases drastically these activities (Kuo and Scheffer, 1964). A causal relationship between vivotoxins and Fusarium wilt of tomato seems therefore so far not to have been established.

Finally, an example of a bacterial toxin. A toxin is produced by *Pseudomonas phaseolicola*, the causal agent of halo blight of bean (Waitz and Schwartz, 1956). Symptom development of the disease suggests action of a toxin. In the centre of the lesion the tissue becomes translucent with a chlorotic halo, and inoculation of the first trifoliate leaf induces water-soaked lesions in these, but also vein-clearing and systemic chlorosis appear in the non-inoculated second and third trifoliate leaves, an indication that the toxin is readily transported from infected to younger uninfected leaves; this does not take place in resistant plants (Waitz and Schwartz, 1956; Patel and Walker, 1963*a*). The toxin is non-specific, although the pathogen is host-specific (Rudolph and Stahmann, 1965, *fide* Goodman, Kiraly and Zaitlin, 1967). In leaves affected by the toxins a substantial increase in several amino compounds, in particular ornithine, has been observed (Patel and Walker, 1963*b*). This indicates that chlorotic halo and systemic chlorosis in halo blight are caused by the toxin-induced interference with the normal amino-acid metabolism.

18 Factors Inherent in the Host Affecting Establishment and Course of Disease

INTRODUCTION (RESISTANCE AND TOLERANCE)

Any infection course is a result of interactions between pathogen and host resulting in different mechanisms of their effects on the host and host population. The inherent factors may be classified according to the manner in which they affect infections. Morphological and physiological properties of the host may directly influence (1) establishment of infection and (2) course of infection. And they may indirectly affect infection by providing conditions for interaction between pathogen and other microorganisms, e.g. by providing exudates that affect the microflora of the environment, particularly the soil microflora (rhizosphere effect); or by providing conditions for establishment of latent infection by other microorganisms which then may affect the interactions between pathogen and host. A complicated system of such different factors operates in any host-pathogen interaction.

There are two aspects of the 'defensive' reaction of host to pathogens or pathogenic factors: resistance and tolerance. *Resistance,* in broader sense, is the act or capacity to escape or exclude infection or efficiently to slow down or suppress either intrusion of an agent into the host, or its spread within the host, or both. *Tolerance* is the act or capacity to endure the harmful effects of a pathogen or pathogenic factor. Susceptibility and resistance are assessed by measuring the degree of disease in terms of direct damage to the host, such as intensity of symptoms, extent of lesions and decay, degree of lowering of viability, etc. Sensitivity, or intolerance, and tolerance are assessed by measuring the effect of a given level of disease on the productive performance of the host, in practice on yield (Caldwell *et al.*, 1958). Two equally infected and equally susceptible cultivars may yield differently because one may endure the same intensity of the disease more than the other.

Flor, Gaines and Smith (1932) showed that in one wheat cv. Hybrid 128, 16 per cent increase of bunt reduced yield 21 per cent (plant infection/yield reduction ratio 1:1.3), in another, Turkey, 31 per cent bunt increase reduced yield only 23 per cent (ratio 1:0.7) which suggests that Turkey is more tolerant. Brönnimann (1969), working with *Septoria nodorum*, compared two tolerant and two intolerant wheat cultivars, testing them by inoculation and by artificial defoliation. Defoliation caused more damage in the tolerant cultivars than did infection, and inversely in the intolerant cultivars; this seems to indicate that in the tolerant cultivar any leaf tissue left uninfected was more efficient in photosynthesis than uninfected leaf tissue in the intolerant cultivars. This interpretation is supported by the observations of Scharen and Krupinsky (1969) who found that photosynthesis measured by means of CO_2 absorption paralleled yields thus providing an additional basis for selecting for tolerance to *Septoria nodorum* infection. In England a very early maturing wheat cultivar, Sterling, appeared to be extremely tolerant, sustaining in contrast to other cultivars no yield loss despite severe infection (Scott, 1973).

The condition of probably any host plant as a suscept changes with age. As introduction to our discussion of the factors operating in pathogen–host interactions, the ontogenetic factors of susceptibility and resistance will first be considered.

18.1 SUSCEPTIBILITY AND RESISTANCE AS RELATED TO ONTOGENESIS OF THE HOST

In the life-cycles of plants, periods of susceptibility are often related to youth and senescence. Cereals and grasses provide examples of hosts passing through very short periods of susceptibility to smut and ergot fungi in certain youth stages of the host. The conditions during these short periods determine the *rate of seed infection*. Wheat and barley are susceptible to loose smut, *Ustilago tritici* and *U. nuda*, only for a few days following fertilisation of the flowers, while cereals and grasses are infected by ergot fungi only before pollination and fertilisation. Thirty minutes after pollination Bahia grass, *Paspalum notatum*, is already immune to *Claviceps paspali* (Burton and Lefebvre, 1948). Male sterile hybrids of Bahia grass, sorghum, wheat and others are more severely infected by ergot fungi, including *C. purpurea*, because fertilisation is not accomplished and the young tissues of the stigma remain in a susceptible condition, see Section 11.4.

Establishment of seedling infecting smuts is limited to a few days after germination of seed, and this short period determines the *rate of seed transmission*. The infective hyphae produced by the sporidia of *U. hordei*, covered smut, invade the coleoptile of barley, which is susceptible only as long as it is unsplit; when the first green leaf has protruded, the possibility for penetration is practically nil, see Section 14.5.

Many seed-borne Fungi Imperfecti affect the host either in its seedling stage (damping-off) or later when the hosts, or parts of the host, are weakened, as particularly in senescence. This holds true for many species of *Alternaria*, such as those affecting crucifers, *A. brassicae*, *A. brassicicola* and others, or many species of *Phoma*, e.g. *P. betae* in beet, or *A. radicina* in carrot, and *P. lingam* in crucifers. When the host plant has passed the seedling stage, infections generally become confined to superficial, scarcely visible necroses, often located at the soil level. Later infections may spread to weakened parts of the host such as dying leaves and petals or any host tissues losing vitality as they do during the maturation of seed; often these conditions give an opportunity for the pathogen to get a foothold in the developing seeds, thus determining the *rate of seed infection*.

Other fungi are less confined to these extreme stages of the life-cycle of the host. *Alternaria zinniae* on *Zinnia* and *A. dianthicola* on *Dianthus caryophyllus* may both produce damping-off but as long as the plants are vigorous they are seldom severely affected. However, as the plants reach the flowering stage severe attacks may develop on leaves, stems and flowers. The diseases produced by *Cercospora beticola* on sugarbeet (Pool and McKay, 1916) and *Cercosporidium punctum* (syn. *Passalora kirchneri*) on dill are other examples of increasing severity of attacks in ageing plants.

Adult plants in full growth are susceptible to a different category of pathogens, including powdery mildews and rust fungi, whereas seedlings and senescent plants are often resistant to these fungi. Snapdragon, *Antirrhinum majus*, is resistant to *Puccinia antirrhini*, snapdragon rust, in the seedling stage. Consequently, teliospores introduced with imported seed packets into snapdragon beds in Rhodesia did not infect the seedlings from the contaminated seed but they did infect adult neighbouring plants, cf. Section 13.8. Here resistance of the seedlings restricts the *rate of seed transmission*.

18.2 MORPHOLOGICAL IMPLICATIONS OF SUSCEPTIBILITY AND RESISTANCE

The capacity of a plant to withstand or evade infection and subsequent damage to any degree can be regarded as resistance in the widest sense of this term. This conception would include 'disease escape' or 'klendusity', which is effective only when the plants are grown in conditions in which they are not exposed, or less exposed, to the pathogen. Obviously, such evasion of attack is not true resistance. However, in discussing from an agricultural point of view the possibilities of reducing damage caused in plants by pathogens, the method of defence, whether it is 'active' or 'passive', is immaterial in so far as it can be utilised in practical efforts of plant sanitation, including breeding for any kind of 'resistance'.

18.2.1 Morphological escape mechanisms

The above view pertains particularly to morphological characters of the host, because growth type may be an important factor in disease escape. A dense closed growth favours development of a humid microclimate thus providing conditions for infection. A stiff dispersed growth counteracts this development. Kidney bean cultivars with a low closed growth are more subject to attacks of *Colletotrichum lindemuthianum* than cultivars of a high stiff growth type.

Barley cultivars having closed flowers (erectum-, zeocritum-barley) are much less affected by *Ustilago nuda* than cultivars with open flowers. If the flowers are opened artificially, they become largely affected (Henning, 1913). Doling (1966*a*) points to the observation that the introduction of Scandinavian barley cultivars with an open flowering habit in the early 1950s coincided with an increase in the severity of loose smut. There is much evidence that the flowering habit of barley is closely related to the size of the lodicules. It is generally assumed that swelling of the lodicules is an important part of the mechanism by which the glumes open for flowering (Pedersen and Jørgensen, 1965). The same is true for wheat and loose smut, *U. tritici* (Tavcar, 1934). Maize cultivars with tight and complete husks escape from maize smut, *U. maydis*, to a greater extent than do cultivars with loose or incomplete husks (Kyle, 1929).

Gramineae with closed flowers are well protected against *Claviceps purpurea*, whereas those with open flowers, for instance barley cv. Hanchen, are more severely affected; the side flowers in the spikes of 6-row barley are of the open type and also much more severely affected than the closed top flowers.

In cotton predominantly the bolls of cultivars with bracts are attacked by *Physalospora rhodina (Diplodia gossypina)* as shown by Luke (1968). After inoculation, Diplodia rot developed in 83 per cent bolls with bracts, but only in 23 per cent of those without bracts. Field experiments supported these observations. High relative humidity within the floral bracts provided conditions for invasion of the parasite into the bolls. Bractless cultivars would assist control.

The growth rhythm of the host is of great importance in disease cycles, where only certain stages or organs are susceptible. The more rapidly such stages are passed, the greater the chance of escaping infection. Quickly germinating wheat cultivars such as Ohio have been found less affected by *Tilletia caries* and *T. foetida* than those which germinate slowly. Rye cultivars which pass more quickly through the flowering period are less affected by *Claviceps purpurea* than slowly flowering cultivars.

18.2.2 Preformed mechanical barriers

The degree of resistance or susceptibility may depend on structural differences in the outer protective tissues of the host, such as the thickness of cuticle and outer walls of epidermal cells, or in the morphological characters of stomata and lenticels influencing the possibilities of the pathogen to penetrate through these openings. Well-known examples are increased resistance of older fruits of tomato against *Alternaria tomato* and *Botrytis cinerea* (Rosenbaum and Sando, 1920; Ainsworth, Oyler and Read, 1938), because the cuticle and epidermal cell walls are thicker in these fruits. Varietal susceptibility of coffee berries, *Coffea arabica*, to *Colletotrichum coffeanum* has been related to physical and chemical differences in the cuticle that make penetration more or less difficult (Nutman and Roberts, 1960).

Ito and Sakamoto (1939–1943) found that resistance of rice to blast, *Pyricularia oryzae*, was positively correlated with resistance to mechanical puncture of leaf epidermis. Similarly resistance to brown leaf spot, *Drechslera oryzae*, has been found to be positively correlated with thickness of epidermal cells of rice leaves (Suzuki, 1934; Chattopadhyay and Chakrabarti, 1957; Mishra and Prasad, 1964). In cotton *Physalospora rhodina (Diplodia gossypina)* is most destructive to bolls younger than 15 days and older than 28 days (Wang and Pinckard, 1971). In the intermediate period of relative resistance the cuticle of the bolls contains maximal amounts of waxes and cutin acids which are inhibitory to spore germination and mycelial growth. The cuticle thickness and quantity of waxes and certain acids are greatest for the intermediate-aged bolls as they approach full size; in this stage the bolls are also resistant to *Rhizoctonia solani* and *Phytophthora parasitica* (Wang and Pinckard, 1973).

Pool and McKay (1916), often quoted in literature, found positive correlation between infection by *Cercospora beticola* in leaves of sugarbeet and the size of the stomata. Penetration of conidial germ tubes occurs only through open stomata. In the less susceptible young leaves they found stomata only half as large as those in older, more susceptible leaves, and penetration was therefore more likely to occur through the larger stomata. However, in recent investigations Ruppel (1972) found no correlation in six sugarbeet cultivars between degree of resistance and size or density of stomata.

18.2.3 Induced mechanical barriers

Morphological structures that reduce or prevent invasion or colonisation of pathogens may develop as a response to infection, in contrast to the normal, preformed morphological structures discussed above (Wood, 1967). Walls of infected cells of leaves or other organs such as stems, fruits and roots thicken as a reaction to infection; the parasite is confined to these cells and does not become intercellular, as in susceptible cultivars. This reaction takes place in cucumber infected by *Cladosporium cucumerinum* (Pierson and Walker, 1954), in oats infected by *Drechslera avenae*, and in flax infected by *Fusarium oxysporum* f.sp. *lini*. It is not unlikely, although not proved, that the increase in thickness of the walls makes these less readily penetrated than normal walls.

Other morphological responses protective against colonisation of pathogens are wound gums which impregnate walls of cells adjacent to infected cells or which fill intercellular spaces, thus forming barriers against further penetration by the parasite. Such reactions are known in rice cultivars resistant to *Drechslera oryzae* and *Pyricularia oryzae* (Kawamura and Ono, 1948; Yoshii, 1957), and in cucumber cultivars resistant to *Cladosporium cucumerinum* (Pierson and Walker, 1954; Hijwegen, 1963). In vascular wilts formation of tylosis is a typical response, prominent in wilts caused by *Fusarium oxysporum* and *Verticillium alboatrum*. There is a more pronounced development of gel plugs in resistant than in susceptible plants of lupin, *Lupinus luteus*, infected by *F. oxysporum* f.sp. *lupini* (Saaltink, 1963), and in resistant tomato plants infected by *F. oxysporum* f.sp. *lycopersici* (Finley, 1950).

18.3 PHYSIOLOGICAL IMPLICATIONS OF SUSCEPTIBILITY AND RESISTANCE

Distinction is made between physiological conditions present in the plant before infection and conditions developed in the plant as a result of infection (Wood, 1967).

18.3.1 Preformed chemical barriers

Resistance may be based on preformed substances that may act directly against the parasite, or indirectly by acting on associated microorganisms. Exudates from the host may affect the pathogen. Root exudates from pea reduce to a certain extent conidial germination, but not hyphal growth, in races of *Fusarium oxysporum* f.sp. *pisi*, pea wilt and pea near wilt, to which the pea cultivar concerned is resistant (Buxton, 1957); the inhibition is more pronounced with exudates from younger than from older plants. It is doubtful that the role of exudates in resistance of pea to the wilt organism is very great (Wood, 1967). Conidial germination of *Cercospora beticola*, leaf spot of beets, has been found to be reduced on leaves of resistant plants and in washings from resistant leaves, as compared with leaves from susceptible cultivars (Kovács, 1955). A classical example of preformed substances providing resistance is the group of substances acting in the red- and yellow-skinned cultivars of onion,

Allium cepa, but not in the susceptible white cultivars. They prevent germination and growth of certain seed-borne fungi including *Botrytis allii*, grey mould or neck rot of onion, and *Colletotrichum dematium* f.sp. *circinans*, smudge of onion, at the surface of the dry, dead scales. The brown to yellow pigments localised in the outer scales are not themselves the cause of resistance. The effect in coloured cultivars depends on fungitoxic substances in the dead cells of dry scales; they are not present in living scales even when these are coloured. Much of the effect can be attributed to the water-soluble fungicidal phenols catechol and protocatechuic acid (Angell, Walker and Link, 1930; Walker and Stahmann, 1955).

Cultivars of chick-pea, *Cicer arietinum*, resistant to 'anthracnose', *Ascochyta rabiei*, have more glandular leaf hairs secreting malic acid than susceptible cultivars. Malic acid inhibits spore germination and retards hyphal growth of the parasite (Hafiz, 1952).

Phenolic substances have been claimed largely to be responsible for the general resistance in higher plants to parasitic bacteria and fungi. In potato plants concentrations of phenols are higher in potato plants resistant to *Verticillium alboatrum* than in susceptible, and in younger than in older, more affected plants (Lee and Le Tourneau, 1958); in rice increase in concentration of polyphenols is correlated positively with resistance to *Pyricularia oryzae*; and in pea concentration of leucoanthocyanin with resistance to *Ascochyta pisi*, *Phoma medicaginis* var. *pinodella* and *Mycosphaerella pinodes* (Clauss, 1961). Sörgel (1952) attributed this resistance to the presence of substances which inhibit growth and sporulation of these fungi in culture. Jacobs and Dadd (1959) have demonstrated that antibacterial substances in the seed coat of sweet peas operate as a contributing factor of resistance to *Corynebacterium fascians*, fasciation or leafy-gall disease, and Kraft (1973) found that pea selections with pigmented seed exuded substances during germination which inhibited sporulation and conidial germination of *Fusarium solani* f.sp. *pisi*, and that these selections were resistant to *Fusarium* infection. Green-seeded susceptible peas were much lower in amounts of phenols exuded than were those with pigmented seed. Various glycosides are considered as significant in resistance, e.g. in maize 6-methoxybenzoxazolinone to *Xanthomonas stewartii*, bacterial wilt. Also, amino acids have been related to resistance, increase leading to increase in resistance, e.g. cucumber to *Cladosporium cucumerinum*, probably by altering the metabolism of plants (Wood, 1967).

18.3.2 Sugar level

The concentration of carbohydrates, particularly sugars, in host tissues, has been considered to play an important role in resistance. Horsfall and his collaborators distinguish between high sugar diseases and low sugar diseases, referring susceptibility to higher or lower levels of sugar in the tissues of the host. In general, older plants with less sugar are more susceptible to low sugar diseases than younger plants which are more susceptible to high sugar diseases. As low sugar diseases are considered early blight of potato and tomato, *Alternaria porri* f.sp. *solani* (Horsfall and Heuberger, 1942; Rowell, 1953), wilts produced by *Fusarium oxysporum* (Waggoner and Dimond, 1957; Patil and Dimond, 1968), spot blotch or seedling blight of cereals and grasses, *Drechslera sorokiniana* (Hsia and Christensen, 1951), leaf spot of finger millet, *D. tetramera* (Vidhyasekaran, 1974), whereas as high sugar diseases are considered rusts (Ibrahim, 1951; Livingston, 1953), powdery mildews (Yarwood, 1934; H. J. Miller, 1952). The sugar level hypothesis has been supported by the fact that growth-regulating substances can induce decrease (2,4-D) or increase (maleic hydrazide) in the sugar content of host tissues and alter susceptibility or resistance according to expectation (Prabhu and Swaminathan, 1968). Thus severity of the above three first-mentioned low sugar diseases were increased by treatment of the host with 2,4-D and reduced by maleic hydrazide (Horsfall and Dimond, 1957; Horsfall and Lukens, 1966). Application of excessive nitrogen fertiliser can induce reduction in sugar concentrations and increased protein synthesis for which sugars are utilised (Vidhyasekaran and Kandasamy, 1971). It is likely that increased sugar concentration in diseased leaves inhibits production of enzymes which degrade cell walls (Vidhyasekaran, 1974).

18.3.3 Induced chemical barriers

In addition to preformed substances discussed above, resistance may depend on toxic substances produced as a reaction to infection. Some substances toxic to the parasite may, however, be produced by itself as staining substances.

Another response to infection is increase in enzyme activity. Enzymes catalyse reactions involving phenols and related compounds. Oxidation of phenols by oxidases forming quinones toxic to microorganisms is one of the processes known to occur and probably involved in wound reactions. These include impregnation of cell walls near the surface of the wound thus making them resistant to most parasites (Wood, 1967). There is evidence for an increase of polyphenoloxidase activity in relation to resistance in pea to *Fusarium oxysporum* f.sp. *pisi* race 2, near wilt (Wilding, 1960) and in tomato to *F. oxysporum* f.sp. *lycopersici* (Schachinger, *fide* Wood, 1967).

In recent years the concept of phytoalexins has attracted growing interest. Much of the earlier work was done by Müller and his coworkers, and an excellent brief review on the topic has been given by Wood (1967) in his textbook on physiological plant pathology. A summary of Wood's description of the concept of phytoalexins follows.

Phytoalexins are toxins formed in host cells in response to certain stimuli, including products secreted from infection hyphae, germinating conidia, and application of substances such as $CuCl_2$ or $HgCl_2$ in suitable concentrations. Often microorganisms that stimulate formation of phytoalexins are those which do not actively parasitise tissues. Production of phytoalexins is essentially a response of living cells; the substances are formed as a direct result of infection and appear as a local response to infection. Tissues of the healthy host are considered usually not to contain phytoalexins. The importance of phytoalexins in disease resistance remains to be established. Briefly, in the definition of Müller (1956), phytoalexins describe antibiotic principles present in hypersensitive host tissues.

A phytoalexin in pea, *pisatin*, has been studied in detail (Cruickshank and Perrin, 1961, 1963*a*, 1963*b*, and others). Pisatin may be produced in pods of pea by inoculation with conidia of *Sclerotinia fructicola* which is non-pathogenic to pea. The substance is fungistatic to a variety of fungi not normally parasitic on pea, as well as on six parasites on this host, but on these only in higher concentrations. Pisatin is not produced by mechanical damage but as a response to drops of $CuCl_2$ or $HgCl_2$ placed on the pods, whereas under these conditions $CaCl_2$ and NaCl are ineffective. Cruickshank (1963*a*, 1963*b*) considers pisatin to be a critical factor in resistance of pea to fungi parasitic in this host.

A closely related phytoalexin is *phaseollin* which is induced by inoculation on French bean, *Phaseolus vulgaris* with conidia of *S. fructicola*, non-pathogenic to this host (Müller, 1958; Cruickshank and Perrin, 1963*b*). Bailey (1973) substantiated that phaseollin is induced when and where bean cells undergo necrosis and show browning as they do in response to incompatible hyphae or, as in his experiments, a virus, tobacco necrosis virus. There seems to be a connection between a rather 'slow' process of cell death and phytoalexin production. Similar inhibitory substances have been induced in rice by *Pyricularia oryzae*, blast of rice, resistant cultivars producing more of the substance than susceptible cultivars (Uehara, 1958) and by *Xanthomonas oryzae*, bacterial blight of rice (Uehara, 1960). Other phytoalexins seem to be induced by species of *Pseudomonas* and *Xanthomonas* in Phaseolus bean (Klement and Lovrekovich, 1961, 1962, 1964). Post-invasion fungal inhibition produced in oats, resistant or susceptible to *Drechslera victoriae* (Nishimura and Scheffer, 1965; Samaddar and Scheffer, 1970), seems not to be due to phytoalexins because the inhibitions affect pathogenic and non-pathogenic forms equally.

Hypersensitivity to infection is of extreme importance, particularly in resistance to obligate parasites. Cells invaded by the parasite, sometimes even in advance of invasion, are killed and therefore become unsuitable substrate for the parasite which commonly lives in living cells. The parasite is confined to a local necrosis and is eventually killed or sealed off. This reaction is equivalent to immunity.

Hypersensitivity has been studied extensively in *Phytophthora infestans* by Müller and his co-workers, in obligate rust and mildew fungi, and in viruses (local lesion reactions). Also non-obligate parasites may meet hypersensitive response in resistant plants, e.g. in resistant cultivars of Phaseolus bean to *Colletotrichum lindemuthianum*, anthracnose, in cotton to *Xanthomonas malvacearum*, bacterial wilt (Thiers and Blank, 1951).

Király, Barna and Érsek (1972) question the phytoalexin theory of resistance. Working with *Phytophthora infestans*, *Puccinia graminis* and *Uromyces phaseoli* they showed that both hypersensitivity reaction and production of phytoalexin, occurring simultaneously, are the effects of the pathogens' death, not the cause. An unknown resistance mechanism includes the release of endotoxins and causes host necroses and phytoalexin production. Van Der Plank (1975) regards phytoalexins as preformed antibiotics protecting the plant against *secondary* infections.

18.4 GENETIC IMPLICATIONS OF SUSCEPTIBILITY AND RESISTANCE

Inheritance of disease resistance in plants follows Mendel's laws. Biffen (1905, 1912) was the first to pronounce this. He demonstrated that the F_2 progeny of a cultivar, highly resistant to yellow stripe rust of wheat, *Puccinia striiformis*, crossed with a susceptible cultivar, segregated in the ratio of three susceptible to one resistant.

Inheritance of resistance to a definite disease may be monogenic, oligogenic or polygenic. The number of genes for resistance varies greatly with host and with pathogen, as does also the mode of inheritance.

Single dominant gene resistance has been established in tomato to race 1 of *Fusarium oxysporum* f.sp. *lycopersici*, Fusarium wilt (Bohn and Tucker, 1940), in pea to pea wilt, *Fusarium oxysporum* f.sp. *pisi* race 1 (Wade, 1929) and near wilt, *F. oxysporum* f.sp. *pisi* race 2 (Hare, Walker and Delwiche, 1949), in sorghum to milo disease, *Periconia circinata*, a widely distributed seed-borne soil inhabitant that produces an exotoxin causing disease in certain cultivars (Bowman *et al.*, 1937; Leukel, 1948; Scheffer and Pringle, 1961), and in barley to loose smut, *Ustilago nuda* and black smut or semi-loose smut, *U. nigra* (Cherewick and Buchannon, 1969). Zaumeyer and Harter (1943*a*) showed that resistance in bean to bean southern mosaic virus is controlled by a single dominant gene. Resistance in some cultivars of onion to pink root, *Pyrenochaeta terrestris*, is controlled by a single recessive gene (Jones and Perry, 1956;

Nichols *et al.*, 1965), and resistance in pea to pea (seed-borne) mosaic virus seems also to be conditioned by a single recessive gene (Hagedorn and Gritton, 1973). The outer-scale resistance of onion to smudge, *Colletotrichum dematium*, is based on three gene pairs that determine the formation of the phenolic colour compounds which inactivate the pathogen (Rieman, 1931; Clarke, Jones and Little, 1944). According to the gene combination, colour of the outer-scale and disease reaction may range from red or yellow and high resistance through pink or cream and intermediate resistance to white and susceptibility.

Inheritance of resistance may not follow the same pattern in different cultivars. In two cultivars of barley resistance to net blotch, *Drechslera teres*, has been found conditioned by two dominant genes (Shands, 1951), in another cultivar by one completely dominant gene, possibly supplemented by one or more modifying genes (Schaller, 1955); three dominant genes were found in barley by Mode and Schaller (1958), and in studies on crosses between three susceptible and seven resistant and semi-resistant cultivars Omar, Selim and Khalita (1970) established three dominant genes for susceptibility and three dominant genes for resistance, out of which were two for semi-resistance.

Resistance to specific pathogenic races of a pathogen can usually be referred to specific genes. Resistance against bean anthracnose, *Colletotrichum lindemuthianum*, race alpha, beta and gamma, is controlled by independent gene pairs, resistance being dominant (Andrus and Wade, 1942). Also resistance to flax rust, *Melampsora lini*, is dominant, based on many distinct independent genes. Flor (1946, 1947, 1954, 1955, 1959) has substantiated a theory of gene-to-gene relationship between resistance in the host and pathogenicity in the fungus, in which pathogenicity is recessive to non-pathogenicity. Each differential cultivar of a list of standard differentials, established by Flor, has a single unique gene for resistance to flax rust. Twenty-six such genes have been identified (Flor, 1955; Flor and Comstock, 1972).

In wheat resistance to bunt, *Tilletia caries* and *T. foetida*, is based on a number of major genes and some minor genes. Each of the dominant genes controls resistance against a certain group of races, e.g. M_1-gene seven races of *T. caries* and four races of *T. foetida*, while gene H controls resistance to eight races of *T. caries* and three races of *T. foetida*, some of which, however, are controlled by M_1. Omar was the first cultivar which combined these two genes for resistance (Holton *et al.*, 1956). Factors for resistance to bunt may be strong or weak, they may be independent or linked, and effective against a single race or against several. By crossing resistant with resistant cultivars there may be transgressive segregation (Fischer and Holton, 1957, who discussed in detail inheritance of factors for resistance to bunt, oat smuts and loose smut of barley and wheat).

While usually specific genes for resistance are related to specific pathogenic races, inheritance of general resistance, or field resistance, or horizontal resistance (discussed in Section 18.5) to non-specific pathogenic races is generally polygenic, though oligogenic inheritance also occurs. This type of resistance may be due to a range of general morphological and physiological characters working together in reducing possibilities for establishment of infection such as thickness of cuticle and epidermis, epicuticular wax coating, size and number of stomata, hairiness of leaves or fruits, closed or open flowers, chemical defences such as in onion against smudge, early maturation determining escape from late season diseases, etc.

Polygenic resistance is more subject to influence by environmental factors such as modifications of host nutrition than is monogenic resistance, which usually is completely stable but which may fail by the appearance of a new pathogenic race.

The importance of environment to the final effect of resistance is relevant to the testing techniques to be applied in breeding programmes. Standard conditions and optimum conditions for development of disease must be provided. In pea wilt monogenic resistance breaks down at high soil temperatures (Schroeder and Walker, 1942; Wells, Hare and Walker, 1949) and in acid soil (Hubbeling and Mathur, unpublished). Resistance to Fusarium wilt in cultivars of cotton, tomato and pea breaks down in the presence of parasitic nematodes (see Section 16.7) and resistance to dry rot, *Phoma lingam*, by cauliflower mosaic virus in cultivars of swede (Smith, 1960; see Section 16.14). In addition to field tests, breeding materials should be tested in plant houses where temperature, light and humidity can be controlled. In any breeding programme standard testing conditions must be provided which can be maintained uniform from season to season. An essential aspect is establishment of 'plant disease gardens' where attempts are made to accumulate presently and potentially important pathogens under favourable conditions for abundant disease development. But here it must be emphasised that such 'nurseries of all evils' must be carefully established and controlled. Sometimes the accusation that the most important distributors of new plant pathogens are plant pathologists appears to be deplorably justified!

18.5 EPIDEMIOLOGICAL IMPLICATIONS OF SUSCEPTIBILITY AND RESISTANCE

Any resistance in crop plants carries implications for the population dynamics of disease development in crops. The character of resistance has direct bearing on the epidemics of the disease concerned.

In the previous sections mechanistic and genetic implications of resistance have been discussed. However, the mechanisms and genetics of resistance are not necessarily directly related to their epidemiological consequences.

There are two types of resistance in plants which may well be placed in the two epidemiological categories which Van Der Plank (1963) has called 'vertical' and 'horizontal'; other authors have used the terms 'specific' (race specific) resistance, and 'general' or 'generalised' (race non-specific) resistance (Stakman and Christensen, 1960). General resistance is almost synonymous with 'field resistance'. Useful discussions on the terminology of resistance have been given by Robinson (1969, 1971) and the Federation of British Plant Pathologists (1973).

Vertical (race specific) resistance operates against specific pathogenic races or pathotypes (i.e. 'vertical pathotypes') but not against others. It may be expressed as nearly complete resistance, immunity, as known in resistance of recent potato cultivars against certain races of potato blight, *Phytophthora infestans*. The response of host to pathogen is often a 'yes or no' response — the effect is qualitative.

The epidemiologic effect of this type of resistance is delay of the onset of the epidemic because of reduction of initial inoculum. *The disease gets a poorer start.*

Horizontal (race non-specific) resistance operates against all pathogenic races of a pathogen. It may be explained as that resistance which is left when a cultivar is attacked by all existing races. The response is a 'more or less' response — the effect is quantitative.

Characteristic of this kind of resistance is a reduction in the rate of increase which by definition includes the process of infection, incubation and reproduction. *The development of the disease is slowed down.*

The terms 'vertical' and 'horizontal' refer to the effect of the two types of resistance in a graphic presentation with a vertical and a horizontal axis (Van Der Plank, 1968; see Fig. 18.1). The advantages of these terms of mathematical origin are that they are free from ambiguity, that they are interrelated, and that their definitions have a mathematical precision. The terms have the virtue of recalling the diagrams vividly! In genetics the two terms are generally covered by the long-used standard terms 'qualitative' and 'quantitative' inheritance.

As mentioned, horizontal resistance is effective against all races and is independent of new races that may appear. But, as shown in Fig. 18.1, this kind of resistance may vary substantially from one cultivar to another, as well as from one host species or genus to another within a wide host range.

Obviously, horizontal resistance is by nature stable while vertical resistance is unstable. On the other hand, usually a higher degree of resistance is obtained by vertical than by horizontal resistance.

There is a certain risk of retrogression in breeding for resistance if in attempts towards vertical resistance the role of horizontal resistance is neglected. By improving vertical resistance one may obtain reduced horizontal resistance. If a cultivar with high vertical resistance but with a low horizontal resistance is crossed with a cultivar in order to improve the vertical resistance of this cultivar, the resulting hybrid may, in addition to better vertical resistance, receive the lower horizontal resistance of the parent cultivar. If the race population of the new cultivar is changed with a breakdown of the vertical resistance, then this cultivar is in a worse position than its predecessor. Loss of horizontal resistance due to negligence in breeding for vertical resistance has been termed the 'Vertifolia effect' because the potato cultivar Vertifolia (in which the vertical resistance is monogenic) provides a conspicuous example of this effect (Van Der Plank, 1968).

Breeding for resistance must take both types into consideration, as is observed in the current breeding of, for instance, rice for resistance against blast, *Pyricularia oryzae*, although little is known about horizontal resistance to blast (Ou, 1972).

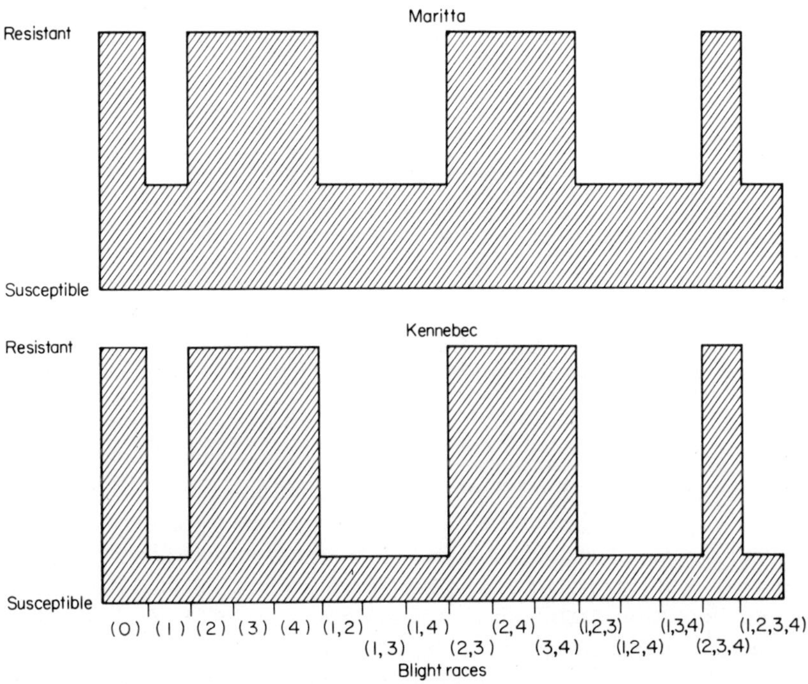

Fig. 18.1. Diagram of the resistance to potato blight, *Phytophthora infestans,* of the foliage of two potato cultivars, Kennebec and Maritta. The resistance is shown shaded to 16 races of the pathogen. Against certain races, e.g. (0), (2), (3) and (4), both cultivars show vertical and complete resistance. Against other races, e.g. (1), (1,2), (1,3) and (1,4), resistance is horizontal. This is small in Kennebec and moderate in Maritta. (Van Der Plank, 1968)

Crop epidemics depend, of course, on the susceptibility of cultivars in the crop concerned. Fields crops commonly involve large populations of plants which are genetically similar; many are clones, and they are usually of the same age and arranged uniformly in fixed positions. If all plants are equally susceptible and exposed to infection, and the relevant factors of weather are known, prognostication of an epidemic can be accomplished. When resistant cultivars are grown, then modifications in the course of the epidemic are introduced.

The rate of increase in any plant disease depends on the susceptibility of the host and the environmental conditions such as temperature and humidity. In the initial phase, when conditions begin to be adequate for infection, the development is usually slow; in the final phase naturally it slows down with the steady decrease of available uninfected host tissue; in the intervening period the rate of increase is highest. Thus, the typical course of an epidemic describes an S-formed curve, as shown in Fig. 18.2 (Van Der Plank, 1968).

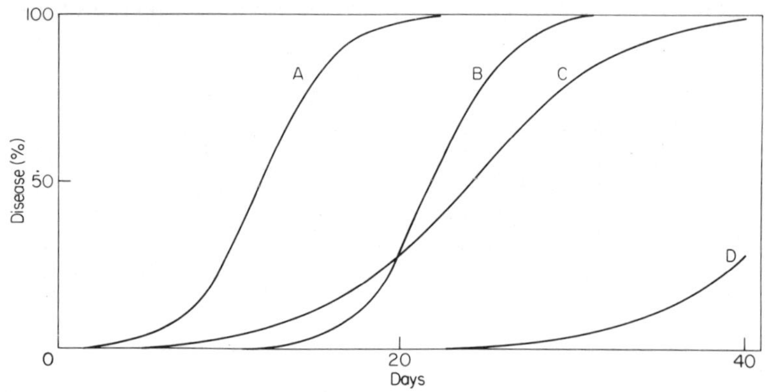

Fig. 18.2. The effect of vertical and horizontal resistance, separate and combined, on development of an epidemic. A, host with no vertical resistance and little horizontal resistance; B, host with vertical resistance and little horizontal resistance; C, host with no vertical resistance but considerable horizontal resistance; D, host with both vertical resistance and considerable horizontal resistance. (Van Der Plank, 1968)

The epidemiological implications of susceptibility and the two types of resistance may be explained by this figure. Curve A represents a host with no vertical resistance and very little horizontal resistance, B a host with vertical resistance and with little horizontal resistance as in A, resulting in delayed infection depending on building up of inoculum of such pathogenic races against which the vertical resistance of the B-host does not work; C is a host with no vertical resistance, as in A, but with a considerable horizontal resistance slowing down development of infection of any pathogenic race; finally, in D, vertical resistance is combined with considerable horizontal resistance, providing both a poor start and a general slowing down of the epidemic.

Two examples of hazards due to crop homogeneity favouring epidemics are Victoria blight susceptible oats which in 1945 covered 80 per cent of the oat acreage in the U.S.A. (98 per cent in Iowa!) and suffered such heavy losses from *Drechslera avenae* that resistant cultivars had to be introduced (Simons and Murphy, 1961); and Texas male sterile cultivars of maize grown on four million acres in the U.S.A. in 1970, when a race of *D. maydis* caused a destructive southern blight epidemic on these cultivars (Moore, 1970) which then had to be replaced by 'normal cytoplasm' cultivars. To counteract such consequences of lack of heterogeneity in self-pollinated crops plants, development of multiline cultivars, consisting of a blend of compatible lines of different genotypes, is a possibility (Jensen, 1952; Borlaug, 1953, 1959).

Van Der Plank (1963) emphasises the importance of vertical resistance as a form of sanitation comparable with other means of sanitation, such as eradicant chemicals, and seed and soil disinfections. These procedures reduce inoculum, thus slowing down initiation of infection. Vertical resistance does not reduce the rate of increase. When races are not controlled by resistance, their reproduction is normal.

Horizontal resistance, on the other hand, reduces the rate of increase — the epidemic progresses more slowly than in susceptible cultivars. The effect of horizontal resistance is comparable with the effect of protective fungicides. A fungicide layer on a leaf surface largely prevents germination of spores and establishment of infection, thus providing reduction of infection rate.

PART III
Principles of Control

Introduction to Part III

By taking precautions against infectious diseases we interfere in the life-cycle of pathogens. Naturally, control should be directed against those phases of their cycle that are most sensitive. A rational combination, weighing the efficiency of one measure against others and utilising their mutual supplementary effects, forms the strategy of an integrated control programme.

First, some important lines of action related to crop chronology are considered in the synopsis below, on the biological functions of control measures.

Biological functions of control measures

1 *Avoidance and elimination of inoculum*
- 1.1 *Quarantine precautions for imported seed*
- 1.2 *Eliminating sources of inoculum in home-grown seed*
  - Use of controlled, health-tested, possibly treated seed
  - Utilisation of vertical resistance
  - Elimination of collateral hosts

2 *Reduction of established inoculum*
- 2.1 *In soil:*
  - Starving-out or weakening of inoculum by crop rotation and fallow
  - Physical and chemical soil treatment (burning, steaming, chemical treatment)
  - Irrigation
  - Deep ploughing
- 2.2 *In collateral hosts and plant residues:*
  - Weed control
  - Adjustment of soil cultivation
- 2.3 *In seed:*
  - Seed processing (cleaning)
  - Seed storage
  - Selection of seed lots (seed health testing)
  - Seed treatment

3 *Slowing down development and spread of inoculum*
- 3.1 *Utilising climatic differences*
  - Geographic location of seed production (appropriately distributed precipitation)
- 3.2 *Influencing microclimate*
  - Planting site (favouring low degree of humidity by ventilation)
  - Plant type (e.g. against too leafy types favouring humidity of microclimate)
  - Planting date (e.g. considering conditions for spore germination)
  - Distance between plants } (favouring ventilation)
  - Staking of plants }
  - Avoidance of overhead watering

3.3 *Impeding spread of inoculum*
Utilisation of horizontal resistance
Adjustment of distance between crops of same host range
Planting date (e.g. considering vector population)
Control of vectors
Critical use of transplanting
Hygienic precautions in handling crops
Short term cultivation of perennial crops

4 *Improving conditions for plant development* (reducing susceptibility)
4.1 *Proper growth place*
Suitable production area and planting site
4.2 *Proper growth season*
Planting date
4.3 *Proper soil and plant management*
Soil reaction
Fertilisers
Soil cultivation
Preparation of seed bed
Planting distance

Next, reviewing the possibilities of control, seed-borne pathogens are tentatively categorised in the following *six epidemiological types*:

(1) *Pathogens depending exclusively or almost exclusively on seed transmission*. Examples are most smuts and ergot fungi of cereals and grasses, *Botrytis anthophila*, anther mould of clover, *Drechslera graminea*, leaf stripe of barley, and *Anguina tritici*, ear cockle of wheat.

(2) *Pathogens being soil or plant residue invaders against which crop rotation is one of the necessary measures*. Different species of *Fusarium*, *Drechslera*, *Alternaria* and *Septoria* which produce a range of diseases in many crops, including seedling rot, seedling blight, root and foot rot, stem rot, leaf spot and leaf blight diseases, head blight in cereals, fruit and seed rot of many crops. Other hosts (weeds, volunteer plants) are often important and largely neglected reservoirs of inoculum. Some soil inhabitants (e.g. *Rhizoctonia solani*) may, on account of pathogenic specialisation, rather be referred to this category of pathogens.

(3) *Pathogens for which vectors play a dominant role*. Many vector-borne viruses including nematode-borne viruses (see Chapter 14, Introduction, and Section 16.8), the viruses of lettuce mosaic and bean common mosaic, and *Erwinia stewartii*, bacterial wilt of maize, which is very dependent on flea beetles (Section 16.1).

(4) *Pathogens in which the perfect stage is important in providing air-borne inoculum very efficient in dissemination*. Examples are *Sclerotinia sclerotiorum* in many crops, *Mycosphaerella pinodes* in pea, *Gloeotinia temulenta* in grasses.

(5) *Pathogens highly dependent on weather conditions for rapid epidemiological spread, and often transmitted from trace amounts of seed-borne inoculum*. Downy mildew, rusts, and certain bacteria, such as those producing diseases in bean, cotton and tomato (cf. Section 32.5).

(6) *Pathogens often or exclusively present in the host as systemic infection often being symptomless*. Important representatives of this group are *Fusarium oxysporum* in a wide range of hosts and differentiated in many *formae speciales* and pathogenic races, *F. moniliforme* in many hosts, including major cereal crops such as rice and maize, *Verticillium albo-atrum*, wilt, and many fungi which are more recognised as producing local lesions than being systemic, such as *Colletotrichum dematium* in soybean, and the bacterium *Xanthomonas campestris* in cabbage; also viruses including those of barley stripe mosaic in barley and wheat, and avocado pear sunblotch in avocado. For all of these pathogens control may be complicated by the frequent presence of symptomless carriers.

Finally, we may briefly conclude that an integrated control programme must be directed against the sources of inoculum and rationally utilise the influence of environment on pathogen and host, and that in selecting measures the type of pathogen as outlined above must be taken into account. In attempting a simple formulation the following *seven principles* are suggested as most essential for *an integrated sanitation programme*.

(1) *Exclusion by quarantine*. The efficiency of quarantine depends entirely on the adopted policy and techniques, and the present situation cries out for improvement, particularly for seed-borne pathogens. Especially there is a great need for specification of pathogens and inspection procedures (Chapter 24).

'Regional' domestic quarantine objects, comprising pathogens having a moderate rate of increase and being typical in forming 'pocket infestations' (Category B, Section 24.3), must be considered. Examples of this type are the lucerne race of stem eelworm, *Ditylenchus dipsaci* (British quarantine provisions), and *Sclerotinia sclerotiorum*; the last pathogen is very efficiently disseminated by air, e.g. from a field, where it has been introduced by seed, miles away to other crops of a region (Subsection 1.3.3).

(2) *'Clean' seed*. Certain *diseases chiefly dependent on seed transmission* have been reduced to minor importance by seed certification schemes and consistent seed treatment. In advanced agricultural countries cereal smuts and ergots have largely been checked although not at all eliminated. In less developed countries where such schemes are lacking, these and similar pathogens are often still destructive factors in food production.

Seed transmission of *soil invaders* is often neglected but is important for three reasons: the pathogens are efficiently inoculated in a field because they are equally spread with the seed; when introduced to a field they survive for years, attacking subsequent crops; and the close association of a pathogen with the seed increases possibilities for infection and establishment of the disease in the crop. This category of pathogens should be regarded as important in certification schemes.

Virulent and rapidly spreading pathogens, carried in seed often only in traces, may be difficult to detect by routine procedures; some have been largely checked by shifting seed production to semi-arid regions and by very strict certification schemes involving field inspection and sophisticated testing procedures, such as have been used for bacterial blights of Phaseolus bean.

Systemic infections that often may persist in a symptomless manner are largely neglected in seed: *Fusarium oxysporum*, *F. moniliforme*, *Verticillium alboatrum, Colletotrichum dematium*, e.g. in soybean (Section 14.1), and many others are readily detectable by laboratory tests while in the seed crop they may not produce specific symptoms or they may cause various degrees of stunt, or on the other hand, 'bakanae' symptoms (*F. moniliforme*) where the plants are tall, but which are difficult or downright impossible to detect in the field. Seed treatment is insufficient, while screening of seed lots by laboratory testing is workable and cheap. Systemic bacterioses, e.g. *Xanthomonas campestris* in crucifers, and viruses, present the same difficulties in field inspection and the same possibilities in laboratory testing.

Pathogens largely disseminated by air are particularly dynamic but can be efficiently checked by screening seed lots: *Mycosphaerella pinodes*, *Sclerotinia sclerotiorum*.

Mechanical seed cleaning may materially reduce inoculum. Ear cockle, *Anguina tritici*, was prevalent in England in the 1920s but is now a rarity because of efficient seed processing by modern equipment. In many developing countries, however, where primitive methods of seed processing are used, the pathogen is common and serious. Contamination of stem eelworm in clover, lucerne and onion occurs in seed from different countries in amounts according to high or low standard of cleaning.

(3) *'Clean' soil*. Crop rotation is imperative and practised in advanced agriculture against pathogenic soil invaders but is not considered in developing countries still based on monoculture production. The requirement of 'clean soil' includes control of weeds and volunteer plants (cf. Subsection 19.2.4). Direct control of contamination accumulated in plant residues may be important: burning of stubbles in grass seed fields in Oregon is the most important factor in controlling blind seed disease, *Gloeotinia temulenta*. Flooding is effective against *Sclerotinia sclerotiorum* and this suggests introduction of lowland rice in crop rotation, wherever appropriate. Steaming of soil is applicable for crops of high value, such as greenhouse crops. Burying inoculum by deep ploughing is used in Oregon against apothecium production by *G. temulenta* and in Japan against *Pyricularia oryzae*, the overwintering inoculum thus being largely checked leaving seed as the main source of inoculum.

(4) *Observation of climate and microclimate*. Selection of climatically

suitable regions for seed production and observing factors pertaining to microclimate, all as outlined above (Synopsis 3.2), cf. Section 19.1.

(5) *Chemical control*. Consistent use of seed treatment, particularly by organic mercury compounds, has greatly reduced certain seed-borne diseases such as smuts, *Fusarium* and *Drechslera* diseases in cereals, especially leaf stripe disease of barley, in advanced agricultural countries but the tendency now is to be more critical in the use of seed treatment, especially those which are poisonous. Mercury compounds have been abandoned in some countries or restricted to use on seeds tested and found to require treatment. The situation points to the need for more attention to seed-borne pathogens in certification schemes, and emphasis on seed treatment of the first generations in the multiplication process — Breeder's Seed, Pre-Basic Seed, Basic (Foundation) Seed — thus restricting the use of the more efficient but more dangerous mercury compounds to the smaller quantities of the first generations. By more severe screening of seed lots seed-borne inoculum is reduced, which eventually means greater efficiency of seed treatment and the possibility of reducing the dosage.

The need for chemical control in seed fields depends, of course, on the prevailing climatic conditions, the need being increased under adverse conditions. A control programme involving many chemical applications is expensive but may be justified for the first generations in seed multiplication in attempting complete control. Particularly in an intensive spraying or dusting programme, rational selection of chemicals, based on diversity, is advisable because of the very specific effects of many formulations of newer chemicals; mixed treatment schedules may expand the 'lives' of chemicals. Control of vectors is helped by growing seed crops when vector populations are low.

(6) *Good conditions for plant growth*. Good establishment of the seedlings is crucial for further development of the crop. Many seed-borne pathogens such as covered smuts, *Fusarium*, *Drechslera*, invade the seedlings during the short pre-emergence phase. Soil conditions during this period can be greatly influenced by agricultural practices. Infection from a certain amount of seed-borne inoculum varies tremendously according to planting date, which is often determined by the temperature and humidity of the soil; soil reaction can be adjusted, depth of ploughing, seeding rate, fertilisers used, and breaking soil down to the appropriate tilth, smaller seeds needing a finer seed bed.

Growing of crops under suitable climatic conditions is fundamental but not always observed. The example of *Fusarium graminearum* in the U.S.A. demonstrates how the pathogen is prevalent in maize in the cooler northern States, in wheat in the warmer southern States, while maize in the southern States (maize belt) and wheat in the northern States (wheat belt) escape the disease, cf. Subsection 15.2.2.

(7) *Resistant cultivars*. Breeding for resistance against some diseases of world-wide importance, in particular rusts in cereals, comprising innumerable races with specific pathogenicity, is based on vertical resistance. However, selection for horizontal resistance ('field resistance') has been utilised consciously or unconsciously as long as agriculture has existed, but in recent years horizontal resistance has become an object in itself for systematic breeding (Section 18.5). More extensively than ever there is an international flow of cultivars and germ plasm throughout the world, and cultivars with specific resistance are introduced from regions with climatic and other environmental conditions different from those of the region of introduction. There is a great need for criticism (not only from the quarantine point when importing materials!) as to the adoption of better yielding cultivars of any crop under conditions different from the place of their origin. Developing countries with their special needs for new or improved crop plants are particularly vulnerable in this respect.

19 Seed Crop Management

19.1 LOCATION OF SEED PRODUCTION

19.1.1 Geographic location, macroclimatic conditions

Seed-borne parasitic fungus flora varies according to climatic conditions. Beet seed grown in Denmark and Holland is considerably more heavily infected by *Phoma betae* than beet seed grown under the drier climatic conditions of Czechoslovakia. In Austria beet seed is much less infected by *Cercospora beticola* in the major regions of Niederösterreich than in the more rainy Oberösterreich. Wenzl (1959*b*) found the average number of conidia carried in seed lots from Oberösterreich to be up to seventyfold the number encountered in seed lots from Niederösterreich.

In the U.K. flax seed grown in the humid districts north and west of the Pennines became more infected by *Colletotrichum lini*, *Aureobasidium lini* and other pathogens, than flax seed from districts south and east of the Pennines (Muskett and collaborators, cf. Muskett, 1950, 1953). In New Zealand seed of crucifers from the semi-arid districts of Central Otago is less infected by black leg, *Phoma lingam*, than seed grown in other regions of the country (Blair, 1952). Cabbage seed, grown in Holland, is more heavily infected by *P. lingam* than cabbage seed grown in Denmark (Neergaard, unpublished).

In the U.S.A. and Canada seed growing of many crops has been transferred from the rainy eastern States to semi-arid regions, such as California and British Columbia. Thus, before 1925 the production of Phaseolus bean seed in the U.S.A. was limited to areas in New England, New York and Michigan, and anthracnose, *Colletotrichum lindemuthianum*, and bacterial blight, *Xanthomonas phaseoli*, were widespread and destructive. Now bean seed is grown mainly in the Middle West and western States, particularly Idaho and California, where the climatic conditions are favourable for production of seed free from bacterial pathogens (Mackie, Snyder and Smith, 1945).

In Canada aerial photography has illustrated, rather dramatically, the importance of using healthy seed — as shown in Colour Plate III (Wallen and Philpotts, 1970). The field, designated 1, is a select registered bean plot from disease-free seed imported from Idaho. The plot concerned contained one small infection focus. Fields 2 and 3 are registered bean plots, grown from 1967 Foundation Seed which was blight free. On the other hand, field 4 was grown from seed which had been produced on the farm for a number of years and a high incidence of seed-borne infection of fuscous blight (*Xanthomonas phaseoli* var. *fuscans*) had resulted. The field shows extensive foci of seed-borne infection and an extended blight pattern.

Since bean seed of cultivars used for processing has been produced in irrigated areas west of the Continental Divide of America, halo blight, *Pseudomonas phaseolicola*, occurs only occasionally in Wisconsin but then in epidemic proportions. Walker and Patel (1964*a*) showed experimentally that in an average Wisconsin crop season as few as 12 infected seeds/acre are sufficient to provide a general epidemic.

Pea seed for gardens and canning formerly grown in the north-eastern States is now produced in Idaho, eastern Washington and California where climatic conditions minimise development of *Ascochyta pisi* (Middleton and Snyder, 1947) and bacterial blight, *Pseudomonas pisi*. Likewise, seed of cabbage, cauliflower and other crucifers is grown in Puget Sound and other semi-arid areas along the Pacific Coast to prevent development of *Phoma lingam* and *Xanthomonas campestris* (Baker and Snyder, 1950).

Seed of *Zinnia elegans*, produced in California, is rarely and if so only weakly infected by *Alternaria zinniae*, whereas seed produced in Italy is so often heavily infected that about 40 per cent of the seed lots inspected for export from Denmark are rejected in spite of a very lenient tolerance (Neergaard, 1965).

Seed produced in semi-arid areas is not necessarily completely disease-free. Leach (1960) surveyed fungi on forage legume seed produced in Oregon. In this State, low summer precipitation and low relative humidities create conditions theoretically ideal for production of 'disease-free' seed. Nevertheless, pathogens of considerable importance were shown to be widespread in seed lots, although present in quite low levels of infection. Thus, in 70 per cent of 193 pea seed lots *Mycosphaerella pinodes* was encountered, in 41 per cent of 112 alsike clover and 66 per cent of 214 crimson clover seed lots *Phoma medicaginis* var. *pinodella (Phoma trifolii)* was present, and 63 per cent of 337 seed lots of hairy vetch were infected by *Ascochyta pisi*.

19.1.2 Planting site, mesoclimatic conditions

Also within smaller areas, indeed on individual farms, the selection of planting site is important. In some countries with cold climates, for instance in northern Europe, so-called frost pockets provide a definite risk for many seed crops in favouring establishment of infections. In Scotland, under exposed conditions of this kind, crops of pea are predisposed to *Pseudomonas pisi* (Noble, 1951). In Denmark clover rot, *Sclerotinia trifoliorum*, in clover, black medick, lucerne and other legumes, is markedly more severe in low-lying situations (Stapel and Bovien, 1943).

In Canada aggravation of bean blight, *Xanthomonas phaseoli*, has been observed in lower parts of a field of white beans, where morning fogs and dews were present, and persisted longer than in any other location of the field (Wallen and Philpotts, 1970), cf. Colour Plate V.

In Denmark seed growing of cabbage and other crucifers is placed primarily on certain small islands and in coast areas where winds prevail and produce ample ventilation of the fields, thus reducing humidity and consequently development of *Alternaria brassicicola* which otherwise is serious to Danish cruciferous seed crops. For the same reason wind-exposed upland is often better suited for seed growing than low-lying fields.

19.1.3 Soil conditions

Low, waterlogged, acid soils often favour development of soil- and seed-borne fungi such as *Claviceps purpurea* (Stapel and Bovien, 1943), *Rhizoctonia solani, Sclerotinia trifoliorum* (Stapel and Bovien, 1943), and *Colletotrichum lindemuthianum*. Consequently, good drainage and suitable pH are imperative even if only to provide conditions for good growth of the mother plants. In India black rot, *Xanthomonas campestris,* and stump rot, *Erwinia carotovora*, following severe infection of black rot in seed crops of cauliflower, were much aggravated in waterlogged areas. In five fields surveyed, 65 per cent of the plants were killed under waterlogged conditions, 6 per cent under non-waterlogged (Patel *et al.*, 1970).

Soil type may be important to development of seed-borne infections. Gassner (1925) found heavy attacks of *Tilletia caries* and *T. foetida* in sandy soils amounting to more than 90 per cent infected plants, while in loam only weak or no attacks were encountered. However, Leukel (1937) reports that sandy soil produces less infections of bunt than loam soil. These conflicting observations indicate that in assessing the influence of soil physical conditions must be related to other edaphic factors.

In Wisconsin, U.S.A., Walker and Snyder (1942) have demonstrated that Fusarium wilt of pea, *Fusarium oxysporum* f.sp. *pisi*, becomes established more rapidly in some soils than in others. In their experiments the percentage of wilted plants steadily increased in successive plantings in sandy loam soil but remained quite low in clay soil. The experimental results substantiated extensive field observations.

19.2 CROPPING

19.2.1 Crop rotation

Many seed-borne fungi, bacteria and nematodes are soil inhabitants or soil invaders. Examples of soil inhabitants are *Rhizoctonia solani* and some species of *Fusarium*, including some species parasitising cereals, grasses and legumes. Against these organisms crop rotation is inadequate.

Soil invaders, on the other hand, may survive for only a limited number of years as resting spores, sclerotia, cysts, etc., or subsist as saprophytes, or saprozoons, in plant residues. Some species of *Fusarium* belong to this category of soil-borne pathogens. Other soil invaders are *Phoma lingam*, black leg, and *Xanthomonas campestris*, black rot, in cabbage; *Colletotrichum lindemuthianum*, bean anthracnose, *X. phaseoli*, common bacterial blight, and *Pseudomonas phaseolicola*, halo blight, in bean.

In countries where severe frost prevails during winter, these organisms seldom persist beyond 2–3 years, and crop rotation of three or four years with immune crops between the hosts usually provides sufficient security against building up of inoculum likely to occur when susceptible plants are cropped closely (Walker, 1969).

According to Chupp and Sherf (1960), in the U.S.A. at least a four-year rotation is needed against *Phoma lingam*, whereas a two-year crop rotation is a sufficient safeguard against *Xanthomonas campestris* (Schaad and White, 1974).

The same authors recommend 3–4-year crop rotations for control of *X. carotae* in carrot, three-year rotations for *Phomopsis vexans* in eggplant, and for *Colletotrichum lindemuthianum* in bean; four years or longer against *Septoria lactucae* on lettuce, *S. lycopersici* on tomato, *S. pisi* and the 'Ascochyta complex' in pea. After pea crops infected by *Pseudomonas pisi*, two years or longer rotations are needed.

Where clovers, lucerne, black medick and kidney bean have been attacked by *Sclerotinia trifoliorum*, there must be at least four years with non-susceptible crops between crops of the species concerned (Stapel and Bovien, 1943). Some pathogens require a much longer rotation to be starved out. To check the red clover race of *Ditylenchus dipsaci* a crop rotation of no less than eight years is necessary (Stapel and Bovien, 1943).

Against building up inoculum of Fusarium root rot in cereals a four-year crop rotation is common practice in northern Europe; this also counteracts other seed-borne diseases such as ergot, *Claviceps purpurea*. The same pattern of crop rotation applies with some reservations to control of foot rot *Fusarium* of legumes.

Fusarium wilt in flax, caused by *Fusarium oxysporum* f.sp. *lini*, is also checked by good rotation practice, growing flax only once in five years (FAO, 1961).

Chupp and Sherf (1960) point out that even long rotation does not control Fusarium wilt of pea, *F. oxysporum* f.sp. *pisi* race 1, but that when five or six years intervene pea crops can be grown successfully for a number of years because of the slow increase of inoculum. Also against Fusarium dry rot of bean, *F. solani* f.sp. *phaseoli*, the usual three-year rotation is insufficient, the disease being continually aggravated until finally bean growing in these fields becomes impossible. The disease can be checked where a six-year or longer rotation is practised.

In wheat districts of the U.S.S.R. teliospores of *Tilletia contraversa*, dwarf bunt, retain their viability and infectivity for 4.5 years. Hence, in infested areas sowing of winter wheat should be discontinued for at least five years (Zakharova, 1964). The spores could still germinate after three years at a depth of 30 cm and the disease occurred up to 25 per cent in wheat grown after an interval of three years (Böning and Von Minckwitz, 1966). In eastern Sweden 2–3 years of continued cropping of wheat have led to catastrophic escalation of dwarf bunt and, as a consequence, in some places farmers replaced wheat by rye (Kolk, personal communication, 1970).

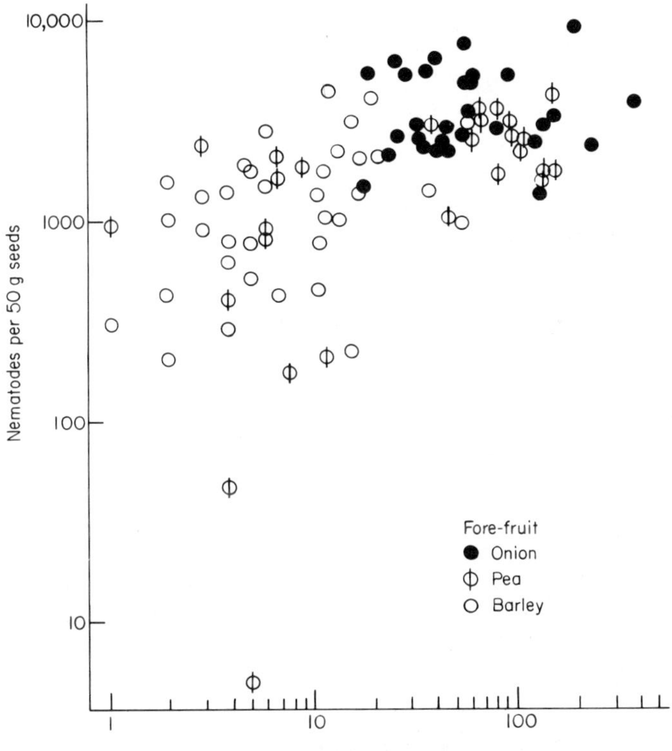

Fig. 19.1. Correlation between numbers of stem eelworm, *Ditylenchus dipsaci*, per 500 g soil before planting and the number of eelworms in the seed harvested from onions in the infested soil. (Seinhorst and Koert, 1971)

Strict rotation of cotton eliminates *Colletotrichum gossypii* and *Ascochyta gossypii*, both prevalent in the U.S.A. (FAO, 1961), and significantly reduces incidence of *Verticillium dahliae* on cotton, increasing yield.

In pea Walker and Hare (1943) found that anthracnose, *Colletotrichum gloeosporioides (C. pisi)*, bacterial blight, *Pseudomonas pisi*, and Mycosphaerella blight, *Mycosphaerella pinodes*, developed in proportion to the number of seasons between pea crops in Wisconsin, U.S.A. The rate of reduction of the two first-mentioned diseases was, however, modified by the factor of annual introduction of seed-borne inoculum, while the incidence of Mycosphaerella blight was modified by the introduction of both seed-borne and air-borne inoculum, the latter as wind-borne ascospores from other fields.

Alternaria may keep alive on dead leaves of the host from one season to another, as demonstrated by Rands (1917) and Van Haltern (1935) for early blight, *Alternaria porri* f.sp. *solani*, in potato and tomato. In Austria trials showed that crop rotation contributes significantly to controlling carrot leaf blight, *A. porri* f.sp. *dauci* (Schmidt, 1965), and in the U.S.A. Chupp and Sherf (1960) indicate three-year or longer rotations as important.

Against stem eelworm, *Ditylenchus dipsaci*, in field crops including red clover and lucerne, rotation with sugarbeet, maize or potato for two years or more is recommended (Walker, 1969). Seinhorst and Koert (1971) correlated the numbers of onion stem nematodes in the soil and in the ensuing seed crop. Onion as previous crop increased, barley decreased the amount of nematodes in the seed (Fig. 19.1).

19.2.2 Fallow

Fallow is likely to reduce inoculum of soil invaders. Intensively cultivated soil in Hong Kong yielded *Fusarium dimerum*, a species frequently encountered in rice seed, whereas fallow soils rarely contained this rice fungus (Chu, 1966). Even *Rhizoctonia solani* considered as a soil inhabitant, although seed-borne races may rather be invaders (cf. Section 1.4), was isolated much more frequently from cultivated soil, 41 per cent, than from fallow, 3 per cent (Chu, 1966). In Australia crown rot of wheat, *F. graminearum*, was less severe in crops following a long fallow (McKnight and Hart, 1966). In India crop rotation of cumin with pearl millet, *Pennisetum typhoides*, increased incidence of Fusarium wilt, *F. oxysporum* f.sp. *cumini*, whereas a summer fallow decreased it (Mathur, Sankhla and Mathur, 1967).

19.2.3 Distance between crops of same host range

Overwintered biennial and perennial seed crops may harbour inoculum that may spread to nearby first-year fields of crops within the same host range.

Perennial leguminous crops should not be grown in close proximity to susceptible annual crops. Some viruses affect many leguminous crops, annual as well as perennial, and may be transmitted from one crop to another by vectors such as *Macrosiphum pisi*; this vector overwinters on clover in northern Europe. Walker and Hare (1943) established the importance of sufficient distance between seed crops and other fields of pea due to the risk of wind-borne dissemination of ascospores of *Mycosphaerella pinodes* from these fields.

Beet rust, *Uromyces betae*, and downy mildew, *Peronospora farinosa*, may readily spread from seed crops to first-year beets (Stapel and Bovien, 1943; Nietzke, 1966).

Particularly important are precautions against spread of inoculum from surrounding fields to areas under production of Certified Seed for which controlled and treated Basic (Foundation) Seed has been used. In Canada the problem of proper isolation of stock seed production areas for barley, free from loose smut, *Ustilago nuda*, has been worked out by legislation (Clark, 1952). Similar legislative precautions have been taken in the German Democratic Republic with respect to stock seed production of wheat and barley from loose smut, *U. tritici* and *U. nuda*, seed crops being rejected if loose smut is encountered in another field within a distance of less than 50 m in the prevailing direction of wind (Kublan, 1952). In Holland Oort (1940*b*) demonstrated that a minimum distance of 100 m from infected fields must be secured for barley grown under certification schemes.

So far as insect-borne virus diseases are concerned it is not possible to state a minimum distance for isolation. In experiments performed in the U.K., new lettuce plots that were sown adjacent to old plots had more than 60 per cent plants infected with lettuce mosaic, whereas crops separated by half a mile of non-susceptible crops had only about 3 per cent at harvest (Broadbent, 1958).

Physical barriers may contribute to isolation of seed crops. In the seed certification scheme of Phaseolus bean in Victoria, Australia, maize or other tall crop barriers, 30 m wide, are used to subdivide bean seed crops over 2 ha in any area submitted to inspection for certification (Bartha, 1963).

Gay, Johnson and Chalfant (1973) showed that a barrier of two rows of pearl millet, *Pennisetum glaucum*, treated with a systemic insecticide disulfoton effective against vectors, significantly reduced infection by cowpea mosaic virus in cowpea plants within the trap-crop compared with non-screened control. A yield increase of about 20 per cent was obtained.

Cohen and Marco (1973) were able substantially to reduce the spread of aphid-transmitted viruses, cucumber mosaic virus and potato virus Y, by trapping the aphids on sticky yellow polyethylene sheets.

19.2.4 Elimination of other host plants

Wild perennial hosts may become important loci of overwintering inoculum where the pathogen does not survive in the soil or on debris of annual crops, and this fact may affect seed production.

From California Houston and Knowles (1949) provided a striking example that wild flora may play a role in maintaining important crop diseases. In two fields which had not been under cultivation for more than 40 years and where no flax crop had been grown for 50 years, extended severe outbreaks of Fusarium wilt. *Fusarium oxysporum* f.sp. *lini,* were encountered in flax crops. Circumstances clearly indicated that the pathogen was not introduced into these fields on the seed. The abundance of inoculum present in absence of flax culture was found due to the presence of 'wild' flax (either *Linum angustifolium* or escape from the domestic flax) growing in the region. Moreover, it was observed that three other diseases, pasmo, *Mycosphaerella linorum*, browning, *Aureobasidium lini*, and anthracnose, *Colletotrichum lini*, occurred naturally in the uncultivated flax.

In 1968–69 Septoria leaf spot of lettuce, caused by *Septoria lactucae*, became a serious concern to seed production of lettuce in New South Wales, Australia, because of an unusually rainy season. The seed, produced especially for freedom from lettuce mosaic, was unsaleable to the U.S.A. because of the infection of the fungus. In the following season the growth conditions were more favourable to the crops which showed no symptoms. However, now the disease developed on wild or prickly lettuce (*Lactuca scariola*) in the growing areas, a new source of inoculum thus having been established — M. V. O'Reilly (Yates & Co.), personal communication, 1969.

Ustilago bromivora, brome grass smut, which is serious to seed production of corn brome grass (*Bromus arvensis*), also attacks soft brome grass (*B. mollis*), a common wild grass along roads, etc., in northern Europe. Stapel and Bovien (1943) recommend cutting the grass in due time in such places. This measure also contributes to control of ergot, *Claviceps purpurea,* and choke, *Epichloë typhina,* both parasitising many wild grasses. In Canada ergot has been successfully controlled by application of a single spray of maleic hydrazide on grasses near cereal crops inhibiting heading and so eliminating the chief source of inoculum (Campbell and Freisen, 1959).

In New Zealand, shepherd's purse (*Capsella bursa-pastoris*) has been found to be a weed host for *Xanthomonas campestris*, black rot of crucifers. The disease occurred in a crop of chou-moellier (*Brassica oleracea* var. *acephala*), which provided a continuous source of inoculum for the weed host (Young, 1969). Many members of the Cruciferae are hosts to *X. campestris,* hence cruciferous weeds should be regarded as potential carriers of the pathogen.

Many weeds are hosts of soil-borne and seed-borne viruses of great economic importance. Shepherd's purse (*Capsella bursa-pastoris*) and chickweed (*Stellaria media*) are both attacked by tomato black ring virus, raspberry ringspot virus, and arabis mosaic virus, all of which are seed-borne in sugarbeet, soybean, and many other crop plants. Dandelion (*Taraxacum officinale*), chickweed (*S. media*), and shepherd's purse (*C. bursa-pastoris*) are frequent carriers of tobacco rattle virus which has more than a hundred species of host plants, and is seed-borne both in the weeds mentioned and in soybean, tobacco and petunia.

Mock cucumber or wild cucumber (*Echinocystis lobata*) serves as a reservoir of viruses of the cucumber mosaic group in Czechoslovakia (Valenta, 1960), and certainly in many other countries. Cucumber mosaic virus, cause of a yellowing disease of outdoor lettuce in Britain, often occurs in several annual weeds common in lettuce fields. These include, according to Tomlinson *et al.* (1970) and Tomlinson and Carter (1970*a*, 1970*b*), chickweed (*Stellaria media*), dead nettle (*Lamium purpureum*), mallow (*Malva sylvestris*), *Cerastium holosteoides* and corn spurrey (*Spergula arvensis*) in all of which the virus is seed-transmitted, and often symptomless; also in groundsel (*Senecio vulgaris*), annual nettle (*Urtica urens*), shepherd's purse (*Capsella bursa-pastoris*), and mayweeds (*Matricaria* spp.). Lettuce mosaic virus can be carried symptomless in *Sonchus asper*, in groundsel (*S. vulgaris*) and in *Chenopodium quinoa*, in all of which it is also seed-borne, and from which it may be transmitted by aphids to lettuce (Ullrich, 1954; Kemper, 1962; Phatak, personal communication, 1972).

Seeds of these weeds may be abundantly present in soil. In Britain *Stellaria media* has been found in numbers of $0.1–97.0\times10^6$ seeds per acre (Roberts and Stokes, 1966). In tobacco fields in Switzerland 1–30 per cent of chickweed plants harboured cucumber mosaic virus (Häni, Pelet and Kern, 1970; Häni, 1971). With seed transmission rates of up to 40 per cent from infected plants as found in Britain (Tomlinson and Carter, 1970*a*) there are considerable chances for infection of susceptible crops, as for example tobacco and lettuce, from this source.

Effective control of a virus disease, though not a seed-borne one, by weed control has been demonstrated by Stubbs, Guy and Stubbs (1963) in Australia. In a vegetable farm sowthistle (*Sonchus oleraceus*), the only known source of necrotic yellow virus and its specific vector, *Hyperomyzus lactucae*, was controlled in an area 430×155 m surrounding a lettuce field. At maturity, 6 per cent of lettuce plants in this field was lost through the virus infection while in a field of a neighbouring farm, where sowthistles were uncontrolled, 76 per cent of the crop was lost through the disease. In this crop the pattern of disease spread was typically marginal from thistles located along one boundary.

Weed control, such as by extensive use of weed killers, may reduce sources of inoculum considerably, even if total eradication of other hosts is too expensive.

19.3 SELECTION OF CULTIVARS

It is obvious that the seed grower is interested in growing seed on plants resistant to any disease and pest which may reduce the output of seed crops. However, distinction may be made between (1) seed-borne pathogens that directly reduce the quality and quantity of seed yield but are of little or no importance to subsequent crops, and (2) those seed-transmitted pathogens which do not exert any material effect on the yields of seed, but may be harmful to subsequent crops — although the two groups of diseases overlap. A disease in the latter category may attack the subsequent crop under climatic conditions different from those of the seed-producing areas.

The seed grower is primarily interested in cultivars resistant to pathogens directly affecting the yield of seed. The consumer, on the other hand, is less concerned with the yield of the seed crop, but is interested in high-yielding crops for consumption. Taking the long view, however, the interests of both groups of growers coincide. If good health control is needed in the interest of crop production for consumption, the economic consequence to the seed grower may be that infected seed is sold at a lower price, or rejected. In such cases productive cultivars susceptible to seed crop diseases carry an extra risk and lead to increase of expense due to control measures. Consequently, the seed grower may rightly require stock seed of resistant cultivars if not otherwise rewarded for the risk.

Gloeotinia temulenta, the blind seed organism of ryegrass, exemplifies seed-borne pathogens involving losses particularly heavy to the seed grower, and thus creating urgent demands for resistant cultivars for seed production. Noble (1951) discussed the consequences entailed by the introduction of some improved, more productive, leafy cultivars or strains of English ryegrass in New Zealand and the U.K. These strains produce a much more uniform stand than the old strains as well as a shorter, more simultaneous flowering. Thus adverse weather conditions in the critical flowering period, when the parasite is active and the host susceptible, became considerably more harmful to the seed-bearing crop, because the possibilities of escape were less. Cultivar S24 was susceptible but has been reselected for resistance, in work carried out in Ulster (Wright and Sproule, 1965).

Breeding for resistance of seed is an important possibility. The rate of seed transmission may vary according to cultivar. Most cultivars of bean transmit bean common mosaic virus at high rates, up to more than 80 per cent, but some at lower, e.g. Beka with a maximum transmission rate of only 15 per cent (Schippers, 1963), cf. Section 13.1; this cultivar may be utilised in breeding for resistance of seed, as may three cultivars, Red Mexican × Pinto, Red Mexican & Small White and Ida Refugee × Pinto which do not transmit the virus through seed (Medina and Grogan, 1961). The rate of seed transmission in lettuce is up to 16 per cent but seeds of cv. Cheshunt Early Giant show immunity (hypersensitivity) thus offering potentials for breeding against seed transmission.

Cultivars of wheat and barley resistant to loose smut provide attractive material for breeding. There are different types of resistance: embryo resistance; embryo susceptibility linked with adult plant resistance (barley cv. Keystone), both types preventing seed transmission; and 'resistance' based on closed flowering providing escape of infection.

Two pathogens, *Alternaria brassicicola* and *Phoma lingam*, important in seed crops of cabbage and cauliflower, depend on humid conditions in the stages of flowering and seed setting. In Denmark early maturing cultivars, harvested under fairly dry conditions at midsummer, largely escape attacks, while late cultivars often are severely attacked because of prevailing humidity during harvest, cf. Section 11.3. In the last decades breeding has resulted in increasing predominance of early maturing cultivars of the different types of cabbage, thus improving control in seed crops.

Another important aspect is resistance in relation to geographical origin. Cultivars and strains originating from different countries may show great differences in resistance. Danish strains of red clover are much more resistant to *Sclerotinia trifoliorum*, clover rot, than particularly early South European strains as well as French and British strains, although otherwise there are considerable differences in resistance between Danish strains (Stapel and Bovien, 1943; Frandsen, 1946). Ylimäki (1969) found marked differences of resistance in clover cultivars grown in different localities of Finland; the resistance was based on the ability of the host to delay the domination of the disease. Red clover of Italian origin has been found to be more susceptible to *Kabatiella caulivora*, scorch or northern anthracnose, in Switzerland than in Italy (Fischer and Gäumann, 1929), red clover of American origin to be more susceptible than red clover of European origin, and south European forms to be more susceptible than north European ones (Stapel and Bovien, 1943).

The problem associated with the origin of the seed is further complicated by the occurrence of pathogenic races, such as those of *Ustilago* spp., *Colletotrichum lindemuthianum* and *Sclerotinia trifoliorum*. The local composition of races may determine the extent to which different cultivars are attacked. Consequently, the origin of stock seed is of utmost importance to the seed grower. This is true also because there is an ever-existing risk of introducing new 'foreign' pathogenic races with the seed.

Resistant cultivars have definite limitations related to pathogenic specialisation, as in the smuts. Race populations frequently change, overcoming host resistance of a cultivar, necessitating a continuous programme of breeding new resistant cultivars.

Resistance through escape may be related to transportation of seed from one climate to another. Different growing places may offer different local possibilities of coincidence of pathogen and susceptible stage of host. Some early cultivars of cowpea escape attacks of wilt and root knot in the U.S.A. because they mature before the season of these pathogens (Stevens, 1960).

19.4 REDUCTION OR ELIMINATION OF SEED-BORNE INOCULUM

19.4.1 Seed certification schemes

19.4.1.1 PRINCIPLES

Seed quality control systems are fundamental in seed production. Properly carried through they provide a reasonable guarantee as to genetic purity of the seed, its purity from seeds of other cultivated plants and weeds and of inert matter, and its viability and freedom from inoculum of pathogens that may reduce the output of the crop. Briefly, they serve to maintain reasonable standards.

Seed certification is a legally sanctioned system for quality control of multiplication and production of seed, involving a series of field and laboratory inspections and evaluations. The entire process of seed production is followed through a definite number of generations, from the small quantities of nuclear stocks through the increasing quantities of the subsequent generations ultimately providing seed for crop production. The classes of Certified Seed are usually designated as in the certification schemes referred to in Table 19.1.

Table 19.1. Classes of Certified Seed

U.S.A. (Association of Official Seed Certifying Agencies)	*Europe* (Organisation for Economic Co-operation and Development and European Common Market)
Breeder Seed	Pre-Basic Seed
Foundation Seed	Basic Seed
Registered Seed	Certified Seed, 1. generation
Certified Seed	Certified Seed, 2. generation

The number of generations from parental material before Basic Seed is strictly limited according to the OECD schemes, and the number of generations of Certified Seed after Basic Seed may vary from one crop plant to others according to technical conditions.

The backbone of certification schemes are the minimum certification standards valid for each class. These standards must necessarily be strict to the maximum degree. In starting a scheme, practical reasons may suggest a setting of somewhat lower standards but these will have to be raised as experience is gained and better equipment becomes available.

Disease tolerances are fixed at the lowest level in Pre-Basic Seed (Foundation Seed) and usually have to be increased in Certified Seed. In principle, tolerances should be zero at least in the important first generations of seed multiplications in which inspections and direct control measures such as seed treatment can be accomplished most thoroughly and most effectively.

In certification schemes two types of seed-borne diseases may be considered — those that are exclusively seed-borne and those that also are soil- or plant residue-borne.

Diseases exclusively seed-borne can be combated, indeed eliminated, by the use of healthy stock seed. Stock seed free from inoculum can be obtained either by elimination of diseased seed crops from seed production through appropriate certification schemes, or by effective seed treatment. The appropriate precautions are largely dependent on the rate of increase of the seed-borne inoculum concerned. Whenever a trace of seed-borne inoculum may give rise to severe epidemics, accepting disease tolerances above zero or relying on the effect of seed treatment are bound to give disappointing results. Bean bacterial blights and bean anthracnose provide examples of seed-borne diseases that spread explosively under favourable climatic conditions from slight traces of inoculum — they must be controlled by appropriate seed certification programmes. On the other hand, seed-borne diseases allowing a slow build up of inoculum requiring perhaps several years may very well be controlled satisfactorily by appropriate seed treatment. Although this treatment may leave slight traces of inoculum, seed treatment year after year may keep down the seed-borne inoculum to a negligible level. This holds true as far as many cereal smuts are concerned, for instance loose smuts of barley and wheat, *Ustilago nuda* and *U. tritici*, and barley leaf stripe disease, *Drechslera graminea*.

Seed-transmitted diseases which are also soil-borne or perpetuated on plant residues, are affected by any reduction of the seed-borne inoculum in proportion to its relative importance in the cycle of the disease. Seed-borne fungi such as *Fusarium nivale*, *F. culmorum*, *F. graminearum*, *F. avenaceum,* and *Drechslera* spp. on cereals and grasses, *Septoria nodorum* in wheat, *Cercospora beticola* in beet, etc., is usually established earlier from seed-borne inoculum than from soil-borne. This initiation of epidemics may be responsible for additional losses in field emergence and for increased severity of later attacks. Precautions directed against seed-borne inoculum of such diseases contribute to slowing down disease development in the parent crop as well as in subsequent crops.

19.4.1.2 FIELD INSPECTION OF SEED CROPS

In Section 31.3 field inspection of seed crops is discussed as a testing procedure. However, ultimately the purpose of field inspection is effective control. Inspection of seed crops should be carried out by well-trained people. In any doubtful case field observations must be followed by laboratory examinations of affected plants.

A minimum of two inspections is needed. This sequence enables, whenever justifiable, roguing of infected plants in the intervening period, initiation of a spray programme and other precautions. The last inspection must be carried out after flowering, and late enough to permit examination of the health condition of the fruits. Examination of 100 pods, siliquas, capsules, etc., in the field equals, quantitatively, examination of several hundred seeds in the laboratory.

If, in previous inspections, attacks of seed-borne fungi and bacteria have been seen, this does not necessarily mean that the seed becomes infected because weather conditions during flowering and maturation of the seed may not be favourable for establishment of infection. On the other hand, it is important to be informed from field inspections that pathogens such as *Colletotrichum lindemuthianum*, bean anthracnose, and *Ascochyta pisi*, leaf and pod spot, have been present in the crop, even if the seed does not become infected, as this may be substantiated later by a laboratory test, because contamination of the seed even occurs during the process of threshing. Based on this kind of information from field inspections, it may be justified to advise seed treatment (Leach and Snyder, 1944), even in spite of a negative issue of a laboratory test. Against such superficial contamination seed treatment is likely to be effective, whereas usually it is an unreliable measure against more well-established, so-called 'deeper' infections of the seed.

Roguing of seed fields against development of seed-borne diseases may have a very limited effect. In California Zink, Grogan and Bardin (1957) investigated the comparative effect of using mosaic-free seed and roguing as control measures for lettuce mosaic. A comparison was made of the effects of roguing commercial lettuce crops by (1) cutting the plant off with a hoe and leaving it in place, (2) oil roguing by a destructive mixture of diesel oil, parathion and a contact herbicide, and (3) roguing the affected plant and one plant on each side in the row by methods (1) and (2). This was done up to three times. Roguing once had practically no effect on the total percentage infection at harvest, while roguing twice reduced infection significantly. The greatest reduction resulted from three roguings with oil, application (2), from 17 per cent for the unrogued to 6 per cent for the rogued. Three hoeings were almost as effective. However, the use of mosaic-free seed was more effective than any form of roguing: one trial grown from mosaic-free seed had 11 per cent infection; commercial seed not rogued 84 per cent; rogued 78–84 per cent.

19.4.2 Seed treatment in seed production

Seed treatment is often taken as a precaution in seed production. In California where the stem canker of nasturtium, *Acroconidiella tropaeoli* (syn. *Heterosporium tropaeoli*), is a disease of importance to seed production of *Tropaeolum majus*, hot water treatment is done every second or third year to counteract build up of inoculum, the rate of increase from generation to generation being approximately as follows (Baker and Davis, 1950*b*): in the first year traces of infection, in the second year about 5 per cent, and in the third year about 95 per cent infection (cf. discussion on rate of re-establishment of seed-borne inoculum, Section 32.2).

In seed certification schemes the main purpose of seed treatment should be to control unnoticed infections or contaminations, particularly in Pre-Basic and Basic (Foundation) Seed, to keep inoculum effectively down to zero. Seed treatment should primarily be used to provide additional security that disease-free seed has been obtained rather than as an attempt to 'repair' more obvious failures in the screening of seed crops for infections. However, the small quantities of the first generations in seed multiplication justify the use of more effective, although more poisonous, materials, such as mercury compounds, for seed treatment, cf. Subsection 22.10.6. Scrupulous use of seed treatment in the early generations largely prevents need of more expensive, more hazardous and less effective operations for the bigger quantities of seed in subsequent generations.

19.5 REDUCTION OF SOIL-BORNE INOCULUM

19.5.1 Burning, steaming or chemical treatment

In northern Europe it has become common practice to burn straw and stubble where cereals are harvested by combine, partly as a precaution against foot rot fungi such as *Fusarium* spp., in reducing the soil- and plant residue-borne inoculum. However, there is experimental evidence that the process is too incomplete to be effective against soil- and stubble-borne pathogens (F. Joan Moore, personal communication, 1974).

In grass seed crops in Oregon, U.S.A., Hardison (1946, 1948*b*, 1953, 1957*a*) and Kreitlow *et al.* (1961) established control of grass seed nematode, *Anguina agrostis*, blind seed disease of perennial ryegrass, *Gloeotinia temulenta*, and ergot, *Claviceps purpurea*, by burning straw and stubble in the field. Adoption of post-harvest burning as a general practice from 1949 is regarded as the most significant factor which led to successful control of the blind seed disease in Oregon (Hardison, 1957*b*). In Australia burning of diseased stubble has been recommended as a control measure for flag smut of wheat, *Urocystis agropyri* (Hamblin, 1921). However, open field burning may have to be banned to reduce air pollution, and substitute control methods should be considered.

Soil disinfestation by baking, steaming or any adequate chemical treatment may prevent soil transmission of diseases to seed crops grown continually under glass, where inoculum may survive on plant debris from one year to another. This is especially important in production of valuable Elite or Foundation Seed of tomato, lettuce and cucumber or certain kinds of flower seed such as *Senecio cruentus* grown under glass in Denmark for later extensive seed production in southern countries. Foundation Seed cabbage produced under glass in Denmark has been found to become infected by *Phoma lingam* from inoculum surviving from season to season in the greenhouse (Andersen, personal communication, 1969).

19.5.2 Flooding — Irrigation practices

In certain regions some diseases and pests may be controlled efficiently by flooding the ground. Sclerotia of *Sclerotinia sclerotiorum* in Florida soils can be killed by flooding for a period of 3–6 weeks, the water being held either continuously or flooded and drained at 3-day intervals (Moore, 1949). Accordingly, Stoner and Moore (1953) pointed out that *Sclerotinia sclerotiorum* may be controlled through summer planting of lowland rice. Under these conditions the sclerotia are completely destroyed, and winter vegetables may be profitably grown in such a crop rotation. This has a bearing on seed crops of vegetables in such regions.

Ocfemia (1924) found infection of *Drechslera oryzae* in lowland cultivars of rice to be prevented by submerging the soil in water to a depth of about 10 cm, provided the water is at 24–28°C. At these temperatures lowland cultivars emerged through water 10–20 cm deep.

Cralley (1957) observed that rice sown in water was much less infected by *Aphelenchoides besseyi*, white tip of rice, than when drilled and flooded later, apparently because the nematodes resume activity and are lost in the water before the seed germinates.

Also smuts may be effectively controlled by flooding the ground. In Egypt, El-Helaly (1939, 1948) demonstrated that kernel smut of sorghum, *Sphacelotheca sorghi,* and flag smut, *Urocystis agropyri,* in wheat are counteracted when light irrigation follows seeding immediately (the *afir* method); infections become higher by irrigation prior to seeding (the *herati* method).

Irrigation methods are also of importance in the development of wheat bunt, *Tilletia caries*, covered smut of barley, *Ustilago hordei*, and millet smut, *Sphacelotheca destruens* (Jones and Seif-El-Nasr, 1938, 1940*a*, 1940*b*). Again the *herati* method gave highest infection, whereas seeding in dry soil and then flooding with water (the *afir* method) gave less infection; however, lowest incidence was obtained by flooding the soil and then broadcasting the seed one hour later (the *mud-sowing* method). This experimental evidence led to a general recommendation of shallow seeding in wet soil.

In Israel the method of irrigation is important to development of two major diseases of tomato, Stemphylium blight, *Stemphylium botryosum* f.sp. *lycopersici*, a destructive disease in the coastal plain and known only in Israel, and bacterial blight, *Xanthomonas vesicatoria*. Rotem and Cohn (1966) established that tomato plots irrigated by overhead sprinkling were more severely affected by these diseases, showing disease incidence higher by 30–80 per cent of both combined, than furrow-irrigated plots. Yields of furrow-irrigated plots exceeded those of sprinkled plots by 30 to almost 50 per cent. Furrow-irrigation did not affect the microclimate prevailing in the field whereas sprinkler-irrigation was associated with increased relative humidity and reduced air temperature. The increased severity of the blights was found to be due to prolongation of the moisture periods by overhead sprinkling. In contrast to this relationship, a third tomato disease which is not seed-borne, powdery mildew, *Oidiopsis taurica*, known to be favoured by a dry environment, developed by 30 to 40 per cent more on furrow-irrigated plots.

Grogan and Kimble (1967) observed that secondary spread of halo blight of bean, *Pseudomonas phaseolicola*, developed more in sprinkler-irrigated plots than under furrow-irrigation, and infections were especially abundant close to the sprinkler outlets.

19.6 ADJUSTMENT OF CULTURAL PRACTICES

Perhaps it may be appropriate here to quote one of the few more critical agricultural authors of the eighteenth century, a period when speculations and mysticism prevailed and all kinds of concoctions were used for grain as 'fructifying liquors' to avert the fate of the crop against all sorts of illness. What Bryant (1783) prompted the farmers of his age, is still worth emphasising to the agriculturist of today who sometimes may be a little too pleased about the ever-rescuing 'concoctions' so abundantly offered him by the modern chemical industry. Bryant recommends the grower 'to lay his old, irrational practice aside; to pay particular attention to a proper culture of the soil; to keep it clean and free from weeds; and to supply it with proper manure'. And he concludes: 'These are the best remedies for diseases in corn; these will contribute most to health and vigour, and make his fields, as well as himself, laugh and sing.' (From Buttress and Dennis, 1947.)

19.6.1 Preparation of the seed bed

The soil structure is of importance to the development of vigorous plants. Mead (1942*b*) found that packing of the soil caused a significant reduction in emergence of barley and an increase in the amount of blight and stunting produced by *Drechslera sorokiniana*, only one demonstration of the fundamental agricultural principle that seed should be sown in well-aerated soil.

Fischer and Holton (1957) point out that in the Pacific Northwest fall ploughing of summer fallow is generally successful as a means of controlling wheat bunt, *Tilletia caries*.

Details in preparation of the seed bed may be of importance to control of diseases. Sowing on ridges with furrows on either side has been shown to have beneficial effect in controlling damping-off of sugarbeet because of better soil drainage thus established in the immediate vicinity of the seedlings, and Dobromysloff (1932) found that bunt in the early growth stages of wheat was favoured on the flat as against ridged rows.

In California, U.S.A., where *Verticillium* has been found most prevalent in the upper layer of cotton soils, the 'rigolen' method of deep ploughing has been suggested to counteract infection. The deepest infected root was found at about 40 cm and the suggested method of ploughing involves the careful inversion of surface soil, usually at the depth of about 30 cm, with soil from below and a winter cover crop of rye was found to help accomplish the inversion by holding the surface soil together (Wilhelm *et al.*, 1967).

19.6.2 Timing and techniques of sowing

Sowing date. The selection of the date of sowing depends primarily upon the temperature requirements of the host and parasite. A general consideration outlining the conditions for development of healthy, more resistant as against weakened, more susceptible seedlings may give some understanding of the principles of precautions to be taken.

In temperate zones the lower temperatures of spring and autumn, i.e. a few degrees Centigrade above the freezing point, are normally favourable to germination of herbaceous plants such as the cereals, grasses, clovers, vegetables, as well as annual and biennial ornamentals, and most species of the wild herbaceous flora.

Under low-temperature conditions the seedlings develop a short, low leaf-rosette or a low, compact cluster of shoots, the plant being provided with a well-developed root system. If, on the other hand, the seed germinates in soil which is already too warm, the germination proceeds too quickly, the hypocotyl becomes long and thin, the plant lax, and the leaves become often abnormally small. Higher temperatures intensify transpiration, bring about earlier lignification of the stem, thus impeding its growth. This forced development of the seedling may eventually lead to premature flowering, the flowers becoming small and weak such as encountered in crops of ornamentals such as *Matthiola incana*, *Callistephus chinensis*, *Nemophila insignis*, and *Zinnia elegans* in Denmark. The end result is often incomplete maturation of the seed crop and, as a consequence, poor quality in seed. Plants impaired in this way often become unduly susceptible to diseases.

Field experiments in Manitoba over a period of four years showed that early sowing of spring wheat considerably reduced root rot caused by *Drechslera sorokiniana* and *Fusarium* spp. and increased yields (Greaney, 1946). In Roumania leaf spot, *Drechslera papaveris*, in poppy, was found also to decrease with early sowing in spring (Rădulescu, Persecă and Docea, 1961). Corresponding to early spring sowing, late sowing of winter crops in autumn is preferable. Dickson (1923) recommends as a precaution against headblight or scab, *Gibberella zeae*, in wheat, sowing in cold soil, late in autumn or early in spring, because the soil temperature most favourable to disease development ranges from 12°C to 28°C. In contradistinction to wheat, maize should be sown in relatively warm soil, i.e. late in spring; the temperature favourable to headblight attacks is lower for this host, 8–20°C (Dickson, 1923).

Maude (1966*b*) established that in England carrot seed crops sown in the autumn became heavily infected by black rot, *Alternaria radicina*, whereas spring-sown carrot grown in fertile soil showed little infection either in the succeeding autumn or after storage.

In India (Jabalpur, Madhya Pradesh) five cultivars of soybean showed clear response in seed quality to planting date (Nicholson and Sinclair, 1973). The seed was planted during the three-month monsoon season with two-week intervals from June 1 to September 2. The number of discoloured seeds and seeds infected by fungi decreased, and the germination capacity increased gradually with delay of planting (Figs 19.2 and 19.3). Seeds from the three first sowing dates produced considerably more seedlings with *Colletotrichum dematium (C. truncatum)* lesions than those from later planting dates. The fungi increasingly involved at earlier planting were, in order of frequency, *Macrophomina phaseolina* and *Sclerotinia sclerotiorum*, followed by *Colletotrichum dematium*, *Sclerotium rolfsii*, *Cercospora* and *Fusarium*, the first three species being largely responsible for reduction of seed germination. During the matura-

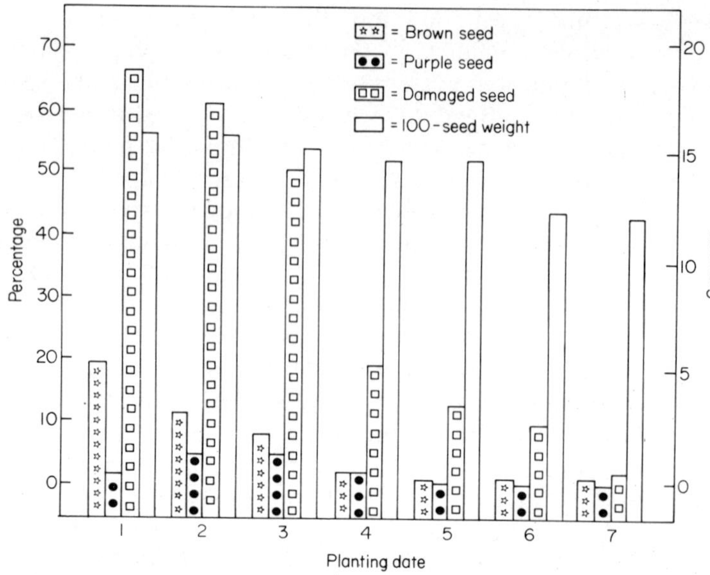

Fig. 19.2. Influence of planting date on seed quality in soybean. Percentage of purple- and brown-stained seed, damaged seed and 100-seed weights of soybean seed harvested from plants sown on seven planting dates (1, June 1; 2, June 16; 3, July 1; 4, July 16; 5, July 31; 6, August 14; and 7, September 2). (Nicholson and Sinclair, 1973)

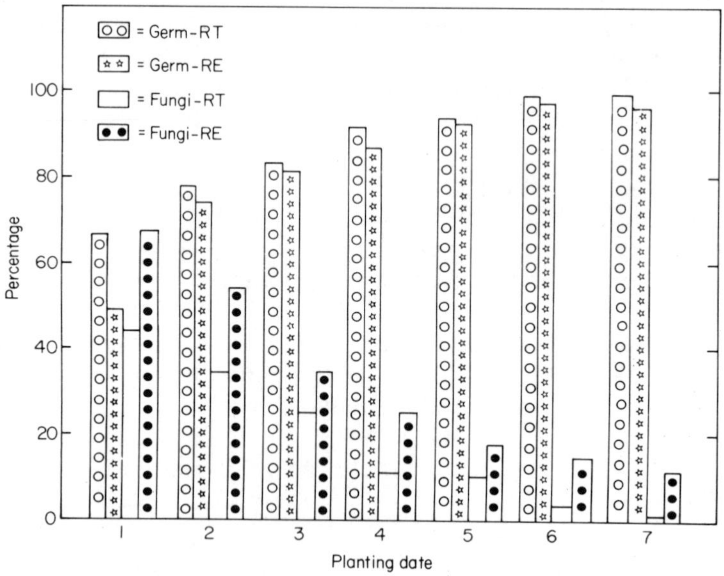

Fig. 19.3. Influence of planting date on seed quality in soybean. Percentage of germination and fungi recovered from surface-sterilised soybean seed from seven planting dates (1, June 1; 2, June 16; 3, July 1; 4, July 16; 5, July 31; 6, August 14; and 7, September 2) and stored either at room temperature (15–45°C) or in a refrigerator (1–15°C). RT, seed stored at room temperature; RE, seed stored in refrigerator before sowing. (Nicholson and Sinclair, 1973)

tion period of plants from the first three planting dates, there were extended periods of rain.

According to a classical contribution by Heald (1918), in the Pacific Northwest of the U.S.A., incidences of bunt in winter wheat are high in fields sown in the 4-week period from mid-September to mid-October, low for fields sown before and after that time. Early sowing allows seedlings to develop beyond the susceptible stage ahead of the infection stage of the smut; late seedling provides conditions for spore germination and subsequent dissipation before the seeds germinate (Fig. 19.4). Observations on sowing time and bunt in winter wheat in Germany are well in agreement (Rabien, 1927).

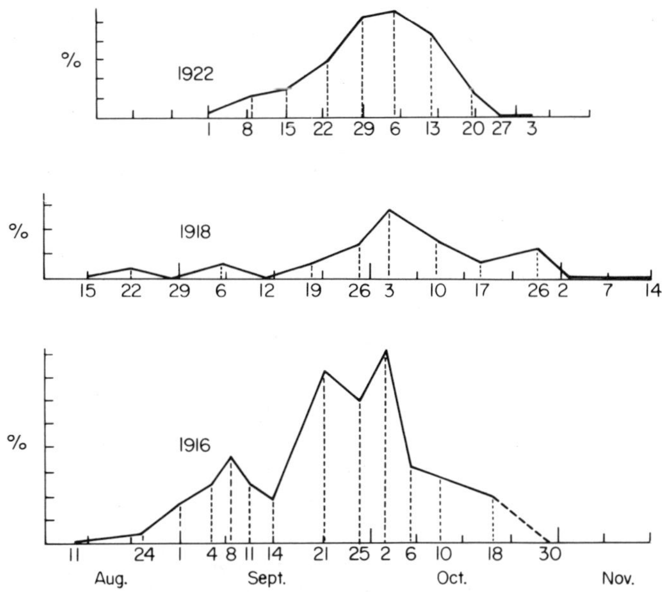

Fig. 19.4. Influence of planting date on percentage of stinking smut in winter wheat for three different years in Washington State, U.S.A. (Heald and Gaines, 1930)

Other earlier German investigations also showed the importance of seeding time in the development of smuts in cereals (Appl, 1915; Caspar, 1926; Moldenhauer, 1927; Bonne, 1931), but the results are conflicting. These conflicting results, however, can be explained by variation in soil factors, particularly soil temperature, in the sensitive stage of seedling and pathogen. For example, Bonne (1931) found for bunt of wheat, *Tilletia caries*, that infection percentages closely corresponded to soil temperature in the period from the 11th to 15th day after sowing. At low temperatures, near an average at 0°C, infection rates were high, at higher temperatures infection rates were low. This explains why for summer wheat there was a sudden decline in bunt percentages from 30 per cent when soil temperatures were about 3°C in the latter fortnight of March, to only 3 per cent bunt in the first fortnight of April, when soil temperatures were 5–6°C, thus indicating that early sowing of summer wheat under these circumstances should be avoided.

In Washington State, U.S.A., results similar to those for bunt in the same state were obtained by Hoffmann and Purdy (1967) for dwarf bunt, *Tilletia contraversa*, in winter wheat, the disease being reduced by very early or very late sowing. Most infection occurs in plants with a few or no tillers before December; the developing tillers were found to be the primary sites of entry. In Sweden Johnsson (1973) found that maximum infection of dwarf bunt occurred when the mean temperature was close to 6°C at 7 a.m. during days 11–50 after sowing, i.e. during the period of host susceptibility. In the island Gotland this mean temperature followed sowing at the end of September; before and after this date the mean temperature was higher and lower, respectively, and attacks decreasing.

In Pakistan early sowing of wheat in autumn at temperatures above 28°C is preferable on account of flag smut, *Urocystis tritici* (Hafiz, 1948).

In contrast to these examples of seedling-infecting smuts the situation is different for the embryo-infecting type. Gassner (1952*b*) and Pedersen (1965) found no influence of sowing time on the amount of loose smut (*Ustilago nuda*) in winter and summer barley, and Neill (1925) and Batts and Jeater (1958*b*) showed that also loose smut of wheat (*U. tritici*) is unaffected by the date of sowing.

Seeding time may be of importance to development of virus diseases. In Germany the time of seeding is an essential factor in development of lettuce mosaic (Kemper, 1962). When lettuce is sown in March and grown to maturity in June, before a large population of green peach aphids, *Myzus persicae*, has developed, the crop becomes much less severely infected than when it is grown later in the season, e.g. in June–August when the vector population normally reaches its maximum.

Depth of sowing. The depth of sowing exerts an influence on germination and development of seedlings which in some respects is similar to that of the time selected for seeding. Proper depth depends upon the size of the seed, the time of seeding and the type of soil. Shallow sowing gives rapidly growing, solid plants and the susceptible phase is shortened. Heavy, crusty soils require shallow sowing.

Investigations based on application of different depths of sowing showed that increase of depth usually leads to increase of smut of the seedling infecting type. This was established by Caspar (1926) and Fomin (1935) for wheat in relation to common bunt, *Tilletia caries*; by Heald and Woolman (1915) and Woolman and Humphrey (1924) in relation to both *T. caries* and smooth-spored bunt, *T. foetida*; by Moldenhauer (1927) for oats in relation to loose smut, *Ustilago avenae*; and by Taylor and Zehner (1931) with regard to covered smut, *U. hordei*.

Kendrick (1956), on the other hand, established that three times as much dwarf bunt of wheat, *T. contraversa*, developed when seed was sown at only 2.5 cm deep, than at 10 cm, and Hoffmann and Purdy (1967) confirmed this.

Tieman (1925) observed in wheat slightly less attack by loose smut, *U. tritici*, by moderately deep seeding than by shallow and deep seeding. Pedersen (1965) recorded great negative influence of excessive depth of sowing on the severity of loose smut in barley, *U. nuda*. An increase in depth from 2 to 4 cm caused only a slight decrease in percentage of attacked plants; by deeper seeding the percentage of attack decreased considerably, as from 40 per cent at 4 cm depth of sowing to 20 per cent at 12 cm depth of sowing. Undoubtedly, deep seeding, leading to reduction of seedling emergence, has a selective effect, infected seeds being slightly smaller and less vigorous; consequently, the ratio healthy plants/attacked plants may be affected. In usual farm practice the seed is not often sown so deep as to exert any influence on the percentage of attack.

Seedling blight of oats, caused by *Fusarium nivale* and *F. culmorum*, is increased by deep planting (Simmonds, 1928; Noble and Montgomerie, 1956; Millar and Colhoun, 1969), and this holds true also for root rot of wheat, caused by *Drechslera sorokiniana* and *Fusarium* spp. (Greaney, 1946).

Slow emergence of barley, with increased exposure to infection, leads to high incidence of barley stripe, *D. graminea* (Gaümann, 1950).

In Oregon, U.S.A., it is recommended, as a control measure against blind seed disease, *Gloeotinia temulenta*, to plant seed of perennial rye-grass more than 1 cm deep (Hardison, 1948*a*, 1957*b*); in this way apothecial production is suppressed (Wilson, Noble and Gray, 1945).

Rate of seeding. The seeding rate may be an important factor where it adversely affects plant growth and favours development and spread of pathogens. Gibson (1956) suggested three ways in which increased sowing density may further spread of damping-off pathogens in seed beds: (1) by giving rise to a dense population of seedlings which would reduce evaporation from the soil, thus producing conditions favourable to spread of damping-off organisms; (2) by increasing the number of seedlings available for infection; and (3) by reducing the average distance between individual seedlings, thus facilitating spread of pathogen from plant to plant. A fourth factor is that of increased inoculum density when seed is infected or contaminated by a damping-off pathogen. Muskett and Colhoun (1947) established that damping-off of flax, caused by *Colletotrichum lini*, is increased by thick sowing.

Heavy seeding of wheat increases incidence of bunt, *Tilletia caries*, in comparison with thin seeding (Milan, 1924; Caspar, 1926; Mourashkinsky, 1932).

In loose smut of wheat, *Ustilago tritici*, the rate of sowing has been shown to have no effect on the percentage of plants infected but a great effect on the percentage of ears infected. In fact, any condition that increases tillering lowers percentage of ears infected (Milan, 1939). Batts and Jeater (1956*b*) explain this effect by the limit in amount of mycelium available for infection of tillers. If few tillers develop, all may become infected; if many tillers develop, the available mycelium may be able to infect only a proportion of them.

Doling (1964) found in barley, by increasing the seed rate from half to four times the normal rate, a heavy reduction in percentage of tillers infected by *Ustilago nuda*. The effect was believed to be due to the competition between healthy and infected seeds or seedlings, similar to the effect on the disease, exercised by excessively deep seeding, shown by Pedersen (1965) and discussed above.

Dense stands of winter rye with limited tillering tend to flower simultaneously with the effect that secondary infection from 'honeydew' conidia of the ergot fungus, *Claviceps purpurea*, is reduced while more widely spaced plants lead to prolongation of the flowering period and hence increased infection (Krebs, 1936; Gaümann, 1950). Another suggested explanation of increased infection is that light is needed for apothecial production (Noble, personal communication).

According to an old practice rice seedlings are usually planted several together. Investigations in Thailand (Piya, personal communication, 1968) have shown that reduction to one plant does not reduce yield because that plant produces tillers. On the other hand, with the greater number of seedlings the inoculum potential of any seed-borne pathogen is increased, the incidence being directly correlated to the number of seedlings. This is, after all, precision seeding which is practicable only when the rice seed, like any other seed, is of dependable quality.

For further discussion on spacing in crops, see Subsection 19.6.6.

19.6.3 Transplanting

Transplanting of seedlings from seed beds is an operation that requires careful consideration. When infected plants are dispersed over large areas, increased potentiality of inoculum is secured. A critical evaluation of young seedlings before transplanting is important.

For many crops the advantages of transplanting must be weighed against the disadvantages. The practice of transplanting pine seedlings in East African nurseries was found by Hocking (1968) to lead to heavier attacks of seedling diseases than direct sowing, possibly due to mechanical damage of plants or spread of inoculum during the operation.

In the U.S.S.R. Nikitina (1966) found in field experiments that transplanting increased tobacco mosaic virus in tomato by 10–15 times, and that cultivation without transplanting increased yield by 19 per cent and improved fruit quality. It is now realised that during the process of transplanting, seed-borne inoculum of the virus which often is present in the seed coat adhering to seedlings, particularly if these arise from freshly harvested seed, easily may be inoculated into the seedling (cf. Broadbent, 1965). But for many years the way in which transmission takes place was not understood. The risk of inoculation can be greatly reduced, however, by spraying plant beds of tomato, as well as tobacco and pepper, with milk just before plants are pulled for transplanting; hands of workers may also be dipped in milk, tobacco mosaic virus being rapidly inactivated by proteins in milk (Crowley, 1958; Hare and Lucas, 1959).

19.6.4 Hygienic precautions during handling of crops

Pathogens may be effectively disseminated in fields by labourers — especially anthracnose and bacterial blight of bean in wet crop. *Septoria lycopersici* and *Alternaria porri* f.sp. *solani* are disseminated in tomato crops on the hands and clothing of tomato pickers (Martin, 1918), and *Septoria apiicola* in celery crops by contact with clothing of labourers (Linn, 1939).

The principal method of spread of many viruses is by handling. Contamination of hands, etc., with infected material should be prevented; frequent washing of hands to remove accidental contamination should be observed. In an experiment, made in Victoria, Australia, to simulate implements brushing leaves of tobacco plants, transmission of tobacco mosaic virus was reduced from 95 to 4 per cent when the contaminated equipment was washed with 10 per cent trisodium phosphate, and to 12 per cent with reconstituted skim milk (Anonymous, 1970). Tobacco mosaic virus is readily spread by workmen weeding tobacco beds; in tobacco fields, one-row wide foci have been observed up to the time of topping because plants are handled row by row; then two rows are topped at a time and the foci become two rows wide (Wolf, 1935, according to Van Der Plank, 1963). In greenhouse crops of tomato the virus is commonly spread during the pruning and tying operations. Some brands of cigars and cigarettes may carry fairly high concentrations of the virus and thus be a source of infection (Johnson, 1937). The virus is also so stable that it can infect after two years on farm equipment such as seed bed cloth and frame boards (Johnson and Ogden, 1929). Another virus, potato virus X, can persist for up to six weeks on clothing (Todd, 1958), and Winther–Nielsen (1972) showed that a tractor which passed from an infected potato field into one free from virus X was capable of infecting plants the first 150 m that it passed through this field.

19.6.5 Precautions when seed crops are grown from plants kept in storage through winter

In cold temperate regions young plants of certain biennial crops, e.g. root crops, are stored in the field through winter in heaps covered with straw and soil to keep out frost and allow roots to breath, and transplanted in spring. This kind of storage often provides favourable conditions for establishment of parasitic fungi which may develop seriously in the subsequent seed crop. In northern Europe this applies to *Typhula betae* on beets overwintered in such pits. Previously, around the turn of the century, when in Denmark carrot seed was grown on plants from roots kept in pits, *Phoma rostrupii* was widespread and destructive. Now the parasite has no significance because plants for seed production remain standing in the field over winter. Another parasite of carrot, *Alternaria radicina,* cause of black rot, a major disease in northern Europe, is worst in crops grown from roots overwintered in pits. This procedure is still followed for growing selected plants for production of Foundation Seed. Also black rot of crucifers, *Xanthomonas campestris*, is greatly favoured by keeping the host in pits over winter.

Careful handling of larger roots for seed production is imperative; they must be kept in sand, well apart from each other, and the temperature must be maintained at a few degrees above freezing point. Diseased roots must be discarded before and after storage. The place of the pits must be shifted every year to prevent parasites such as *Phoma lingam* and *Xanthomonas campestris* in crucifers from infecting subsequent crops.

In technically advanced countries maintenance of proper conditions for overwintering plants is provided by well-equipped storehouses where temperature and humidity can be controlled.

19.6.6 Ventilation of seed crops, microclimatic conditions

Proper ventilation is extremely important to seed crops frequently attacked by bacterial and fungal diseases. Seed production should be avoided in enclosed localities where stagnant, damp air promotes splash-borne fungi such as *Colletotrichum lindemuthianum* in bean (Schaffnit and Böning, 1925). Close to the edge of a wood or to a windbreak, aeration may be hampered. Observations made in Denmark by the author illustrate the effect of shelter leeward on disease development. A seed crop of *Godetia hybrida*, placed adjacent to a high windbreak, showed markedly decreasing severity of a devastating attack of *Alternaria godetiae* with increasing distance from the windbreak. Also in an onion crop, attacked jointly by *Peronospora destructor* and *Pleospora herbarum*, the disease gradually decreased with increasing distance from a windbreak.

Distance between plants within a crop has a bearing on development of infectious diseases, increase in distance usually leading to decrease of infection. Spread of inoculum from plant to plant is reduced, admission of light to lower leaves improved, and better aeration provides quicker drying after rain and dew.

Ample spacing of rows in cabbage seed crops reduces substantially attacks of *Alternaria brassicicola* and other *Alternaria* spp. in Denmark. Schorch or northern anthracnose of clovers, *Kabatiella caulivora,* is counteracted by reducing the density of the stand (Stapel and Bovien, 1943); also for seed crops susceptible to grey mould, *Botrytis cinerea,* it is imperative to keep sufficient distance between rows.

Narrow strips of low-growing field crops, such as pea or bush bean, alternating with higher seed crops of dill, provide favourable conditions for ventilation of the latter thus preventing serious outbreaks of *Cercosporidium punctum* (Neergaard and Helmbring, unpublished).

However, wide spacing may have the opposite effect. In Germany Ballarin (1950) observed in opium poppy substantial increase in infection and damage of *Drechslera papaveris* with increased spacing, and plants at the outskirt of crops were more heavily attacked than those inside the crop. Wider space favours luxuriant growth, prolongs vegetation, promotes branching, and hence retains the susceptible stage (Grümmer, 1955), entailing aggravation of infection in capsules and reduction in seed yields, while in healthy crops wider distance between plants increases yields.

Lodging in stands counteracts aeration and favours development of infections, e.g. blight, *Pyricularia oryzae*, in rice; the new short- and stiff-stemmed cultivars are resistant to lodging thus tolerating higher dosages of fertilisers. Regnault (1973) showed by extensive investigations in France that lodging crops of rape, *Brassica napus*, were much more damaged by *Sclerotinia sclerotiorum* (21–38 per cent being seriously attacked) than standing crops, including those in which the plants had been supported (1–4 per cent seriously attacked). Increased spacing has a similar influence on the disease. Steadman, Coyne and Cook (1973) found reduced rates of infection by *S. sclerotiorum* in Phaseolus bean by wider row spacing, and in the U.S.A. it is well known that bean crops grown on low ground surrounded by woods, which reduce air circulation, suffer the greatest losses (Zaumeyer and Thomas, 1957).

19.6.7 Adjustment of soil reaction

Fusarium wilt of cotton, pea, tomato, soybean, and other crops, incited by different pathogenic races of *Fusarium oxysporum*, is substantially favoured by acid soil, and conversely, largely checked in alkaline. By raising the soil pH, wilts of this type can be reduced or controlled.

Edgerton and Moreland (1920), Sherwood (1923), Scott (1924), Fisher (1935), Jones and Woltz (1968) and Jones and Overman (1971) demonstrated that Fusarium wilt of tomato, *F. oxysporum* f.sp. *lycopersici* races

1 and 2, can be controlled by adjusting the soil pH. Jones and Overman showed in Florida, U.S.A., that by increasing the soil pH from 6.0 to 7.0 or 7.5, yields of tomato, affected by the disease, are increased as much as by applying any of the soil fumigants while the pH remained at 6.0 (Fig. 19.5). They pointed out that it would be economically more feasible to

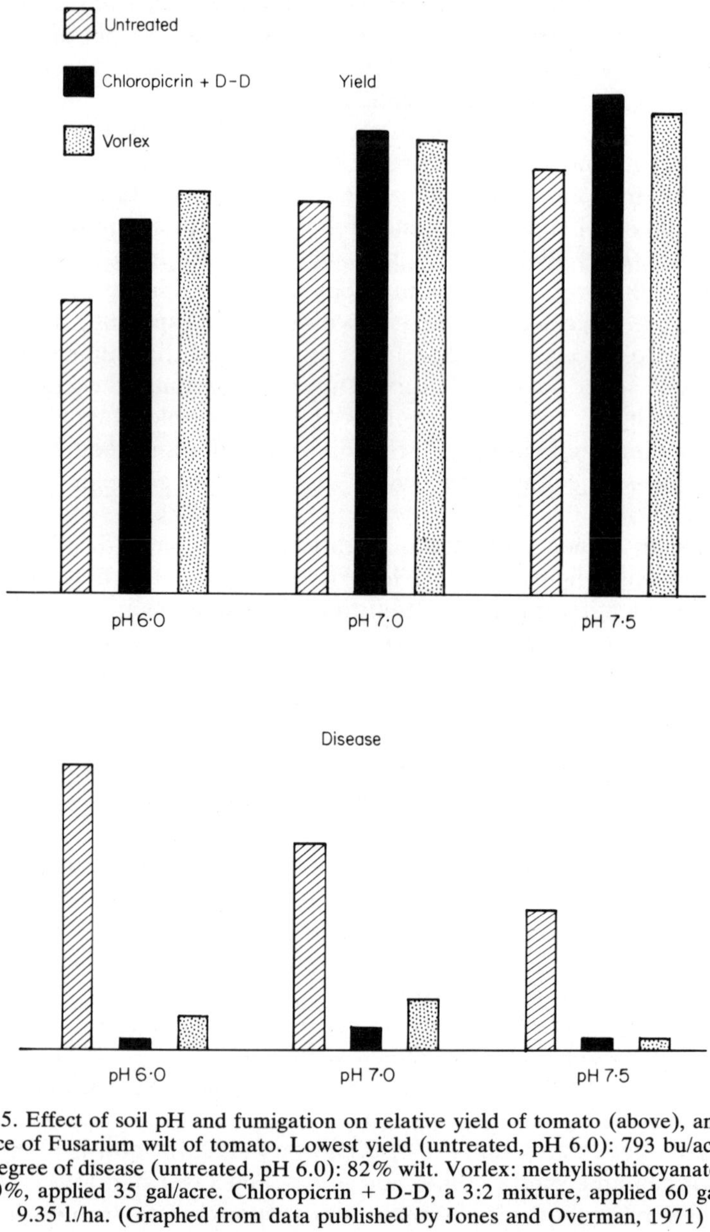

Fig. 19.5. Effect of soil pH and fumigation on relative yield of tomato (above), and on the incidence of Fusarium wilt of tomato. Lowest yield (untreated, pH 6.0): 793 bu/acre. Maximum degree of disease (untreated, pH 6.0): 82% wilt. Vorlex: methylisothiocyanate 20% + D-D 80%, applied 35 gal/acre. Chloropicrin + D-D, a 3:2 mixture, applied 60 gal/acre = 9.35 l./ha. (Graphed from data published by Jones and Overman, 1971)

raise the soil pH than leave it at a low pH and fumigate. Soil fumigation would cost 250–300 dollars/acre, whereas liming of sandy soil from pH 6.0 to 7.0 or 7.5 would amount to 30–50 dollars/acre. However, a combination of both procedures gives additional profit.

In Florida, as in many other places, Verticillium wilt, *Verticillium alboatrum*, is another menace to tomato cultivation. The disease is considerably enhanced by a soil pH of 7.5 (Subsection 15.2.3) and is difficult to control even by fumigation. Therefore, according to Jones and Overman, a pH of 7.0 would be preferred in order possibly to avoid Verticillium wilt.

19.6.8 Use of fertilisers

All the complexity of fertilising any crop is involved in growing seed crops. The effect of the components of fertilisers on predisposition of hosts as well as on the development of established infections is not least in seed crops of a complex nature. Physicochemical conditions of soil, soil type as well as climatic conditions greatly influence the effect of fertilisers. Some typical relationships may illustrate the versatility of the subject.

Holton and Heald (1941) have reviewed the influence of manuring on development of *Tilletia caries* and *T. foetida* in wheat. From many conflicting results in this review it may be concluded in agreement with Fischer and Holton (1957), that according to one investigator nitrogen tends in general to reduce bunt infection, and potassium and phosphoric acid tend to increase it, while according to another investigator potash and superphosphate at the rate of 32,000 kg to the hectare reduced amount of infection! According to Tapke's review (1948) bunt of wheat is increased by potassium and usually by phosphoric acid and reduced by nitrogen. Calcium cyanamide seems definitely to reduce the extent of infection.

In Australia Forster and Vasey (1929) found that applications of lime and farmyard manure, either combined or independently, increased infection of flag smut of wheat, *Urocystis agropyri*, possibly due to changes in soil reaction, the pH being increased from 6 to 8. Millikan (1939) thoroughly studied the influence of fertilisers on development of flag smut. In sand culture experiments deficiency of calcium, and to some extent deficiency of nitrogen and potassium, resulted in reduction of the disease, while deficiency in phosphorus increased severity. Corresponding field experiments were much more variable but he was able here to correlate the first appearance of flag smut about nine weeks after germination with an optimum concentration of calcium accumulated in the host plant.

In barley and wheat, infected by loose smut, the growth of both mycelium and host follows the same time schedule and is subjected to the same effects of growth factors and, hence, the development of attack will not normally be affected by external conditions. Pedersen (1965) investigated the effect of extreme exhaustion of plant nutrients from the soil in attacks of *Ustilago nuda* in spring barley. In 1961 and 1962 he placed his field trials in plots which had received exactly the same treatment continually since 1899, and the soil which had received no nitrogen, phosphorus, potassium and calcium for more than 60 years had no modifying influence on the amount of loose smut.

Cotton wilt, *Fusarium oxysporum*, f.sp. *vasinfectum*, is known often to be aggravated in potassium-deficient soil (Walker, 1969).

Excessive application of nitrogenous fertiliser increases the susceptibility of rice plants to blast, *Pyricularia oryzae* — the size, particularly the width, of the lesions are increased (Kozaka, 1963; Sridhar, 1969, 1970; Sakamoto, 1970). This is a critical factor in growing high-yielding rice cultivars, farmers being urged to apply large doses of nitrogenous fertilisers to get maximum yields. Brown leaf spot, *Drechslera oryzae*, which usually prevails under poor farming conditions, may become severe both by deficiency and excess of nitrogen (Chattopadhyay and Dickson, 1960), and also bacterial blight, *Xanthomonas oryzae*, becomes severe with high nitrogen dosages (Have and Kauffman, 1972; Ou, 1972).

However, it has been established that it is generally the inorganic form of nitrogen available to host or pathogen that affects disease severity rather than the amount of nitrogen. Some diseases are increased by nitrate nitrogen and decreased by ammonium nitrogen, while other diseases are affected inversely.

The form of nitrogen in soil depends greatly on soil conditions. The rate of biological mineralisation of organic nitrogen to inorganic ammonium nitrogen and its subsequent nitrification are influenced by physical factors such as pH, soil type, nitrogen concentration, source of nitrogen, oxygen tension, temperature, salt concentration, previous crop and agricultural chemicals, as reviewed by Huber and Watson (1974). From this very useful review on nitrogen form and plant disease some examples of seed-borne diseases may be listed — see below.

The severity of the following diseases are *reduced by nitrate nitrogen* and increased by ammonium nitrogen:

Northern leaf blight of maize, *Drechslera turcica*
Fusarium stalk rot or wilt of maize, *Fusarium moniliforme*
Fusarium wilt of tomato, *Fusarium oxysporum* f.sp. *lycopersici*
Fusarium wilt of cotton, *Fusarium oxysporum* f.sp. *vasinfectum*
Fusarium wilt of pea, *Fusarium oxysporum* f.sp. *pisi*
Root rot of bean, *Fusarium solani* f.sp. *phaseoli*
Blast of rice, *Pyricularia oryzae*

In contrast, these diseases are *increased by nitrate nitrogen* but decreased by ammonium nitrogen:

Fruit and root rot of tomato, *Colletotrichum gloeosporioides*
Bacterial canker of tomato, *Corynebacterium michiganense*
Diplodia stalk rot of maize, *Diplodia zeae*
Bacterial wilt of maize, *Erwinia stewartii*
Verticillium wilt of tomato, *Verticillium alboatrum*

Farmyard manure may act as an important source of inoculum for some seed-borne diseases. Thus transmission may occur through roots of crucifers contaminated by *Phoma lingam* and *Xanthomonas campestris*, and introduced with farmyard manure into seed crops of the hosts. Oospores of *Sclerospora graminicola* and *S. sorghi* may be disseminated through farmyard manure — the oospores remaining infective in the dung (Safeeulla, 1971). Starr and Kercher (1969) established that sheep manure may contain bean halo-blight-producing bacteria, *Pseudomonas phaseolicola*.

19.6.9 Short-term cultivation of perennial seed crops

In perennial seed crops the amount of inoculum and development of diseases may increase from year to year, resulting in ever-decreasing yields and quality. In Denmark seed crops of cocksfoot were previously grown for four to five years and, as a consequence, often the crops were seriously attacked by choke, *Epichloë typhina*, which is seed-transmitted in red fescue, and yellow slime disease, *Corynebacterium rathayi*. Now only short-term crops are grown, and these diseases have lost their importance. Seed of cocksfoot should not be grown beyond two years on the same plants.

In Oregon, U.S.A., losses from blind seed disease, *Gloeotinia temulenta*, have been reduced to a negligible level by following a strict control programme, in which ploughing after second seed harvest was an essential part before field burning was generally practised (Hardison, 1957).

In the U.S.A., Frosheiser (1970) assayed 56 seed lots of lucerne and found 48 samples infected with alfalfa mosaic virus, also that incidence of infected seeds increased with increased age of the stands of mother plants when the seed was harvested, from 0–1 per cent in seeds from 0–1 year stands to 1–6 per cent in seeds from 6–8 year stands. Also crops infected by wilt, *Verticillium alboatrum*, show increase in incidence in each successive year and may frequently in Britain have no value after three years. Even harvests after the second year may be unprofitable (Isaac and Lloyd, 1959).

19.6.10 Breaking the life-cycle of the pathogen

Control of grass seed nematode, *Anguina agrostis*, in fodder grasses can be obtained by prevention of seed formation, thus breaking development of the nematode. This can be accomplished by clipping, use of maleic hydrazide to prevent floral formation, or by pasturing of an entire crop to prevent floral formation (Couch, 1973); cf. Subsection 19.2.4 on control of brome grass smut and other pathogens in wild grasses, using the same techniques.

19.7 CHEMICAL PROTECTION OF SEED CROPS

19.7.1 Purpose and requirements

Depending on climatic conditions and presence of pathogens, pesticides may have to be applied, as a regular practice, to make seed production profitable.

A programme of spraying or dusting seed crops is directed against three major groups of pests — fungi, bacteria and virus vectors — the last group comprising primarily aphids and leaf hoppers.

The primary incentive is, of course, to avoid yield loss and thus to be of direct benefit to the farmer.

Beyond this purpose, however, a control programme in seed production should in principle aim at complete prevention of even slight traces of any seed-borne pathogen of importance, even though this control may bring no direct benefit to the grower through the immediate cash value of the seed lot produced. Rather than merely being satisfied by precautions against losses in production of seed, emphasis must be laid on adopting strictly hygienic principles, thus safeguarding against later development of diseases appearing in the progeny of the seed produced. This holds true particularly in the production of primary sources of seed, Foundation Seed and First Generation of Certified Seed, but certainly this approach is relevant to all seed production including any Certified Seed. Appreciation of this view will increase as techniques in detecting even weak incidences of seed-borne infections are improved, and as our knowledge is widened on the mechanism of seed transmission and the related development of epidemics.

19.7.2 Economic benefit of direct control

Innumerable trials with many different crops suffering from epidemic diseases have shown that fulfilment of a well-planned spray programme may substantially improve the yield of seed, quantitatively and qualitatively. This has been well demonstrated in experiments with Bordeaux mixture, the results of which still hold in principle.

Some outstanding older records of yield increases with copper sprays may still be improved upon because of avoiding the phytotoxic effects of copper in crops such as bean, carrot and cabbage.

In northern Europe celeriac has been protected for decades by copper compounds against devastating attacks of *Septoria apiicola*, early blight of celery, the classical procedure being 1–3 applications of Bordeaux mixture. In Danish experiments (Jørgensen, 1934) one application gave 60–80 per cent increase of seed yield, and in Germany three applications led to an increase of 25–115 per cent (Ellsmann, 1936; Wenk, 1939). Dillon Weston and Cronshey (1947), by several applications of Bordeaux mixture to seed crops, obtained a high germination capacity of the seed (98 per cent) with 0.5 per cent *S. apiicola* as compared with 84 per cent infection in untreated. The plants were severely attacked when the control measures were initiated.

In German experiments (Hähne, 1942), 2–3 applications of Bordeaux mixture against bean halo blight, *Pseudomonas phaseolicola*, gave a considerable increase in yield of green pods, the production of seed was increased by 35–100 per cent, and seed-borne infection considerably reduced. Complete control was not obtained and cannot be expected. A modern spray programme is likely to include antibiotics, but emphasis must be laid on exclusion through an effective seed health certification programme.

In Denmark and Scotland *Alternaria brassicicola* may cause heavy losses of seed, both quantitatively and qualitatively, in cabbage and other cruciferous crops, and make seed production unprofitable. Two to three applications of Bordeaux mixture have in some years increased yield by up to 70 per cent, while the quality of the seed in terms of germination capacity and health condition is considerably improved (Nielsen, 1933). Bordeaux mixture is still included in spraying programmes, e.g. in the U.K. (Maude and Dudley, 1971).

19.7.3 Materials

Hundreds of compounds are available for crop protection as sprays, dusts, fumigants, systemics and other categories of remedies. Reference is made to descriptions and discussions of this variety of chemicals and their effects as protectants or eradicants in handbooks, monographs and reviews such as those published by Horsfall (1945, 1956), Horsfall and Dimond (1951), Sharvelle (1961), Rose (1963), Torgeson (1967, 1969), Commonwealth Mycological Institute (1968), Van Der Kerk (1969), Nene (1971), Martin (1972, 1973) and Billings (1974), and annually by many countries operating schemes of approval of crop protection chemicals.

Sulphur and *copper* are the classical *fungicides* and have been more widely used than any other. For the last two or three decades there has been a general shift from the old protectants, including lime sulphur, flotation and micronised sulphur, Bordeaux mixture and insoluble copper, to the modern types of fungicides such as organic metallic compounds, and other organic compounds, including those of *ethylenebisdithiocarbamate* (nabam, dithane, zineb, maneb and others), *zinc* and *ferric dimethyldithiocarbamate* (zerlate, fermate), *bisdimethylthiocarbamoyl disulphide* (Arasan, Nomersan, thiram, etc.), *trichloromethylmercapto* compounds (captan), *triphenyltin* compounds, and other types. In this list must be included also the *antibiotics*. Still, however, the classical sulphur and copper fungicides are not at all totally superseded; they are still in use even in technically advanced countries, although perhaps particularly in developing countries.

Sulphur and copper fungicides have a broader range of effect than the new organic compound fungicides which are often more effective within a narrow range of pathogens. Some of the modern fungicides are effective where, for instance, Bordeaux mixture fails.

A modern spray programme takes into consideration the more specific applicability of modern fungicides in combining different materials, either as a mixture to achieve a wider range of effectiveness or scheduled at different times in the timetable of a control programme, some being preferred at one stage, others at another stage of the sensitivity of crops. Varied application of pesticides in a control programme is important also because pathogens and pests may develop 'tolerance' or 'resistance' to compounds used for some years.

Examples of the limitations in use of fungicides are maneb, ziram and captan that are effective against anthracnose fungi, as in cucurbits *Colletotrichum orbiculare (C. lagenarium)*, and tomato (*Colletotrichum gloeosporioides*), whereas copper is practically worthless against this category of fungi.

In a spray programme against bacterial spot of tomato (*Xanthomonas vesicatoria*), recommended in South Florida by Cox (1966), materials known to be effective include neutral copper, streptomycin and maneb. Copper plus maneb is applied from the time that the cotyledons are fully expanded to the second true leaf stage, but streptomycin is not recommended for this growth stage purely for reason of economy since because of limited leaf area most of the material will simply be lost. Streptomycin is used until plants are approximately 45 cm tall, together with the other fungicides, but not during prolonged dry periods as streptomycin is more effective during humid periods.

As in fungicides, development of new *insecticides* and other pesticides has led to new types based on organic compounds. The classical nicotine and arsenical compounds are now largely replaced by organic *thiophosphorus* compounds (including methyldemeten, diazinon, malathion, methylparathion, parathion), *pentachlorodiphenylethane*, or DDT, now banned, *hexachlorocyclohexane* (lindane), and many others.

The modern insecticides have in common with modern fungicides that they are more specific in their effects, the classical insecticides having a broader range of effect.

The importance of using proper insecticides against vectors such as aphids and leaf hoppers, for control of viruses in seed crops, is obvious. Vectors such as aphids in lettuce and bean must be killed as they enter a crop. Good results have been obtained by Broadbent, Burt and Heathcote (1956) and Broadbent, Heathcote and Mason (1958) by use of fortnightly applications of persistent and systemic insecticides against aphids acting as carriers of potato viruses. It was shown, however, that when infective aphids land on a sprayed crop they can infect plants before they die. Total prevention of introduction of virus in the crop concerned could not be obtained, but a delay in the process of increasing infection from year to year could be obtained. Systemic insecticides (dimethoate, phorate, disulphoton), applied in granular form to the soil at planting, kept aphids from colonising the plants for three months (Burt, Broadbent and Heathcote, 1960; Broadbent and Burt, 1962).

Tu and Cheng (1970) showed the importance of using insecticide in combination with fungicides in controlling cotton boll black rot, caused by *Diplodia gossypina*. There is a close relationship between infection by this disease and injuries caused by various chewing insects such as cotton bollworm, *Heliothis zeae*, spiny bollworm, *Earias fabia*, pink bollworm, *Pectinophora gossypiella*, and cotton stainer, *Dysdercus megalopygus*.

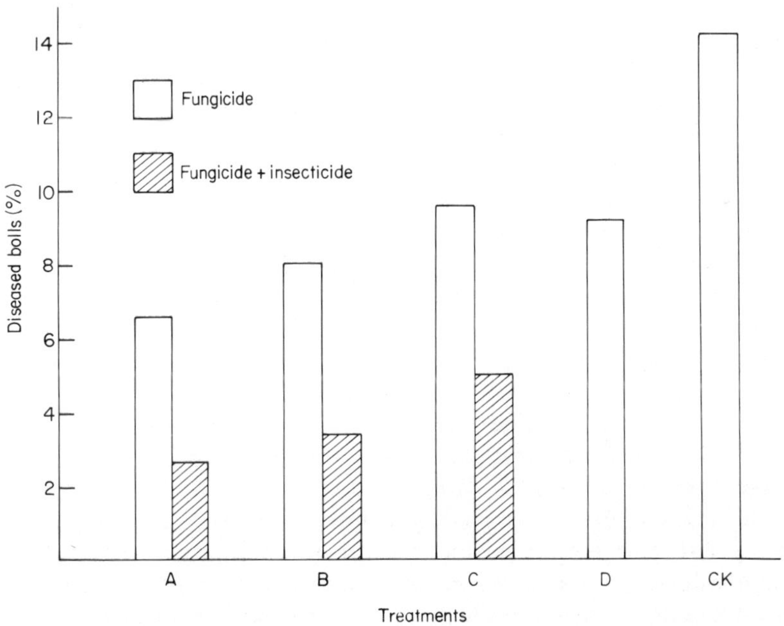

Fig. 19.6. Per cent cotton bolls infected after three applications of different fungicides with and without addition of an insecticide. A, Difolatan; B, Polyram; C, Maneb; D, Mercury–Bordeaux mixture; CK, untreated. (Tu and Cheng, 1970)

One of three fungicides — (A) Difolatan, (B) Polyram (a mixture of zineb and thiram) and (C) Maneb — as shown in Fig. 19.6 was mixed with the insecticide, EPN E.C. (45 per cent *O*-ethyl-*O*-p-nitrophenyl phenylphosphonothioate) already widely used in Taiwan for control of insects on cotton. Fig. 19.6 shows substantial increase in control of the disease by application of the fungicide–insecticide treatment, as compared with use of fungicide alone. Mercury–Bordeaux mixture, indicated as D, was used for comparison as a standard fungicide, commonly used in Taiwan.

A parallel example is the use of an insecticide in sprays or dusts against early blight of tomato, *Alternaria porri* f.sp. *solani*. In controlling flea-beetles, the insecticide also helps to reduce infection of the disease (Chupp and Sherf, 1960).

Nematode vectors are important in the life-cycle of a number of serious seed-borne viruses designated NEPO-viruses (Cadman, 1963). Hence, control of these vectors to prevent transmission must be considered. In controlling transmission of tomato black ring and raspberry ringspot viruses by treatment of the soil with D-D (dichloropropene-dichloropropane) Murant and Taylor (1965) obtained a 95 per cent kill of the nematode, *Longidorus elongatus*. Harrison, Peachey and Winslow (1963) reduced by 99 per cent a population of the vector of Arabis mosaic virus, *Xiphinema diversicaudatum*, by using D-D and methyl bromide.

Other effective nematicides are ethylene dibromide (EDB), chloropicrin, 1,2-dibromo-3-chloropropane, sodium methyl dithiocarbamate and 3,4-dimethyl-1,3-5,2H-tetrahydrothiadiazine-2-thione. These are sold singly or in mixtures under various trade names. Chupp and Sherf (1960) have expounded possibilities of nematode control.

19.7.4 Diseases and timing of chemical applications

The factors to be considered in setting up a programme of field application of chemicals are: the type of pathogen, its rate of spread, effect of damage and sensitivity to chemicals; the type of crop, its form and growth and sensitivity to chemicals; the climatic conditions in relation to development of diseases; the chemicals available and their cost of application.

The sprays and dust commonly used are protectants; however, eradicants, and particularly systemics, are becoming important. The pathogens concerned are those disseminated by air, by air plus rain splashes, etc., or by insects or other vectors. This actually implies that chemical protection of seed crops is directed primarily against epidemic diseases. These may be quite different; the increasing rate varies inherently from one pathogen to another, is dependent on climatic conditions and, for some pathogens more than others, on the growth stage of the crop. All of these facts have bearing on the strategy to be adopted.

There is usually a definite connection between the type of disease and the type of pathogen involved. A discussion on principles of setting up a spray (or dust) programme may therefore appropriately be based on categories of pathogens. Those dealt with below are extensively seed-borne.

Bacterial diseases. Many of these diseases spread very quickly, and provided that conditions of temperature and humidity are adequate they may develop rapidly on all growth stages of the host, from young seedlings to maturing plants. Under epidemic conditions frequent applications are necessary. In recent years antibacterial antibiotics, e.g. streptomycin, have been included in spray programmes but they still have limited use; they must be accepted by government authorities before they can be applied legally to food plants. Certain fungicides, basic copper, maneb and others, may be used.

In seed production complete prevention of infection is imperative. Against bean bacterioses weekly applications of streptomycin are advised; formerly Bordeaux mixture was recommended but this fungicide reduces yield as well as bacterial blights (Chupp and Sherf, 1960).

The following four quite distinctive types of fungus disease are common in seed crops:

(1) *Alternaria diseases*, which are predominantly epidemic in a definite period during the growth season, usually not reaching severe proportions throughout the season. Expanding epidemics during active growth may develop from *Alternaria porri*, purple blotch of onion, *A. porri* f.sp. *dauci*, Alternaria blight of carrot, and *A. porri* f.sp. *solani*, early blight of tomato. For all of these, 5–10 day intervals, dependent on weather conditions, must be considered, the fungicides commonly used being maneb, ziram and zineb, and at least for early blight also captan. The applications may have to begin when seedlings are quite young, 7–10 cm tall for carrot, and for tomato 4–6 weeks after the plants are set into the field (Chupp and Sherf, 1960).

For other alternarioses, such as those caused by *A. brassicae* and *A. brassicicola* in cabbage which become harmful particularly as the host matures, in the stage of senescence, two applications after flowering, or three, one before and two after, give adequate protection under weather conditions prevailing in Denmark (Stapel and Bovien, 1943).

(2) *Anthracnoses and blights*, caused by various species of *Colletotrichum*, *Phoma* and *Phomopsis*, and often dominant in Phaseolus bean, eggplant, pepper, soybean, spinach and tomato, produce deep lesions on stems, petioles and leaves as well as heavily sporulating fruiting bodies of a quite persistent nature. This makes any control most difficult once the disease has become established. To obtain the necessary complete prevention many frequent applications are needed. Where conditions favour disease, sprays must begin as soon as the plants break through the soil.

Usually applications at 5–7–10-day intervals with ziram, ferbam or maneb are effective against anthracnoses on the hosts mentioned. Foliar applications of soybean with benomyl, once at mid-flower or at late pod, or once at both stages, has been shown to be effective against internally seed-borne infections, including those of *Phomopsis sojae* (Ellis *et al.*, 1974).

(3) *Downy mildews* (*Peronospora* spp., and other genera) are typically epidemic diseases usually having an extremely high rate of increase. When effectively seed-borne, they are transmitted as mere traces but these may act like a spark bringing about an explosion. Some mildews may at any rate be very harmful to seed crops, particularly beet, onion and tobacco, less to spinach and cabbage, and to flower seed crops such as *Antirrhinum*, *Cheiranthus*, *Clarkia* and *Papaver*. Bordeaux mixture has been in common use but is now being replaced by organic compounds, chiefly maneb, zineb and ferbam. Against serious types of mildew spraying begins with the appearance of the first true leaf and continues throughout the season at weekly or 5-day intervals, according to conditions.

(4) *Septoria diseases*, many of which are epidemic and of great economic importance, being major diseases on many grasses and cereals, on celery, celeriac, lettuce and tomato, and on various flower crops, including *Callistephus*, *Digitalis* and *Phlox*. Bordeaux mixture is the classical remedy being replaced mainly by zineb and maneb. Timing of application must follow the pattern for epidemic fungus diseases, beginning with young seedlings, and continuing at 5–7-day intervals, according to conditions of weather and crop.

In England a spray programme for wheat was evolved by Jenkins and Morgan (1969) to assess crop loss due to *Septoria nodorum* and *S. tritici*. They obtained partial control of combined attacks of these fungi in winter wheat with frequent applications of tank-mix zineb. In a complete spray programme, spraying at about 10-day intervals was continued from late tillering stage, growth stage 5 according to Feeke's scale (1941), until the crop was mature, at growth stage 11.3 a total of 11 sprays being applied. The full spray programme gave very satisfactory control of the *Septoria* diseases and increased grain yield by about 36 per cent. Most of this increase was accounted for by the increase in size and weight of the grain, and undoubtedly this was due to the fact that premature death of photosynthetic leaf tissue was largely prevented by the treatment (Fig. 19.7). When only two late sprays were applied, at growth stage 10.5 (Feeke), just after ear emergence, and 10 days later, the yield increase was only about 10 per cent.

Other fungal diseases. During the last decades fungicides have been increasingly used to control cereal diseases. In developing countries cereal fields are often small and spraying is practicable without much difficulty. Spraying or dusting of rice fields against rice blast, *Pyricularia oryzae*, and brown spot, *Drechslera oryzae*, is a common practice. Phenylmercuric compounds are usually preferred (Ou, 1972) but also dithiocarbamates are recommended against these pathogens.

For seed production of cereals, use of fungicides is profitable whenever serious bacterial and fungus diseases occur. For the highest quality seed a consistent spray programme against seed-borne fungi and bacteria may be justified in spite of considerable costs. This holds particularly for seed production at the basic levels, Foundation Seed and Certified Seed, but only if the improved quality and health are recognised and attract a higher price.

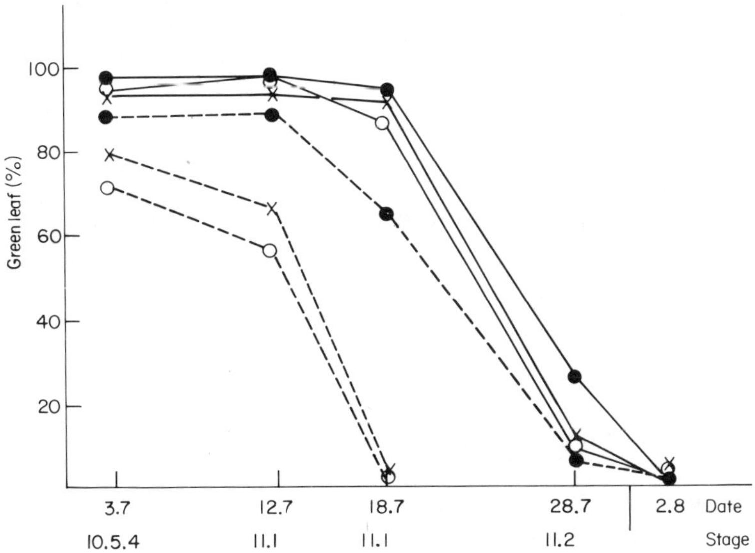

Fig. 19.7. The effect of zineb sprays on the percentage green leaf area on leaf 1 (flag leaf) and leaf 2 (the leaf below the flag leaf) of winter wheat, cv. Cappelle Desprez, sown in the previous autumn. Leaf 1——; leaf 2----; complete spray programme •; late spray ×; and water o. (Jenkins and Morgan, 1969)

Blind seed disease, *Gloeotinia temulenta*, in perennial ryegrass, *Lolium perenne*, can be controlled effectively by a single application of benomyl at the soil surface over overwintered infected seeds. In Oregon, U.S.A., the treatment, made in November, December or January, eliminated apothecia during April–May (Hardison, 1970, 1972), but in Australia an application just before flowering did reduce the formation of apothecia but did not completely control the ascospore inoculum (McGee, 1971).

Virus diseases. Complete prevention of virus infection by effective control of vectors is not to be expected. Flight of aphids and other vectors may continue over a long period of time. Non-persistent viruses are acquired by aphids in a minute's feeding and transmitted to a healthy plant in an equally short time. When infective aphids feed on an insecticide-sprayed crop they can infect plants before they die.

Although insecticides, applied to vectors, may not prevent introduction of a virus into a crop they may substantially reduce the development of infection. In field tests in Israel Lobenstein, Alper and Levy (1970) found that oil emulsions applied to pepper from emergence were highly effective against two stylet-borne viruses on pepper, cucumber mosaic virus, seed-borne in this host, and potato virus Y, the active vector being predominantly *Myzus persicae*. In two consecutive years, oil spraying suppressed virus incidence almost completely in autumn nurseries during periods with high aphid population. Spraying at 3–7-day intervals in the nursery stage was highly effective. When spraying was terminated after transplanting into field, practically no further decrease in yields of green peppers was obtained. While early infections in the nursery caused a complete failure of the crop, protection afforded by oil sprays resulted in yields of 5,000–6,000 kg of green pepper from 1,000 m^2. In Germany 3–5 per cent maize oil (Hein, 1965) and 1–3 per cent groundnut oil, 4–6 weekly applications (Jaeger, 1966) have been used against *M. persicae* to control lettuce mosaic. The latter spraying programme resulted in 43 per cent reduction of virus spread. In California, seed stocks of lettuce free from lettuce mosaic are produced in isolated fields with frequent applications.

19.8 PROTECTIVE INOCULATION

In order to minimise losses from virus infections inoculation of a weak or attenuated strain of the virus concerned has been suggested (Broadbent, 1964; Paludan, 1968, 1973, 1975).

Rast (1972) succeeded in producing a mild strain, probably a mutant of tobacco mosaic virus. This strain causes symptomless infection on tomato and tobacco cv. Samsun in contrast to the normal parent strain. In cross-protection tests the symptomless mutant, inoculated into tomato seedlings, protected the plants against the parent strain as well as against other severe strains. The mutant was introduced into commercial practice in 1969 and tried with encouraging results by many tomato growers in Holland. Also in France promising results have been achieved (Migliori *et al.*, 1972) and in the U.S.S.R. yield in treated plants has been increased by 20–30 per cent (Vlasov, 1972).

This interesting method is somewhat controversial for obvious reasons. Mutations often occur in viruses and may result in increased virulence; pre-inoculated plants may be infected by some other virus and the combination then established may exercise a synergistic effect such as is known from certain types such as tomato streak originating from combined virus infections.

19.9 PRECAUTIONS TO BE TAKEN AT HARVESTING AND PROCESSING OF THE SEED

The treatment of the seed crop after reaping may be crucial to the quality in respect of the germination capacity and health condition of the seed. In northern Europe, e.g. Denmark and the U.K., drying of late-maturing seed crops should take place on seed racks to allow maximum aeration. Thus in Denmark it has been established that germination of radish seed from racks was significantly superior to germination of seed from siliques left on plants laid in swath during the period of weathering (Troelsen, 1947). Seed-bearing plants or flowers lying on the ground are easily infected by soil-borne fungi such as *Rhizoctonia solani*, a risk reduced by drying on racks or by use of fungicide-treated canvas sheets, or plastic sheets.

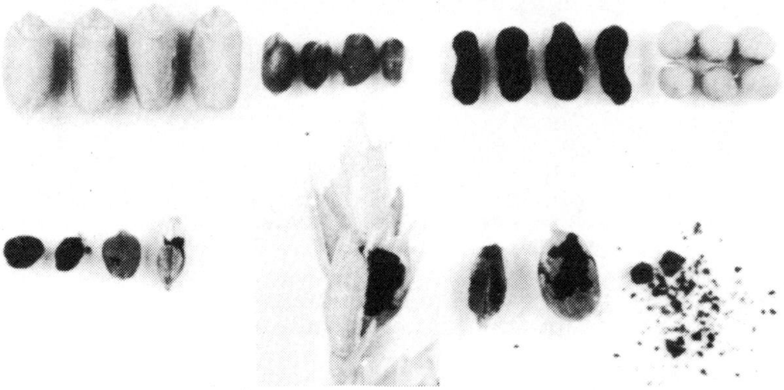

Fig. 19.8. Comparison of nematode galls or ear cockle, centre above, with normal kernels of wheat, and with small bunt balls, below left, and bunt balls of almost normal size of kernels, below right; centre below, part of a wheat spike infected by bunt. Above right, sclerotia of *Sclerotinia sclerotiorum* from a seed lot of *Brassica napus,* seeds of which are shown on the extreme right. Magnification approx. ×3.

During the operation of threshing there are ample possibilities of contamination of the seed from vegetative parts of plants.

In Chapter 2, the importance of damage due to threshing is discussed. The risk of damage on this account is dependent on the condition of the seed at the moment of the operation, particularly with regard to the water content of the seed. If the seed is too dry, the seed coat is more or less brittle and may easily be injured with ensuing increased risk of subsequent invasion of microorganisms and of reduced tolerance to chemical seed treatment. If the seed is too soft due to high water content, the effect of threshing may be internal lesions, which later result in abnormal seedlings, whereas the seed coat usually remains undamaged. The speed of rotation of the threshing machine should be regulated with due consideration of the kind of seed.

Leguminous seed is particularly sensitive. Different types of threshing machine may work more or less gently. In Denmark the so-called 'swede-thresher' is preferable for sensitive seeds as against the so-called 'beater-thresher' (Throelsen, 1947).

Fig. 19.9. Seeds (Cremocarps) or coriander (*Coriandrum sativum*): on the right normal healthy seeds, the vittae (ribs) clearly visible, on the left seeds hypertrophied due to stem gall disease *Protomyces macrosporus*. Approx. ×4

Proper cleaning reduces inoculum of many seed-borne pathogens, particularly those accompanying the seed as sclerotia (*Claviceps purpurea*, *Sclerotinia sclerotiorum*, see Fig. 19.8, etc.), galls (*Anguina tritici*, *A. agrostis*), Fig. 19.8, deformed seeds (*Protomyces macrosporus*), Fig. 19.9, or in the form of fructification organs in plant debris, e.g. *Gaeumannomyces graminis* (syn. *Ophiobolus graminis*) in wheat, *Melampsora lini*, flax rust, and *Septoria linicola*, pasmo disease, in flax, *Corynebacterium insidiosum*, bacterial wilt, in lucerne, or as free nematodes in the dry tissues of plant debris such as *Ditylenchus dipsaci*, stem eelworm, in onion, lucerne and red clover.

Ergots (*Claviceps purpurea*) in rye are about the same size as the grain and cannot be entirely removed by sifting through screens. However, the specific gravity of the sclerotia is 1.10–1.15, while that of rye grain is 1.30–1.35. Niemann (1956*a*) found that about half of the ergots could be removed by a horizontal airblast, but this procedure involved loss of 15 per cent of the grain.

In 1966, samples of winter rape from W. Schleswig-Holstein in Germany contained 432 sclerotia/kg of *Sclerotinia sclerotiorum* on an average or 0.3 per cent by weight. By proper mechanical cleaning most of the sclerotia could be removed, only 0.002 per cent by weight being left, namely those similar to the seed in size (Schlösser, 1968).

In Britain a London Company, Hurst Gunson, Cooper & Tabor, has developed an electronic machine named 'Sortex', separating discoloured, diseased seeds from normal seeds by electronic processes. The procedure has been successfully used for pea seed visibly infected by *Ascochyta* spp. and other fungi, and especially against anthracnose, *Colletotrichum lindemuthianum*, on white Phaseolus beans. The procedure can be used for any kind of distinctly discoloured seed, such as soybean seed discoloured due to infection by soybean mosaic virus (Colour Plate IV).

Nematode galls and ergots can be removed mechanically or by immersion of the seed in a 20 per cent saline solution, the seed being carefully stirred. Chu (1945) described a special machine 'wheat nematode eliminator' successfully used in China.

The importance of proper cleaning of lucerne seed as a measure of reducing concomitant seed-borne inoculum of stem eelworm, *Ditylenchus dipsaci,* is well illustrated by the data, in Table 19.2, secured from an infested field in Utah, U.S.A. (Thorne, 1961).

Table 19.2. Effect on *Ditylenchus dipsaci* by cleaning lucerne seed (Thorne, 1961)

Type of seed screenings	*Nematodes* (per lb)
Recleaned seed ready for market	236
Thresher-run seed before cleaning	1,088
Fine seed passing through screens	1,134
Recleaned screenings	1,170
Fine screenings	16,629
Blowings from cleaning plant	4,398

The size of seed is often substantially reduced as a result of infection. The weight of chickpea, *Cicer arietinum*, from pods infected by *Ascochyta rabiei* has been found to be only 40 per cent of the weight from healthy pods, essentially due to poor development of the seed, cf. Table 19.3 (Halfon-Meiri, 1970). Dunleavy (1971) found that seeds of soybean, that were encrusted by the oospores of *Peronospora manshurica,* were from 12 to 26 per cent lighter than healthy seeds from the same seed lot.

Table 19.3. Effect of *Ascochyta rabiei* on development of chickpea seed. Distribution of seeds from healthy and diseased pods according to size (Halfon-Meiri, 1970)

Size of sieve hole (mm)	*1,000-seed weight* (g)	*Different fractions in 500 g of seed* (per cent) Healthy pods		Diseased pods	
		Sample A	Sample B	Sample A	Sample B
8	360	70.0	80.7	11.5	11.9
7	280–360	26.0	11.0	32.5	42.7
6	190–280	3.5	8.3	35.3	28.4
<6	190	0.5	0.0	20.7	17.0

By proper cleaning and grading, removing small, poorly developed, shrivelled seed, the infection percentage can be materially reduced. This holds true for instance for *Septoria linicola* in flax (Sackston, 1950), *Alternaria brassicae* in rape (McDonald, 1959), *Phoma lingam* in cabbage (Lloyd, 1959*a*) and *Phoma medicaginis (Ascochyta imperfecta)* in lucerne (Kernkamp and Hemmerick, 1952). In supplementing the last reference, infected trefoil seed may even be larger than the majority of the healthy seed, being formed earliest. This type of infected seed, named 'brown flats' in commercial practice, can be removed mechanically (Mary Noble, personal communication, 1969).

Small grains of wheat yield a higher infection percentage of *Ustilago tritici* than larger grains (Tiemann, 1925; Taylor, 1928), and similar results have been obtained with *Drechslera graminea* in barley (Genau, 1928). Krull *et al.* (1966) found 8 per cent *U. nuda* in the largest grains and 27 per cent in the smallest grains of barley. Corresponding results were obtained by Lavery (1965) for two cultivars of barley, and Kuznetsova (1971) reduced the incidence of loose smut by sowing large barley seed. However, for loose smut the difference in grain size is not sufficient to allow mechanical removal.

Drechslera graminea has been known to be transmitted from an infected seed lot of barley to another healthy one, the inoculum having been retained in the grader. This observation from practice was confirmed experimentally by using the same diseased seed lot. The interval between the successive processes was several months (Mary Noble, personal communication, 1967). This raises the question of what other contaminations, such as bunt, may take place during the processing of seed. In one incidence in Lesotho 20 bags of a healthy wheat lot had to be discarded due to contamination of bunt or stinking smut, *Tilletia foetida*. The contamination took place in a threshing-machine in which a heavily contaminated wheat lot had just been treated (personal information during a visit to Lesotho, 1970). Obviously, proper disinfestation of machinery during processing must be considered.

Grogan and Kimble (1967) and Guthrie (1970) found evidence that bean seed may be contaminated with halo blight, *Pseudomonas phaseolicola*, during threshing, cleaning and other operations after the seed is harvested. They demonstrated that seedlings may become infected from such surface seed-borne contaminations.

Virus infections may reduce the size and weight of seed. A virus that produces a ringspot of tobacco causes the seeds to be smaller and lighter (Marcelli, 1955), and similarly Middleton (1944) reports that light, poorly filled, deformed seeds of squash, *Cucurbita pepo*, proved to carry a significantly higher percentage of squash mosaic virus (0.96 per cent out of 14,492 seeds) than well-filled, heavy plump seed (0.14 per cent out of 31,866 seeds). Paguio and Kuhn (1974) found increasing seed transmission of groundnut (peanut) mottle virus with decreasing seed size of groundnut, 0 per cent through seeds sized above 8 mm, 1.6 per cent through seeds sized 5–8 mm, and 2–3.6 per cent through seeds sized less than 6.5 mm.

Shrivelled seeds of two barley cultivars susceptible to barley stripe mosaic showed 69 and 56 per cent infection, while large plump seeds only 2 and 16 per cent, respectively. In cowpea small shrivelled seeds of three cultivars showed transmission of 20–50 per cent cowpea mosaic virus, whereas bold smooth seeds did not yield any infected plants (Phatak and Summanwar, 1967).

The size of seeds, whether infected or not, may also be of importance as a predisposing factor. Underdeveloped small seeds germinate usually at an abnormally slow rate, and the seedlings are exposed to infection for a longer period than rapidly growing seedlings.

20 Management of Seed Storage

20.1 STORAGE CONDITIONS AND MOISTURE CONTENT OF SEED

The longevity of seed kept in storage is predominantly dependent on the moisture content of the seed, and on the temperature and relative humidity in the store room. Seed being markedly hygroscopic its moisture content varies considerably with the degree of the atmospheric humidity. Toole (1942) and Harrington (1972) indicate approximate percentages of moisture content in different kinds of seed as related to different degrees of relative humidity of the air (Table 20.1). Seeds high in oil content will have a much lower moisture content than seeds high in protein and starch content, as seen by comparing oil seeds such as flax and soybean with cereals, e.g. rice and wheat.

Table 20.1. Relation between the relative humidity of the air to approximate percentage of moisture content in seed (combined from Toole, 1942; Harrington, 1972)

Seed	*Relative humidity*					
	30	45	60	65	75	80
Bean	6.5	8.5–9.0	11.0	12.0	14.0	16.0
Cabbage	4.5	6.0	7.0	8.0	9.0	10.0
Carrot	6.0	7.0–7.5	9.0	10.0	11.5	12.5
Celery	—	9.0	—	11.0	12.0	13.5
Flax	5.5	6.5	8.0	—	10.0	—
Lettuce	5.0	6.0	7.0	8.0	9.0	10.0
Pea	7.0	8.5	11.0	—	14.0	—
Soybean	6.5	7.5	9.5	—	13.0	—
Spinach	8.0	9.5–10.0	11.0	12.0	13.0	14.5
Tomato	7.0	8.0	9.0	10.0	11.0	12.0
Barley	8.5	10.0	12.0	—	14.5	—
Maize	8.5	10.5	12.5	—	14.5	—
Rice	9.0	10.5	12.5	—	14.5	—
Rye	8.5	10.5	12.0	—	15.0	—
Sorghum	8.5	10.5	12.0	—	15.0	—
Wheat	8.5	10.0	11.5	—	14.5	—

High moisture content in the seed provides conditions favourable for spread and establishment of fungi and bacteria in the seed and for establishment of an escalating process involving increase in respiration, in water content, in temperature of the seed, and causing accumulation of toxic waste products, resulting in a rapid decline of its vitality with ultimate loss of its germination capacity. Naturally, higher temperatures in the store room will enhance this process. When seeds are kept at temperatures of about 20°C, the relative humidity of the atmosphere must not exceed 60 per cent, while at temperatures between 4°C and 10°C, which are favourable for storage of most kinds of seed, it should be maintained at a level of preferably 50 per cent, not exceeding 70 per cent.

Certain seed-borne fungi are sensitive to reduced moisture content of the seed. This holds true for downy mildews of maize. Seeds at 30 and 18 per cent moisture content, infected by *Sclerospora sorghi*, develop systemically infected plants, while infection is prevented by drying to 9 per cent and storage for 40 days, irrespective of initial seed moisture (Jones, Leeper and Frederiksen, 1972). When maize seed, infected by *S. sacchari*, is dried to 20 per cent moisture content, transmission is prevented (Chang, 1970); transmission of *S. maydis* is prevented by air-drying maize seeds before planting (Semangoen, 1970), and *Sclerophthora macrospora*, crazy top, is transmitted at lower rates from fully mature infected maize seeds than from immature seeds (Ullstrup, 1952).

Harrington, of the University of California, has given a summary of the effects of moisture on seed in storage (according to Mullett, 1969). The processes occur as follows at the indicated levels of seed moisture:

above 45 per cent: germination occurs
above 18–20 per cent: heating may occur
above 12–18 per cent: mould grows on and in seed
below 8–9 per cent: little or no insect activity
below 4–8 per cent: sealed storage is safe

In any arrangement for drying seed it is essential to secure proper ventilation during the process. Moving air will dry material at a much lower temperature than is needed by direct heating. In following this principle often rather simple seed dryers may be used, with oil, coke or wood burners as the heat source and a simple fan.

During the drying process temperature should be kept at 32°C, until the moisture content is below 18 per cent; then it may be raised to 37°C until a moisture level of 10 per cent is reached; the final drying can take place at 42°C but not higher (Mullett, 1969).

In evaluating the moisture content of a seed lot, such as a lot of grain in bin, it must be emphasised that there may be a considerable range of moisture content within a lot. A range of 1 per cent is commonly found between samples from any carload or truckload, and in bins of grain that have been stored for a few weeks a range of 2–3 per cent or more is not unusual. For safe storage the highest moisture content in any portion of a given lot must be known, since the moisture content is usually different from place to place in a bulk of stored grain. Thus there is a great storage risk in mixing grain lots of different moisture content to get an average presumed to be safe for storage (Christensen and Kaufmann, 1969).

20.2 INACTIVATION OF INOCULUM BY AGEING DURING STORAGE

Some seed-borne pathogens are outlived by their host. This implies that seed transmission of the parasites concerned in year-old seed is either negligible or may be entirely prevented by storage of the seed under suitable conditions for a certain period.

20.2.1 Seeds

In estimating this principle of control, consideration should be given to the fact that the life-span of seed from a commercial point of view is considerably shorter than its biological life-span, since high germination capacity must be maintained. Good storage conditions, as usually provided in seed warehouses, greatly increase the longevity of seeds but often also the life duration of seed-borne pathogens. Crocker (1938), Barton (1961) and Roberts (1972) reviewed the literature on seed preservation and longevity, and Christensen and Kaufmann (1969) published a monograph on grain storage in relation to damage by storage fungi.

From the point of view of *maximum longevity of seed*, as we may designate the longevity in terms of the longest duration of viability of any seed in a bulk, many kinds of seed have a life-span over decades, a few over centuries, under optimum storage conditions. According to old terminology, seed was classified based on longevity under optimum conditions (Ewart, 1908): (1) *microbiotic* whose life-span does not exceed 3 years; (2) *mesobiotic* whose life-span ranges from 3 to 15 years; and (3) *macrobiotic* whose life-span ranges from 15 to more than 100 years. Obviously these classes overlap.

The meaning of these terms, which may be applied to seed-borne microorganisms as well, is only relative in as much as any reference to these classes is dependent on the optimum conditions of storage, and these may steadily be improved as techniques improve.

In practical seed trade and agriculture it is not the theoretical life-span of the longest living seed of a bulk, the maximum longevity, that matters. It is rather the life duration of a seed lot in terms of the level of germination capacity and seedling vigour in the sense of soil penetration power acceptable as satisfactory from a commercial and agricultural point of view. This 'qualified' life-span we may call the *optimum longevity of seed*. It is not only the ultimate germination percentage of seed that is important but, for many kinds of seed, the speed with which they germinate. In a laboratory test of, for instance, seed of cabbage and other crucifers, the germination energy or 'speed' is recorded after three days of incubation, as against the total percentage obtained after completion of the test, four days later. Slowly germinating seeds usually lead to uneven stand and poor development. The important relationship between seed age and plant vigour, which is not necessarily reflected in the germination capacity recorded in the laboratory, may be illustrated by the following examples.

Dungan and Koehler (1944) established a distinct influence of age of maize seed upon field performance in terms of yield of grain. Yield reduction was caused by both reduced stand and weak plants from old seed. Five-year-old seed yielded 13.5 per cent less than one-year-old seed (Fig. 20.1). Crocioni (1934) observed that plants from old seed of *Triticum, Brassica*, *Medicago* and *Trifolium pratense* were retarded in growth and less resistant to adverse conditions, plants from old seed of wheat and cabbage being smaller and their yield decreased. Filutovicz and Bejnar (1954) recorded that 1–3-year-old seed of sugar beet germinated more uniformly than 5–7-year-old seed, but nevertheless there was no reduction in the quality of beets produced from the old seed.

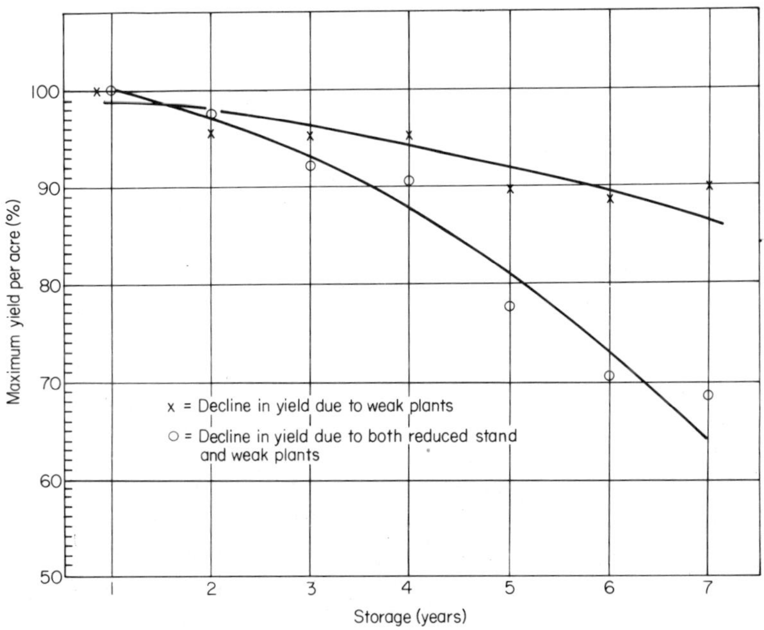

Fig. 20.1. Reduction in yield of grain from old maize seed. Reduction in plant vigour as a factor. (Dungan and Koehler, 1944)

Table 20.2. Plant vigour in relation to seed age. Total yields for seed lots of different age, of tomato, pepper and lettuce. Summary of data from Barton (1961)

Lot	*Age of seeds*	*Yields in ounces*		
		Tomato fruits	Pepper fruits	Lettuce plants
A	Fresh	3,819	829	—
B	7–8 months	3,705	753	560
C	13 years, in containers* at room temperature	1,845	730	641
D	Fresh	3,755	—	—
E	7–8 months	3,589	—	574
F	13 years, dried, in sealed containers, −5°C	3,919	—	658

*Tomato and pepper seed kept in open, lettuce in sealed, containers.

Barton compared plant quality and yield from (A) freshly harvested seed, (B) 7–8-month-old seed, and (C) 13-year-old seed of pepper, tomato and lettuce. A summary of the results (Barton, 1961) is given in Table 20.2. There was a marked decrease in planting value of the old seed stored in open containers (tomato, pepper) in terms of yield, whereas no depreciation of yield was encountered in plants from seed stored with reduced moisture content in sealed containers (tomato, lettuce). Pepper fruits from plants raised from stored seed were of poorer quality, being considerably smaller in size.

Seeds weakened by storage under unfavourable conditions are sensitive to any adverse condition. Injury caused by seed treatment, whether physical or chemical, has direct relation to the viability of the seed such as shown by Leach (1943) for tomato seed. Older seeds of low vigour suffered much more by the treatment than younger seeds of higher vigour (Fig. 20.2.).

In practice, seeds in large quantities such as grain seed are generally stored under the natural conditions available, even in most of the technically advanced countries. This means that the optimal longevity of seed, as experienced by farmers and the seed trade, varies according to local climatic conditions which range from the damp atmosphere and high temperatures of humid tropics to the conditions of cool, moderately dry temperate regions.

The usual optimum longevity of seed under fairly favourable storage conditions, as provided by seed companies in Europe, is about 3 years for most kinds of seed. In Denmark, an important seed-exporting country, this holds true for vegetable seed, some differences existing between the different categories. Stored seed lots are tested for germination every year, their quality being evaluated on basis of a conventional minimum of germination 'energy' or 'speed', a standard dealt with above. As seed becomes older, the germination capacity, particularly the germination

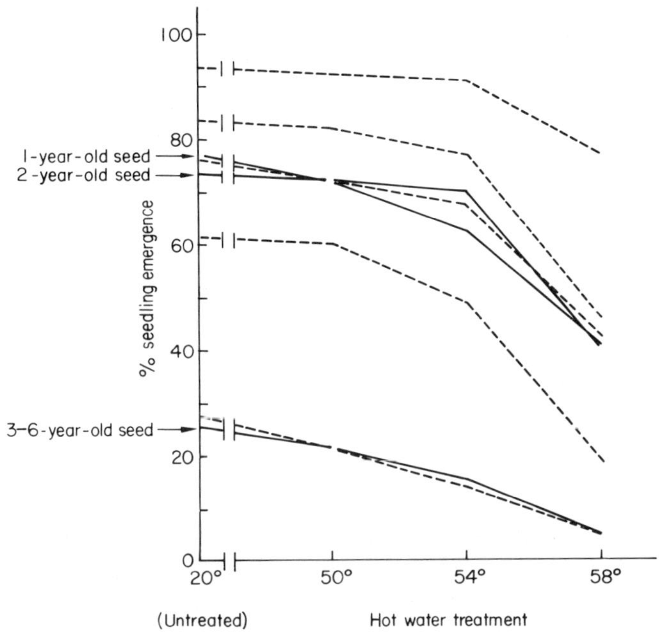

Fig. 20.2. Relationship of age (——) and viability (----) of tomato seeds to their tolerance of hot water treatment at three temperatures for 30 min. The five viability groups, according to per cent emergence in the untreated (soaked in water, 20°C for 30 min), are from above: 100–90; 89–80; 79–70; 69–50; and below 50. (Data from Leach, 1943)

energy, is gradually reduced, and the number of abnormal seedlings increases. This means uneven field emergence and stand, increased weakness in adverse conditions, and ultimately reduced yield.

Usually, as in experience of the seed company J. E. Ohlsens Enke, Copenhagen, seed lots of cabbage and related species are not sold beyond a storage period of three years. For radish, carrot, tomato, cucumber and many other kinds of seed this maximum period of storage applies. Celery and celeriac may keep in good condition for about five years. On the other hand, seeds of onion and leek when freshly harvested hold a water content of 12–14 per cent, must be dried down to 8 per cent moisture content and placed in plastic bags, before storage beyond the year of harvest. Some kinds of seed show better germination after some period of storage — spinach after 2–3 months, celery and celeriac after one year. A prerequisite for good keeping quality is that the seed has been well grown and harvested under favourable conditions, securing adequate maturation of the seed and resulting in good germination energy.

Now, drying process and packing of seeds in metal cans, and vacuum packing, have improved their optimum longevity during storage and transportation under adverse conditions.

The optimum longevity of cereals is similar to that of most vegetables — about 3 years. According to Vilmorin (Crocker, 1938) the 'average' life of seeds of barley, oats and wheat is 3 years, maize 4 years, rye and timothy 2 years. These indications are based on experiences in France, but they hold true for most temperate countries. Cereals in these countries are usually not kept beyond 1–2 years; in the tropics rice, wheat and maize not beyond 1 year (for rice, cf. Cabral and Ferreira, 1974).

20.2.2 Fungi

Krout (1921), Campanile (1926) and Mujica (1943) suggested control of *Septoria apiicola* in celery by inactivation of the parasite during storage of seed. They found the fungus to be incapable of producing viable conidia when the seed had been stored for 2–3 years. However, the longevity of the parasite is variable. Stirrup and Ewan (1931) demonstrated that the fungus may survive five-year-old seed, 70 per cent infected seeds producing 24 per cent infected plants; celery usually germinates well up to four years (Krout, 1921) or even five years (J. E. Ohlsens Enke, personal communication). The viability of the parasite is checked by special laboratory tests (Hewett, 1962; Sheridan, 1963) in conjunction with seed health testing of celery and celeriac. Maude (1963) found in such tests, performed by streaking exudate of conidia on prune lactose yeast agar, that the proportion of seeds yielding viable spores dropped from 82 to 2 per cent in 8 months from harvest. These results were confirmed by insertion of seeds in celery petioles.

Colhoun and Muskett (1948) investigated the longevity of seed-borne fungi in flax. They established that the longevity varies from one seed lot to another within the following limits: *Colletotrichum lini* 26–69 months, *Aureobasidium lini* 16–55 months, *Botrytis cinerea* 16–40 months, and *Phoma* sp. (probably *P. exigua* var. *linicola*) 27–43 months. The longevity of the pathogens was found to some extent to be correlated with the severity of infection. The more heavily the seed was infected by *A. lini*, the longer the fungus survived in the seed. Moreover, it was found that even by long storage complete inactivation of the seed-borne fungi was not obtained without considerable reduction of the germination capacity of the seed. As a rule the pathogens outlived their host.

The life-span of *Botrytis cinerea* in seed is comparatively short. An investigation of 354 seed lots of *Calendula*, *Callistephus*, *Centaurea*, *Clarkia* and *Godetia* showed that the fungus survived until about two years after the harvest of the seed (Neergaard, 1950). According to Colhoun and Muskett (1948), the fungus may survive longer in flax seed than in the kinds of flower seed mentioned, when stored under usual storage conditions by a seed company in Denmark. In the U.K. the fungus survived in flax up to 40 months, but was considerably weakened after 2 years of storage. For most kinds of seed 2 years of storage are sufficient to free seed from heavy infection of *Botrytis cinerea*. The longevity is largely dependent on storage temperature. Van Der Spek (1965) found a rapid decline of vitality of the fungus in flax seed. The parasite died after only 23 days at 34–35°C and 75 per cent relative humidity. *B. anthophila* in red clover, on the other hand, may survive 6 years of storage at 5 per cent relative humidity, in air-tight containers (Narkiewicz-Jodko, 1974).

Red clover seed, infected by *Kabatiella caulivora*, scorch or northern anthracnose, showed after 9 months an infection percentage of 0.3 per cent against 92.5 at the beginning of the storage (Minyaeva, 1951).

Perennial or Italian ryegrass, infected by *Gloeotinia temulenta*, blind seed disease, is free from infection after storage for 2 years (Wilson, Noble and Gray, 1945; Hardison, 1948*b*), and hence in Oregon, U.S.A., planting only disease-free seed or seed more than 2 years old (Hardison, 1948*b*) is the recommended alternative.

Shands (1937) found that *Fusarium graminearum* could live in barley up to 27 months, Christensen (1963*b*) recorded 20 months, *F. culmorum* and *F. avenaceum* were viable for even a shorter period of time (Shands, 1937), whereas Ponchet (1966) recorded survival of *F. nivale*, although in low percentage, in wheat after 42 months. Dungan and Koehler (1944) recorded that, in maize, *F. graminearum* died out completely in 2 years, most of the seed being free of the pathogen already after 15 months. On the other hand, some of the seeds still harboured viable *F. moniliforme* after 8 years, while *Acremonium strictum* (*Cephalosporium acremonium*) died after 7 years. When maize is properly stored, satisfactory yields can still be obtained after storage for 4 years (Dungan and Koehler, 1944). *F. oxysporum* in seeds of red clover could still be isolated after 6 years of storage in air-tight containers at 5 per cent relative humidity of the air (Narkiewicz-Jodko, 1974).

Kuribayashi (1928) established that conidia of *Pyricularia oryzae* on rice seeds kept under dry conditions at room temperature lived more than 1 year, Suzuki (1930) that the fungus survived storage of the seed for 2 years. Distinction must be made between categories of inoculum. While seed-borne conidia of *P. oryzae* may survive for 1–2 years, ultimately in a condition of heavily reduced germination capacity, resting mycelium may survive in the tissues of the seed, predominantly in the seed coat, for a much longer period, of at least 4 years.

Cercospora beticola provides another example of two types of inoculum. The pathogen is harboured in beet seed, for instance sugarbeet, as resting mycelium and as conidia. Schmidt (1938) found that the fungus died out after storage for 1 year. Takimoto (1923) showed, however, that the conidia, kept under dry conditions, were still alive after 16 months and that seed transmission may occur through them. Aebi and Rapin (1954) recommend control of the pathogen by use of two-year-old seed but

Wenzl (1959*a*) established that the parasite could survive as mycelium for 2.5 years in the seed, while the conidia on the seed had died long before. Lehman (1952) found that another species of *Cercospora*, *C. kikuchii*, in soybean died out after about 2 years of storage.

Another beet parasite, *Phoma betae*, was shown to survive in the seed for 5 years (Newton and Bosher, 1946). Lloyd (1959*a*) found that *Phoma lingam* could survive for only 17 months in cabbage seed, but Neergaard (1969) observed development of pycnidia on seed of various lots of cabbage, sown on blotter after the seed had been stored in a commercial warehouse for nearly 4 years.

Phoma valerianellae, a pathogen on *Valerianella locusta* producing a seedling rot, survived 33 months of storage when the infected seed was kept in paper bags at room temperature. The longevity could be prolonged when the seed was kept in glass containers with silica gel at 20°C, and even better at 10°C (Rohloff, 1968*a*).

Selenophoma bromigena, common leaf and stem spot of smooth bromegrass, *Bromus inermis*, survived as conidia for 2–7 years when the seed was kept in closed polyethylene bags at ±10°C but not for more than 18 months in cotton bags at 18–24°C. When stored at controlled humidities for 16 months the survival of the spores appeared directly related to per cent relative humidity. No viable spores were found in seed samples stored at relative humidities above 50 per cent (Smith, 1970*a*).

Another pycnidial fungus, *Ascochyta pisi*, was found viable in 9-year-old seed of *Vicia faba* (Sprague, 1929), in 7-year-old seed of hairy vetch, *V. villosa* (Crosier, 1939), while the fungus died out in pea seed after 7 years (Wallen, 1955). However, most seed lots of hairy vetch and pea were found to retain their germinability at a high level after the parasite was eliminated after only 2–3 years of storage (Crosier, 1939), cf. Fig. 20.3. *Phomopsis leptostromiformis* remains viable more than 3 years in seed of *Lupinus luteus* (Strukchinskas, 1971).

Machacek and Wallace (1952) established that *Septoria nodorum* may survive in wheat seed until 7 years after harvest.

Many parasitic fungi, perhaps most of them, are so persistent in seed that they are capable of outliving their hosts. Many species of *Alternaria* and *Stemphylium* commonly encountered in seed are long-lived under storage of the seed, usually at least 5–6 years (Neergaard, 1945), i.e. just as long or rather longer than most kinds of seed can be kept still good for sowing. In cabbage seed lots *Alternaria brassicicola*, *Ulocladium consortiale* and *Stemphylium botryosum* were viable after storage for over 7 years, and *A. tenuis* after storage for over 6 years (Neergaard, 1969). Neergaard (1941) found that surface-borne conidia of *A. brassicicola*, washed from cabbage seed, which 1 year after harvest had been submitted to combined irradiation by a quartz lamp and a Sollux sun therapy lamp

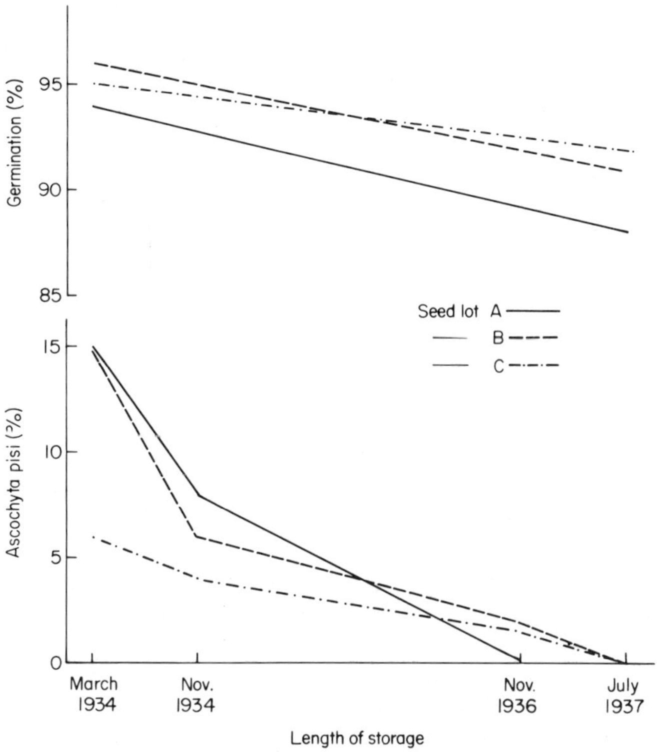

Fig. 20.3. Comparison of longevity of *Ascochyta pisi* with longevity of seed hairy vetch, *Vicia villosa,* in three seed lots. (Data from Crosier, 1939)

for 1 hour and since kept in sealed containers, were viable 8.7 years after harvest; non-irradiated seed carried only dead conidia.

In barley Christensen (1963*b*) in America found *A. tenuis* viable after storage for 6 years and that in general *A. tenuis* lost viability in barley kernels somewhat more rapidly than did *Drechslera sorokiniana*. However, in Canada Machacek and Wallace (1952) established that *A. tenuis* survived 10 years of storage in both wheat and barley.

Drechslera sorokiniana has been found definitely to be long-lived. The resting mycelium can survive for 5 years (Mead, 1942*a*), but the fungus may still survive in seed of barley and wheat after 10 years (Machacek and Wallace, 1952); in other investigations on seed lots of barley from various localities the fungus was found to survive for 6 years (Christensen, 1922, 1963*b*). Machacek and Wallace (1952) found *D. avenae* in oat seed and *D. teres* in barley seed to survive after storage for 10 years, Leukel, Dickson and Johnson (1933) that *D. graminea* was viable and still infective after 5 years and Shands (1937) that it survived after 10 years. Paxton (1922) established that in vegetative material stored for 16 years the fungus survived as dormant mycelium, whereas Isenbeck (1930) recorded that the germination of conidia, kept at 0–5°C for 36 months, was only 20 per cent, an indication that the conidia of this fungus possibly are shorter living than the dormant mycelium. In seed of *Phleum nodosum*, refrigerated at −20°C, *D. phlei* remained viable for 7 years (Smith, 1970*b*). Suzuki (1930) found that *D. oryzae* can survive in seed of rice for 4 years at least.

Arndt (1953) established that cotton seed stored at 1°C for 13.5 years was still capable of germinating and that *Colletotrichum gossypii*, anthracnose, and *Fusarium moniliforme* had survived and could infect seedlings. *Colletotrichum gloeosporioides* (*C. hibisci*), tip blight of kenaf (*Hibiscus cannabinus*), has remained viable for 31 months in seed (Sy and Lo, 1958). Weimer (1952) found that *Colletotrichum gloeosporioides*, anthracnose, survives in blue lupin, *Lupinus angustifolius*, for about 18 months in storage.

The longevity of fungi is, of course, quite dependent on storage conditions. Lutey and Christensen (1963) found substantial reduction in percentage of field fungi such as *Alternaria* (undoubtedly *A. tenuis*), *Drechslera*, and particularly *Fusarium* in barley kernels kept for a few months at 14 per cent moisture content at 20°C. Kept under these conditions for 24 weeks all *Drechslera* and *Fusarium* died out, and *Alternaria* was reduced from 80 per cent in the original seed to 10 per cent. The germination capacity of the seed remained high, above 90 per cent. When the moisture content was reduced to 12 per cent, the field fungi mentioned were all present, although in low percentages, after 53 weeks of storage at 20°C. On the other hand, grain stored at 16 per cent moisture content and 20°C and 30°C was invaded by storage fungi before the field fungi died, and the germination percentage of the seed decreased rapidly.

Downy mildews are usually short-living; mycelial infection of *Sclerospora sorghi* in sorghum seed may endure storage for less than 2 months (Jones, Leeper and Frederiksen, 1972), and other downy mildews die when the seed is air-dried before planting, cf. Section 20.1. However, oospores of *Peronospora manshurica* on soybean seed can germinate after 8 years of seed storage (Pathak, Mathur and Neergaard, unpublished), and Wester, Goth and Drechsler (1966) established transmission of *Phytophthora phaseoli* in overwintered seeds of lima bean only if oospores were present.

Smut fungi lose their viability slowly. Resting mycelium of loose smut of barley, *Ustilago nuda*, has been found viable after up to 11 years (Porter, 1955*a*), thus clearly outliving the host. The fungus was still pathogenic after 11 years in the seed, and was capable of forming promycelium, but not sporidia, hence no new generation of spores could perpetuate the fungus. Fezer (1959) investigated over a period of 4 years the survival of loose smut of barley and found for 29 cultivars of barley, differing in susceptibility, an average of 33 per cent in 1956, the year after the mother plants were inoculated by the air-blast method, and 34 per cent in 1959. He concluded that there is no evidence of difference in survival time between barley seed and the smut harboured in it, during the first 4 years following harvest. Russell (1961), on the other hand, in Saskatoon, Canada, found in 15 stocks of barley seed sown each year, a decline of loose smut in the growing crop from 14 per cent at the start of his studies to 3 per cent. The initial smut percentage was approximately equal to the percentage of infected embryos found in the seed. He established that the reason for the decline was the successively larger proportion of infected seeds that failed to germinate, and that the parasite, although dormant, was able to kill the host as it became weakened by ageing. As it appears, dormant mycelium of the loose smut fungi of barley and wheat is long-living. Teliospores of *Ustilago nuda*, on the other hand, have been found non-viable after only a few months when stored *in vitro* at room temperature (15–16°C). However, when stored at −2–0°C 14-year-old spores were still able to infect flowers and produce smutted heads.

Tilletia caries, wheat bunt, has been found to survive as teliospores in wheat seed stored for more than 11 years and the disease was established in plants from 9-year-old contaminated seed (Caspar, 1926). When kept in the laboratory the fungus may survive for more than 18 years (Fischer, 1936; Kendrick, 1954), and *T. foetida* survived after 25 years in herbarium material (Fischer, 1936). Differences in longevity between pathogenic races of *T. caries* and *T. foetida* have been recorded. It made little difference whether teliospores were in balls or powder (Kendrick and Holton, 1960). The teliospores of *Neovossia indica* (*T. indica*), Indian bunt of wheat, collected in bunted grains, have been found to survive for 4–5 years at most (Mathur and Ram, 1963).

Fischer (1936) established, in a study of herbarium specimens of 77 species of many genera of smut fungi, longevity of a remarkable length for several species: *Sphacelotheca sorghi* 10 years, *Ustilago hordei* 23 years, *U. avenae* 13 years, *U. bromivora* 10 years; and in *U. crameri*, allegedly 64 years old, 1 per cent of the spores were still capable of germinating! Fischer established as a general rule the long life-span of smut fungi, particularly in the *Tilletiaceae*. Tarr (1962) and Ramakrishnan (1963) indicate that the teliospores of *Sphacelotheca cruenta* retain viability for 4 years and Noble (1933) that teliospores of *Urocystis agropyri* survive at least for 10 years, when kept dry.

There seems, however, to be at least one exception from the general long life-span of smut fungi. Kreitlow (1947) suggests control of *Ustilago striiformis* on *Phleum pratense* by storage of the seed for one year.

Little is known about longevity of the few seed-borne rust fungi. Agarkov and Assaul (1963) found that teliospores of *Uromyces betae*, beet rust, maintained viability for nearly 2 years in store houses. Prasada and Chothia (1950) observed two morphologically similar types of teliospores

Table 20.3. Recorded maximum longevity of seed-borne fungi — Summary of review

	Taxonomic category	*Fungus*	*Host*	*Longevity* (years)
Mostly short-living (microbiotic)	Moniliaceous Hyphomycetes	Acremonium strictum	Zea mays (systemic!)	7
		Botrytis anthophila	Trifolium pratense (systemic!)	6
		Botrytis cinerea	Many hosts	2 – 3
		Cercospora beticola	Beta vulgaris	2 – 3
		Cercospora kikuchii	Glycine max	2
		Fusarium avenaceum	Hordeum vulgare	2
		Fusarium culmorum	Hordeum vulgare	2
		Fusarium dimerum	Oryza sativa	2 – 3
		Fusarium graminearum	Zea mays	<2
		Fusarium moniliforme	Gossypium (systemic!)	13.5
			Zea mays (systemic!)	8
		Fusarium oxysporum	Trifolium pratense (systemic!)	6
		Gloeotinia temulenta, mycelium	Lolium perenne	2
		Pyricularia oryzae, conidia	Oryza sativa	2
		Pyricularia oryzae, mycelium	Oryza sativa	4
	Uredinales	Puccinia carthami	Carthamus tinctorius	1.5
		Uromyces betae	Beta vulgaris	>2
Fairly long-living (mesobiotic)	Melanconiales	Colletotrichum gloeosporioides	Hibiscus cannabinus	3
		Colletotrichum gloeosporioides	Lupinus angustifolius	1 – 2
		Colletotrichum gossypii	Gossypium	13.5
		Colletotrichum lini	Linum usitatissimum	6
		Kabatiella caulivora	Trifolium pratense	1
	Sphaeropsidales	Ascochyta pisi	Pisum sativum	7
			Vicia faba	9
			Vicia villosa	7
		Phoma betae	Beta vulgaris	5
		Phoma exigua	Phaseolus vulgaris	2 – 3
		Phoma lingam	Brassica oleracea	4
		Phoma valerianellae	Valerianella locusta	3
		Phomopsis leptostromiformis	Lupinus luteus	3
		Selenophoma bromigena	Bromus inermis	1.5 – 7
		Septoria apiicola	Apium graveolens	2 – 5
		Septoria nodorum	Triticum aestivum	7
	Ascomycetes	Claviceps purpurea	Many hosts	>13
		Diaporthe phaseolorum	Phaseolus lunatus	3
		Protomyces macrosporus	Coriandrum sativum	8
		Sclerotinia sclerotiorum (sclerotia)	Many hosts	7

Table 20.3 *continued*

	Taxonomic category	*Fungus*	*Host*	*Longevity* (years)
Mostly long-living (some macrobiotic)	Dematiaceous Hyphomycetes	Alternaria brassicicola	Brassica oleracea	7 – 8
		Alternaria longissima	Oryza sativa	>5
		Alternaria tenuis	Brassica oleracea	6 – 7
			Hordeum vulgare	10
			Triticum aestivum	10
		Aureobasidium lini	Linum usitatissimum	4 – 5
		Drechslera avenac	Avena sativa	10
		Drechslera graminea	Hordeum vulgare	10
		Drechslera oryzae	Oryza sativa	>4
		Drechslera phlei	Phleum	7
		Drechslera sorokiniana	Hordeum vulgare	10
		Drechslera teres	Hordeum vulgare	10
		Phaeoisariopsis griseola	Phaseolus vulgaris	<1
		Stemphylium botryosum	Brassica oleracea	7 – 8
		Trichoconis padwickii	Oryza sativa	>4
		Ulocladium consortiale	Brassica oleracea	7 – 8
	Ustilaginales	Sphacelotheca cruenta	Sorghum spp.	4
		Sphacelotheca sorghi	Sorghum spp.	10
		Tilletia caries	Triticum aestivum	7 – 18
		Tilletia foetida	Triticum aestivum	25
		Tilletia indica	Triticum aestivum	5
		Urocystis agropyri	Triticum aestivum	10
		Ustilago avenae	Avena sativa	13
		Ustilago bromivora	Bromus	10
		Ustilago crameri		64
		Ustilago hordei	Hordeum vulgare	23
		Ustilago nuda, mycelium	Hordeum vulgare	11
		Ustilago striiformis	Phleum pratense	<1
		Ustilago tritici, mycelium	Triticum aestivum	5

in *Puccinia carthami*, one short-living, for a few months only, and ready to germinate soon after formation, the other requiring a resting or maturation period of about a half-year. Calvert and Thomas (1954) established that the teliospores on seed stored for about 18 months in an unheated building in Texas were infective.

Contradictory results on individual fungus species may occur. These discrepancies are probably mainly due to differing environmental conditions. On the other hand there is evidence enough to make some fairly clear distinctions with reference to different natural groups of fungi (cf. Table 20.3).

(1) *Moniliaceous Hyphomycetes*. Obviously, more or less hyaline fungi characterised by thin-walled conidia ('hyalospores'), mainly *Hyphomycetes* such as *Botrytis*, *Cercospora*, *Gloeotinia* (conidial stage) and most

of *Fusarium* spp. considered, are usually short-lived; they may often be controlled by storage of the seed for a suitable period of time. However, noteworthy exceptions in this group are *Fusarium moniliforme* and *Acremonium strictum* both of which may invade their host systemically. The location and condition of these fungi within the tissue of the seed may account for their protracted longevity.

(2) *Dematiaceous Hyphomycetes*. Fungi that may be placed at the other extreme by the characters of strong pigmentation and thick conidial walls ('phaeospores'), as exemplified by species of *Drechslera*, have been found to be quite long-lived, according to various authors. A special case is the genus *Aureobasidium*, the thick-walled arthrospores of which are coloured, whereas the thin-walled yeast-like, budding conidia are hyaline and not likely to be very persistent.

(3) *Coelomycetes*. Most of the species within *Sphaeropsidales* and *Melanconiales*, as represented in the list, appear to be persistent. This is possibly due to protection of spores by the fruiting bodies, or to the ability of the fruiting bodies of some of these fungi to produce fresh conidia after years of storage.

(4) *Ustilaginales*. Some smut fungi are long-lived, both in the form of resting mycelium, deeply embedded in the tissue of the seed, and as surface-borne pigmented spores.

(5) *Resting mycelium*. Resting mycelium, especially if thick-walled, protected within the tissue of seed, usually the seed coat, has a much better chance of survival than spores borne unprotected on the surface such as those of *Cercospora beticola*. Hyaline conidia of *Pyricularia oryzae* may keep alive usually for only about a year, whereas resting mycelium may survive for at least 4 years, dry sclerotia of *Sclerotinia sclerotiorum* for 7 years (Chupp and Sherf, 1960), and sclerotia of *Claviceps purpurea* for more than 15 years (Clark, 1972).

The species of the host, in particular the conditions provided by the anatomical and physiological character of the seed, is likely to influence the longevity of fungi. Direct comparative studies in this respect are lacking but it appears that cotton seed provides particularly favourable conditions, so that *Colletotrichum gossypii* and *Fusarium moniliforme* maintain their vitality and virulence up to 13.5 years.

Also the condition of spores at the time of harvest has bearing on their longevity. Sampson (1928*a*) found in her studies on the viability of the spores of oat smut fungi that the state of maturity largely determines the longevity. Fischer and Holton (1957) refer to the general belief that 'in all the smut fungi the fully mature spores maintain their viability for a much longer time than do less mature spores of the same species or even of the same collection'.

Further studies under controlled optimum storage conditions and due consideration of the condition of seed and pathogens at the time of harvest are needed. The condition of the fungus in and on the seed during storage should be specifically investigated. In this connection inherent differences between the kinds of seed harbouring a definite pathogen should be taken into account. Determination of survival should include testing of the virulence and sensitivity to seed treatments.

20.2.3 Bacteria

Bacteria borne on the surface of seed may survive for a period of 2–3 years. This holds true in the case of *Xanthomonas vesicatoria* on tomato (Gardner and Kendrick, 1921*a*, 1923; Mickail, Bishay and Farag, 1970). Xanthomonads are relatively resistant because of their capacity to form a mass or 'gum'. Gardner and Gilbert (1921) observed that *Pseudomonas lachrymans* on cucumber could survive storage for more than 20 months but less than 32 months. Graham (1953*a*) found that *P. glycinea* and *P. tabaci* on seed of soybean still were alive after storage for 6 months but not after storage for 18 months, while *Xanthomonas phaseoli* var. *sojense*, bacterial pustule disease of soybean, in the same seed lot survived for 30 months and produced primary infection on seedlings when the seed was sown in soil. Nicholson, Sinclair and White (1973), however, established survival up to 24 months for *P. glycinea* in soybean.

According to Chupp and Sherf (1960) *P. phaseolicola* lives longer than 2–3 years on the seed, and Schuster and Sayre (1967) isolated virulent *Corynebacterium flaccumfaciens* var. *aurantiacum*, *Xanthomonas phaseoli* and *X. phaseoli* var. *fuscans* from 15-year-old seed of Phaseolus bean, and *C. flaccumfaciens* var. *violaceum* from 8-year-old seed stored at 10°C. Basu and Wallen (1966) established that *X. phaseoli* var. *fuscans* was viable in bean seed for 3 years at temperature ranging from −20°C to 35°C. Johnson and Murvin (1925) found that *Pseudomonas tabaci* could survive 2 years' storage of tobacco seed. Hunter and Brinkerhoff (1964) established long term survival of *X. malvacearum* in and on cotton seed, the pathogen being viable after more than 4.5 years. *Xanthomonas oryzae* has been isolated from about 12 per cent of rice seed, kept for two years and four months (Soepriaman, Mathur and Neergaard, unpublished). Habish and Hamad (1971) recorded that *X. sesami* on inoculated seed of sesame, *Sesamum orientale*, survived on about 10 per cent of the seeds for up to 16 months; this was much longer than the period of survival in moist soil (5 months) and dry soil (a little more than 7 months). Chupp and Sherf (1960) inform that *Pseudomonas apii* lives on and in celery seed for one year after harvest.

Corynebacterium insidiosum, bacterial wilt of lucerne, was still viable in lucerne seed stored at 21–27°C for about 3 years, and the pathogen was still capable of infecting roots of lucerne (Cormack, 1961). He found that lucerne straw was a better medium for survival of the bacteria than seed; they survived storage for 10 years in infected lucerne hay. In wheat *Corynebacterium tritici* has remained viable for at least 5 years in the galls of *Anguina tritici* (Mathur and Ahmad, 1964).

Table 20.4. Recorded maximum longevity of seed-borne bacteria — Summary of review

Pathogen	*Host*	*Longevity* (years and months)	
Corynebacterium flaccumfaciens	Phaseolus vulgaris	24	
— var. aurantiacum	Phaseolus vulgaris	15	
— var. violaceum	Phaseolus vulgaris	8	
Corynebacterium insidiosum	Medicago sativa	3	
Corynebacterium tritici	Triticum aestivum	5	
Pseudomonas apii	Apium graveolens	1	
Pseudomonas glycinea	Glycine max	2	
Pseudomonas lachrymans	Cucumis sativus	>1	8
Pseudomonas phaseolicola	Phaseolus vulgaris	2–3	
Pseudomonas tabaci	Glycine max	0	6
Pseudomonas tabaci	Nicotiana tabacum	>2	
Xanthomonas malvacearum	Gossypium hirsutum	4	8
Xanthomonas oryzae	Oryza sativa	2	4
Xanthomonas phaseoli	Phaseolus vulgaris	15	
— var. fuscans	Phaseolus vulgaris	15	
— var. sojense	Glycine max	2	6
Xanthomonas sesami	Sesamum orientale	1	4
Xanthomonas translucens	Triticum aestivum	3	
Xanthomonas vesicatoria	Lycopersicon esculentum	>2	

Klykov (1945) found, in surface-disinfected wheat kernels macerated in sterile water, the following quantities of living *Xanthomonas translucens*: at the beginning of the investigations 12,600,000 bacteria per gramme grain, after storage for 6 months 870,000 bacteria per gramme, and after storage for 3 years 1,000 bacteria per gramme. For another cultivar he found 860,000, 181,000 and 4,000 bacteria, respectively, while seven other cultivars which contained from 690,000 to 890,000 bacteria per gramme grain in the beginning were free of bacteria after a period of 3 years.

However, the record of established survival of bacteria on seed seems to be 24 years of persistence of *Corynebacterium flaccumfaciens* in bean seed recorded by Burkholder (1945) in the U.S.A. (see Table 20.4).

20.2.4 Viruses

Some viruses may survive in seed for only a few months, others may persist for many years (Table 20.5).

Embryo-borne viruses usually show considerable longevity. Bean common mosaic virus persists for many years in bean seed; Polak and Chod (1967) found it to be transmitted after 6 years, and Pierce and Hungerford (1929) that it was infective even after 30 years in the seed.

Table 20.5. Recorded maximum longevity of seed-borne viruses — Summary of review

Disease	*Host*	*Longevity* (years and months)	
Alfalfa mosaic	Medicago sativa	5	
Barley stripe mosaic	Hordeum vulgare	6	6
	Triticum aestivum	3	
Bean common mosaic	Phaseolus vulgaris	30	
	Glycine max	2	
Bean southern mosaic	Phaseolus vulgaris	0	7
Bean western mosaic	Phaseolus vulgaris	3	
Broad bean stain	Vicia faba	1	
Cowpea (aphid-borne) mosaic	Vigna sinensis	2	
Cucumber mosaic	Stellaria media	1	6
	Phaseolus vulgaris	2	3
Echtes Ackerbohnenmosaik	Vicia faba	1	
Hop nettlehead	Humulus lupulus	2	
Lychnis ringspot	Lychnis divaricata	9	
	Silene noctiflora	1	9
Prune dwarf	Prunus cerasus	4	6
Prunus necrotic ringspot	Prunus pennsylvanica	6	
Raspberry ringspot	Capsella bursa-pastoris	6	
	Stellaria media	6	
Sowbane mosaic	Chenopodium murale	14	
Squash mosaic	Cucurbita pepo	3	
	Cucumis melo	5	
Tobacco mosaic	Lycopersicon esculentum	9	
Tobacco ringspot	Glycine max	6	
	Nicotiana tabacum	5	6
	Petunia violacea	0	6
Tomato black ring	Capsella bursa-pastoris	6	
	Stellaria media	6	
Tomato bunchy top	Physalis peruviana	1	2
	Solanum incanum	3	3

Bean (western) mosaic virus retained its full activity after 3 years of storage (Skotland and Burke, 1961). In *Vicia faba* both broad bean stain virus and Echtes Ackerbohnenmosaikvirus kept their infectivity for 1 year in seeds stored at 15°C (Cockbain, Cook and Vorrai-Ura, 1973), and in cowpea seeds, stored in paper bags at 'room temperature' (25–40°C) cowpea (aphid-borne) mosaic virus was infective after 2 years, although the infection percentage was reduced by about one-half (Khatri and Chohan, 1972). In lucerne seeds, 20 per cent carrying alfalfa mosaic virus, there was no significant decrease during 5 years of storage (Frosheiser, 1970, 1974). On the other hand, bean southern mosaic virus is not very persistent in Phaseolus beans, keeping infective for less than 7 months, but a strain of the virus, 'bean mosaic virus 4A', did survive for more than 7 months (Zaumeyer and Harter, 1943*b*).

Tobacco mosaic virus has been found to decline rapidly on tomato seed after storage for 1 year, but some virus could still be recovered after 3 years (Alexander, 1960), and Broadbent (1965) established that the virus may persist for 9 years in the endosperm.

McNeal, Mills and Berg (1961) found no loss of viability or virulence of barley stripe mosaic virus in wheat seed after storage of the seed for 3 years, and Scott (1961) detected by serological means the virus in barley seed stored for 6.5 years.

Sowbane mosaic virus was still transmissible from seed of *Chenopodium murale* stored for 14 years (Bennett, 1969). Hop nettlehead (hop infectious sterility) virus remained infective for more than 2 years in hop seed (Blattny and Osvald, 1954).

Lychnis ringspot virus showed no decrease in seed of *Lychnis divaricata* after 9 years of storage (Bennett, 1969); in *Silene noctiflora* the virus was still active after almost 2 years (Bennett, 1959).

The infection percentage of squash mosaic virus decreased from 28 to 93 per cent in freshly harvested melon seed to 3–5 per cent in 3-year-old seed (Rader, Fitzpatrick and Hildebrand, 1947), but could still be recovered from seeds stored for 5 years (Middleton and Bohn, 1953). According to Middleton (1944) squash mosaic virus remained viable in 3-year-old seed of squash, maintaining a constant percentage of seed transmission. In cantaloupe, however, seed transmission generally declines with storage of the seed. Powell and Schlegel (1970) found that seven seed lots which showed 11–31 per cent transmission in 1967 showed no transmission in 1969, and that in only one out of thirteen seed lots the percentage of virus transmission remained essentially unchanged.

Tomato black ring virus and raspberry ringspot virus were retained in seeds of *Capsella bursa-pastoris* and *Stellaria media* for 6 years (Lister and Murant, 1967). In seeds of *S. media* cucumber mosaic virus remained infective after storage at room temperature for at least 1 year and 6 months (Häni, 1971), in seeds of Phaseolus bean for at least 2 years and 3 months (Bos and Maat, 1974).

Tobacco ringspot virus was still present in tobacco seeds after 5.5 years of storage (Valleau, 1939), while in seeds of *Petunia violacea* survival of the virus was recorded after 6 months of storage (Henderson, 1931). In soybean seed stored for 6 years the infection percentage of the virus remained almost unchanged (Laviolette and Athow, 1971).

McClean (1948) reported activity of tomato bunchy top virus in seed of *Physalis peruviana* after storage for 3 years, in seed of *Solanum duplosinuatum* for 1 year and 6 months, and in seed of *S. incanum* after 6 years.

Gilmer (1964) recovered prune dwarf (cherry sour yellows) virus from cherry seed stored for 54 months at 4°C. He points out that storage of infected cherry seed is a practical research procedure for maintaining the virus protected against other viruses, and the virus is recoverable at any season. Fulton (1964) found infection of Prunus necrotic ringspot virus constant in seed of *Prunus pennsylvanica* after 4 years of storage at 2°C, while after 6 years the infection percentage had dropped materially with little loss in seed viability.

20.2.5 Nematodes

Nematodes may outlive seed for many years. *Anguina tritici,* ear cockle, in wheat survives as larvae in the gall for 14 years, and these are, according to Raeder (1954), still capable of infecting seedlings; Fielding (1951) found them viable after storage for 28 years.

Goffart (1951) established survival of *Ditylenchus dipsaci* in seed of lucerne for 5 years, and in oats for 8 years.

Surviving surface infestation of *Aphelenchoides besseyi*, white tip of rice, for 2 years on rice has been recorded by Todd (1953) and Todd and Atkins (1958), while Yoshii and Yamamoto (1950) found the nematode still living after 3 years.

Epps (1968) found infective larvae in cysts of soybean cyst nematode, *Heterodera glycines*, carried in seed bags stored for 22 months.

20.2.6 Conclusions

Any consideration of control of short-lived seed-borne pathogens by protracted storage must be related to the optimum longevity of the seed concerned, and this is the maximum period of storage in seed trade practice. Under conditions prevailing in commercial warehouses in temperate zones this longevity is for most cereals and vegetables up to 3 years; for a few kinds of seed, such as celery, up to 5 years. In tropical zones the period of storage does not usually exceed 1–2 years, except in specially built warehouses.

Some pathogens die within these periods of time. This holds true for *Kabatiella caulivora* in red clover seed; *Botrytis cinerea,* important in flax and different kinds of flower seed; *Cercospora beticola* in beet seed; *Fusarium graminearum* in barley and maize seed; *Gloeotinia temulenta* in seed of ryegrass; and *Phaeoisariopsis griseola* in bean seed; and, particularly, downy mildew fungi. These pathogens have been established as essentially microbiotic and may be controlled by storing the seed for 3 years or less, with few exceptions.

Storage for 2–3 years may substantially reduce, but not completely remove, inoculum of some important pathogens — *Ascochyta pisi* in pea after 2–3 years, *Septoria apiicola* in celery and celeriac after 1 year, or more often after 3–4 years.

The longevity of a certain pathogen may vary considerably from one seed lot to another and may be a short period of time in some seed lots.

Many pathogenic fungi may survive the seed of their hosts by many years: most fungi belonging to *Dematiaceae* of *Hyphomycetales*, including many species of *Drechslera*, and some of *Alternaria*; several forms of

Sphaeropsidales, including *Ascochyta*, *Phoma* and *Septoria*; some of *Melanconiales*, defined as species of *Colletotrichum*; and apparently most species of smut fungi, including *Tilletia* and *Ustilago*. This is also true of some systemic fungi, *Acremonium strictum* and *Fusarium moniliforme* in seeds of different hosts.

Some kinds of seed seem to provide better conditions for long-term survival of pathogens than others, cotton seed being an example, with infections of *Colletotrichum gossypii* and *Fusarium moniliforme* viable after 13–14 years.

Seed stored for 3 years outlives most species of surface-borne bacteria. Internally borne bacteria, such as *Xanthomonas phaseoli* and *Pseudomonas phaseolicola* in bean seed, and *Corynebacterium insidiosum* in seed of lucerne, may keep alive for 3 years or more, while *C. flaccumfaciens* in bean seed may survive for even more than 20 years.

Some viruses, certainly, are outlived by their hosts and remain infectious for only a few months but most embryo-borne viruses may be expected to be very persistent in seed, such as those infecting seed of leguminous crops and stone fruit trees.

Certain nematodes may outlive their hosts, *Anguina tritici* by almost 30 years.

It must be emphasised that longevity of pathogen as well as of seed depends on storage conditions, which vary greatly. As a general conclusion, control of seed-borne inoculum by ageing of seed must be regarded as unreliable.

21 Seed Treatment, Procedures and Equipment

21.1 DEVELOPMENT OF CHEMICAL TREATMENT

Seed treatment is probably the cheapest and often the safest method of direct plant disease control. Relatively small quantities of plant material are handled, and in principle there is every possibility that all the seeds are in fact treated.

Early history. Seed treatment has been used for centuries. About 1660, salt brine was accidentally discovered to be effective against bunt of wheat (Table 21.1). The procedure of brining, which remained common practice for about a century (Buttress and Dennis, 1947), originated from a cargo of wheat that sank near Bristol (Tull, 1733); being unfit for bread the grain was purchased by farmers and sown; all other wheat in England, it is said, was smutty that year. Tull, however, realised the uncertainty of this experience, as he remarks, '. . . our drowned Wheat might possibly be Foreign, and then might not have been Smutty the next year, tho' it had been soak'd in the Sea-Water' (from Buttress and Dennis, 1947).

At the end of the 18th century Abbé Tessier, in France, tested many chemical compounds to improve wheat seed, without affecting agricultural practice (Woolman and Humphrey, 1924). After Prévost (1807) had demonstrated the effect of copper sulphate against bunt of wheat this fungicide was adopted although not widely used, until Kühn (1873) showed its effectiveness when applied to seed. A lime bath after treatment was recommended to avoid injury to the seed (Woolman and Humphrey, 1924) and in effect this treatment was a precursor of Bordeaux mixture.

Twentieth century. Formaldehyde was the next major chemical to be used as an effective and cheap fungicide against many kinds of seed-borne fungi, particularly smuts (Geuther, 1895; Bolley, 1897). It is less injurious to seed than copper sulphate.

Copper carbonate was the first dry-treating material used in control of wheat smut (Darnell-Smith, 1910, 1915, 1917). According to McCallan (1967), by 1925 it was estimated that in the U.S.A. one-tenth of all the

Table 21.1. Three centuries of germicidal seed treatment.

1660	Salt brine discovered as effective against bunt of wheat
1807	Copper sulphate discovered as effective against bunt of wheat
1873	Biological establishment of copper sulphate for seed treatment against smut
1888	Hot water treatment against cereal smuts introduced
1895	Formaldehyde introduced for seed treatment
1909	Biological establishment of hot water treatment specifically for loose smut of barley and wheat
1910	Copper carbonate introduced as dust for wheat smuts
1912	Organic mercurials discovered for seed treatment
1920–30	Copper carbonate dust generally used Liquid mercurials used on a small scale
1930–40	Mercurial dusts and some non-mercurial dusts introduced
1940–60	Wettable dusts, slurries and fluid concentrates introduced and used on a large scale
1960–70	Large-scale use of mercurials, increasing use of non-mercurials Systemic fungicides introduced, applicable against loose smut of barley and wheat
1970–	General banding of mercurials and replacement of these by non-mercurials

seed wheat was treated with copper carbonate dust to control bunt.

The modern era of fungicides and seed treatment started with the introduction of the organic mercury compounds. Riehm (1913, 1914, 1915) recommended treatment of wheat seed with mercury chlorophenol, after investigations during 1911–1912 by Wesenberg had established its suitability as a seed treatment (Wesenberg, 1938). The first product containing this compound was marketed by Bayer's in Germany under the trade name Uspulun, and in 1920 a new German product, Germisan, was marketed. These and similar preparations now replaced copper sulphate for seed treatment. However, various other copper compounds, such as basic copper sulphate, red copper oxide, copper oxychloride and particularly copper carbonate, have remained in use for dry seed treatment.

A new category of mercury preparations based on ethylmercuric chloride later appeared and included a preparation marketed in 1926 under the name of Ceresan. New Improved Ceresan, based on ethylmercuric phosphate and introduced in the U.S.A. in 1933, is highly volatile and controls smut at about one-fourth the application rate of the original Ceresan (Fischer and Holton, 1957).

Post-war organic mercurials. During and after World War II other preparations of mercury compounds followed, such as phenylmercuric urea (e.g. Leytosan and Agrox), phenylmercuric acetate (Mersolite), cyano(methylmercuric)guanidine — all dry treatments. Liquid preparations of the two latter are used and of 2-methoxyethylmercuric acetate, e.g. under the trade name Mercuran.

Ceresan M, ethylmercuric *p*-toluene sulphonanilide, appeared in 1948 and largely replaced the previous formulations. Panogen is a Swedish preparation originally based on methylmercuric dicyandiamide; Semesan is based on hydroxymercuric chlorophenol; and Aagrano, produced in the Netherlands, is based on methoxypropylmercuric bromide.

Post-war non-mercurial organic fungicides. Since World War II there has been a vigorous development of non-mercurial organic fungicides applicable to both seed and growing plants. The first fungicides of this type appeared in the 1930s when, in 1934, Tisdale and Williams took out U.S. patent rights for *thiram* (tetramethylthiuram disulphide, or bis(dimethylthiocarbamoyl)disulphide), a derivative of dithiocarbamic acid:

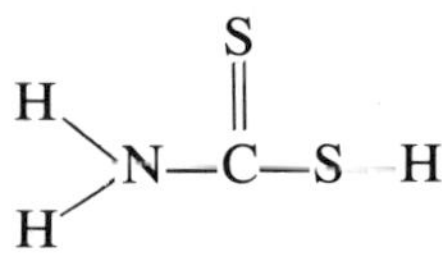

Other dithiocarbamate compounds followed, such as *ferbam* and later *zineb*, see Section 21.5(8). Other organic compounds from this period are *chloranil* and *dichlone*, see Section 21.5(9), *captan*, see Section 21.5(12), and various *polychlorobenzenes*. These were developed during the war, probably as a consequence of the war-time shortage of organic mercury preparations, see 21.5(10).

Origin and development of hot water treatment. Hot water treatment has been the standard method for controlling embryo-infecting loose smuts of barley and wheat for about 75 years. The procedure was established by the Danish seedsman J. L. Jensen, in 1888, who proposed the method for controlling all cereal smuts. However, these have different life-cycles, and although wheat bunt and oat smut could be controlled by soaking the seed for 5 min in water held at 55°C, chemical treatment is easier and effective. For this reason hot water treatment was little used until the life-cycle of loose smuts of barley and wheat was explained, the diseases understood as specific types of infections, and the method taken up on a new theoretical base and elaborated specifically for these smuts (Appel and Gassner, 1907*a*; Freeman and Johnson, 1909; Appel, 1909). According to the modified method, the grain of both cereals was presoaked for 5 h, then wheat was soaked for 10 min in water held at 54°C, and barley for 15 min in water held at 52°C. The method was generally adopted as a standard, although various modifications were developed some of which have been reviewed by Fischer and Holton (1957). The treatment is used as a routine for foundation seed of barley in New Zealand and other countries. Hot water treatment of cereals must be followed by treatment with a mercury compound because useful antagonists have been removed by the former procedure.

21.2 AIMS AND REQUIREMENTS

The aim of seed treatment is to prevent infection of the seedlings and the subsequent crop. The mechanism of the effect may vary; the inoculum in or on the seed may be killed directly by the treatment itself, or later during the germination of the seed and pathogen.

Treatment of infected seed may thus be curative but should also protect the seeds from attack by organisms that are present in the soil.

21.3 TYPES OF SEED TREATMENT PROCEDURES

A considerable range of physical and chemical procedures have been brought into use for disinfection or protection of seed against pathogenic fungi, bacteria and, to some extent, nematodes. The efficiency of these procedures is highly dependent on the equipment used, often on the mechanical principles adopted for chemical treatment of seed. Seed treaters range from simple containers, adapted for treatment of small batches, to high capacity continuous treaters for industrial processing.

There are two main types of processes — physical and chemical. These two principles may be combined but there are few examples of this having been done.

21.3.1 Physical procedures

(1) *Hot water treatment*. Sometimes, after presoaking, the seed is held in water at a definite temperature, often a little above 50°C and close to the maximal temperature of both pathogen and host for a definite period of time. Heat appears to be the effective agent.

Hot water treatment is commonly used against loose smut, *Ustilago nuda* and *U. tritici*, in barley and wheat; black leg, *Phoma lingam*, and black rot, *Xanthomonas campestris*, in crucifers; *Septoria apiicola* in celery (Walker, 1923, 1969); bacterial blight, *X. incanae*, in stock (Baker, 1956); parasitic nematodes in rice, and stem eelworm, *Ditylenchus dipsaci*, in lucerne (Edwards, 1932).

The treatment is also effective for seed of onion, tomato and some other vegetables.

(2) *Anaerobic water treatment*. The seed is soaked in water and held under anaerobic conditions for a time. For instance, barley was efficiently treated against loose smut by being soaked in water at 24°C and at 28°C for 6 h and then drained and stored in air-tight containers for 42 and 31 h, respectively, with one-half of the container filled with the grain (Hebert,

1955). Another efficient procedure is described by Ujević and Dias (1970). The procedure reduces the hazard of injury to seed germination from overheating.

(3) *Dry heat treatment*. This has been little used, although different combinations of temperature have given promising results against certain pathogens: *Colletotrichum gossypii* in cotton (Lehman, 1925) and *Puccinia antirrhini* in snapdragon (Yarwood, 1948; Baker, 1952; J. Walker, 1954), while attempts against other fungal pathogens gave negative results (Racovita, 1957). For pretreatment in seed health testing of oats positive results were obtained by Malone (1962).

Dry air treatment of tomato seed at 70°C for 3–4 days is partly effective against tobacco mosaic virus — treated seed lots must be tested before being accepted as virus-free (Broadbent, 1965). Lettuce mosaic virus can be inactivated by dry air treatment of lettuce seed for 80–120 days at 55°C without material reduction of the germination capacity of the seed (Kristensen, 1970).

(4) *Solar heat treatment* (solar energy method). In countries with a hot climate, grain (wheat and barley) is presoaked in water for 4–5 h in the shade or in a room and then dried on ground in thin layers in sun for 1 h, usually at noon. The procedure has proved effective against loose smut of barley and wheat (Luthra and Sattar, 1934; Luthra, 1941; Kozhewnikova, 1953; Bedi, 1957) but has been used only to a limited extent in practice in India, particularly in the northern areas such as the Punjab, where the procedure is applied mainly in May or June, and in Pakistan.

The procedure is also used against smuts on sorghum in Burma, India and Tanzania (Tarr, 1962) but certainly fungicidal treatment is preferable for this type of smut.

(5) Aerated-steam treatment. The principle is to treat dry seed in moving vapour, by pressure flow. Different types of equipment have been designed — a recently developed type is discussed below in Section 21.4(11). Baker (1969*b*) mentions the following successful treatments: beet against *Phoma betae*, 56°C/20 min, after humidification; china aster against *Alternaria tenuis* and *Botrytis cinerea*, 53°C/30 min, non-humidified; cabbage against *Alternaria brassicae* and *Phoma lingam*, 56°C/30 min, after humidification; parsnip against *Itersonilia pastinacea*, 45°C/30 min, non-humidified; stock against *Fusarium oxysporum* f.sp. *matthiolae*, 54°C/15 min, non-humidified; and zinnia against *Alternaria zinniae*, 57°C/30 min, non-humidified.

21.3.2 Biochemical procedure

(6) *Aerobic fermentation*. As a regular procedure for separating tomato seeds from the macerated fruit pulp, the mixture is fermented for 96 h at 20°C; this treatment is effective against the bacterial canker organism owing to the increase in acid content of the matrix. As a derivative procedure, freshly extracted seed may be disinfected by soaking in 0.8 per cent acetic acid for 24 h (Walker, 1969). If, however, alkali is used for seed extraction, virus infection is not killed.

21.3.3 Chemical procedures

(7) *Steep treatment*. The seed is soaked in a fungicidal liquid solution or suspension, for a definite period of time and then drained and dried.

(8) *Sprinkle treatment*. The seed is sprinkled with a fungicidal liquid, solution or suspension, left damp with this for a definite period of time, and then dried.

(9) *Dust treatment*. The seed is mixed with a dust fungicide in a mixing device until the seed is well coated. Usually about 2 g of mercurials per kg grain is used. Handling hazards must be considered.

(10) *Slurry treatment*. The seed is mixed with a dust fungicide in a special treater in which small calibrated amounts of liquid (5–20 ml./kg grain) are added, thus forming a soup-like slurry to ensure coating without undue wetting.

(11) *Wet treatment* (Benetzungsbeize, according to Gassner, 1927*a*). The seed is mixed with a relatively small amount of concentrated liquid (about 100–300 ml./kg grain). There is no recovery of liquid after treatment, but drying is necessary.

(12) *Quick-wet (or short-wet) treatment*. The seed is mixed with a relatively small quantity of a concentrated, usually volatile liquid fungicide (about 20–40 ml./kg grain), ensuring good coating. The treated seed usually requires no drying, at least not when it is sown within a few days.

(13) *Oil-fungicide (Panogen) treatment* (Zade, 1943). The seed is mixed with a slightly volatile mercurial germicide in an oily suspension in extremely small ('ultralow') quantities (1–3 ml./kg grain). No subsequent drying is needed. The process is used also with slightly volatile pesticides other than Panogen, e.g. chlorinated hydrocarbon insecticides (Lindström, 1961).

(14) *Fumigation*. The seed is treated in an air-tight container or chamber for a definite period of time by use of volatile material, a fungicide, nematicide (or insecticide).

(15) *Pelleting*. Application of a sticker prior to mixing the seed with a dust which may contain pesticides, fertilisers or other suitable materials. The seed is coated first with methocel, a cellulose acetate sticker in dilute solution, and is then agitated with the dust in such a way that each seed becomes a pellet containing an outer zone of protectant (Walker, 1969). Sometimes seed is treated with a fungicide before pelleting, such as sugar beet seed which is usually steeped in ethylmercuric phosphate before the process.

The seeding rate is decreased when pelleted seed is used; consequently there is a bigger demand for high quality seed than when natural seed is used.

21.4 SEED TREATMENT EQUIPMENT

The basic principle in any chemical seed treatment is mixing seed and chemical in order to obtain adequate coating and, for some treatments, adequate penetration.

(1) *The shovel method*. The simplest implement for treating seed is the shovel. The seed is piled on a grain-tight floor, fungicide dust is scattered over the pile and mixed by portions until even coating of the seed appears to be obtained. The procedure may also be used with liquids as in application of formaldehyde to oat seed (Koehler, 1935). The method is primitive, not very efficient and not without danger to the operatives.

Cherewick and Popp (1951) made a survey of the effect of seed treatments in actual use in Canada during the years 1947–1949. At that time 55 per cent of the farmers in Manitoba used a treating machine; the others used shovel, a loading auger or other less effective methods of application. Farmers who treated seed of oats and barley with a machine had an average of 0.72 per cent smut in their fields, whereas those who used the more primitive methods had an average of 1.32 per cent.

Simple home-made devices for application of liquids and dust have been extensively used and are still in common use. In developing countries the use of such simple treaters may appropriately be encouraged.

(2) *Stationary upright gravity treater*. In its simplest form this usually home-made treater for small quantities of seed consists of a cylinder, usually a metal oil drum, containing a hopper with a funnel-shaped base through which a regulated flow of grain falls in the apex of a screen cone to allow dust to pass through, and from this cone into a further inverted cone from which it passes through a chute into bags (Dillon Weston and Taylor, 1945; Machacek, Russell and Tyner, 1950). This gravity-feed type

(Fig. 21.1) is not so efficient in producing an even mixing as the rotational drum or barrel described below. Dillon Weston and Taylor (1945) compared the mechanical principles of these two treaters, assessing them in controlled experiments with bunt of wheat, *Tilletia caries*. They found the principle of treatment by rotation clearly superior to that of gravity-feed operation.

The *Minnesota seed grain treater* for dry treatment applications (Moore, 1937) is a popular small-scale gravity treater with a capacity of about 40 bushels per hour (Purdy, 1967).

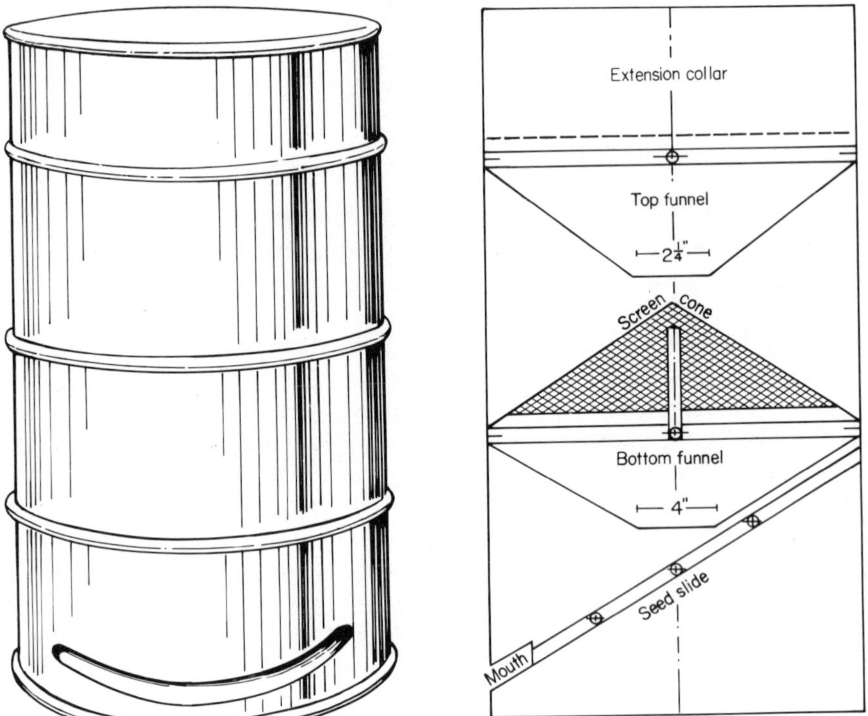

Fig. 21.1. Stationary gravity treater. (Redrawn from Machacek, Russell and Tyner, 1950)

(3) *Hand-rotated treater*. This is a revolving drum or barrel, as exemplified in Oldenberg's barrel (Fig. 21.2). The container is usually provided with baffles inside to increase agitation. Home-made treaters are often made from metal oil drums. The container may be placed in a diagonal position by which further agitation in rotating is obtained. The rotary treater should be turned slowly for about 50 revolutions (cf. Dillon Weston and Taylor, 1945) for the seed to tumble from one end of the container to the other, so as to give even distribution of the agent.

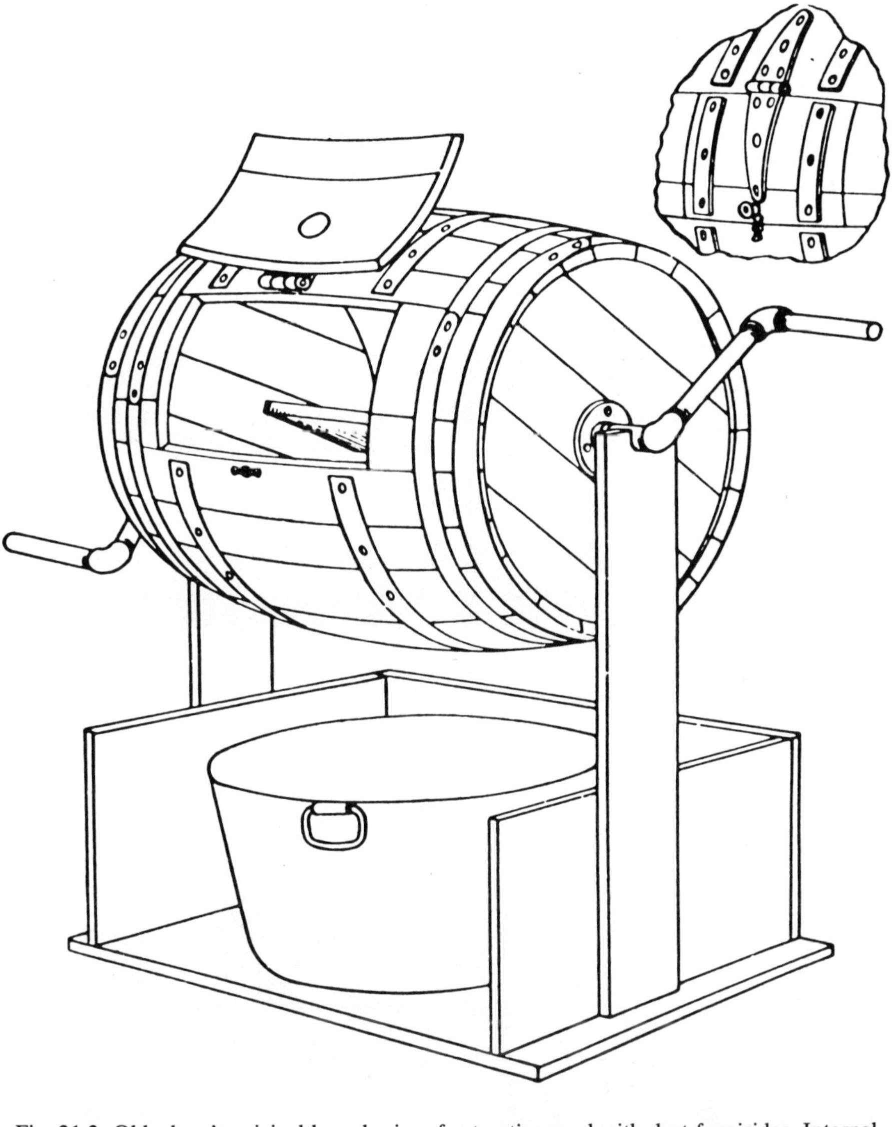

Fig. 21.2. Oldenberg's original barrel mixer for treating seed with dust fungicides. Internal baffles, one of which can be seen through the open shutter, increase agitation of seed with dust. Nowadays an oil drum is more likely to be used. (Leukel, Martin and Lefebvre, 1951)

The rotary drum or barrel is suitable for small seed lots of small grain and in general of seed that is not easily injured. For large-seeded legumes and similar kinds of seed, sensitive to damage, this procedure might be unsatisfactory. The treater lends itself well to small farming and is widely recommended in developing countries.

A hand-rotated treater, adapted for short-wet treatment has been described by Nagel (1927).

(4) *Smaller power-rotated treaters*. Several modifications following the principle of agitating by rotation, and driven by motor, are in use. Small or standard models of cement mixers are used frequently in treating seed. Rickard and Moeller (1970) described a roller-type mill for treating seed, applicable both for powders and liquids (Fig. 21.3).

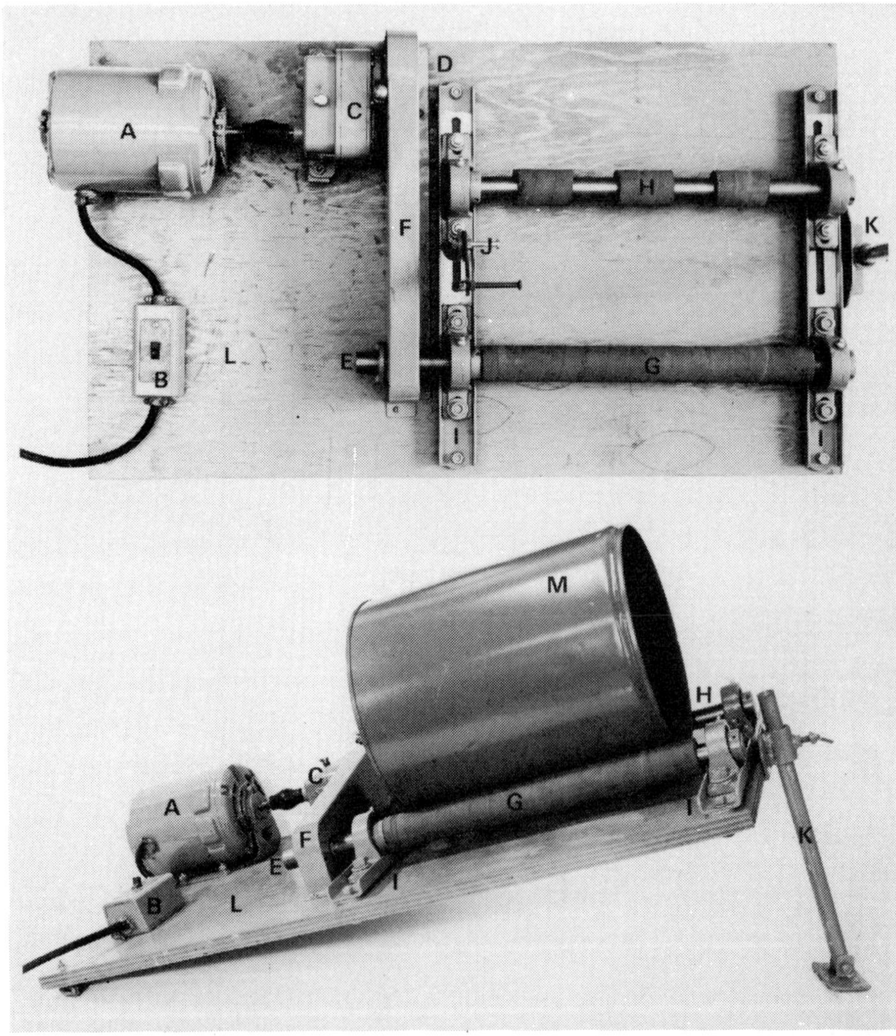

Fig. 21.3. Modified roller-type mill for treating seeds and soil. A, 1/6 hp electric motor; B, on–off switch; C, variable speed transmission; D, drive sprocket; E, roller sprocket; F, chain guard; G, driving roller; H, idling roller; I, roller adjustment tracks; J, container stop; K, adjustable leg; L, plywood platform; M, 5-gallon container. (Rickard and Moeller, 1970)

(5) *Continuous operation dust treaters*. This type of seed treating machine was the first to be developed after World War I as a consequence of the wider use of copper carbonate in wheat against bunt. The seed is mixed with the dust in different types of chambers, in rotating drums or by gravity through chutes provided with baffles (Hurst *et al.*, 1936). Accurate metering of copper carbonate dust is not needed as excess dust will fall off into the bottom of the sack or bulk bin, and underdosages can be observed as insufficient coverage of seed with the chemical and can easily be checked at the discharge end of the treater (Hurst *et al.*, 1936). When organic mercurial dusts are used accurate metering is imperative; the relatively low rates of applications make it difficult to estimate adequate covering by eye. Hurst *et al.* (1936) developed a fungicide feeder with accurate metering, thus providing the first well-calibrated seed treating machine.

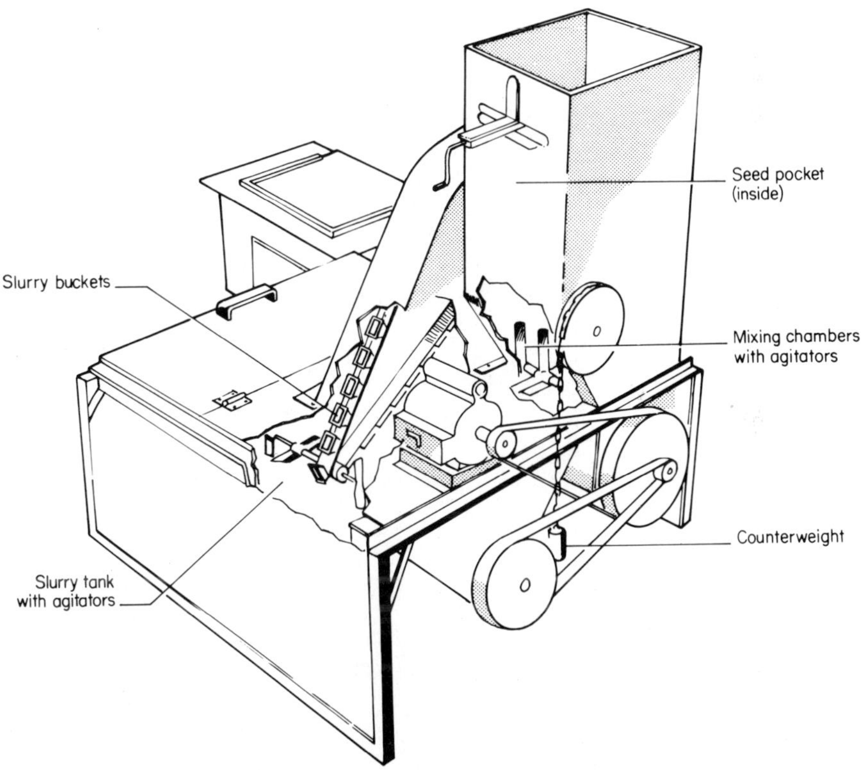

Fig. 21.4. Slurry treater, which is equipped with a seed gate that controls the seed flow pattern to the weighing or seed dump pan. A constant dump weight for a given kind of seed is obtained with proper seed gate setting. The mixing chamber is fitted with an auger-type agitator that mixes and propels the seed to the bagging end of the mixing chamber. (E.I. du Pont de Nemours & Co., Inc., Seed Treating Manual, 1966)

(6) *Continuous operation slurry treaters*. Several brands of this treater are available. Large-scale application of dust treatment to seed has inevitably serious disadvantages — dustiness around the treater and, consequently, danger to operators of continual operation. Most dusts in use cause irritation to skin or to the mucous membrane of the respiratory tract. To overcome these difficulties wettable powders were developed for slurry application. The dust is applied as a heavy suspension in water to the surface of the seed. The machine (Fig. 21.4) is provided with seed and slurry metering devices which mix the seed and fungicide in calibrated amounts from small buckets drawn on an endless chain. Slurry machines are adaptable to all types of seed and rates of treating up to 2,200 l. (600 bushels) per hour. Bagging can be done directly because a drying process is unnecessary. This type of treater is still the most suitable one for many kinds of seeds, although the procedure may give difficulties because of slurry settling in the machine.

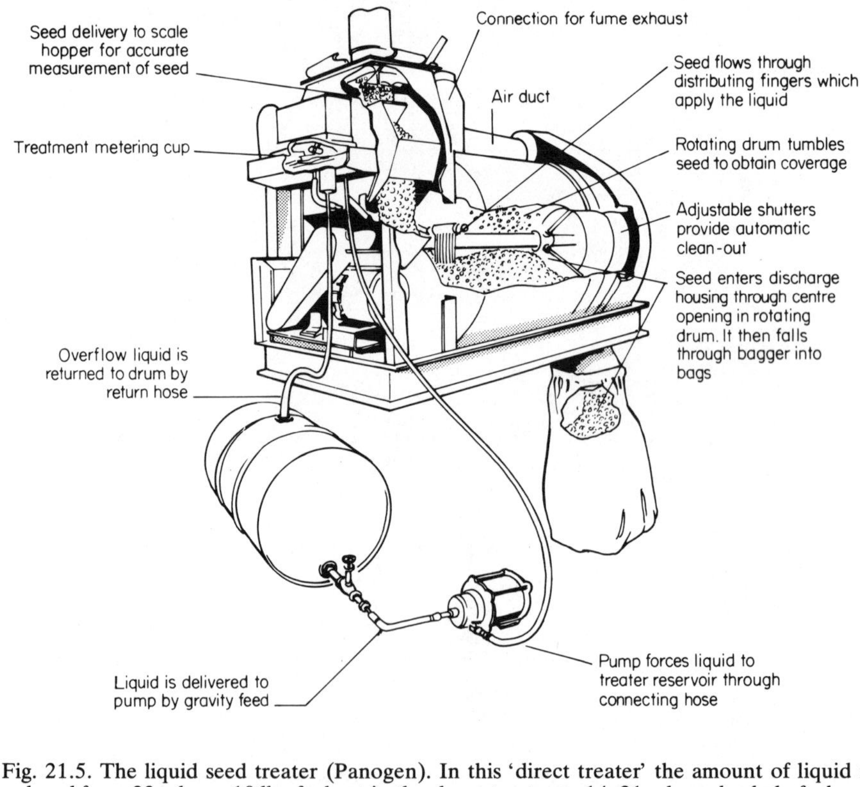

Fig. 21.5. The liquid seed treater (Panogen). In this 'direct treater' the amount of liquid is reduced from 23 ml. per 10 lb of wheat in the slurry treater to 14–21 ml. per bushel of wheat. This small quantity is suitable only for volatile liquid mercurials where complete disinfection is obtained through fume action rather than through complete seed coverage. (E.I. du Pont de Nemours & Co., Inc., Seed Treating Manual, 1966)

(7) *Continuous operation liquid treaters*. Special treaters have been designed for direct (undiluted) application of liquid preparations such as mercury preparations in concentrated formulations, e.g. Panogen and Ceresan L, in an ethylene glycol or similar carrier. The machine (Fig. 21.5) is provided with an adjustable hopper to control seed flow, calibration devices for metering seed and fungicide, and an inclined drum with baffles at the discharge and to control the time the seed is retained in the rotating drum mixing chamber (Purdy, 1967). The capacity of this type is near 2,200 l. (600 bushels) per hour, equal to the slurry treater. Different models designed for different kinds of seeds are produced — one well-known type, of Swedish origin, under the trade name Panogen Seed Treater, shown in Fig. 21.5. An exhaust fan is included in the seed treater to remove dust and chaff from the drum, assuring cleaner treated seed. Also, excess fumes are removed to promote a safer as well as more pleasant work area. The capacity of Panogen Treater Model E is 5–20 tons of wheat and barley per hour, 3–15 tons of oats per hour, and 2–5 tons of beet seed per hour (Casco, personal communication, 1970). The capacity of Panogen Model VS-60C Cotton Seed Treater is up to 8 tons per hour of mechanically delinted cotton seed, and up to 2,500 l. (700 bushels) per hour of small grains (Casco, personal communication).

(8) *Continuous operation mist treater*. This type is a further development of the slurry and liquid treaters in an attempt to obtain complete coverage of the seed by less volatile preparations. The best-known brand of these machines is the Mist-O-Matic seed treater (Fig. 21.6). In addition to seed and fungicide devices the machine is provided with rapidly spinning discs that break the fungicide solution into small droplets, the dispersal of which ensures complete and uniform coverage when the machine is properly adjusted and used.

(9) *Continuous operation spray treater*. In this treater the fungicide suspension is introduced through spray nozzles into the mixing chamber with the seed. This treater type was originally developed for treatment of sugarbeet seed (Kepner and Leach, 1949). Various types of the spray application method have been reviewed by Purdy (1967), including application of undiluted liquid formulation, suspensions of wettable powders, and dust applications in association with application of water to aid the retention of dust by the seed. The capacity is near 2,200 l. (600 bushels) of small grains per hour.

Another development of this treater type is the 'Plantector' seed treater, devised and designed by Plant Protection Ltd of England. 'Plantector' will treat barley, wheat, oats, field beans, peas and delinted cotton seed at rates up to 8,700 l. (2,400 bushels) per hour. The treater can be fitted with liquid- and powder-feeding attachments which may operate separately or

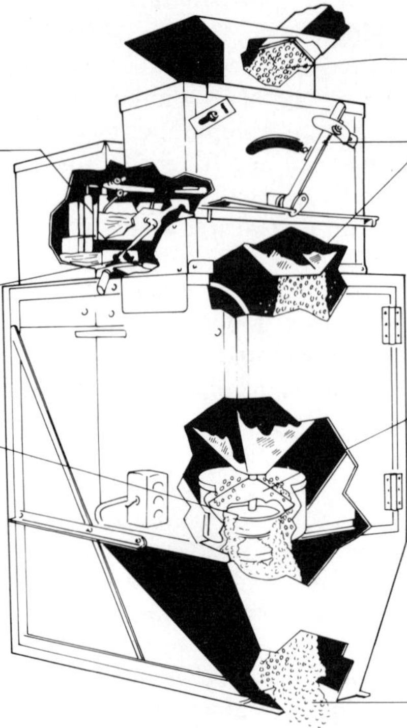

Fig. 21.6. Mist-O-Matic seed treater. Applies the treatment as a mist to give instantaneous and complete coverage of the seed without the necessity of seed mixing. Quantities as small as 44 oz. material per bushel may be applied and still give good seed coverage. There are several models of the treater, of which the M-1 has become more or less the standard. (E.I. du Pont de Nemours & Co., Inc., Seed Treating Manual, 1966)

in combination, and the treater can be used to apply slurries. The operating mechanism is shown in Fig. 21.7(A). The feed for the seed is operated by a paddle rotating in an enclosed chamber and the seed and treating materials are mixed by the action of an auger. The seed is conveyed up through the mixing chamber to the bagging-off spouts. The powder dispenser, Fig. 21.7(B), consists of a broad rubberised endless slowly-moving belt which forms the floor of the hopper. The powder is taken off the belt by a rotating brush and a fine cloud of dust settles on the seed. The powder feeder must be calibrated before treatment to achieve the desired application rate. The liquid dispenser, Fig. 21.7(C), consists of a special pump driven from the metering paddle shaft. The pump is adjustable over a wide range of application rates. Liquid is sucked direct from a seed dressing drum and delivered to a special spray jet mounted in the seed hopper. Liquid is sprayed, at low pressure, into the base of the mixing chamber at the same time as the seed. The liquid feeder must be

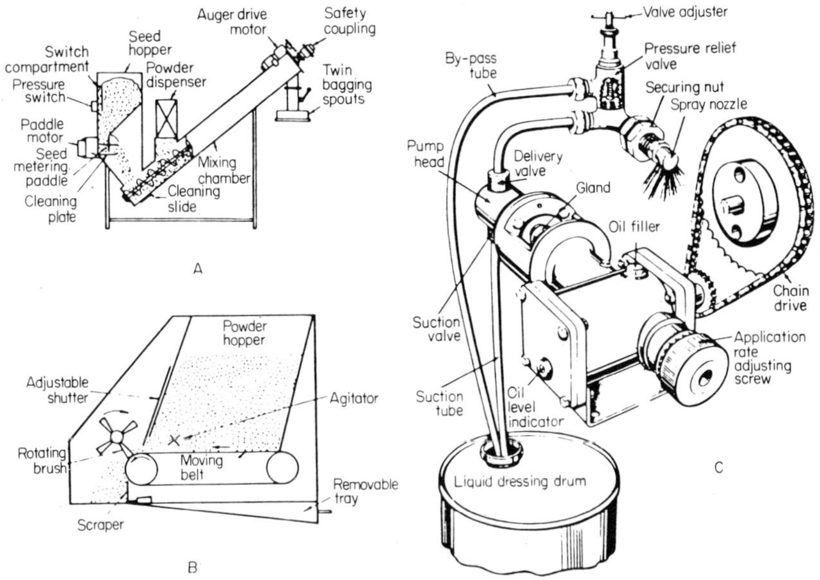

Fig. 21.7. The 'Plantector' seed treater. A, operating mechanism—sectional working programme; B, powder dispenser—sectional working programme; C, liquid dispenser—liquid circuit diagram. (Plant Protection Ltd, ICI, Instruction Manual for the 'Plantector' seed treater)

calibrated to the desired application rate. In the slurry technique the machine dispenses liquid and powder separately but simultaneously to the seed. No hand measuring or agitation of water and powder is required; it is only necessary to replenish the two containers periodically. (Information from Instruction Manual for the 'Plantector' seed treater, Mark 3, received 1970.)

(10) *Hot-water treaters*. Equipment for the application of hot-water treatment ranges from the simple tank type for dipping bags of seed to large-scale types, including electric hot-water treaters (De Vilmorin, 1949; Shurovenkov and Volkov, 1952). An example of a fairly large-scale hot-water treater is one devised by the Engineering Faculty of the University of Saskatchewan, Canada, used in a certification scheme for production of loose smut-free barley seed in Canada (Clark, 1952). The seed is treated in half-bushel lots in two-bushel, loosely woven jute bags, tied at the top. It is presoaked for a period of 5 h in water at 21°C. The bags are then attached to an endless belt that slowly passes through a tank of water at 52°C. It is held in this tank for 11 min. A false bottom in the tank, which moves in the opposite direction to the carrier belt, forces the water through the loose barley so that it is all equally and evenly heated. The

seed is cooled quickly by immersing the bags in cold water for a few minutes, then placing it on a draining rack for a short time before moving it to the dryer. The dryer consists of a rectangular trough, measuring 20×2×2 in. (about 51×5×5 cm), with wire-covered openings in the top, over which the bags of seed are spread. Warm air, controlled to 35°C, is forced into the trough and thus up through the seed. The capacity of the hot water unit is 110 l. (30 bushels) per hour and of the dryer about 35 l. (9 bushels) per hour. The operating rate of the latter varies with atmospheric conditions. Webb's, the English seed firm, produces hot-water treaters on a commercial scale.

Fig. 21.8. Equipment for automatic temperature and flow-rate control of aerated steam for treatment of seed. Flowmeter at left, pneumatic thermostat at top with two gauges for pneumatic control. Needle valves in air and steam lines at bottom. Output at top rear, connected to a micro seed treater. (Courtesy of Kenneth F. Baker)

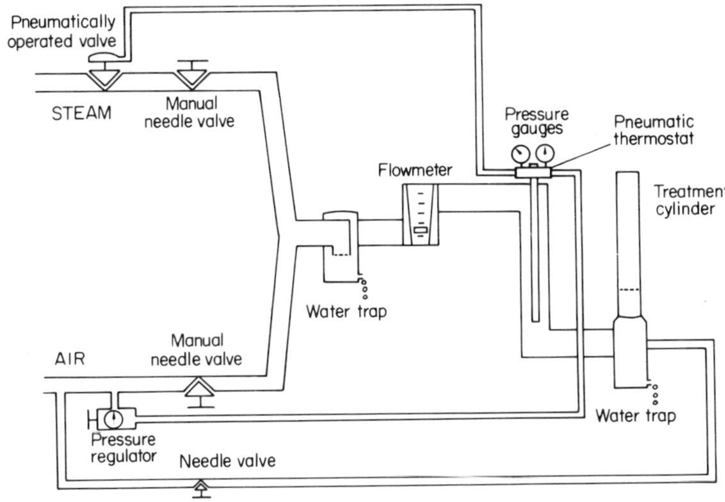

Fig. 21.9. Diagram of seed treatment equipment shown in Fig. 21.8. The treatment cylinder (Fig. 21.10) has a water trap built in, eliminating the separate output trap shown. The pressure regulator maintains air pressure at the necessary 15 lb/in.^{2} in the thermostat line. The needle valve in the steam line acts as a micrometer for the pneumatically operated valve. The needle valve in the air-bypass line is used in the cooling phase, when the steam-air mixture is shut off; it is also used for minor temperature correction, during treatment, for such factors as changing heat or wetting. (Baker, 1969*a*)

(11) *Aerated-steam treaters*. Vapour-heat treatment has been used on a limited scale for some decades, particularly for control of insects and mites in fruits, vegetables, strawberries, bulb crops and ornamentals (Latta, 1932; Smith and Goldsmith, 1936), and later for seed (Miller and McWhorter, 1948). The different types of equipment have been critically reviewed by Baker (1969*a*), who developed a very promising pressure-flow steam treater specially designed for seed treatment. He established that successful aerated-steam treatment of seed requires that the vapour be forced through the mass of material, the seed must be treated and cooled rapidly and precisely to avoid injury, and seed treated in steam chambers should be held neither in trays, cloth bags nor in plastic mesh bags, because heat penetration is poor under these conditions. Mixing air with steam dilutes it and thus lowers its temperature to any desired level. Baker's aerated-steam treater (Figs 21.8 and 21.9) is automatically temperature-regulated in the steam line; aerated steam is passed down and through the seed column in a double-walled 'Plexiglass' input cylinder (Fig. 21.10), the inner wall being heated to prevent condensation. The flow rate of aerated steam is sufficient to permit heating through the whole seed bulk at once rather than in progressive layers. The treater supplies a source of compressed air capable of delivering 20 ft^3/min without overheating. The flow rate is such that the temperature of the vapour

at the output during initial heating is not more than 1.5°C lower than at the input.

In various respects aerated-steam seed treatment has advantages over hot-water seed treatment. Seed treated by aerated steam is much drier than that treated with hot water. At the conclusion of treatment the steam is turned off, the seed dried by air flow and the temperature quickly lowered by evaporative cooling. Germination is usually not reduced as much as with hot water. Temperature control is easier and more accurate with aerated steam than with hot water, and the margin of safety between effective pathogen control and excessive seed injury is generally greater than that with hot water and dry hot-air treatments. However, seed used for aerated-steam treatment should be fresh, of high vigour and germination, and free from mechanical injuries.

Fig. 21.10. 'Plexiglass' cylinder for aerated treatment of seeds. Vapour enters outer cylinder through tube at left, spirals upward against outer wall (dropping condensate to bottom, where it drains through the tube at the right), passes into the inner cylinder at the top, down through the seeds it contains, and out through the hole near the bottom. The wall of the inner cylinder is heated, preventing condensation. The unit is loaded and unloaded by loosening the turnbuckles and removing the top. A thermometer or thermocouple is inserted through the rubber stopper in the opening in the top. (Baker, 1969*a*)

21.5 CATEGORIES OF SEED TREATMENT CHEMICALS

(1) *Inorganic copper compounds*. Copper sulphate was the first fungicide established on a scientific basis for seed treatment. Prévost (1807) showed that germination of bunt spores could be inhibited by this compound, and Kühn (1859) established a definite basis for the general adoption of copper sulphate as standard treatment for control of wheat bunt.

Copper carbonate was the first dry-treating material used in control of different smuts in cereals, including bunt and flag smut of wheat, and kernel smuts of sorghum (cf. Fischer and Holton, 1957). Other dry copper compounds have been used for seed treatment, including basic copper sulphates prepared by reacting copper sulphate with sodium hydroxide; red copper oxide (cuprous oxide, introduced as a seed protectant by Horsfall, 1932*b*); copper phosphate; copper oxalate; and copper oxychloride — all for control of smuts (Leukel and Livingston, 1945). Seed treatment with inorganic copper compounds is now almost obsolete.

(2) *Inorganic sulphur compounds*. While this category of compounds has been used extensively for protection of growing plants in the form of lime sulphur and wettable sulphurs, these chemicals have not been found applicable for seed treatment. Sulphuric acid is used in delinting cotton seed and is effective against such cotton pathogens as *Colletotrichum gossypii* (*Glomerella gossypii*), anthracnose, *Verticillium* spp., wilt, and *Xanthomonas malvacearum*, angular leaf spot.

Various types of powdered sulphur, as finely powdered as possible to ensure that they stick to the seed, may control smuts. Sulphur is cheap, non-poisonous, and is used to some extent for seed treatment, mainly in tropical countries. It is, for instance, fairly effective against kernel smut of sorghum.

(3) *Inorganic mercury compounds*. Mercuric chloride (corrosive sublimate) was introduced by Hiltner (1915) as a seed disinfectant against *Fusarium nivale* in rye, and commonly used under the commercial name Fusariol. It is still rather extensively used, particularly in Germany and the U.S.A., for seed treatment of many kinds of seed, chiefly seeds of vegetables and ornamentals, e.g. *Callistephus chinensis* against *Fusarium oxysporum* f.sp. *callistephi* (Baker, 1953); on *Capsicum* against *Xanthomonas vesicatoria*; *Cheiranthus cheiri* against *Alternaria cheiranthi*; watermelon against anthracnose, *Colletotrichum orbiculare*; stock, *Matthiola incana* against *Alternaria raphani*; and lucerne and black medick against spring black stem, *Phoma medicaginis (Ascochyta imperfecta)*.

(4) *Inorganic silver compound*. Silver nitrate is occasionally used for seed treatment, notably in Rhodesia on tobacco seed against angular leaf spot, *Pseudomonas angulata*, and other diseases, see also Johnson and Murvin (1925).

(5) *Other inorganic materials*. To remove virus from tomato seed, treatment with 10 per cent trisodium phosphate for 10 min has been found effective (John and Sova, 1955). Other procedures are extraction of seeds from ripe fruits to which concentrated hydrochloric acid has been added, 10 ml. to 10 kg pulp, or soaking the seed in 12 per cent solution for 1 h (Broadbent, 1965).

(6) *Formaldehyde*. This colourless gas, dissolved in water, was introduced in Germany in 1895 as a seed disinfectant by Geuther (1895) and by Bolley (1897) in the U.S.A. It has since been extensively used efficiently against fungi, e.g. smuts, and is also effective against bacteria. However, it is injurious to many kinds of seed at effective dosages, although less so than copper sulphate which it largely replaced. Formaldehyde is available on the market as formalin, a 34–37 per cent solution of the gas in water. The treatment consists of steeping the seed for a defined period of time in usually 1 per cent of commercial formalin. Its popularity has declined because of its phytotoxicity and its fumes disagreeable to eyes and breath. Attempts to provide preparations for dry treatment to utilise its volatile nature have failed (Fischer and Holton, 1957).

(7) *Organic mercuric compounds*. Several hundred organic mercuric compounds have been tested for use in plant disease control, many as plant protectants, and the great majority of them for seed disinfection. The first compounds contained much more mercury than those being used now, with 18 per cent in the first preparation, Uspulun, as against 1–2 per cent in recent products, 0.6–1.5 per cent in powder preparations and 0.6–2.0 per cent in liquid preparations.

The organic mercuric compounds consist of pure or substituted aliphatic or aromatic radicals such as methyl-, ethyl-, phenyl-, or methoxyethyl radicals (R), and as inorganic or organic acid group (X), according to the basic formula R—Hg—X.

Gassner (1951*a*) investigated 100 different mercuric compounds and established that it is the R radicals that are of importance to the efficiency of the compounds as plant protectants. The percentage of mercury in a compound is not an index of its fungicidal effect. Dillon Weston and Booer (1935) found that, in organic mercuric compounds, R—Hg—X, with an acidic X radical, the fungicidal power decreases with an increase in the molecular weight of radical R.

Some alkyl compounds, particularly methyl- and ethyl- compounds, have shown outstanding properties as plant protectants; cf. Section 22.4 as to degrees of phytotoxicity of mercury salts.

Examples of some successful compounds are listed in Table 21.2, with indication of the vapour pressure of some of them according to a review by Ulfvarson (1969), defining the different degrees of volatility.

To achieve efficiency against inoculum protected by host tissues or inoculum that has not been brought into physical contact with the fungicide, a certain degree of volatility of the compound is necessary. Methyl- and ethylmercuric compounds are volatile enough to kill smut teliospores and barley leaf stripe fungus mycelium beneath the hulls of grains. Alkylmercuric compounds generally have a greater volatility than aryl- and alkoxyalkyl derivatives. The anion modifies the volatility considerably. According to Lindström (1961) the ideal saturated vapour concentration is between 100 and 10,000 $\mu g/m^3$. Too high a volatility, above 10,000, leads to loss of fungicidal effect during treatment and also to handling hazards.

However, the development of liquid preparations after World War II and application of new techniques as described above (quick wet treatment), in specially designed machines, reduces hazards due to exposure to fumes and eliminates the dust problem.

Storage for some time of seed treated with mercury disinfectants before seeding increases the effect of the treatment. Cherewick and Popp (1951) found, in their survey of seed treatments in actual practice in Manitoba, Canada, that 35 per cent of the farmers stored their treated seed 3 days or less before seeding and had an average of 1.7 per cent smut, whereas 65 per cent allowing 4 days or more for post-treatment storage had only 0.27 per cent.

In recent years there is a definite trend away from use of mercuric compounds, in view of the risk involved to personnel and to animals. Legislative precautions have been taken in various countries to limit or eliminate the use of these compounds. This aspect is further considered in Subsections 22.10.5 and 22.10.6.

(8) *Organic sulphur compounds*. Since the introduction of dithiocarbamic acid derivatives in the 1930s as foliage fungicides, carbamates have come increasingly into use in plant protection, some for seed treatment. Thiram (tetramethylthiuramdisulphide) is used extensively by the slurry method under the trade names Arasan and Tersan against damping-off and foot rot in bean and pea, seed-borne *Verticillium* in lucerne, and damping-off in maize. Maneb (manganese ethylene-1,2-bisdithiocarbamate) is used in place of mercuric compounds against *Drechslera graminea* in barley, *Ustilago avenae* in oats, and *Tilletia caries*, *Septoria nodorum* and *Fusarium nivale* in wheat (Ponchet and

Table 21.2. Some organic mercury compounds used for seed treatment

*Commercial brand (older and newer)**	*Cation*	*Anion*	*Vapour pressure at 20°C* (μg/m³) Ulfvarson (1969)
	methyl mercury	acetate	75,000
Aabiton	methyl mercury	benzoate	2,000
	methyl mercury	bromide	94,000
	methyl mercury	chloride	94,000
Chipcote	methyl mercury	cyanide	
Panogen 8; Panogen 12, 15, 42	methyl mercury	dicyandiamide	300
	methyl mercury	hydroxide	10,000
Ceresan L	methyl mercury	2,3-dihydroxy-propyl mercaptide methyl mercury acetate	
Ortho LM	methyl mercury	8-hydroxyquinolate	
	methyl mercury	iodide	90,000
Chipcote	methyl mercury	nitrite	
Tillantin 12	methyl mercury	*para*chlorobenzoate	
	methyl mercury	propandicarboxylate	
Ceresan M	methyl mercury	toluene sulphonanilide	
	methyl mercury	toluenesulphonate	15,000
Aagrano	ethyl mercury	bromide	7,000
Ceranit 61; Granosan; Ceresan 2%; Betoxin F	ethyl mercury	chloride	8,000
	ethyl mercury	iodide	9,000
New Improved Ceresan	ethyl mercury	phosphate	
Lunasan	ethyl mercury	thiuronium chloride	
Ceresan MDB; Ceresan M	ethyl mercury	toluene sulphonanilide	
Tillantin dry; Mersolite; Uspulun secco; Ceresan Berk Leytosan	phenyl mercury	acetate	17
Setrete	phenyl mercury	ammonium acetate	
Ceredon Special	phenyl mercury	chloride	5
Germisan; Uspulun powder	phenyl mercury	pyrocathechin	
Agrox; Leytosan	phenyl mercury	urea	
Panogen Metox; MEMA; Mercuran; Tillantin 6, 12	methoxyethyl mercury	acetate	2
Ceranit 6, 12, 63, T, TH, VF; Ceresan Universal-Nassbeize; Agallol; Aretan; Certosan	methoxyethyl mercury	chloride	2,600
Tillantin-S, -T; Tillantin 1875; Abavit-Neu; Ceresan Univ.-Trockenbeize	methoxyethyl mercury	silicate	
Corrosive sublimate	mercury	chloride	330
	metallic mercury		14,000

* Name commercially used for a definite compound may be different from country to country.

Ventura, 1963; Darpoux, Ventura and Cassini, 1968). Thiram soak treatment (0.2 per cent for 24 h, then dried) is effective against the 'Ascochyta complex' in pea, *Septoria apiicola* and *Phoma apiicola* in celery, *P. lingam* and *Alternaria brassicicola* in cabbage, *P. betae* and *Colletotrichum dematium* in red beet, *Alternaria porri* f.sp. *dauci* and *A. radicina* in carrot and *A. zinniae* in zinnia (Maude, 1966*a*, 1966*b*, 1973; Maude and Keyworth, 1967; Maude, Vizor and Shuring, 1969); the procedure is practicable particularly for small bulks of high value seeds.

While organic mercuric compounds have a broad range of efficiency (cf. the German commercial name 'Universal-Beizmittel'), organic sulphur compounds and many other organic fungicides of various categories have a narrower range of efficiency and may in some cases be quite specific. They are less poisonous to man and warm-blooded animals than the mercury preparations.

(9) *Quinones*. Two compounds of this group are outstanding fungicides for seed treatment: Chloranil (tetrachloro*para*benzoguinone) and dichlone (2,3-dichloro-1,4-naphthoguinone), both among the first organic compounds recognised as fungicides (Cunningham and Sharvelle, 1940; Sharvelle, Young and Shema, 1942; Ter Horst and Felix, 1943).

Chloranil is used under the trade name Spergon as a dust or slurry for seeds of clover, lucerne, maize, sorghum (against kernel smut), peanuts and vegetables. Dichlone is used under the trade name Phygon for seed treatment of maize, rice, sorghum (against kernel smut), and for many vegetables. Both fungicides are particularly effective against different types of seed rot and seedling blight. They are less phytotoxic and less effective against smuts than the organomercurial compounds. Dichlone is toxic to nitrogen-fixing bacteria of the Leguminosae.

(10) *Polychlorobenzenes*. Hexachlorobenzene (HCB, not to be confused with the insecticide BHC which can be phytotoxic to seed) is effective against bunt of wheat and used for control of this disease under the trade names Anticarie, Sanocide, No-Bunt, Smut-Go, and others, and also against dwarf bunt. Treated grain is poisonous (cf. Table 22.6 and Subsection 22.10.5). Pentachloronitrobenzene (PCNB), sold as Brassicol, Terrashlor, Tritisan, and other trade names, is also used for the control of bunt. PCNB and, particularly, HCB seed treatments are effective against seed-borne and soil-borne spores of common bunt.

(11) *Methyl bromide*. This is an odourless, non-flammable gas of low (3.6°C) boiling point. It is used for fumigation of seed against nematodes, e.g. 1 g/50 l. for 16 h against *Aphelenchoides besseyi*, white tip of rice; against *Ditylenchus dipsaci*, stem eelworm, in clover, lucerne, and onion seed; and against soybean cyst nematode, *Heterodera glycines*. It is not

fungicidal and may be phytotoxic, especially to certain cultivars of rice.

Methyl bromide is widely used for fumigation against storage insects of seed and grain. It is dangerous to the fumigation operator, and only personnel specifically trained in its use should carry out fumigation with this material. Gostick (1970) reviewed the effect of methyl bromide on seed germination. The resistance of seed to fumigation damage decreases with increasing moisture content. Defining dosages of fumigation in terms of concentration time product (CTP) according to international standard, the maximum safe moisture contents are given in Table 21.3 for most seeds considered above. Gostick's review contains corresponding information about many other kinds of seed.

Table 21.3. Safe maximum moisture content of different seeds for given CTPs of methyl bromide. (Data from Gostick, 1970)

Product	*Temperature of treatment* (°C)	*CTP* (mg h/l.)*	*Maximum moisture content of seed tolerated* (%)
Clover	10	1,600	12
	18–21	600–700	11
	21	1,600	12
	32	480	11
Lucerne	10	1,600	12
	18–21	600–700	11
	21	1,600	12
	32	480	11
Onion	18–21	600–700	11
Rice	10	1,200	12
	18–21	600–700	11
	21	1,600	12
	32	480	10

*Most of these dosages are probably much higher than needed for nematode control.

CTP for a given pest is defined as the product of concentration×time required to obtain 99 per cent mortality. To determine the actual concentration which is altered in the course of treatment due to certain factors, e.g. decreasing the initial dosage, it is required that atmospheric samples be taken for analysis from various points inside the chamber during fumigation. Special instruments testing this by thermal conductivity are available.

Hague (1968) controlled a heavy infestation of *Ditylenchus dipsaci* in seed of shallot, *Allium ascalonicum*, which had a moisture content of 10 per cent, by a CTP of 856 mg h/l. The germination capacity of the fumigated seed was not reduced, whereas the germination speed was reduced. CTP could be increased to 1,284 mg h/l. without further effect on the germination.

(12) *Heterocyclic nitrogen compounds*. The most outstanding of this category of fungicides is captan (*N*-trichloromethylmercapto-4-cyclohexene-1,2-dicarboximide), marketed as Orthocide 75 and applied as dust or slurry to seed of maize, rice, sorghum, pea, bean and lima bean, being particularly effective as a protectant against seed-rotting organisms. It has a broad range of fungicidal action. As seed treatment it has also been found effective against bacterial wilt of maize, *Erwinia stewartii*. Another compound of this group is dyrene (2,4-dichloro-6-chloranilino) more recently developed as an agricultural fungicide. It controls cereal seed-borne fungi (*Drechslera*, *Fusarium*, *Rhizoctonia*) on aerial plant parts and on turf grasses, but has not been established as a seed treatment. Finally, thiabendazole(2-(4'-thiazolyl)benzimidazole), a systemic fungicide, is effective against seed-borne bunt of wheat, covered smuts of oats and barley (Hoffmann, 1970), and *Septoria nodorum* in wheat (Darpoux, Ventura and Cassini, 1968); it has been registered as a seed treatment in the U.S.A. against *Drechslera oryzae* in rice, as well as for barley, cotton, maize and oats.

(13) *Antibiotics*. Several metabolic products of microorganisms have shown remarkable properties for seed treatment, being effective against pathogenic fungi and particularly bacteria not combatable by other pesticides. Streptomycin can control halo blight (*Pseudomonas coronafaciens*) in oats, silvering of red beet (*Corynebacterium betae*) in beet, angular leaf spot (*P. lachrymans*) in cucumber, halo blight (*P. phaseolicola*) in bean, and bacterial leaf spot (*P. sesami*) in sesame (Thomas, 1959*a*; Knösel, 1965, 1967). Combinations of streptomycin and oxytetracycline are on the market as Agrimycin and are used against various bacterial diseases (Thirumalachar, 1968).

Other antibiotics effective against smuts, e.g. actidione against bunt of wheat and covered smut of oats (Henry *et al.*, 1951; Henry, Millar and Peterson, 1952), helixin B against bunt of wheat, oat smut and covered smut of barley; and cycloheximide as a seed treatment against rust, *Puccinia carthami*, on safflower (Zimmer, 1965); rimocidin and pimaricin are effective against *Ascochyta pisi* when the seed is soaked in a solution for 24 h, although the parasite is not completely eradicated (Dekker, 1957). Blasticidin-S which acts as a systemic fungicide is extremely effective as a spray against blast fungus, *Pyricularia oryzae*, in rice but highly resistant strains of the fungus are known (Huang, Misato and Asuyama, 1964) and it is toxic to animals and humans. It is losing ground to the new antibiotic kasugamycin (Dekker, 1971). Kasugamycin is used on a large scale as a spray against rice blast; used as a 2 per cent wettable powder on seed it provides protection against the disease in the field beds for a month (Dekker, 1971). The Indian antibiotic 'aureofungin' is fairly effective as a seed treatment against *Pyricularia oryzae* and *Drechslera oryzae* in rice

(Thirumalachar, 1967; Rahalkar and Neergaard, 1969) and against *Ustilago hordei* in oats and *Drechslera graminea* in barley (Dharam Vir and Raychaudhuri, 1968, 1969, 1971). In Canadian investigations aureomycin killed *Xanthomonas campestris*, black rot, in cabbage (Sutton and Bell, 1954).

A serious drawback to the use of antibiotics in agriculture is the stable chemical nature of these compounds in food materials. Antibiotics may remain intact in food material from treated plants and animals, and when continually consumed in small amounts the effect of these antibiotics against human pathogens may decrease so that a dangerous resistance is developed. Consequently, government authorities are adopting a cautious policy in many countries, and antibiotics must be accepted by the authorities before they can legally be applied to food plants.

(14) *Miscellaneous compounds* (*systemic fungicides*). Various systemic fungicides have recently been developed. Carboxin, trade name Vitavax (2,3-dihydro-5-carboxanilido-6-methyl-1,4-oxathiin), and benomyl under the trade name Benlate (methyl-1-(butylcarbamoyl)-2-benzimidazolecarbamate) are effective against loose smut of barley and wheat, carboxin being slightly superior in barley. Both are highly effective against bunt of wheat, seed-borne as well as soil-borne, and effective against flag smut of wheat, covered smut of barley and oat smuts. Pyracarbolid, closely related chemically to carboxin, has a similar range of activity but is slightly more potent, e.g. controlling loose smut (Stingl, Härtel and Heubach, 1970). Benomyl controls also a wide range of Fungi Imperfecti and *Ascomycetes*, such as *Ascochyta*, *Colletotrichum*, *Fusarium*, *Phoma*, *Septoria* and *Sclerotinia*. Accordingly, Bollen and Fuchs (1970), in extensive *in vitro* studies, found that most *Ascomycetes* and the *Hyphomycetes–Blastosporae–Phialosporae–* and *Aleurosporae* were usually highly sensitive to benomyl, whereas *Oomycetes* and *Hyphomycetes–Porosporae* (*Alternaria, Drechslera*) were resistant. Indicative of the remarkable fungicidal properties of Benlate is its capacity of eliminating in pea seeds high infections of *Ascochyta pisi*, known to resist many seed disinfectants (Maude and Kyle, 1970).

There are two categories of systemic seed treatments: (1) those killing or preventing development of fungi arising from seed infection or seedling infection systemic in the plant, and (2) those preventing development of aerial inoculum. In addition to the effect on loose smut, seed treatment of barley with benomyl has given marked control of powdery mildew (*Erysiphe graminis*) during the establishment of the crop, an effect due to the systemic translocation of the fungicide in the seedling (Evans, Rickard and Whitehead, 1969).

Shift from mercurial to non-mercurial seed treatments. During the past 5–10 years there has been a dramatic shift in practice from mercurial to non-mercurial seed treatments. Although organomercurials still offer the most effective and versatile control of many seed-borne diseases, it is now generally realised that precautions against hazards in use of poisons in agriculture must be given more attention (Subsections 22.10.5 and 22.10.6). Particularly, some of the systemic formulations are becoming important as seed treatments.

Combination of seed treatments. Combination of materials is one of the ways to overcome difficulties arising in using the more expensive selective systemic fungicides now becoming more important because of the general banning of mercurials. By combining a cheap fungicide having wide-range effects with a more selective one, the overall cost is reduced, compared to a higher dosage of the systemic. Using a fungicide with a much wider range of activity, the chances for development of resistant strains are reduced. Combinations often also offer the advantage of synergistic effects. Mills and Wallace (1968, 1970) demonstrated marked differential action of fungicides on seed-borne flora of barley, wheat, buckwheat and oil seed, cf. Section 22.5 and Fig. 22.10. Djerbi and Hafsa (1972) showed that for treatment of wheat against loose smut, *Ustilago tritici*, a combination of 75 per cent of the very effective Quinolate V-4-X (50 per cent carboxin, 15 per cent copper oxyquinoleate) and 25 per cent of either maneb or thiram is equally effective and reduces the price of the treatment by 25 per cent; and other combinations may further reduce costs.

Annual lists of pesticides. Many countries making wide use of plant protection chemicals, including seed treatments, now publish annual lists of those approved for efficiency as well as safety to personnel. In view of the rapidly changing situation, these lists should be consulted. Reference is also made to the compilations of reports dealing with tests of fungicides and nematicides, published annually by the American Phytopathological Society (e.g. Zehr *et al.*, 1973).

22 Mechanism and Practical Use of Chemical Seed Treatment

INTRODUCTION

In Fig. 22.1 the sequence of the main factors operating in germicide seed treatment is presented. Briefly the course of action may be outlined as follows.

The seed is first *mixed* with the disinfectant which is mechanically *distributed* in the form of dust, wettable powder, slurry or liquid. The material *covers* more or less the surface of the seeds and *adheres* to the surface. During and long after the mechanical application the chemical *diffuses* in the seed bulk, *penetrating* between and into the seeds. This diffusion and penetration may take place in liquid state, or it may occur in gaseous state exercising a *vapour action*. Some of the chemical is *resorbed* by the tissues of seeds and seed-borne organisms which respond by specific *reaction*. This reaction may continue for some time, depending on the effective compounds of the chemical, the kinds of organisms involved and the storage conditions of the seed.

In the field the material *adhered* and *resorbed* on and in the seed becomes subject to the same conditions as those of the germinating seed. *Dilution* and *diffusion* of the chemical take place in the soil surrounding the seed and seedling and in the seedling itself; adsorption to soil particles and colloids occurs, and this reduces the chemotherapeutical effect according to the properties of the chemical and the soil. By *absorption* of water the dormant phase of seed and seed-borne organisms are brought into the vegetative phase, and the seedlings and microorganisms may again *resorb* compounds of the disinfectant and respond to renewed *reaction*.

The exact *time of action* of seed germicides, whether they act during treatment and storage of the seed or after sowing, depends chiefly on the method of application. The principles of this relationship are shown in Table 22.1. Between chemicals, however, there are considerable differences in the mode of action under given conditions of application. Some are more volatile than others and exercise an effect before sowing. Some are more easily absorbed by seeds and pathogens than others. Chemicals that are not easily absorbed may leave more material adsorbed to the surface of the seed. This residue provides conditions for a later germicidal effect after sowing, while germicides that are easily absorbed may not exercise such an effect.

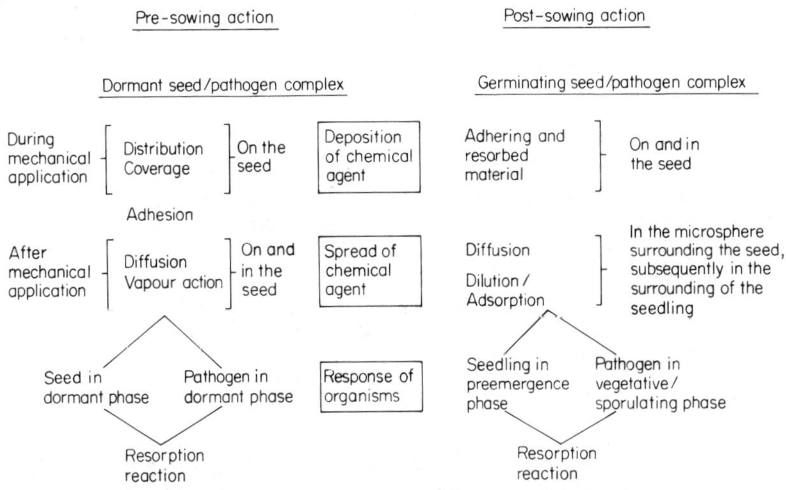

Fig. 22.1. Factors operating in the mechanism of chemical seed treatment

22.1 PRE-SOWING ACTION OF GERMICIDES

22.1.1 Distribution and coverage

The mechanism of chemical seed treatment is governed by the physical form of the agent and the mixing process in the treater. When the seed is soaked in a dilute solution, complete distribution of disinfectant and uniform and complete coverage of seed is obtained, provided the seed remains long enough in the solution to secure complete wetting of the bulk. Because of the necessity of a subsequent drying operation, application of dust in special seed treaters was introduced.

The problem of insufficient coverage arose with the use of dusts because of the relatively low doses applied of some phytotoxic fungicides. Copper carbonate could be used in excess doses, 2 oz per bushel, without damage and still provide complete coverage of all seeds, but mercurial fungicide powders at the rate of 0.5 oz per bushel does not ensure complete coverage of all seeds (Purdy, 1967), especially as applied by certain machines.

Even slurry applications of organic mercurials fail to cover seeds completely and may leave some untreated. Purdy (1958) showed, on the basis of colour distribution, that about 25 per cent of the seeds in commercially

Table 22.1. Action of seed germicides before and after sowing, according to method of application

	Method of application	*Deposit on treated seeds ready for sowing*	*Pre-sowing (primary) effect*	*Post-sowing (secondary) effect*
Liquids	Steep treatment: soaking in solution or suspension	Little or none. Some adsorption may occur in some preparations	Practically all effect during soaking, some reaction of seed and pathogens during storage	No effects, or little effect from adsorbed material
Liquids	Slurry treatment: dust plus liquid in soup-like slurry	Considerable amount, according to adhesiveness of the material	Mainly effect during treatment and storage through diffusion and vapour action	Some or considerable effect
Liquids	Quick-wet treatment: concentrated usually volatile liquid	Material adjustable to the level of concentration applied	Mainly effect during treatment and storage through diffusion and vapour action	Some or considerable effect
Dusts	Dust treatment: seed mixed with dry dust	Considerable amount depending on adhesiveness and adequate concentration applied	Practically no effect except by vapour action from volatile materials	Practically all of the effect occurs under influence of soil moisture
Dusts	Pelleting	Large	No effect	All of the effect
Vapour	Fumigation: seed subjected to action of vapour	None	All of the effect during treatment	No effect
	Hot water treatment	None	The effect exclusively during treatment	No effect

treated seed were inadequately covered. However, the volatile nature of the mercurial fungicides, especially those based on methyl mercury, theoretically, offset the lack of complete coverage. Nevertheless, in the opinion of Purdy (1967): 'Without complete coverage, control is also incomplete. There is nothing associated with seed-treatment fungicides and their application that will offset the lack of complete and uniform coverage.'

Zade (1943) developed a liquid disinfectant, known as Panogen, to be used with an 'ultralow' amount of liquid — in fact, a tenth or less of the quantity of liquid used in the short- or quick-wet treatment developed by Gassner (1927*a*) and Nagel (1927). Consequently, the Panogen-procedure does not produce any essential change in moisture content of the seed. But because of the extremely small amount of liquid disinfectant used, uniform distribution remains a problem (De Ong, 1953).

During the mechanical process of seed treatment, contamination may be redistributed. Smut spores in cereals may move and settle in relatively larger proportions in the furrow of the kernel and in the hairs. Powder chemicals, however, move along with the spores (Nagel, 1927).

22.1.2 Adhesion

Effectiveness of treating materials is to some extent related to their power of adhesion which is influenced by the texture of the dust, the type and condition of the seed, and the method of application (Fischer and Holton, 1957). Good adhesion reduces handling hazards and loss of fungicide, e.g. in sowing.

Adherence is related to distribution. At Rothamsted Station, Lord (personal communication, 1972) found that in general the better the adherence to seed the more difficult it is to get good and uniform distribution between individual seeds. Undoubtedly this is an effect of immediate local adhesion (Lindström, 1958).

Adherence is a quality of the material itself. Metallic oxides are very adhesive, including cuprous oxide. Salts of metals adhere less well, such as copper sulphate (Horsfall, 1932*b*). Mercuric chloride and silver nitrate are retained to some extent by sweet pea seed after repeated washing and soaking in water, while chloramine-T is readily released from seed treated with it (Dadd and Jacobs, 1958).

Mills (1972) investigated the rate of adhesion of five seed fungicides. He related the maximum amount of material retained to gross surface of seeds, type of seed surface (Fig. 22.2) and adhesion characteristics of the fungicides (Fig. 22.3), and defined the amount of adhesion per unit surface area of seed. The seven crop seeds included in the investigations were characterised by the number of seeds per bushel (=35.24 l.) and the gross surface area in m^2 per bushel, relative to that of wheat in percentage, indicated as follows:

barley, *Hordeum hexastichon*:	64×10^8 seeds/bu: 34.6 m^2 (=65%)
buckwheat, *Fagopyrum sagittatum*:	115×10^8 seeds/bu: 40.3 m^2 (=75%)
crambe, *Crambe hispanica*:	211×10^8 seeds/bu: 46.4 m^2 (=87%)
flax, *Linum usitatissimum*:	491×10^8 seeds/bu: 68.7 m^2 (=128%)
rape, *Brassica napus*:	555×10^8 seeds/bu: 55.5 m^2 (=104%)
sunflower, *Helianthus annuus*:	16×10^8 seeds/bu: 22.7 m^2 (=42%)
wheat, *Triticum aestivum*:	105×10^8 seeds/bu: 53.6 m^2 (=100%)

The seeds were maintained at a standard moisture grade.

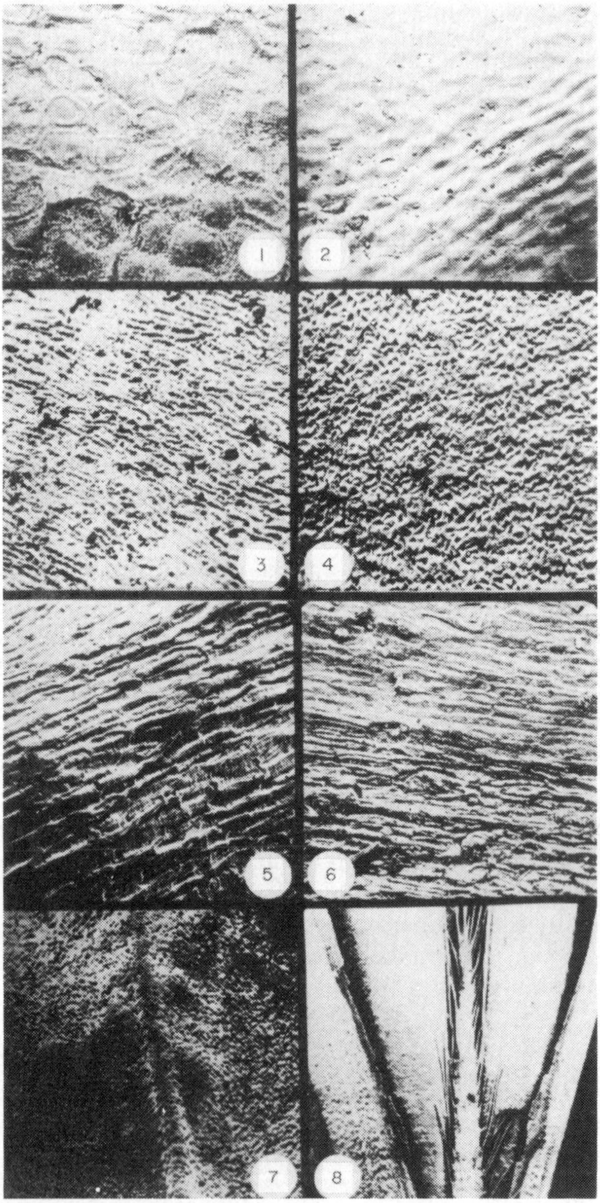

Fig. 22.2. Details of seed surfaces to which fungicides adhere. (1) Crambe cv. Indy (pericarp); (2) flax cv. 'Noralta' (testa); (3) buckwheat cv. Tokyo (pericarp); (4) rape cv. Target (testa); (5) wheat cv. Manitou (pericarp); (6) sunflower cv. Peredovik (pericarp). All ×160. (7), (8) barley cv. Conquest: (7) dorsal view of midportion of lemma; (8) ventral view of lower portion of seed showing parts of lemma, palea and rachilla. Both ×40. (Courtesy of J. T. Mills)

The following five seed treatment fungicides, formulated as dusts, were tested for differential adhesion:

Arasan 75	(bis(dimethylthiocarbamoyl)disulphide), 75.0% (thiram);
Benlate T	(methyl 1-(butylcarbomyl)-2-benzimidazole carbamate) (benomyl 30.0%+thiram 30.0%);
Manzate D	(ethylenebis(dithiocarbamato)manganese), 80.0% (maneb);
Panogen Px	(methylmercury dicyandiamide), 0.9%;
Vitaflo DB	(5,6-dihydro-2-methyl-1,4-oxathiin-3-carboxanilide) (Vitavax 40.0%+thiram 40.0%)

Seed treatment formulations differ in volatility, diluent, sticker, other additives, particle size and tendency to form aggregates. The maximum quantities adhering to a defined area of seeds were determined for the above formulations, and details in the structure of the seed surfaces were observed. Irrespective of crop the five fungicides ranked as follows in capacity to adhere to seeds: in increasing order, Panogen Px < Vitaflo DB < Arasan 75 < Manzate D < Benlate T. The fungicides adhered to the crop seeds increasing progressively in the following order: sunflower, barley, buckwheat, wheat, flax and rape. Crambe could not be so rated because the effects of the fungicides on this crop were not parallel to those on the other crops. The adhesion to rape seed was particularly strong because it has a high gross surface area per bushel and a very rough surface, Fig. 22.2(4). Manzate D and Benlate T were more friable and had fewer larger aggregates of particles, Figs 22.3(4) and 22.3(5), than the other three fungicides, hence greater adhesion. Panogen Px, Fig. 22.3(1), depends for effectiveness on its volatility, and adhesion is therefore not a factor limiting its use.

Hilgendorf (1928) found that for various fungicide dusts the length of contact is important; 2–5 min is required for full adhesion in wheat and rye; he found that kernels with smooth surfaces take up more dust than those with a wrinkled surface.

Most seed treatment dusts are reduced to particles less than 10 μm. Burchfield and McNew (1950) established average radii of 2.3–9.5 μm in four dichlone formulations successfully used for protection of pea seed. Heuberger and Horsfall (1939) showed that fungitoxicity of cuprous oxide increases with decrease of the particle size, and Somers and Thomas (1956) found increase in toxicity against *Alternaria tenuis* down to a particle size of 6 μm; below this size there was no further increase. Also, specific gravity is important. Dounine and Simsky (1932) recorded greatest adhesiveness in dusts with a relatively low specific gravity.

The effectiveness of volatile material is also dependent on particle size, particularly in mercury compounds.

Oil improves adhesiveness of dusts (Ginsburg, 1927), mineral oil being preferable to glyceride oils.

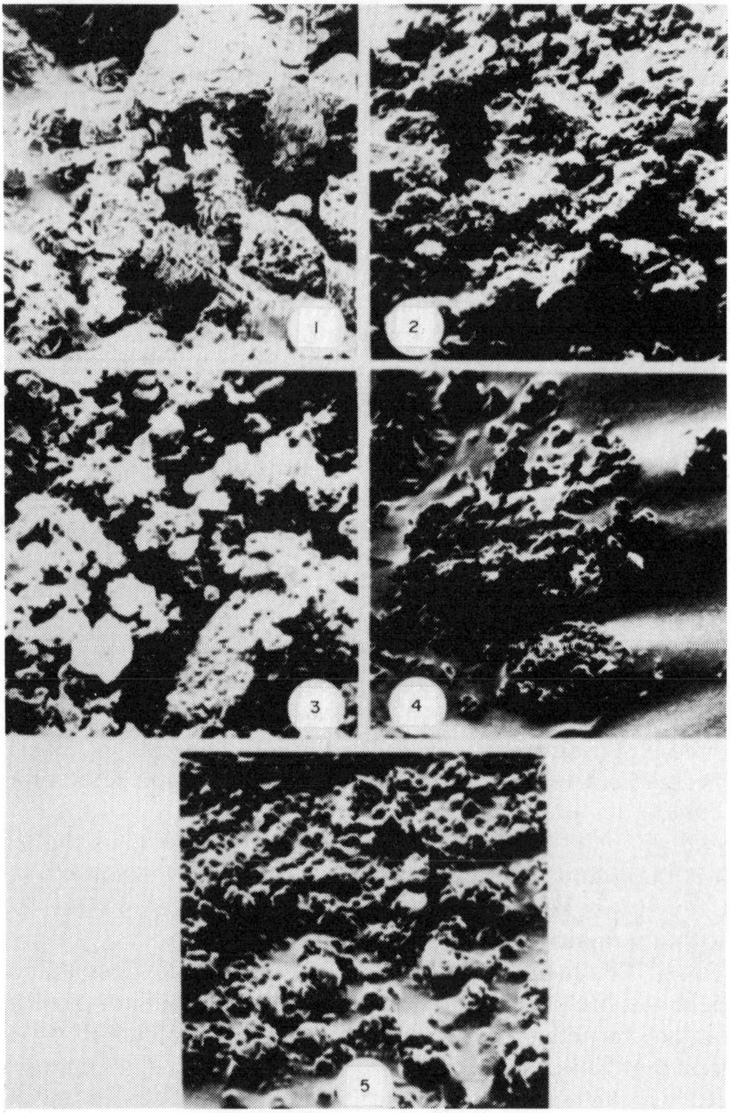

Fig. 22.3. Stereoscan electromicrographs of seed treatment materials. Increasing rate of adhesion in order 1 to 5: (1) Panogen Px, ×2,000; (2) Vitaflo DB, ×2,500; (3) Arasan 75, ×2,000; (4) Manzate D, ×2,400; (5) Benlate T, ×2,400. (Courtesy of J. T. Mills)

Lindström (1958) compared the adhesion of the original oily liquid mercury fungicide, Panogen, and a commercial organomercurial dust, Betoxin TD, and found the liquid to be far superior; the adherent portion after application of a jet of air and washing procedures was 97 per cent for the liquid, but only 47 per cent for the dust. Smart (1963) confirmed this rate of retention, 97 per cent for Panogen, after washing of wheat and barley grain, and the retention on wheat of only half of the material of one dry formulation. However, in the cases of three other dry treatments examined, two on barley, one on wheat, only 10–15 per cent was washed off; the two formulations applied were based on phenylmercuric acetate and ethylmercuric chloride combined.

22.1.3 Diffusion and vapour action

There are two possibilities in the mode of action of seed germicides. The effect may be brought about by diffusion of liquid or by vapour action. Some materials are volatile, some are not. As shown in Table 21.2 some organomercurials are volatile, e.g. the ethylmercuric compounds and particularly the methylmercuric compounds which have a high vapour pressure. Other organomercurials have practically no vapour pressure, such as phenylmercuric compounds, e.g. Germisan and Uspulun powder. The vapour pressure of some of the alkylmercuric compounds is so high that the weight loss due to evaporation may be quite substantial. According to Gassner (1951*b*), the weight of the effective substance of methylmercuric compounds, exposed to open air at room temperature for 48 h, is reduced by more than 40 per cent, ethylmercuric compounds by about 10 per cent. Hexachlorobenzene is an example of a non-mercurial volatile seed germicide.

Coverage of volatile mercurials occurs in two steps: first, mechanical distribution in the treating equipment, secondly, redistribution by evaporation and resorption of the compound (Lindström, 1958). Evaporation and condensation of mercurials take place simultaneously from and on to the surface of seeds. Transference from surfaces rich in the material to surfaces poor in this material, where it is resorbed, is continuous until equilibrium is reached. By this vapour action distribution of the deposits obtained initially is materially improved and may in part compensate for poor mixing.

Lindström (1958) established that quick distribution in the first moments of the mixing process is important because of the immediate effect of local adhesion and adsorption. He found that satisfactory distribution of Panogen by evaporation was obtained 2 h after mixing, and that further improvement at increased storage time was so small that it was of no practical importance, see Fig. 22.4.

The initial distribution is negatively influenced by rapid resorption at a high moisture content, cf. Subsection 22.1.4. Lindström (1958) found that treatment of wheat with a radioactive non-volatile thallium nitrate solution gave very poor distribution on high moisture seed, while the initial distribution was very good on seed of lower moisture.

The effectiveness of vapour action is reduced by aeration. Arny and

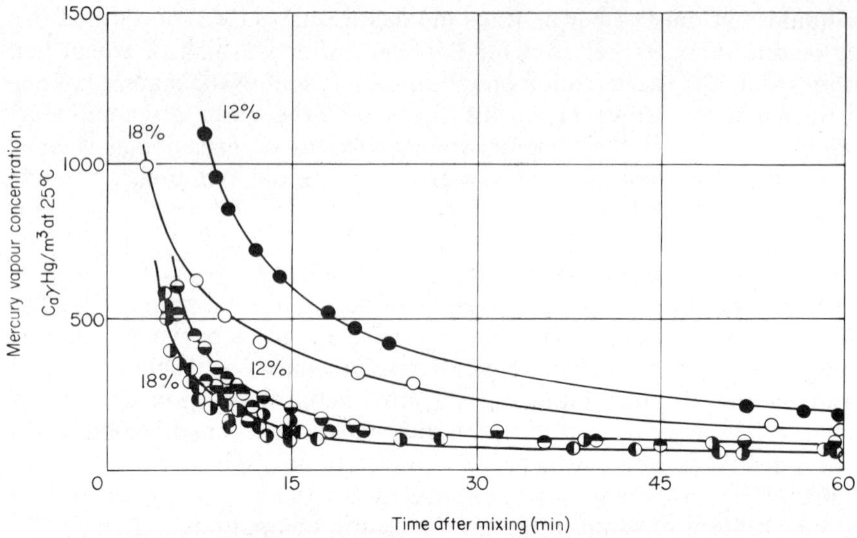

Fig. 22.4. Mercurial derivative in the atmosphere above treated seeds in a closed system as a function of time after seed treatment. For determination of mercurial vapour concentration airstream passed above the seeds. Two upper curves: Betoxin F, liquid (0.8 per cent mercury, solubilised ethylmercuric chloride); two lower curves: Panogen, liquid (0.8 per cent mercury, cyano (methylmercuric) guanidine), two formulations combined in each curve. The indicated percentages refer to the moisture content of the treated seed. (Lindström, 1958)

Leben (1954) and Arny (1962) reported that the vapours alone of various organic mercurials, particularly Panogen 15, considerably reduced bunt of wheat and Drechslera seedling blight (*Drechslera sorokiniana*) in barley, and eliminated Victoria blight (*D. victoriae*) in oats. Treated and non-treated seeds were mixed, and the vapour diffused from the former to the latter (Arny, 1962). If treated seeds were exposed to the air for several days before they were mixed with non-treated seeds, the effectiveness was greatly reduced. If they were stored in closed containers, the length of storage time before mixing had no marked effect. When treated seed is stored in cotton bags or paper bags from which vapours escape, little or no increase in the effectiveness of the materials against smut is obtained (Purdy, 1956*b*).

Discussing techniques of testing fungicides for seed treatment Leukel (1953) emphasised the necessity of storing of seed in closed (metal or glass) containers. He referred to the experience that if paper envelopes are stored in contact one with another, the fumes from a volatile mercurial fungicide in one envelope will affect the seed in the adjoining envelope and if this envelope contains an untreated check sample, it will then be useless for that purpose. Jørgensen (1966) demonstrated such decrease of available mercury in two organomercurials, Tillantin 1875 and Panogen 8, after storage of the treated seed in small paper bags, as measured in bioassays.

22.1.4 Resorption and reaction

Many seed-borne pathogens are harboured within the tissues of the seed, particularly the outer layers, the husk, pericarp and seed coat, hence successful treatment is largely dependent on the capacity of the germicide to penetrate. Relatively few pathogens infect more deeply situated tissues such as those of the embryo. It is well known that the protective layers, when physically intact, constitute efficient barriers against entry of most organisms, as well as chemicals, into the deeper layers of seeds, while mechanical injuries provide access for increased attacks of invading fungi, and increased damage by toxic chemicals.

The defence mechanism of pericarp and seed coat is mechanical as a physical impediment to organisms and chemicals, but also physicochemical, mainly in being selectively permeable to chemical substances in solution. Brown (1932) demonstrated that the mucilaginous testa (seed coat) of wheat forms a semi-permeable system around the wheat grain, and this holds true also for barley (Brown, 1907; Tharp, 1935), oats and rye (Brown, 1907).

Brown (1932) has distinguished between two types of absorption by wheat grains: (1) absorption by certain water solutes to which the membranes are freely permeable as by water, e.g. iodine, and (2) absorption of solutes for which the membranes have only slight permeability, e.g. silver nitrate. Brown found that absorption of the last category takes place through the micropyle to the outer cuticle of the testa from which it diffuses across the testa, and after partial breakdown of the semipermeability system, proceeds into the endosperm. The physicochemical properties of the solute are important, distinction being made between electrolytes and non-electrolytes; also the effect of the osmotic pressure of the solutions is vital (Brown, 1932).

It is well established that dry grains are relatively impermeable to some substances, somewhat more permeable to others, and readily absorb yet others. They are relatively impermeable to sulphuric acid, nitric acid, sodium chloride, sugars and polyhydric alcohols, somewhat permeable to mercuric chloride, acetic acid and trichloroacetic acid, while they readily absorb water, phenol, iodine, aniline and acids generally (Caldwell, 1961).

Netolitzky (1927) emphasised that in most seeds with endosperm, physicochemical protection is provided primarily by the inner cuticle of the seed coat that surrounds the endosperm. Only the micropyle through which the pollen tube penetrates is originally lacking in cuticular protection; this opening, however, closes tightly after fertilisation. At the opposite position of the seed, the chalazal region, there is also originally an opening in the inner cuticle for passage of nutrient for the growing endosperm and embryo, but this opening also closes as the seed matures. The inner cuticle of the seed coat is the most important barrier against penetration of chemicals such as germicides in moderate concentrations, even if penetration through the outer cuticle should succeed (Netolitzky, 1927).

In seeds of legumes (*Phaseolus*, *Vicia*) the situation is different. The endosperm is negligible; the embryo provides food storage and is surrounded by a cuticle beneath which is an aleurone layer; the seed coat has practically no cuticular protection. The semi-permeable protection is provided mainly by the cuticle of the embryo (Netolitzky, 1927).

Brown (1940) investigated permeability of seed coat membranes, as exemplified by *Cucurbita pepo*, to gases, and established that the inner of the two membranes controls gaseous exchange.

Kennedy (1961) demonstrated the protective effect of the seed coat of pea of moisture content 14.2 per cent against chloropicrin gas. The fungicidal effect of 9 h of fumigation against *Ascochyta* spp. in quartered seeds of pea was similar to that obtained when whole seeds were fumigated for 96 h.

By treating wheat seeds with radioactive formulations of Panogen, Lindström (1958) determined the depth which the mercurial of Panogen penetrates in dry wheat kernels. When labelled mercurial was absorbed, emission of beta particles was reduced while gamma radiation remained practically constant. The reduction of beta activity could be determined by the beta to gamma ratio and was found to be due to diffusion of mercurial. By increasing storage time up to one month, the maximum reduction of beta activity was attained. This was an indication of the maximum degree of absorption (Fig. 22.5). At this time the mercurial was situated behind a layer which, in conformance with the reduction of the beta activity, could be determined (by conversion of the weight of the absorbing layer to its depth) to be about 100 μm deep, corresponding to the thickness of the bran in wheat kernels. From this observation there is

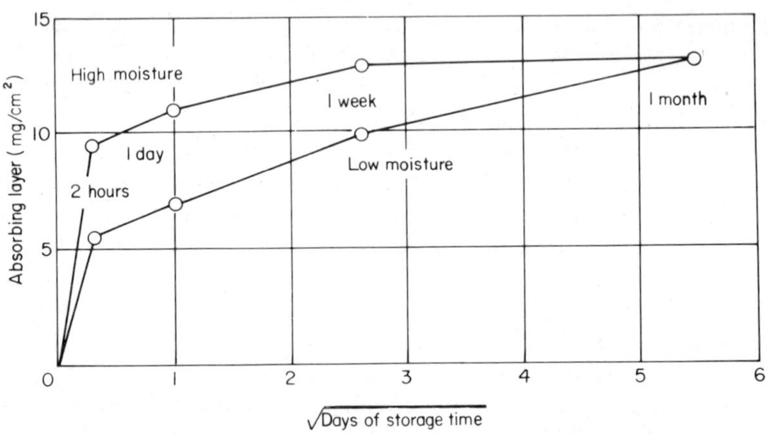

Fig. 22.5. Resorption of mercurial in kernels of wheat as a function of time and moisture content. High moisture, 18 per cent; low moisture, 12 per cent. (Lindström, 1958)

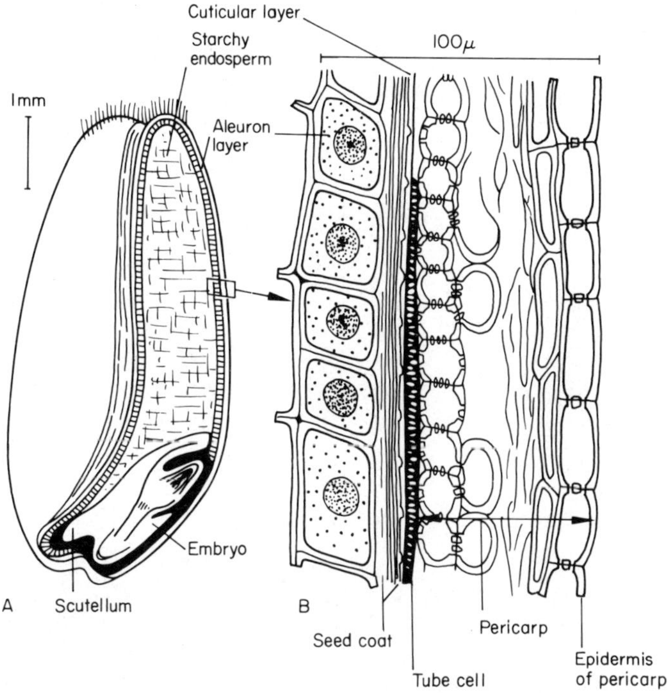

Fig. 22.6. A, Caryopsis of wheat, *Triticum,* and B, parts of its pericarp in longitudinal section. (Esau, 1960)

strong evidence that the mercurial ultimately becomes located in the deeper parts of the bran, including the pericarp and the nucellar and aleurone layers, see Fig. 22.6. Lindström (1959) established by chemical analysis that no mercurial is present within the starchy endosperm.

Arny and Leben (1954), in bioassays made on caryopses of barley from which the hulls had been removed after exposure to vapour action of Panogen, found that part of the vapour had penetrated beneath the hull. In Fig. 22.7 the amount of fungicide taken up by barley seed in the surface layers during air-tight storage has been converted into equivalents of fungicidal effect in terms of ounces of fungicide per bushel of grain. The vapours alone were effective in controlling completely Victoria blight of oats, *Drechslera victoriae*.

Lehman (1943) exposed cotton seed to vapours from ethylmercuric materials and recorded reduction in seedling blight and seed decay as a result of the treatment.

Seeds are more reactive at high moisture content levels than at lower levels, as shown in Figs 22.4 and Fig. 22.5, in which absorption of mercurials in wheat kernels at two moisture levels is compared. According to Hinton (1959) the moisture content in the bran layer of wheat is about 2

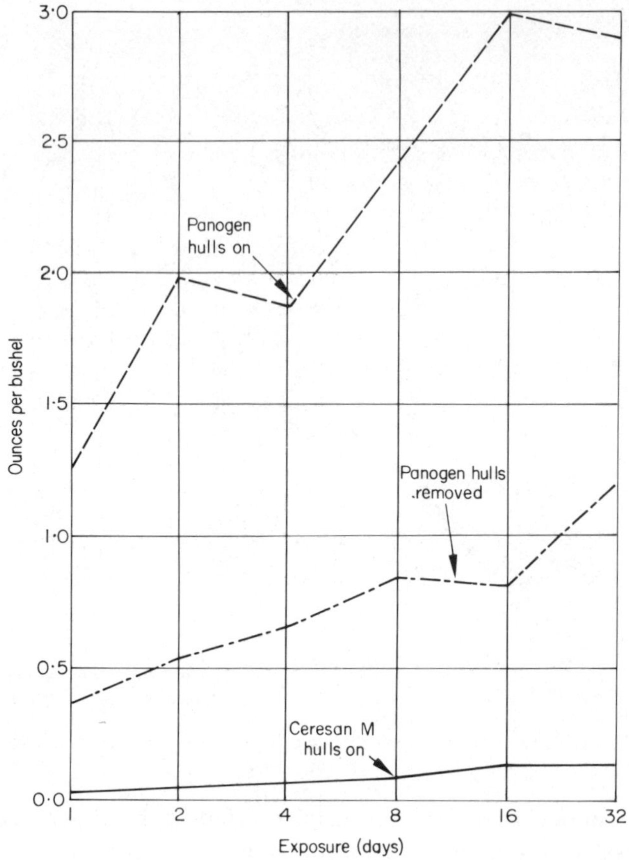

Fig. 22.7. The amount of fungicide on barley seed when seed was exposed to Ceresan M or Panogen vapours for various times in sealed crocks. Determinations made by bioassay of seed in the surface layer and expressed in terms of ounces per bushel. After the exposure period, hulls were removed from some seeds and only the caryopses assayed. (Arny and Leben, 1954)

per cent lower than in the endosperm, at the moisture levels dealt with here. Also, Gorringe (1950) found in his critical studies on sorption of metallic mercury vapour by wheat, that the rate of sorption increased with increased moisture content as well as higher temperature.

Copper is absorbed by pea seeds particularly in the seed coat, and is also localised in the embryonic radicle but only slightly absorbed in the cotyledons and plumule bud (Wain and Wilkinson, 1945).

Not much is known on the nature of the physiological reaction of seeds and pathogens to chemical seed treatment. According to various authors, quoted by Foster (1947), and his own investigations, copper fungicides cause injury to seeds of crucifers, cucumber and pea, while in adequate dosages they consistently exert beneficial effects on seeds of beet, egg-plant, pepper, spinach and tomato. Foster found evidence that the nature of the enzyme systems responsible for respiration of the seeds determines whether the action of copper is beneficial or injurious. The seed category reacting unfavourably to copper requires the presence of sulphhydryl groups in one or more of its enzymes, and these enzymes are inhibited by copper. The second category of seeds contains the enzyme tyrosinase which requires the presence of copper ions, and hence might be benefited rather than injured by moderate amounts of copper. Foster, however, points to the danger of oversimplification in this classification into two distinct groups. Many kinds of seeds may contain both copper-sensitive and copper-containing enzymes.

In wheat kernels, about 90 per cent of total protein is present in the nucellar and aleurone layers. These are situated just below the pericarp and, as mentioned previously, are likely to be tissues in which mercury is resorbed and accumulated after long exposure to vapour action. Sulphhydryl proteins are claimed to react with the mercurial fungicides in blocking reactions (Lindström, 1961).

The reaction of seed to overdoses of mercurials seems to be chromosomatic and comparable to the effect of colchicine. The hypertrophy of the coleoptiles of cereals produced in such cases is due to formation of polyploid cells and consequent arresting of growth (Fig. 22.8). The multinucleate cells of hypertrophied organs contain both very small and very large polyploidal nuclei as a result of incomplete mitosis (Sass, 1937; Purdy, 1956*a*).

Fig. 22.8. Phytotoxic reaction of wheat seedlings due to treatment of the seed with mercurials. Symptoms: stunted roots and hypertrophy of coleoptiles which may assume almost global form (lower row, left). (Courtesy of Jacques Ponchet)

Kostoff (1941) demonstrated this type of abnormality induced by ethylmercuric chloride in seeds of different crop plants and observed the mitosis to be affected in pea, rye and wheat. Gassner (1951*a*) established this effect by treatment of wheat grains with methyl-, propyl-, butyl- and cyclohexylmercuric compounds.

The principal mode of action of germicides on seed-borne pathogens is by diffusion of liquid and subsequent resorption. Surface-borne smut spores, e.g. of bunt of wheat or covered smut of barley, are largely killed by organomercurials in this way. The situation is different when oats are treated against loose smut, *Ustilago avenae*. This parasite is largely harboured between the glumes and pericarp as teliospores or resting mycelium. This somewhat protected inoculum may be reached by gases rather than by diffusion of liquid. Against this pathogen the almost non-volatile mercuric compounds have little effect and the ethylmercuric compounds with higher vapour pressures are better but not fully effective. The best control is given by methylmercuric compounds with very high vapour pressures, enabling penetration of the gases through and between the protective tissues (Gassner, 1951*b*).

Formalin is also effective against loose smut of oats just because of the penetrating power of the gaseous formaldehyde.

The direct mode of action of fungicides on fungi has been studied, particularly in their effects on spores. Fungicides may be fungistatic at low dosages, fungicidal at higher, with varying threshold values for dosages and time of effect for the transition from the arresting to the killing effect.

Copper compounds in lower effective dosages are predominantly fungistatic. Copper is chemically adsorbed by smut spores, being partially combined with phosphoric acid and the albumenous substances diffusing from the spores. When bunt spores are removed from the copper solution, washed free from adhering copper by treatment with 0.1 per cent hydrochloric acid, the spores will germinate (Bodnár and Terényi, 1925). The amount of copper adsorbed depends on both the compound and its copper content and on the length of exposure. Copper sulphate has a lower adsorption value than copper ammonium sulphate (Bodnár and Terényi, 1930). Hecke (1902, 1903) showed that the smut spores take up the same amount of copper, about 1 per cent of their weight, whether exposed for 5 min or 2 h, at concentrations of copper sulphate varying from 0.25 to 3 per cent. While copper sulphate may be washed off from the spore wall, copperammonium ions are absorbed into the spores of *Ustilago avenae* when the exposure time is longer than 15 min; in this instance the action of copper is not reversible, an indication that absorption has occurred (Terényi, 1930).

As indicated in Section 22.4, the activity of organomercurials depends chiefly on the organic radical, while the anion is of practically no importance (Gassner, 1951*a*). The toxic effect of mercuric salts on bunt spores, *Tilletia caries*, was shown by Bodnár and Terényi (1926, 1932) to be influenced very much by the particular salt used, and they found that the spores only could take up lipophilic compounds such as the mercuric halides. More mercury ions are necessary than copper ions to prevent spore germination. Mercury ions are insoluble in lipoids, whereas molecules of mercuric choride and mercuric bromide are lipoid-soluble; they are absorbed, combine with albumen and kill the spores. Matsui *et al.* (1962) have shown that in conidia of *Drechslera oryzae*, soaked in a 0.1 per cent solution of mercuric chloride, mercury is concentrated first on membranes of the nucleus and other organelles, then in the matrix of the nucleus and in the cytoplasm.

Mills (1970) studied the effect of different fungicides, including Dithane M45 (zinc-coordinated manganous ethylenebis dithiocarbamate), Panogen Px (methylmercury dicyandiamide), and Polyram (zinc-activated polyethylene thiuram disulphide, 80% formulation), on conidia of *Drechslera sorokiniana*, obtained from treated infested barley seeds. He found that the most likely path of entrance of fungicide is through the germination pores, as the two ends are the first parts to become affected; the conidium has a firm exospore with two terminal germination areas, and affected conidia were observed to produce lateral germ tubes instead of the normal terminal.

22.2 POST-SOWING ACTION OF GERMICIDES

Gassner (1927*b*) was the first to make clear distinction between primary, i.e. the pre-sowing, and secondary, i.e. the post-sowing, effect of germicides upon seeds and pathogens. He emphasised that dry dust application is based exclusively on the effect of chemicals after sowing under influence of the soil moisture. Moreover, that by the steeping and short-wet applications the action, through diffusion and absórption, occurs during the treating process, partly during storage in the short-wet treatments. After sowing there is little or no effect of the chemical left on steeped seed. By short-wet application there is a secondary effect of germicide because it is present after sowing in more or less efficient doses, encrusted on the seed.

Different chemicals vary in their capacity to exert primary and secondary effects. Gassner (1927*b*) showed that steeping of wheat in Germisan, an organomercurial, and subsequent inoculation of the seed with bunt spores reduced disease incidence appreciably, whereas there was very little or no such secondary effect when the seeds had been soaked in equal doses of two other mercurials, Uspulun and Segetan-Neu. Gassner (1927*b*) found, on the other hand, that the primary spore-killing effect was higher in the two last preparations than in Germisan. He concluded that Germisan is less easily absorbed and therefore adsorbed in somewhat greater proportions to the surface of the seed. Because of its mode of action, Germisan was found preferable to the others for the short-wet technique.

Naturally, manifestation of *primary* effect of seed treatments to germination of seed can only be recorded after sowing. Damage to germination of seeds that have been treated in hot water or formaldehyde is a primary effect. A secondary effect is excluded in these cases because there can be no adsorption of chemical and hence no residual material left.

The *secondary* effect of seed treatment by dry dust application of fungicides is well known already from the first metallic dusts, copper carbonate and basic copper sulphate. Both of these improve emergence and stand by combating soil-borne fungi. The widely used organomercurial dusts are applied on seeds in much smaller quantities than the classical copper compounds, and the secondary effect is limited, e.g. in their control of soil-borne inoculum of wheat bunt, *Tilletia caries*.

Hexachlorobenzene is effective against both seed-borne and soil-borne inoculum of bunt, including soil-borne dwarf bunt, *Tilletia contraversa* (Siang and Holton, 1953; Zobrist and Thiollière, 1954). However, used only as a seed treatment, this fungicide gives good but still incomplete protection against seedling infection by the soil-borne spores of the fungi. Hexachlorobenzene has a low vapour pressure, and the vapours produce a zone free of smut spore inoculum around the germinating seed, so that the seedling is not infected in its early development. A few of the seedlings that have grown beyond this zone may yet become infected (Purdy, 1957*a*). Organomercurials, and particularly the metallic dusts mentioned, act by diffusion of the liquid formed after dissolution of the dusts in the soil.

The mode of action of systemic dust fungicides is another example of purely secondary effect. Thapliyal and Sinclair (1971) demonstrated, by using labelled and unlabelled fungicides as seed treatments of soybean, that benomyl (Benlate) and chloroneb (Demosan) initially tended to localise in the cotyledons, while carboxin (Vitavax) did not. Subsequently, according to radioautographs and bioassays, chloroneb became localised in cotyledons, lower hypocotyl and roots of the seedling, carboxin was distributed uniformly throughout the seedling, with higher concentrations in the epicotyl, whereas benomyl moved only into the epicotyl. The mode of redistribution of these fungicides defines their applicability as seed treatments: chloroneb has potential for control of root and lower-stem infections, while benomyl and carboxin would be more effective against pathogens invading the epicotyl tissues. Both benomyl and carboxin have been established as effective against the loose smuts of barley and wheat, the infection of which is located in the embryo, and most effectively in the plumule bud.

Jacobsen and Williams (1971) reported that infection of *Phoma lingam*, located in the seed coat and embryo of broccoli, can be completely controlled by benomyl if applied as a soak for 24 h at 30°C, but not if used as a dust. Hence, control of seed-borne inoculum is satisfactory through the primary effect, but not through the secondary only.

Thiram is used for seed treatment as a powder or as a soak treatment in a colloidal suspension. Many kinds of vegetable seeds are treated by this chemical. It is non-volatile, consequently the secondary effect is provided by the liquid. Thiram soak is more than a question of simple diffusion; the thiram seems to be concentrated by the passage into the seed. It is effective against a considerable range of seed-borne pathogens (Maude, 1973, cf. Section 21.5(8); see also later, in Tables 22.2 and 22.3) as well as soil-borne fungi such as *Pythium* and *Phytophthora*. In the last capacity it is acting as a so-called 'seed protectant', in the first as a 'seed disinfectant' — a fact that may provide an argument against the use of this often ambiguous distinction.

22.3 SEED TREATMENT AND TYPE OF HOST

Seed treatment procedures must be appropriate to the type of seed to be treated. For instance, steeping is not workable for flax seeds because of the mucilaginous hygroscopic seed coat, and powder fungicides largely fail to adhere to its smooth surface coat when dry; flax therefore requires a heavier dosage than most other seeds. When, for small seed lots, an overdosage of dry treatment is corrected by subsequent sifting, some seeds such as hairy ones may retain so much material that it becomes phytotoxic; the dosage can be reduced by diluting the powder in talc or similar inert matter (Hahn, 1970).

The glumes of Gramineae largely prevent fungicide powders reaching spores of smuts and other fungi; thus volatile mercurials are most effective.

The physical requirements for treatment vary according to type of seed. Nagel (1927) found for the short-wet procedure that to obtain a certain effect of organomercurials, 2–3 l. of water per 100 kg grains was required for wheat and rye, 2.5–3.5 l. for barley, and 3–4 l. for oats.

Special seed processing procedures may essentially influence the outcome of germicide treatment. For treatment of cotton seed the selection of delinting technique, either acid-delinting or machine-delinting, may be important. Mechanical delinting may injure the seed.

Differences between cultivars may be important to selection of seed treatment procedures. An old example is the reaction of two cultivars of sorghum to steeping of the seed in 2 per cent copper sulphate. 15 min of soaking were sufficient for one cultivar but 30 min were needed for the other because of the glumes being closely attached (Robertson, 1927, 1928).

22.4 SEED TREATMENT AND PHYTOTOXICITY

Doses lethal to seed lead to complete failure of germination, sublethal doses to reduction of germination as well as to abnormalities in the seedlings. These may resemble mercury-injured wheat and maize seedlings with symptoms as described in Subsection 22.1.4 (Sass, 1937; Purdy, 1956*a*), see Fig. 22.8.

The reaction of seeds is influenced very much by their previous harvest and storage conditions. Phytotoxicity by mercurial seed treatment may be increased if the moisture content of the grains has been raised unduly near maturity of the seeds (Chambers, 1961). Differences in sensitivity to chemical treatment between seed lots of the same cultivar and with the same moisture content are certainly due to such often unknown circumstances.

Gassner (1951*a*) investigated *dosis toxica* (see Section 23.1) of wheat seed in relation to organic mercury compounds applied as dry powders, and found that the highest concentration of mercury tolerated is largely related to the cation of mercury salts as follows:

butylmercuric salts	10 ppm
propylmercuric salts	12 ppm
cyclohexylmercuric salts	26 ppm
phenylmercuric salts	37 ppm
ethylmercuric salts	56 ppm
methylmercuric salts	72 ppm
methoxyethylmercuric salts	74 ppm
mercuric chloride salts	380 ppm

Some fungicides are toxic to certain kinds of seeds, beneficial to others. According to Leukel (1953), Cuprocide is injurious to seeds of crucifers, lima beans and peas but beneficial to lettuce (cf. discussion on the effect of copper in Subsection 22.1.4), which in turn is injured somewhat by thiram (Arasan) and ferbam (Fermate); zinc oxide is injurious to peas but is highly beneficial to crucifers and spinach.

Cultivars may react quite differently to chemicals. Injury to sorghum seed, for instance, depends largely on cultivar. Cultivars of the Leoti sorgo group have been shown to suffer considerably more from mercury injury than Sharon Kafir cultivars (Leukel, 1943; Leukel and Livingstone, 1945). Hsi (1957) classified sorghum cultivars according to susceptibility to injury by seed treatment with Panogen 15, applied at rates from 0.75 to 6 oz per bushel. Kafirs and kafir-milo hybrids were more susceptible than sorgo, milo and hegari types. Four cultivars were tolerant, seven were moderately tolerant and seven susceptible to injury from Panogen seed treatment. Tolerant cultivars could withstand the highest dose, susceptible cultivars reacted by some injury even at the lowest rate of application.

Methyl bromide fumigation against white tip nematode, *Aphelenchoides besseyi*, is applicable in some cultivars of rice but is injurious to others (Noble and Richardson, 1968).

Gassner (1951*a*) established that the method of application has consistent influence on the level of *dosis toxica* (cf. Section 23.1) of organomercurials in wheat. When different mercurials with the same composition and the same mercury content are compared, *dosis toxica* is lowest, i.e. most toxic, at the steeping procedure, higher when dry dust is applied, and highest in the short-wet procedure. Many categories of mercurials used by the soaking procedure were found impracticable for dry dust treatment. Methylmercuric acetate, for example, was found excellent for steeping but useless for dry dust treatment.

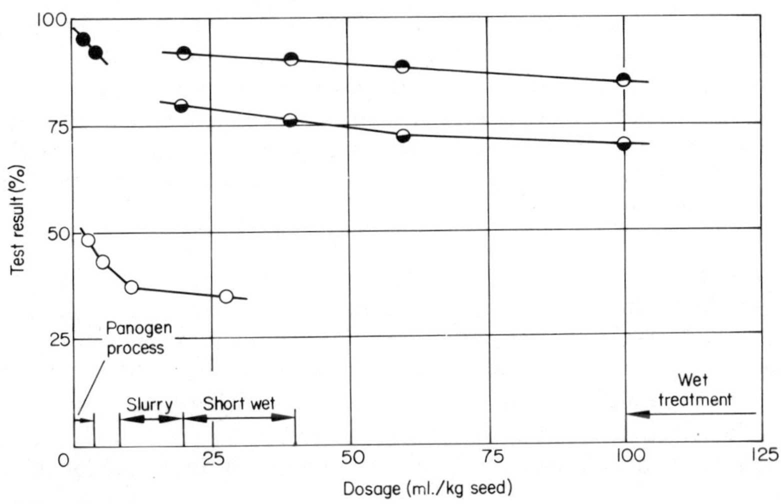

Fig. 22.9. Influence of the liquid dosage on biological performance. Evaluation of data from tests where constant amounts of mercurial have been applied with varying amounts of the liquid carrier (Lindström, 1961):

- ○ Field emergence of vegetable seeds (tomato, cucumber, beet, radish, bean, maize), according to Wallen (1953). Emergence in check 30.5%.
- ◒ Rate of germination ('Triebenergi') for wheat treated Segetan-Neu, according to Gassner (1927*b*)
- ◓ Vitality ('Triebkraft') observed in the test reported by Gassner (1927*b*).
- ● Emergence of Panogen treated oats in a laboratory test conducted by Markkula, cf. Lindström (1961).

In Fig. 22.9 Lindström (1961) has summarised extensive data obtained by various investigations on the depreciating influence of increasing liquid dosages in organomercurials upon emergence of vegetable and cereal seeds. Or inversely, mercurial damage is reduced with decreased liquid dosages.

Seed treatments in which incompatible materials are combined induces phytotoxicity. Jones and Collins (1971) found a consistent phytotoxic interaction with all combinations of phenylmercuric ammoniumacetate (as used in the Mist-O-Matic process) and benomyl or carboxin, the systemic fungicides all used in standard doses to wheat. There was no such interaction when either of the two systemic fungicides was combined with Panogen 15 or Ceresan PL (methylmercuric acetate) at standard rates.

22.5 SEED TREATMENT AND TYPE AND AMOUNT OF INOCULUM

Every seed treatment chemical has a distinct range of therapeutic effect in terms of the pathogens against which it is effective. The old seed germicides, such as copper sulphate, copper carbonate, mercuric chloride and formaldehyde, as well as the newer organic mercurials act on a greater number of pathogens than the most recent non-mercurial organic compounds which are more selective. In Table 22.2 examples of the effectiveness of seed treatment materials to categories of pathogen are given. Table 22.3 surveys some important seed treatment procedures in use for important crop plants.

Although a certain fungicide may be effective against a certain category of pathogens, the degree of effectiveness, the dosage needed, may differ between the members of such a category. Different species of smut fungi may react differently to a certain fungicide, and the differences in reaction of the pathogens may not be consistent from one fungicide to another. This is exemplified in Table 22.4, in which Kišpatić (1944) has recorded threshold concentrations (*dosis curativa*, see Section 23.1) and their relative values of four fungicides against the spores of four species of smut fungi.

Any pesticide may be regarded as selective. Mills and Wallace (1968, 1970) demonstrated differential action of fungicides upon fungi on barley seed. The selective action operates on pathogens as well as saprophytes (Fig. 22.10). The effect may be selective inhibition or selective stimulation and may work on antagonistic microorganisms, thus affecting the microbiological balance.

There is generally, within certain limits, a clear correlation between the amount of seed-borne inoculum and the dosage of chemical required.

Table 22.2. Major types of seed treatment pesticides in their use against seed-borne pathogens. (Copper carbonate, used particularly in developing countries, is now mostly replaced by new fungicides) *Combined with hot water

	Pathogen		*Copper carbonate*	*Silver nitrate*	*Mercuric chloride*	*Thiram*	*Organomercurials*	*Hexachlorobenzene*	*Pentachloronitrobenzene*	*Captan*	*Benomyl*	*Carboxin*	*Streptomycin*	*Methylbromide*	*Hydrochloric acid*
Oomycete	Phytophthora	cryptogea				×				×					
Fungi Imperfecti, Ascomycetes	Alternaria	brassicicola				×	×								
		godetiae					×								
		porri f.sp.dauci				×									
		radicina			×	×	×								
		zinniae				×	×			×					
	Ascochyta	fabae				×									
		pisi				×				×	×				
	Colletotrichum	graminicola	×				×								
		gloeosporioides					×								
		orbiculare			×										
		lindemuthianum									×				
		lini				×	×								
		dematium				×									
	Drechslera	avenae				×	×								
		graminea				×	×								
		oryzae					×								
		sorokiniana					×								
		teres					×								
	Fusarium	graminearum					×								
		moniliforme				×	×			×					
		oxysporum (Callistephus, tomato)			×										
		spp.				×					×				
	Phoma	apiicola				×									
		betae				×	×								
		exigua v.linicola				×	×								
		medicaginis			×	×									
		lingam				×	×*								
	Pyricularia	oryzae					×								
	Septoria	apiicola				×									
		drummondii								×					
		nodorum				×									
	Verticillium	alboatrum (lucerne)				×									
	Rhizoctonia	solani				×	×								
Ustilaginales	Sphacelotheca		×			×	×								
	Tilletia		×				×	×	×		×	×			
	Urocystis		×				×	×							
	Ustilago (embryo infection)									×	×	×			
	Ustilago (seedling infection)		×				×			×	×				
Bacteria	Corynebacterium	(Beta, tomato)											×		
	Pseudomonas	(bean, Cucurbita, oats, soybean, tobacco)		×									×		
		(Hibiscus)			×										
	Xanthomonas	(Brassica, Cyamopsis)											×		
		(Capsicum)			×										
Nematodes	Aphelenchoides	(rice)												×	
	Ditylenchus	(clover, lucerne, onion, rice)												×	
Virus	Tobacco mosaic virus	(tomato)													×

Table 22.3. Seed treatments against pathogens of important crop plants; most procedures have been established in general practice. Some treatments against soil-borne pathogens (e.g. causing damping-off) have been included, as well as some heat seed treatments. The table is not exhaustive. (Some references: Maude, 1968; Nene, 1971; Marsh, 1972)

Cereals (dosage per 100 kg)		
Crop	*Pathogen*	*Treatment*
Barley	Drechslera graminea, D.teres, D.sorokiniana	organomercurials, 100 – 400 g or slurry 250 g/5 l; maneb; carboxin 150 g; mancozeb, 200 g
	Fusarium culmorum, F.graminearum, F.nivale	benomyl or carboxin 100 g (some effect); Neo-Voronit 200 – 300 g; thiabendazole 160 g (effective against internal infection)
	Ustilago hordei	organomercurials 250 g, slurry 250 g/5 l; carboxin, or pyracarbolid 100–150 g; thiabendazole 120 g
	Ustilago nuda	carboxin, or thiabendazole 100 – 120 g; Panogen-Metox 200 ml; hot water
Maize	Diplodia maydis	chloranil 200 g
	Drechslera maydis, D.turcica	carboxin (some effect)
	Fusarium graminearum	thiram 200 – 250 g
	Fusarium moniliforme	organomercurials
	Ustilago maydis	organomercurials but ineffective against soil contamination; thiram
Millet, sorghum	Cercospora sorghi	thiram; chloranil
	Gloeocercospora sorghi	thiram; chloranil
	Drechslera spp.	organomercurials 250 g; ferbam 250 g
	Fusarium moniliforme	organomercurials 250 g
	Sphacelotheca cruenta, S.sorghi and other spp.	organomercurials 250 g; elemental sulphur 500 g (fairly effective); chloranil 200 g; thiram 250 g; carboxin 250 g (moderate effect)
	Pyricularia grisea	organomercurials 200 g
Oats	Drechslera avenae	organomercurials 250 g but fail against mercury-resistant strains; maneb or carboxin 150–200 g; mancozeb 200 g
	D.sorokiniana, D.victoriae	organomercurials 300 g
	Fusarium graminearum, F.nivale	organomercurials 300 g; Neo-Voronit 200 – 300 g
	Septoria avenae	try organomercurials 300 g
	Ustilago avenae, U.hordei	organomercurials 250 g or slurry 250 g/5 l; benomyl 100 g; carboxin 100 – 200 g; pyracarbolid 10 g; thiabendazole 120 g; Panogen-Metox 300 ml
Rice	Drechslera oryzae	organomercurials 250 g; thiabendazole; thiram 200 – 250 g; aureofungin
	Fusarium moniliforme	organomercurials 250 g
	Pyricularia oryzae	organomercurials 200 g; benomyl 400 g; thiabendazole; aureofungin; kasugamycin
	Trichoconis padwickii	tepid water 16 h + water 54°C/15 min
Rye	Fusarium nivale	organomercurials 200 g; Neo-Voronit 200 – 300 g; thiram 200 g; benomyl 100 g
	Urocystis occulta	organomercurials 200 g; benomyl 100 g; carboxin 150 – 200 g; thiabendazole 120 g; Neo-Voronit 200 – 300 g

Table 22.3 *continued*

Cereals (dosage per 100 kg)		
Crop	*Pathogen*	*Treatment*
Wheat	Alternaria triticina	water 4 h + water 52 – 54°C/10 min suggested
	Drechslera sorokiniana	organomercurials 100 – 200 g; carboxin–thiram 250 g
	Fusarium culmorum, F.graminearum, F.nivale	organomercurials 100 – 200 g; thiabendazole 160 g; thiram 250 g; Neo-Voronit 200 – 300 g
	Septoria nodorum	organomercurials 200 g; thiabendazole; maneb. Difficult to obtain satisfactory control; benomyl 100 g
	Tilletia caries, T.contraversa, T.foetida	hexachlorobenzene 330 g; PCNB 200 g; benomyl 100 g; carboxin 150 – 225 g; thiabendazole 120 g; Neo-Voronit
	Urocystis agropyri	carboxin 75 g; oxycarboxin 30 g; pyracarbolid 50 – 100 g
	Ustilago tritici	carboxin 75 g; carboxin 0.2%/6 h; oxathiin or pyracarbolid 100 g
Industrial plants (fibre plants, oil plants) and fruit trees (dosage per 100 kg)		
Beet, sugarbeet	Cercospora beticola	organomercurials 600 g
	Phoma betae	organomercurials 600 g (partly effective); aer-steam 56°C/30 min; thiram 600 g
	Uromyces betae	organomercurials 600 g
	Corynebacterium betae	streptomycin
Cotton	Colletotrichum dematium, C. gossypii	organomercurials 250 g
	Rhizoctonia solani	quintozene 230 g; PCNB 250 g; thiram 250 g
	Xanthomonas malvacearum	mercuric chloride-iodide (abavit), short-wet
Flax	Alternaria lini, A.linicola	organomercurials 250 g
	Botrytis cinerea	thiram 0.2% 30°C/24 h; organomercurials 300 g
	Colletotrichum lini	thiram 150–250 g
	Phoma exigua var. linicola	thiram 150–250 g
Groundnut	Aspergillus flavus, A.niger	thiram 200–250 g; organomercurials 200–250 g
Rape, turnip, colza	Alternaria brassicae, A.brassicicola, A.raphani	organomercurials 0.5% soak 30 min
	Phoma lingam	see cabbage
	Rhizoctonia solani	organomercurials 0.5% soak 30 min; water 50° C/30 min
Safflower	Puccinia carthami	thiram 200–250 g
Sesame	Pseudomonas sesami	streptomycin 0.1%, reduces infection
	Xanthomonas sesami	Abavit B 7 g
Tobacco	Cercospora nicotianae	thiram 250 g
	Damping-off	thiram 250 g
Citrus	Phytophthora citrophthora, P.nicotianae var.parasitica	water 51–52°C/4–10 min, then arasan, and stored in polythene sacs at 4–7°C
	Xanthomonas citri	hydrogen peroxide, 1 part + 2 of water/10 min
Coriander	Protomyces macrosporus	thiram 250 g (+ broadcasting on soil 8 kg/acre)
Avocado	Phytophthora cinnamoni	water 49–52°C/30 min

Table 22.3 *continued*

Ornamentals (dosage per kg)		
Crop	*Pathogen*	*Treatment*
Esch-scholtzia	Acroconidiella eschscholtziae	water 52°C/30 min
Matthiola	Alternaria raphani	try thiram 0.2% 30°C/24 h or thiram 2.5 g
	Fusarium oxysporum	aer-steam 54°C/15 min
	Xanthomonas incanae	water 54–56°C/10 min
Zinnia	Alternaria zinniae	water 52°C/30 min; aer-steam 57°C/30 min; thiram 2.5–3 g; thiram 0.2%/24 h
	Rhizoctonia solani	water 52°C/30 min

Vegetables (dosage per kg)		
Bean	Colletotrichum lindemuthianum	benomyl; thiabendazole 2 g
	Damping-off	captan 2 g; PCNB 2.5 g
	Corynebacterium flaccumfaciens Pseudomonas phaseolicola Xanthomonas phaseoli	streptomycin or other antibiotics 0.02% 15 – 30 min only partly effective but possibly effective against trace contamination
Cabbage, cauliflower	Alternaria brassicae, A.brassicicola	thiram 0.2% 30°C/24 h; thiram 2.5 g; organomercurials 3 g
	Plasmodiophora brassicae	thiram 2–2.5 g
	Phoma lingam	thiram 0.2% 30°C/24 h; benomyl 0.2% 30°C/24 h; thiabendazole 0.2%; aer-steam 56°C/30 min
	Xanthomonas campestris	antibiotic, e.g. 0.3%/30 min; water 50°C/25 min (not always effective), then thiram
Carrot	Alternaria porri f.sp.dauci	thiram 0.2% 30°C/24 h
	A.radicina	thiram 0.2% 30°C/24 h
	Cercospora carotae	dichlone fairly effective
	Xanthomonas carotae	water 50°C 15–20 min, then captan 5 g
Celery, celeriac	Phoma apiicola	thiram 0.2% 30°C/24 h
	Septoria apiicola	thiram 0.2% 30°C/24 h
Chickpea	Ascochyta rabiei	thiram 1 g
Chilli, pepper	Cercospora capsici	organomercurials 2.5 g; PCNB 2.5 g
	Damping-off	organomercurials 2.5 g; PCNB 2.5 g
	Rhizoctonia solani	water 52°C/30 min (against seed-borne inf.)
Lettuce	Septoria lactucae	water 46°C/30 min but germination reduced
Onion	Urocystis cepulae	thiram 2.5 g, encrust seed using methylcellulose sticker
Pea	'Ascochyta complex'	thiram 0.2% 30°C/24 h; benomyl 3–5 g; thiabendazole; carbendazin
	Damping-off	captan 2.5 g; thiram 2.5 g
	Pseudomonas pisi	streptomycin (slurry) 2.5 g
Red beet	Colletotrichum dematium f.sp. spinaciae	thiram 0.2% 30°C/24 h; thiram 6 g; organomercurials 6 g
	Phoma betae	thiram 0.2% 30°C/24 h; thiram 6 g; organomercurials 6 g

Table 22.3 *continued*

	Vegetables (dosage per kg)	
Crop	*Pathogen*	*Treatment*
Soybean	Cercospora kikuchii	thiram 2.5 g; chloranil 2 g
	Colletotrichum glycines, C.dematium	try thiram 2.5 g
	Peronospora manshurica	try thiram 2.5 g
Spinach	Pythium	thiram 5 – 10 g
Tomato	Damping-off	organomercurials 2.5 g; thiram 2.5 g; PCNB 2.5 g
	Didymella lycopersici	try thiram 5 g
	Fusarium oxysporum	mercuric chloride 0.05%/5 min
	Phomopsis phomoides	thiram 5 g
	Corynebacterium michiganense	streptomycin 0.04%/30 min
	Xanthomonas vesicatoria	hexachlorophene 5 ppm
	tobacco mosaic virus	trisodium phosphate 10%/10 min

Table 22.4. Threshold concentrations (*dosis curativa*) of four seed fungicides tested for their effect on spores of four species of smut fungi (Kišpatić, 1944)

*Seed fungicide**	*Tilletia caries*	*Ustilago hordei*	*Ustilago avenae*	*Sphacelotheca destruens*	
Abavit	0.015%	0.005%	0.005%	0.01%	Absolute value
Ceresan	0.025%	0.020%	0.02%	0.05%	
Fusariol	0.020%	0.015%	0.02%	0.045%	
Germisan	0.022%	0.006%	0.003%	0.04%	
Abavit	60	20	20	40	Relative values. 0.025% = 100
Ceresan	100	80	80	200	
Fusariol	80	60	80	180	
Germisan	88	24	12	160	
Abavit	100	33	33	67	Relative values. Values for bunt control = 100
Ceresan	100	80	80	200	
Fusariol	100	75	100	225	
Germisan	100	27	14	182	

*Abavit Nassbeize 3330a; Ceresan Nassbeize U564; Fusariol Nassbeize 2115a; Germisan Nassbeize Retorte 3659 I.

Heavily infected or contaminated seed requires more fungicide. The higher the spore load of smut fungi the higher are the concentrations of fungicide needed to obtain complete control; but there is a limit to the quantity which the seed can retain!

In Danish experiments, increasing percentages of loose smut in barley required increasing dosages of Vitavax to eliminate infection (Nød-degaard, Hansen and Rasmussen, 1970).

In Japan Hino and Furuta (1968), treating rice seed samples with an organomercurial, obtained a low disinfection rate in samples with an average embryo infection of 9 per cent *Fusarium moniliforme* and increasing control in seeds with decreasing percentages of embryo infection.

Heavily smut contaminated sorghum seed requires increased dosages of fungicide as compared with slightly contaminated seed (Uppal and Malelu, 1928; Uppal and Desai, 1931).

When the spore load of smut fungi, or rate of infection of *Fusarium*, *Drechslera* and other seed-borne pathogens is very low, seed treatment may not be necessary especially if antagonistic antibiotic-producing organisms are also present.

Hansen (personal communication, 1972) compared the effect of four dosages of organomercurials on three infection levels of barley leaf stripe disease, *Drechslera graminea*, Fig. 22.11. Increasing dosages are required to control increasing levels of infection.

The consistent use through decades of a 'subdose' of organomercurial seed treatment of cereals in Denmark, half of the standard dosage used in other European countries, pertains to the problem of adjusting dosage. The implications of this policy of using a low dosage in relation to human and animal environment are discussed in Subsections 22.10.5 and 22.10.6.

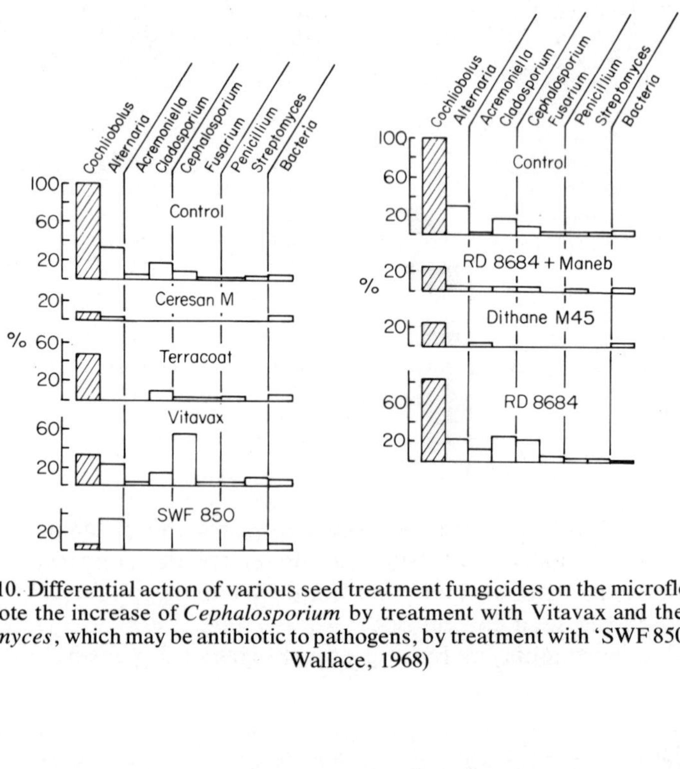

Fig. 22.10. Differential action of various seed treatment fungicides on the microflora of barley seed. Note the increase of *Cephalosporium* by treatment with Vitavax and the increase of *Streptomyces*, which may be antibiotic to pathogens, by treatment with 'SWF 850'. (Mills and Wallace, 1968)

Another problem pertaining to dosage is integration between seed treatment and other procedures of control, including biological control. If infected seed is sown in disinfested soil, seedling diseases often become more prevalent as damping-off by *Rhizoctonia solani* in different crops when Fenaminosulf ('Dexon') has been used to control *Pythium* (Garren, 1963). Any change in the antagonistic flora, in soil or in seeds, may influence dosage of seed treatment. Hot water treatment of cereals used against loose smut changes the seed-borne antagonistic microflora and may increase the necessity of subsequent chemical treatment.

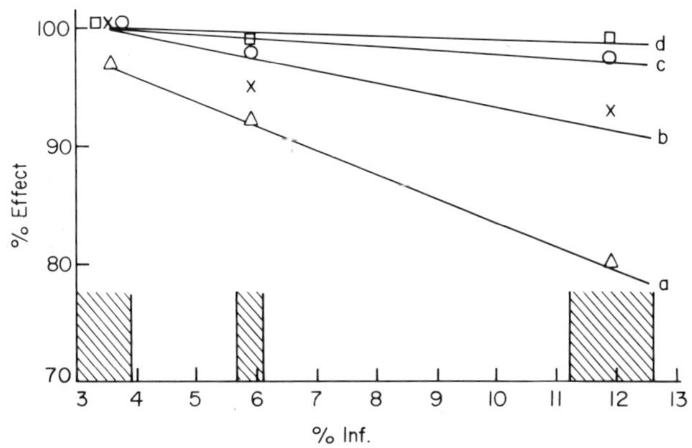

Fig. 22.11. The diminishing effect of dosage of organomercurials on increasing infection levels of *Drechslera graminea,* barley leaf stripe disease. Field experiments (total 98) at Statens Plantepatologiske Forsøg, Lyngby, Denmark, 1965–1971:

Dosages: a = 1/4 normal d. = 0.25 g/kg (26 exp.)
b = 1/2 normal d. = 0.50 g/kg (23 exp.)
c = 1/1 normal d. = 1.00 g/kg (26 exp.)
d = 2/1 normal d. = 2.00 g/kg (23 exp.)

The dosage indicated as 'normal' is in accordance with Danish standard, cf. Subsection 22.10.5. Levels of infection (indicated by hatched columns): 3.0–3.9%; 5.7–6.1%; 11.2–12.6%.

(Knud E. Hansen, personal communication)

22.6 SEED TREATMENT AND VARIABILITY OF PATHOGEN

Pathogens may vary substantially from biotype to biotype in their reaction to pesticides. Weltzien (1957*a*) investigated 20 accessions of *Tilletia caries* from various parts of Germany for their resistance to Ceresan-Trockenbeize (methoxyethylmercuric silicate). He found constant differences in the relative germinability of the spores after treatment with an underdosage of Ceresan, 1 g per kg seed, cf. Fig. 22.12.

Kuiper (1965) reported failure of hexachlorobenzene to control bunt of

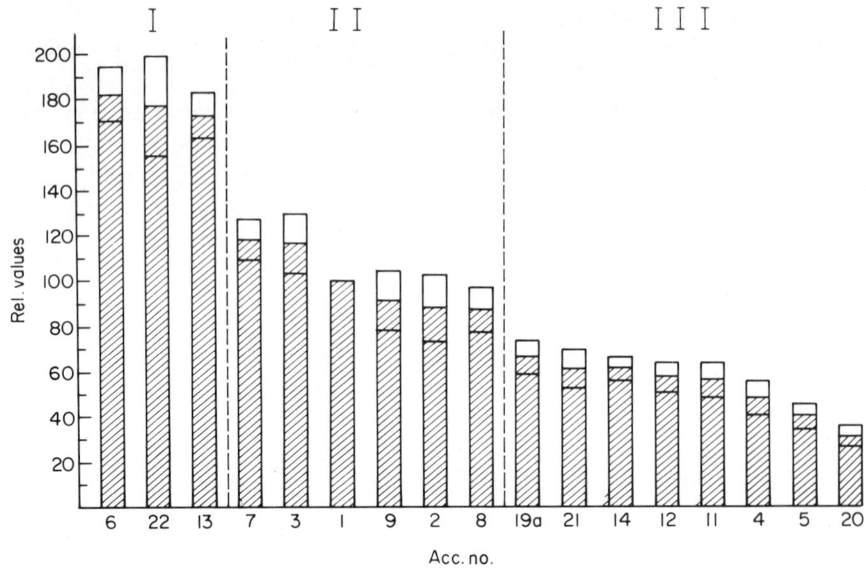

Fig. 22.12. Relative mean values for spore germination of 17 accessions of *Tilletia caries* (hatched columns) with the standard deviation indicated above the transverse line, Accession No. 1 being the standard of measurement, equal to 100. Distinction is made between three groups of resistance to the fungicide. Treatment: 1 g Ceresan-Trockenbeize per kg grain. (Weltzien, 1957*a*)

wheat, *Tilletia caries* and *T. foetida*, in Victoria, Australia; two biotypes of *T. foetida* have been found to differ only in their susceptibility to this fungicide, one being controlled, the other virtually uncontrolled (Anonymous, 1967).

Some biotypes of *Rhizoctonia solani* are resistant to pentachloronitrobenzene, and these are also more pathogenic to plants (Shatla and Sinclair, 1963). Buxton, Sinha and Ward (1962) observed apparent development of tolerance of spruce seedling pathogens to formaldehyde. In Taiwan Chien (1964) found that resistance of *Pyricularia oryzae* to mercuric chloride is coupled with increased virulence to rice.

The natural differences between biotypes of a pathogen may range from sensitivity to a high degree of resistance against a germicide. By using, year after year, a certain chemical, resistant biotypes may be selected, thus rendering use of the germicide inefficient for practical control. A dramatic example is the penicillin-resistant *Staphylococcus* that plagues hospitals.

The most striking example among plant pathogens from recent years on variation of resistance or tolerance to a type of fungicide, is development of mercury-resistance of *Drechslera avenae (Pyrenophora avenae)* in oat seed. Noble *et al.* (1966) were the first to report that some strains of this parasite, in Scotland, were resistant to mercury compounds and this was confirmed by Old (1968). Malone (1968) recorded that the fungus was resistant in 25 of 84 infected oat seed samples of Northern Ireland but susceptible in 34 other samples, and also Old (1968) recorded mercury-resistance in Ulster oat seed. Mercury-resistance of the oat parasite was furthermore reported from New Zealand (Sheridan, Tickle and Chin, 1968; Sheridan and Vachrabhorn, 1968), England and Wales (Dickens and Sharp, 1970), Denmark (Dharam Vir, Mathur and Neergaard, 1970) and the U.S.A. (Crosier, Waters and Crosier, 1970). Also, in Holland strains of the fungus showing increased tolerance to organic mercuric compounds have been isolated (Limonard, 1968). Resistance was only to the compounds which contained an aryl radical and not to those with an alkyl radical (Dekker, 1969). Sheridan (1973) found in New Zealand that all samples of British-grown oats examined were infected with mercury-resistant *Drechslera avenae*, whereas incidence and levels were still low in New Zealand but appeared to be increasing. He considered import of seed treated with organo mercurials to be potentially dangerous because of introduction of new highly resistant strains of the fungus.

Greenaway and Cowan (1970) found in their studies on 26 isolates of *Drechslera avenae* two distinct classes, susceptible and resistant. Conidia from susceptible gave rise only to susceptible colonies, while conidia from resistant isolates gave both susceptible and resistant colonies. Growth of many resistant isolates was stimulated by mercury concentrations between 1 p.p.m. and 10 p.p.m., whereas susceptible isolates succumbed to 5–10 p.p.m.

In 1970 benomyl suddenly failed to control *Botrytis cinerea* in cultures of cyclamen in Dutch nurseries. Mycelial growth of the resistant strain on potato dextrose agar containing 1,000 μg/ml. benomyl was less inhibited than an isolate of the normal form of *B. cinerea* with 0.5 μg/ml. of the fungicide. The resistant strains also showed increased resistance to three other systemic fungicides, methyl-thiophanate, furidazol and thiabendazole (Bollen and Scholten, 1971). Benomyl-resistant strains of *B. cinerea* from nursery crops of lettuce and tomato were also found in England (Miller and Fletcher, 1974). A similar development of resistance to benomyl has been encountered in *Cercospora beticola*. Regular spraying of sugarbeet crops in northern Greece and in Texas, which had given excellent results in previous years, suddenly failed to be effective. The resistance probably developed locally as a consequence of consistent, frequent and uniform application of this selective compound (Georgopoulos and Dovas, 1973; Ruppel and Scott, 1974). Dekker (1973) provided a review on acquired resistance of fungi to systemic fungicides.

Established natural differences in germicide resistance between biotypes indicate that increased resistance of a pathogen may be due to selection of resistant biotypes. Another mechanism may be by mutation incited by the chemical concerned. Christensen (1946) suggested that increased tolerance of *Fusarium graminearum* to mercuric chloride and certain other fungicides is due to mutation. Mutated types failed to revert to susceptible types. He observed that mercuric bichloride, ethylmercuric phosphate and malachite green in nutrient media were decidedly mutagenic to *Colletotrichum lini*, *Fusarium graminearum*, *Drechslera carbonum* and *D. sorokiniana* (Stakman and Christensen, 1953). Bartels-Schooley and MacNeill (1971) showed that induced mutants of *Fusarium oxysporum* f.sp. *melonis* were resistant to three systemic fungicides — benomyl, thiabendazole and furidazol.

22.7 SEED TREATMENT AND CONDITION OF THE SEED

It is a common experience that seed of poor germination (poor 'vigour') suffers more from chemical seed treatment than seed of high germination power. Weakness of seed may be due to many factors, some of which are physiological immaturity of the seed, mechanical damage such as that caused by careless handling during threshing, etc., and high moisture content. Infection by seed rot-producing organisms, even slight infection, may also provide avenues for penetration of chemicals to the deeper

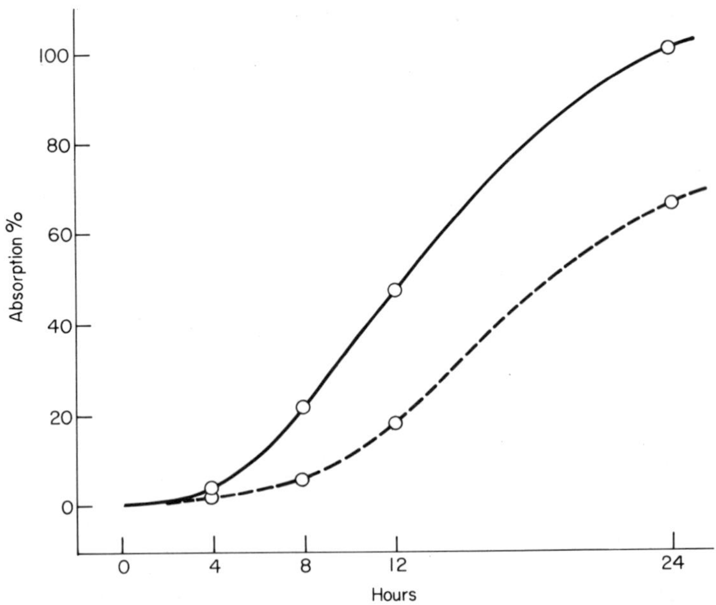

Fig. 22.13. Absorption of water in necrotic bean seeds. severely infected by *Colletotrichum lindemuthianum* (—) and in healthy seeds (----). (Grümmer and Mach, 1955)

layers of the seed, thus favouring conditions for phytotoxic effects. Lindström (1958) found, in his radioactive studies on resorption of mercurial compounds in intact and damaged wheat kernels, that damaged kernels carried at least 50 per cent more chemical per unit weight than intact kernels. Similarly, antibiotics have been found to be much more easily absorbed in damaged than in intact seeds — patulin and streptomycin in bean (Grümmer and Mach, 1955), rimocidin and pimaricin in pea (Dekker, 1957).

The permeability of the seed coat of bean is substantially increased by necroses produced by *Colletotrichum lindemuthianum* (Grümmer and Mach, 1955), cf. Figs 22.13 and 22.14.

Dusting of dry and superficially moist wheat with an organomercurial led to increase of 15 per cent abnormal seedlings in the moist seed, partly due to increase of chemical retained by moist seed (Brett and Dillon Weston, 1941). The dose of methyl bromide required to kill white tip nematode (*Aphelenchoides besseyi*) in rice varies inversely with moisture content of the seed; seed containing 14 per cent moisture is more likely to be damaged than seed of a lower moisture level (Cralley, 1952; Todd and Atkins, 1959).

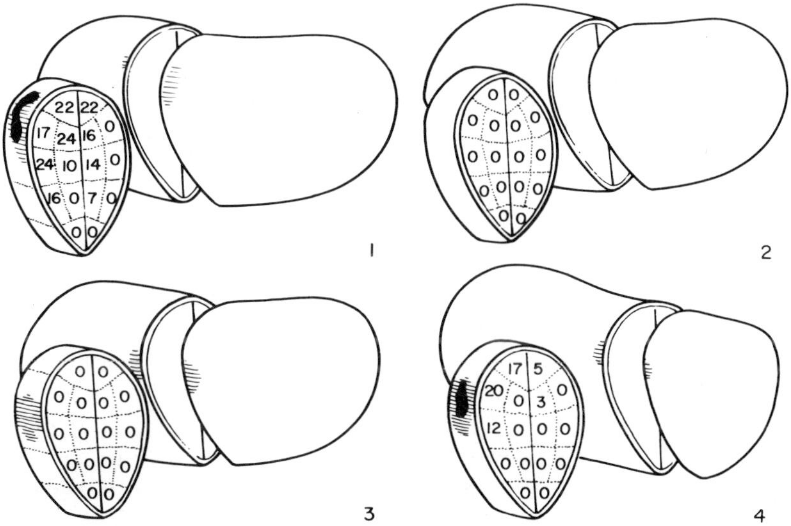

Fig. 22.14. (1) Absorption of 1 per cent solution of patulin in bean seed, severely damaged by *Colletotrichum lindemuthianum*—the figures on the individual pieces of the slice indicate average zone of inhibition, in mm, recorded in a bioassay with *Bacillus subtilis* as test organism; (2) absorption in healthy bean seed; (3) absorption in weakly affected bean seed; (4) absorption in moderately affected bean seed. (Grümmer and Mach, 1955)

22.8 STORAGE OF TREATED SEED

Storage of treated seed may be beneficial by increasing the germicidal effect, or detrimental by increasing phytotoxicity. The major environmental factors operating before and during storage are collated, and their actions against the *vitality* of seeds and pathogens (defined in Fig. 22.15) are shown in Fig. 22.16.

Pre-harvest and pre-storage environmental factors are within the control of the seed grower and merchant: proper handling can be undertaken during harvest and processing of the seed, the moisture content of seed can be regulated, contaminated or infected seed can be treated to eliminate or reduce inoculum.

Storage conditions can generally also be controlled, suitable levels of humidity and temperature established, insects controlled, vapour action of germicides maintained for a protracted period and the rate of respiration reduced. Conditions favourable for longevity of seed are usually also favourable for longevity of pathogens. The inherent longevity of both must be considered (Chapter 20).

Storability of treated seed is naturally related to the phytotoxic properties of the germicide used. Among organomercurials some alkylmercuric compounds are highly phytotoxic (Gassner, 1951*a*), cf. Section 22.4; so it is not surprising that Johannes (1961) was able to establish consistently bad effects of ethylmercuric compounds on grains during storage.

Vapour action of volatile germicides may operate under suitable storage conditions, the course of action being dependent on the level of vapour pressure, cf. Table 21.2. The gaseous components of Panogen are absorbed by seeds in closed storage during the first 2 h after treatment (Lindström, 1958); thereafter there is only a slight increase of absorption and fungicidal effect (Subsection 22.1.3 and Figs 22.4 and 22.5). Hedén and Ulfvarson (1963) found practically complete efficacy of this vapour action of Panogen only 20 min after treatment of wheat and oats. Pesticides with low vapour pressure operate in storage over long periods of time. Conditions of storage where gases escape from the treated seed reduce vapour pressure.

The condition of seed at the time of treatment is of decisive importance in phytotoxicity. Seed poorly matured (example 2 in Fig. 22.16) with high moisture content, or mechanically injured (Roane and Starling, 1958) (example 3 in Fig. 22.16), is highly reactive to toxic materials. However, if storage conditions are suitable, detrimental reactions may be lessened.

Schlösser (1954) found that mechanically produced monogerm sugar-beet seed suffered from storage when treated with a standard dosage of an organomercurial, Germisan. The germination capacity of the seed was reduced substantially after 2 years of storage, while the normal seed in the clusters from which it was produced was practically unaffected; the seed

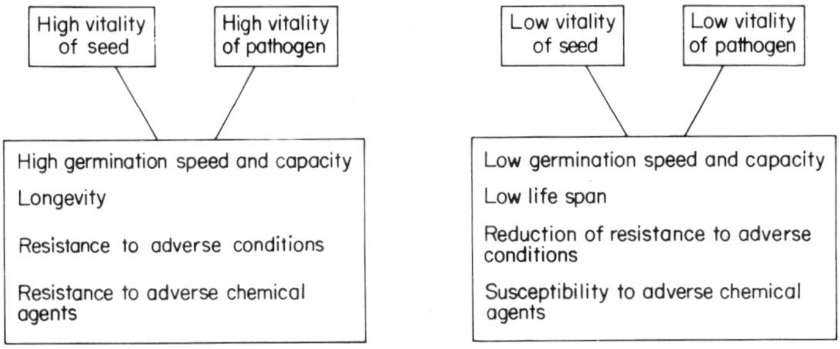

Fig. 22.15. Definition of vitality

had lost its protective cover by the processing. Later, Eifrig (1961) again compared the effect of storage on normal seed and mechanically produced monogerm sugarbeet seed, because, as he emphasised, equipment and procedure for this kind of processing had been much improved in recent years. He found little difference in germination capacity between treated normal seed and treated monogerm seed after 3–4 years of storage and concluded that treated, mechanically produced monogerm seed of good standard under suitable conditions can very well be kept for at least 3 years. Recently, Krexner (1971) recorded decrease of germination capacity in treated seed as well as untreated mechanically produced monogerm seed after 2 years of storage, as compared with the normal seed from which it was produced. However, when the seed was kept under dry conditions the decrease was less in treated than in untreated seed. He found 1 year of dry storage of treated natural as well as processed seed to be safe.

The reaction of seeds to chemicals in storage is greatly influenced by temperature and humidity (Fig. 22.5). At higher levels of humidity even liquid diffusion may take place, increasing germicidal and phytotoxic effects. Some pesticides, particularly certain ethylmercuric compounds, reduce germination of wheat seed after storage at moderate or higher temperatures, 20°C or higher, cf. example 4 in Fig. 22.16 (Koehler and Bever, 1956).

The investigations by Krexner (1971), discussed above, provide an example of how increased water content during storage may lead to heavy reduction of germination. This was true particularly of mechanically produced monogerm seed when treated with Germisan.

The time factor is important. In Canada an extensive survey of farmers' practices in treating seeds showed that the effect of fungicides on loose and covered smuts of oats and covered smut of barley increased substantially with increase of storage time from 3 days or less to 4 days or more (Cherewick and Popp, 1951), cf. Section 21.5(7). Machacek (1954) reported that wheat treated with organomercurials against bunt, should be

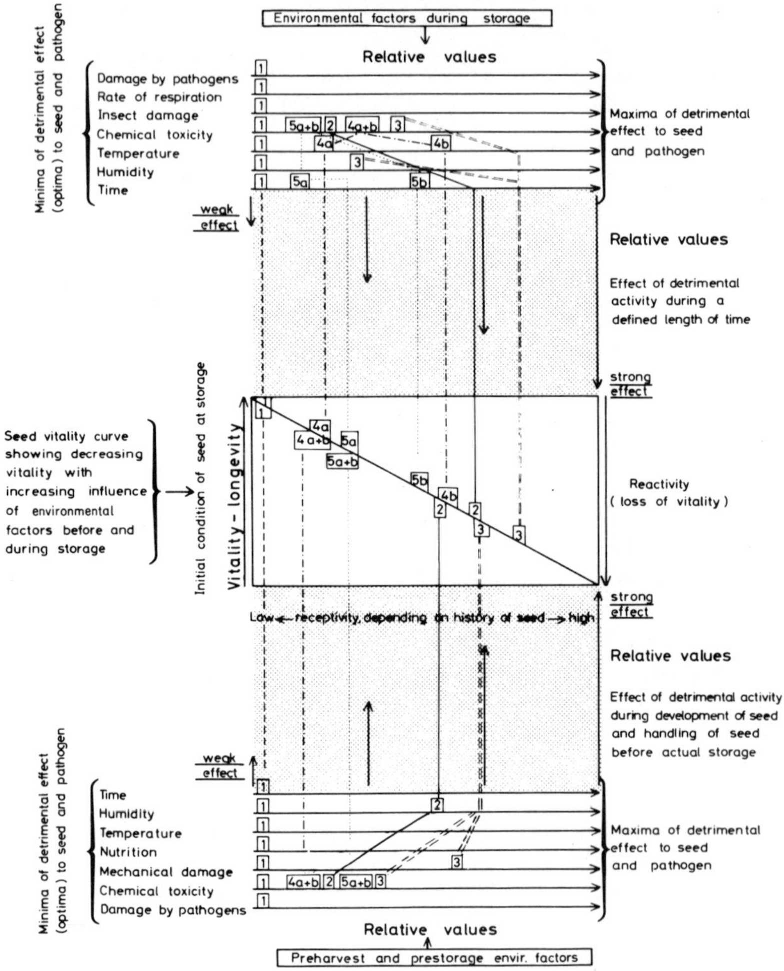

Fig. 22.16. Major environmental factors effective on seeds and seed-borne pathogens during storage of treated (and untreated) seed. Examples of effects are indicated by the numbers referred to below. When placed below the vitality curve in the middle of the figure, the numbers indicate the relative vitality of the seed at the beginning of storage; above the curve, after storage. The relative values of operating factors and of vitality are applied, in each example, independently of the other examples.

Examples of effects indicated in the chart

(1) Seed of high vitality due to optimum field conditions and good handling of the harvested seed and with low moisture content was stored reasonably soon after harvest and exposed to low relative humidity of air, suitable low temperature, and other optimum conditions. The vitality was maintained practically unaffected.

(2) Seed, produced under unduly humid field conditions near maturity of the crop, was treated with a standard dosage of an organomercurial, cf. Section 22.4 (Chambers, 1961). The

seed was injured, because it was more reactive to chemicals than normal, and a post-effect of the volatile fungicide further reduced the vitality during storage, even though the seed was kept under good storage conditions.

(3) Because seed was harvested under very dry conditions threshing caused considerable mechanical damage to the fragile seed coat (pericarp). The seed was treated with an organomercurial at standard dosage. The vitality was reduced by toxic effect operating through the breakages in the seed. Storage of the seed under suboptimal conditions, with rather high relative humidity, allowed the toxicity of the chemical further to reduce its vitality.

(4) Seed of good quality was treated with an ethylmercuric compound without damage, part of the seed was stored under suitable conditions, at low temperature (4a), another at a higher temperature, above 20°C (4b). After storage, the seed from the low temperature was unaffected, whereas the seed kept at high temperature had lost some of its vitality.

(5) Seed of good vitality was treated with an overdosage of Panogen which had no notable effect on its vitality. It was stored under suitable conditions for two months without reduction of vitality (5a), but after a longer period the seed showed reduced vitality (5b).

stored for 4 days or more, whereas storage was not needed for the full effect of hexachlorobenzene (HCB) or pentachloronitrobenzene (PCNB). Overdosage of Panogen on cereal grains did not impair germination after 2 months' storage but caused injury after longer periods, see example 5 in Fig. 22.16 (Andrén, 1945). This slowly developing effect may apply to many other kinds of seed and volatile germicides.

Attempts have been made to treat seed against storage fungi but all experience points to such treatment having a quite limited effect. While germicides may very well reduce inoculum of ordinary field and storage fungi, they do not fully eliminate them. If storage conditions favour development of storage fungi, as do unduly high levels of temperature and humidity, the concentrations used in standard treatments have negligible effect. Christensen and Kaufmann (1969) have given a short critical review of literature pertaining to tests with fungicides for the specific purpose of controlling storage fungi.

22.9 SEED TREATMENT AND FIELD CONDITIONS

The effects of chemical seed treatment vary substantially according to field conditions. Any adverse effect of treatment to seeds may be enhanced by field conditions unfavourable for emergence and seedling development. Also, the therapeutic effect of germicides is greatly influenced by the edaphic factors. Effect of seed treatment chemicals established under laboratory conditions, such as effect on spore germination or mycelial growth on artificial media, often gives no direct indication of the therapeutic effect under natural conditions. This is especially the case with the new fungicides of the benomyl type. Chemicals are adsorbed to soil particles and soil colloids, and hence their toxic effect is greatly reduced against pathogens and hosts. This reduction of effect may be determined as the ratio between the *in vitro* and the *in vivo* effect in soil or a standard substitute for soil, and is termed *adsorption factor* (Mulder, 1943). For instance, Gottlieb and Siminoff (1952) found for an antibiotic, chloromycetin, that 60 times as much of the antibiotic was required to inhibit *Bacillus subtilis* in soil as was required in broth.

The adsorption factor varies according to the chemical used. Gottlieb and co-workers (Gottlieb and Siminoff, 1950, 1952; Siminoff and Gottlieb, 1951; Gottlieb, Siminoff and Martin, 1952; Martin and Gottlieb, 1952) established that the alkaline antibiotic, streptomycin, is readily adsorbed on to soil colloids, whereas neutral and acidic antibiotics, such as actidione and clavacin respectively, are relatively unaffected.

Mulder (1943) pointed out that adsorption of phenylmercuric compounds used in the older type of seed treatment fungicides is approximately ten times greater than that of the later ethylmercuric compounds.

Soil reaction may be decisive to chemical injury from treated seed. Spinach and beet seed treated with chloranil have been damaged when they were planted in alkaline soil but not in acid soil (Burchfield, 1960).

22.10 PRACTICAL USE OF SEED PESTICIDES

22.10.1 Amounts of seed pesticides used

Satisfactory statistics on usage of seed pesticides are available for very few countries but the examples of usage given below are believed to be indicative of the actual situation.

The majority of fungicides used for seed treatment are organomercurials. Gayner (1962) estimated that in 1958 about 300 tons of mercury were used for the preparations of fungicides for world agriculture. It is likely that the major part of this production is used for seed treatment.

The greatest quantity of fungicides for seed treatment is used for cereals and cotton. In technically advanced agricultural countries most of the cereal-growing areas are sown with treated seed. In the period 1947–1951 60 per cent of the major categories of crop seed in Canada, which were assumed to be in need of treatment, were actually treated, the amount of chemicals used being about 500,000 kg annually. This estimate of the percentage was based on the assumption that all of oats, barley, rye and flax, and half of the wheat sown, required treatment (Machacek, 1953). Ordish and Mitchell (1967) estimated that Canada used about 8 tons of mercury per year for seed treatment of grain crops, covering 17.7 mill. ha. In the period 1940–1965 about 80 per cent of all cereal seed sown in Denmark has been treated, most of it with organomercurials; this means that the seed for about 1.2 mill. ha is treated every year (Stapel, 1966*b*). The percentage of 80 in Denmark has remained stable until the last 2–3 years, during which there has been a definite decrease due to the restrictions in use of mercurials, and also as a consequence of the gradually improved health standard obtained by the many years of seed treatment (Nøddegaard, personal communication, 1975). In 1965, 80 per cent of cereal seed was treated with organomercurials in Sweden, 70–80 per cent in Austria, Britain and the Netherlands, 65 per cent in Eire, 50–60 per cent in Federal Republic of Germany, Hungary, Switzerland and Turkey, 35 per cent in France (Granhall, 1965) and, in the period 1953–1965, 77 per cent in Finland (Blomqvist, 1970*a*); but also in these countries there has been a decrease in the last few years. Crosier, Waters and Crosier (1970) reported that, in 1967, 97 per cent of oat seed lots sold in New York State had received a slurry treatment; the percentage of treated oats decreased slightly in 1969 and 1970. In most developing countries, on the other hand, only a small part of seed for sowing is treated, but by development of certification schemes and multiplication centres, increasing proportions of seed are being treated, until recently mostly with organomercurials. Small farmers in Africa and Asia often use cheaper and safer preparations of older types, copper carbonate and sulphur dust, and recent restrictions on the use of mercury in Europe, Canada and the U.S.A. will certainly also be introduced elsewhere; see Subsection 22.10.6.

According to Ordish and Mitchell (1967) about 80 per cent of the world cotton area is grown from treated seed, and an amount of 540 metric tons of materials were used. About 25 per cent of this amount was based on copper, 75 per cent on organomercurial preparations.

22.10.2 Organisation of seed treatment

At first, seed was treated by individual farmers. Now, in all advanced agricultural countries there is a well-developed system of seed certification schemes, mostly incorporating seed treatment. Seed treatment is also carried out by professional enterprises operating on contract for many farmers. In Denmark, in 1958, about 1,100 such stations treated cereal seeds in conjunction with the processes of drying, cleaning and grading — now this number is much less. In Sweden the number of registered seed-treating stations is about 1,000. They operate under the supervision of the Government Seed Testing Organisation (Esbo, personal communication, 1970). In some countries large seed companies now treat seed of some field crops and vegetables, either as a general practice or on buyer's request. In addition to the major crops mentioned above, seeds of flax, sugar beets and peas are commonly treated commercially — see also Subsection 22.10.6.

22.10.3 Technical achievements of seed treatment in practice

In practice seed treatment is only partly effective. The preparations used may not be fully effective under all conditions, or the technical performance of the treatment itself may not be up to expected standard.

In Denmark systematic investigations during four years showed that 20–30 per cent of commercial treatment of cereal seeds was highly inefficient, largely because of inadequate, partly old, obsolete treating equipment (Stapel, Skou and Martins, 1961). In Wisconsin, U.S.A., Arny (1952) bioassayed the amount of Ceresan M on oat samples treated by 20 processors using different makes of treating machines. Undertreatment was frequent, and amounts applied highly variable. Most samples carried less than the recommended 0.5 oz per bu.

In the international comparative seed health testing schemes, organised by the Plant Disease Committee of the International Seed Testing Association, practically all commercially treated seed lots tested showed infections, the percentages varying from station to station according to seed testing conditions (see Subsection 29.5.4.7 on *Septoria nodorum* in wheat seed).

Depending on the rate of increase of the pathogen concerned, the effect of residual infection in treated seed may be from unimportant to disastrous. Residual infection of *Fusarium nivale* that has been commonly found surviving in cereal seeds treated with organomercurials in Sweden may be decisive as to whether the crop will be profitable or not (Olofsson, 1971).

22.10.4 General economic benefit of seed treatment

It is unquestionable that chemical seed treatment has greatly reduced losses from cereal smuts, blight and root rot fungi, some major diseases of cotton, seed-borne and soil-borne damping-off pathogens, and some systemic diseases such as barley leaf stripe disease, *Drechslera graminea*, and wilt caused by *Fusarium moniliforme* in maize, rice and many other crops. In many countries consistent use of seed treatment, irrespective of the actual level of seed-borne inoculum to be controlled, has been carried out for many years as an insurance against increase in development of inocula.

Extensive investigations on the economy of seed treatment were carried out in Denmark, 1931–1949, and in Sweden, 1933–1963. Results obtained were assessed by Stapel (1966*b*) and are summarised in Table 22.5.

There was an important difference in the planning of the experiments in the two countries, hence the great difference in the level of yield increase. In Denmark, seed assumed to be of good quality and to carry little seed-borne inoculum was used. In Sweden, on the contrary, infected or con-

Table 22.5. Experiments on the economy of seed treatment in Denmark and Sweden (Stapel, 1966*b*)

Cereal	*Denmark 1931–1949* (Olsen, 1950) Number of experiments	Increase of yield in crops from treated seed (kg/ha)	*Sweden 1933–1963* (Granhall, 1963) Number of experiments	Increase of yield in crops from treated seed (kg/ha)
Wheat	56	90	92	620
Rye	131	90	100	300
Barley	129	100	113	380
Oats	118	60	127	170
Total and average	434	80	432	about 350

taminated seed lots were used consistently, such as wheat with bunt, oats with loose smut, barley with leaf stripe disease, and rye with foot rot (*Fusarium* spp.). These seed lots were regarded to be representative of such a low health condition of seed that was usual before modern seed treatment was generally adopted, and which might recur, if the seed were left untreated (Andrén, 1964). The profit obtained through treatment of cereals in Denmark, on the basis described above, amounts to 40–45 mill. Danish kroner (Stapel, 1966*b*). The cost of seed treatment, 5 D.kr. per ha, amounts to 7–8 mill. D.kr. (E.P., 1965).

If we calculate the potential profit obtained by treatment of seed, in the condition it would be if seed treatment had not been carried out for many years, sown on the total area, about 2.4 mill. ha used for cereals in Sweden, then the annual gross profit achieved through seed treatment is 325–350 mill. D.kr. On the same basis the gross profit in Denmark is 150–175 mill. D.kr. These figures are probably conservative estimates.

22.10.5 Handling hazards — poisoning of wildlife, domestic animals and man

The dangers of poisoning by certain seed treatment materials are well recognised, and precautions are generally recommended and adopted to meet them (Table 22.6). The mercury compounds are by far the most dangerous group, copper compounds being much less poisonous.

Wheat treated with copper carbonate is in fact often used for livestock feed (Fischer and Holton, 1957), although this disposal certainly is not recommended. Other domestic animals, such as poultry and sheep, have been fed with copper carbonate-treated wheat without any detrimental effect (Richardson, 1925; Keast, 1943).

In recent years there has been a trend away from organomercurials to

Table 22.6. Major hazards of poisoning that may be caused by seed germicides

Chemical	*Effect of poisoning*	*Precautions*
Basic copper sulphate Copper carbonate	Extreme nausea and vomiting if inhaled or swallowed	Use dust mask or respirator
Formaldehyde	The gas is pungent, affecting eyes and respiratory organs	Treat seed in a well-aired place. Do not breathe fumes
Hexachlorobenzene	When consumed with grain, hexachlorobenzene is stored in the body. It causes blistering and epidermolysis of the skin, liver injury, in some patients osteomyelitis and arthritis, especially of the fingers; sub-normal temperature, weight loss, muscle atrophy. In Turkey poisoning of estimated 3,000 cases caused 10 per cent mortality	Treated grain should be made readily distinguishable from untreated grain to prevent consumption of treated grain. Bags should be adequately labelled with warnings
Mercury compounds, both in solid, liquid and gaseous state. The volatile compounds, particularly methylmercuric compounds, also ethylmercuric compounds, are dangerous when operators are continually exposed.	Organic mercurials are easily and rapidly absorbed in the lungs and via the gastrointestinal tract, and by skin. Some compounds are strong vesicants, producing oedema by contact in solutions or as gas (Dermatitis venenata). More serious is damage to the central nervous system. Symptoms are tiredness, conjunctivitis, muscular pains, headache, dizziness, decrease of libido, loss of appetite, flatulence, diarrhoea, later loss of ability to co-ordinate the action of muscles. 85–95% of excretion occurs via the faeces which assume a characteristic mercury odour. Higher doses may be lethal (Gassner, 1951*b*; Ulfvarson, 1969)	Do not breathe dusts or fumes of the chemical or let it come in contact with skin, especially if skin is moist. For continual work use overall and hood; roll sleeves down, use rubber gloves. Use dust mask; if exposed to volatile materials, use respirator with air-line. If the skin is contaminated, wash immediately with soap and water, also after treating the seed and before eating, drinking or smoking. At sowing operations with treated seed, observe the same precautions. Prevent consumption of treated grain! Bags should be adequately labelled with warnings
Thiram	Causes irritation of skin and mucous membranes	Protect mouth and nose. Wash hands after handling the material
Any poisonous chemical	The effect of cumulative poisons taken in over a longer period must be taken into account, mercury compounds and hexachlorobenzene being the most important examples. Certain less poisonous chemicals when in combination can be dangerous Treat any unknown chemical as poisonous until it is proved to be safe!	Secure treating machines against escape of dust and fumes into working rooms. Proper ventilation. Treated seed should not be used for feed or food, or mixed with untreated seed and used for this purpose. Sacks used for treated seed should not be used for feeding stuffs. In any case do not wash out empty containers in streams as animals may drink from them

less poisonous chemicals. This is only partly due to the direct handling hazards involved in the use of organomercurials. There is a growing recognition of the fact that mercury is accumulated in animals which eventually may contain poisonous doses dangerous if they enter human food. Danger of this sort has become acute through industrial pollution. An example of this disastrous effect is accumulation of mercury in the fish stock of the Baltic. The increasing content of mercury had led to serious concern as to consumption of fish from this sea. Alarm in Danish newspapers and on radio and television about this situation, which may have been somewhat over-dramatised, resulted in a serious decline of fish consumption in Denmark and even the closing down of some shops.

The problem of the effect of mercury used for seed treatment is more specific and the amounts brought into the fields very limited. By the usual standard of 200 g organomercurial per 100 kg grain containing 1–2 per cent mercury, only about 4–8 g/ha is introduced in the fields. In Denmark, where a standard dosage of only 100 g/100 kg seed is used, no danger to wildlife has so far been recorded. In Sweden, where the higher dose is standard, many cases of poisoning and death of grain-eating birds such as woodpigeons, rooks, pheasants, partridges and various small birds have been encountered. The cases of poisoning are frequent at the time of sowing, and more frequent in spring than in autumn (Hansen, 1965).

In addition to this direct effect on birds, a series of chain processes is started. Mercury is accumulated in eggs; as a consequence the fertility is reduced, and eventually also the population of the birds. Moreover, mercury is transferred and accumulated in birds of prey which feed on grain-eating birds. Swedish investigations have revealed considerable concentrations of mercury in dead wild birds and also in freshwater fishes and hens' eggs. Although there may be other important sources of this extensive distribution of mercury in animals, there is little doubt that feeding on treated seed is an important source (Granhall, 1965). Obviously, these experiences have an impact on the policy in the use of organomercurials for seed treatment. The implications are discussed in Subsection 22.10.6, which follows.

The disastrous effect of organomercurials used for seed treatment in Sweden, in contrast to the situation in Denmark, may be due to a combination of factors. As mentioned, the dose usually applied in Sweden, is, as in other European countries, twice that used in Denmark. Furthermore, however, Danish grain is dry-treated, whereas Swedish seed has been treated for some years by the oily methylmercury suspensions, predominantly Panogen which is a Swedish invention. As we know, there is much higher retention of the fungicide by seeds in this liquid application than there is by dust treatment. Furthermore, the increasing use of certain machines for sowing has resulted in a substantial and undesirable increase of kernels remaining on top of the soil and hence they are exposed to birds. Thirdly, in Sweden at the time of sowing in spring there is very limited access for birds to natural sources of food.

When mercury-treated seed is by negligence used directly for food, the results can be catastrophic. Incidences from recent years due to ingestion of flour or wheat seed treated with methyl- and ethylmercury compounds are known from Iraq, Pakistan, Guatemala (Bakir *et al.*, 1973) and from Afghanistan (Sadeq, personal communication, 1973). In 1960 in Iraq, thousands were affected by methylmercury poisoning, 370 being admitted to hospital. In Guatemala during the wheat-growing seasons 1963, 1964 and 1965, 45 people were affected and 20 died. A similar outbreak occurred in Pakistan in 1969. Most serious was the 'epidemic' of methylmercury poisoning in farmers and their families in Iraq in 1971–1972. A total of 6,530 cases of poisoning were admitted to hospitals, and there were 459 hospital deaths attributed to methylmercury poisoning. People became exposed to the poison when they ate home-made bread prepared from wheat seed treated with the fungicide. Flour samples analysed showed a mean content of 9.1 μg methylmercury per gram. Consumption of meat and other animal products obtained from livestock fed by treated barley contributed to accumulation of the poison in the patients. There were also cases of suckling infants poisoned through maternal milk, and it was established that hazardous amounts of methylmercury can enter the foetus *in utero*. Distribution of treated seed began in September 1971, and the rate of admission to hospital increased in early January 1972 to several hundreds each day. Relevant to this fatal development of mass poisoning is the fact that the latent period between dose and onset of symptoms may have given the farmers a false sense of security — some gave treated grains to their chickens for a few days and observed no harmful effects, and humans eating the contaminated bread did not have symptoms for weeks or months. Another reason for false security was that the grain was coloured with a brownish-red dye which could largely be removed by washing, giving the impression that also the poison had been removed.

As indicated in Table 22.6, hexachlorobenzene has caused catastrophic mass incidences of poisoning when treated grain has been used for bread. In Turkey there were, in the 1950s, an estimated 3,000 cases, with 10 per cent mortality (FAO/WHO, 1974).

22.10.6 Policy in use of seed pesticides

In recent years the use in plant protection of certain kinds of mercury compounds has been prohibited in some countries. The stability of the compounds in nature is essential. Methyl- and ethylmercuric compounds are quite stable and are the primary sources of pollution. Restrictions against these compounds were introduced in Sweden in 1965 (Stapel, 1966*a*), Denmark in 1968 (Bonnevie *et al.*, 1969), and in Finland in 1969 (Blomqvist, 1970*a*). Methoxyethylmercuric compounds, on the other hand, are rapidly decomposed in soil (Booer, 1944) and are still accepted for seed treatment, for example in Scandinavia, U.S.A., Canada. Phenylmercuric compounds are also unstable in soil (Booer, 1944; Kimura and Miller, 1964) and are still largely in use.

The development of cereal seed treatment practices and legislation in Sweden during recent years provides an interesting example of public policy towards seed pesticides. Legislative safeguards have taken into consideration the impact of seed treatment practices on the environment and on public health.

As a consequence of the situation in Sweden restrictions on the use of these compounds were prescribed by a governmental decree of 15 October, 1965. The restrictions are limited to spring wheat, spring barley and spring oats, covering 2.5 million acres, or one-third of the arable land of Sweden (Esbo, personal communication, 1970). Mercury treatment of seed of these categories is allowed only if the need for treatment has been certified by an official seed testing station. The autumn cereal crop covers about 0.6 million acres and is not subject to these restrictions. Seed for export may be treated without a licence. The need for treatment is assessed by laboratory seed health testing, and increasing incidence of seed-borne pathogens referred to one of six classes, 0–5. Licences are granted for classes 3–5. For class 3, treatment is allowed and recommended; for classes 4 and 5, treatment is strongly recommended and, if a certificate is required, treatment is mandatory.

In the six years 1965–1971, 23, 13, 11, 9, 3 and 15 per cent of the spring cereal seeds were assessed to require treatment, the need fluctuating according to harvest conditions (Esbo and Fritz, 1972). During the period 1966–1973 less than one-third of spring cereal seeds were treated with mercuric compounds, as against 80–90 per cent before the restrictions were introduced (Olofsson and Kolk, 1973).

What has been the effect of adopting this principle of treating seed according to established need? In the period 1966–1971 the content of mercury in Swedish birds has been reduced to a level comparable with that existing before mercurials were used for seed treatment. Also, the mercury content in animal agricultural products has been reduced. The success is, however, partly due to the shift, imposed by legislation, from alkylmercuric compounds, used before 1966, to alkoxyalkylmercuric compounds which are more rapidly decomposed (Olofsson, 1971).

Another factor is the increase in the foreseen use of non-mercurial and non-poisonous compounds. Voronit (furylbenzimidazol and hexachlorobenzene), a dust, has been tested in Sweden since 1965, and Neo-Voronit (furylbenzimidazol and dimethyldithiocarbamate), a liquid preparation, since 1966. Olofsson (1971) reports that, on average, the increase of yield obtained in six years of testing rye and wheat treated with Voronit is higher than that obtained by organomercurials. The same holds true for

Neo-Voronit used for rye, while for wheat the effect is approximately the same as obtained by organomercurials. Olofsson recommends that the organomercurials for winter cereals should be replaced by the non-mercurials, if these continue to show good performance. Voronit is not fully effective against some pathogens in barley and oats, and these crops therefore still have to be treated with organomercurials. This is a reflection of the fact that the traditional compounds of sulphur, copper and mercury are non-specific, while the newer types of organic substances are more selective, being more effective on a narrower range of pathogens.

The Swedish principle of treating seed according to need, as established

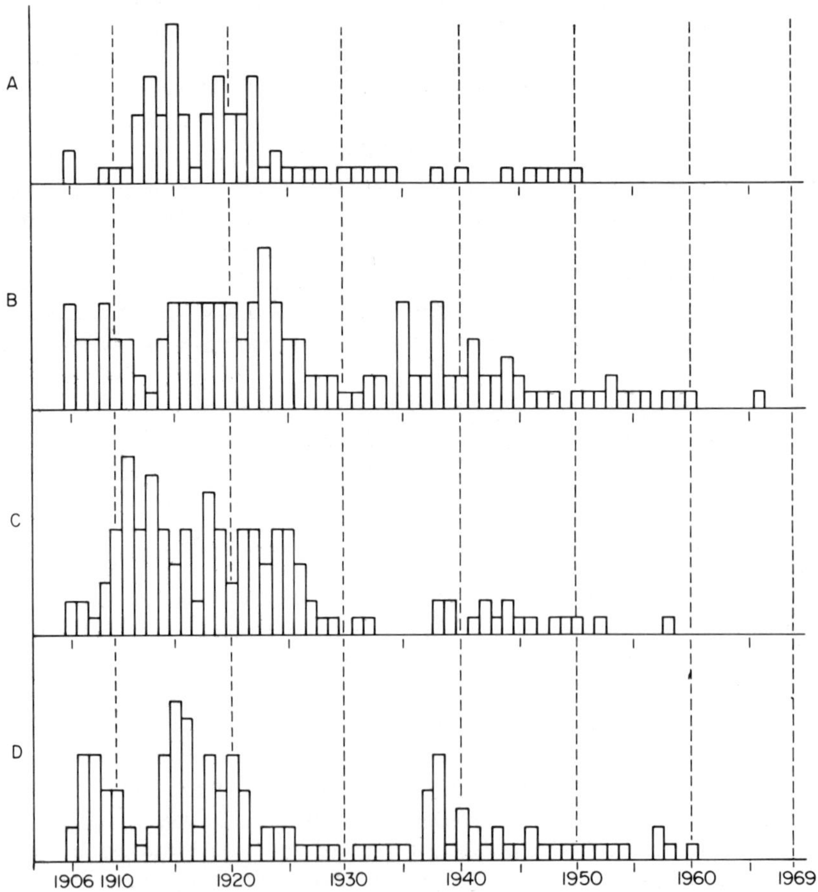

Fig. 22.17. Relative severity of attacks of four seed-borne diseases in Denmark 1906–1969. Decrease of severity is due to seed treatment with organomercurials from the middle of the 1920s. Assessment of distribution and severity according to Monthly Reports on Plant Diseases in Denmark 1906–1969. From above: A, loose smut of oats (*Ustilago avenae*); B, barley leaf stripe disease (*Drechslera graminea*); C, stem smut of rye (*Urocystis occulta*); D, bunt of wheat (*Tilletia caries*). (Stapel, 1972)

by health testing of the seed, may be evaluated against the Danish principle of general treatment. In Denmark all cereal seeds are treated but with a lower standard dose than in Sweden. As a consequence, seed-borne diseases are maintained on a comparatively low level (Fig. 22.17). On the other hand, untreated seed, if in categories 0–2 according to the Swedish practice, may lead to increase of inoculum within a generation or two, as indeed happened for barley leaf stripe disease and covered smut (Esbo and Fritz, 1972).

For the latter view it may be argued (1) that practically any seed treatment will leave some inoculum unaffected and thereby possibilities for its subsequent increase, (2) that tolerances for the classes of health condition mentioned above may be regulated in such a way that the risk of too rapid increase of inoculum from untreated seed is duly taken into consideration, and it may be pointed out, (3) that organised seed health testing in principle is a more sound basis for using treatment because unexpected heavy infections are detected, while the overall costs and hazards of seed treatment are reduced.

At any rate, emphasis should be made on treating the first generations in seed multiplication. According to regulations recently introduced in Denmark (Nøddegaard, 1973; Miljøministeriet, 1973), only approved and registered stations are permitted to use mercuric compounds for seed treatment of cereals, and treatment with such materials is principally restricted to nuclear seed stocks, including Selected Seed ('Pre-Basic Seed'), Foundation Seed ('Basic Seed') and Registered Seed ('Certified Seed, First Generation').

23 Testing Seed Germicides and Treated Seed

23.1 DEFINITION AND INTERPRETATION OF GERMICIDAL EFFECT

Screening of seed germicides must be based on their capacity to destroy pathogens and on their toxic effect on the seeds. The concept of *dosis curativa*, *dosis toxica* and *dosis tolerata* was adopted by German plant pathologists (Gassner, 1923, 1926, 1951*a*; Nagel, 1925, 1927), who followed traditions from pharmacology, originating in Ehrlich's work on chemotherapy of human syphilis. *Dosis curativa* is the lowest dosage necessary to destroy a given inoculum, *dosis toxica* the threshold dosage causing injury to the seed, and *dosis tolerata* the highest dosage which the seed can support without injury. The *chemotherapeutical index* is the *dosis curativa* divided by the *dosis tolerata* or *dosis toxica*. Naturally, attempts are made to find disinfectants with a low *dosis curativa* and a high *dosis tolerata*. The index of such preparations shows a low fraction of 1.

The chemotherapeutical index is particularly useful when different chemicals are compared by laboratory testing. Gassner (1951*a*) evaluated, in his classical investigations on organomercurials, 100 compounds defining their respective index. To define the quality of germicides on this basis, the testing conditions must be accurately standardised. Under field conditions, environmental and other factors that cannot be controlled alter the performance of the germicides in the growing crop to such an extent that extrapolation of results from laboratory tests on the seed are not necessarily conclusive. The *dosis toxica* and *dosis tolerata* may also vary with cultivar, condition of the seed and conditions of storage.

Another tool of interpretation, provided by American workers (Wilcoxon and McCallan, 1939), for assessing differences between germicides is the *dosage response curve* and the *related lethal dosage 50* (LD 50), or *effective dose 50* (ED 50), both terms of dosage indicating the dosage effective on 50 per cent of a pathogen population or disease control. The relation of dosage to 50 per cent is preferred for statistical reasons. By plotting on logarithmic-probability paper, the sigmoid (S-formed) curve of the diminishing return, which is the usual type of curve obtained on arithmetical paper, is converted to a straight line. The straight line allows interpolation. The slope of the dosage response curve shows clearly differences between germicides as well as differences between conditions of application of one germicide. The slope of the straight curve indicates the effect obtained by increase or decrease of dosage. In practice, reduction of a defined dosage on seeds may occur by loss of material due to poor retention or volatility, or by adsorption to soil after sowing or in other ways. If the straight curves of two fungicides cross, especially if their LD 50s are similar, one is superior to the other at some doses but inferior at others. By defining a certain fixed dose such as *dosis curativa* or LD 50, a merely static expression of effect is achieved on the assumption of certain narrowly defined testing conditions.

Both principles of interpreting testing results, the chemotherapeutical index and the dosage response curve, have been used widely in screening seed germicides. They have produced a bulk of useful data on germicides and procedures of seed treatment.

Leaving the principles of screening seed germicides, a partly different approach is taken in assessment of the practical performance of established seed treatments. Standard laboratory tests provide data to assess whether a seed treatment is up to a definite standard or not. Some of the techniques of such tests are briefly summarised.

23.2 LABORATORY TESTING

23.2.1 Evaluation of dye coverage

Usually, seed germicides contain a dye for colouring to indicate that the seed has been treated with germicide. The dye distribution may be evaluated by determination of the dye content of individual seeds by extraction and colorimetric analysis. However, this procedure has proved erratic because of the strong adsorption of the dye to seed (Lindström, 1961).

It is better to evaluate dye distribution by visual examination. The seeds may be classified into heavily, medium and weakly coloured, etc.

Distribution of the dye is independent of the actual distribution of the germicide, as in highly volatile fungicides.

23.2.2 Chemical analysis of fungicide coverage

Small quantities of fungicides in treated seed can be determined by different conventional chemical methods as well as by the now usually adopted chromatographic and photospectrometric methods. The principles of chromatography and photospectrometry are given in standard textbooks on organic chemistry and biochemistry (e.g. Roberts and Caserio, 1965; White, Handler and Smith, 1968; Vejlby and Dich, 1971).

Recent work at Rothamsted, England, in which the insecticidal compounds of combined treatments can be very accurately measured by thin layer or gas chromatography, is also giving information on distribution of mercury.

23.2.3 Bioassay of germicide distribution

Seeds are placed on agar plates on which propagules of a test organism have been equally spread. A zone of inhibition around seeds treated with the germicide indicates the germicidal effect of the dosage carried by individual seeds. The diameter of the zone of inhibition may be measured and compared with the diameter of zones produced by a correctly treated sample. In principle, this testing procedure has been used by Heatley (1944), Mead (1945), Machacek (1950), Leben and Keitt (1950), Arny (1952), Lockwood, Leben and Keitt (1952), Arny and Leben (1954), Kovacs (1963), Jørgensen (1966) and many other workers. Often *Glomerella cingulata* has been used as the test organism, also *Penicillium purpurogenum* (Machacek, 1950), *P. atrovenetum* (Thapliyal and Sinclair, 1971), *Drechslera sorokiniana* (Mead, 1945; Stapel, Skou and Martins, 1961), as well as species of *Curvularia*, *Myrothecium*, *Phoma*, *Sarcina* and *Stemphylium* (Crosier, Waters and Crosier, 1970).

Jørgensen (1966) has used *Corynebacterium rathayi* as test organism; results can be recorded after only 24 h incubation. Kovacs (1963) found the bacterium to be susceptible to organomercurials. Conditions of incubation must be carefully standardised. Jørgensen found that increasing depth of the agar layer, increasing concentrations of the agar in the medium and increasing concentrations of the bacteria in the medium, decreased the size of the inhibition zones.

Test organisms should be selected critically, many fungicides being effective to a small number of pathogens. Hexachlorobenzene is specific for wheat bunt fungus and cannot be tested on *Glomerella cingulata* or most other fungi. In this case the bunt fungus must serve as test organism, as described below.

The bioassay technique has been used successfully for studies on the translocation of systemic fungicides in seedlings, different parts of treated seedlings being assayed (Thapliyal and Sinclair, 1971).

The inhibition zone bioassay has its limitations. The inhibition obtained indicates only the secondary effect of the germicides, not the primary effect which may have been in operation by liquid diffusion and by vapour action. Consequently, the zone of inhibition may decrease substantially when seed has been stored for some time after treatment, particularly when the seed, as in the examples shown in Fig. 22.5, has been treated with a volatile liquid (Panogen), but less with a dust preparation (Tillantin 1875). With increased moisture content seeds become more reactive,

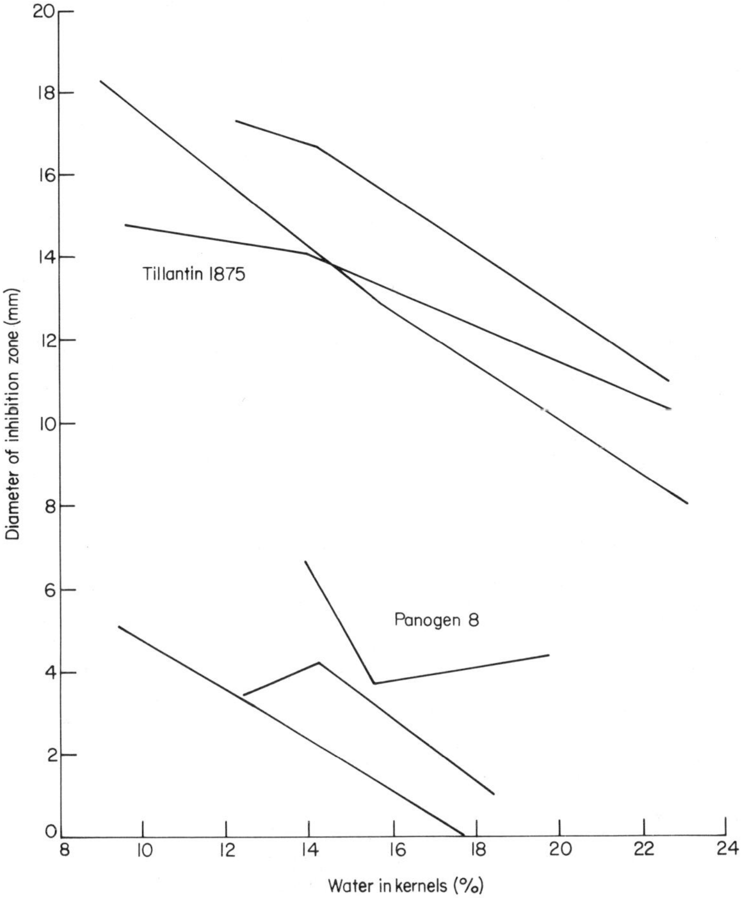

Fig. 23.1. The diameter of the inhibition zone produced by two organomercurials, a dust preparation (Tillantin 1875) and a liquid preparation (Panogen 8), around barley kernels with different levels of water content. (Jørgensen, 1966)

absorption of the germicides increases (Fig. 23.1) and the response in the bioassay decreases.

Fungicides for smut control have been tested by using wheat bunt fungi as test organisms. The medium used for germination of the smut spores is finely sieved, air-dry soil placed over a layer of vermiculite or a similar substance of good water-holding capacity. On the moistened substrate a standard dose of spores is dusted evenly over the surface. The treated seeds are planted and incubated. A dark halo surrounding each seed appears in contrast to the white mycelial cover produced by the spore germination. The diameter of the zone indicates the degree of effect (Pichler, 1935; Zogg, 1954). Again this is a measure of the secondary effect.

Gassner (1944) tested the primary effect of fungicides on the wheat bunt fungus. According to his procedure, still in use, wheat kernels are artificially contaminated with a standard dose of bunt spores and subsequently treated. The kernels are planted in finely sedimented moist soil with the embryo end up and above the surface (Fig. 23.2) in order to delay germination. After incubation at 10–12°C for 5–6 days, at 12–15°C for 4 days the seeds have germinated and are removed. After a further 2–3 days spore germination is completed. Around the holes left by the removed kernels the sporidia and white mycelial mats can be seen under low power microscope, wherever kernels are untreated or insufficiently treated. For further precision of records, incidences of fungus growth may be graded, e.g. into classes 0–4. In Turkey, Saydam, Copcu and Öğüt (1972) found, in testing fungicides against *Tilletia foetida*, Gassner's technique to be unsatisfactory; by using $Ca(NO_3)_2$ added water agar medium results were obtained that corresponded with those from experimental field tests.

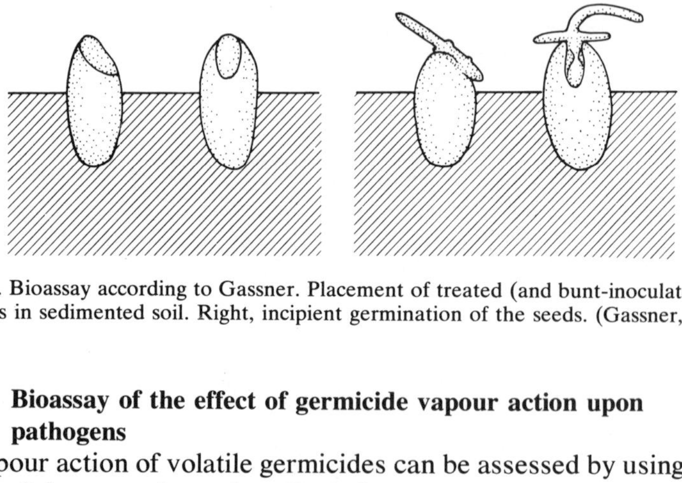

Fig. 23.2. Bioassay according to Gassner. Placement of treated (and bunt-inoculated) wheat kernels in sedimented soil. Right, incipient germination of the seeds. (Gassner, 1944)

23.2.4 Bioassay of the effect of germicide vapour action upon pathogens

The vapour action of volatile germicides can be assessed by using modifications of the procedures described above.

The principle is to exclude liquid diffusion of germicide to the test organism so that only the effect of the vapour action is recorded. Treated seed may be placed in one of the halves of a plate of agar or of sedimented fine soil (as used by Nagel, 1927). The plate is divided across the middle by a gap made by pressing a slide into the medium, then removing the slide leaving an air space between the two halves. Any effect from the material on the treated seed upon the test organism across the air-space must be ascribed to the action of vapour.

Gassner (1951*b*) followed a different procedure. A standard suspension of wheat bunt spores is spread in parallel strips on plates of fine sedimented soil, kept superficially dry. Then a slurry of the germicide is stroked across in a 4 mm broad strip by a soft brush near the middle of the inoculated plate. A gap across the middle of the plate, parallel with the germicide strip, is made. After 6–10 days of incubation at 12–15°C the spores have produced strips of mycelial felt that can be observed macroscopically. In the half where a non-volatile fungicide has been applied there is a distinct inhibition zone along the stroke of the fungicide, indicating diffusion, whereas none has passed into the other half. A volatile fungicide, on the other hand, shows the effect of vapour action across the air-space, by more or less suppressing spore germination in the untreated half. Moreover, the inhibition zone along the stroke of fungicide is more diffuse, thus providing further indication of the vapour action, and in contrast to the sharp limit of the inhibition zone produced by non-volatile fungicides. Gassner found that phenylmercuric compounds were not active across the air-space provided by the test but that alkylmercuric materials, especially ethyl- and methylmercuric compounds, were highly active across it (cf. Table 21.2).

Vapour action can also be assessed by measuring the uptake of fungicidal vapour by seed in the agar diffusion bioassay. Untreated seeds are exposed for differing periods to only the vapour of the chemical to be tested, and then bioassayed. Arny and Leben (1954) studied the vapour action of different organomercurials on oats naturally infected by *Drechslera victoriae*, cf. Subsection 22.1.4. Fig. 23.3 shows the vapour effect of 0.77 ml. Panogen on increasing quantities of oat seed after exposure for 16 days in sealed jars, as established by the bioassay. The penetration of fungicide through seed layers of various depths was also assessed. At the surface level the dosage conveyed by the gas was equivalent to 0.4 oz per bushel used directly; 1 cm down, 0.2 oz; 2 cm down, 0.1 oz; 3 cm down, trace; and 4 cm down and below, none.

23.2.5 Evaluation of germicidal effect on natural seed-borne microflora

Crosier, Waters and Crosier (1970) reported from New York State, U.S.A., that in routine examinations of oat samples for fungi and for adequacy of chemical treatment, the presence of *Alternaria tenuis* and *Epicoccum purpurascens* forming conidia on the seed indicated that less than half of the recommended amount of mercurial was present on the seed. For all oats grown along or near the Atlantic seaboard the natural occurrence of *Alternaria tenuis* was used in a macrobiologic test for seed treatment, and until 1966 consistently supported the microbiologic assay conducted with *Glomerella* and other test organisms. If the two mentioned fungi were absent and red-coloured areas on the seeds evident, it was certain that the seed was adequately treated against smuts and to a lesser extent against scab, *Fusarium graminearum*. From 1967, however, *Drechslera avenae* was found in such seed samples that previously would have been assessed to be adequately treated; this was due to the estab-

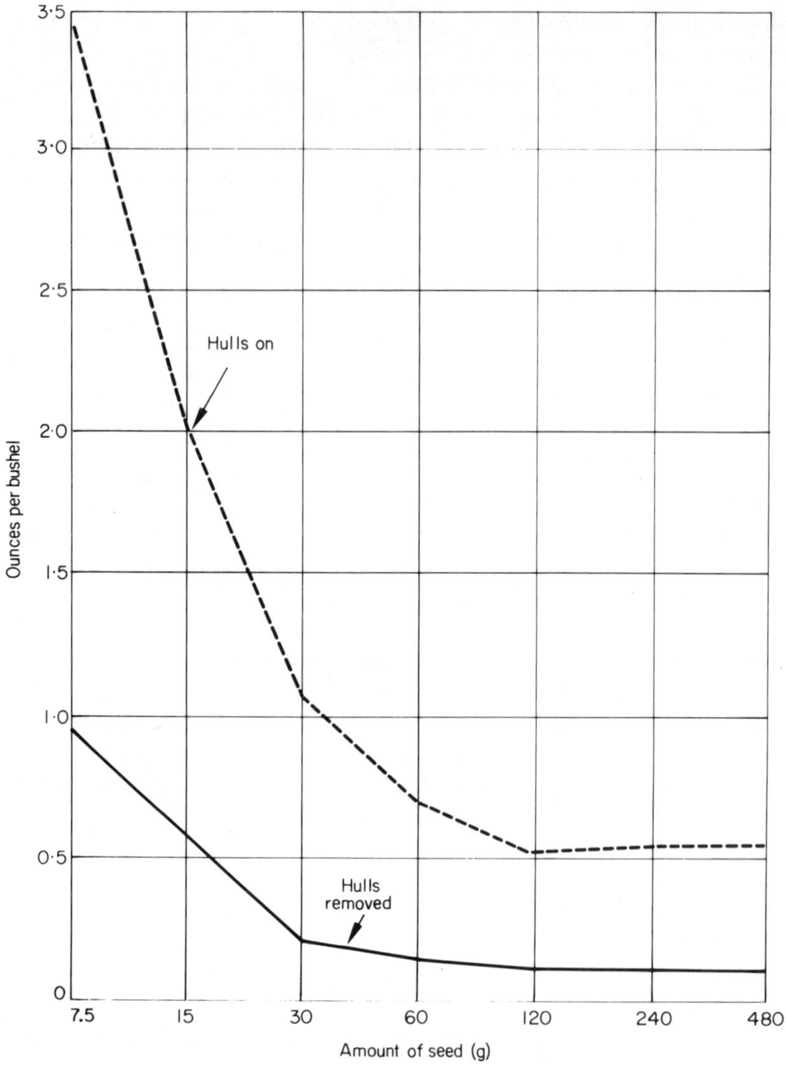

Fig. 23.3. The amount of fungicide on oat seed when various quantities of seed were exposed to a fixed amount of Panogen vapour for 16 days in sealed jars. Determinations were made by bioassay of seed drawn from the surface layer, and results expressed in terms of ounces of the chemical per bushel. After the exposure period, hulls were removed from some seeds and only the caryopses assayed. (Arny and Leben, 1954)

lished resistance of this pathogen to organomercurials, as discussed in Section 22.6.

Danish sugar beet seed lots treated with organomercurials before export and submitted to health testing according to quarantine regulations of importing countries for issuance of Phytosanitary Certificates, may equally well be assessed for efficiency of treatment by a macrobiological test. Untreated seed is consistently contaminated by *Alternaria tenuis* which is absent in adequately treated seed lots. In these, presence of the red dye provides additional evidence of treatment (Neergaard, unpublished).

23.2.6 Evaluation of germicidal effect on development of seedling diseases

The blotter method is the simplest procedure for testing seed treatments against seedling diseases. The procedure is valuable for preliminary screening of germicides and for preliminary assessment of dosages needed. Results are not fully valid because the soil adsorption factor is absent in this technique. In screening work a special standard technique for assessing the adsorption factor was developed by Mulder (1943), and now standard composts containing known proportions of humus are available for such tests.

The seedling symptom tests (Section 28.5) are applicable for testing emergence and health condition of treated seed with special regard to seedling diseases. The procedures more commonly used are the Hiltner brickstone method, the sand method and the standard soil test. The standard disinfection method (Subsection 28.5.5), used in Norway and Sweden, provides assessment of the need of treating the seed.

23.2.7 Evaluation of phytotoxic effect

The simplest procedure for assessing phytotoxicity of seed treatment chemicals is to test the treated seed for germination on moistened blotter. Lethal doses of the chemical may quite suppress germination, less toxic doses may (1) reduce percentage of germination, (2) retard germination, although perhaps eventually without reduction of the percentage of germination (Fig. 23.4), and (3) produce toxic symptoms as exemplified by the polyploidy malformations caused by mercury in seedlings (Fig. 22.8).

Phytotoxic effect of vapours can be tested according to Gassner (1951*b*) by germinating wheat seeds on blotter in a petri dish which is then placed in a larger dish, around which the germicide to be tested is placed. The effect varies from complete inhibition of the germination of the kernels to none. Injured seedlings may show pronounced symptoms of polyploidy when volatile organomercurials are tested.

In any test on phytotoxicity it is decidedly preferable to use pathogen-free seeds, as the treatment can be partly taken up by the pathogens and symptoms caused by the chemical confused by those produced by the pathogens.

Phytotoxicity shows up more readily in a blotter test than in sand, brick grit (Fig. 23.4) or especially soil with humus, because the adsorption factor is eliminated. The results of tests using more natural substrates usually correspond better with performance under field conditions.

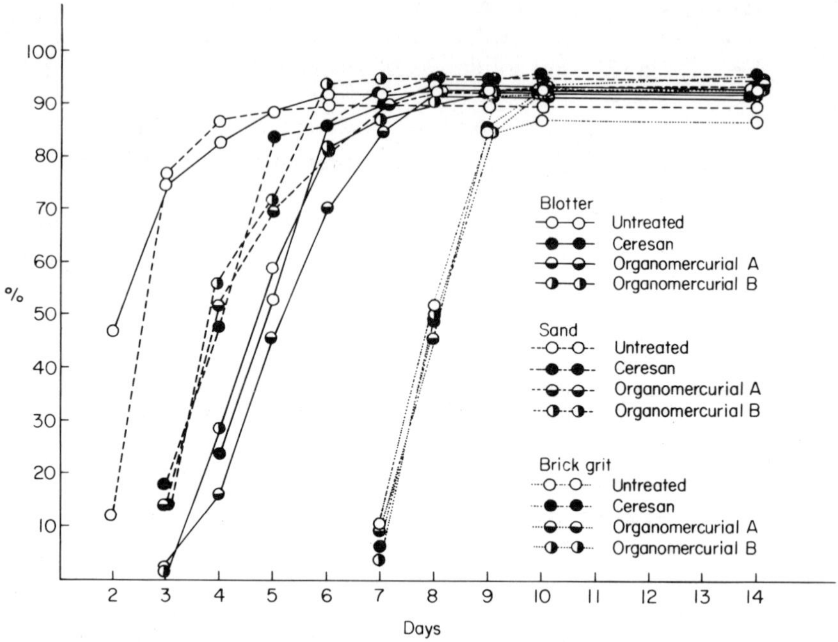

Fig. 23.4. Influence of three organomercurials on the rate of germination of barley on blotter, in sand and in brick grit. (Graphed from data published by Roth, 1959)

In Fig. 23.5, Gassner (1951*a*) has demonstrated the importance of taking the 'speed factor' in germination into account. Two facts are apparent. (1) With the less toxic phenylmercuric acetate, after 12 days of germination there are no differences in the effect of doses up to 500 g/100 kg grain, whereas there is a clear effect after 5 days of germination even at the lowest doses, 50–100 g/100 kg. (2) The effect of increasing doses in the two tested types of organomercurials shows roughly almost identical response in terms of germination speed, whereas the two compounds are distinctly different in their response of germination capacity, as recorded after 12 days. Rate of germination (germination speed) is a more sensitive measure of phytotoxicity than germination capacity.

23.2.8 Testing for efficiency of commercial seed treatment

For testing commercially treated seed for efficiency of treatment *in vitro*, any workable bioassay may be used. However, tests leading to sufficiently conclusive results pertaining to field emergence and freedom from seedling diseases are greatly preferable (Subsection 23.2.6). For this purpose the basis of the assessment should be a comparison of seed treated with the amount of material recommended by the manufacturer or the germicide-approving authorities, with control samples known to have been correctly treated.

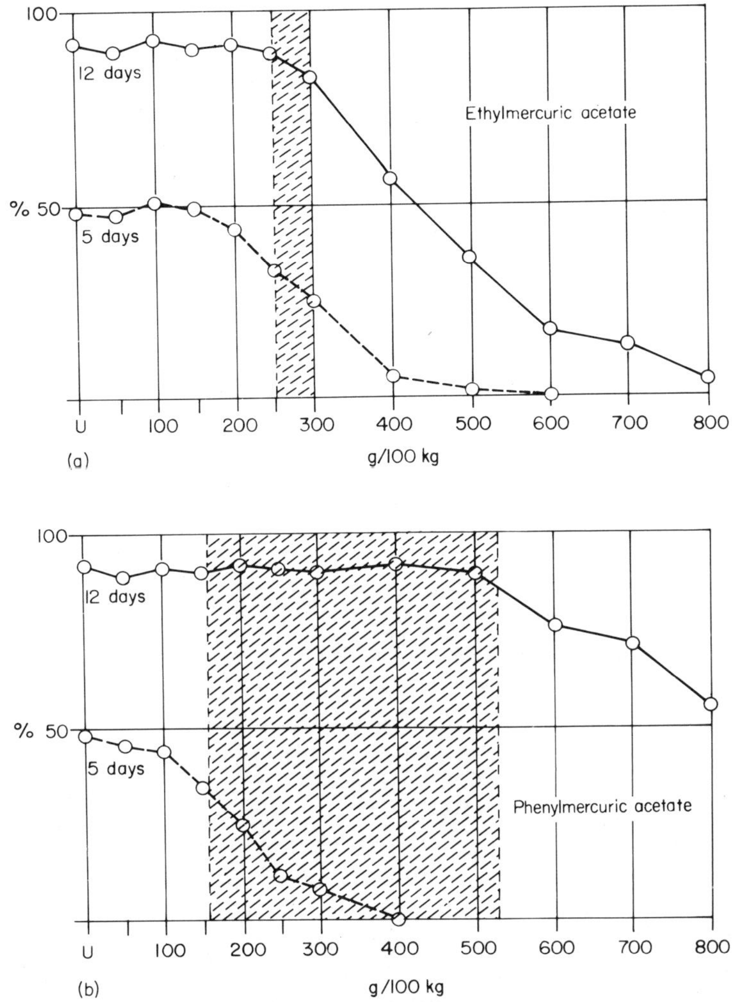

Fig. 23.5. Germination rate of wheat, sown in sand, as affected by treatment with ethylmercuric acetate (a) and phenylmercuric acetate (b), both preparations with 2 per cent mercury content. The hatched areas indicate the distance between the *dosis* causing distinct hampering of germination and the *dosis* inducing the first reduction of the ultimate percentage of germination. (Gassner, 1951*a*)

23.3 FIELD TESTING

The procedure and purpose of field testing is to subject uniformly treated seeds to a variety of natural conditions sufficiently divergent to allow a general evaluation of their field performance, and to carry through comparisons between seed treatments under comparable conditions. Briefly, to follow the principles of diversity and uniformity in field testing, and to establish efficiency of seed treatment under natural conditions.

23.3.1 Diversity in field testing

In the U.S.A. and in Canada, as in many other countries, extensive comparative schemes on field testing of seed treatments have been organised. Results from different parts of individual States or from different States are collated in State-wide or country-wide accounts, e.g. Pinckard and Ivey (1971), Wallace (1971) and Mills (1973). Tests on individual fungicides and procedures are reported from a series of years. Ideally, information on such schemes should be accumulated under consideration of a range of climatic conditions, soil types, time and technique of sowing, edaphic factors, biotic factors, methods of cultivation, cultivars grown, as well as with consideration of the condition of the treated seed, the type and amount of seed-borne inoculum, the history of the seed, including its pre-harvest history, the period and conditions of storage of the seed and, not least, the actual practical performance of the seed treatment purporting to be an agreed standard under test. In a well-organised co-ordinated scheme the procedure of treating the seed is centralised, and duly drawn representative samples of the treated seed are distributed to the co-operators.

23.3.2 Uniformity in field testing

Any system of field testing contains many uncontrolled variables which reduce the reliability of the results. By using replications, four or more in each individual trial, it is assumed that statistical treatment of the data will overcome difficulties in assessment of the results due to variation from these factors. Undoubtedly, field testing has largely been based on statistical probability not sufficiently supported by experimental technique.

Pinckard and Ivey (1971), who have reported nation-wide co-ordinated cotton seed treatment trials in the U.S.A., describe the variability of methods used by various co-operators across the cotton belt, including the method of planting, rate of seeding, size of test plot and final spacing of the plants — all arrangements so far left to be settled by individual co-operators. The methods of planting range from hand-dropping seeds to the use of animal-drawn one-row planters and tractor-drawn, one- and four-row planters. Arndt (1950) found differences as great as 25 per cent in the rate of seeding by mechanical planters, in spite of efforts to calibrate them equally.

Variation due to such inconsistencies in the rate of seeding and depth of planting, which usually occurs with use of conventional planting methods, obviously reduces the value of seed treatment evaluation tests. In order to eliminate such variation Pinckard and Ivey designed and used in their field tests in Louisiana a 'planting plate' which ensures that the number of seeds planted in each individual replication is exact and constant and that they are planted at a uniform depth. This simplified field-planting procedure has other advantages: the land required for the tests is reduced, seedling counts are easier because of the uniform planting pattern, and photographs of representative replications can be made and filed.

It is important that sowing time and soil type be selected so that conditions are favourable to disease development, such as soil temperature and moisture conducive to infection. For cereals, this means a humid but sufficiently well-aerated soil to allow germination of spores. Leukel (1953), an American expert in seed treatment experimentation, made the following list of temperatures conducive to infection of cereal diseases:

Bunt of wheat	5–10°C
Barley leaf stripe disease	8–15°C
Barley covered smut	10–20°C
False loose smut of barley	15–20°C
Smuts of oats	18–22°C
Kernel smuts of sorghum	24–30°C

However, periods favourable to infection cannot be predicted with certainty, and in their absence significant field data are not obtained.

23.3.3 Seed materials for testing seed treatment

There are two types of material to be considered for evaluation of curative effects of germicides: contaminated seed and infected seed.

Cereal seeds contaminated with covered smuts of the seedling infecting type belong to the first category: bunts of wheat, covered smut of barley, etc. Natural contamination takes place by any handling of seed that contains smut balls, particularly during threshing and processing. The contamination is purely mechanical, and consequently such contaminated seed lots can easily be produced artificially for seed treatment experiments. Healthy seed lots of any desirable standard or condition can be inoculated with fixed amounts of inoculum in the form of weight units of spores per weight units of grains. Results from tests using such material must not be taken to apply to any other.

The second type of material, infected seeds, is more critical from the point of view of providing material appropriate for seed treatment experiments. For any realistic evaluation, the seed material to be used must be representative of natural infection: loose smut of barley and wheat, pathogens infecting the pericarp, testa and even embryos.

In experimental work with loose smut of barley and wheat the importance of using seeds with natural infection is well understood. There is hardly any alternative. But for many other seed-borne pathogens the line of least resistance is often followed and seed is artificially contaminated with spores or mycelium of some pathogenic species of *Fusarium* or *Drechslera*, on the assumption that fungicides can thus be evaluated. Experiments conducted in this way are not really different from *in vitro* experiments, and they lead to erratic results. Attempts to inoculate seeds by using vacuum techniques usually add little to reproduction of natural infection. If seed lots with natural infection cannot be made available, infected seeds should be provided by inoculation of growing plants under suitable conditions and allowing the natural development of the seed. Inoculation of mature and perhaps stored seeds cannot be expected, even under conditions suitable for mycelial invasion, to reproduce the natural condition of tenacious resting mycelium in the tissues of the seed, well established behind the protecting outer layers of testa and pericarp.

23.3.4 Criteria for field evaluation of seed treatments

Field performance of seed treatments is usually evaluated in terms of seedling emergence, control of disease and plant growth and increase of yield.

Seedling emergence is recorded by counting the number of seedlings and by evaluating seedling vigour. Harvested seedlings may be classified according to size or weight. Symptoms of phytotoxicity or infections are recorded. Usually two records are taken, e.g. in the U.S.A. 11–14 days and 33–34 days after sowing is a standard for cotton (Pinckard and Ivey, 1971).

Disease control assessment at the seedling stage is often carried out in pots, flats or beds in the greenhouse, where soil conditions can be checked and results more easily recorded. Final tests, however, should be carried out under field conditions. Number of seedlings infected by damping-off, blights, etc., are recorded. Diseases at later stages — leaf spot diseases, blights, stem rots, root rots, etc. — should be recorded according to standard scales for disease assessment, such as those collated under the aegis of the F.A.O. (Chiarappa, 1970), and by percentages. Systemic diseases, such as barley leaf stripe disease, wheat bunts and other cereal smuts, are also recorded as percentage of infected plants, or smuts as number of smutted spikes per unit of area.

Increase of yield, as produced by seed treatment, is the conclusive criterion of effectiveness. For critical evaluation, treated disease-free, or practically disease-free, seed should be included in the trials along with untreated seed. This makes it possible to distinguish the effect of fungicides on soil-borne pathogens, i.e. the seed-protectant qualities, from the effect on seed-borne inoculum.

PLATES III, IV and V

4
1
2
3
FG
FC
FA
FA
FA
FG
FA
FC
FA

Plate III (*opposite, top*). Colour infrared aerial photograph showing bean fields near Hensall, Ontario, infected with bacterial blight (1, 2, 3, 4). (Courtesy V. R. Wallen.) The foci of infection in these fields can be identified by the darker spots. Select registered bean plots originating from imported disease-free seed are shown in 1, 2 and 3. A severely infected bean field is shown in 4. The seed used in this field was grown on the farm for a number of years, resulting in a subsequent build up of seed-borne infection. Scale 1:3,600. (Wallen and Philpotts, 1970)

Plate IV (*opposite, bottom*). Top row (left): Soybean seed with black and red mottling of the seed coat, due to infection of the mother plants by two strains of soybean mosaic virus–normal seeds (above) have no mottling. Top row (right): Seed rot and discoloration in maize seeds due to *Diplodia maydis*.

Middle row (left): Rice seed with discoloration due to *Drechslera oryzae* and *Trichoconis padwickii*; the pink tinge in some of the spots indicates presence of the last pathogen. Middle row (right): Soybean seed with discoloration caused by *Cercospora kikuchii*, purple seed stain.

Bottom row: Agar plate test on oxgall-potato dextrose agar, showing different species of *Fusarium*: FA, *F. avenaceum*, FC, *F. culmorum*, FG, *F. graminearum*; other fungi are present. Bottom row (left): Colony characters seen from above. Bottom row (right): The reverse of the dish showing characters of the submerged mycelium.

Plate V (*below*). Aerial photograph of a field of white beans near Hensall, Ontario, Canada, taken in August 1968 by Ektachrome 8443 colour infrared. The darker spots consist of the bean blight, *Xanthomonas phaseoli*, the red striping is rows of green bean plants, the white striping between the plants shows the soil. Scale 1:500. (Courtesy V. R. Wallen)

Ground surveys revealed that the heavily infected area at the right was on average 3 ft lower than the remainder of the field; in the heavily infected area, heavy morning fogs and dews were present and persisted longer than in any other location in the field.

Due to grey-tone gradation difficulties in black-and-white infrared photography, it is not possible to interpret the exact extent of the bacterial blight. A more reliable interpretation is obtained by using visible colour infrared photographs as shown here. (Wallen and Philpotts, 1970; see also Wallen and Philpotts, 1971)

24 Quarantine for Seed

INTRODUCTION

There is hardly a more efficient means of spreading plant diseases over long distances than by transportation of seed, and practically no limit as to how far seed may be sent. Any seed-borne inoculum surviving from harvest to time of sowing may be safely conveyed from any part of the world to another, irrespective of geographical barriers such as oceans and mountains. This is true for any part of plants transported over long distances, such as vegetative plant material for propagation. But from several points of view there is a particular risk attached to seed.

Seeds are small units easy to handle. Anybody can send a handful across the world in an envelope. Badly infected seed may look healthy.

Seed may harbour pathogens of considerable longevity. Seed-borne inoculum may be present at such a low incidence that it is not present in a sample otherwise regarded as being representative of the consignment. Laboratory testing, even when applicable, would not reveal infections in such cases, and other precautions such as field inspection and post-entry growing-on tests must be applied.

Seed infected at a level which would be present in a representative sample of a reasonable size may be tested efficiently in the laboratory. Implementation of this method of inspection, applicable for certain types of seed-borne diseases, requires, however, special arrangements and personnel that are still largely unavailable in most countries.

The nature of seed transmission of diseases, taken as a whole, is most complex. The precautions to be taken in connection with international movement of seed require a special policy, by which certain specific provisions are imperative. This will be discussed in the following.

24.1 INTERNATIONAL SPREAD OF SEED-BORNE DISEASES DUE TO EXPORT AND IMPORT OF SEED

The value of plant quarantine cannot be experimentally demonstrated, but there are numerous examples of destructive pests and diseases that have spread and caused disastrous losses in hitherto uninfested areas because no effective precautions were taken. Obviously, the examples given below are but a few fairly well-established incidences (numbered data refer to the numbers in Fig. 24.1).

(1) In 1942, bacterial canker of tomato (*Corynebacterium michiganense*) was introduced into the U.K. from the U.S.A. (Mary Noble, personal communication, 1972). The disease occurred then on greenhouse crops in two localities in Sussex, and in the subsequent year in a further 10 localities (Moore, 1946*a*). In Ireland the disease occurred for the first time in 1947 on plants raised from imported seed (McKay, 1948).

(2) According to Noble (1951), bacterial blight of soybean (*Pseudomonas glycinea*) was introduced with seed into Scotland, where in the late 1940s attempts were made for the first time to grow this crop. These efforts failed, mainly because the crop was heavily attacked under Scottish conditions. The seed was imported from Sweden where the disease was unknown until the 1940s, when the disease was believed to have been introduced on seeds of new American cultivars.

(3) In Wisconsin, U.S.A., introduction of black rot of crucifers (*Xanthomonas campestris*) has often been encountered on plants from seed imported from Europe (Monteith, 1921). In 1961 the pathogen caused losses in kale and cauliflower crops as a new plant disease in Portugal, from seed imported in July 1961 from France (Pinto-Ganhão, 1962). In recent years the disease has developed in India on cauliflower

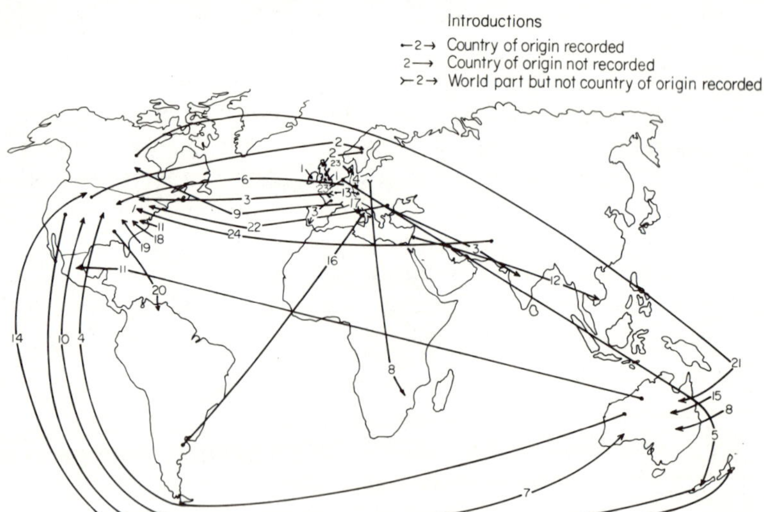

Fig. 24.1. International spread of diseases by seed. Numbers are references to examples in the text.

and other crucifers from seed imported from Europe, particularly from the Netherlands.

(4) The yellow slime disease, or Rathay's disease of cocksfoot, *Corynebacterium rathayi* on *Dactylis glomerata*, had been reported from Germany only once, in 1932, until it was encountered in 1968 in all of six replications of a Danish cultivar 'Adefa' in an experimental field near Husum in the Federal Republic of Germany. The severe outbreak of the disease developed from seed introduced from Denmark and sown three years earlier. None of other 25 foreign cultivars of *D. glomerata* was infected in spite of favourable infection conditions (Schlösser, 1969). In 1945 the disease was encountered for the first time in the U.S.A. on a strain of cocksfoot, imported from New Zealand, in all probability introduced with the seed, three replications of the strain showing diseased panicles (Hardison, 1945).

(5) In 1970, outbreaks of common bacterial blight of bean, *Xanthomonas phaseoli*, and fuscous blight, *X. phaseoli* var. *fuscans*, occurred for the first time in New Zealand. The diseases were traced back to the use of bean seed imported from the Netherlands in 1969 (Watson, 1970).

(6) *Puccinia malvacearum* is an early example of rust fungi spreading by seed from one part of the world to another. Farlow (1886) referred the sudden appearance in 1885 of this parasite in the U.S.A. to introduction from Europe of seed of *Malope trifida* (*M. grandiflora*). Undoubtedly the seed was contaminated in the form of debris of capsules carrying the sori.

Yet another rust, *Puccinia coronata* (*P. beckmanniae*), crown rust, was introduced to Australia on seed of *Beckmannia eruciformis*, imported from the U.S.A. (McAlpin, 1906, according to Orton, 1931); and in view of the strict quarantine in New South Wales (7), Australian plant pathologists found indirect evidence that also race 2 of snapdragon rust (*Puccinia antirrhini*), known in North America and Rhodesia only, arrived at Australia as viable spores on seed imported from North America; the first record was made in October 1952 in New South Wales (8), from where it spread within a year to all other states of the Commonwealth except Western Australia, and in 1953 this race of the rust was encountered in New Zealand, carried by winds across the Tasman Sea (J. Walker, 1954; cf. also Close and Moar, 1973). The rust was introduced to Rhodesia in seed packets of snapdragon (Hopkins, 1940; cf. Section 13.8, and Section 24.2, interception in New Zealand).

(9) Beet rust (*Uromyces betae*) was encountered in Canada on beetroot grown from seed imported from Europe (Newton and Peturson, 1943); the seed concerned carried, abundantly, sori of uredospores as well as teleutospores.

(10) In all probability, bunt of wheat (*Tilletia caries*) was introduced to California from Australia at about the middle of the past century. John Bidwell (1860), who was familiar with grain cultivation in the Sac-

ramento Valley since 1841, refers to the first appearance of wheat bunt in 1854 in this locality, in a field where Australian seed had been used.

(11) Another serious wheat disease, flag smut (*Urocystis agropyri*), was introduced from Australia to Mexico with wheat intended for milling, but which was used for sowing. As a consequence, the authorities immediately issued quarantine regulations to prevent further spread of the disease from the regions where infected seed had been used (Borlaug, Harrar and Stakman, 1946). In the U.S.A. the disease occurred for the first time in 1919 in a wheat field in Madison county, Illinois, undoubtedly introduced on imported wheat, although the source of infection was not discovered (Humphrey and Johnson, 1919).

(12) Loose smut of wheat, *Ustilago tritici*, was detected as the first record in Laos on wheat cv. Florence, imported from Israel (Reddy, 1970).

(13) Onion smut (*Urocystis cepulae*) is believed to have been introduced, in 1924, to Switzerland on seed from France (Müller-Thurgau and Osterwalder, 1924).

(14) Blind seed disease of rye grass (*Gloeotinia temulenta*), which is one of the most serious diseases of fodder grasses in Oregon, U.S.A., was introduced to this State on seed from New Zealand about 1940 (Hardison, 1948*a*).

(15) According to Magee (1935), downy mildew of beets (*Peronospora farinosa*), was introduced into Australia on imported seed which carried an abundance of oospores of the fungus. *Peronospora aestivalis* is reported in one case (16) to have been introduced to Italy on seed of lucerne, imported from Argentina (Campbell, 1922), and the first record of downy mildew of pea (*Peronospora viciae*) in Italy (17), since 1881, was on plants raised from seed imported from Holland (Ciccarone, 1952).

(18) During World War I, *Botrytis ricini*, cause of a serious blight on castor bean, *Ricinus communis*, was introduced to the U.S.A. (Stevenson, 1926), and *Sclerotium oryzae*, a serious rice pathogen, was also introduced to the U.S.A. (19) with seed (Stevenson, 1926).

(20) Zonate leaf spot of sorghum (*Gloeocercospora sorghi*) came to Venezuela in 1949 on seed of *Sorghum vulgare* grown in the U.S.A. (Ciccarone, 1949). This was the first report of the pathogen in South America.

(21) One of the most serious flax diseases, pasmo disease, *Septoria linicola*, was established for the first time in Australia on plants grown from seed imported from Canada (Millikan, 1948).

(22) Wernham (1942) reported a case of introduction of *Epichloë typhina* on seed of *Festuca rubra* from Hungary to Pennsylvania, U.S.A.

(23) Barley stripe mosaic virus is one of several important seed-borne quarantine objects of viruses. The virus has been isolated only twice in Great Britain, from barley seed imported from France (Kassanis and

Slykhuis, 1959; Watson, 1959), and from *Hordeum zeocriton*, grown from seed from Denmark (Catherall, 1972).

(24) According to Leppik (1964*b*) squash mosaic virus was introduced to the U.S.A. by seed from Iran and disseminated in Iowa by cucumber beetles. After several years of intensive work the disease was eradicated and disease-free seed produced.

(25) In Canada, *Ascochyta rabiei* was introduced with seed but its origin could not be determined; it was destructive in yield trials of various cultivars of chickpea in 1973, this incidence of the disease being the first in North America (Morall and McKenzie, 1974).

24.2 INTERCEPTION OF SEED-BORNE DISEASES

Quarantine control measures properly applied in practice provide many examples of successful interception. These are practically always reported from controlling agencies of importing countries. There is usually little access to information about inspected seed consignments condemned by exporting countries because of seed-borne diseases, at least as far as clear reference to the intended country of importation is concerned. Examples of interceptions are given below (numbered data refer to the numbers in Fig. 24.2).

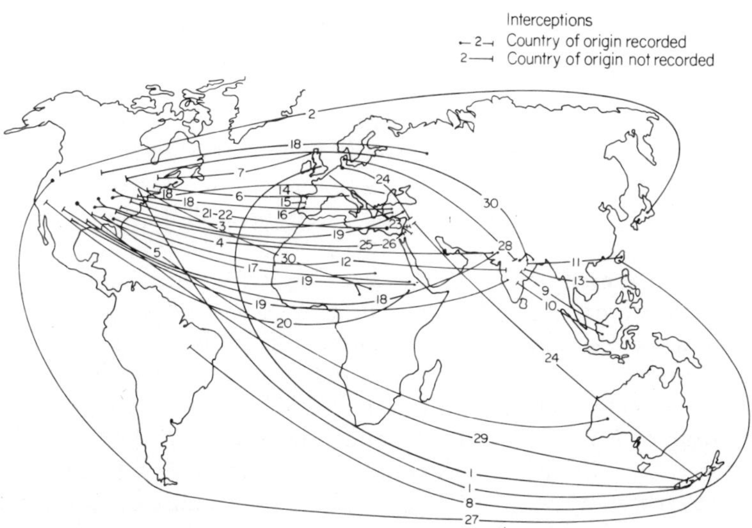

Fig. 24.2. Interception of seed-borne pathogens. Numbers are references to examples in the text.

(1) In New Zealand, *Puccinia antirrhini*, rust of snapdragon, on *Antirrhinum majus* was intercepted in 1953 and 1957 on seed imported from the U.S.A. and the U.K. (Anonymous, 1954).

(2) A rust on bush clover (*Lespedeza* sp.), *Uromyces lespedezae-procumbentis*, was intercepted in the U.S.A. on seed from Japan (Anonymous, 1942).

(3) In the U.S.A., *Tilletia pancicii* on *Hordeum* sp. from Turkey was intercepted in 1952, and *Neovossia indica*, bunt of wheat, from Afghanistan (4) (Locke and Watson, 1955). Moreover, *Septoria tritici*, speckled leaf spot, was intercepted on debris and grain of wheat from Australia (5) (Pollack, 1945), *Coniothyrium hellebori* on seed of *Delphinium* from Italy (6) (Limber, 1962), and *Phoma lineolata* on *Larix* from Scotland (7) (Limber, 1962).

In 1969 a consignment of a new cultivar of rice, Taichung, was imported to Brazil, São Paulo State, from Taiwan (8). The crop, 5.5 ha, showed a serious attack of the bakanae disease, *Fusarium moniliforme*, a new rice disease for Brazil. By order of the Brazilian Plant Protection Service the whole crop was destroyed and the area involved put under quarantine. To secure eradication of the inoculum, cereals were not allowed to be grown for 5 years, but only bean and other crops assumed not to be susceptible (Cardoso, personal communication, 1971).

(9) In seeds of *Hevea brasiliensis*, rubber, from Malaya, *Botryodiplodia theobromae*, die-back or collar rot, and *Phomopsis heveae*, stem die-back, were intercepted at Madras quarantine station in India (10) (Srivastava, 1956).

A further two important pathogens were intercepted in India on soybean, at the import seed health testing laboratory of the Indian Agricultural Research Institute, New Delhi: *Colletotrichum dematium*, anthracnose, seedling blight, on seeds from Taiwan (11), and *Myrothecium roridum*, leaf spot, on seeds from the U.S.A. (12) (Lambat *et al.*, 1969). Moreover, *Aphelenchoides besseyi*, white tip of rice, was intercepted in 1968 on rice seed imported from the Philippines (13) (Nath and Sethi, 1970).

In Portugal, three major plant diseases have been intercepted by growing-on tests carried out in greenhouse as part of post-entry control (Branquinho de Oliveira, personal communication, 1963): tobacco blue mould, *Peronospora tabacina*, was intercepted several years before the disaster in 1958 (14), when the disease spread in four years from the U.K. to the whole of Europe and the Mediterranean countries (Fig. 24.3); the second serious disease intercepted was bacterial wilt of lucerne, *Corynebacterium insidiosum* (15), the third one being bacterial wilt of maize, *Erwinia stewartii* (16).

In growing-on tests for post-entry inspection of seed at the N. Central Regional Plant Introduction Station in Ames, Iowa, the following seed-

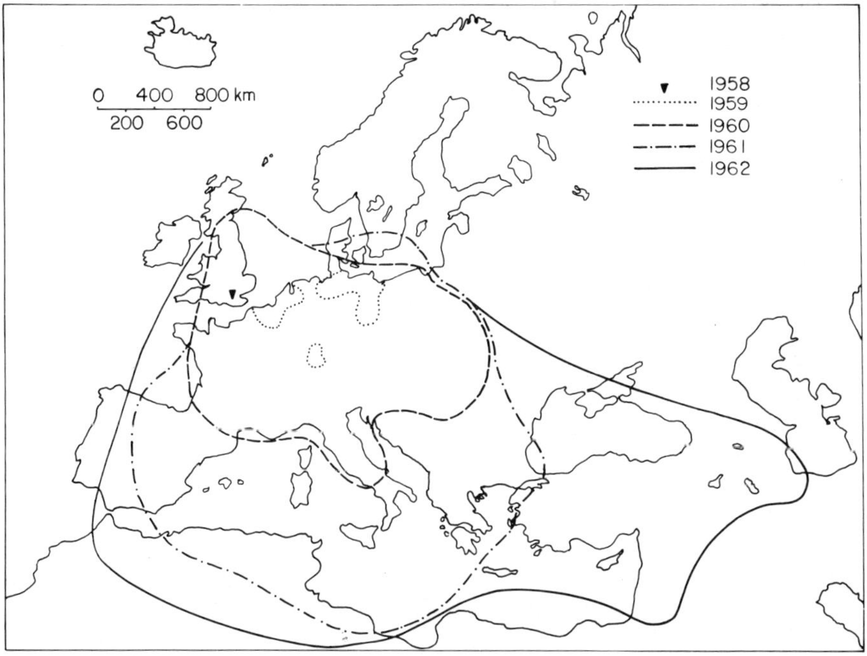

Fig. 24.3. Spreading of tobacco blue mould, *Peronospora tabacina,* 1958–1962, from one source of inoculum in England. (Corbaz, 1964)

borne diseases, foreign to the U.S.A., were intercepted (Leppik, 1962*a*, 1962*b*):

(17) *Ascochyta pisi*, evidently a pathogenic race new to the U.S.A., detected on seed of *Lathyrus sativus* and *L. cicera* from Ethiopia.

(18) Downy mildew, *Plasmopara halstedii*, on sunflower, *Helianthus annuus*, evidently a new pathogenic race, reported from eastern Europe and found on plants grown from seed from Turkey, Pakistan, the U.S.S.R. and Roumania (Leppik, 1962*b*).

(19) *Cercospora traversiana*, leaf spot, on fenugreek, *Trigonella foenum-gracum*, a destructive leaf spot, known in some eastern European countries, common in the Near East and India, but not yet reported from North America. It was found on seed samples from the Near East, India and Ethiopia.

(20) Finally, a mosaic on *Melilotus suaveolens*, in plants from seeds imported from Ethiopia.

In all, this post-entry inspection revealed ten seed-borne diseases, including pathogenic races of three species of fungi, all of them new to the U.S.A.; namely, one introduced from Afghanistan, three from Ethiopia, one from China, three from Europe, two from India, two from Pakistan and two from Turkey.

In Israel, many seed lots, all provided with the International Phytosanitary Certificate, have been intercepted due to quarantine pathogens. In 1959 an oat seed lot with *Drechslera victoriae* (21), and in 1961 a maize seed lot with *Diplodia maydis* (22), both from the U.S.A., were refused entry. In 1968 and 1969 seed lots of bean with *Colletotrichum lindemuthianum* from France (23), in 1968, 1969 and 1970 ryegrass seed lots from Holland and New Zealand (24) with *Gloeotinia temulenta*, and in 1969 and 1970 cotton seed lots from the U.S.A. with *Colletotrichum gossypii* and *Xanthomonas malvacearum* (25), (26), were intercepted. None of these pathogens has been encountered in Israel. For the period considered, a total of 35 quarantine-certified seed lots was rejected but many more lots were intercepted because of other infections of importance to crop production in Israel (Halfon-Meiri, personal communication, 1972).

(27) In New Zealand, race T of *Drechslera maydis*, southern leaf blight, a potential maize pathogen in the U.S.A., was intercepted in two small packages of popcorn, produced in the States (Scott, 1971).

(28) In 1971, 6,000 kg sugar beet seed from Denmark was intercepted and consequently destroyed in India because of *Heterodera schachtii*, present as concomitant contamination (Sethi, 1971, unpublished, according to Lambat, personal communication, 1972). This nematode has not, so far, been encountered in India.

(29) In New Zealand, squash mosaic was recorded for the first time in Honeydew rockmelon plants grown from seed imported from the U.S.A. It was not detected in the field (Thomas, 1973).

(30) Ten out of 169 small seed samples of sorghum collected in Africa and India and sent to the U.S.A. for breeding purposes were intercepted by the Plant Quarantine Station, Washington, D.C., because they were infected by long smut, *Tolyposporium ehrenbergii*; the infection ranged from 8 to 100 per cent. This smut is widespread in Africa and Asia but does not occur in the U.S.A. (Mabry and Lightfield, 1974).

24.3 PRINCIPLES OF SETTING UP QUARANTINE REGULATIONS

To some extent seed-borne diseases have been taken into account in the import regulations of most countries but the precautions so far taken are of an extremely sporadic nature. Only a few serious diseases have been blacklisted in quarantine provisions; many are neglected, perhaps mainly for the following reasons.

(1) Seed-borne inoculum of many diseases is inconspicuous and cannot be revealed by direct inspection of the dry seed, nor by examination of the dry seed under a binocular microscope.

(2) Routine seed health testing methods have been developed rather slowly, being internationally standardised and more widely brought into use only in the last 25 years.

(3) For many important seed-borne diseases, laboratory routine seed health testing methods are not yet available.

(4) Seed-borne inoculum of many diseases is present in very small amounts, such as one infected seed in many thousands, or as debris mixed with the seed. For this reason the sampling procedures commonly used may not be sufficient. Such diseases may be of utmost importance in quarantine calling for special precautions (downy mildews, certain viruses and bacteria, etc.). Nevertheless, they are largely neglected as such.

(5) The policy followed by many countries in setting up quarantine regulations with regard to seed-borne diseases is not precise. Often no specification of requirements is provided, and a mere 'general' seed health certificate is required with no further guidance for the exporting country. Loose formalities of a certificate are often preferred to factual reports. There is a great need for re-assessment of the situation by authorities and inspectors within this field.

Distinction should be made (Neergaard, 1972) between (a) dangerous pathogens and pests which do not occur in the importing country but which, when introduced, may spread very quickly (Category A); (b) pathogens and pests that do not occur or have a limited distribution in the importing country, and which have a moderate rate of increase, their introduction into an uninfested area or field being of particular importance to the growers concerned (Category B); and (c) pathogens and pests that occur generally in the importing country and which may reduce the planting value of the seed (Category C).

Only the first two categories may justifiably be regarded as proper subjects for quarantine regulations, whereas the third category may be considered as such under certain circumstances. This category relates to the quality of the seed, the criteria of evaluation being comparable to those pertaining to properties of seed quality such as germination capacity and purity.

For the first two groups, which appropriately may be classified as 'prohibited organisms', a tolerance of zero must be assumed. For other pathogens and pests which reduce the quality of the seed for sowing or consumption the purpose of import regulations may be to ensure that the seed is of good planting quality. For such cases, specification of tolerances may be acceptable, such as the two per cent bean common mosaic permitted by Australia. However, indication of tolerances must be based on a prescribed inspection procedure or testing method.

The question of seed health testing methods for quarantine was taken up by one of the regional plant protection organisations (cf. Appendix I of this chapter), the European and Mediterranean Plant Protection Organisation (EPPO), in a Working Party held in London on July 1965 with participation by representatives of the International Seed Testing Association (ISTA). Since 1957 the Plant Disease Committee of this association has organised comparative seed health testing schemes and annual international workshops dealing with the results of these testing activities. As a result, some methods have been standardised and included as recommendations in the International Rules for Seed Testing (ISTA, 1966*a*). A brief description of the most important testing procedures based on these standardised methods is included in the report of the mentioned EPPO Working Party (EPPO, 1966).

It is obvious that the value of a plant health certificate issued for seed is entirely dependent on the testing procedure applied. Hence, prescriptions on definite testing methods or inspection procedures to be applied should be given in quarantine regulations dealing with seed-borne organisms. Until definite testing methods are generally agreed upon and put into use it is desirable to give explicit information in the certificate on the method used; according to the International Rules for Seed Testing such information shall be given on the Orange International Seed Lot Certificate. Also, in issuing the FAO Model Phytosanitary Certificate (Rome Certificate), some countries introduce information on actual testing procedures under the certificate's heading of Additional Declaration. The certificates are given in Appendix II of this chapter.

The specification of pathogens and pests is handled in different ways in existing quarantine regulations. Some countries provide a separate list of hosts that must be inspected and a separate list of pests and diseases that must be sought. As a consequence, it is left to the certifying agency of the exporting country to decide on the host range to be inspected for the diseases and pests included in the regulations concerned. It is much preferable that the provisions clearly state *which* plant species should be examined for *which* disease or pest, because some pathogens may be serious on one host but unimportant on others. Important steps toward more specification in quarantine provisions for seed have been made by publication of two annotated lists of dangerous seed-borne pathogens (EPPO, 1954, 1959).

This point of view is reflected in quarantine provisions pertaining to *Verticillium alboatrum*, which is regarded as a quarantine object in seed of only *Medicago sativa* by the U.S.A. and in seed of only *Lycopersicon esculentum* by Greece, and on *Ditylenchus dipsaci*, which is regarded as a quarantine object in seed of only *Medicago sativa* by Great Britain — other hosts are disregarded (Neergaard, 1970*c*), the race in question being limited to lucerne. In quarantine regulations specification of races restricted to certain hosts under consideration would be helpful. Some countries issue detailed host lists of important quarantine objects, while in the provisions of other countries only the main host is indicated.

As it appears, there is an urgent need for more precisely formulated requirements on seed-borne diseases and pests in quarantine regulations of most countries, with clear indication of the plant species to be inspected, the organisms (diseases) to be observed, and the inspection procedure (testing method) to be used.

This leads us to formulate the following principles, to be taken into account in setting up regulations for quarantine.

The main principle to be observed is specification of pathogens and inspection procedures. Any specification must take into consideration a clear differentiation between the three above-mentioned categories of pathogens discussed further below.

(1) *Pathogens which develop particularly dynamic epiphytotics* (Category A). These include many downy mildew fungi, certain rust fungi and many bacteria, including certain wilt-producing bacteria. Pathogenic races of certain widely distributed pathogens should be considered, as exemplified by the numerous races of *Pyricularia oryzae* in rice, still neglected in the quarantine provisions of practically all of the rice seed importing countries.

Many pathogens of this category may be present in seed only as a trace. For this reason sampling of the consignment and subsequent laboratory testing are inadequate. For such pathogens the exporting country must be required to ascertain that *the pathogen concerned does not occur in this country* or, where applicable, in the region in which the seed has been produced. When there is a special need to accept small consignments of seed in the country of importation, for instance for the purpose of introducing new cultivars, or for breeding, post-entry inspection may be carried out as growing-on tests in quarantine greenhouses, effectively equipped to prevent escape and contamination from outside (Sheffield, 1968).

(2) *Pathogens of a moderate rate of increase*, and which may have a restricted distribution in the importing country (Category B) for which inspection of the seed in the exporting country by testing representative samples according to laboratory procedures, usually involving incubation of seed, may be accepted. The procedure to be used should be specified in the provisions of the importing country. This usually applies to pathogens of both the second and the third category, discussed below.

(3) *Pathogens of importance only to the planting value of the seed* (Category C) are not true quarantine objects. For this category sampling and laboratory testing according to standard procedures are usually applicable, and the provisions concerned would form a parallel to those import regulations valid in certain countries that require minimum standards for germination capacity and purity, and issuance of the related

International Orange Certificate. Also, for this category of pathogens due specification of organisms, diseases and inspection procedures should be provided by the importing country.

In summarising, quarantine regulations may include the following items:

I. *Specifications*
Specification of pathogens and inspection procedures.

II. *Export control*
(Regulations related to the controlling agency of the exporting country.)

(1) *Certification of complete freedom from infestation*, in the country, or in a defined growing area of the seed crop concerned. For this purpose, issuance of an international standard phytosanitary certificate according to the provisions of the importing country.
(2) *Field inspection of the seed crop* to ascertain freedom of infection (or accordance to requirements of tolerances) in this crop. Issuance of an international standard phytosanitary certificate according to the provisions of the importing country.
(3) *Inspection of the seed lot* by drawing representative samples and testing these according to standard procedures approved by the importing country. Issuance of an international standard phytosanitary certificate according to the provisions of the importing country.
(4) *Seed treatment* whenever practicable. This provision should be considered with great reservation. It may be an additional precaution against quarantine object; seed treatment should never be regarded as a complete safeguard in itself.

III. *Import control*
(Regulations related to the controlling agency of the importing country.)
These may appropriately include the following provisions:

(1) *Embargo*. Total prohibition of importation of certain kinds of seed, or restricted, under the conditions of issuance of an *import permit* previously granted by the authorities of the importing country for individual consignments. The importer must request authorisation to import, usually from the Director of Agriculture or the Chief of the Plant Protection Service.
(2) *Inspection of the seed lot* by drawing representative samples and testing these in accordance to standard procedures, such as incubation tests.
(3) *Post-entry growing-on tests* of small seed consignments of special importance, cultivated from seed to mature seed-bearing plants in adequately equipped quarantine greenhouse units, as discussed below. The seed grown under these conditions should be tested and again found free from any infection. It may be released for cultivation in the open but the plants grown from the seed may be still required to be inspected before ultimate release.

24.4 PRINCIPLES OF APPLYING QUARANTINE METHODS

Plant quarantine legislation currently in force in various countries and territories to prevent introduction of important seed-borne diseases and pests are, generally speaking, of the following five types of quarantine action: (1) *embargo* and *import permit*; (2) *inspection* (field inspection, laboratory testing, etc.) *in the exporting country* before shipment of the consignment; (3) *seed treatment* for diseases and pests that with reasonable certainty can be effectively eliminated by disinfestation or fumigation; (4) *post-entry growth inspection* of the importing country in closed quarantine; and (5) *certification*. The Model Phytosanitary Certificate as defined by the International Plant Protection Convention of 1951, adopted in Rome under the aegis of the FAO (FAO, 1951) has been accepted by practically all countries requiring a phytosanitary certificate attesting freedom of seed from pests and diseases. According to the Convention: 'Each contracting Government undertakes not to require consignments of plants intended for planting or propagation imported into its territories to be accompanied by phytosanitary certificates inconsistent with the model set out in the Annex to this Convention' (cf. Appendix II to this chapter, Phytosanitary Certificate).

24.4.1 Pre-entry control

24.4.1.1 IMPORT FROM DISEASE-FREE AREAS

The reliability of certifying for freedom of areas from diseases is dependent on the effectiveness of the disease surveys made in the countries concerned. The Commonwealth Mycological Institute, Kew, England, issues distribution maps of plant diseases based on surveys and individual records. These maps are under continual revision, and they provide information essential to elaboration of quarantine regulations and execution of quarantine control. However, even the information provided by the maps is of a rather general nature and can only be as dependable as the survey work on which they are based. Details on distribution of pathogens within areas designated on the maps as infested may vary from widespread occurrence to local incidences ('pockets'), or they may represent merely one or a few isolated records. Although detailed information is given in the bibliography of each map, such details must be disregarded on the maps but yet be of vital importance to quarantine policy and practice of issuing phytosanitary certificates.

In countries where quarantine objects occur in restricted areas, effective surveying is particularly important, and here careful inspection of the seed crops concerned may be an acceptable prerequisite of issuing certificates.

24.4.1.2 FIELD INSPECTION IN THE EXPORTING COUNTRY

Standard procedures complying with the regulations of the importing country must be followed. Approval of domestic or international standard seed certification schemes (such as the OECD seed certification scheme) on the part of the importing country with a view to seed-borne diseases, may contribute substantially to standardisation and simplification of certification procedures. For further reference, see Section 31.3.

24.4.1.3 LABORATORY TEST OF SEED IN EXPORTING COUNTRIES

In the International Rules for Seed Testing (ISTA, 1966*a*), precise directions on procedures of drawing representative samples of seed lots of specified types of seed have been given. The minimum requirements contained in these directions must be strictly followed.

As to mixtures of seeds from two or more lots, either from the same country or different ones, the EPPO/ISTA Working Party of 1965, mentioned above, gave the following advice: 'If . . . the lots are separate when the application for a test is made, the various lots should be tested separately. Then afterwards, a test should be made to ensure that the subsequent mixing is done properly having regard to any different standards revealed by the original examinations and finally, the lots must be resealed to preserve their identities.'

For any laboratory testing of seed for quarantine it is fundamental to adopt standardised procedures. Such are available for a comparatively small number of pathogens but certain incubation procedures (blotter method, agar plate method) may be adopted generally for a large range of pathogens, apart from those for which these procedures have been definitely shown to be workable.

Brief surveys of seed health testing methods for quarantine control have been published as Appendices I and II in the above-mentioned Report of the EPPO/ISTA Working Party (EPPO, 1966). Reference is made, moreover, to Chapter 29, dealing with various standard incubation procedures.

24.4.1.4 SEED TREATMENT

The Report of the EPPO/ISTA Working Party (EPPO, 1966) states, concerning seed treatment: 'The authority carrying out the examination is keenly interested in knowing whether the seeds have received any special processing such as treatment with hot water, chemicals or fumigants. The receiving countries are also interested in such information, since a repetition of certain treatments could have an adverse effect on certain seeds and moreover, there are often dangers to wild life if certain chemicals have been used.'

There are special chemical testing procedures available to assess whether certain treatment of seed, such as with any organic mercury compound, has actually been carried out. The presence of seed protectants on seed may also be demonstrated in a biological test. Standardisation of such procedures should be considered and, in fact, some are commonly used internationally in investigations on seed treatment, cf. Section 23.2.

A post-treatment health test, to ascertain that the treatment has been effective, may be required.

24.4.2 Post-entry control

24.4.2.1 CLOSED QUARANTINE AND PRODUCTION OF PATHOGEN-FREE SEED

Closed quarantine provisions are directed against serious diseases that may have escaped inspection in the country of origin, or perhaps were symptomless in a particular cultivar imported. The principles of closed quarantine procedures have been described by Sheffield (1968), based on experiences gained in East Africa where a plant quarantine station has been in operation for about 40 years (Fig. 24.4). A few main points of the basic requirements and the function of a post-entry quarantine station must be emphasised.

Strict scientific supervision by specialists in plant pathology is the primary essential of such a station. All plants under quarantine should be grown in controlled conditions, usually under glass, the conditions suited to the plants concerned should be provided and consequently there should be an experienced horticulturist on the staff. Small separate glass-house units (275×305×366 cm has been found to be an appropriate size) are preferable in order to ensure good control of atmospheric conditions

Fig. 24.4. Glasshouse units for closed post-entry quarantine control at East African Plant Quarantine Station, Muguga, Kenya. (Courtesy of E. J. Guthrie)

within each chamber and to reduce the risk of contamination. The floor of the chamber should be of concrete with holes for drainage, and the chamber should be provided with an insect-protected induction fan, insect-protected ventilators, overhead sprinklers to provide a fine mist, overhead lighting with time switches, and thermostatically controlled heating units. All controls for temperature, lighting, watering, humidity and ventilation should be placed externally to reduce the need for personnel to enter the growth chambers. Each unit must be provided with a gauze door or screen and surrounded by a water-filled trough to reduce entry of insects. Facilities must be available for the sterilising of soil before and after use, of tools, flower pots, and for complete destruction of diseased material.

From the small consignment of seed accepted for closed quarantine a few seeds should be sown and grown through to maturity. If the plants remain healthy after having been constantly inspected and indexed for pathogens, seed can be harvested from them and be released. By using only part of a small consignment of seed, the process can be repeated if all resultant plants of the first or later batches of tested seeds are diseased.

In Rhodesia a quarantine glasshouse has been adapted so that tobacco seed, if necessary, can be imported even from countries where blue mould, *Peronospora tabacina* (syn. *P. hyoscyami*) occurs. The plants are grown under individual polythene covers, a spore proof filter is placed over an extractor fan and only seed from healthy plants is released (Whellan, 1965).

Indexing for seed-borne viruses that remain symptomless in the resultant plants is an important part of the quarantine procedure. Certain indicator plants, such as spp. of *Chenopodium*, *Nicotiana* and *Petunia*, should be included as they are all susceptible to infection of a very large number of viruses.

The purpose of a quarantine station is to remove the risk of introducing pathogens with new cultivars. Destroying all such material enforced by conventional quarantine regulations would mean a serious loss of promising new crops. Valuable foreign seed consignments, such as germ plasma for breeding purposes, is not lost even if found to be infected and detained in quarantine. Special methods and appropriate techniques can be applied to such plant material for the production of seed free from dangerous pathogens (Leppik, 1968).

Seed disinfection may be effective for slightly infected seed lots, but to ascertain complete freedom from dangerous infections the treated seed must be planted in closed quarantine as described above. Particularly, risk may be expected from systemic pathogens including viruses which may show up only on mature plants. All plants showing symptoms must be eliminated before neighbouring plants can be infected. Seed from healthy plants must be grown for checking.

Under certain circumstances, when only a few original plants are available and more simple procedures are not workable, pathogen-free stocks may be obtained by meristem tip cultures as described by Dimock (1962), Baker and Phillips (1962) and Braverman (1975).

24.4.2.2 LABORATORY TESTING OF SEED SAMPLES

As an extra safeguard the importing country should check the health condition of a seed consignment that has been provided with the 'blue stamp' of a Phytosanitary Certificate, issued by the exporting country. Standard procedures such as those recommended in the International Rules for Seed Testing would improve the conditions for obtaining comparable test results. The EPPO/ISTA Working Party on Health Certification of Seeds for Export (EPPO, 1966) may again be quoted: 'A universal adoption of these Seed Testing Rules would go far in the elimination of the situations which arise when examinations in the receiving country do not confirm the health certificate accompanying the seeds. They arise not by bad faith but nearly always through the two plant protection services concerned working to different tolerance levels and using different techniques. If one service performs a ''visual examination'' on dry seed and the receiving service conducts a critical test, the room for difference is obvious.'

24.4.2.3 FIELD PLOT INSPECTION

Whenever observations in closed quarantine cannot be made because of insufficient capacity of the quarantine inspecting agency or plant introduction station, field observations provide the sole possibility of detecting foreign pathogens imported by seed. Leppik (1964*a*) emphasises that, even in cases where plants were kept under constant surveillance in greenhouses, new pathogens were first detected later in the field. This holds true for smuts, the systemic mycelium of which generally grows with the host but without producing outwardly visible characteristics, the first reliable symptoms developing on ripening fruits or on ageing stems or leaves.

24.5 PRACTICE IN ADOPTING AND APPLYING QUARANTINE PROVISIONS FOR SEED

24.5.1 Requirements of quarantine control in various countries

More than 100 countries have issued quarantine import regulations for seed. About 50 of these countries require a general plant health certificate for all or nearly all kinds of seed (Neergaard, 1970*c*). Some countries issue a list of prohibited seed-borne pathogens, which implies some specification for a few species of crop plants. However, as mentioned above, many countries do not indicate any seed-borne quarantine objects at all. As a consequence of this lack of specification and precision the major part of quarantine seed health testing carried out at Plant Protection Services in various seed exporting countries, such as that of Denmark, is left entirely to the interpretation of the export controlling agency. This applies both to pathogens to be looked for and methods to be used.

In 1964–1965 the Plant Protection Service in Copenhagen issued the Rome Certificate for about 12,000 seed lots exported to 81 different countries. The United Kingdom and the Netherlands are other European countries with a considerable seed export and charged with similar responsibilities in issuance of seed health certificates.

According to an admittedly incomplete review (Neergaard, 1970*c*), specification on seed-borne pathogens to be observed in inspecting seed for export has been given in the plant quarantine provisions of at least 32 countries. The specifications include a total of 16 species of bacteria, 72 species of fungi, 11 viruses, 6 species of nematodes, as well as 31 species of insects, chiefly storage pests, a total of 49 specified genera or families of host plants having been considered.

No indications as to prescribed seed health testing methods have been noticed in the quarantine regulations referred to in the above-mentioned review. Some countries, however, do prescribe inspection of the seed crop concerned during active growth for bacterial and virus diseases, and certain other diseases.

In general, field inspection, carefully accomplished, must be a minimum requirement before freedom from these categories of diseases can be declared in a certification. For a few of these diseases reliable laboratory procedures are now available (for example, for bean common mosaic and lettuce mosaic) and may supplement or replace field inspection.

Indications on permissible *tolerances* are given in the quarantine provisions of a few countries. Australia indicates 2 per cent and New Zealand 1.5 per cent bean common mosaic as the maximum infection permissible, Turkey 0.1 per cent lettuce mosaic, Hungary 3 per cent *Corynebacterium flaccumfaciens* in bean, and 2 per cent *Xanthomonas malvacearum* in cotton — the tolerances allowable for the two last-mentioned severe quarantine objects seem difficult to justify. For pathogens specified in quarantine regulations without such indications, a tolerance of zero must be assumed.

24.5.2 Current practice in issuing seed health certificates

(1) *Certificates.* There are two international standard certificates that may be used for documentation on the phytosanitary condition of a seed lot: (A) the *Orange Certificate* of the International Seed Testing Associa-

tion, and (B) the FAO Model Phytosanitary Certificate, the so-called *Rome Certificate*. They serve different purposes.

(A) The *Orange Certificate* (Appendix II) is the international standard certificate in use for certification of the *quality* of a seed lot and for seed trade transactions in general. Issuance of this certificate must be based on analysis of a representative sample drawn from the seed lot concerned and examined according to the prescriptions laid down in the International Rules for Seed Testing (ISTA, 1966*a*)*. The so-called Blue Certificate is issued on samples which have not been drawn officially.

In practice seed is tested mainly for germination capacity and purity and often also for moisture content and trueness to variety, but only rarely for health condition. According to the International Rules for Seed Testing a detailed test for the health condition of seed samples is required only when requested by the owner of the seed. This implies that a seed lot may be classified as of excellent quality, with a high germination capacity and no impurities, although the subsequent crop may be destroyed owing to serious seed-borne infection. Actually, there has been very little demand for a detailed test for health condition, and in some countries none at all. It is true that an Orange Certificate provided with an ostentatious display of Latin names of fungi, perhaps of little or no consequence to the development of the crop, could do more harm than good to the seed trade, because of the risk of wrong interpretation. The possibility of such unreasonable cases, which certainly could be prevented by adequate regulation, should not impede the proper evaluation of disease potentialities that really menace the crop.

(B) *The Rome Certificate* (Appendix II). In contrast to the ISTA Certificate, the FAO Model Phytosanitary Certificate, the so-called Rome Certificate, is a specific *quarantine* certificate issued as an obligatory precaution against introduction of seed-borne diseases from one country to another. The certificate indicates, without any specification, unless introduced under the heading of 'Additional declaration', that the lot in question has been found 'to be substantially free from injurious diseases and pests; and that the consignment is believed to conform with the current phytosanitary regulations of the importing country'. The value of this document is, of course, quite dependent on the testing procedure applied. A certificate based on an improper technique is not worth a dishonoured cheque!

The wording of the existing Rome Certificate, as used for seeds, has been criticised from various angles. The text does not include certain necessary information, particularly as to the method used. A special quarantine certificate for seed may eventually be preferable. However, at

*The certificate, see Appendix II, contains on the reverse side the *Declaration of the Signatory*.

the Plant Protection Service in Denmark indications under 'Additional declaration' of the method used was introduced from 1963 as, for example, the following text indicating the one most frequently applied: 'The seed investigated has been tested by the blotter method, including microscopic examination after incubation.'

The importing country may require a certain specification under an 'Additional declaration', often that a certain disease has not been found in the seed lot in question, or has not been encountered in the exporting (seed-producing) country, that the seed has been disinfected (indication on disinfectant and procedure may be included), or that the seed crop has been inspected during active growth — another important implication of seed health testing.

The issue of Rome Certificates is required not only by official authorities. Once in a while special requests come from exporting firms as part of their agreement with importing firms. Such requests may include some undue demands for 'Additional declarations' that are not required by the authorities of the importing country and, moreover, they may lead to demands for certificates for seed lots to be imported to countries that do not require the Rome Certificate for the seed concerned, or indeed for any kind of seed at all. At the Plant Protection Service in Denmark certificates are issued even in such cases, but the following explicit indication is provided under 'Additional declaration': 'The current phytosanitary regulations of the importing country are not known to include provisions pertaining to the consignment.'

In some cases merchants are surely entitled to ask for freedom or low tolerances even when no transmissible disease is involved, e.g. hollow heart of peas. Seed testing stations should provide this service rather than the plant protection services.

Another difficulty is connected with the re-export of seed, in that it may have been produced under climatic conditions quite different from those of the re-exporting country and of the country where it is to be sown. There is almost no limitation as to the geographic origin of seed exported from any one country, and hence the technical demands as to the testing for seed-borne diseases, carried out in one country, is markedly international in scope.

(2) *Practice of testing seed for quarantine*. More than 100 countries have published regulations for quarantine export and import control. Only rather few countries carry out many seed health tests for export control for quarantine. For several decades this has been carried through in Denmark, Holland and the United Kingdom (England).

There is information from two countries about the relative number of seed lots rejected in quarantine export control due to seed-borne diseases. In England, according to W. C. Moore (1946*a*), 30,000 seed lots were examined for seed-borne diseases in connection with export quarantine control during the period 1925–1943, mainly by direct inspection of seed under low power microscope without incubation of the seed prior to the examination. Of the total, 650 seed lots or 2.2 per cent were rejected for export. The causes of rejection were *Septoria* in *Apium graveolens* and *Petroselinum crispum*; *Phoma betae* and *Uromyces betae* in *Beta vulgaris*; *Sclerotinia sclerotiorum* in *Helianthus annuus*; *Colletotrichum lindemuthianum* and *Pseudomonas phaseolicola* in *Phaseolus vulgaris*; *Ascochyta pisi* and marsh spot in *Pisum sativum*. Later, during recent years relatively many seed lots of *Lolium perenne* have been rejected on account of blind seed disease, *Gloeotinia temulenta*.

Denmark has a considerable seed export and an extensive control in this connection. Neergaard (1965) has given a rather detailed survey over 11 years of the quantities of seed controlled, including the number of seed lots examined and rejected, the number of certificates issued and the number of importing countries. In the period 1954–1965 a total of about 11,000 Rome Certificates were issued. In the period 1953–1965, 5,840 seed lots of vegetables were examined mainly by incubation methods, 8.5 per cent of which were rejected; 1,400 seed lots of ornamentals, 15.7 per cent rejected; 980 forest seed lots, 0.6 per cent rejected; 14,470 seed lots of cereals and field crops, 0.5 per cent rejected — making a total of 22,690 seed lots with an average of 3.5 per cent seed lots rejected.

A list of pathogens on account of which these rejections have been made has been published (Neergaard, 1965). The list comprises 41 species of host plants and a total of 41 seed-borne diseases and pests, some of which occurred in several hosts.

Both in the United Kingdom and in Denmark the causes of rejections were primarily pathogens which would be classified as belonging to the third category ('C'), as discussed in Section 24.3, i.e. those that presumably occur generally in the importing country and may cause damage to the seed or reduce its planting value. The rejections were almost exclusively made for consignments bound for countries whose quarantine regulations had no specifications on the pathogens to be considered. Because of this it was left entirely to the controlling agency to make decisions about pathogens to be considered and maximum disease tolerances to be applied. Reference is made to the author's previous discussions of some basic principles to be considered in setting up such tolerances (Neergaard, 1962*a*, 1962*b*).

Of the 41 seed-borne diseases and pests that caused rejection of seed lots in Denmark, only a few are among those that have been included in the specifications of seed-borne quarantine objects in the regulations of some countries. They are the following: *Corynebacterium rathayi* in *Dactylis glomerata*, *Gloeotinia temulenta* in *Lolium perenne*, bean common mosaic in *Phaseolus vulgaris*, lettuce mosaic in *Lactuca sativa*, and

Phoma lingam in *Brassica oleracea* and other *Brassica* spp.

A picture of current practice in seed health testing for quarantine is, however, incomplete, if no mention is made of the fact that in some seed exporting countries Rome Certificates are issued regularly from one day to the next for seed consignments, actually without any previous inspection of the seed. There is a great need of better understanding, on the part of some responsible plant quarantine officers and some authorities, as to what serious consequences may arise from adopting such a policy of freely issuing certificates, which are assumed by the importing countries to be safeguards in accordance with the International Plant Protection Convention.

Apart from detection of seed-borne diseases and pests in export control of seed consignments, there are other problems connected with quarantine seed certification. One has been defined in the Report of the EPPO/ISTA Working Party of 1965 (EPPO, 1966) as follows:

'It is well-known that certain types of seeds move from country to country in the normal course of trade before they find their final destination. This movement gives rise to a number of plant health problems, particularly when seed merchants are reluctant to disclose the origin of the seeds which they are seeking to market abroad. The FAO Certificate calls for this important information and yet often this section is left blank on the application form submitted to the seed tester. Thus, neither the exporting nor the receiving plant protection service can be on their guard against the pests and diseases which are associated with the parts of the world in which the seeds were grown. Nor, under these circumstances, can they be assured that the seeds were produced in countries where the authorities take precautions to limit infection. Furthermore, the mixture of seed lots of different origin has an important bearing on the issue of a certificate and this kind of information is vital to proper certification.'

Another problem of importance is that of seed treatment, such as treatment with hot water, chemicals or fumigants. The EPPO/ISTA Working Party emphasised the need 'for all plant protection services to insist that all application forms for an export plant health certificate should include a declaration of the origin of seeds and a clear statement of any special treatment given to them, although not necessarily in great detail'.

In view of the need to keep the FAO Certificate as simple as possible and free from additional statements, there was, however, no unanimity among the members of the Working Party as to whether this information should be stated on the Certificate itself. It should, however, be noted that it is recommended for inclusion on the Certificate by the Inter-African Phytosanitary Commission.

With respect to seed lots treated against infection and contaminations, it should be clear that no declaration on such a treatment should be taken as a guarantee that the seed lot concerned is free of diseases and pests.

At the Plant Protection Service in Copenhagen, in a considerable number of cases cabbage seed lots that have been rejected on account of *Phoma lingam* have been submitted to hot water treatment and then retested, and sometimes again rejected because the pathogen could still be detected in the treated seed lot. In seed destined for certain countries, *P. lingam* must not occur and mere reduction of the seed-borne infection is not enough to allow the honest issue of a Rome Certificate; the seed must be entirely free from the parasite as far as detectable by the test.

There are many other examples of treated seed lots still containing viable infection; consequently, treated seed must be tested on its own merit.

A third important item studied by the EPPO/ISTA Working Party of 1965 concerned declaration by the exporter of information concerning the inspection for domestic purposes of seed crops during active growth. The Report says:

'There can be no doubt that a tester is provided with much valuable phytosanitary information if he knows that the seeds have passed through the safeguards of an appropriate domestic certification scheme. Indeed, if the technical demands of importing countries could be linked to such schemes it might even be possible to meet them by requiring exporters to supply seeds from certain categories of certified stocks. But having resource to the detailed field reports of seeds grown in domestic certification schemes, while technically useful, could lead to commercial complications, for all countries exporting seeds do not have domestic certification schemes. Even when these schemes do exist, the regulations governing them are seldom the same and thus no comparable information is available on potential causes of rejection. We are, however, of the opinion that, for some organisms which can only be detected in the field, any seed which is likely to be carrying such pathogens obnoxious to the prospective importing country should be required to be derived from a recognised and appropriate domestic certification scheme and that the would-be seed exporter should be required to declare the identity of the seed in relation to such a scheme and supply details to the plant protection service when applying for an export certificate.'

Appendix I Regional Plant Protection Organisations

EPPO	European and Mediterranean Plant Protection Organisation
IAPSC	Inter-African Phytosanitary Council
OIRSA	Organismo Internacional Regional de Sanidad Agropecuaria
SEAPPPC	Plant Protection Committee for the South-East Asia and Pacific Region
NEPPC	Near East Plant Protection Commission
CIPA	Comité Interamericano de Protection Agricola
OBSA	Organismo Bolivariano de Sanidad Agropecuaria
CPPC	Caribbean Plant Protection Commission

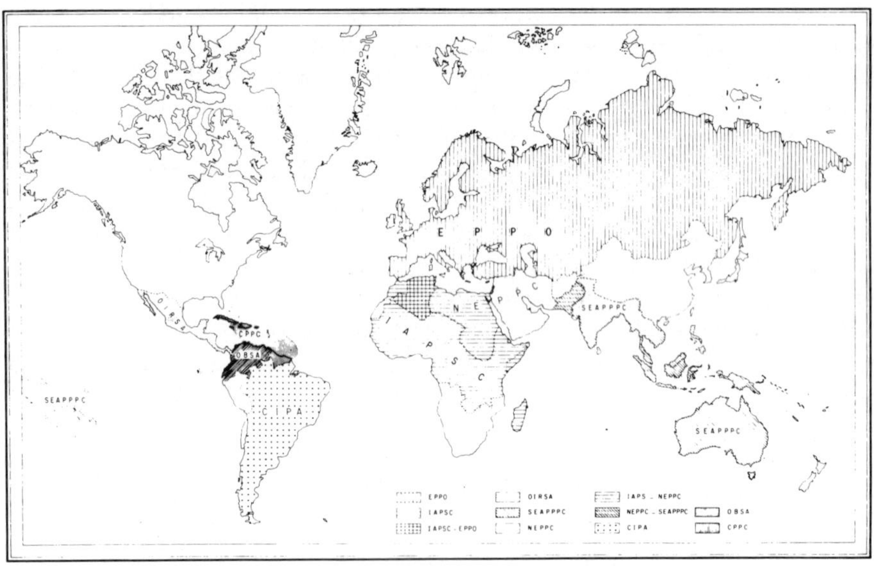

Fig. A.1. Map of regional plant protection organisations. (Reproduced from Caresche *et al.*, 1969)

Appendix II International Certificates Applicable for Certification of the Health Condition of Seed

Certification for plant quarantine

(1) *Rome Certificate*. The international standard quarantine certificate, set up in accordance with the FAO Model Phytosanitary Certificate. English version of the certificate as used by the Danish Plant Protection Service.

Certification for quality of seed

(2) *Orange International Seed Lot Certificate*. The international standard seed quality certificate, issued by the International Seed Testing Association, and to be used in accordance with the International Rules for Seed Testing. Front page and back of the certificate (front page, 1976 edition; back page, 1966 edition).

(3) *Green International Seed Lot Certificate*. A modification of the Orange Certificate, used for seed lots duly sampled and sealed in one country, but tested and certified in another country (1966 edition).

(4) *Blue International Seed Sample Certificate*. Used in cases where the Orange or Green Certificates cannot be used, the result reported referring only to the sample tested (1966 edition).

The back of the Green and Blue Certificates carries a text identical with that of the Orange Certificate, except for the keywords 'Green' and 'Blue' and a few pertinent details.

Note on the reproduction of the three ISTA certificates

For technical reasons these certificates have not been photomechanically reproduced from the original certificates; they have been composed using a typography resembling as closely as possible that applied in the original certificates (reproduction by permission of the International Seed Testing Association).

The certificates may be used *only* in the form issued by ISTA and only by authorisation of this organisation, from which they must be purchased.

0

Plant Protection Service
DENMARK

Phytosanitary Certificate

OF THE INTERNATIONAL
PLANT PROTECTION CONVENTION 1951

A Nr. 39302

ORIGINAL

This is to certify

that the plants, parts of plants or plant products described below or representative samples of them were thoroughly examined on .... by ....

an authorized officer of the Plant Protection Service, Denmark, and were found to the best of his knowledge to be substantially free from injurious diseases and pests; and that the consignment is believed to conform with the current phytosanitary regulations of the importing country both as stated in the additional declaration hereon and otherwise.

Fumigation or disinfection treatment:

Date: .... Treatment: ....

Duration of exposure: .... Chemical and concentration: ....

Additional declaration:

Issued by Statens plantetilsyn, Gersonsvej 13, 2900 Hellerup, ...

STATENS PLANTETILSYN

Erling Jacobsen

ERLING JACOBSEN
Head, Government Plant Protection Service

Description of the Consignment

Name and address of exporter: ....

Name and address of consignee: ....

Number and description of packages: ....

Distinguishing marks: ....

Origin: ....

Means of conveyance: ....

Point of entry: ....

Quantity and name of produce: ....

Please turn over

Botanical name: ....

Stamp of Station
Cachet de la Station
Stempel der Station

ORANGE INTERNATIONAL SEED LOT CERTIFICATE
BULLETIN ORANGE INTERNATIONAL DE LOT DE SEMENCES
INTERNATIONALER ORANGER BERICHT ÜBER EINE SAATGUTPARTIE

(See back — Voir au verso — Rückseite beachten)

STATED BY APPLICANT – INFORMATIONS DU REQUERANT – ANGABEN DES ANTRAGSTELLERS

Without responsibility of the station — Sans responsabilité de la station — Ohne Verantwortung der Saatgutprüfstelle

Name of applicant
Nom du requérant . . . . :
Name des Antragstellers

Species, cultivar (variety), weight of lot etc.
Espèce, cultivar (variété), poids du lot etc. . . . . . :
Art, Cultivar (Sorte), Gewicht der Partie usw.

OFFICIAL INFORMATION – INFORMATIONS OFFICIELLES – AMTLICHE ANGABEN

Issuing station
Station qui délivre le bulletin . . . . :
Ausstellende Saatgutprüfstelle

Sampling and sealing agency
Organisme d'échantillonnage et de plombage . . . :
Probenahme- und Plombierungsstelle

Official marks of lot
Marques officielles du lot . . . . :
Amtliche Kennzeichnung der Partie

Official seal of lot
Plomb officiel du lot . . . . :
Amtliche Plombe der Partie

Number of containers Nombre de contenants Anzahl der Behälter	Date of sampling Echantillonnage effectué le Datum der Probeziehung	Date sample received Echantillon reçu le Eingangsdatum der Probe	Date test concluded Analyse terminée le Datum des Prüfungsabschlusses	Test number No. de l'analyse Untersuchungs-Nr.

ANALYSIS RESULTS – RESULTATS DE L'ANALYSE – UNTERSUCHUNGSERGEBNISSE

SPECIES — ESPECE — ART:
Latin name — Nom latin — Lat. Name

PURITY — PURETE — REINHEIT			GERMINATION — KEIMFÄHIGKEIT						MOISTURE CONTENT (wet basis) TENEUR EN EAU (poids humide) FEUCHTIGKEITSGEHALT (Frischeinwaage) %
% Weight — % en poids — % Gewicht				% Number — % en nombre — % Anzahl					
Pure seeds Semences pures Reine Samen	Inert matter Matières inertes Unschädl. Verunreinigungen	Other seeds Semences d'autres plantes Andere Samen	Number of days Nombre de jours Anzahl der Tage	Normal seedlings Germes normaux Normale Keimlinge	Hard seeds*) Graines dures*) Harte Körner*)	Fresh ungerminated seeds Graines fraîches non germées Frische nicht gekeimte Samen	Abnormal seedlings Germes anormaux Anomale Keimlinge	Dead seeds Semences mortes Tote Samen	
1	2	3	4	5	6	7	8	9	10

Kind of inert matter
Nature des matières inertes
Art der unschädl. Verunreinigungen

Species Latin names Espèces Noms latins Arten Lat. Namen	Other seeds Semences d'autres plantes Andere Samen

OTHER/AUTRES DETERMINATIONS – WEITERE UNTERSUCHUNGSERGEBNISSE

(See also back — Voir aussi au verso — **Siehe auch Rückseite**)

Place and country — Localité et pays — Ort und Staat	Date — Datum	Signature — Unterschrift
*) Only for seeds of / Seulement dans le cas de / Nur für Samen von { Gossypium, Hibiscus, Leguminosae }	N = { Not tested / Non déterminé / Nicht geprüft } — Tr. = { Trace / Spuren }	See declaration on back — Voir déclaration au verso Siehe Erklärung auf der Rückseite

DECLARATION OF THE SIGNATORY — DECLARATION DU SIGNATAIRE — ERKLÄRUNG DURCH DEN UNTERZEICHNETEN

I certify that sampling, sealing and testing have been carried out in accordance with the International Rules for Seed Testing of the ISTA and that the tests have been made at an Official Station authorized by the International Seed Testing Association to issue International Certificates.

Je, soussigné, certifie que l'échantillonnage, le plombage et les essais ont été effectués conformément aux Règles Internationales de l'ISTA pour les essais de semences dans une Station Officielle autorisée, par ladite Association Internationale d'Essais de Semences, à délivrer des Bulletins d'Analyse Internationaux.

Ich bestätige hiermit, dass Probeziehung, Plombierung und Untersuchungen in Übereinstimmung mit den Internationalen Vorschriften der ISTA für die Prüfung von Saatgut von einer staatlichen Saatgutprüfstelle durchgeführt worden sind. Diese Saatgutprüfstelle ist von der Internationalen Vereinigung für Saatgutprüfung zur Ausstellung von Internationalen Saatgut-Untersuchungsberichten ermächtigt.

ADDITIONAL OBSERVATIONS — OBSERVATIONS COMPLEMENTAIRES — ZUSÄTZLICHE BEMERKUNGEN

DEFINITIONS:

Number determinations — endorsements of test results have the following meanings:
Complete test. An examination of the full quantity of seed shown in Table 2A, column 5, of the International Rules for Seed Testing in which all other species found to occur are reported.

Limited test. An examination of the full quantity of seed shown in Table 2A, column 5, of the International Rules for Seed Testing in which only those species designated by the applicant are reported.

Reduced test. An examination of a quantity of seed less than that shown in Table 2A, column 5, of the International Rules for Seed Testing but in which all other species found to occur are reported.

The above terms should be used separately or, where appropriate, in combination.

Dénombrements — les termes utilisés dans l'expression des résultats d'analyse ont le sens suivants:
Essai complet. Essai pour lequel la totalité de l'échantillon d'analyse prescrit dans la colonne 5 du Tableau 2A des Règles Internationales pour les Essais de Semences est utilisé pour la recherche des semences de toutes les autres espèces.

Essai limité. Essai pour lequel la totalité de l'échantillon d'analyse prescrit dans la colonne 5 du Tableau 2A des Règles Internationales pour les Essais de Semences est utilisé pour une recherche limitée aux seules espèces nommément désignées par le requérant.

Essai sur échantillon réduit. Essai pour lequel une quantité de semences inférieure à celle prescrite dans la colonne 5 du Tableau 2A des Règles Internationales pour les Essais de Semences est utilisée pour la recherche des semences de toutes les autres espèces.

Les termes ci-dessus doivent être utilisés séparément ou, le cas échéant, conjointement.

Zahlenmässige Bestimmungen — die Bestätigungen der Untersuchungsergebnisse haben folgende Bedeutung:
Vollständige Untersuchung. Eine Untersuchung der gesamten in Tabelle 2A, Spalte 5 der Internationalen Vorschriften für die Prüfung von Saatgut angegebenen Saatgutmenge, bei welcher alle anderen darin gefundenen Arten berichtet werden.

Begrentze Untersuchung. Eine Untersuchung der gesamten in Tabelle 2A, Spalte 5 der Internationalen Vorschriften für die Prüfung von Saatgut angegebenen Saatgutmenge, bei welcher lediglich die vom Antragsteller bezeichneten Arten berichtet werden.

Abgekürzte Untersuchung. Eine Untersuchung einer geringeren Saatgutmenge als in Tabelle 2A, Spalte 5 der Internationalen Vorschriften für die Untersuchung von Saatgut angegeben, bei welcher jedoch alle anderen darin gefundenen Samen berichtet werden.

Die obengenannten Ausdrücke sollen getrennt voneinander, oder, wo angängig, in Verbindung miteinander verwendet werden.

CERTIFICATES — BULLETINS — BERICHTE

The certificates should be typed and if the result of a determination is negative it should be entered as —0—, otherwise the results should be entered as —7— or —10—. If a determination has not been made the words "Not tested" (abbreviated "N") should be entered in the space provided.

No attention should be paid to a certificate unless it contains the stamp of the issuing station and the signature of the director or his assignee as well as the test number and date of issue. No certificate should be accepted which shows signs of amendment, alteration or erasure.

If a different result is found and confirmed in a subsequent test the issuing station cannot be held liable for compensation.

The analysis results on the Orange International Seed Lot Certificate relate only to the seed lot as a whole at the time of sampling.

Un Bulletin doit être, autant que possible, dactylographié. Si le résultat de certaines déterminations est négatif, l'espace prévu pour l'indication du résultat doit être ainsi rempli: —0—, sinon, les résultats doivent être ainsi indiqués: —7— ou —10—. Si on n'a pas effectué certaines déterminations, l'espace prévu pour l'indication du résultat doit être rempli par les mots: "Non déterminé" (en abrégé "N").

Doit être considéré comme sans valeur un Bulletin où ne figurent pas le No. d'analyse, le cachet de la Station qui l'a délivré, la date de sa délivrance, et la signature du directeur ou de son délégué.

Un Bulletin présentant des signes de surcharges ou de corrections, ou des ratures, ne doit pas être accepté.

Si un résultat d'analyse différent est ultérieurement obtenu et confirmé, la Station qui a délivré le Bulletin ne peut être tenue pour passible d'une indemnité.

Der Untersuchungsbericht soll mit Maschine geschrieben sein. Wenn das Ergebnis einer Analyse negativ ist, soll die entsprechende Rubrik wie folgt ausgefüllt werden: —0—. Alleinstehende Ziffern in folgender Weise: —7— oder —10—. Nicht durchgeführte Analysen werden in der betreffenden Rubrik durch die Worte "Nicht geprüft" (abgekürtz "N") gekennzeichnet.

Atteste, in denen Untersuchungsnummer, Datum der Probeziehung, Siegel der Station oder Ausstellungsdatum und Unterschrift des verantwortlichen Leiters fehlen, sind ungültig. In gleicher Weise ungültig sind Atteste, welche Ausbesserungsspuren, Abänderungen oder Radierungen erkennen lassen.

Wird zu einem späteren Zeitpunkt ein von den vorstehenden Werten abweichendes Resultat ermittelt und bestätigt, so kann die Saatgutprüfstelle nicht haftbar gemacht werden.

Stamp of Station
Cachet de la Station
Stempel der Station

INTERNATIONAL SEED LOT CERTIFICATE
BULLETIN INTERNATIONAL DE LOT DE SEMENCES
INTERNATIONALER BERICHT ÜBER EINE SAATGUTPARTIE

(See back — Voir au verso — Rückseite beachten)

STATED BY APPLICANT – INFORMATIONS DU REQUERANT – ANGABEN DES ANTRAGSTELLERS

Without responsibility of the station — Sans responsabilité de la station — Ohne Verantwortung der Saatgutprüfstelle

Name of applicant
Nom du requérant . . . . :
Name des Antragstellers

Species, cultivar (variety), weight of lot etc.
Espèce, cultivar (variété), poids du lot etc. . . . . :
Art, Cultivar (Sorte), Gewicht der Partie usw.

OFFICIAL INFORMATION – INFORMATIONS OFFICIELLES – AMTLICHE ANGABEN

Issuing station
Station qui délivre le bulletin . . . . :
Ausstellende Saatgutprüfstelle

Sampling and sealing agency
Organisme d'échantillonnage et de plombage . :
Probenahme- und Plombierungsstelle

Official marks of lot
Marques officielles du lot . . . . :
Amtliche Kennzeichnung der Partie

Official seal of lot
Plomb officiel du lot . . . . :
Amtliche Plombe der Partie

Number of bags Nombre de sacs Anzahl der Säcke	Date of sampling Echantillonnage effectué le Datum der Probeziehung	Date sample received Echantillon reçu le Eingangsdatum der Probe	Date test concluded Analyse terminée le Datum des Prüfungsabschlusses	Test number No. de l'analyse Untersuchungs-Nr.

ANALYSIS RESULTS – RESULTATS DE L'ANALYSE – UNTERSUCHUNGSERGEBNISSE

SPECIES — ESPECE — ART:
Latin name — Nom latin — Lat. Name

PURITY — PURETE — REINHEIT				GERMINATION — KEIMFÄHIGKEIT						MOISTURE CONTENT (wet basis) TENEUR EN EAU (poids humide) FEUCHTIGKEITSGEHALT (Frischeinwaage) %
% Weight — % en poids — % Gewicht					% Number — % en nombre — % Anzahl					
Pure seeds Semences pures Reine Samen	Inert matter Matières inertes Unschädl. Verunreinigungen	Other crop seeds Semences d'autres plantes cultivées Fremde Kultursamen	Weed seeds Graines de mauvaises herbes Unkrautsamen	Number of days Nombre de jours Anzahl der Tage	Normal seedlings Germes normaux Normale Keimlinge	Hard seeds*) Graines dures*) Harte Körner*)	Fresh ungerminated seeds Graines fraîches non germées Frische nicht gekeimte Samen	Abnormal seedlings Germes anormaux Anomale Keimlinge	Dead seeds Semences mortes Tote Samen	
1	2	3	4	5	6	7	8	9	10	11

Kind of inert matter
Nature des matières inertes
Art der unschädl. Verunreinigungen

Species Latin names Espèces Noms latins Arten Lat. Namen	
	Other crop seeds Semences d'autres plantes cultivées Fremde Kultursamen
	Weed seeds Graines de mauvaises herbes . . : Unkrautsamen

OTHER/AUTRES DETERMINATIONS — WEITERE UNTERSUCHUNGSERGEBNISSE

(See also back — Voir aussi au verso — Siehe auch Rückseite)

Place and country — Localité et pays — Ort und Staat	Date — Datum	Signature — Unterschrift
*) Only for seeds of / Seulement dans le cas de / Nur für Samen von { Gossypium, Hibiscus, Leguminosae } — N = { Not tested, Non déterminé, Nicht geprüft } — Tr. = { Trace, Spuren }		See declaration on back — Voir déclaration au verso Siehe Erklärung auf der Rückseite

D — No. 3 G.I.C.

Stamp of Station
Cachet de la Station
Siegel der Station

INTERNATIONAL SEED SAMPLE CERTIFICATE
BULLETIN INTERNATIONAL D'ECHANTILLON DE SEMENCES
INTERNATIONALER BERICHT UBER EINE SAATGUTPROBE
(See back — Voir au verso — Rückseite beachten)

STATED BY APPLICANT – INFORMATIONS DU REQUERANT – ANGABEN DES ANTRAGSTELLERS

Without responsibility of the station – Sans responsabilité de la station – Ohne Verantwortung der Saatgutprüfstelle

Name of applicant
Nom du requérant . . . . :
Name des Antragstellers

Species, cultivar (variety), etc.
Espèce, cultivar (variété), etc . . . . :
Art, Cultivar (Sorte) usw.

Weight of lot — Poids du lot Gewicht der Partie	Number of bags — Nombre de sacs Anzahl der Säcke	Date of sampling — Echantillonnage effectué le Datum der Probeziehung

OFFICIAL INFORMATION – INFORMATIONS OFFICIELLES – AMTLICHE ANGABEN

Issuing station
Station qui délivre le bulletin . . . . :
Ausstellende Saatgutprüfstelle

Seal of sample — Plomb de l'échantillon — Plombe der Probe	Date sample received — Echantillon reçu le — Eingangsdatum der Probe	Date test concluded — Analyse terminée le Datum des Prüfungsabschlusses	Test number — No. de l'analyse Untersuchungs-Nr.

ANALYSIS RESULTS – RESULTATS DE L'ANALYSE – UNTERSUCHUNGSERGEBNISSE

SPECIES — ESPECE — ART:
Latin name — Nom latin — Lat. Name

PURITY — PURETE — REINHEIT				GERMINATION — KEIMFÄHIGKEIT						MOISTURE CONTENT (wet basis) TENEUR EN EAU (poids humide) FEUCHTIGKEITSGEHALT (Frischeinwaage) %
% Weight — % en poids — % Gewicht					% Number — % en nombre — % Anzahl					
Pure seeds Semences pures Reine Samen	Inert matter Matières inertes Unschädl. Verunreinigungen	Other crop seeds Semences d'autres plantes cultivées Fremde Kultursamen	Weed seeds Graines de mauvaises herbes Unkrautsamen	Number of days Nombre de jours Anzahl der Tage	Normal seedlings Germes normaux Normale Keimlinge	Hard seeds*) Graines dures*) Harte Körner*)	Fresh ungerminated seeds Graines fraîches non germées Frische nicht gekeimte Samen	Abnormal seedlings Germes anormaux Anomale Keimlinge	Dead seeds Semences mortes Tote Samen	
1	2	3	4	5	6	7	8	9	10	11

Kind of inert matter
Nature des matières inertes . . . . :
Art der unschädl. Verunreinigungen

Species Latin names Espèces Noms latins Arten Lat. Namen	
	Other crop seeds Semences d'autres plantes cultivées Fremde Kultursamen
	Weed seeds Graines de mauvaises herbes . . : Unkrautsamen

OTHER/AUTRES DETERMINATIONS — WEITERE UNTERSUCHUNGSERGEBNISSE

(See also back — Voir aussi au verso — Siehe auch Rückseite)

Place and country — Localité et pays — Ort und Staat	Date — Datum	Signature — Unterschrift
		See declaration on back — Voir déclaration au verso Siehe Erklärung auf der Rückseite

*) Only for seeds of / Seulement dans le cas de / Nur für Samen von { Gossypium, Hibiscus, Leguminosae } | N = { Not tested / Non déterminé / Nicht geprüft } | Tr. = { Trace / Spuren }

No. 2 B.I.C.

PART IV
Seed Health Testing Methods

25 Objectives of Seed Health Tests

25.1 BASIC REQUIREMENTS

Routine methods to be adopted for seed health testing must fulfil the following demands of efficiency and economy:

(1) A test must give reliable information pertaining to field performance and quarantine requirements.
(2) The results must be *reproducible* within statistical limits.
(3) The *time*, *labour* and *equipment* for carrying through a test must be kept within economic limits.
(4) In tests requiring incubation the results must be made available quickly.

25.2 OBJECTIVES OF SEED HEALTH TESTS

Ultimately, the objectives of seed health testing are related to the actual policy towards seed improvement, seed trade and plant protection. The testing technique will depend invariably on the importance of the pathogen carried in the seed, and the disease potential assigned to this pathogen in a given situation.

The principal seed health tests and some related factors are summarised in Table 25.1, according to objectives of certification.

Testing for quarantine. Closed quarantine in greenhouses especially equipped for post-entry growing-on tests is used for examining particularly valuable seed material in order to exclude dangerous pathogens. This seed health testing procedure is very expensive and is limited to valuable seeds (Subsection 24.4.2.1).

For seeds to be certified by an exporting country as 'substantially free' from a particular pathogen, a sufficiently large representative sample must be taken and the testing procedure must be sufficiently sensitive to reveal even traces of infection. The agar plate test is more sensitive than the blotter test for many fungi, e.g. for *Fusarium oxysporum* f.sp. *lini* in

Table 25.1. Seed health testing in plant disease and seed quality control

Principle of disease control	*Purpose of seed health testing*	*Principal tests applicable*	*Certificate*
Exclusion of inoculum. Health checked before release	Testing for quarantine	Growing-on test in closed quarantine Indicator test	Post-entry control (FAO International Phytosanitary Certificate)
Exclusion of inoculum, sampling being accepted	Testing for quarantine	Blotter test Agar test Growing-on test Indicator test Embryo count test Serological test	FAO International Phytosanitary Certificate
Sanitation: Eradication or reduction of inoculum, tolerances being accepted	Improvement of seed stocks in certification schemes	Blotter test Agar test Growing-on test Indicator test Embryo count test Serological test	National organisations of Certification Schemes, ISTA Orange Certificate
	Testing of planting value of seed for farmers	Blotter test Agar test (Growing-on test) (Indicator test) (Seedling-symptom test) Embryo count test Serological test	ISTA Orange Certificate. For export seed lots often: FAO International Phytosanitary Certificate
Sanitation: Controlled eradication of inoculum	Testing for advisability of seed treatment, or of treated seed	Blotter test Agar test Seedling-symptom test	National certification on advisory or obligatory basis
Quality control of produce	Evaluation of quality of grain, etc., for consumption	Agar test (e.g. for storage fungi) Blotter test	National certification on advisory or obligatory basis

flax. To meet demands of importing countries that seed shall be free from specified pathogens, special selective procedures may be preferable rather than the versatile, but less sensitive standard tests. Testing by indicator plants may provide a high degree of security as to freedom from infection, and such indicator tests may be established as a regular part of national certification schemes. Seed certified under such schemes may be accepted without further testing by the inspection services for quarantine purposes.

A third category of pathogens observed in inspection of seeds for quarantine are those seed-borne pathogens that have not been specified in the quarantine regulations of importing countries requiring a general phytosanitary certificate. These pathogens, of very varying importance, are not quarantine objects in the strict sense, but rather pathogens affecting the planting value of individual seed lots. Interpretation of results, especially when the testing techniques are not specified, may vary substantially from country to country. Therefore, tolerances may be accepted at the discretion of the certifying agency of the exporting country. For this category of pathogens, versatile standard procedures are appropriate. In fact, today most quarantine seed inspection is devoted to these kinds of pathogen (Neergaard, 1965). The tolerances traditionally accepted by inspection services vary, and international standards have not been agreed upon (Neergaard, 1962*a*; EPPO, 1966).

Testing for certification schemes. The purpose of seed health certification schemes for improvement of seed stocks is elimination of pathogens; consequently, testing procedures applied in this connection must aim at maximum recording. This suggests use of selective procedures. Indicator tests capable of revealing traces should also be taken into consideration, particularly for virus diseases. For many fungi the agar plate test may reveal traces more readily than the blotter test. But this is not always so, as exemplified by *Pyricularia oryzae* in rice. The agar seedling-symptom test for *Xanthomonas campestris* is a sensitive selective procedure suitable for seed health certification schemes for cabbage and other crucifers.

Testing for evaluation of planting value. In assessing the planting value, a seed health test is comparable with a germination test. The aim of both is to forecast field emergence and to some extent the health of the mature crop. However, assessing the effect of pathogens is extremely complex because of innumerable and often unpredictable factors in the field. High correlation between laboratory results and field performance may be difficult to obtain from tests of a few seed lots, but clearer when many test results are subjected to statistical analysis.

Nevertheless, the correlation between laboratory and field results depends on two major factors: the testing procedure, and the transmission and rate of increase of the pathogen.

The first problem, selection of methods predicting field performance in the seedling stage rather than maximum results in laboratory, has been discussed by De Tempe (1963*a*, 1963*b*) and by Limonard (1968). Appropriate selection of methods depends primarily on local conditions. 'The presence or absence of the pathogen is in many cases more important than exact figures' (Limonard).

The second major problem, that of transmission and multiplication rate, is subject to all the factors affecting epiphytotics. Most simple in this respect are seed-borne systemic diseases such as barley loose smut. There are examples of complete correlation between percentage of seed infection and percentage of yield reduction for these diseases. This is indeed an ideal situation for disease forecasting. Another such disease is barley leaf stripe, *Drechslera graminea*, for which the ratio between percentage of infected seeds and percentage of yield reduction is close to 1:1. Pathogens that cause true 'explosive' epidemics constitute the other extreme and depend entirely on climatic or other conditions, e.g. bacterial blights of bean, carrot, cotton and rice. One or two infected seeds in ten thousand may give rise to total loss of the crop. For such pathogens there should be zero tolerance. Between these extremes there is a large range of pathogens in which the rate of increase varies according to potentialities and conditions.

Testing for advisability of seed treatment. The outcome of a test presents three possibilities: (1) the seed is suitable for sowing without treatment; (2) the seed can be used after treatment according to prescription; and (3) the seed is unsuitable for sowing. For many years a testing scheme for peas has been carried through in Ottawa, Canada, according to these three choices. In recent years, an extensive scheme for testing cereals for seedling pathogens with assessment of the need for seed treatment has been carried out in Sweden. In that country, treatment of seeds with mercuric compounds is permitted only when specified by the Government Seed Testing Station. The Swedish procedure, based on the seedling-symptom test, was introduced in order to counteract the hazards to wildlife brought about by extensive use of mercury compounds for seed treatment.

Testing of treated seeds. There is a growing interest in checking the health condition of seeds treated by chemicals or other means. The experience gained through the international comparative seed health tests organised by the ISTA Committee on Plant Diseases, with commercially treated seeds, has demonstrated the justification for this kind of test.

Obviously, testing of treated seeds is particularly important in connection with inspection of seeds for export in accordance with quarantine regulations. Modification of standard techniques must be considered, especially extension of the period of incubation.

Testing seeds for storage quality or for feeding. The quality of seeds is highly dependent on the presence and dominance of certain microorganisms which may be detrimental to its keeping qualities as well as its value for feeding and human consumption. The role of storage fungi is discussed in Chapter 7. Testing is vital in predicting storage behaviour of grain lots, and determining the extent of actual damage. The tests should include examination for moisture content and for injuries of pericarp and seed coat (Subsection 7.6.2). Injuries may provide invasion sites for storage fungi.

Testing seeds for resistance of cultivar. Seeds of cultivars resistant to a certain pathogen may be tested by the indicator test to confirm that the seed is, in fact, of the suggested resistant cultivar (Section 30.1).

Most of the routine methods in use have been developed empirically. The first seed health testing laboratory was founded in 1919 at the Government Seed Testing Station in Wageningen. Since 1957 the Committee on Plant Diseases of the International Seed Testing Association (ISTA) has organised annual international comparative seed health testing schemes, based on selected methods used by all participants according to directions. Furthermore, annual workshops are organised, during which the results are discussed and duplicate samples submitted to joint laboratory studies of the participants, who are familiar with the samples and the technique from their co-operative home studies. These activities have contributed in no small measure to standardising some methods and to clarifying problems and difficulties pertaining to the techniques as well as to problems of personal skill and practice.

The methods, which are briefly discussed and evaluated in Chapters 27–31, may be grouped into the following principal categories:

(1) Direct inspection of the seed with unaided eye, hand lens or under low power stereoscopic microscope. The examination may be assisted by such simple procedures as cutting through individual seeds to see insects, submerging seeds in water drops or placing them for a few hours or one day on moist filter paper to obtain swelling of pycnidia, etc.

(2) Microscopic examination of suspensions obtained by the washings from the seed.

(3) Microscopic examination after application of a clearing and/or staining procedure.

(4) Examination of seeds and seedlings after incubation of the seed sown on blotter, within blotting paper, on agar media or any other essentially sterile media.

(5) Examination of symptoms developed on seedlings grown in soil, sand or similar material.

(6) Examination based on growing-on tests, carried out in greenhouse, environment-controlled chamber or in the field.

(7) Tests based on biochemical reactions (serological tests).

(8) Tests performed as bioassays (indicator test, phage plaque test).

(9) Tests under field conditions.

25.3 EQUIPMENT

25.3.1 Basic equipment

(1) *Microscopes*. It is imperative to provide good microscopes: *stereoscopic binocular microscopes* with magnification up to 60× providing stepwise increase, rather than the zoom system, experience having shown the advantage of knowing the precise magnification, and fitted with the two lamps (Subsection 29.5.5.1 and Fig. 29.4). These lamps should provide cool light, and the light should come from two sides to avoid shadows on seeds. Also, a good hand lens is indispensable, and a very useful aid is a magnifier as used by a watchmaker. Good *binocular compound microscopes* for higher magnification are needed for detailed studies of microorganisms. They should have a fitting for a camera, and a drawing apparatus should also be available.

(2) *Incubation chambers*. These can be either compact standard units purchasable from different firms or they may be arranged as walk-in chambers with thermostatic controls including cooling and heating (Fig. 29.3). The latter type if preferable. Seed health testing should be performed at standardisation incubation temperatures, 20°C, for many tropical crops at 25–28°C.

(3) *Light*. To stimulate sporulation the above chambers should be fitted with near ultraviolet light tubes (Philips Black Light lamps TL 40W/08), 2 tubes hanging horizontally 20 cm apart from each other at a distance of 40 cm above the petri dishes according to international standard. If black light lamps are not available, cool white fluorescent daylight tubes (for example Philips TLF 40W/34 De Luxe) have proved quite efficient (Sub section 29.5.4.6). Light periods can be controlled by automatic time switch. A 12 h light exposure each day is, at present, international standard.

(4) *Containers*. Seeds can be incubated in any kind of container, petri dishes being in general use. If the incubation room is fitted with near ultraviolet lamps, plastic or pyrex petri dishes and containers should be used, as only they allow near ultraviolet light to pass.

(5) *Glasshouse or controlled environment growth room* is required to test for presence of certain organisms that are not detected on seedlings. Various pathogens on imported seed material, including obligate parasites and viruses, may be detected under these conditions.

(6) *Media*. Clean blotters (absorbent paper or paper towelling), preferably produced and packed under sterile conditions, are needed; for the agar plate method, potato dextrose agar and malt extract agar are suitable standard media.

(7) *Additional equipment*. A centrifuge operating at about 3,000 rev/min, and a mechanical shaker, should be available for the washing test to obtain suspensions of surface-borne spores, etc., by a standardised

procedure. A haemocytometer to be used for counting of spores per unit of liquid as the basis for calculating the numbers of spores per seed or per gramme of seed. An autoclave for preparing sterile media. A vacuum counter for quick counting of seeds to be sown on blotter (Fig. 28.2). Small sundry items: glass slides, cover slips, beakers, flasks, petri dishes, culture tubes, forceps, pencils for marking on glass and wet paper; plastic cups may be used for working samples of seed.

25.3.2 Special equipment

Special equipment is needed for sampling (Chapter 26) and for special tests, such as for the whole embryo count procedure for loose smuts of barley and wheat (Section 27.4), for the washing and sedimentation procedure for detection of stem eelworm (Section 27.3), for the standard soil test, see Fig. 28.8, and perhaps for field trials (Section 31.2). In particular, special equipment is needed for bacteriological and virological investigations, and manuals dealing with techniques of identifying bacteria and viruses should be consulted; e.g. for bacteria those of Stapp (1958), Dowson (1957), Commonwealth Mycological Institute (1968), Harrigan and McCance (1966), and for viruses those of Bawden (1964), Corbett and Sisler (1964), Fritzsche *et al.* (1972), Klinkowski (1974) and Phatak (1974), and for both categories, Society of American Bacteriologists (1957).

26 Sampling

Practically any health test of a seed lot is based on examination of a sample which usually is an extremely small portion of the seed lot. The procedures applied in drawing a representative sample and in dividing this into working samples for analysis are therefore fundamental in order to obtain uniform, accurate and reproducible results. The accumulated experience from practice and research has been laid down in the International Rules for Seed Testing (ISTA, 1966*a*, 1976). In drawing samples for seed health tests, these rules should be followed strictly. For details on sampling techniques, reference is made to the chapter on sampling of the rules. The following is a summary, partly in quotations, of this chapter.

The seed lot. For any test carried out in accordance with the International Rules for Seed Testing, the maximum size of a seed lot of agricultural and horticultural seed is 20,000 kg for seeds of the size of wheat or larger, and 10,000 kg for seeds smaller than wheat.

For tree seeds the maximum is 5,000 kg for the size of *Fagus* spp. or larger, and 1,000 kg for seeds smaller than *Fagus* spp.

A seed lot should be homogeneous but this ideal can never be quite reached. In spite of careful bulking and mixing it is impossible in practice to obtain a completely homogeneous seed lot. Uniformity of a seed lot refers to any attribute that can be measured in a test, e.g. purity, number of weed seeds, germination.

The seed sample. Several individual samples are drawn from different parts of the bulk. Each of these probes is called a *primary sample*. These are combined to make the *composite sample*, which usually is much bigger than required, and must be reduced. The properly reduced sample is called the *submitted sample,* which is the sample submitted to the testing station.

From the submitted sample a reduced sample must be obtained for the actual test. This subsample is called the *working sample*.

Techniques of drawing submitted samples. 'In drawing samples, approximately equal amounts of seed shall be taken from each container (bag, etc.) sampled, or from each place in such container, or, when sampling seed in bulk (bins, cars, wagons, etc.), from each location sampled.' . . . 'Suitable sampling instruments must be used when sampling in bulk and in containers. Samples may also be drawn directly from the seed stream during the cleaning or bagging operation with a suitable instrument, by hand or automatically.'

In the case of seeds which flow with difficulty, such as chaffy seed, it is sometimes necessary to draw the sample by hand.

The sampling instruments described in the International Rules for Seed Testing are (1) the stick trier or sleeve-type trier, and (2) the Nobbe trier.

The *stick* or *sleeve-type trier* 'consists of a hollow brass tube inside a closely fitting outer shell or sleeve which has a solid pointed end. The tube and sleeve have open slots in their walls so that when the tube is turned until the slots in the tube and sleeve are in line, seeds can flow into the cavity of the tube, and when the tube is given a half turn the openings are closed. The tubes vary in length and diameter, being designed for different kinds of seeds and various sizes of containers, and are made with or without partitions.

'The stick triers may be used for most seeds with the exception, perhaps, of some very chaffy species. Up to a certain diameter of tube it may be used through sack walls of coarsely woven jute or other similar material. When the trier is removed the point should be run across the hole a couple of times in opposite directions to pull the threads together and close the hole. Closed paper bags may also be sampled by puncturing the bag and afterwards sealing the hole with a special adhesive patch.'

The *Nobbe trier* 'is made in different dimensions to suit various kinds of seeds. It is a pointed tube, long enough to reach the centre of the bag, with an oval hole near the pointed end. The total length of the instrument should be approximately 500 mm, including a handle of about 100 mm and a point of about 60 mm, leaving about 340 mm to penetrate into the bag, which should be sufficient to reach the centre of all types of bags. For cereals the internal diameter of the tube should be about 14 mm, but for clover and similar seeds 10 mm is sufficient. The Nobbe trier is suitable for sampling seed in bags but not in bulk.

'The sampling should be varied from top, middle and bottom of the bags. To sample the bottom of standing bags they may be raised off the floor and placed on top of other bags. The holes made in bags by the trier may be closed as described for the stick trier.'

Sampling during the cleaning process. 'For sampling seed during the cleaning process the sampling container shall be of such construction that the entire cross-section of the seed stream is uniformly sampled and the seeds entering it cannot bounce out again. The container may be moved through the seed stream either by hand or automatically.'

Sampling by hand. 'In certain cases and for certain species, especially chaffy, non-freeflowing species, hand sampling is sometimes the most satisfactory method. However, it is difficult by this method to sample deeper than about 40 cm. This will, of course, mean that it is impossible to obtain samples from the lower layers in bags and bins. In such cases the seed sampler may take special precautions such as requesting that some bags be emptied or partly emptied to facilitate sampling, and then be refilled. When sampling is done by hand great care should be taken to keep the fingers tightly closed about the seeds so none may escape.'

Sampling intensity. 'For seed lots in bags or containers of similar size the following sampling intensity shall be regarded as the minimum requirement: Up to 5 containers: sample each container and always take at least 5 primary samples. 6 to 30 containers: sample at least 1 in every 3 containers, but never less than 5. 31 containers or more: sample at least 1 in every 5 containers, but never less than 10. One should choose the bags or containers to be sampled at random and vary the sampling from top, middle, and bottom of each container. From larger containers one should take larger primary samples.'

Seed in bulk should be sampled at random locations and the samples should be drawn from varying depths. The sampling intensity should be as for seed lots in bags, etc., i.e. up to 500 kg, at least 5 primary samples (except for very small lots); further up to 3,000 kg, 1 primary sample for each 300 kg, but not less than 5 samples; and further up to 20,000 kg, 1 primary sample for each 500 kg, but not less than 10 primary samples.

Minimum weights of submitted samples. The International Rules for Seed Testing give prescriptions for minimum weights of submitted samples, and reference is made to a list of agricultural, horticultural and tree seeds,with indications of sample minimum weights. Examples:

1,000 g Most cereals. *Arachis hypogaea, Corylus avellana, Glycine max, Fagus sylvatica, Gossypium* spp., *Helianthus annuus, Lupinus* spp., *Phaseolus* spp., *Pinus pinea, Pisum sativum*
900 g *Prunus avium, Sorghum vulgare*
500 g *Beta vulgaris, Prunus serotina*
400 g *Oryza sativa*
250 g *Sorghum sudanense, Trifolium subterraneum*
200 g *Sinapis alba*
150 g *Capsicum* spp., *Cucumis melo, Cucumis sativus, Pennisetum glaucum, Solanum melongena*
100 g *Brassica napus, Brassica oleracea, Pastinaca sativa*
70 g *Sesamum indicum*
50 g *Malus* spp., *Medicago lupulina, Medicago sativa, Melilotus officinalis, Rosa* spp., *Trifolium pratense*
40 g *Brassica chinensis, Cichorium endivia, Petroselinum crispum, Picea abies*

25 g *Agrostis* spp., *Alnus glutinosa, Larix* spp., *Nicotiana tabacum, Phleum pratense, Poa* spp.
15 g *Betula* spp., *Lycopersicon esculentum*
5 g *Eucalyptus* spp. (small-seeded), *Populus* spp., *Salix* spp.

Marking and sealing the seed lot. 'At the time of sampling seeds for the issuance of an official Certificate (Orange International Seed Lot Certificate, ISTA; Rome Certificate, FAO) all bags must be marked or labelled to show a lot identification.' (See Fig. 26.1.)

'The bags must then be sealed by the authorised sampling and sealing agency with an official recognised seal and in such a way that it is impossible to open the bags unless the seal is destroyed. When the bags are tied by hand, both ends of the twine must be sealed together; and when the bags are sewn, the loose end of the stitching must be sealed. No sampled lot or part of a lot may be left unsealed.'

Marking, sealing, packing and sending the sample. 'Each sample must be marked in a way that establishes the connection between lot and sample. An additional label may be made for the lot and attached to the sample or put into it. The submitted sample packet should be provided with a printed form on which the necessary information can be filled in. For an orange or green certificate the sample must be sealed.'

Techniques of obtaining representative working samples. 'It is very important that every effort be made to ensure that the working sample is truly representative of the submitted sample.'

According to the International Rules one of the following methods must be used in obtaining the working sample: (a) mechanical divider method, (b) random cup method, (c) modified halving method, (d) spoon method.

There are several types of mechanical dividers, some of which are described in the rules: conical divider (Boerner type), soil divider, centrifugal divider (Gamet type). However, there is an important disadvantage of using mechanical dividers in connection with seed health testing: contamination may occur from one sample to subsequent ones, and safe cleaning after each sampling is too cumbersome to be practicable. For working samples in seed health testing it is advisable and convenient to use plastic cups of size according to size of seeds; to prevent contamination between samples the cups should be discarded immediately after use for one sample. For small working samples plastic dishes may be used, a clean unused one for each sample.

Reference is made to the International Rules for Seed Testing and Annexes (ISTA, 1976) for details of procedures.

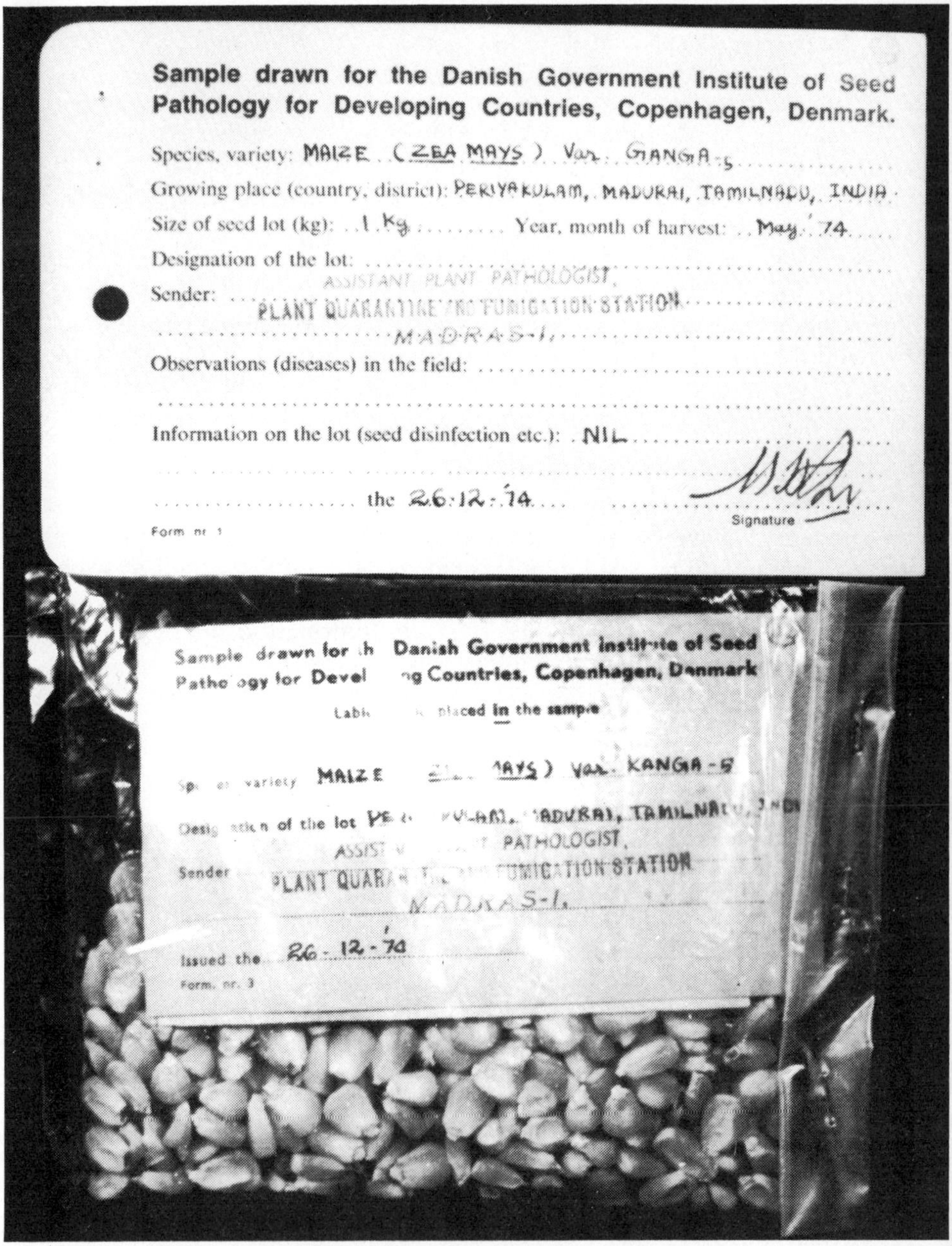

Fig. 26.1. Proper labelling of seed samples is imperative for identification of the seed lot. An 'outer label' is recommended for full information, and an 'inner label' for emergency use if the main label is lost

Considerations on working samples for health tests. For an examination of the health condition of a seed lot there is usually need for a minimum of two special tests: (1) examination of a dry sample for admixtures of infected plant debris, sclerotia, cysts, etc., and (2) examination of seeds and seedlings after incubation of the working sample.

For the first test a working sample of the size fixed in the International Rules for purity tests should be drawn. For the incubation test, usually a minimum of 400 seeds shall be counted at random into replicates of 100, 50 or 25 seeds.

For tests in which low permilles of infection are considered the working sample must, of course, be sized accordingly, such as in the tests for loose smut, lettuce mosaic, etc.

Selective sampling may be justified in testing for certain pathogens. Ryder and Johnson (1974) demonstrated a procedure by which heavy and light portions of lettuce mosaic-infected lettuce seed samples can be separated: vertical airstream concentrates the infected seed in the light portion. This method permits smaller samples to be tested for the virus than does the standard procedure.

TESTING PROCEDURES

Working Sheets of Testing Procedures
In the *Handbook on Seed Health Testing*, published *seriatim* by the International Seed Testing Association, Series 3 comprises sheets, each dealing with test procedures for one pathogen on one host. They constitute an important source of information for selection of test type.

27 Examination of Ungerminated Seed

27.1 INSPECTION OF DRY SEED

(1) *Procedure*. The dry seed is examined for impurities, classified partly as 'inert matter' in the International Rules for Seed Testing, such as plant debris, sclerotia, galls, insects, smut balls, etc., also for symptoms such as discoloration, malformation and similar indications of infection, including fruiting bodies of fungi, resting hyphae on the surface of the seed, spore or bacterial masses on the seed, as well as mechanical damage and, if required for the test, fungicidal residues. To carry out such examinations in a proper way, a low power stereoscopic microscope, with magnifications up to 50–60 times, is required, as well as good light for the observations. 'Inert matter' should also be incubated either on blotters or agar media, and examined after a standard period of incubation (Sections 28.1 and 28.2).

(2) *Application*. In seed testing stations the procedure is closely related to testing for purity; inert matter, provided from this test, should be specially examined as part of the seed health test. Sclerotia of *Sclerotinia sclerotiorum*, *Botrytis cinerea*, *Claviceps purpurea*, and other species of *Claviceps*, *Typhula* and other genera. Galls of nematodes, particularly species of *Anguina* in grasses and wheat, bunt balls and smutted kernels, free insects or other indications such as excrements originating from them, are the most important entities encountered.

Still another object should not be disregarded in applying this procedure: plant residues mixed with the seed and often present in the so-called inert matter of the purity test. The immense number of pycnidia of *Septoria linicola* on plant debris with flax seed, amounting to 50–75 per mm^2 containing conidia up to a total of about one million per mm^2 is an example already mentioned. The occurrence of stem eelworm in seed lots of red clover, lucerne and onion is the result of inefficient cleaning of the bulk, thus Palo (1962) found no nematodes in or on seed from heavily infected plants of lucerne, but great numbers in the inert matter removed from the seed lots concerned. Commercial seed lots of the species affected by stem eelworms are now usually well cleaned in technically advanced countries, only a small amount being left with the seed: 0.05 per cent in French, 0.03 per cent in American and Canadian seed of lucerne imported into Denmark, according to Statsfrøkontrollen (1963). But the fact that concomitant contamination occurring in plant debris may account for seed transmission does show the importance of testing not only the 'pure' seed but also adherent or loose plant debris and other entities mixed with it.

Dry seeds may show symptoms in varying degrees, due to necroses or discoloration from stains produced by microorganisms, or in the form of fruiting bodies or masses of spores or bacteria. In seeds of bean, pea and other Leguminosae, brown spots often indicate *Colletotrichum lindemuthianum* and other anthracnose fungi, or *Ascochyta pisi*, *Mycosphaerella pinodes* and other parasites in pea. Bacterial blights, *Xanthomonas phaseoli*, in bean seeds are sometimes detectable by the presence of the 'varnished' masses of bacteria, particularly in the hilum area. Blackening of the germ of cereals, caused by storage fungi, is of great importance to grading of wheat and other cereals in the U.S.A. (Christensen, 1957), black point (*Alternaria tenuis*) in wheat in India (Dharam Vir, personal communication, 1970), and 'pecky rice', i.e. kernels discoloured due to saprophytes and weak parasites (cf. Neergaard, 1970*b*), are of considerable importance as well to grading of the cereals. Pycnidia of *Septoria* on celery, celeriac and parsley, and of *Phoma betae* on the clusters of beet and mangold, are commonly seen on dry seeds.

Examination of dry seed for certain infections may be easier under ultraviolet light. The type of screened lamp used in seed laboratories for distinguishing between annual and perennial ryegrass in blotter tests, and the black light tubes used in seed health testing for incubation tests, are suitable. *Ascochyta pisi* causes a yellow-green fluorescence on pea, *Pseudomonas phaseolicola* (cf. Parker and Dean, 1968) and *Xanthomonas phaseoli* produce a bluish-white fluorescence on white, cream coloured or buff bean seed, and *Septoria nodorum* a greenish fluorescence on wheat seed. However, fluorescence does not always work in these cases. Wharton (1967) found 112 of 542 fluorescing seeds of white-seeded cultivars of Phaseolus bean being infected by *Pseudomonas phaseolicola*, but only 6 of 495 non-fluorescing seeds from the same samples. He recorded that 64–68 per cent of infected seeds were of the fluorescent type. Taylor (1970) confirmed these results. Naoumova (1957) reported detection of loose smut, *Ustilago tritici*, in wheat by a bluish fluorescence in ultraviolet light, but at the Institute of Seed Pathology (1970, unpublished) no correlation was found between fluorescence and infection of loose smut in wheat.

Lamps of greater output in 'watts' may incite a more vivid fluorescence response (Parker and Dean, 1968). It should be emphasised that the presence or absence of fluorescence, in itself, is not assurance of either the presence or absence of infection. The fluorescence procedure is supplementary to examination of dry seed in ordinary light and may be regarded as useful for a quick estimate of the level of infection of the mentioned pathogens in pea, bean (Anselme and Champion, 1962) and wheat.

When seeds are placed in water or lactophenol, observation of pycnidia of *Septoria* and other fungi is facilitated. Also ryegrass investigated for blind seed disease, *Gloeotinia temulenta*, and flax seed for *Aureobasidium lini*, are placed in water drops to provide conditions for examination of spores exuding from the seeds — in the case of ryegrass the glumes may be teased loose to enable spores to get free but often this is unnecessary. Certain nematodes may be detected by placing seeds in water, for instance *Aphelenchoides besseyi* and *Ditylenchus angustus* in rice.

The occurrence of insects and the implication of injuries due to mechanical damage such as threshing injury are discussed in Chapter 2. Particularly, mechanical damage may also be important in relation to establishment and seed transmission of pathogens.

(3) *Limitations*. Direct inspection provides quick information, does not require much equipment and may readily be applied in seed testing stations in conjunction with purity tests without much additional labour, provided that it has been well practised. It is a procedure indispensable for examination of seed lots for contents of sclerotia and other material regularly removed from working samples submitted to germination tests or health tests requiring incubation. Thus the procedure is a necessary supplement to the incubation methods. Preliminary examination of samples from unprocessed material of seed lots may give information on the presence of underdeveloped, shrunken, diseased seeds that should be removed by efficient grading of the seed lot.

However, apart from sclerotia and similar entities, only few diseases are adequately detectable by direct inspection. Although anthracnose fungi and *Ascochyta* spp. of Leguminosae produce symptoms on the seeds, investigations based on development of the fungi through suitable incubation are imperative to provide full information. Weak incidences of such fungi may not show external signs of infection on the dry seeds.

Finally, certain fungi adequately detectable by direct inspection, such as *Septoria* on celery, celeriac and parsley, may actually be dead. An additional viability test is therefore required, for which agar media or test plants are used. This procedure is followed as a routine test at the Official Seed Testing Station in Cambridge to fulfil a statutory requirement that *Septoria* infection of celery and celeriac 'must be declared', since such may originate from lots that have been treated by hot water, or stored beyond the usual longevity period of the parasite.

27.2 EXAMINATION OF SUSPENSIONS OBTAINED FROM WASHINGS OF SEED

(1) *Procedure*. The seed is shaken for a fixed period of time in water to which a detergent may be added, or in alcohol, a fixed amount of seed being applied with a fixed amount of the fluid. An automatic shaker facilitates the work and makes standardisation easier. The suspension obtained may be examined directly, or the suspended spores may be concentrated by centrifuging (Safronova, 1936; Cherewick, 1944, 1948; Russell, 1946), evaporating or filtering the fluid. The concentrated remainder may then be diluted in a known quantity of fluid and examined under the microscope. A haemocytometer may be used for this purpose (Rice, 1939). In using this procedure several mounts should be made. Rice found, by the chi-square test, that counting the number of bunt spores in a quarter of the field of five different mounts gave a much better estimate of spore load per wheat kernel than by counting the number of spores in the whole field or one mount. He developed a formula for computing the spore load per kernel. Since 4 mm^3 are examined, that is five quarters fields of 0.8 mm^3 each, the volume examined is 1/250 of 1 ml. or 1/1250 of 5 ml., the latter being the quantity of water used for each 100 kernels. The formula then becomes

$$N = \frac{(x)\ 1250}{(0.85)(100)} = (x)(14.7)$$

in which x is the number of spores in 4 mm^3 of spore suspension, 0.85 a correction factor indicating that about 85 per cent of the spores were washed off in 50 agitations applied, and N the number of spores per kernel. The correction factor may be altered according to agitation procedure. Instead of using a haemocytometer the spores may be collected on filter paper, which is made transparent by xylol so that the spores can be counted under the microscope (Gentner, 1929). Many variants of the procedure have been used or are still in use (Beck, 1920; Bredemann, 1923; Heald and Boyle, 1923; Crosier, 1926; Cherewick, 1947; Keding, 1952*b*, Kietreiber, 1952; Schoen, 1974); reviews of different relevant procedures have been given by Dounine and Michailow (1931), Zillig (1932) and De Tempe (1963*c*). Wenzl (1960) described the procedure applied for detection of *Cercospora beticola* in beet. He also used the method for testing the ability of the fungus to produce conidia on seed clusters as a criterion of the presence of viable mycelium.

The amount of spores in a suspension may also be determined on a spectrophotometer; the turbidity of the suspension varies according to the number of spores. After calibration this procedure is much quicker than procedures based on counting.

(2) *Application*. The technique has been largely used for quantitative determination of the spore load of smut fungi in cereals but can also be used for determination of any fungus adhering to the seed surface, such as *Alternaria*, *Stemphylium*, *Drechslera*, *Fusarium*, *Pyricularia* (rice) and others. In addition to the spore count at lower magnification, it may be necessary to identify spores to species at high magnification, for instance where closely related smut fungi are present together. In the case of *Tilletia caries* and *T. foetida*, as against *T. contraversa*, a preparation in diluted Indian ink reveals the thick slime layer surrounding the spores of the latter species. The procedure is used in France for detection of oospores of *Peronospora valerianellae* in seed of *Valerianella locusta* and of *P. farinosa* in seed of spinach (Champion, 1975, personal communication).

(3) *Limitations*. The method is adequate for such contaminations that are present exclusively as externally loosely borne spores, provided that these spores lead to field infection. In *Ustilago avenae* and *U. hordei* it has been claimed (Gassner, 1953) that only the spores which have penetrated between the glumes and the caryopsis are responsible for infection of the seedlings, whereas spores borne on the surface are not, and consequently the amount of these spores is not an adequate expression of the inoculum potential. The same is true in the very many seed lots in which mycelial infection occurs side by side with spore contamination by the same pathogen. *Pyricularia oryzae* is regularly present in rice seed both as loosely borne conidia and as mycelial infection. The conidia lose their germination capacity within the first year after harvest; there is no correlation between conidial contamination and mycelial infection, so spore counts of this parasite are therefore not reliable for more than preliminary information on freshly harvested seed lots (Neergaard, Lambat and Mathur, 1970).

Where heavy spore loads can be shown to lead to material infection in the field, or where they can be correlated to regularly simultaneously occurring mycelial infection leading to field infections, the method is applicable. But in general, with the important exception of certain smut fungi, the method is not so much misleading as failing to give full information. Consequently, it must be supplemented or replaced by suitable incubation procedures as described below.

27.3 EXAMINATION AFTER WASHING OF SEED, AND SEDIMENTATION

(1) *Procedure* (Seinhorst, 1950; Nielsen, 1964, cf. Rao, 1972*a*). A working sample of 50–100 g of seeds of clover, lucerne or onion, suspected to be

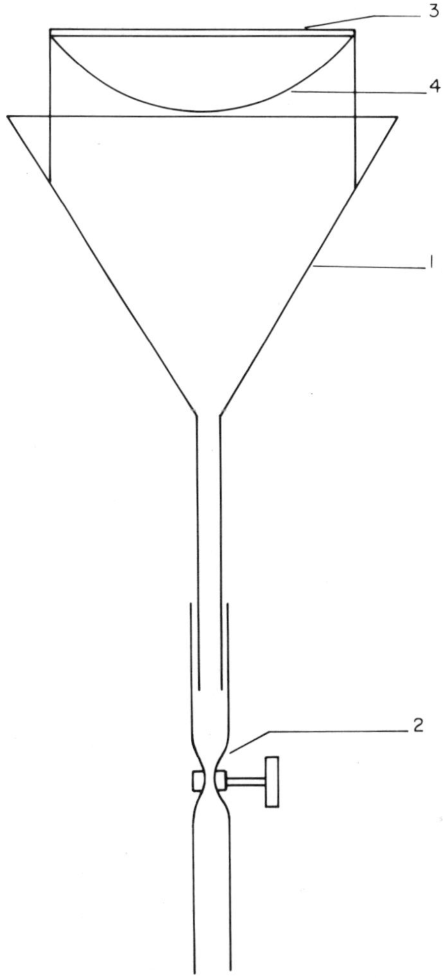

Fig. 27.1. 'Baermann' hopper for driving off nematodes. (1) Glass hopper; (2) rubber tube with screw clip; (3) metal ring, supported by three legs on the inside of the hopper, and on which a bent wire gauze has been placed; (4) the wire gauze, the lowest point of which just touches the water surface in the hopper. The water may be supplied by continuous spraying through one or several nozzles. (Nielsen, 1964)

contaminated with the stem eelworm, *Ditylenchus dipsaci*, is placed on filter paper just above a hopper of which the outlet is closed with a tap. During a fixed period, usually 24 h, the seed is submitted to a constant, fine water spray. The water penetrates between the seed, which is nevertheless open to the effect of the air, and the superfluous water flows off over the edge of the hopper. Under these conditions of moisture and air the eelworms move through the filter paper and sink into the water of the outlet just above the tap. The eelworms may, after the period of the treatment, be collected in a few drops of water and identified, cf. Fig. 27.1.

(2) *Application*. The procedure has so far been used only for detection of stem eelworm in the three kinds of seed mentioned: clover, lucerne and onion.

(3) *Limitations*. The eelworms are present very often only in relatively few pieces of plant debris, unevenly distributed in the seed lot. A major part of contamination is removed by the cleaning process. Large working samples are needed, but may still not be sufficient to provide reliable data.

27.4 THE WHOLE EMBRYO COUNT METHOD

The staining technique was applied to cereal loose smut mycelium by Skvortzov (1937) who dissected out the embryos, macerating them in sodium hydroxide and staining with aniline blue. Later, many workers modified this procedure, mainly by developing chemical methods of separating the embryos so that many seeds could be tested on a routine basis.

(1) *Procedure*. A working sample of at least 2,000 seeds is usually taken. The seeds (barley, wheat) are soaked overnight at 22°C in 10 per cent NaOH. Next morning the seeds are shaken with warm water and poured through three sieves, as detailed below, so that all the embryos are retained by the final sieve having the smallest holes. The embryos are then transferred with a small pipette to a 100 ml. beaker. Any embryos retained within the glumes are extracted manually, and added to those already recovered. Excess water is drained off and lactophenol added, the beaker placed in a water bath and heated slowly until all the embryos have cleared. A modification of the procedure is used for wheat.

With slight modifications this procedure has been developed by Wöstmann (1942), Yablakova (1944), Simmonds (1946), Russell (1950), Popp (1951), Pedersen (1956*a*) and Rennie (personal communication, 1972). Pedersen found it preferable to place the seeds in boiling water and boil them for 5 min, because boiling and maceration permit the scutellum to tolerate the subsequent treatment with sodium hydroxide better than if Simmonds' procedure is followed. After rinsing in hot water the seeds are boiled in 5 per cent sodium hydroxide and retained at the boiling point for 5 min. After draining off and rinsing, the seeds are left for 7–9 h at 55°C in a 2.5 per cent solution of sodium hydroxide, to which is added 2.5 per cent formalin. This last part of the procedure has a hardening effect on embryo and scutellum. The embryos are transferred to alcohol, for 2 h in 50 per cent, then for 2 h in 70 per cent, and finally they are left overnight in 95 per cent. They may safely remain in this solution for several weeks and can be examined when, 1–2 days before the analysis, they are transferred to glycerine to render them clear and translucent.

At the Seed Testing Station in Edinburgh the sample is soaked for 24 h in 1 l. of 5 per cent solution of sodium hydroxide at 20–22°C, in a container of at least 2 l. capacity. The sample is then transferred to a modified Fenwick can (as for nematology) in which the water inlet is at the base and the top is open. In this can the sample is continuously washed in a stream of water at 50–70°C to separate the embryos from the endosperm and to allow the embryos to rise and pass over the lip of the can with the flow of water — Fig. 27.2(1). The extraction process is continued until all

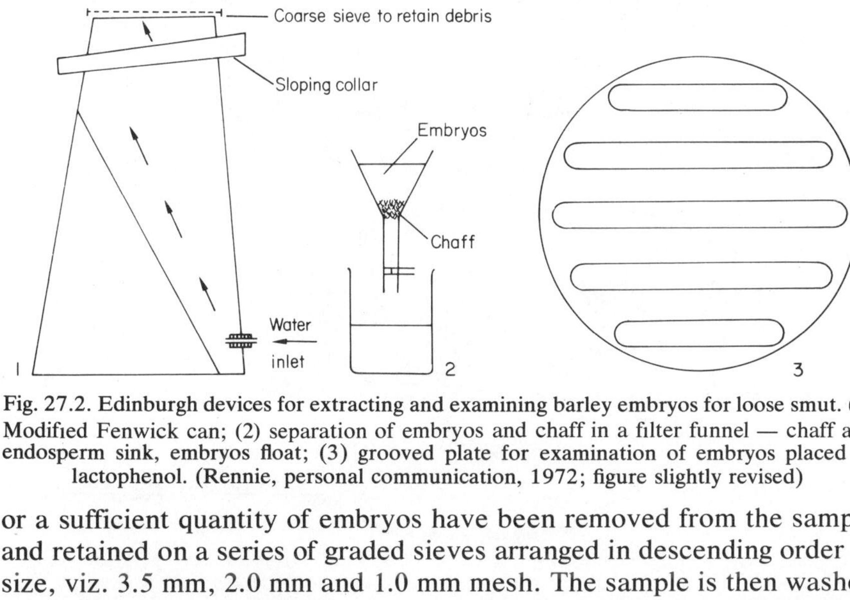

Fig. 27.2. Edinburgh devices for extracting and examining barley embryos for loose smut. (1) Modified Fenwick can; (2) separation of embryos and chaff in a filter funnel — chaff and endosperm sink, embryos float; (3) grooved plate for examination of embryos placed in lactophenol. (Rennie, personal communication, 1972; figure slightly revised)

or a sufficient quantity of embryos have been removed from the sample and retained on a series of graded sieves arranged in descending order of size, viz. 3.5 mm, 2.0 mm and 1.0 mm mesh. The sample is then washed into a wire basket or muslin cloth and dehydrated in ethyl alcohol for 2 min. When the sample is transferred to a filter funnel, Fig. 27.2(2), and a lactophenol/water solution added, the embryos float and the chaff and endosperm tend to sink and can thus run off. Should the chaff be slow to separate a little more water can be added to the solution. Clearance is obtained by boiling in lactophenol for 10–20 min; the embryos are poured with the lactophenol into grooved perspex plates 1.5 mm thick and so designed that the embryos easily settle in rows, Fig. 27.2(3), the plates fitting into plastic disposable petri dishes.

The embryos are examined with a stereoscopic microscope, 20–40× magnification, for the presence of mycelium which may vary from two or three fine hyphae to complete infection of the scutellum, the light coming from below. It may be advisable occasionally to use a higher power microscope to check identification.

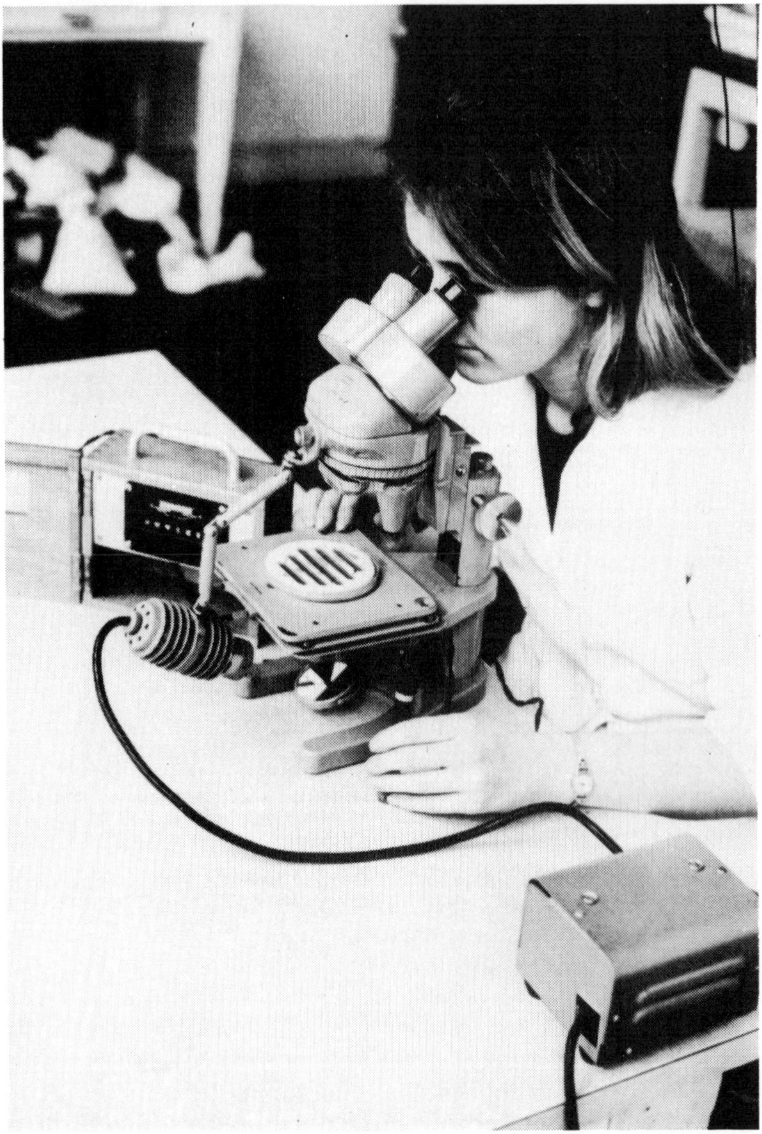

Fig. 27.3. Examination of embryos for loose smut through low power microscope, provided with light from below. Number of embryos is recorded by electric counter; the handpiece in the left hand is knocked on the table for speed, and the figures are registered on the counter left of the microscope. Infected embryos are recorded manually. (Courtesy of Department of Agriculture, Agricultural Scientific Services, Edinburgh, Scotland)

(2) *Application*. The method has been established as very reliable for detecting percentages of loose smut of barley and wheat, those being very close to counts obtained in field tests (Russell, 1950; Russell and Popp, 1951; Pedersen, 1956*a*; Marshall, 1959; Kavanagh and Mumford, 1960; Morton, Tool and Hagen, 1960; Russell, 1962). Morton, Tool and Hagen (1960) obtained for 295 barley samples a correlation coefficient of 0.946 between the results of the embryo test and the percentage of smutted heads that grew from the samples. Pedersen (1956*a*) and Russell, Sallans and Cannon (1961) point out the possibility of cutting down labour by practising subsequential analysis, dividing the working sample into several subsamples and discarding the remainder when the tolerance level is reached. According to Rennie (personal communication, 1972), the time required for extracting embryos is reduced to about 5 min if a sufficient supply of hot tap water is available. A further 10 min is required for the separation and cleaning processes and the examination can be completed within 15 min. When two analysts work together, 20 samples can be tested in 4 h (Fig. 27.3). If large numbers of tests are carried out the cost of chemicals is very low, probably less than £0.5 sterling (50 n.p.).

(3) *Limitations*. Distinction should be made between embryo resistant cultivars and adult-plant resistant cultivars (Batts and Jeater, 1958*a*). In

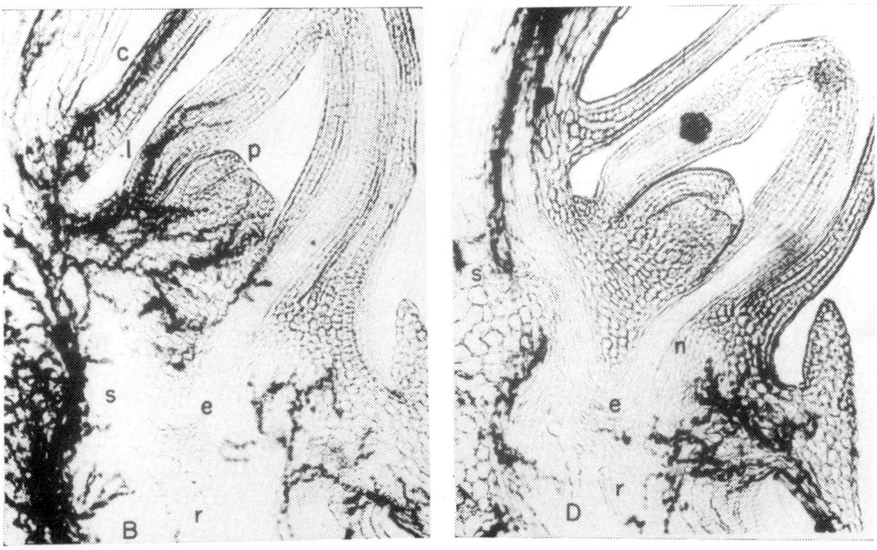

Fig. 27.4. Typical distribution of loose smut mycelium (*Ustilago tritici*) in wheat embryos of adult-plant susceptible and resistant cultivars, as seen in sectioned embryo tissue (×85). Left: section of embryo of a susceptible cultivar, showing mycelium in plumular bud (p), primary foliage leaf (l), coleoptile (c), scutellum (s), primary root (r) and embryonic axis (e). Right: section of embryo of a resistant cultivar with mycelium only in scutellum (s), primary root (r), coleoptile node (n) and embryonic axis (e). (Popp, 1959)

Canada, the embryo count method is no longer used for the barley cv. Keystone, because it is mature-plant resistant, although extremely embryo susceptible. Also, in barley cv. Jet infection in the embryo is not transmitted to the seedling (Noble and Richardson, 1968), nor in the European cultivar Emir (Hewett, 1972). Popp (1959) found that in wheat the location of the mycelium in the embryo is important in relation to amount of disease developed in the crop; only infection within the plumule bud is correlated with infection in growing plants, cf. Fig. 27.4. Pedersen (1956*a*) pointed out that *Ustilago nigra*, which can hardly be distinguished from *U. nuda* in field tests, may be a source of error, leading to discrepancies between results obtained from the embryo test for barley and those recorded in field tests.

28 Incubation Tests I: Procedures

28.1 THE BLOTTER TEST

(1) *Procedure*. The blotter test is a combination of the *in vitro* and the *in vivo* principles of investigation. The seeds are sown in petri dishes, perspex boxes or other suitable containers on moistened absorbent paper (blotter), usually three layers to provide enough moisture for duration of the test (Fig. 28.1). If a light-coloured fungus like *Verticillium* is concerned, black or grey blotters may be used to facilitate observations. The seeds are placed at a fixed distance according to the size of the seed. A vacuum counter (Fig. 28.2) may be used. The seeds are incubated for a fixed period of time, usually one week, at a fixed temperature, for many

Fig. 28.1. Preparing substrate, and sowing, according to the blotter method

kinds of seed at 20°C±2° (the standard temperature used for most seed germination tests).

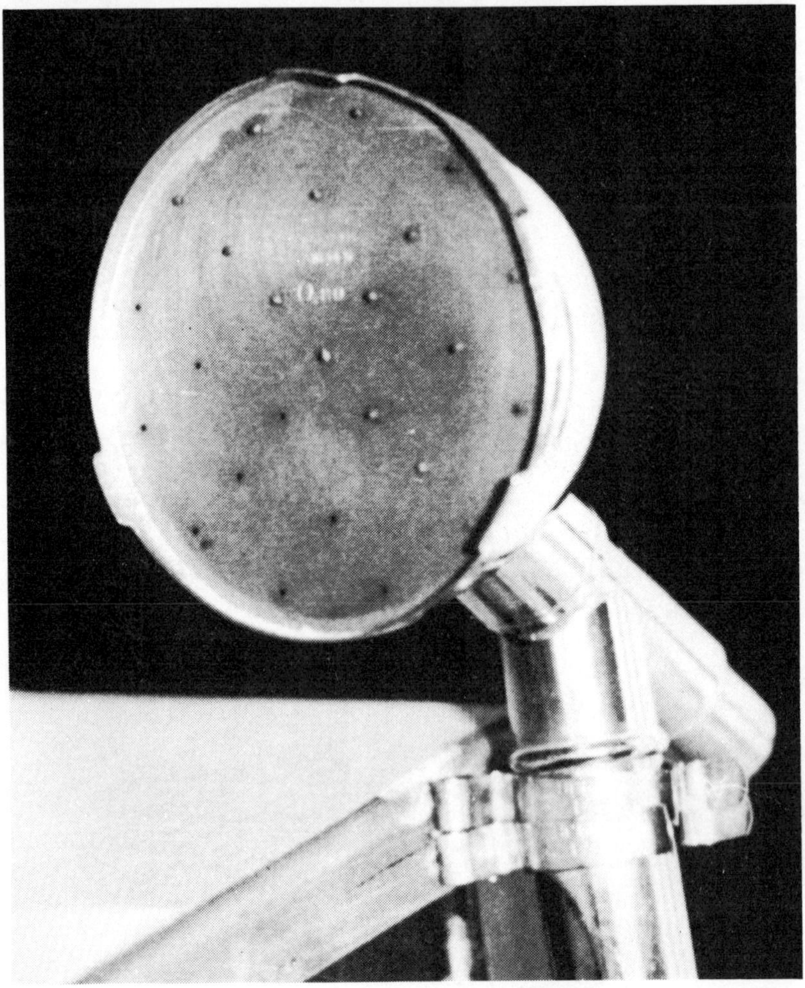

Fig. 28.2. One hundred seeds adhere to this counting head of a vacuum counter. Heads are interchangeable and may be provided with holes of different sizes and in different numbers

Pieces of plant debris, and other 'inert matter' that may have been separated during examination of the dry seeds, are plated at reasonable distance in separate dishes and incubated. Often the period of incubation for such material is shorter than for the seeds concerned.

The incubators should be provided with equipment for near ultraviolet radiation, which has been shown to stimulate sporulation in many seed-borne fungi, as discussed in Subsection 29.5.4.6. A near ultraviolet light–darkness cycle of 12/12 h has been adopted as a standard procedure (ISTA, 1966*a*). Recording is made by a low power stereo-microscope with up to 50 or 60× magnification, as described in Section 29.6. A stereo-microscope enables observation of the fungi as developed on their host *in situ*, undisturbed, in a condition of natural growth. This principle of observation is the great merit of the blotter method. Different pathogens may require different periods of incubation as well as differences in other conditions of incubation; thus the temperature may range from about 12°C for a part of the incubation period for seeds from temperate areas, to about 28°C for fungi on crops grown in warmer subtropical or tropical areas (Fig. 28.3). The infection percentage is based on presence of the pathogen as well as on signs of disease in the seedlings.

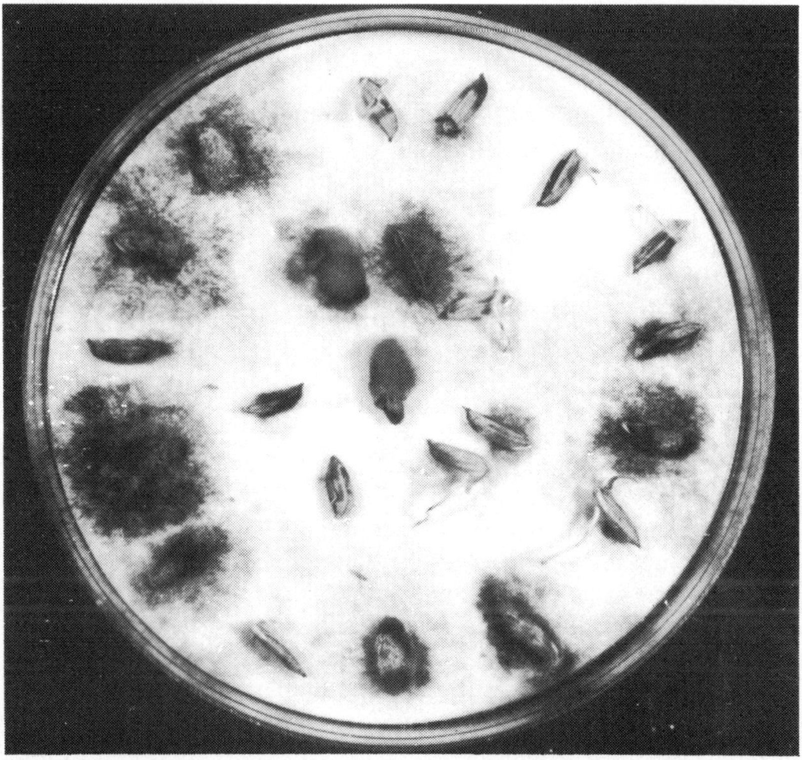

Fig. 28.3. The blotter test. Rice seed incubated for 8 days at 28°C and 12/12 h NUV/dark cycles. Most seed covered by dark colonies of *Drechslera oryzae*

The essence of recording in blotter tests is quick identification of habit characters, each species displaying characteristic features such as the form, length and arrangement of conidiophores; the form, size, septation, colour, chain formation, etc., of conidia, and their arrangement on the conidiophores, appearance of spore masses, characters of mycelium, density of colony, etc. — see Colour Plates I and II (*Fusarium*) and Figs. 5.13–5.15 (*Drechslera*).

In order to facilitate recording, the germination of seed tested by the blotter method may be retarded or stopped by using a 0.1–0.2 per cent water solution of the sodium salt of 2,4-dichlorophenoxyacetic acid (the herbicide 2,4-D) instead of pure water (Neergaard, 1969). This procedure is widely used in testing seed of crucifers for *Phoma lingam*. The procedure should be adopted critically as 2,4-D may have a weakly fungistatic effect, e.g. *Alternaria brassicicola* is delayed in development on cabbage seed by 1–2 days at the level of 0.2 per cent. Also, in this case of course there are no seedling symptoms for extra guidance.

For certain kinds of seed which usually are heavily contaminated with saprophytes, pretreatment is necessary — 1–3 per cent chlorine is often used, and a proprietary product, Chloros, has been recommended.

(2) *Application*. The 'blotter method' is widely used where regular seed health testing is carried out, especially in circumstances where the agar test is impracticable. It is applied to all kinds of seeds, including cereals, vegetables, ornamentals and forest seeds. The method may be designated as a hybrid between the moist chamber procedure used in plant pathology and the germination test used everywhere in seed technology. The method combines the advantages of *in vitro* investigations with those of *in vivo* observations.

(3) *Limitations*. The blotter method provides excellent conditions for development of mycelial growth of many Hyphomycetes, for conidial sporulation of many Fungi Imperfecti, and for development of signs of infection produced by pathogenic forms of these fungi in the seedlings sprouted during the incubation. These advantages describe also the limitations of the method. Fungi, whether belonging to Fungi Imperfecti or to other taxonomic groups of fungi, which do not ordinarily produce mycelium or sporulate profusely under the conditions of the blotter method, may readily be overgrown by more vigorous fungi. Certain genera of fungi are frequently encountered in the blotter test, e.g. *Alternaria, Ascochyta, Botrytis, Cercospora, Cladosporium, Colletotrichum, Curvularia, Drechslera, Fusarium, Myrothecium, Phoma and Stemphylium*.

The occurrence of pathogenic bacteria in the blotter test is very rare, *Corynebacterium rathayi* in cocksfoot being one of the few exceptions. Many important seed-borne pathogens are not revealed in the blotter test, or they are present only inconspicuously. Among Fungi Imperfecti, *Ramularia* and *Verticillium* are poor in mycelium, and within Oomycetes several species of *Peronospora* may be present as mycelial infections but have never been observed to develop their imperfect state on seeds. Rarely, oospores are found, as in the cases of *Sclerophthora macrospora* on oats and pearl millet (*Pennisetum glaucum*) and *Peronospora manshurica* on soybean, and they are detectable even without incubation of the seed. Finally, the blotter method favours development of saprophytes that very often seriously impair recording by quick-growing parasitic fungi. Among the most troublesome contaminators in this respect are *Alternaria tenuis*, *Penicillium expansum*, *Rhizopus stolonifer* and other species of *Rhizopus*, *Mucor* spp., and at higher incubation temperatures, *Aspergillus niger*, and other species of *Aspergillus*. Such contaminations may be checked by surface disinfection. However, within these limitations the blotter method is extremely useful for routine seed health testing.

28.2 THE AGAR PLATE TEST — EXAMINATION OF COLONIES DEVELOPED FROM SEEDS PLATED ON AGAR MEDIA

(1) *Procedure*. The seeds are plated in petri dishes on a suitable agar and spaced according to the size of the seed. Malt extract agar or potato dextrose agar are the media most commonly applied. In testing for storage fungi, agar media with high salt concentrations such as 18 per cent NaCl must be used, necessary for detection of *Aspergillus restrictus*. Prior to plating, a pretreatment of the seed may be applied to prevent profuse development of saprophytes. Plating on malt extract agar without pre-treatment was used for routine seed health testing of flax in Ulster, and is known as the Ulster method (Muskett and Malone, 1941). But often pre-treatment is required and an internationally accepted procedure is soaking for 5 min in 1 per cent sodium hypochlorite and plating immediately without washing the treated seed.

The seeds are incubated at a fixed temperature, mostly about room temperature (20±2°C) for an incubation period of from five to eight days (Fig. 28.4).

As in the blotter method, plant debris and other 'inert matter' found in examination of the dry seed is plated and incubated under the same conditions as the seed concerned.

The principle of recording in the agar plate test is macroscopical examination of fungus colonies. An experienced analyst, familiar with the colony characters of the fungi present in certain seeds, can identify and count colonies by naked eye without hesitation, checking both sides of the plates. Muskett and co-workers in Northern Ireland first used this method in a large-scale testing scheme for health condition of 25,000 flax seed lots in 15 years (Muskett and Malone, 1941).

There is an extensive literature on cultural characters of fungi grown in

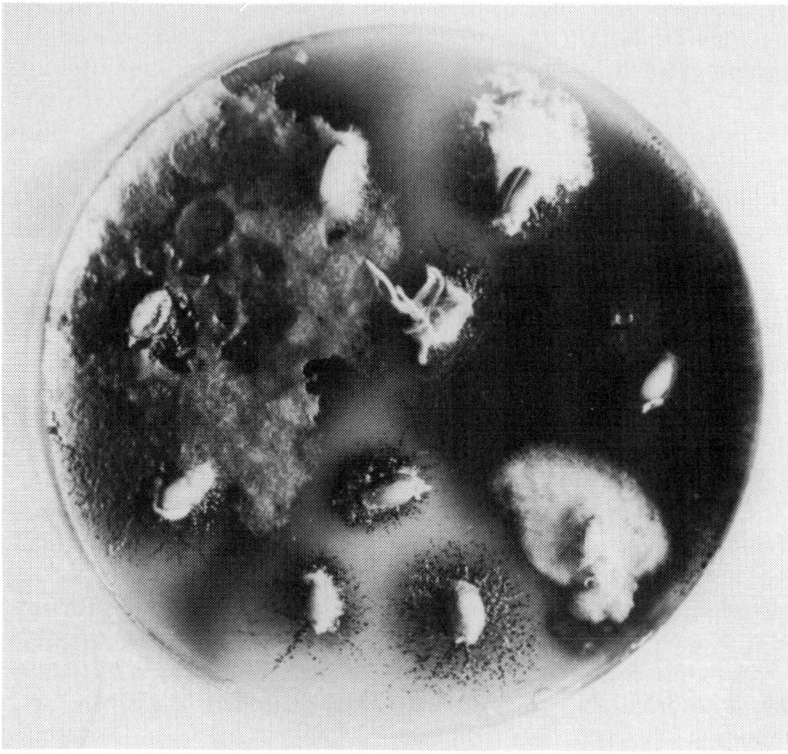

Fig. 28.4. The agar plate test. Brönnimann's method. Wheat seeds incubated at 15°C under continuous artificial daylight for 7 days. Abundant formation of pycnidia around infected seed, covered by white mycelium. Dark colonies of *Alternaria tenuis* and *Epicoccum purpurascens*

pure culture on agar media. Special problems arise, however, when seeds are planted on agar, because here the colonies, often several from one seed, represent a complex of fungi masking and perhaps inhibiting each other, some sectoring within others, etc. There is much need for further study of the behaviour of such combinations of associated fungi under the conditions of the agar plate test, and on the development of selective media.

A special agar plate procedure for detection of *Phoma betae* in beet has been proposed by Mangan (1971). Seeds, untreated or chlorine protected, are sown on water agar and the fungus is identified, after 7 days' incubation, by its characteristic 'holdfast'-like structures or hyphal clumps visible under compound microscope (Fig. 28.5). Sharp (1974) found untreated seed preferable (compare, however, Lysle D. Leach's utilisation of pretreated against untreated seed for ecological investigations — see Section 11.3).

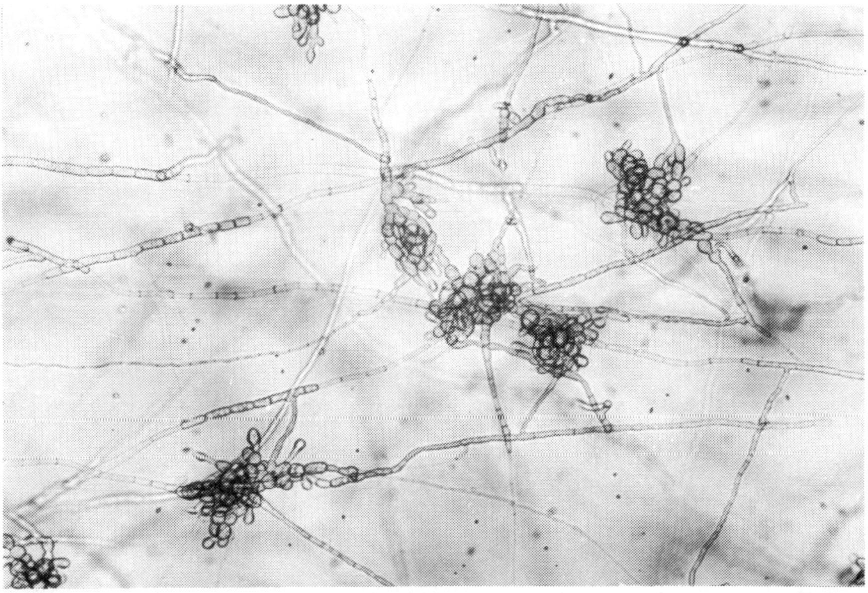

Fig. 28.5. 'Holdfast'-like structures or hyphal clumps of *Phoma betae* as seen in Mangan's water agar plate procedure. Magnification ×200

(2) *Application*. The agar plate method is applicable to those kinds of seed in which saprophytic species do not materially impair quick identification of the pathogens, a difficulty often overcome by pretreatment. The procedure is preferable where the blotter method does not provide adequate conditions for development of mycelial growth, sporulation or symptoms of the pathogens on seedlings and seeds, but when, on the other hand, characteristic colonies are developed on nutrient agar. As mentioned above, the agar plate method has been widely used for flax and with pretreatment for oats in Northern Ireland; also in Canada for peas, for which the procedure has been found preferable to the blotter test in the international comparative testing schemes of ISTA.

(3) *Limitations*. Much regular practice with each kind of seed is necessary to acquire and maintain experience. When the method is used only occasionally, the analyst will hesitate in distinguishing cultural characters, and frequent checking by time-consuming microscopic examinations will be needed. On the other hand, if the test is well adapted for the seed concerned, a skilled analyst is capable of carrying out a test more quickly by this method than by the blotter method. However, in the agar plate test slow-growing fungi may be more or less suppressed by vigorously growing fungi. Before one of the methods is preferred to the other, careful comparative evaluation is necessary whenever the number of tests of a

certain kind of seed is sufficiently high to render consideration of the agar plate method desirable. This is also a rather expensive test because of the amount of agar and petri dishes required.

28.3 THE FREEZING METHOD

This method (Fig. 28.6), as originally proposed by Limonard (1966), is a modification of the blotter method.

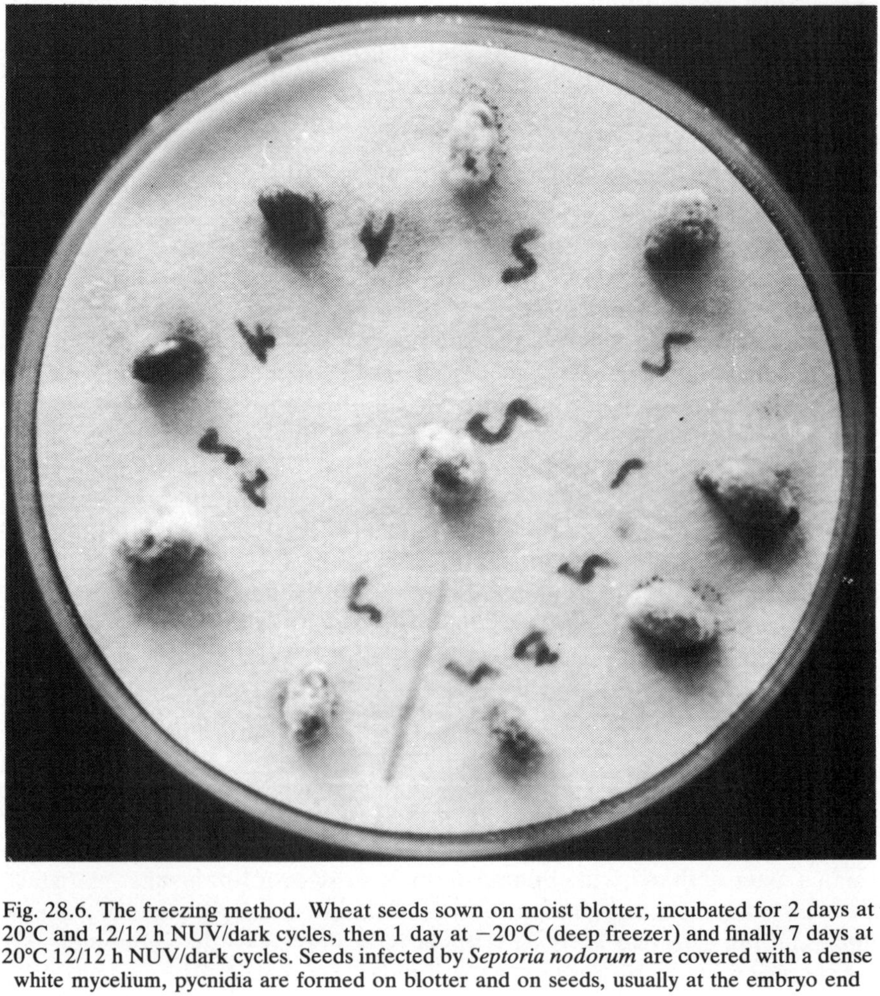

Fig. 28.6. The freezing method. Wheat seeds sown on moist blotter, incubated for 2 days at 20°C and 12/12 h NUV/dark cycles, then 1 day at −20°C (deep freezer) and finally 7 days at 20°C 12/12 h NUV/dark cycles. Seeds infected by *Septoria nodorum* are covered with a dense white mycelium, pycnidia are formed on blotter and on seeds, usually at the embryo end

(1) *Procedure*. The seed is sown on blotter. Cereals are incubated at 10°C for three days to induce germination, then at 20°C for two days; other kinds of seed at 20°C for four days. The seedlings are then frozen overnight at −20°C and subsequently incubated at 20°C in near ultraviolet light for another five to seven days. The dead seedlings provide a substrate for profuse development of fungi such as *Fusarium* and *Septoria* in cereals, *Phoma* in cornsalad and beet. Bacterial antagonism can be eliminated by adding an antibiotic, such as terramycin, to the blotters.

To overcome secondary infections from seed to seed due to the properties of filter paper, Limonard (1968) proposed a modification of the procedure. The blotters are replaced by perspex plates, 10×26 cm, 6 mm thick, provided with 75 holes of about 15 mm diameter and a few millimetres deep. Liquid can be placed in each hole, together with one seed (*perspex-freezing method*).

(2) *Application*. The percentages of infection obtained have been found to be close to those obtained on agar media, the dead seed being comparable to an agar medium, without any host resistance operating as in the blotter test. The method has been in satisfactory use for some time at the Seed Testing Station in Wageningen in routine testing of commercial samples of cereals and grasses for *Drechslera* and *Fusarium* and of various small-seeded vegetables. The procedure has been found very suitable for testing of maize for fungi (Singh, Mathur and Neergaard, 1974*a*) and cabbage, in particular for *Phoma lingam*.

(3) *Limitations*. The procedure has the advantage of using a more natural medium than agar and requiring less labour. The usual blotter test has, however, the advantage of enabling evaluation of the effect of pathogens on the host, this possibility being eliminated in the freezing method. The sporulation of certain pathogens may be favoured by this procedure (e.g. *Alternaria porri* on onion, as indicated by Limonard) as compared with other incubation procedures. Further investigations, including comparison between the different methods, should be made to evaluate the merits of the freezing method.

28.4 THE WATER AGAR PLATE METHOD FOR DETECTION OF BACTERIA

This testing procedure, developed at the Institute of Seed Pathology by Srinivasan, Neergaard and Mathur (1973) for the detection of *Xanthomonas campestris* in seed of cabbage and other crucifers, has only the agar medium in common with the ordinary agar plate method. In this agar plate procedure the medium serves for germination of the seed and development of the seedlings, not for development of colonies.

(1) *Procedure*. The seeds are soaked in an antibiotic solution such as aureofungin at a concentration of 200 ppm for three or four hours. After draining off the excess of antibiotic solution, the seed is plated out on agar (1.5 per cent), 25 seeds per petri dish, and incubated at 20°C in darkness. Recordings are made after 12 days. Diseased seedlings show delayed germination and stunted growth. The hypocotyl and cotyledons are yellowish and pulpy, the seedling collapsing on the agar surface. From infected parts a yellowish bacterial ooze can be seen under the compound microscope. Other seedlings show V-shaped marginal black necroses which also yield bacterial ooze. Seedlings showing collapse due to systemic infection can be readily recognised in sharp contrast to healthy seedlings. Identification of the pathogen may be confirmed by transference of bacterial ooze to a medium selective for *Xanthomonas* (Kado and Heskett, 1970), Medium D 5. This medium contains 10 g cellobiose, 3 g K_2HPO_4, 1 g NaH_2PO_4, 1 g NH_4Cl, 0.3 g $MgSO_4$ $7H_2O$ and 15 g agar per litre.

(2) *Application*. The procedure has been successfully used for a number of seed lots of cabbage and cauliflower and has been found to provide fairly reproducible results of infection percentages based on 400 seeds per working sample. It still remains to establish a further evaluation of the method, based on tests carried out with a great number of seed samples of different origin.

(3) *Limitations*. The procedure is easy and quick, and has been found to be reliable in showing natural symptoms of the disease on the seedlings. The method is acceptable as a qualitative test and further investigations may establish the procedure as a quantitative test. If standardised, it would prove useful for seed health certification schemes.

28.5 SEEDLING SYMPTOM TESTS

In the blotter and agar plate methods the medium used is entirely artificial. An approach to establishing more natural conditions is obtained by sowing the seed in autoclaved soil, gravel, sand or similar material. Under standard conditions of temperature and moisture, infected seeds and seedlings may develop symptoms comparable with those developed under field conditions. For certain pathogens the procedures of this category provide information pertaining to field performance of the seed lot in relation to seed-borne seedling diseases.

28.5.1 Hiltner's brickstone method

(1) *Procedure* (Hiltner, 1917). As a medium, sterile crushed brickstone with a maximum grain size of 3–4 mm (Fig. 28.7) is used. Ground brickstone has, because of its capillarity, high waterholding capacity and is rather heavy. When applied for small-grain cereals, 100 kernels are sown well spaced in each container and covered with 3 cm of the medium. Prior to sowing, the medium is moistened with water so that later watering is not needed. The containers are placed in darkness at room temperature and the results recorded after two weeks. Kietreiber (1971) recommends rather to keep a lower incubation temperature, 10–12°C, to favour seedling attacks, especially *Fusarium*; in this case results should be recorded after three weeks. Here the seedlings have to penetrate a considerable mechanical barrier in high humidity similar to adverse field conditions. All seeds and seedlings are removed from the medium to enable close examination of the roots and coleoptiles for signs of disease and also to enable recording of those seedlings that have been incapable of penetrating through the medium.

Fig. 28.7. Hiltner test, carried out in a medium of sterile granulated brickstone. Oat seedlings tested for *Fusarium nivale*. (Department of Agriculture, Agricultural Scientific Services, Edinburgh, Scotland)

(2) *Application*. The method has been widely used at seed testing stations in many countries, and is still in use in Germany, Sweden, Norway (upon request), Egypt and other countries. It can be used for testing treated seed, especially cereals, and is also used for peas and beans in Germany.

(3) *Limitations*. The procedure requires an incubation period of at least two weeks as against seven days in the blotter method. It is a field performance test giving information on seedling symptoms, rather than a test for the presence of pathogens which may attack later and which are more readily identified in the blotter and agar plating tests.

28.5.2 Sand methods

(1) *Procedure*. Sterile, fine-grained sand, well wetted, is used as a medium. The seeds are covered by a 3–4 cm deep layer of water-saturated brickstone gravel (simplified pot method, according to Eggebrecht, 1949) or by a 2–3 cm deep layer of coarse-grained sand (Derlitzki, 1918). For small-grained seed, such as those of *Abies* and *Pinus*, the layer of coarse-grained sand should be 1 cm deep only. The water supply is, according to the prescriptions of these methods, also strictly standardised, but the amounts needed may vary according to the containers used.

(2) *Application*. The methods are used at seed-testing stations in Germany for cereals (the simplified pot method) and for agricultural, horticultural and forest seeds. Mead, Broadfoot and Machacek (1950) investigated the method applied for detection of *Drechslera*, *Septoria* and *Fusarium* in wheat, oat and barley, using non-sterile soil and washed non-sterile sand contained in cardboard boxes.

(3) *Limitations*. The method has the same advantages and disadvantages as the Hiltner method. The procedure is rather time-consuming — forest seed are examined as late as 30 days after sowing (Eggebrecht, 1949). On the other hand, information pertaining to field performance is provided.

28.5.3 Standard soil method

This procedure has been elaborated at the Swedish Government Seed Testing Station (Karlberg, 1974).

(1) *Procedure*. Basic material for the medium is a uniform soil mixture, 'Frühstorfer P-soil' composed of 4 parts of clay and 6 parts of peat with fertiliser added, after the prescriptions of the German professor Frühstorfer (Karlberg, 1974). The soil is sold in Sweden in plastic bags for use in gardens and greenhouses. The medium for the test is prepared by mixing 3

Fig. 28.8. Standard soil test. (Courtesy of Helmut Kolk)

parts of P-soil, 1 part of sterilised sand and 0.6 parts of water for 5 min to get uniformity; plastic multipot-trays are used. The multipots are filled, the tray wiped off and the seed is planted, see Fig. 28.8. A thin polyethylene bag is pulled over the tray to keep the moisture throughout the incubation. In Sweden, cereals are kept at 10°C for two weeks for development of *Fusarium nivale*, and subsequently one week at 20°C for *Septoria* and *Drechslera*. The symptoms are then easy to observe, and the percentage of attacked seedlings can be recorded. Counting takes place after 2–4 weeks, depending on the kind of seed and temperature. Every seedling is picked up and examined. After use the soil is sterilised by heating at 95°C for 20–30 min and can be used again.

(2) *Application*. In Sweden the method is adopted for routine testing of cereals for germination and health, and for testing various kinds of seed which in germination on blotter exhibit phytotoxic symptoms resulting from insecticide treatment. These symptoms are suppressed in the soil test.

(3) *Limitations*. The method is fairly well standardised by the use of multipot trays and a uniform standard soil mixture and is applicable as a rational and work-saving procedure on a larger scale. By growing every seedling separately the risk of contact infection among the seedlings is almost eliminated. The conditions of the test are closer to natural field conditions and when samples show phytotoxic or disease symptoms, evaluation of seedlings is more accurate in the soil test than on artificial media. On the other hand, only symptoms of seedling-parasitising pathogens are revealed, whereas to obtain specification of other pathogens the standard tests on artificial media are preferable.

28.5.4 Use of other media essentially sterile

Other media, such as china clay, may be used for the same type of test as provided in the above seedling symptom tests. In evaluating such media as to their suitability for examination of seedling symptoms, it is important to consider whether the particles of the medium concerned can easily be removed from the roots of the seedlings and that the colour of remaining particles does not cover symptoms to be investigated. China clay meets these requirements.

28.5.5 Standard disinfection methods

(1) *Procedure*. One of the methods discussed above is used. A working sample is divided and a sample treated in the laboratory with a commercial product according to a standard procedure. The degrees of disease developed in appropriate tests, especially a soil test, from the untreated or treated subsamples are compared. The method is regularly used in the seed testing stations in Norway and Sweden.

(2) *Application and limitations*. The procedure provides information on planting value related to seed-borne inoculum and on the advisability of suitably treating the seed lot.

28.5.6 Test tube agar method

(1) *Procedure*. The seed is sown in test tubes of 16 mm diameter, one seed per tube, on water agar, 10 ml. per tube, and incubated at 20°C or another suitable temperature according to pathogen and host, under artificial daylight tubes, 12/12 h cycle; to retain moisture the tubes are covered, in groups, by aluminium foil which is removed when the seedlings have reached the cover. The length of incubation is determined by the time needed for development of seedlings and symptoms. For barley tested for

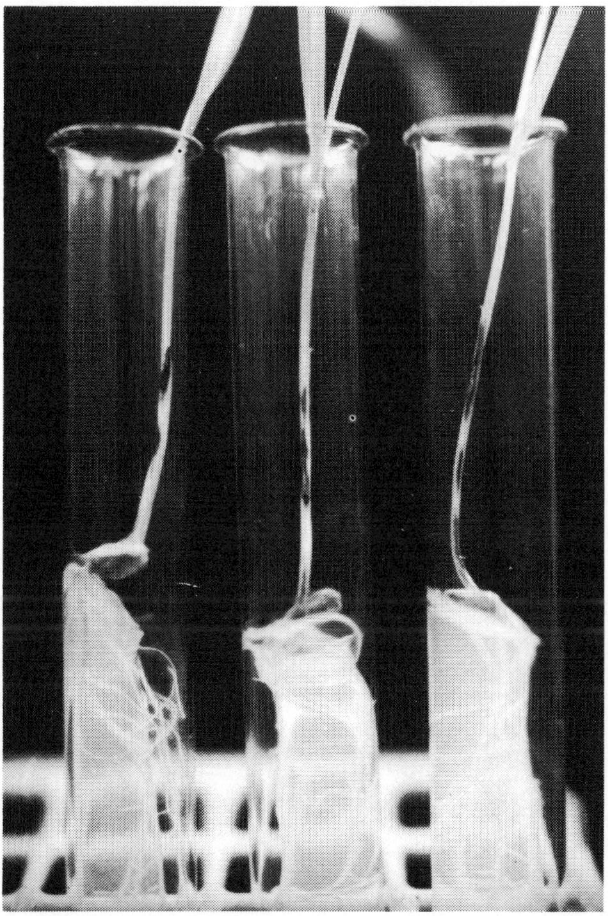

Fig. 28.9. Test tube agar method. Wheat seedlings showing symptoms produced by *Septoria nodorum*. 14 days' incubation.

Drechslera graminea, *D. teres*, *D. sorokiniana* and other pathogens producing seedling symptoms, 10 days' incubation is suitable; for wheat tested for *Septoria nodorum*, 14 days' incubation (Khare, Mathur and Neergaard, in print) — Fig. 28.9.

Water agar may be replaced by a nutrition or selective medium, or nutritional matter may be supplied later, after the development of the seedling. Also, blotter and water may be used as substrate (cf. Neergaard, 1945).

(2) *Application and limitations*. The procedure has recently been developed at the Institute of Seed Pathology and has been used for barley, wheat and legumes. The procedure allows better development of seedlings and symptoms than the blotter test, and symptoms can be easily studied being visible on roots as well as on the green parts; neighbour infection is excluded. When well organised, the procedure is probably not very time-consuming but it does require labour for preparation of medium and cleaning of tubes. Before international acceptance, the procedure would have to be tried out in international comparative tests. The procedure should prove useful in testing for quarantine of valuable material because healthy seedlings can be transplanted for further inspection and for production of seed.

29 Incubation Tests II: Factors Affecting Test Results

29.1 SELECTION OF TEST TYPE

An incubation test for detection of seed-borne pathogens has, as its objectives, the determination of (1) the *kind of inoculum*, (2) the *amount of inoculum*, (3) the *distribution of inoculum* within a seed lot (according to sampling procedures), and (4) the *vigour of inoculum* at the time of testing.

The *relative* vitality and virulence of pathogenic fungi may be measured in terms of (1) mycelial growth rate (agar test), (2) mycelial growth rate plus sporulation and ability to produce symptoms (blotter test), (3) ability to reduce field emergence and to produce seedling diseases (soil test, sand test, Hiltner tests, etc.), (4) ability to produce disease in seedlings grown from the seed under test, and (5) ability to induce disease in indicator plants.

There are five major types of tests based on incubation of seed sown on a suitable medium. They may be defined as follows:

(1) *Tests based on growth of pathogen only, as in the standard agar plate test.* Inoculum may be recorded as percentage of incidence, and further determined by measuring the growth rate of the colonies.

(2) *Tests based on growth of seedling and pathogen.* These are effected in the standard *blotter test* in which both growth rate and virulence of the inoculum can be studied. Inoculum may be recorded as infection percentage, and this value may be supplemented by records on infection intensity in terms of classified standard degrees of infection.

(3) *Tests based on growth of seedling and development of symptoms.* This type of test, which may be termed a *seedling-symptom test*, has a special, restricted purpose. The test forecasts the effect of seedling-infecting pathogens on field emergence of the seed lot concerned, rather than giving a detailed specification of the pathogens involved. For this purpose seeds are sown in *soil, sand or a semisterile inert medium* such as brickstone chips (Hiltner method), perlite, vermiculite, etc. This procedure

often indicates the vigour of germination of the seed as well as the total of inoculum potentials of seed-borne seedling pathogens. The method is excellent for testing treated seed.

(4) *Tests based on growth of the host beyond the seedling stage, and on development of symptoms*. This procedure, usually called a *growing-on test*, involves cultivation of plants to mature stages to allow development of those diseases that appear on mature plants only. The procedure is often used for post-entry quarantine control when it is the only method available. It is applicable for detection of downy mildews and certain systemic diseases not detectable in the seedling stage of the host, including those produced by fungi that infect the vascular system, and virus diseases. In closed quarantine for important pathogens, the procedure is to grow plants to maturity for production of seed which may be released after proper checking for health condition (Sheffield, 1968).

(5) *Tests based on growing of indicator plants and inoculation by pathogens harboured within the seed being tested*. This type of procedure which may be termed an *indicator test*, is often used for detection and identification of seed-borne viruses and for checking viability of pathogens, as in the test for *Septoria apiicola* in celery (Hewett, 1962, 1968).

29.2 FACTORS INVOLVED IN THE CONDITION OF SEED

The condition of the seed at the time of testing is important to the outcome of any culture test aiming at detection of pathogens. Low vigour of seed, whether due to premature harvesting, poor maturation or some physical or physiological damage, may favour development of saprophytes and substantially influence the competition between fungi harboured in the seed. Mechanically injured seeds offer entry points for saprophytes as well as rot-producing pathogens.

Ageing of seeds may induce a shift in the seed–parasite relationship because of differences in longevity. The surface-borne microflora of the seed, particularly when antagonists are present, may have a crucial effect on the development of pathogens in the test. *Botrytis cinerea* may occur in high percentages in newly harvested seed of many species, but infections are eliminated or strongly reduced after only one year of storage (Neergaard, 1950). Although infections of *Ascochyta pisi* in pea and vetch are much reduced after one or two years' storage this parasite may still be viable after 7 or 8 years (Sprague, 1929; Crosier, 1939; Wallen, 1955). On the other hand, decline in antagonistic fungi in seeds as a consequence of ageing may lead to an increase in percentage of pathogens, as observed for *Drechslera avenae* in oats (Malone, 1962; Limonard, 1968) and *Septoria nodorum* in wheat (Brönnimann, 1968; Limonard, 1968).

Finally, treatments of seed by fungicides, hot water or other means may lead to problems which could require the use of special testing techniques, cf. Subsection 29.5.4.7.

The condition of seeds can be improved in various ways to meet the technical requirements of a test. For certain kinds of seed which often are heavily contaminated by surface-borne saprophytes, chemical or other pretreatment may be necessary to prevent profuse growth of contaminants which can hamper development and recording of pathogens. As a mild disinfectant, a solution of sodium hypochlorite, containing approximately 1 per cent w/w available chlorine, is generally used for cereals, beans and peas tested by the agar plate method (Section 28.2).

Pretreatment of cruciferous seed by an antifungal antibiotic, e.g. aureofungin, is part of the procedure for detection of *Xanthomonas campestris* (Srinivasan, Neergaard and Mathur, 1973) (Section 28.4).

Pretreatment in a solution of the sodium salt of 2,4-dichlorophenoxyacetic acid may be used to suppress germination of seeds, thus facilitating recording of pathogens. Hagborg, Warner and Phillips (1950) first applied 2,4-D in agar medium for testing bean seeds for anthracnose, then Neergaard (1956AR), testing cabbage seed for *Phoma lingam* by the blotter test, added 2,4-D to the water (Section 28.1).

29.3 AMOUNT OF INOCULUM PER SEED, AND THE FACTOR OF ONSET

In agar tests the colony diameter of a given fungus usually varies considerably from seed to seed, even under constant environmental conditions. Such variation in growth rate reflects differences in the inoculum harboured in individual seeds. Two principal factors are involved: the amount and vigour of inoculum and the site of infection or contamination.

Malalasekera and Colhoun (1969), using wheat seeds artificially inoculated with *Fusarium culmorum*, correlated the amount of inoculum in terms of spore load per seed with the growth rate of mycelium on agar. The correlation between the common logarithm of spore load and colony size was highly significant, likewise the correlation between log spore load and disease recorded in terms of reduced dry weight of infected seedlings and mean percentage of pre-emergence death of seedlings was highly significant for seeds which were inoculated with logarithmically increasing spore loads. Various loads per 10 g of seeds yielded the following colony diameter on an agar medium: 2.9 cm by 10^2 spores, 5.4 cm by

10^4 spores, 6.5 cm by 10^6 spores and 7.2 cm by 10^7 spores (cf. Fig. 17.4). The relation of spore load to colony diameter on agar medium thus has a direct bearing on agar plate procedures for measuring similar seed-borne inoculum potential.

Mycelial infection of seeds is much more common than spore contamination. Seed coat, pericarp and glumes are common sites of infection for many seed-borne fungi in the imperfect phase, such as *Drechslera sorokiniana* in wheat and barley (Weniger, 1923; Mead, 1942*a*), *D. oryzae* in rice (Nisikado and Nakayama, 1943; Fazli and Schroeder, 1966), and *D. graminea* in barley, particularly in the apical part of the grain (Genau, 1928). *Drechslera avenae* is carried as dormant mycelium between the glumes and the pericarp of oats (Rathschlag, 1930). For all these fungi, considerable variation in the colony diameter on agar media is well known. This indicates differences in amount and vitality of the mycelial inoculum.

Obviously, the exact site of the inoculum within the tissues, i.e. whether it is located in the outer or inner layers, or confined to a particular area such as the apical part of a grain, has a bearing on the emergence of a fungus from the seed. For example, some colonies of *Drechslera* spp. in cereals become visible after only two days of incubation, whereas others take four or five.

In blotter tests, pathogenic fungi such as *Fusarium moniliforme* located deep within the tissues of rice and maize seeds may visibly grow and sporulate after a comparatively long time, cf. discussion of this in Subsection 29.5.4.7. Late development was demonstrated by Kucharek and Kommedahl (1966) who found much higher percentages of *Fusarium moniliforme* when halved kernels of maize rather than whole kernels were plated on potato dextrose agar. For example, 24 and 1 per cent, 20 and 0 per cent, were reported for two seed lots with halved and whole kernels, respectively. In rice, this fungus has been found to be harboured mainly in the embryo (Vidhyasekaran, Subramanian and Govindaswamy, 1970). In maize, Manns and Adams (1923) and Koehler (1942) found all parts of the kernel invaded by the mycelium, but in recent investigations internal infections were confined to the hilum region (Gemawat, Mathur and Neergaard, unpublished).

In the blotter test, differences similar to those in the agar test occur in the development of colonies on and around individual seeds, and in the intensity of infection in individual seedlings. Incidence may range from one or a few spores or spore chains of *Alternaria brassicicola* on cabbage seeds and *Drechslera* spp. on cereal seeds, to a dense, sporulating colony covering the entire seed and producing conspicuous symptoms on the seedling. Such differences in the amount of inoculum per seed account for discrepancies in test results from station to station because the conditions of incubation may vary and suppress weak incidences, and from person to person because slight incidences are missed to a greater or lesser extent by individual analysts.

Eventually it may be possible to estimate individually the inoculum on each seed and record it according to a defined scale. Such a procedure would considerably increase the precision of results which at present usually record only infection percentage.

29.4 RESPONSE OF PATHOGEN TO TEST CONDITIONS

The results of seed health tests depend on the reaction of pathogens to the conditions of incubation. In the standard laboratory procedures, test conditions are such that certain groups of pathogens are found, namely those which are capable of producing sufficient mycelial growth and/or sporulation on agar media or of producing mycelial growth, sporulation and symptoms, if any, on seedlings in the blotter test. Thus, the standard laboratory seed health testing procedures are applicable for detection of many genera in the Fungi Imperfecti responsible for seedling diseases, certain foliage diseases, certain types of rot and necroses of stems, roots and fruits, and a few wilt diseases. These pathogens include species of *Alternaria*, *Ascochyta*, *Botrytis*, *Cercospora*, *Colletotrichum*, *Curvularia*, *Drechslera*, *Fusarium*, *Myrothecium*, *Phoma*, *Stemphylium*, *Verticillium*, and others (cf. examples given in Table 29.1). On the other hand, many important seed-borne fungal pathogens such as downy mildews and smut fungi are not detectable by the standard methods mentioned.

The response of pathogens to a selected testing procedure naturally determine its applicability (Table 29.1). Thus in the agar plate method the rate of mycelial growth is decisive. In Table 29.2, types of seed-borne fungi detectable by this procedure are listed, grouped according to rate of mycelial growth, roughly characterised as slow, medium and rapid.

Table 29.1. Applicability of culture test procedures to different categories of seed-borne pathogens and diseases*

Category of seed-borne pathogen/disease	*Culture test procedure*† (1)	(2)	(3)	(4, 5, 6)
Physiogenic diseases:				
Boron deficiency	blotter			
Cavitation (hollow heart)	blotter			
Manganese deficiency	blotter			
Molybdenum deficiency			seedl.-sympt.	
Zygomycetes:				
Rhizopus arrhizus (cotton)	blotter?	agar?		
Rhizopus stolonifer	blotter	agar		

Table 29.1 *continued*

Category of seed-borne pathogen/disease	*Culture test procedure*† (1)	(2)	(3)	(4, 5, 6)
Oomycetes:				
Phytophthora capsici, P. infestans, etc.				growing-on
Phytophthora phaseoli				very wet sand
Plasmopara halstedii, P. tabacina, P. arthuri, etc.				growing-on
Sclerospora graminicola, Sclerophthora macrospora, etc.				growing-on
Ascomycetes, cf. also Fungi Imperfecti, (with few exceptions always as conidial state):				
Cochliobolus geniculatus, C. miyabeanus, C. setariae, C. spicifer, C. sativus, C. victoriae	blotter	agar		
Diaporthe phaseolorum		agar		
Diplodia gossypina, D. macrospora, D. maydis	blotter	agar		
Gibberella fujikuroi, G. zeae	blotter	agar		
Glomerella cingulata	blotter	agar		
Glomerella gossypii		agar		
Griphosphaeria nivalis	blotter	agar	seedl.-sympt.	
Khuskia oryzae	blotter	agar		
Leptosphaeria avenaria, L. nodorum	blotter	agar		
Mycosphaerella ligulicola, M. pinodes	blotter	agar		
Mycosphaerella melonis	blotter			
Pleospora björlingii, P. herbarum, P. calvescens	blotter	agar		
Pyrenophora avenae, P. graminea, P. teres, P. tritici-repentis	blotter	agar	seedl.-sympt.	
Sclerotinia sclerotiorum (mycelium, sclerotia)	blotter	agar		
Trichometasphaeria turcica	blotter	agar		
Fungi Imperfecti (incl. conidial state of most of the Ascomycetes listed above):				
Acremonium strictum	blotter			
Acroconidiella eschscholtziae				
Alternaria brassicae, A. brassicicola, A. cinerariae, A. dianthi, A. dianthicola, A. lini, A. linicola, A. longissima, A. porri (incl. f.sp. dauci, f.sp. solani), A. radicina, A. raphani, A. sesami, A. tenuis, A. zinniae, etc.	blotter	agar		
Alternaria triticina		agar		
Ascochyta chrysanthemi, A. gossypii, A. pisi, A. rabiei, etc.	blotter	agar		
Aureobasidium lini		agar		
Botryodiplodia palmarum	blotter			
Botrytis allii, B. byssoidea, B. cinerea, etc.	blotter	agar		
Cercospora beticola, C. kikuchii	blotter	agar		
Colletotrichum dematium (incl. f.sp. circinans, f.sp. spinaciae, f.sp. truncatum) C. gloeosporioides, C. godetiae, C. lindemuthianum, C. lini	blotter	agar		
Corynespora cassiicola	blotter	agar		

Table 29.1 *continued*

Category of seed-borne pathogen/disease	*Culture test procedure*† (1)	(2)	(3)	(4, 5, 6)
Curvularia cymbopogonis, C. eragrostidis, C. geniculata, C. lunata, C. oryzae, C. pallescens, C. trifolii, etc.	blotter	agar		
Drechslera avenacea, D. avenae, D. catenaria, D. dictyoides, D. graminea, D. halodes, D. longirostrata, D. oryzae, D. papaveris, D. phlei, D. poae, D. rostrata, D. setariae, D. sorokiniana, D. teres, D. tritici-repentis, D. turcica, D. victoriae, etc.	blotter	agar	seedl.-sympt.	
Didymella lycopersici, imperf. state	blotter	agar	seedl.-sympt.	
Epicoccum purpurascens	blotter	agar		
Fusarium avenaceum, F. culmorum, F. equiseti, F. graminearum, F. nivale, F. scirpi, F. semitectum, F. solani and others	blotter	agar	seedl.-sympt.	
Fusarium moniliforme, F. oxysporum	blotter	agar		growing-on
Itersonilia pastinacae	blotter	agar		
Kabatiella caulivora	blotter	agar		
Macrophomina phaseolina	blotter	agar		
Myrothecium leucotrichum, M. roridum, M. verrucaria, etc.	blotter			
Nigrospora oryzae	blotter	agar		
Pestalotiopsis funerea	blotter			
Phoma bellidis, P. betae, P. exigua, P. herbarum, P. medicaginis and others	blotter	agar		
P. lingam	blotter			
Phomopsis sojae		agar		
Phyllosticta antirrhini	blotter	agar		
Pleiochaeta setosa	blotter		seedl.-sympt.	
Pyricularia oryzae	blotter			
Ramularia cyclaminicola				growing-on
Rhizoctonia solani	blotter	agar		
Selenophoma donacis				growing-on
Septoria apiicola				indicator, muslin test (Sheridan, 1963)
Septoria glycines, S. nodorum	blotter	agar		
Stemphylium botryosum	blotter	agar		
Trichoconis padwickii	blotter	agar		
Truncatella hartigii	blotter		seedl.-sympt.	
Verticillium alboatrum, V. dahliae	blotter			growing-on
Basidiomycetes:				
Marasmius perniciosus				growing-on
Thanatephorus cucumeris	blotter	agar		
Bacteria:				
Corynebacterium michiganense				growing-on
Pseudomonas angulata				growing-on
Pseudomonas lachrymans	blotter			growing-on (vermiculite)
Pseudomonas glycinea, P. phaseolicola, P. solanacearum				growing-on

Table 29.1 *continued*

Category of seed-borne pathogen/disease	*Culture test procedure*† (1)	(2)	(3)	(4, 5, 6)
Xanthomonas campestris				agar seedl.-sympt.
Xanthomonas oryzae, X. phaseoli, X. vesicatoria				growing-on
Viruses:				
Arabis mosaic, avocado pear sunblotch, barley stripe mosaic, bean common mosaic, bean southern mosaic, cineraria mosaic, cowpea mosaic, cucumber mosaic, lettuce mosaic, mungbean mosaic, pea seed-borne mosaic, prunus necrotic ring-spot, raspberry ringspot, soybean mosaic, soybean stunt, squash mosaic, tobacco mosaic, tobacco rattle, tobacco ringspot, tomato black ring, tomato ringspot, tomato spotted wilt				growing-on, indicator

*For details on hosts, occurrence, control, authorities for Latin names, synonyms, etc., consult Index.

†Names of test procedures, abbreviated above: (1) blotter, (2) agar plate, (3) seedling-symptom, (4) growing-on, (5) indicator, (6) other methods (specified).

Table 29.2. Growth rate of fungi developed from seed on standard agar media at 20°C

Growth rate	*Organism*
Slow: Diameter below 2 cm after 7 days	Aureobasidium lini Cercospora oryzae Cladosporium cladosporioides Gloeotinia temulenta Pyricularia oryzae Septoria apiicola Septoria nodorum Verticillium alboatrum
Medium: Diameter 2–6 cm; usually 3 ± 2 cm, after 7 days	Alternaria (e.g. A. tenuis auct., A. porri, A. brassicae, A. brassicicola, etc.) Colletotrichum (all or most spp., e.g. C. graminicola, C. lini, C. dematium) Curvularia (all or most spp.) Drechslera (all or most spp., e.g. D. oryzae, D. rostrata, D. sorghicola, D. sorokiniana, D. teres) Fusarium (most spp., e.g. F. avenaceum, F. culmorum, F. equiseti, F. oxysporum, F. semitectum, F. solani) Phoma (all or most spp., e.g. P. betae, P. exigua, P. glomerata) Stemphylium botryosum Trichoconis padwickii Ulocladium (all or most spp.)

Table 29.2 *continued*

Growth rate	*Organism*
Rapid: Diameter more than 6 cm; often about 10 cm after 7 days	Botrytis cinerea Epicoccum purpurascens Fusarium graminearum Fusarium nivale Fusarium poae Nigrospora oryzae Rhizoctonia solani Rhizopus stolonifer Sclerotinia sclerotiorum Sordaria fimicola Trichoderma viride

In practice, slow-growing fungi will usually be overgrown by others and therefore often recorded in too low percentages. On the other hand, rapidly growing fungi such as *Botrytis cinerea* which also begin growth quickly are detectable after a period of incubation shorter than standard; after this period they will have overgrown other pathogenic fungi, which cannot then be properly recorded.

29.5 FACTORS OF VARIATION IN INCUBATION TESTS

In attempts to establish standard conditions of incubation, all factors promoting growth, fructification and development of symptoms of pathogens must be considered. These factors are dependent on the fungi under test and must be closely controlled according to a defined standard. Lack of one essential factor in a procedure may lead to its complete failure as a standard test. The principal factors to be considered are the following:

1. Variation due to sampling
2. Storage conditions of the sample
3. Conditions of pretreatment
4. Conditions of incubation
   - 4.1. Container
   - 4.2. Spacing of the seeds
   - 4.3. Medium
   - 4.4. Temperature
   - 4.5. Humidity (particularly in the blotter test)
   - 4.6. Light
   - 4.7. Time (period of incubation)
5. Recording
   - 5.1. Conditions of recording (sources of light, magnification used)
   - 5.2. Distinction (personal)
   - 5.3. Early recognition
   - 5.4. Estimate (between infection/contamination and healthy/non-contaminated)

29.5.1 Variation due to sampling

As discussed above (Chapter 26), any health testing of seed lots must be based on a representative sample. The uniformity of a seed lot may be less with regard to infections than to germination capacity or purity. Little has been done so far to investigate this aspect of sampling. Also, the working sample must be drawn in accordance with the prescriptions of the International Rules for Seed Testing.

29.5.2 Storage conditions of the sample

The longevity of some seed-borne parasites and saprophytes is more restricted than that of the host. Some seed-borne microorganisms die within a few months of storage, others survive the seeds by many years. The storage conditions of the sample after it has left the warehouse of the seed company are of great importance. The conditions during transportation and in the laboratory of the testing agency may lead to testing results which do not reflect the actual condition of the seed lot as maintained in the warehouse. Also, comparison between results obtained by different stations on the same seed lot may be confused if the samples submitted to them have been stored under different conditions. The ISTA Committee on Plant Diseases therefore recommends referee seed samples to be stored at a temperature not higher than 10°C. Plastic bags may be used to prevent desiccation.

29.5.3 Conditions of pretreatment

The purpose of pretreatment is to remove contamination of competing organisms, mainly saprophytes, which impair recording. The treatment must bring about a mild disinfestation but should not interfere with the development of pathogens under test. It may be carried out as a chemical or physical treatment (dry heat treatment, as proposed by Malone, 1962). Immersion of the seed for 10 min in a solution of sodium hypochlorite containing approximately 1 per cent w/w available chlorine and letting the surplus liquid drain off, is a commonly used and usually satisfactory procedure. A somewhat stronger treatment is immersion in a solution of one of the commercial mercury compounds, followed by thorough washing of the seeds to remove any trace of the fungicide that may interfere with the development of the pathogens. In order to obtain internationally comparable results it is imperative strictly to follow a standard procedure of pretreatment.

29.5.4 Conditions of incubation

29.5.4.1 CONTAINER

There is a great variety of containers in use for the blotter test, including glass containers and plastic (perspex) containers of different size. According to the experience of the International Comparative Seed Health Testing Schemes, organised by the ISTA Committee on Plant Diseases, the type of container is of no direct importance. However, plastic containers transmit light with the wavelengths stimulating sporulation of fungi and have been recommended by the Sixth International Workshop on Seed Pathology in preference to glass containers except Pyrex (ISTA, 1964*a*).

29.5.4.2 SPACING OF SEEDS

Spacing is dependent on the size of the seed and the growth rate of colonies from each seed in relation to the length of the incubation period. Formulae have been worked out for calculation of sowing distance when the gross diameter of the pathogen colony at the date of recording is known, or that of the area of infected seeds/seedlings involved, and the maximum infection percentage to be excepted for that pathogen (Neergaard, preliminary note in Condensed Version of the discussions at the Sixth International Workshop on Seed Pathology, Lisbon, 1963 (ISTA, 1964*b*)). The possibility of neighbour infection must be considered. In recording *Phoma lingam* and other pathogens that may spread on the blotter, infecting surrounding seeds and seedlings, only centres of infection should be counted. *Botrytis cinerea* is known to spread within a container so quickly that secondary infection of seedlings occurs at some distance from the primary foci. Rapidly spreading fungi such as *Rhizoctonia solani* and *Sclerotinia sclerotiorum* can hardly be recorded as per cent incidence, because detection of originally infected seeds at an early date in the test is difficult. One seed per container would be the ideal, and some suggestions to apply the 'multipot' principle used in nurseries have been made. However, as a laboratory procedure for use in blotter tests or agar plate tests, the multipot system has presumably little prospects.

29.5.4.3 MEDIUM

The type of filter paper for the blotter test seems to be of subordinate significance, but in the agar plate test the selection of medium is of great importance.

Two nutrient agar media, malt extract agar (MEA) and potato dextrose agar (PDA), are in common use in seed health testing. The International Rules for Seed Testing recommend 2 per cent MEA in the agar plate method for testing of cereals, legumes and flax. In the International Comparative Seed Health Testing Schemes, organised by the ISTA Committee on Plant Diseases, PDA has been adopted for many kinds of crop seeds, including cereals, mainly because this medium has been found preferable for recognition of *Fusarium* spp. In order to obtain maximum uniformity of media, standard brands of PDA and MEA are used for comparative tests — Difco 0013-0 (final pH 5.6 at 25°C) and Difco MEA 0012-01 (final pH 4.6 at 25°C) have been adopted.

Although malt preparations not having the same diastatic value may lead to distinct differences in appearance and growth rate of fungi, the counts may not vary substantially from one batch of media to another. Malone (1959) tested oats heavily contaminated by *Drechslera avenae* on five brands of MEA and found that the percentage and type of colonies produced on the media were quite comparable.

The amount of agar medium per dish, i.e. depth of the agar plate, should be standardised. Brancato and Golding (1953) found that the depth of medium influences amounts of mycelium and spores produced, whereas the diameter of the colony is not affected. Leach (1962*b*) demonstrated that depth of medium affects the quality and intensity of transmitted ultraviolet and visible radiations. Absorption increases with decrease in wavelength as well as increase in depth of medium. Shorter wavelengths and higher light intensities caused pycnidia of *Ascochyta pisi* to develop deeply submerged in the medium.

Variation of the pH of a medium may be important for development of cultural characters and, hence, to identification and recording. For the comparative testing schemes it has been generally accepted that changes in the pH of the standard nutrient agars should not be made.

Table 29.3 illustrates colour reactions due to development of pigments in *Fusarium* spp. at different hydrogen-ion concentrations. Reagents may, however, be added to change the pH and obtain a colour reaction specific to the fungus species under consideration.

Table 29.3. Colour of diffusible pigment of three *Fusarium* species, developed at different hydrogen-ion concentrations in a solution of dextrose in combination with certain inorganic salts, and in 2 per cent dextrose agar with the same inorganic salts. Colour designations according to Ridgway's *Color Standards and Nomenclature,* Washington, 1912. (Data from Sideris, 1925)

	Fusarium culmorum		*Fusarium moniliforme*		*Fusarium solani*	
pH	Dextrose solution	Dextrose agar	Dextrose solution	Dextrose agar	Dextrose solution	Dextrose agar
3.0	Begonia rose*	—	Flesh	—	Apricot orange	—
4.0	India red	Hydrangea pink	Hydrangea pink	Hydrangea pink	Nopal red	Maroon red
5.0	Victoria Lake	Vinaceous pink	La France pink	La France pink	Maroon dark	Nopal red
6.0	Colourless	Purplish pink	Colourless	Vinaceous pink	Colourless	Cinnamon purple
7.0	Colourless	Purple	Colourless	Vinaceous purple	Colourless	Greenish purple
7.5	Colourless	Purplish blue	Colourless	Lilac	Colourless	Greenish blue

*Reference is made to Rayner's (1970) *Mycological Colour Chart,* in which Ridgway's colour names are listed in alphabetical order, enabling translation into modern notations; e.g. Begonia rose becomes 2.5 R/5.8/120.

Joaquim, Mathur and Neergaard (in preparation) found that pH of the medium may greatly influence infection percentage in the blotter test and mycelial growth (colony diameter) in the agar plate test. Optimum for detection of *Drechslera oryzae* on blotter was between pH 7 and pH 9, with a peak at pH 8, and for *Trichoconis padwickii* between pH 4 and pH 6, with a peak at pH 5. In the agar plate test, optimum for mycelial growth on potato dextrose agar was at pH 7.9 for *Drechslera oryzae*, at pH 7.3 for *Fusarium moniliforme* and at pH 5.3 for *Trichoconis padwickii*.

Distilled water should be used in preparing the media because tap water may vary considerably from place to place. It has been brought out (ISTA, 1966*a*) that the tap water in Wageningen, Holland, is very high in manganese; in the U.K. the water in the Edinburgh area contains less nitrate than that at Cambridge and to achieve uniformity with Cambridge KNO_3 is added to the water in the Scottish seed testing station for germination of certain grass seeds (Noble, 1971). A monograph has been published on standard waters (Ashworth and Crozier, 1972) in which recommendations are made for preparation of standard waters used for pesticidal formulations; a FAO survey of naturally occurring water is given.

Selective agents may be added to agar media. Papavizas (1967) evaluated a number of media that had been suggested for selective isolation of fungi from soil, and recommended 18 media for selective isolation of *Fusarium* by the soil dilution plate method. After preliminary testing of three of these media for seed health testing, Limonard (1968) found peptone–PCNB agar (Nash and Snyder, 1962) to be most suitable. This medium (modified) is prepared from Difco peptone with the addition of certain salts, terramycin (50 ppm) and 1,000 ppm of pentachloronitrobenzene (PCNB), and two per cent agar. This modified medium proved highly selective for various species of *Fusarium* in cereals. The average percentage recorded on PDA increased twice or three times on peptone-PCNB agar in seed samples of wheat, rye, barley and oats. Peptone-PCNB agar requires no sterilisation, and incubation may be in the dark or in light. The medium may prove useful for further development of a test selective for *Fusarium*.

In order to stop the growth of fast-growing fungi and to obtain more reliable counts, growth-retarding agents are commonly used for isolation of soil organisms. Two of these agents are rose bengal (Martin, 1950) and oxgall (Littmann, 1947; Miller, Peers and Neal, 1951). Limonard (1968) found 2,000 ppm oxgall in PDA to be the most satisfactory of these, facilitating recording of *Septoria nodorum* in wheat; see Colour Plate IV, oxgall PDA used for detection of *Fusarium* spp. Rose bengal is workable but causes confusion in identification and recording because it colours colonies of fungi and the medium red.

The use of selective fungicides in agar media for seed health testing has not been put into practice. Bollen and Fuchs (1970) tested a large number of pathogenic fungi incubated at 22–25°C on PDA, to which was added different concentrations of benomyl. This fungicide is highly selective, exercising a pronounced fungistatic effect, distinctly more to certain groups of fungi than to others. With dosages of 0.1–1.0 ppm the following fungi were suppressed: *Sphaeropsidales (Ascochyta, Phoma), Moniliales-Blastosporae (Botrytis, Cladosporium), Moniliales-Phialosporae (Aspergillus, Fusarium, Penicillium, Phialophora),* whereas much higher dosages, 100–500 ppm, were needed to check the following groups of fungi: *Oomycetes, Zygomycetes, Moniliales-Porosporae (Alternaria, Curvularia, Drechslera, Stemphylium),* and exceptionally one species of *Phoma*: *P. betae*. This selectivity may offer possibilities for establishing a procedure for detection of seed-borne species of *Phytophthora*, e.g. those known to occur in bean, *Citrus*, pepper, soybean, tomato (cf. Noble and Richardson, 1968), provided that proper nutrient agar media for this type of fungus are used, such as a selective medium proposed by Castillo, Fernández-Northcote and Fernández-Northcote (1974).

In the blotter test, paper is used to provide high humidity during incubation. The type of blotter used, within the range of ordinary commercial brands available, seems to be of minor importance.

Sand and particularly soil as media for seedling-symptom tests and growing-on tests are more difficult to standardise internationally. Within individual countries fairly good standardisation, based on local materials, may be obtained. Such tests may work satisfactorily for local disease problems, when the need for international standardisation is less urgent. However, studies on this subject are very desirable and should be carried out through international comparative tests based on different local materials, but under otherwise standardised incubation conditions. The growing-on test for detection of lettuce mosaic, devised by Rohloff (1962), is an example of a well-standardised version of this type of test. The composition of the medium used in this test has been specified: peat with 0.5–2.0 g $CaCO_3$ and 0.5–1.0 g complete fertiliser per litre.

29.5.4.4 TEMPERATURE

Incubation tests are greatly influenced by temperature which affects germination, growth, reproduction and other activities of organisms. The temperature response curve, defining the minimum, optimum and maximum temperature, and a more or less wide optimum temperature region, is specific for each of the different life processes of an organism, although some of the curves may more or less coincide. The temperature curve is usually skewed to the right, particularly for fungi with a high optimum, the distance between optimum and maximum being shorter than the distance between optimum and minimum. The curve for mycelial growth may differ from the curves for sporulation, germination and infection. The temperature response curve for development of symptoms in one host may differ distinctly from the curve for infection of another host by the same pathogen. Temperature affects the morphology of fungi and conditions for normal typical development sometimes are confined to a rather sharply defined range.

These facts imply that selection of incubation temperature for incubation tests depends largely on the pathogen to be detected as well as the testing procedure. A test based on mycelial growth may require temperatures other than those for seedling growth and development of symptoms.

In agar tests for fungi the incubation temperature must be related to the optimum for mycelial growth as well as for sporulation.

In Table 29.4 a survey is given of the principal temperature requirements for mycelial growth of some mesophilic seed-borne fungi detectable in agar tests. It is important to define the approximate optimum temperature range rather than the exact peak of the temperature curve. Taking into consideration the two more important standard temperatures adopted in the International Rules for Seed Testing for agar tests (ISTA, 1966*a*), it appears that the optimum temperature for all fungi listed in Table 29.4 coincides with either the lower standard temperature adopted in seed health testing, 20°C (18–22°C), or the higher standard temperature, 28°C (26–29°C), or with both of these.

Table 29.4. Optimum temperature region for mycelial growth of some seed-borne fungi on agar media. The establishment of the precise peak of the temperature curve depends on the experiment temperatures actually having been tested around the true optimum temperature

Optimum temperature group (°C)	*Fungus*	*Approx. peak of optimum* (°C)	*Approx. region of optimum* (°C)	*References*
	Phoma apiicola	16		Bennett, 1921
	Cladosporium herbarum	18	15–21	Jaarsveld, 1940
	Alternaria cinerariae	20	16–24	Neergaard, 1945
	Fusarium nivale	19–20	16–24	Tu, 1929; Dahl, 1934; Endo, 1963; Ponchet, 1966
<18–20	Geniculodendron pyriforme	20	15–21	Salt, 1974
	Drechslera victoriae (one isolate), cf. 30°	20		Wilson and Murphy, 1964
	Colletotrichum lindemuthianum	20	18–22	Schaffnit and Böning, 1925
	Mycosphaerella pinodes	20	18–25	Sattar, 1934
	Stemphylium botryosum	22	18–26	Neergaard, 1945
	Fusarium lateritium	20–24	20–24	Massie and Peterson, 1968
	Fusarium tricinctum	22–23	18–24	Seemüller, 1968
	Septoria linicola		18–24	Krüger, 1940
22–23	Alternaria brassicae	22–23	19–25	Neergaard, 1945; Taber *et al.*, 1968
	Botrytis allii, B. byssoidea, B. squamosa	20–25	20–25	Walker, 1926

Table 29.4 *continued*

Optimum temperature group (°C)	*Fungus*	*Approx. peak of optimum* (°C)	*Approx. region of optimum* (°C)	*References*
22–23 *continued*	Phoma destructiva	21–24		Chupp and Sherf, 1960
	Alternaria linicola	23	20–27	Neergaard, 1945
	Drechslera tritici-repentis	23		Ammon, 1962
	Fusarium avenaceum	22/27/28	20–29	Tu, 1929; Chi and Hanson, 1964
	Botrytis cinerea	23–24	18–26	Lauritzen, 1932
	Alternaria raphani	23–25	15–25	Neergaard, 1945; Taber *et al.*, 1968
	Verticillium alboatrum	22–25	20–27	Edson and Shapovalov, 1920; Brinkerhoff, 1969
	Phoma betae	24	18–29	Chupp and Sherf, 1960
	Colletotrichum lini	24	19–26	Krüger, 1940
	Alternaria dianthi	24	20–27	Neergaard, 1945
	Penicillium expansum	24	21–25	Jaarsveld, 1940
	Sclerotinia sclerotiorum	24	20–27	Lauritzen, 1932; Bunschoten, 1933; Bedi, 1962
	Rhizopus stolonifer, cf. 27°	24	23–26	Lauritzen and Harter, 1925
	Fusarium graminearum	24/27	22–28	Jones, 1924; Dickson, 1923; Tu, 1929
	Alternaria brassicicola	24–25	19–30	Neergaard, 1945; Taber *et al.*, 1968
	Rhizoctonia solani	21/25/27	20–29	Jaarsveld, 1940; Endo, 1963; El Zarka, 1965; cf. Sherwood 1970
24–26	Fusarium coaeruleum	25	20–26	Edson and Shapovalov, 1920
	Epicoccum purpurascens	25	20–26	Neergaard and Saad, 1962
	Phoma medicaginis var. pinodella	25	20–27	Sattar, 1934
	Drechslera dactylidis	25	22–27	Ammon, 1962
	Fusarium poae	25	22–27	Wollenweber and Reinking, 1935; Ammon, 1962; Seemüller, 1968
	Ascochyta pisi	25	20–28	Leach, 1962b, Sattar, 1934
	Fusarium solani, cf. 28°	25	20–29	Edson and Shapovalov, 1920
	Drechslera sorokiniana, cf. 28–30°	25		Wilson and Murphy, 1964
	Drechslera avenacea, D. rostrata	25		Ammon, 1962
	Drechslera avenae	25	20–28	Ravn, 1900; Ammon, 1962
	Alternaria triticina	25	19–30	Prabhu and Prasada, 1966
	Drechslera graminea	25/28/29	22–30	Ravn, 1900; Nisikado, 1929; Ammon, 1962; Chandwani and Munjal, 1963
	Drechslera siccans	25	22–30	Ammon, 1962
	Alternaria tenuis, cf. 28°	25–26	20–30	De Haan, 1937; Neergaard, 1945; Neergaard and Saad, 1962
	Drechslera teres	25–26	20–30	Ravn, 1900; Ammon, 1962; Singh, 1962
	Alternaria dianthicola	26	21–28	Neergaard, 1945
	Colletotrichum dematium f.sp. circinans	26		Walker, 1969

Table 29.4 *continued*

Optimum temperature group (°C)	*Fungus*	*Approx. peak of optimum* (°C)	*Approx. region of optimum* (°C)	*References*
27 – 28	Fusarium culmorum, cf. below	27	20–30	Tu, 1929; De Haan, 1937
	Drechslera tuberosa, D. dictyoides, D. phlei	27		Ammon, 1962
	Alternaria zinniae	27	23–28	Neergaard, 1945
	Myrothecium roridum, M. verrucaria	27	24–30	Cunfer, Graham and Lukezic, 1969
	Rhizopus stolonifer, 24°	27	22–29	Harter and Weimer, 1923
	Drechslera turcica	27	23–30	Nisikado, 1927
	Fusarium sporotrichioides	27–28	22–29	Seemüller, 1968
	Fusarium fusarioides	27–28	24–30	Seemüller, 1968
	Fusarium moniliforme	27–30	27–32	De Haan, 1937
	Alternaria macrospora (cotton isolate)	27–30	20–35	Nisikado, Kimura and Miyawaki, 1943
	Alternaria porri (incl. f.sp. solani, f.sp. dauci)	27–28	22–31	Klaus, 1941; Neergaard, 1945
28 – >30	Drechslera oryzae	28	18–32	Ocfemia, 1924; Neergaard and Saad, 1962
	Alternaria radicina	28	22–31	Lauritzen, 1926; Neergaard, 1945
	Drechslera sorokiniana, cf. 25°	28–30		Wollenweber and Reinking, 1935; Wilson and Murphy, 1964
	Fusarium culmorum		24–32	Simmonds, 1928
	Alternaria tenuis (bean isolates), cf. 25 – 26°	28	20–30	Saad and Hagedorn, 1970
	Fusarium oxysporum	28	20–29	Chi and Hanson, 1964
	Fusarium oxysporum f.sp. lycopersici	28	24–31	Edson and Shapovalov, 1920; Jones, 1922
	Fusarium oxysporum f.sp. cepae			Wollenweber and Reinking, 1935
	Fusarium oxysporum f.sp. vasinfectum	28?	25–30	Wollenweber and Reinking, 1935
	Fusarium oxysporum f.sp. pisi	28?	27–30	Wollenweber and Reinking, 1935
	Fusarium oxysporum f.sp. niveum	28–30	25–32	Sleeth, 1934; Wollenweber and Reinking, 1935
	Trichoconis padwickii	28	25–29	Mathur and Neergaard, 1970

Table 29.4 *continued*

Optimum temperature group (°C)	*Fungus*	*Approx. peak of optimum* (°C)	*Approx. region of optimum* (°C)	*References*
	Pyricularia oryzae	28	25–29	Neergaard and Saad, 1962
	Fusarium solani, cf. 25°	28–30	20–32	Tu, 1929; Chi and Hanson, 1964
	Diplodia maydis	28–30		Walker, 1969
28–>30 *continued*	Drechslera victoriae (3 isolates), cf. 20°	30		Wilson and Murphy, 1964
	Drechslera maydis	30	23–33	Nisikado, 1927
	Drechslera setariae	30		Nisikado, 1929
	Macrophomina phaseolina	30/35		Mathur and Sackston, 1963
	Aspergillus fumigatus	37		Rennerfelt, 1941

However, the response of fungi to temperature in agar tests is less simple than it appears from these data. Several modifying factors operate in the response of individual species involved in an agar test. Different growth rates of fungi on a seed, the time needed for different species to grow out from their location in the seed to the agar, and antagonistic interactions may all be decisive for the outcome of competition between organisms harboured in one seed.

These facts account for differences in temperature curves based on mycelial growth of pure cultures on agar media, such as potato dextrose agar, compared with percentage of diseases from seed sown on the same medium. Neergaard and Saad (1952) found that for *Epicoccum purpurascens* on rice the optimum temperature for mycelial growth was 25°C, but for percentage of incidence observed it was 19°C. For *Alternaria tenuis* the optimum for mycelial growth was also 25°C but for the maximum percentage incidence it was 28°C. On the other hand, for *Pyricularia oryzae* the optimum temperature was 28°C for both mycelial growth and percentage incidence. Finally, for *Drechslera oryzae* there was a broad optimum temperature range of 18–30°C with no definite optimum peak, for both mycelial growth and percentage incidence. Nevertheless, the development of colonies of *D. oryzae* within this optimum temperature region depends largely on the rest of the mycoflora associated with the same seed. The growth rate of *D. oryzae* is much reduced at lower temperatures, 19–22°C, when it is growing in association with other fungi, such as *Alternaria tenuis* and *Epicoccum purpurascens*. These fungi exercise dominance of mycelial growth within the same optimum temperature region (Fig. 29.1).

Differences in the optimum temperature may exist between biotypes, such as in those which originate from different geographic regions; the peak of the temperature curve for mycelial growth of three different

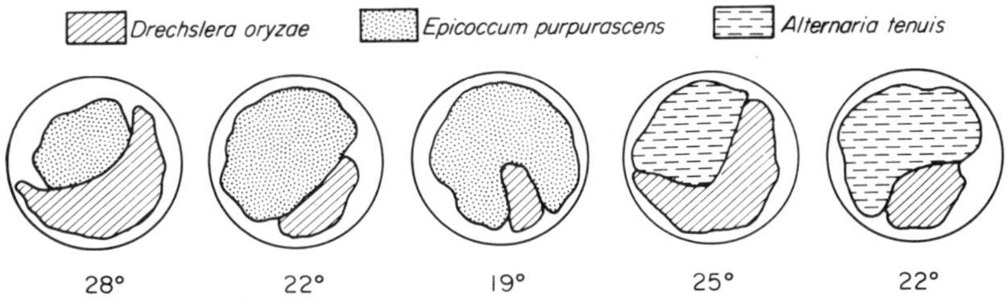

Fig. 29.1. Diagram of interaction between *Drechslera oryzae* and *Epicoccum purpurascens*, and *Drechslera oryzae* and *Alternaria tenuis* on potato dextrose agar at different temperatures. Note dominance of *Drechslera* at 25°C and 28°C, and dominance of *Epicoccum* and *Alternaria* at lower temperatures. (Drawn from photographs; Neergaard and Saad, 1962)

biotypes of *Drechslera teres* was 20°, 25° and 30°C, respectively (Singh, 1962). Other examples are *Alternaria tenuis*, *Drechslera graminea*, *Fusarium avenaceum, F. graminearum, Macrophomina phaseolina* and *Rhizoctonia solani* (see Table 29.4).

The morphology of fungi may be greatly influenced by temperature. In cultures of *Drechslera dictyoides* on agar medium, conidia formed at 21°C or higher are small and atypical (Vargas and Wilcoxson, 1969) and, according to Nisikado (1929), the conidia of *D. turcica* became broad fusiform at 15–20°C, slender at higher temperatures, and irregular in shape above 30°C. Conidia of *D. maydis*, southern blight of maize, produced at 15°C, are straight cylindrical, and with a rise of temperature they become curved to one side. At 25–30°C they assume a crescent shape, and at 33°C they become irregular. Conidia of *D. turcica* become gradually smaller with increasing temperature up to 27–30°C, while in *D. maydis* they enlarge notably at higher temperatures.

Colony characters may change according to temperature. *Epicoccum purpurascens* colonies show transitional shades of colours from light rose at 15°C through wine red at 22°C to yellowish-orange at 28°C (Neergaard and Saad, 1962).

Testing at high temperatures reveals thermophilic fungi, some of which are important in storage. Flannigan (1970) detected on agar media considerable percentages of *Absidia corymbifera* and *Aspergillus fumigatus* in barley, oats and wheat after incubation at 37°C; *Absidia corymbifera* and *Mucor pusillus* were dominant at 50°C.

In general practice, seeds are incubated on blotter and agar at the same temperature, usually 20°C, or 28°C for some tropical seeds. A direct comparison of percentages of fungi from rice sown on blotter and potato dextrose agar, respectively, at a range of temperatures from 9°C to 32°C, has been made by Neergaard and Saad (1962). The results (Table 29.5) illustrate how the optimum temperature for development of a fungus species in incubation tests may shift according to type of medium.

Table 29.5. Optimum temperature for percentages recorded for five major seed-borne fungi of rice on blotter and potato dextrose agar. (Data extracted from Neergaard and Saad, 1962)

Fungus	*Blotter*		*Potato dextrose agar*		
	Optimum temp. for per cent recorded (°C)	Highest percentage recorded	Optimum temp. for mycelial growth (°C)	Optimum temp. for per cent recorded (°C)	Highest percentage recorded
Alternaria	**22**	37	25	**28**	22
Cladosporium	**19**	20	18	**19**	2
Epicoccum	**25**	16	25	**19**	53
Drechslera	**25**	23	18 – 30	**25; 28**	34; 34
Pyricularia	**28**	26	28	**22; 25; 28**	9; 8; 9

It appears then that adoption of a fixed standard temperature for seed health testing must of necessity be a compromise.

Incubation at two temperatures may, however, be needed to provide conditions required by different fungi. Kolk and Karlberg (1971) compared the standard blotter incubation at 20°C with parallel tests at 10°C, alone or in combination with 20°C, for part of the incubation period. They found that the standard procedure, 20°C for 7 days, was suitable for *Drechslera sorokiniana* (barley) and for *D. avenae* (oats), while incubation at 10°C for 14 days was best for *Fusarium nivale* (rye), *F. avenaceum* (barley), and *Septoria nodorum* (wheat). Incubation at 10°C for 4 days, followed by 20°C for 4 days, was also suitable for *F. nivale*, as well as for *D. avenae*, *F. avenaceum* and *S. nodorum*. Finally, incubation at 10°C for 3 days followed by 20°C for 7 days was suitable for both *D. avenae* and *D. sorokiniana*.

The optimum temperature for mycelial growth and for sporulation of pathogenic fungi is often quite different from the optimum temperature for production of disease. For example, the optimum temperature for mycelial growth of *Drechslera dactylidis* and *D. siccans* is 25°C, but for leaf infection of their hosts, *Dactylis glomerata* and *Lolium multiflorum*, respectively, the optimum temperature is 20°C (Ammon, 1962). For development of disease produced by *Fusarium nivale* in seedlings of wheat, Millar and Colhoun (1969) found an optimum soil temperature of 6°C, while the optimum temperature for mycelial growth is 20°C (cf. Table 29.4).

Many seedling blights and rots of cereals are favoured by lower temperatures and an extended emergence period. This is emphasised under unfavourable conditions such as deep sowing in heavy soils. Based on these facts, cereals have been tested in Sweden in soil (in recent years in multipots), at 10°C for 2 weeks, to provide the cold climate conditions for infection by fungi such as *Fusarium nivale*, plus one week at 20°C for fungi such as *Septoria* and *Drechslera* (Karlberg, 1974). Soil temperatures of 10–15°C are most conducive to development of seedling symptoms caused by *Septoria nodorum* (Baker, 1970).

29.5.4.5 HUMIDITY

Humidity has been less subjected to systematic investigation than perhaps any of the other incubation factors. Testing methods in general use, including the few accepted standard procedures, are based primarily on a variety of established practices, but there has been little discussion on improving standardisation of humidity. Discrepancies encountered in international tests have not been regarded as due to differences in humidity during incubation, except where excess water in the blotter has resulted in the 'wet blotter effect', investigated extensively by Limonard (1968). Excessive 'wet blotter' leads to substantial reduction in the percentage of incidence of certain fungi but not of others. This effect has been shown by Limonard to be primarily related to bacterial antagonism. Very wet, medium wet and moderately wet blotter were compared as to infection percentages obtained in testing for different pathogens. When seeds of carrot and cabbage were placed on moderately wetted blotter, considerably higher infection percentages of *Alternaria radicina*, *A. porri* f.sp. *dauci* and *A. brassicicola*, respectively, were obtained than when the seeds were placed on blotter containing more moisture. Very wet blotters reduced the infection percentages strikingly. Similar results were obtained with *Botrytis cinerea* on flax. The effect of high humidity may be due to seeds and seedlings being covered by a water film which disfavours mycelial growth, sporulation and development of seedling symptoms. However, development of antagonistic seed-borne bacteria has also been shown to be responsible for suppressing these fungi and the 'wet blotter' effect can be counteracted by the addition of antibacterial antibiotics (Limonard, 1968). On the other hand, no influence of excessive water was observed on *Drechslera* spp. and *Fusarium* spp. in cereals, *Phoma betae* in sugarbeet and *Phoma lingam* in cabbage.

It has become an established procedure to dip the blotter in the water and let it drip off before it is placed in the container.

The humidity of the incubator is often not adequately controlled in routine testing, and the blotter may dry out prior to recording. Containers may not always be equally tightly closed, ventilators or lamps may cause desiccation of plates, particularly blotters. Even small fluctuations of the temperature in the incubator may produce variation of the humidity of the air with unexpected consequences. It should be borne in mind that larger seeds and vigorously sprouting seeds use much more water than do smaller seeds and those that do not germinate. For this reason alone considerable differences in the supply of humidity for the development of fungi are bound to follow.

In the blotter test the problem is primarily to keep the blotters adequately wet throughout the period of incubation, preferably under such conditions that further supply of water during this period is not needed. Water may be continually provided by a wick, as in the Copenhagen germinator. In petri dishes increasing number of layers of blotters per dish increases the amount of water available, and reduction of the number of seeds per container, especially larger seeds, such as those of cereals, naturally decreases the consumption of water for germination. Evaporation can be reduced by placing small containers like petri dishes in polythene bags, and in larger incubators humidity may be maintained by providing water for continued evaporation through wicks.

Kolk and Karlberg (1971) compared different moisture amounts in blotters (basic weight 400 g/m^2) in petri dishes ranging from 15 to 35 ml. per plastic dish (diameter 17 cm, height 4 cm, with a glass cover). In Sweden approximately 23 ml. water are usually used in the blotter test of cereals with 50 kernels sown per dish. The following optimum amounts of water per dish were found for seven seedling pathogens:

Fusarium avenaceum (barley)	25 ml.
Fusarium nivale (rye)	20 ml.
Drechslera avenae (oats)	20 ml.
Drechslera sorokiniana (barley)	25 ml.
Drechslera graminea (barley)	15 ml.
Drechslera teres (barley)	25 ml.
Septoria nodorum (wheat)	25 ml.

These optima are well in agreement with the usual practice in the ordinary blotter method, except for *Drechslera graminea* which apparently requires a remarkably low moisture content.

For the agar plate procedure, usually 2 per cent nutrient agar is used. The water content of this medium and the usual level of the relative humidity of the atmosphere in the laboratory or incubator, provide conditions which are practically never suboptimal for mycelial growth at the temperatures and within the length of incubation generally used. The situation may be different under extremely dry conditions such as prevailing in some tropical and subtropical regions. The relative humidity of the atmosphere in a laboratory in New Delhi (checked by two hygrographs) has been found occasionally to drop to only 5 per cent (Phatak, personal communication, 1971)! The diameter of covered agar plate cultures is little influenced even at a very low relative humidity (cf. Brancato and Golding, 1953).

There is a definite need for research to elucidate the technical possibilities of regulating moisture in the atmosphere and media during incubation of culture tests.

29.5.4.6 LIGHT

In seed health testing, incubation in darkness has been widely used for decades but Leach (1962*a*, 1963*a*, 1963*b*, 1964, 1967*a*, 1967*b*) has now demonstrated that many fungi, saprophytes as well as parasites, show increased sporulation when subjected to radiation in the blue and ultraviolet range of wavelengths. Using monochromatic radiation, he found that the near ultraviolet (NUV) region of the spectrum, 3,200–4,000 Å (320–400 nm), is very efficient in inducing sporulation, that much of the visible light is ineffective, and that the far ultraviolet is inhibitory or even lethal at high dosages. Because of this work, black light fluorescent lamps which emit light mainly at wavelengths near 3,650 Å (365 nm), and cool white daylight fluorescent lamps which emit some NUV light, have become standard equipment in seed health testing (Fig. 29.2).

Both light quality and light/darkness cycles markedly influence sporulation of many fungi. By using monochromatic light, Leach (1962*a*, 1962*b*, 1963*b*) found for *Ascochyta pisi*, *Alternaria chrysanthemi* and *Pleospora herbarum* that only ultraviolet radiation less than 370 nm is effective in stimulating sporogenesis, while visible light, including wavelengths in the blue region of the spectrum, is ineffective. He found for various fungi, including *Alternaria porri* f.sp. *dauci* and *Ascochyta pisi*, that less energy is necessary to stimulate conidial production at shorter wavelengths of ultraviolet radiation.

Some light-sensitive fungi sporulate when exposed continuously to near ultraviolet radiation, but others require a subsequent period of darkness to complete spore production (Barnett and Lilly, 1950; Lilly and Barnett,

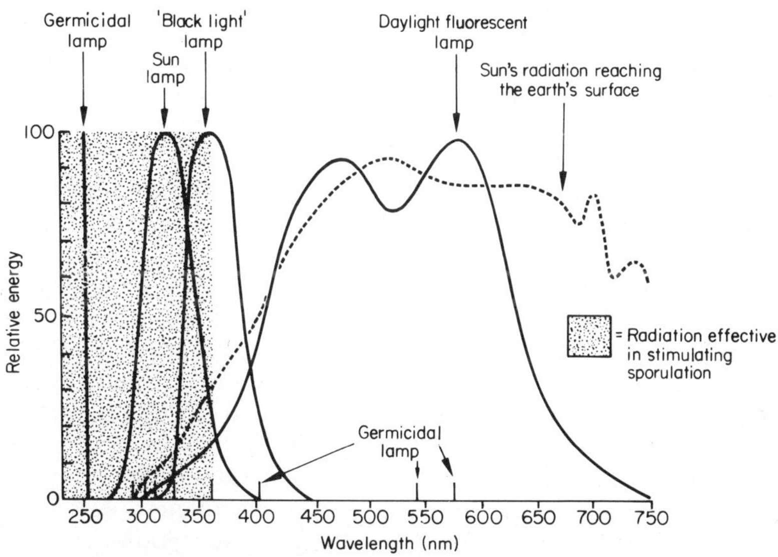

Fig. 29.2. A comparison of radiation emitted by different lamps, solar radiation reaching the earth's surface and the wavelengths of radiation effective in stimulating sporulation in some fungi. (Leach, 1967*b*)

1951; Witsch and Wagner, 1955; Aragaki, 1961; Leach, 1961; Lukens, 1963). In the last-mentioned group, termed 'diurnal sporulators' (Leach, 1967*a*), the mechanism of sporulation is divided into two distinct phases — an 'inductive phase', resulting in formation of conidiophores, and a second, 'terminal phase', or conidial phase which results in the formation of conidia (Leach, 1967*a*). Aragaki (1962) established that in *Alternaria tomato* the terminal phase is actually inhibited by light and this has been confirmed for other diurnal sporulators. Also, in *A. triticina*, sporulation is inhibited by light; maximum sporulation occurs in continuous darkness (Prabhu and Prasada, 1965).

Diurnal conidial sporulation has been established for *Alternaria porri* f.sp. *dauci* (Witsch and Wagner, 1955; Leach and Trione, 1966; Leach, 1967*a*; Zimmer and McKeen, 1969), *A. porri* f.sp. *solani* (Lukens, 1963), *A. tomato* (Aragaki, 1962), *Drechslera graminea* (Houston and Oswald, 1946), *D. oryzae* (Leach, 1961), *D. teres* (Onesirosan and Banttari, 1969), *Fusarium nivale* (Sanderson, 1970), *Peronospora tabacina* (Cruickshank, 1963*c*) and *Pleospora herbarum* (Aragaki, 1962; Leach, 1967*b*). All radiations from 240 to 600 nm will inhibit the terminal phase, the spectral regions 280 and 480 nm being particularly effective (Aragaki, 1962; Leach, 1967*b*).

There is an important interaction of light and temperature which significantly affects sporulation. For *Drechslera graminea* a diurnal fluctuation of both light and temperature is necessary (Houston and Oswald, 1946). This holds true also for *Alternaria porri* f.sp. *dauci* (Leach, 1967*a*), *A. porri* f.sp. *solani* (Lukens, 1966), *A. tomato* (Aragaki, 1961; Leach, 1967*a*) and *Pleospora herbarum* (Leach, 1967*a*, 1967*b*). Formation of conidiophores is stimulated by light (near ultraviolet wavelengths) at relatively higher temperatures, whereas formation of conidia requires lower temperatures and darkness.

It seems that for many diurnal sporulators the temperature critical to the effect of alternating light and darkness lies somewhere between 20°C and 25°C.

Aragaki (1961) demonstrated that after irradiation with wavelengths inhibitory to conidial formation, a period of darkness is necessary to induce sporulation of *Alternaria tomato* at a temperature above 23°C but not at lower temperatures, 6–23°C, where conidia are formed under continuous irradiation. In *A. porri* f.sp. *dauci*, conidial formation in the dark occurs three times faster at 18°C than at 25°C, and if the temperature is 23°C or below, formation of both conidiophores and conidia occurs under continuous irradiation; in continuous darkness this fungus does not sporulate (Zimmer and McKeen, 1969). For a third diurnal sporulator, *Drechslera teres*, the optimum temperature for sporulation in cultures exposed to 12 h of light alternating with 12 h of darkness is 21°C, while the maximum temperature for sporulation is approximately 25°C. No spores are produced in continuous darkness, and only a few in continuous light (Keeling, 1966; Onesirosan and Banttari, 1969). Finally, sporulation of *Stemphylium botryosum* is inhibited by NUV at 25°C and 28°C, but not at 21°C (Leach, 1968). Obviously, for these fungi a standard testing temperature of 20°C±2 is preferable to ensure sporulation.

But some fungi have a low optimum temperature. In *Drechslera dictyoides* conidia are formed at 17°C and 21°C but not at 27°C. However, at 21°C and above the spores are small and atypical (Vargas and Wilcoxson, 1969). The optimum temperature for sporulation of *Fusarium lateritium* is 16°C; more spores are produced in constant light than alternate light and darkness; in constant darkness cultures yield no spores (Massie and Peterson, 1968). *F. nivale* sporulates best at 17°C in constant NUV or at 15°C in NUV alternating with darkness (Leach, 1967*a*).

Other fungi are less dependent on cycles of light and temperature within the optimal range of these factors.

Septoria nodorum sporulates under continuous light or diurnal light–dark cycles but not in constant darkness. In Brönnimann's agar plate procedure (Brönnimann, 1968) continuous artificial daylight is used for incubation. Calpouzos and Lapis (1970) tested the influence of these alternatives of light condition at two temperatures, 15°C and 25°C, on different media and found that this fungus formed pycnidia more frequently at 15° than at 25°C. The difference in sporulation was particularly conspicuous at lower suboptimal light intensities.

The sporulation response to alternating periods of light and darkness, particularly when accompanied by high and low temperatures respectively, agrees with the reproductive behaviour of many pathogenic fungi which under natural conditions show a peak of spore dissemination in the morning after a night of spore production.

As shown in Subsection 29.5.4.4, the effect of temperature upon radial growth of fungi on agar media may be quite different from the effect of temperature upon percentage incidence on these fungi as found in the agar test. Hence, the incubation factors should be examined for their direct effect on the practical tests in terms of percentage incidence. Contaminated seed in a culture test are communities of developing organisms — hosts to a range of microorganisms, some of which are dominant, while others are suppressed by a complex of physical and biological conditions.

Howell (1964) increased counts with sporulating *Drechslera avenae* in oat seeds in the blotter test by 7 days of continuous NUV followed by 1 day of darkness, as compared with 8 days of continuous darkness. The average percentage for three infected samples was 46.3 and 8.7 per cent, respectively. De Tempe (1968) compared the effect of NUV in a 12/12 h cycle with continuous darkness on two pathogens of carrot in the blotter test, incubated at 20°C for 10 days. The infection percentages of *Alternaria porri* f.sp. *dauci* and *A. radicina* were three times higher under

NUV than in darkness. He also compared the effects of 12 h daily periods of NUV with cool white daylight tubes ADL (artificial daylight), upon *A. radicina*. Although there was some reduction of the infection percentage under ADL, he found this lighting rather satisfactory. In earlier investigations De Tempe (1959, 1964) similarly increased the percentages of *Phoma exigua* var. *linicola* in flax, *P. betae* in beet, *P. medicaginis* in small-seeded legumes, *P. lingam* in cabbage, *Drechslera avenae, D. teres* and other species of *Drechslera* in cereals, with the use of NUV lamps.

Kang, Neergaard and Mathur (1972) investigated the effect of NUV and ADL on percentages of seed-borne fungi of rice on blotter. When the seed was incubated for 8 days at 20°C in complete darkness, the percentage recorded of *Curvularia lunata*, *Drechslera oryzae*, *Epicoccum purpurascens*, *Fusarium semitectum* and *Pyricularia oryzae* was much reduced. No infection by *Trichoconis padwickii* was recorded after incubation in darkness, as against about 40 per cent after 8 days of 12 h daily under NUV. For most fungi, 12 h daily irradiation was found to be clearly superior to the effect of 4 h or 8 h, supporting the adoption of a standard 12 h NUV. Furthermore, by comparing the two standard incubation temperatures, 28°C and 20°C, the counts were found practically equal for *Drechslera oryzae* and *Pyricularia oryzae* under 12 h NUV/12 h dark, whereas counts of *Trichoconis padwickii* were considerably increased at the higher temperature. Chuaiprasit, Mathur and Neergaard (1974) substantiated the standard procedure of 12 h NUV for detection of the three pathogens in blotter tests.

The following three sets of incubation conditions have been compared as to their influence on recorded percentage of common seed-borne fungi of blackgram (*Phaseolus mungo*), greengram (*P. aureus*), soybean, wheat, and particularly rice (Agarwal, Mathur and Neergaard, 1972; Mathur and Neergaard, 1972): 20°C NUV, 28°C NUV and 28°C ADL, all under 12 h of daily irradiation. The level of incidence recorded, based on average of records for a number of seed lots of each species, was practically equal, thus supporting the selected standards of incubation conditions for laboratory seed health testing, and moreover supporting the substitution of black light fluorescent tubes (NUB) by cool white daylight fluorescent tubes (ADL) when NUV tubes cannot be obtained as in some developing countries. Fungi investigated included *Colletotrichum dematium* (blackgram, soybean), *Curvularia lunata* (blackgram, greengram), *Drechslera oryzae* (rice), *Epicoccum purpurascens* (rice, wheat), *Fusarium semitectum* (blackgram, greengram, rice), *Macrophomina phaseolina* (blackgram, greengram) and *Pyricularia oryzae* (rice). *Trichoconis padwickii* on rice yielded higher percentage incidence at the higher temperature, and also under NUV rather than ADL.

Light often has a marked effect on the morphology of fungi. Snyder and Hansen (1941) showed that conidia of *Fusarium* of different species produced in light were consistently larger and invariably had more septa than those developed in the dark. Cultures kept in darkness showed consistently greater mycelial growth than those kept in light. In *Ascochyta pisi* the conidia developed in darkness are constricted, whereas in NUV light they are cylindrical and non-constricted; pycnidia from non-irradiated colonies on MA at 21°C are approximately twice the diameter of those from irradiated colonies (Leach, 1962*b*). Conidia of *Drechslera maydis* that are curved at 20°C and 28°C in NUV and ADL become, in the dark, straight, considerably smaller and with fewer septa (Singh, Mathur and Neergaard, 1974*a*).

Pigmentation in fungi is greatly influenced by the presence or absence of light. Irradiation such as by NUV usually stimulates strong pigmentation (Leach, 1967*b*).

Fig. 29.3. Incubation of seed sown on blotter or agar media in plastic petri dishes and kept at a standard temperature(often 20°C) and irradiated by near ultraviolet light (12 h per day), from tubes placed 40 cm above the dishes

The experience gained from these irradiation studies on increased sporulation and improved possibilities of identification in seed health testing practice has been most encouraging. A 12 h NUV/12 h dark cycle has been adopted as a general standard procedure for testing seed for fungal pathogens (ISTA, 1966*a*). Philips Black Light fluorescent lamps (NUV) which emit radiation in a continuous spectrum from 320 to 420 nm have been recommended for this purpose. Also, lamps from Osram, Sylvania, General Electric and other manufacturers may be used. The lamps are obtainable in 20 W, 40 W and 80 W and are usually interchangeable with daylight fluorescent tubes and require no special wiring. As established above, cool white daylight fluorescent lamps can be used when NUV/360 lamps are not available.

When space is available the best arrangement is to place two 40 W lamps horizontally 20 cm apart, and at a height of 40 cm above the dishes (41 cm above the seeds) to be irradiated. This arrangement will adequately cover an area approximately 60 by 120 cm (cf. Fig. 29.3).

29.5.4.7 TIME (INCUBATION PERIOD)

The length of the incubation period prescribed for any organism under test is determined by its rate of growth. However, in most routine testing, efforts are made to obtain easily readable results and maximum numbers.

The length of incubation depends on other factors influencing the rate of development of the organisms involved. Temperature is particularly important. At incubation temperatures lower than the standard 20°C, in which 7 days' incubation is usually required for adequate detection of most fungi, incubation must be prolonged to obtain approximately the same development of colonies (Table 29.6).

Table 29.6. Length of incubation period at 20°C, 15°C and 10°C necessary to attain approximately the same colony diameter of each of five seed-borne fungi of wheat plated on PDA. (Data from Limonard, 1968)

Fungus	*Number of days*			*Approximate colony diameter in the three tests* (mm)
	20°	15°	10°	
Alternaria tenuis	7	10	15+3*	18
Epicoccum purpurascens	7	8	15+2	17
Fusarium avenaceum	7	10	15+1	26
Fusarium graminearum	7	9	15+3	36
Fusarium nivale	7	8 – 9	15+1	35

*The number of days added are obtained by extrapolation

Limonard (1968) found an approximately equal percentage of each of the five fungi listed in Table 29.6 and of *Septoria nodorum* after 7 days on PDA at 20°C, 10 days at 15°C and 14 days at 10°C*.

The period of incubation selected for the standard culture tests reveals most of the fungi detectable by these procedures. In practice there is a tendency to take for granted that a practically complete record of fungi has been obtained after completion of the standard test, i.e. in the 8th day for the blotter test. However, investigations based on records taken after incubation for additional days or weeks revealed important fungi that otherwise would have been missed.

Incubation of rice on blotter beyond the standard period of incubation has consistently yielded an increasing percentage of *Trichoconis padwickii* and *Alternaria longissima* after 13 days incubation at 20°C and 10 days at 28°C (Mathur and Neergaard, 1971). *Fusarium moniliforme* in maize and rice appears very slowly in the blotter test at 20°C. Extension of the period of incubation up to three weeks or more increases counts in maize (Gemawat, Mathur and Neergaard, unpublished), and up to 15 days in rice (Ram Nath, Neergaard and Mathur, 1970). Derbyshire (1961) described a blotter procedure for detection of *Didymella lycopersici* in tomato seed. She emphasises the importance of incubating the seed for a period of three weeks to ensure full development of all the inoculum to the pycnidial stage.

For certain pathogens a period shorter than the standard is preferable. In rice, *Pyricularia oryzae* in a blotter test is easier to record after 4–5 days, and *Verticillium* can be easily detected after 2 days of incubation, whereas after 7 days this fungus may be completely overgrown.

The effect of seed treatment is often limited to mere reduction of the inoculum potential in terms of both reduced infection percentage and reduced potential of inoculum in individual seeds. This is reflected in reduced growth rate of fungi still alive in treated seed. In order adequately to reveal such infections, the period of incubation must be extended. In the international comparative seed health tests of 1968 (Neergaard, 1968), the following increase in percentage of *Septoria nodorum* in commercially fungicide treated wheat seed, tested on potato dextrose agar at 20°C, was reported from the participating station of Paris: 14 per cent after 7 days, 29 per cent after 8 days, 38 per cent after 10 days. Another station (Beltsville) reported on the same treated seed sample: 10 per cent after 7 days and 25 per cent after 12 days (the counts obtained from the untreated sample varied among 13 stations from about 60 to 75 per cent).

*These and similar observations suggest that, for many fungi in temperate climates, the product of *temperature* (Celsius) within the optimal and suboptimal temperature range for mycelial growth and the *length of incubation* (days) required for obtaining standard records in *agar plate tests* is a constant. Investigations to substantiate this point are needed.

29.5.5 Recording

29.5.5.1 WORKING CONDITIONS (SOURCE OF LIGHT, MAGNIFICATION USED)

The personal element in recording is probably responsible for the main part of the discrepancies encountered in the international comparative tests. Lighting during examination must be adequate; two sources of light from different angles ensure that some parts are not left in shade (Fig. 29.4). Lamps with a strong drying effect should be avoided — small lamps, preferably 'cold tube' with concentrated light projecting, are preferable. A stereoscopic low power microscope with a series of magnifications, from 10 to 60, readily changeable, is necessary for routine work.

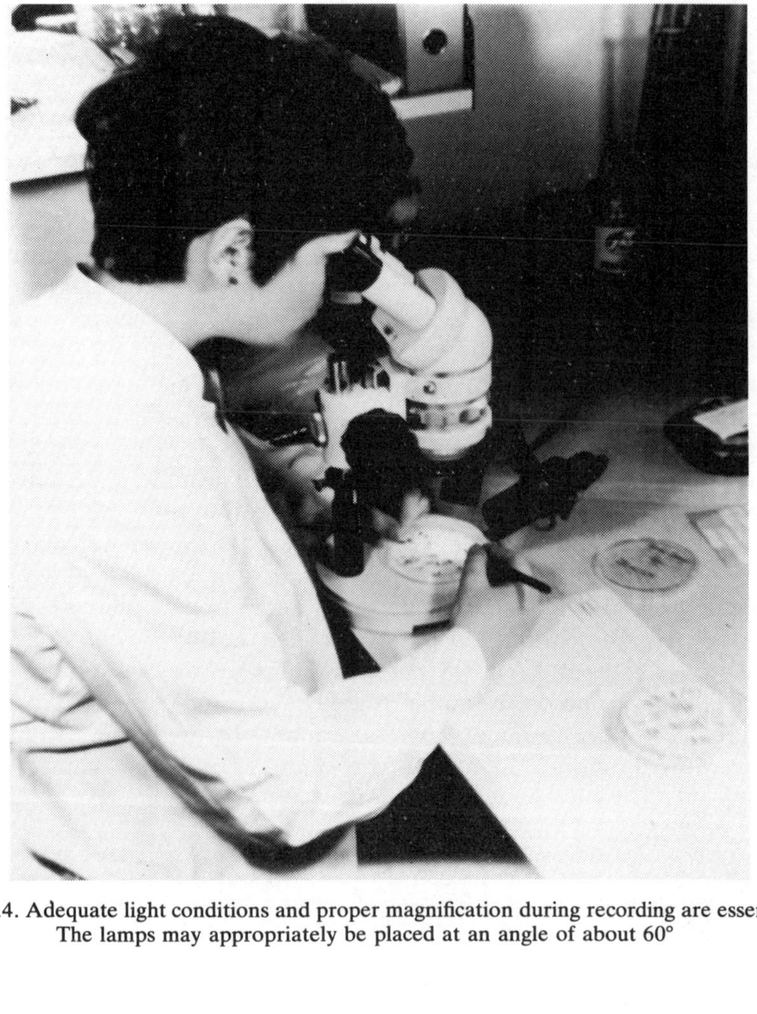

Fig. 29.4. Adequate light conditions and proper magnification during recording are essential. The lamps may appropriately be placed at an angle of about 60°

29.5.5.2 DISTINCTION — REPRODUCIBILITY OF TEST RESULTS

The capacity of distinguishing between fungi or symptoms of similar appearance may be a matter of talent, but lack of practice has been found to be a major cause of discrepancies between well-equipped stations. In the International Comparative Schemes some stations, whose results in the first tests of their participation were not in accordance with the majority, have improved in a few years to such a degree that their results were well in agreement with those obtained in stations with well-trained personnel. Joint laboratory studies of duplicate referee samples at the international workshops have contributed substantially in reducing the negative effect of this unavoidable personal factor. A few testing results from stations of less experience as compared with stations of more experience are shown in Figs 29.5–29.8, which mainly serve to exemplify reproducibility of the blotter and agar tests for detection of some seed-borne pathogens (Neergaard, 1970*a*).

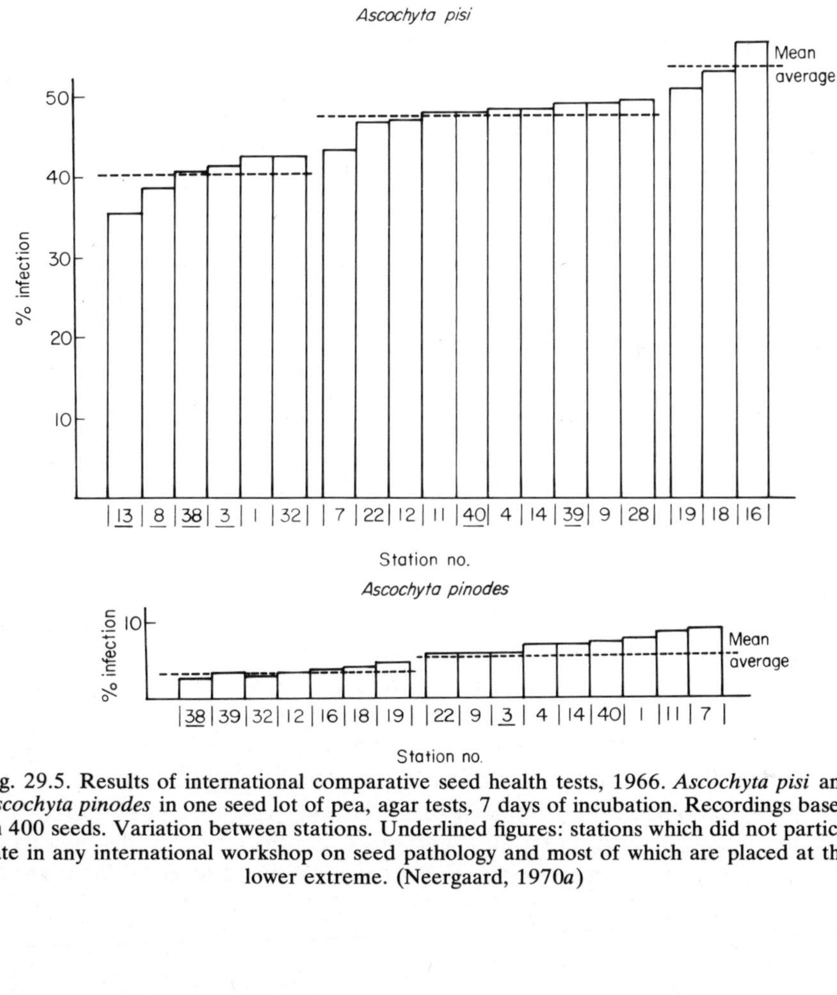

Fig. 29.5. Results of international comparative seed health tests, 1966. *Ascochyta pisi* and *Ascochyta pinodes* in one seed lot of pea, agar tests, 7 days of incubation. Recordings based on 400 seeds. Variation between stations. Underlined figures: stations which did not participate in any international workshop on seed pathology and most of which are placed at the lower extreme. (Neergaard, 1970*a*)

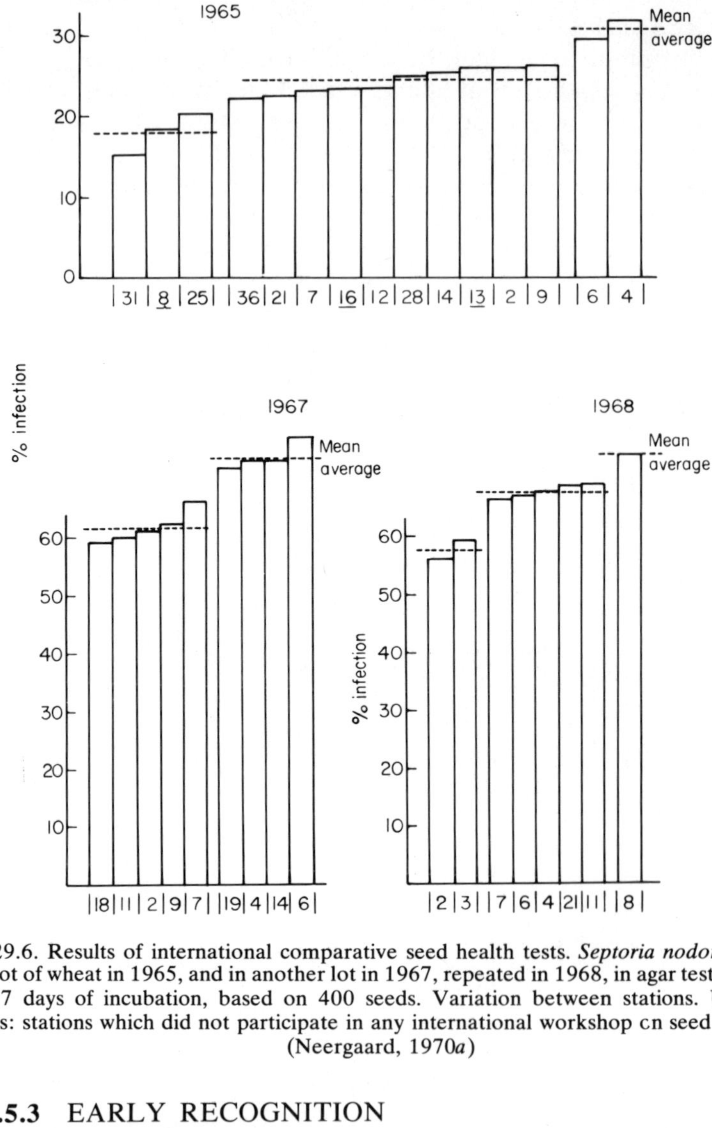

Fig. 29.6. Results of international comparative seed health tests. *Septoria nodorum* in one seed lot of wheat in 1965, and in another lot in 1967, repeated in 1968, in agar tests, recorded after 7 days of incubation, based on 400 seeds. Variation between stations. Underlined figures: stations which did not participate in any international workshop cn seed pathology. (Neergaard, 1970*a*)

29.5.5.3 EARLY RECOGNITION

It is important to develop skill in recognising weak infections as soon as possible, the first faint signs, for instance characteristic mycelial appearance, before sporulation. Thus, 30 years ago testing cabbage for *Phoma lingam* in Copenhagen required a period of incubation of 3 weeks, but this was later reduced to 10 days, and indeed the characteristic silvery mycelium and pycnidial primordia may even be apparent to the skilled analyst after 6 days, although often confirmation of the record at the 10 days count is necessary. Early recognition of a pathogen enables quicker information for seed companies that certain seed lots are not fit for export. The fact that saprophytes may cover up pathogens and symptoms is another reason for recognising, as early as possible, signs indicating the presence of pathogens.

29.5.5.4 ESTIMATE

The decision as to what is to be accepted as healthy and what rejected as diseased is a vital part of the test. There are clear cases that leave no doubt. Often, however, there is room for variation in interpretation. The presence of *Alternaria brassicicola* in cabbage may range from profuse development of conidia on the seed and surrounding blotter to a single chain of conidia on an apparently healthy seed.

The question arises, where to draw the line. Are weak incidences secondary contaminations? Is a weak facultative parasite, e.g. *Pleospora herbarum*, to be included in seed health test reports? Are faint symptoms on seedlings recognised as indications of perhaps important pathogens, for instance the brownish knots produced by *Septoria nodorum* on the

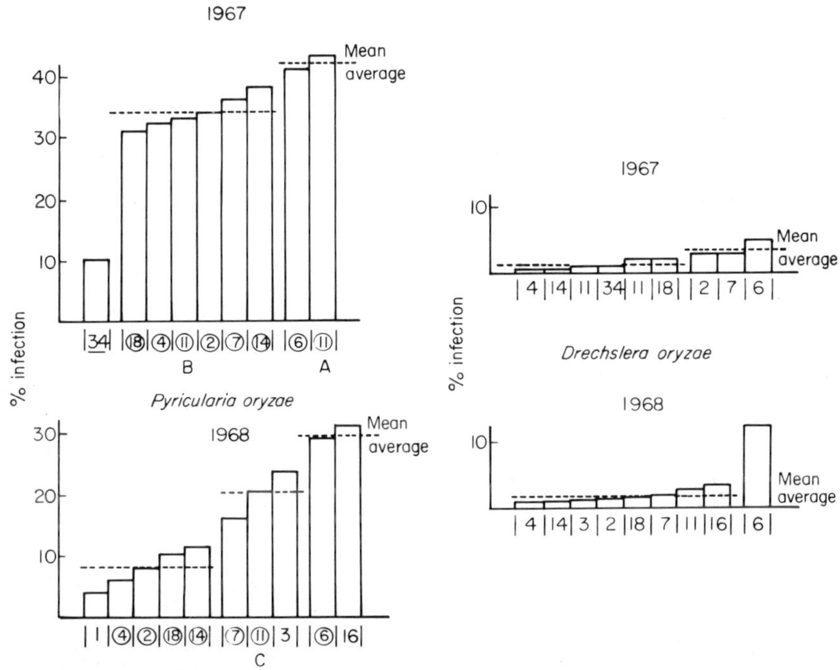

Fig. 29.7. Results of international comparative seed health tests. *Pyricularia oryzae* and *Drechslera oryzae* on one seed lot of rice, 1967 and repeated in 1968, on blotter, recorded after 7 days of incubation, based on 400 seeds. Variation between stations. Underlined figure: station which did not participate in any international workshop on seed pathology and the result disagreeing with those of the more experienced stations. Note discrepancies in 1968 in contrast to 1967, due to weakness of the fungus to adverse conditions after one year of storage. Station A, B, C is identical with the one dealt with in Fig. 29.8.

coleoptile of many wheat cultivars? Such symptoms are readily overlooked if not specifically accepted as important indications. Only permanent international co-operation between specialists with experience can build up a uniform practice of assessment.

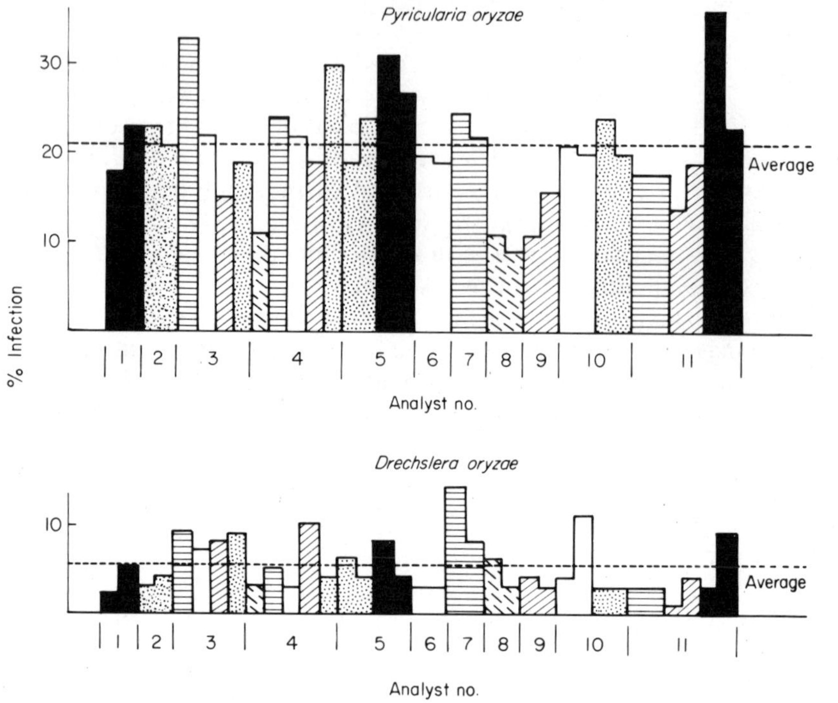

Fig. 29.8. Within-station variation. Results of seed health tests of the same rice bulk sample as that included in the international tests reported in Fig. 29.7, the same two pathogens and same method, 1968, recordings based on 100 seeds. Different signatures of the columns refer to different dates of testing. Note agreement between the eleven analysts of the station and more variation from one date of testing to another. This variation is due to variation in the incubation conditions from one testing period to another. Compare Fig. 29.7 for similar variation from one station to another. (Neergaard, 1970*a*)

30 Bioassays and Biochemical Procedures

30.1 INDICATOR TEST — THE INOCULATION METHOD

(1) *Procedure*. Inoculum in contaminated or infected seeds may be used in various ways to produce symptoms on healthy seedlings or mature plants thus used as indicator plants.

Wilson (1935) used green bean pods as suscepts to detect inoculum of the halo blight bacterium, *Pseudomonas phaseolicola*, but also bean seedlings may be used. In the U.S.A., Saettler (1971, 1973, 1974), Andersen and Young (1973) and Schuster and Coyne (1975) found that inoculation by hypodermic injection of indicator plants with material from seeds slightly infected by *Xanthomonas phaseoli* or *X. phaseoli* var. *fuscans*, is a test more sensitive than serological techniques. One-pound samples of non-seed-treated beans from each lot are pretreated in 2.6 per cent sodium hypochlorite for 15 min and rinsed in sterile water; 1,200 ml. sterile water is added and the sample incubated at room temperature for 18–24 h. From the remainder of the liquid, injections are made into the primary leaf node of 10-day-old Manitou bean seedlings. Positive reaction is in the form of large lesions followed by systemic necrosis. The procedure proved very efficient in a bean certification scheme. Thyr (1969), using artificially infected seed, found indications that the indicator test might be used also in assaying tomato seed for the bacterial canker pathogen, *Corynebacterium michiganense*; one infected seed per thousand could be detected. However, establishment of a procedure by testing inoculated seed is not conclusive; the procedure must be assessed by using naturally infected or contaminated seed.

Marrou and Messiaen (1967) described a method of testing lettuce for lettuce mosaic virus by inoculation on *Chenopodium quinoa* — infection in this indicator host is systemic, while in the highly sensitive *C. amaranticolor* only local lesions are produced; the symptoms in *C. quinoa* are specific and easy to identify. The seed is tested in groups of 700 seeds for commercial certification, 300 for Foundation Seed, and for this purpose a total of at least 3,000 seeds is required, numbers increasing according to decreasing permitted infection. Quantz (1957, 1958) provided a plate test for rapid detection of bean common mosaic. Expressed sap from bean

Table 30.1. Indicator hosts for some seed-borne viruses. Symptoms, such as local lesions, may be characteristic for individual viruses and thus indicative for identification. Responses, positive or negative, may vary according to virus strain and host variety. L, local; S, systemic; no, none. (Data compiled mainly from Smith, 1972 and Phatak, 1974)

	Alfalfa mosaic	Arabis mosaic	Asparagus latent	Barley stripe mosaic	Bean common mosaic	Bean yellow mosaic	Broad bean mottle	Broad bean stain	Cherry leafroll	Clover red vein mosaic	Cowpea aphid-borne mosaic	Cowpea mosaic (beetle)	Cowpea ringspot	Cucumber mosaic	Lettuce mosaic	Lychnis ringspot	Mungbean mosaic	Pea early browning	Prunus necrotic ringspot	Raspberry ringspot	Soybean mosaic	Squash mosaic	Tobacco mosaic	Tobacco rattle	Tobacco ringspot	Tomato black ring	Tomato bushy stunt	Tomato ringspot	Tomato spotted wilt
Chenopodium album			L	L					L							L				L	L								
C. amaranti-color	LS	S	LS	L	no	L	no	L	LS	L	L		L	L	L	L	L	L	LS	LS	no	L		L	L	L		L	L
								no														no							
C. quinoa		LS	LS	L	L	L	no		L	L	L		L	L	LS		L		LS	L	L	no			L	L		L	
					no																no								
Cucumis sativus, cotyl.		L			no	L	no		LS		no		L	S	no		no	L	LS		no	S		L	LS	S	L	L	
Cyamopsis tetragonoloba				L	L	L	L				L		no		no		L		L		no	no							
Datura stramonium					no	no	no				L		L	LS	no		no				no	no	L		S		LS	S	S
Dolichos lablab											L	L																	
D. uniflorus				no	L	no	L				LS		L		no		L				LS	no							
Glycine max				no	no	LS	LS				S		S	no	no		no				S	no							
Gomphrena				L		L				L	L			L	L														
globosa	LS	L	L	no	no	no	no	no					no	no	no		no	LS		LS	no	L				L			
Nicotiana glutinosa	L		L	no	no	no	no				no		LS	LS	no		no			L	no	no	L		L		L		LS
N. tabacum 'Samsun'			LS	no	no	no	no				no		L	LS	no		no				no	no							
Petunia hybrida		L																		L					S	L	S		L
Phaseolus mungo				no	no	no	LS				no		S	S	no		L				no	no							
P. vulgaris prim. leaves	L		L	no	LS	LS	LS				L		LS	LS	L		L	L		S	no	no		L	LS	L	L	L	
P. vulgaris								S																					
'Can. Wonder'								L																					
P. vulgaris 'Pinto'											L	L											L						
P. vulgaris 'Prince'								L												L									
P. vulgaris								S																					
'Tendergreen'								L																					
P. vulgaris 'Topcrop'					L																								
Pisum sativum				no	no	S	S	S		S	L		S	S	no		no	S			no	no							
Vigna unguiculata	L			no	no	LS	S				S		LS	S	no		LS	LS			no	no			LS		L	L	

seedlings is inoculated on to fully-opened primary leaves of a sensitive cultivar of *Phaseolus vulgaris* cv. Topcrop. The leaves are kept on moist blotters in closed petri dishes at 30–32°C under illumination. In 2–3 days discrete, easily distinguishable dark to reddish-brown spots appear. Results are clear at dilutions of the virus up to 1:1,000. However, Bos (1971) and Phatak (1974) have reported that the test is not always reliable.

The indicator test is essential for virus identification; the test is based on response (local or systemic infection) in selected standard indicator test hosts. Some important indicator test hosts for seed-borne viruses are listed in Table 30.1.

Moreover, the indicator test is used to confirm resistance of certain cultivars to certain pathogens. Suspension of spores or conidia are applied to young seedlings, for instance bean seeds are germinated in paper rolls at 10° or 20°C, inoculated 3 days after sowing by soaking of the seedlings for 1 h in a suspension of conidia of nine pathogenic races of *Colletotrichum lindemuthianum* and, after further 5 days of incubation at 22°C, symptoms are recorded at levels from 0 to 6 according to severity (Champion *et al.*, 1973).

(2) *Application and limitations*. For identification of viruses this method is a standard technique. The procedure is little used in routine seed health testing. For detection of bacteria and certain viruses, however, the method offers possibilities that should be considered for development of routine procedures. Pods of Leguminosae may probably be suitable as suscepts, but also detached leaves have proved to be usable in routine testing, for instance detached leaves of *Chenopodium amaranticolor* as indicator suscepts for carnation viruses. The technique may possibly be developed for detection of seed-borne infections such as already described for bean common mosaic. In preliminary investigations detached leaves of *Nicotiana glutinosa* and *N. tabacum* cv. Xanthi have been found to be workable suscepts for testing tomato seed for tobacco mosaic virus (Kristensen, unpublished; Phatak, 1974). For the purpose of testing for mere presence or absence of lettuce mosaic virus in a stock of seed, the procedure of inoculation on *Chenopodium quinoa* is undoubtedly more suitable, because easier and quicker, than the growing-on test procedure for detection of this virus (Rohloff, 1960, 1962), discussed below (Section 31.1). However, this is only true without reservation if indicator plants are always available for the test. Even if not, suitable test seedlings of *C. quinoa* can be raised ready for inoculation in 11–12 days under constant fluorescent daylight, at 25°C, and the reaction of inoculations can be recorded after a further 5–8 days, making a total of 17–20 days — slightly less than the growing-on test procedure according to Rohloff (1968*b*) who compared them. Kimble *et al.* (1975) found, by comparison of the two procedures, that the indicator method requires less

labour and greenhouse space, and hence is less expensive. The indicator test may be an important supplement to other routine tests, such as serological and bacteriophage tests, to meet situations where new strains of a pathogen may appear as it occurred with *Xanthomonas phaseoli* in Canada in 1961 (Sutton and Wallen, 1967); cf. also the limitation of using the bacteriophage technique for *X. oryzae* (Section 30.2), suggesting the possibility of using the host indicator test for this pathogen. In the U.S.A. the indicator test for *X. phaseoli* (Saettler, 1971, 1973) has been used since 1971, according to requirements for certification in the Michigan bean seed programme (Copeland, Wijandi and Adams, 1974).

In France the seedling resistance test is used in testing commercial seed for confirmation of resistance in a cultivar of sunflower against downy mildew, *Plasmopara halstedii*, and in flax cultivars for confirmation of resistance against flax rust, *Melampsora lini*, and in testing breeder's seed of Phaseolus bean for resistance against anthracnose, *Colletotrichum lindemuthianum* (Champion, 1975, personal communication).

30.2 THE PHAGE-PLAQUE METHOD

(1) *Procedure*. This procedure is designed for detection of bacterial pathogens (Katznelson and Sutton, 1951; Sutton and Katznelson, 1953; Katznelson, Sutton and Bayley, 1954), originally to find *Xanthomonas phaseoli* and *Pseudomonas phaseolicola* in bean seed, later *P. pisi* in pea seed. 250 g bean seeds are surface sterilised in 2 per cent sodium hypochlorite solution for 10 min and macerated for 5–7 min in 1 l. sterile nutrient broth in a blender. The sample is incubated for 24 h to permit multiplication of bacteria present. Portions of 10 ml. are removed aseptically to sterile flasks and a suspension containing 4,000–5,000 phage particles is added. Samples of 0.1 ml. of this mixture are immediately, and after 6–12 h, spread on plates seeded with appropriate indicator bacteria. Each phage particle forms zones of lysis or plaques, and these are counted. The presence of the homologous bacteria, either of the two above-mentioned bean pathogens, is indicated by a significant increase in the number of phage particles in the second plating. Ednie and Needham (1973) described a testing procedure for trace infection of *Xanthomonas phaseoli* var. *fuscans* in bean. Seed samples of 1.6 kg (3.5 lb), pretreated in 1 per cent sodium hypochlorite for 10 min and rinsed in sterile water, are wet ground, the slurry produced is vigorously agitated and then incubated for 2 h, and series dilutions are plated on agar plates, incubated at 27–28°C for 5–10 days; finally isolates from *Xanthomonas*-like colonies are phage typed, the polyvirulent bacteriophages Pg 60 for *X. phaseoli* and Pg 176 for *X. phaseoli* var. *fuscans* (Sutton and Wallen, 1967) being

used. The method allows detection of one infected seed in 10,000 as well as internal infection of seeds in representative samples from field with one infected plant in 10,000. A similar procedure by which the seeds, however, were ground to a flour for extraction of the bacteria, was described by Taylor (1970) for *Pseudomonas phaseolicola*.

(2) *Application*. The method has not been used widely for routine testing but in connection with research projects. It is very sensitive. Ercolani (1968) determined infection percentages of *Xanthomonas vesicatoria* and *Corynebacterium michiganense* in tomato seeds using this method. Ednie and Needham's above procedure, supplemented with host-plant infectivity assay (injection), is used for routine testing of Foundation Seeds in Canada (Ednie, personal communication, 1973).

(3) *Limitations*. The method is applicable where great numbers of seed samples are submitted for routine testing, but the procedure requires co-operation of specialists seldom available in routine seed health testing laboratories. It calls for accurate knowledge of the pathogens and concomitant organisms. The sensitivity of the test may decrease for bean seed substantially contaminated with saprophytic bacteria. Such contaminations may occur when seed is harvested under adverse conditions, wherein the pods would be in contact with the soil under moist conditions for a longer than normal period of time (Ednie and Needham, 1973). It is an advantage that the procedure does not require expensive laboratory equipment and, provided workable bacteriophages for a definite pathogen are available, training of personnel on the identification of isolates is readily accomplished. Ou (personal communication, 1969) found that bacteriophages attacking *Xanthomonas oryzae* are very specific to strains of the pathogen and that only a combination of several strains of bacteriophages would provide for a fairly workable test, but complete failures due to the specificity of the bacteriophages cannot be excluded.

30.3 SEROLOGICAL METHODS

Serological testing techniques are widely used in human and veterinary pathology, and have been adopted also to a considerable extent for detection and identification of plant pathogens, particularly viruses and bacteria.

The techniques are based on the following physiologic process. When introduced into an animal, an organism causes formation of substances called *antibodies* in the blood. Also, dead cells or complex proteinaceous cell products injected into an animal cause formation of antibodies. The substances inciting production of antibodies are called *antigens*. Blood from an animal that has been inoculated with an antigen contains the antibodies in the globulin fraction of the serum, and this antibody-containing serum is termed *antiserum*. Serological tests are based on *in vitro* reactions between antigens and antibodies. These reactions may occur in the form of agglutination or precipitation.

A description of the principal serological tests is given in *Plant Pathologist's Pocketbook* (Commonwealth Mycological Institute, 1968), by Wetter (1965) and by Ball (1974). Serological tests are commonly used in diagnosing virus diseases in connection with certification schemes of vegetatively propagated crop plants such as potato, carnation, apple and stone fruit trees.

Serious approaches have been made towards adopting serological techniques for seed-borne viruses and bacteria. References to some contributions are given below but without attempt to give technical details.

(1) *Procedure*. For routine testing of seed for bacteria or viruses an antiserum for each specific pathogen must be provided. The antiserum may appropriately be placed in plastic containers and kept frozen at −30°C. Guthrie, Huber and Fenwick (1965) compared, for detection of *Pseudomonas phaseolicola*, four commonly used procedures: the slide agglutination test, the tube precipitin test, the micro-precipitin test and the gel double-diffusion test, cf. Fig. 30.1. The seed to be examined was surface sterilised, cultured for 36 h in sterile distilled water, and the supernatant fluid tested with antiserum. Traces of inoculum could be found in very large bulks of seed; for instance, four infected seeds in 160,000 could be detected with 95 per cent accuracy if 20 4 lb samples were tested. On the other hand, the detection of 24 seeds in 160,000 seeds would require only 10 samples of 1.25 lb each (Guthrie *et al.*, 1965).

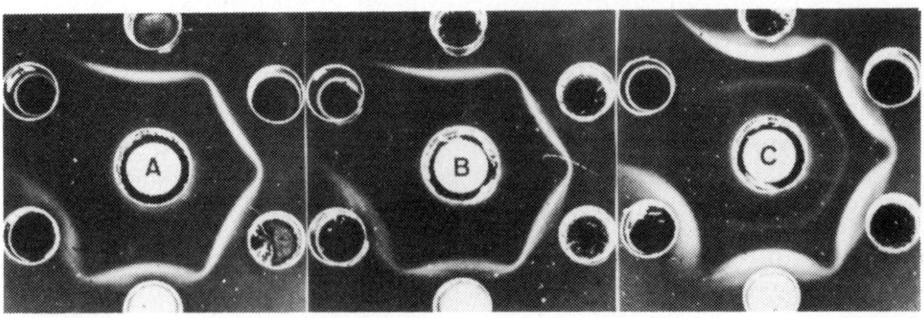

Fig. 30.1. Ouchterlony agar double-diffusion comparison of antigen–antisera reactions of races 1 and 2 of *Pseudomonas phaseolicola.* (A) Race 1 antiserum; (B) and (C) race 2 antisera. The peripheral wells, beginning at the top and moving clockwise, contain antigen for race 2, race 1, race 2, race 1, race 2 and *P. fluorescens.* (Guthrie, 1968)

Hamilton (1965) has developed a rapid and accurate serodiagnostic procedure for detection of barley stripe mosaic virus in barley seed. Extracts from single embryos are made and submitted to the Ouchterlony agar-gel double-diffusion test.

Hamilton described fully the procedure, which can be applied to large-scale screening of seed lots. The immunoprecipitates can be observed easily after two days of incubation. He found excellent correlation between the average percentage infection detected by the embryo test, by the serological assay of seedlings, and by visible inspection of seedlings. The embryo test appeared to be more reliable than the seedling inspection method. This latter method is unsatisfactory because under certain environmental conditions some seedlings from infected seeds fail to exhibit symptoms distinct enough to be classified as diseased. Recently an improved technique has been taken into use — the 'radial-diffusion technique'. In this, immunospecific reactions were assessed 24–36 h after embedding pieces of the primary leaf directly into a 0.5 per cent agar matrix containing antiserum and an appropriate virus dissociating reagent. The test has a 10-fold greater sensitivity than the conventional double-diffusion test (Slack and Shepherd, 1975).

Also the tomato strain of tobacco mosaic virus can easily be detected in tomato by the Ouchterlony agar-gel double-diffusion test. Subsamples of the seed (0.1–1 g) are macerated with a small quantity (0.5–5 ml.) of normal saline solution. An antiserum is serially diluted with saline solution, and controls consist of extracts of known virus-free seed in saline solution. Diffusion is carried out in Difco Bacto agar in plastic petri dishes, incubated at room temperature in moist chambers. Precipitation is visible, within 16 h, as characteristic opaque white bands between antiserum wells and wells with extracts of infected seed. The dilution of the antiserum depends on the titre of the antibodies of a particular sample but a dilution of 1:4 has been found suitable in preliminary investigations (Phatak, 1974).

Another serological technique, the latex flocculation (agglutination) test, known in human pathology, appears very promising (Phatak, 1974). The principle in this test is to adopt the large particles of the biologically inert polystyrene latex (0.81 μm diam.), sensitised with the globuline fraction of virus antisera. When these particles are suspended in a solution containing a homologous antigen, flocculation takes place after short incubation, the precipitin reaction occurring on the surface of the indicator particles which clump together.

The procedure has been used with soybean mosaic virus. Cotyledons and plumules from seedlings, developed in an ordinary blotter test for fungi, are separated and macerated in small quantities (0.5–5 ml.) of normal saline solution (0.85 per cent NaCl in distilled water). A drop (0.05 ml.) of the extract is mixed with a drop (0.025 ml.) of sensitised latex suspension, preferably on a glass cavity-slide. The slides are incubated at about 20°C in a moist chamber to avoid evaporation. Observations are made, after 1 h, through a low power stereo binocular microscope (6–50 × magnification) with reflected light and a black background. Frequently, strong positive reactions can be seen by the unaided eye. If soybean mosaic virus is present in the test drops, even at low concentrations, latex particles aggregate in groups, thereby making the suspension heterogeneous. Concentration of the virus determines the extent of flocculation of the latex particles; the reaction may not occur at very high and very low concentrations. In controls, made by extracts from known healthy seedlings, the particles remain uniformly dispersed. Sometimes a longer incubation period of some hours is required. Also extracts made directly from seed can be used. Batches of, for example, 10 soybean seeds are macerated with 15 ml. saline solution. To the extract 2 ml. ethyl alcohol is added for partial clarification. After standing for 30 min the solution is centrifuged at about 3,000 rev/min for 20 min. The supernatant is used for the flocculation test. Control is provided from known healthy seeds. In less than 2 h positive reactions can be seen in the test drops.

An elegant procedure is the fluorescent antibody technique. The antiserum is conjugated to a fluorescent dye and the complex, the fluorescent antibody, is then used as stain of material from a natural environment. Antigen–antibody reaction can be visualised by fluorescence microscopy (Schmidt, 1973). This immunofluorescence technique has been used for detection of *Pseudomonas phaseolicola* in bean seed (Coleno, 1968).

(2) *Application*. The techniques have not yet been widely used in seed health testing. In Chile the Ouchterlony test has been used regularly in post-entry screening of imported melon seed lots; in Montana, U.S.A., used in a seed certification programme of barley against barley stripe mosaic (Davis and Wallace, 1967).

(3) *Limitations*. Provided that antiserum is available, serological methods offer great possibilities, perhaps particularly for detection of viruses. The procedure is quick and the results usually accurate. Seedlings to be tested for viruses could be made available from the blotter test. Antisera for specific seed-borne viruses and bacteria could be produced and distributed internationally from a few centres. Even so, the degree of specificity of antisera from plant pathogens to be considered has to be checked to avoid failure when an antiserum is used for seeds infected with a pathogen of a serologically different strain. Usually, however, different strains of a virus react almost identically to one antiserum of the virus.

The type of serological procedure to be used depends on type of virus. The Ouchterlony gel double-diffusion test can be used for isometric viruses and for tobacco mosaic virus but not for filamentous viruses for which the latex test may be used. Naturally, serological procedures can be used only for specified viruses, whereas the indicator test must be used to reveal any of a broader range of viruses.

31 Inspection of Plants beyond the Seedling Stage

31.1 GROWING-ON TEST PROCEDURES

(1) *Procedure*. Certain categories of seed-borne diseases require a longer period of incubation to be detected than is provided by the usual incubation procedure, the blotter method, the agar plate method and the seedling symptom tests. This is true particularly for bacterial and virus diseases. The plants must be grown beyond the seedling stage before symptoms can develop. The seeds are sown in autoclaved soil in suitable containers and submitted to suitable conditions of temperature and humidity, often in a glasshouse or controlled environment cabinet, protected against contamination. A controlled environment chamber, as in the standard procedure developed for lettuce mosaic by Rohloff (1960, 1962), is an ideal of the procedure. A test can be completed in 16–21 days by sowing the seed in peat with 0.5–2.0 g $CaCO_3$ added per litre and 0.5–1 g complete fertiliser, and incubating at 25°C under constant fluorescent daylight (8×40 W tubes for each 0.6 m^2, 30 cm above substrate surface). Afanasiev (1956) described a test for barley stripe mosaic that may be conducted in a greenhouse. Barley is sown in flats or polypots, in units of 100 seeds. The first reading is made in the three-leaf stage, the second and final reading about 10 days later. Hampton, Sill and Hansing (1957) established that high light intensities (10,000 foot-candles) and temperatures of 21–30°C are optimum conditions for symptom expression and that symptoms were poor and recorded infection percentage lower at low light intensities (3,000 foot-candles) even at optimum temperatures. Seed health certification schemes of barley for stripe mosaic based on this technique are carried through successfully in Montana and Kansas where this cereal is extensively grown (MacWithey, Mills and Afanasiev, 1957; Hampton *et al.*, 1957).

Also, systemic infections of fungi may be detected in growing-on tests. Porter (1944) indicates that fusarioses in cabbage, flax, tomato and watermelon are detectable with 'reasonable accuracy' by this procedure. *Colletotrichum fuscum* on *Digitalis lanata* is difficult to detect directly on seeds, but the disease is easily detectable when the seeds are sown in dishes in the greenhouse and recorded after 12 weeks (Thomas, 1951).

(2) *Application*. In quarantine control the growing-on procedure is of the greatest importance, as exemplified in the Portuguese quarantine import inspection of seed samples for breeding purposes. Three major pathogens, which were not known to occur in Portugal at the time of inspection, were encountered: *Peronospora tabacina* from tobacco seed, *Corynebacterium insidiosum* from lucerne seed and *Erwinia stewartii* from maize seed (Branquinho de Oliveira, personal communication, 1963). Similar experiences concerning alien races of parasitic fungi and strains of viruses have been reported from the Regional Plant Introduction Station at Ames, Iowa (Leppik, 1962*a*). The method has been highly technically developed at the East African Plant Quarantine Station at Muguga in Kenya, where seed is grown under closed quarantine in aseptically operated greenhouse units. Seeds are harvested from the plants grown under these conditions, and retested for health in the laboratory before they are released for final surveillance under field conditions (Sheffield, 1968).

(3) *Limitations*. In routine seed health testing as now carried out at seed testing stations and plant protection services, particularly in Europe, much emphasis is laid upon laboratory incubation methods. The importance of growing-on tests is underestimated and the procedure little used in routine testing probably because it does require considerable space, greenhouse facilities or environmental controlled rooms, and more time than laboratory procedures. Nevertheless, particularly for plant quarantine inspection, the method is indispensable for detection of such important seed-borne diseases which may spread rapidly in new areas owing to the great rate of increase of their inocula (certain bacterial and virus diseases, certain epiphytotic pathogens such as downy mildew, for instance *Peronospora*, and certain rusts). Many of these pathogens are not detectable by available laboratory procedures, and hence the growing-on procedure must be adopted. On the other hand, many seed-transmitted pathogens may be carried in a symptomless manner or may only produce vague symptoms such as slight stunt or bakanae effects. Wilt organisms such as *Fusarium moniliforme*, *F. oxysporum* and *Verticillium alboatrum* are often symptomless in many hosts, also *Colletotrichum dematium* in soybean, *Xanthomonas phaseoli* in bean, *X. campestris* in cabbage, *X. vesicatoria* in tomato, etc., cf. Section 14.1; finally many viruses, e.g. raspberry ringspot virus, tomato black ring virus in strawberry (Lister and Murant, 1967) and sunblotch virus in avocado (Wallace and Drake, 1962), and often they are symptomless in dependence of cultivar and virus strain, as in bean common mosaic virus in bean (Hubbeling, 1972). For detection of fungi, incubation procedures, including the blotter method, are usually workable; for bacteria several procedures may be considered including the phage-plaque and the

serological tests (Sections 30.2 and 30.3), and for viruses the indicator test (Section 30.1) may work, provided that suitable indicator hosts are available.

31.2 FIELD TRIALS

For detection of certain seed-borne pathogens laboratory routine testing procedures or other methods may be regarded as unsatisfactory or impracticable for various reasons. In such cases an after-control of the seed lots concerned may be carried out by examination of representative samples grown in field trials. The procedure is explained below, as exemplified by the field trial scheme for barley, used in the seed certification schemes for cereals in Denmark.

(1) *Procedure*. Each sample is sown in four widely separated plots. The trials are located in two different farms at different localities, two plots being sown in each farm on two different dates.

The plots are 9 m long, 6 rows being sown in each plot with a distance of 25 cm between the rows — altogether 54 m linear rows. The plots are separated, lengthwise, by a single row of spring wheat, and crosswise by a 1 m path — Fig. 31.1(4) and (5). Small hand sowing machines are used for sowing — Fig. 31.1(2) and (3). There are normally 3,000 plants in each plot, which makes a total of 12,000 plants in the four plots.

When the plants have emerged and before tillering, Fig. 31.1(4), their number is recorded by counting a 1 m row in four different places in each plot. By multiplying the count of the four samples by 13.5 ($13.5 \times 4 = 54$, total of linear metre) the total of plants in the plot concerned is calculated. The health control of barley includes leaf stripe (*Drechslera graminea*) and smut. In Denmark only loose smut (*Ustilago nuda*) is of importance. Stripe is recorded by going through the whole plot. The controller straddles above a row examining the neighbouring row. The infected plants are pulled up and a single spike taken, while the remainder of the plant is discarded. After examination of the whole plot the number of spikes are recorded and thus the total of infected plants.

Loose smut is recorded in the same way but later, when the records of stripe have been completed; this is at the boot stage or just after.

Principles, as well as general and special technical rules for field plot examination, are given in the chapter 'Determination of genuineness of species and cultivar' in the International Rules for Seed Testing (ISTA, 1966*a*). The rules include directions on the procedure of planting field plots of cereals, pulses, oil plants, herbage plants, root crops and certain other crops.

Fig. 31.1. Field trials in Denmark and Sweden. (1) Distribution of seed samples in the field; (2) sowing apparatus for equidistant sowing in rows; (3) operation of sowing; (4) newly germinated plots; (5) and (6) barley plots prior to recording of diseases and harvest

(2) *Application*. The control technique described above is used extensively in Denmark, Norway and Sweden, in conjunction with checks on cultivar identity. The same method has been adopted in other countries such as Portugal, where the procedure forms a part of the national seed certification scheme for rice.

(3) *Limitations*. One minor difficulty in this technique lies in the fact that some of the plants are lost during the period from the first record of emerged plants to the final records of infected plants. Consequently the disease percentages recorded become unrealistically low, if no correction is made. Normally about 80–90 per cent of the original number of plants is expected to remain at the time of the final records but sometimes insect

damage, etc., may further reduce the plant number. Nevertheless, a correction factor has not been adopted in the Danish scheme.

It is an obvious disadvantage that disease records are usually made after the seed lot concerned has already been used for sowing. This disadvantage does not fully apply to Foundation and Registered Seed tested in field trials. If, in temperate climates, trials are carried out during winter in greenhouses, results can be made available before sowing in spring.

For certain crops this procedure is in some respects greatly preferable to field inspection of seed crops, because a direct comparison can be made simultaneously between a large number of seed samples, growing under uniform conditions. Field inspection of seed crops, on the other hand, shows variation with the conditions of growth from place to place, and the observations cannot possibly be carried through with the same care as in field trials.

31.3 INSPECTION OF SEED CROPS

For many seed-borne diseases, field inspection of the seed crop is the most effective control procedure, with rejection of infected seed lots before harvest. For any progress in control of seed-borne diseases this method is invaluable.

(1) *Procedure*. In practically any field inspection scheme for seed crops two inspections are regarded as the minimum. The first inspection is carried out when the plants are green, in full growth, while the second takes place when they have reached full maturity, during flowering or soon after flowering, during the period of seed maturing.

Details in the techniques of examining the crop for infection depend of course on the kinds of crop and diseases involved. Whenever the whole crop cannot be examined plant by plant, proper sampling of the field to be investigated is imperative. In setting up regulations for field inspection precise directions for sampling the field and examining the plants should be given. Usually, infections are recorded as percentage of plants or as the number of plants per unit of area. For some diseases other measurements may be applied, such as in a certification scheme used some 20 years ago in New York State for tomato: 'Any field showing more than 5 per cent of the fruit affected by potato late blight or buckeye rot (Phytophthora footrot) at any one inspection shall be rejected'.

The inspection technique has been worked out fairly well in details for cereals in many countries. In the German Democratic Republic, for instance, crops for Registered Seed of barley and wheat must be inspected according to the following procedure: loose smut should be recorded within 100 steps in each direction, with a total of five records (Kublan, 1952). Usually the plants found within an arm's length of either side are observed. In Holland, barley, oats and wheat must be inspected by examining four samples of the field each of 100 m^2. Field inspection for smut has been discontinued in many places in favour of the laboratory embryo test, especially where permitted infection is of the order of 0.2 per cent.

In the Organisation for Economic Co-operation and Development scheme for the varietal certification of cereal seed moving in international trade (OECD, 1969), a system for sampling ears or panicles is described. The procedure may serve as an example on sampling for cereals and seed crops of grasses:

> . . . a convenient sample size is provided by an area which will include about 10,000 ears or panicles. Under average conditions in North West Europe, 20 sq. metres would fulfil this requirement; it is usually convenient to look at a sample 1 m wide×20 m long, walking across the direction of sowing. . . . The number of sample areas which the inspector can look over in each crop will be limited by the time available. In general, it is considered to be satisfactory if the number of areas examined include 100,000 ears or panicles — usually 10 areas of 20 sq.m. each in North West Europe. This will give a good estimation of the varietal and species purity. It is necessary to make an estimate of the number of ears or panicles per hectare so that the numbers of impurities detected can be expressed as a proportion of the crop as a whole. This can be done by counting the number of ears or panicles in a 1 metre length of row within each of the 10 larger sample areas. The population per hectare can then be calculated from the formula:
>
> $$P=\frac{1{,}000{,}000M}{W}$$
>
> where P is the ear or panicle population per hectare,
> M the mean number of ears or panicles per metre length or row,
> W the width between rows in cm.
>
> 'Ready reckoner' tables can easily be produced for a range of values of M and W to avoid calculating the population for each field. If it is desired to look more closely at ears or panicles than is possible in a 20 sq.m. sample area, the 1 metre length of row in which the plant counts were made can be examined in greater detail.

In England, for instance, the normal standard for Field Approval for infection of loose smut is one smutted ear in 10,000 ears. Lower standards are applicable to susceptible cultivars of both wheat and barley but these are not more than one in 2,000 in commercial crops and not more than one in 5,000 in crops intended to further multiplication (Doling, 1966*b*).

Popp (1947) demonstrated a special technique that provides an easy and precise record of wheat bunt in the spikes. These are harvested, when most of them are mature. Before examination they are soaked in water, either boiling or almost boiling, remaining soaked overnight, or 24 h in water, to which a detergent is added, the water being kept at room temperature. By this procedure the smutted grains swell and become darker, and the glumes become semi-transparent.

(2) *Application*. All seed certification schemes are primarily based on field inspection of the seed crops. Seed certification schemes in operation all over the world have contributed in no small measure to the dramatic rise in agricultural production during the last half-century. By following minimum seed certification standards, disease levels are kept down to a minimum, and if the schemes operate successfully, disease tolerances may gradually be reduced towards the ideal level — a tolerance of zero. The importance of certification schemes in control is further discussed in Subsection 19.4.1.

In North America the leading body for the organisation of certification schemes is the International Crop Improvement Association, founded in 1919, and comprising seed certifying agencies in most of the States of the U.S.A. and all of the states in Canada. The proposals in regulations (Minimum Seed Certification Standard), issued by this organisation, serve as guidance for setting up certification schemes in the individual States of the U.S.A. and Canada. Such schemes are in operation with bean (anthracnose, bacterioses, mosaic), mungo bean (anthracnose, bacterioses, mosaic), southern pea (Fusarium wilt), barley (loose and covered smut), oats (loose and covered smut), wheat (loose, stinking, covered and other smuts, dwarf bunt), small grains (flag smut, loose smut, ergot and other diseases), sorghum (head smut and kernel smut), sudan grass (head smut and kernel smut), rice (all seed-borne diseases), brome grass (head smut and kernel smut), other grasses, including blue grass, tall fescue, fine fescue, orchard grass and wheat grass (all of them: smut), flax (all seed-borne diseases), tomato (bacterioses, Fusarium wilt, *Septoria lycoperisici* and others), and watermelon (Fusarium wilt). Similar schemes are in operation in Belgium, Holland, Germany, the United Kingdom and many other countries. For vegetable seed crops, certification schemes are organised in Holland — bean (halo blight, mosaic), pea (Fusarium wilt), carrot (black rot); in the Federal Republic of Germany — bean (anthracnose, halo blight), celery (*Phoma apiicola*, *Septoria apiicola*), pea (*Ascochyta*); and in Australia particularly successful schemes have been in operation since the 1930s in New South Wales for bean against bacterioses and anthracnose, and in Victoria (Henderson, 1948) for bean (anthracnose, bacterial blight, halo blight), for pea (*Ascochyta*, *Septoria*, *Pseudomonas pisi*) and for tomato (against bacterial wilt and Fusarium wilt).

(3) *Limitations*. The procedure is expensive in terms of labour and transportation, but may often be combined with inspection for other pur-

poses, especially for trueness to variety now often required under legislation governing Plant Variety Rights; also as a general guidance to the farmer on cultural practices. Some diseases, however, may occur so late in the development of the crop that they may not be detected during such a field inspection. This certainly holds true for symptomless carriers. Laboratory tests may here often complete the information obtained by field observations.

Field inspection is often assumed to be more reliable for any disease than laboratory inspection. However, for many diseases a laboratory test is preferable for specific identification. Thus, the two major bacterial diseases of bean, halo blight and common bacterial blight or its variant fuscous blight, cannot be readily distinguished in the field, and if field inspectors are not able to distinguish between them then field inspection is useless for quarantine control.

Weather or cultural conditions strongly affect development of symptoms. Grogan and Kimble (1967) found field inspection for halo blight 'practically worthless' in beans grown under furrow irrigation in a dry climate as in California. When seed contaminated with *Pseudomonas phaseolicola* was grown with sprinkler irrigation, infection was usually apparent, but with furrow irrigation detection of disease was extremely difficult. Under these conditions infected plants could sometimes be detected in the first trifoliate leaf stage, but not after the plants were full grown. The authors state: 'We believe that a more meaningful criterion for certification is to test for seed transmission after seed processing including seed treatment has been completed.'

Symptoms may vary greatly according to weather conditions thus reducing certainty in recording. For example, *Septoria apiicola*, the late blight organism of celery and celeriac, produces dry grey or brown leaf spots under moderately humid conditions, whereas extremely moist conditions lead to development of soft rot symptoms with extended dark soft lesions, quite different from the generally known disease picture described in textbooks. Conversely, under dry conditions black rot of crucifers, incited by *Xanthomonas campestris*, may be modified from the well-known soft rot lesions to atypical dry leaf spots. Under certain growth conditions many pathogens may not develop symptoms at all, the host acting as a symptomless carrier (cf. Section 14.1).

Loose smut in cereals can be counted in the field in certain conditions but inspection agencies have found by bitter experience that at inspection time the bare heads without any smut must be found. In these circumstances and taking into account that infection may come in from outside the crop, there is no doubt that for most cultivars the laboratory embryo test is much more reliable for this disease than any field inspection.

PART V
Assessment of Seed-borne Inoculum

32 Forecasting Losses from Seed-borne Diseases and Assessing Disease Tolerances for Seed Health Testing

INTRODUCTION

Seed health testing of crop seeds for sowing purposes is intended to measure seed-borne inoculum potential. Assessments of the economic implications of seed-borne inocula in terms of possible crop losses should be an integral part of such testing activities. Expressed in another way: how much may the farmer have to pay in crop loss for a definite infection percentage found by testing his seed lot? The information available to answer this question varies extremely widely from one seed-borne disease to another and ranges from total ignorance to well-established facts of loss of profit by the farmer.

In setting tolerances for seed health testing, often 'guestimates' still supersede facts and realities. In the following sections various factors in their relation to development of disease and crop loss will be discussed. The following factors have to be considered:

I. *Biological considerations (infection biology)*
  1. Measuring seed-borne inoculum potential: testing methods
  2. Amount of infection or contamination: inoculum potential
     (1) Percentage infected/contaminated seeds
     (2) Individual intensity of infection in terms of load of inoculum per seed, and vitality of the pathogen, respectively the number of pathogenic propagules per weight unit of seed
     (3) Inoculum density: infection percentage or number of pathogenic propagules × seeding rate
     (4) Relative frequency of seed infection or contamination
  3. Within-species variation of virulence (variation from biotype to biotype, pathogenic races)
  4. Within-species variation of susceptibility of the host (varietal resistance)
  5. Antibiosis
  6. Infection probability
     (1) The quarantine aspect
     (2) Degree of transmissibility from seed to seedling
     (3) Spread of infection in the field: rate of increase
     (4) Probability of soil contamination and infection of subsequent crops

7. Relative importance of other means of transmission
8. Possibility of disinfecting the seed: reduction or removal of the seed-borne inoculum

II. *Economic considerations (loss evaluation)*

9. Crop value (average depending on region, country, year)
   (1) Values converted to hkg barley or wheat
   (2) Values converted to monetary units (dollars at some date?)
10. Degree of damage (loss as percentage of crop value)
11. Total rejection if for export and fails to meet quarantine requirements
12. Costs of control measures against seed-transmitted diseases

32.1 MEASURING SEED-BORNE INOCULUM POTENTIAL

Any indication of seed-borne inoculum must refer to a specified testing method, because the amount of seed-borne inoculum detected varies substantially according to the testing technique. An established standard procedure must therefore be followed and at least for international work comparisons of seed-borne inoculum potentials must be based on data provided by the same testing technique.

The amount of inoculum may be expressed in terms of (1) *percentage* of infected or contaminated seeds; (2) the *load* of pathogen, for instance (a) *weight* of sclerotia in a unit weight of the seed, or (b) the *number* of pathogen units, e.g. sclerotia in a unit weight or a definite number of seeds, or (c) as spores per individual seed as obtained in suspensions from washing seeds; (3) the *degree* of severity of infection as developed according to a standard procedure and recorded according to a standard scale; and, finally, (4) the *viability* of the inoculum, for instance the germination capacity of spores, the longevity of the pathogen.

32.1.1 Infection percentage and index of infection

The sensitivity of a method should be considered. For example, the agar plate method yields a higher percentage or gives a stronger manifestation of certain fungus pathogens than the blotter method or direct inspection. An extreme, although not a unique, example — cabbage seed infected by *Alternaria brassicicola* tested by the blotter method — showed 6.5 per cent infection, but 93 per cent on malt extract agar; in another test, the figures were 2 and 73 per cent, and by 'naked eye' none!

In rice, the spore count method does not reveal mycelial infection of *Pyricularia oryzae,* and hence the number of conidia is misleading as a quantitative measurement of inoculum potential, and particularly so if the germination capacity of the spores is disregarded (Neergaard, Lambat and Mathur, 1970). Moreover, for this parasite the agar plating technique yields percentages much lower than the blotter test, owing to its slow rate of growth on agar media and suppression of the colonies by other fungi (Neergaard and Saad, 1962).

In the incubation methods the results are usually given as infection percentages. Per cent is a simple mathematical notion implying uniform units. But the incidences of a pathogen encountered in individual seeds are very variable, and hence an infected seed is not necessarily a simple unit of inoculum potential. It may be so, as in loose smut of barley. In most cultivars of barley an infected embryo may be accepted as an almost invariable unit of inoculum potential in spite of varying amounts of mycelium present, each seed harbouring such an infection developing into an infected plant. But *Alternaria brassicicola* is quite different in this respect. Individual seeds show a great range of infection degrees, reflected in striking differences between infection percentages obtained in the blotter test and those obtained in the more sensitive agar plating test. In recording infection percentages of this pathogen in blotter tests a simple indexing system may be adopted, for instance 1 for 'weak' infection, 2 for 'medium' and 3 for 'severe' infection — 2 and 3 based on macroscopic inspection (Neergaard, unpublished). According to a similar principle Aulakh, Mathur and Neergaard (1974) established three infection grades of *Drechslera oryzae* in rice seed, and Singh, Mathur and Neergaard (1974*a*) three infection grades of *Drechslera maydis* in maize seed; the grades were positively correlated to severity of seedling losses.

In agar plate tests, colony diameter of a pathogen around individual seeds may indicate the potential of inoculum in each of these. Malalasekera and Colhoun (1969) have shown that colony diameter of *Fusarium culmorum* around individual seeds on an agar medium is positively correlated with inoculated spore load per seed and, moreover, that spore load per seed is positively correlated with severity of disease in wheat seedlings in the field. This suggests a positive correlation also between colony diameter of the pathogen on agar and disease severity in the field and may be utilised in agar tests in indexing for the potential of seed-borne inoculum; cf. Section 17.3 and Fig. 17.4.

This kind of differentiation in recording incidences of infection or contamination may contribute to more accurate measurements of seed-borne inoculum potentials and closer correlation between laboratory records of inocula and field observations of disease development.

In order to investigate the degree of correlation between seed-borne inoculum and disease development in the field, comparisons should be made within a range of inoculum potentials:

(1) as found between different seed lots of the same host cultivar, containing different degrees of infection or contamination; or

(2) it may be established by treating one seed lot with increasing dosages of seed treatment, thus leaving different degrees of seed infection or contamination;

(3) it may be established by diluting a heavily infected seed lot using

increasing rates of seed from a healthy lot of the same cultivar; or

(4) it may be established by increasing dosages of artificial inoculum in the same seed lot.

In all these cases the inoculum must be determined according to a standard seed health testing procedure.

32.1.2 Inoculum density

The amount of seed-borne inoculum actually available in the field is dependent not only on its proportion in the seed lot but eventually on the amount of infected seed that is brought into the soil. The seeding rate multiplied by the infection percentage or any adequate conversion, and the amount of disseminated propagules, such as sclerotia or cysts, per unit of area, defines the actual amount of inoculum with the seed in the field, or the inoculum density.

Smaller seeds are in most cases sown in greater number per unit of area than are larger seeds. Consequently infection percentages are not always directly comparable between different species of the host.

Factors of cultivation practice also have a bearing on the amount of seeds needed per unit of area. For crops raised in seed beds the seeding rate may be very high (e.g. for tobacco), for forage crop seed sown in mixtures the seeding rate may be relatively low. Soil conditions may be decisive: in Denmark *Festuca pratensis* is sown in meadows and on boggy soil at a rate of up to 12 kg per hectare, but on normal not-too-dry agricultural soil the seeding rate may be down to 3 kg per hectare (Pedersen, 1942).

In Table 32.1 selected important species of crop plants are listed. The seed weight (according to United States Department of Agriculture, 1952), the approximate average quantity of seed usually sown per hectare (calculated according to basic figures given by Pedersen, 1942) and the seeding rate in terms of million seeds per hectare are given for each species. All are kinds of seed that in usual practice give dense populations, such as forage crops or seeds sown in seed beds for later transplanting or thinning.

A formula applicable for calculation of infection percentage as related to a given inoculum density, on the basis of seeding rate, is given in Table 32.2. As examples, the infection percentage, corresponding to a tolerance of 100,000 inoculated loci per hectare, has been calculated for each species included in the table.

However, sowing in rows needs further consideration, since by this practice the inoculated loci are not distributed at random on the area. At any given seeding rate increasing distance between rows results in increasing inoculum density within the rows. This fact should be taken into consideration in assessing disease tolerances.

Table 32.1. Seed weight and approximate normal seeding rate of selected important agricultural crop plants

Family	*Species*	*s* (number of seeds per g)	*q* (kg per ha)	*s* × *q*	*r* (1/1,000 sq) (approx. average seeding rate as million seeds per ha)
Chenopodiaceae	Beta vulgaris	58	16 – 25	928 – 1,450	1
Cruciferae	Brassica campestris	536	4 – 5	2,144 – 2,680	2
	Brassica napus	346	5	1730	2
	Brassica nigra	1,256	3 – 5	3,768 – 6,280	5
Gramineae	Avena sativa	28	180	5,040	5
	Hordeum vulgare	30	160 – 180	4,800 – 5,400	5
	Secale cereale	40	170	6,800	7
	Triticum aestivum	25	180 – 200	4,500 – 5,000	5
	Dactylis glomerata	1,441	3 – 6	4,323 – 8,646	6
	Lolium multiflorum	500	3 – 6	1,500 – 3,000	2
	Lolium perenne	500	4 – 12	2,000 – 6,000	4
	Phleum pratense	2,500	2 – 5	5,000 – 12,500	9
	Poa pratensis	4,800	2 – 6	9,600 – 28,800	19
Leguminosae	Medicago lupulina	586	12 – 14	7,032 – 8,204	8
	Medicago sativa	500	20 – 25	10,000 – 12,500	11
	Trifolium pratense	600	10 – 15	6,000 – 9,000	8
	Trifolium repens	1,500	2 – 6	3,000 – 9,000	6
	Vicia sativa	19	100	1,900	2
Linaceae	Linium usitatissimum				
	fibre flax	178	130 – 150	23,140 – 26,700	25
	oil flax	178	80 – 90	14,240 – 16,020	15
Solanaceae	Nicotiana tabacum	15,625	5 (for seedbeds	78,125	78
Umbelliferae	Daucus carota	826	4 – 5	3,304 – 4,130	4

If 10 kg *Festuca rubra* (1,200 seeds per gramme)* and 10 kg *Poa pratensis* (4,800 seeds per gramme), both with 10 per cent *Drechslera*-infected seeds, are sown each on one hectare, four times more *Drechslera*-infected

*All seed weights are according to United States Department of Agriculture (1952): Manual for testing agricultural and vegetable seeds. *Agricultural Handbook No. 30,* Washington, D.C., 445 pp.

seeds are introduced to the soil with *Poa pratensis* than with *Festuca rubra*. A certain percentage of *Fusarium* in *Poa* (12 kg seeds per hectare, 4,000 seeds per gramme), gives about 8.5 times as many inoculated loci per hectare than the same percentage of *Fusarium* in rye (170 kg per hectare, 40 seeds per gramme).

Table 32.2. Calculation of infection percentage, as related to a given inoculum density, on the basis of seeding rate

$$p = \frac{I}{10qs} \text{ (Neergaard, 1962}a\text{)} \qquad (1a)$$

where I = number of inoculated loci per ha (inoculum density)
p = infection percentage
q = quantity of seeds as kg per ha
s = number of seeds per g
may be simplified for calculation as specified in the heading above,

$$p = \frac{i}{r} \qquad (6)$$

where i = per cent of one million inoculated loci per ha, or number of inoculted loci per m² $= \frac{I}{10{,}000}$

r = seeding rate as million seeds per ha, $= \frac{qs}{1{,}000}$

p as above.

Examples:
(6), A: A tolerance of I is fixed at 100,000 inoculated loci per ha or i at 10. Hence the following tolerance infection percentages:

Species	*Seeding rate* r	*Tolerance inf. percentage when* $p = \frac{10}{r}$
Beta vulgaris	1	10
Brassica campestris	2	5
Brassica napus	2	5
Brassica nigra	5	2
Avena sativa	5	2
Hordeum vulgare	5	2
Secale cereale	7	1.4
Triticum aestivum	5	2
Dactylis glomerata	6	1.7
Lolium multiflorum	2	5
Lolium perenne	4	2.5
Phleum pratense	9	1.1
Poa pratensis	19	0.5
Medicago lupulina	8	1.3
Medicago sativa	11	0.9

Table 32.2 *continued*

Species	*Seeding rate* r	*Tolerance inf. percentage when* $p = \frac{10}{r}$
Trifolium pratense	8	1.3
Trifolium repens	6	1.7
Vicia sativa	2	5
Linum usitatissimum		
fibre flax	25	0.4
oil flax	15	0.7
Nicotiana tabacum	78	0.1
Daucus carota	4	2.5

(6), B: A tolerance of *I* of a pathogen of tobacco is fixed at 2,000,000 inoculated loci per ha or *i* at 200 (200 per m^2). Hence tolerance infection percentage 200/78 = approx. 2.6%.

(6), C: A tolerance of *I* of a pathogen of *Beta vulgaris* is suggested at 2,000,000 inoculated loci per ha. Hence tol. inf. percentage 200/1 = 200%, i.e. 2× the indicated normal seeding rate for this crop in using 100% infected seed would be acceptable with the suggested tolerance.

Formulae for calculation of the average distance between inoculated loci in rows and the related infection percentage are given in Table 32.3.

In this purely statistical approach to the effect of inoculation ratio in coherent crop populations, the economic significance of individual crops and diseases has been disregarded. However, this line of thought stresses the importance of seeding rate and inoculum density as factors in assessing disease tolerances.

Formulae applicable for calculation of inoculum density on the basis of infection percentage, and vice versa, are given in Table 32.4.

The problem of inoculum density is important in relation to tolerances with regard to sclerotia and similar structures to be counted per weight unit, and generally speaking the conception of 'inoculum density' as a criterion and as a comparable measurement of inoculum is particularly important as far as forage crops, cereals and seeds to be sown in seed beds are concerned. On the other hand, the 'infection percentage' retains its full validity for crops in which each plant keeps its individuality, such as bean, pea, maize and castor bean.

Tolerances for sclerotia, for instance ergot, are often given as number of sclerotia per kg of seeds, but this measure is not directly comparable from species to species. If, taking into account the number of seeds per gramme for each species, we adopt a fair standard tolerance — for example, as Neergaard suggested as a tolerance for seed inspected in Copenhagen, 100 sclerotia per million seeds for ergot — the following approximate number of sclerotia per kg seeds would be obtained: *Poa pratensis* 500, *Festuca rubra* and *Dactylis glomerata* 125, *Lolium perenne* 50, and rye 4 sclerotia per kg seeds.

This amount of sclerotia would give, for each of the four last-mentioned species, roughly about 1,000 sclerotia per hectare as an acceptable amount of inoculum, but with the important reservation that the smaller the size of the seed, the higher the seeding rate and the number of sclerotia sown. Thus, *Poa pratensis* at the above-mentioned tolerance would give 5,000–6,000 sclerotia per hectare at the normal seeding rate. In setting up tolerances a correction according to seeding rate may be introduced. In adopting such a correcting factor in order to obtain a standard

Table 32.3. Calculation of infection percentage on the basis of inoculum density as related to sowing in rows

$$d = \frac{10{,}000}{\sqrt{I}} \qquad (7)$$

where d = average distance in cm between inoculated loci
I = number of inoculated loci per ha (inoculum density)

$$d_v = \frac{d^2}{v} \qquad (8)$$

where d_v = average distance in cm between inoculated loci in rows
v = distance in cm between rows
d as above.

In assessing infection percentages as related to the distance between inoculated loci in rows as against the distance between inoculated loci from seed sown equally distributed, a

$$\text{correction factor} = \frac{d}{v} \qquad (9)$$

may be used. This factor may be of practical importance when $d/v < 1$ and is of proportionally increasing importance as d/v is decreasing.

Examples:

(8) A seed lot of beet with I = 1,000,000 of *Phoma betae* or d = 10 cm (or with r = 1, see Table 32.2, p = 100%) is sown in rows with a distance of 50 cm between the rows. Hence

$$d_v = \frac{100}{50} = 2 \text{ cm}$$

(9) In using the correction factor for sowing in rows,

$$\frac{d}{v} = \frac{10}{50} = 0.2\ (< 1)$$

The infection percentage corresponding to d_v = 10 cm would be 0.2 × 100 = 20% (or with r = 1, see Table 32.2, I = 200,000).

tolerance of 1,000 sclerotia per hectare, the tolerance of the above-mentioned species would produce the following approximate numbers of sclerotia per kg seeds: *Poa pratensis, Festuca rubra* and *Dactylis glomerata* 100; *Lolium perenne* 60, rye 6. In this calculation, however, the size of sclerotia, varying from one host plant to another, and by this the number of stromata, perithecia and ascospores produced per sclerotium, is disregarded. Obviously, from this point of view the calculated tolerances are relatively lenient to big-seeded crops such as rye.

Table 32.4. Calculation of inoculum density on the basis of infection percentage and vice versa

$$I = q \;.\; 1{,}000s \;.\; \frac{p}{100} = 10\,qsp \qquad (1)$$

where I = number of inoculated loci per ha (inoculum density)
q = quantity of seeds as kg per ha
s = number of seeds per gramme
p = infection percentage.

$$i = \frac{qsp}{1{,}000} \qquad (2)$$

where i = number of inoculated loci per m²
q, s and p as above,

hence, for calculation of infection percentage on the basis of inoculum density,

$$p = \frac{I}{10qs} \qquad (1a)$$

and

$$p = \frac{1{,}000i}{qs} \qquad (2a)$$

Examples:

(1) Rye with 10% *Fusarium:* 10 . 170 . 40 . 10 = 680,000 inoculated loci per ha or

(2) $\frac{170 \;.\; 40 \;.\; 10}{1{,}000}$ = 68 inoculated loci per m².

(1) *Poa pratensis* with 3% *Drechslera:* 10 . 12 . 4,800 . 3 = 1,728,000 inoculated loci per ha or

(2) $\frac{12 \;.\; 4{,}800 \;.\; 3}{1{,}000}$ = 172.8 inoculated loci per m².

(1a) *Poa pratensis* with I = 100,000 inoculated loci per ha.:
$\frac{100{,}000}{10 \;.\; 12.\; 4{,}800}$ = 0.17% or

(2a) i = 10 gives: $\frac{10{,}000}{12.4{,}800}$ = 0.17%.

Formulae applicable for calculation of inoculum density on the basis of number of pathogenic units (e.g. sclerotia) per million seeds or per kg seeds, and vice versa, are given in Table 32.5.

32.1.3 Surveying — estimates on frequency and amount of seed-borne inoculum

In risk calculations we are interested in totals of seed-borne inocula and their distribution. To obtain an all-over picture of the distribution of inocula of a pathogen in an area, in a country or in the world, the frequency of occurrence in seed in terms of the proportion of seed lots infected from year to year in one or many seed-producing areas must be considered. Much information is available, but scattered throughout the literature, and there is a need to collate the relevant valuable data. Table 32.8 is an attempt to compare a few well-known seed-borne diseases with

Table 32.5. Calculation of inoculum density on the basis of number of pathogenic units per million seeds or per kg seeds and vice versa

$$I = qk \qquad (3)$$

where I = number of inoculated loci per ha (inoculum density)
q = quantity of seeds as kg per ha
k = number of pathogenic units (e.g. sclerotia) per kg seeds.

$$I = q \,.\, 1{,}000s \,.\, \frac{m}{1{,}000{,}000} = \frac{qsm}{1{,}000} \qquad (4)$$

where s = number of seeds per gramme
m = number of pathogenic units per million seeds
I and q as above,

hence

$$k = \frac{sm}{1{,}000} \qquad (5)$$

$$m = \frac{1{,}000k}{s} \qquad (5a)$$

$$k = \frac{I}{q} \qquad (3a)$$

and

$$m = \frac{1{,}000I}{qs} \qquad (4a)$$

Examples:

(4) Rye with 100 sclerotia per million seeds (m = 100):
$\frac{170 \,.\, 40 \,.\, 100}{1{,}000}$ = 680 sclerotia per ha (= I) and

(5) $\frac{40 \,.\, 100}{1{,}000}$ = 4 sclerotia per kg (= k).

(3a) Rye rendering 1,000 sclerotia per ha:
$\frac{1{,}000}{170}$ = 5.88 or approx. 6 sclerotia per kg (= k).

regard to their inoculum potentials and distribution of these in terms of per cent infected seed lots or the corresponding estimates.

The following conversions of terms of estimates into terms of per cent infected lots (Table 32.6), and terms of infection percentages (Table 32.7), or other standard terms, respectively, are suggested.

Table 32.6. Scale of frequency of occurrence of a seed-borne disease in seed lots of a given host

Degree of frequency	*Percentage infected seed lots*
Extremely frequent	above 85 of all seed lots inf.
Very frequent	70 – 85 of all seed lots inf.
Frequent	30 – 70 of all seed lots inf.
Rather frequent	15 – 30 of all seed lots inf.
Infrequent	5 – 15 of all seed lots inf.
Very infrequent	0.1 – 5 of all seed lots inf.
Extremely infrequent	below 0.1 of all seed lots inf.

Table 32.7. Scale of infection percentages encountered in seed lots of a given host

Degree of infection	*Inoculum potential* (percentage infected seeds)	
	Alternaria brassicicola in cabbage (%)	Lettuce mosaic in lettuce (maximum: about 17 per cent) (%)
Extremely high	above 85	above 15
Very high	70 – 85	10 – 15
High	30 – 70	3 – 10
Rather high	15 – 30	1.5 – 3
Rather low	5 – 15	0.5 – 1.5
Low	0.1 – 0.5	0.1 – 0.5
Extremely low	below 0.1	below 0.1

These conversions may be convenient if precise statistical information is not available. In applying this range a fair estimate may be placed in one of the adjoining groups and still be valuable as a preliminary appraisal.

The same range of estimates may be applied to inoculum potentials in terms of infection percentage, but the basic percentages within each class must vary according to (1) the maximum infection rate expected to be recorded for a definite seed-borne disease, (2) economic significance of a certain level of inoculum, and (3) the testing method applied. Examples may illustrate such differences pertaining to the basic figures, in evaluation of the severity of different seed-borne inocula.

Table 32.8. Tentative estimates of occurrence and amounts of inoculum potentials of some seed-borne diseases and suggested 'seed infection/yield reduction' ratios. SI, seed infection; YR, yield reduction

Pathogen and host	*Degree of frequency* (per cent seed lots)	*Inoculum potential*	*SI/YR ratio*	*References*
Alternaria brassicicola, cabbage	Extremely frequent, above 85% in Denmark	High to extremely high, up to 100% in Denmark		Neergaard
Alternaria zinniae, Zinnia elegans	Extremely frequent in Denmark	High to very high in Denmark		Neergaard
Anguina tritici, ear cockle, wheat	Infrequent, 3 – 9%, in 1920s, but extremely infrequent, 0 – 0.2% in 1950s, in England	Extremely low in England in 1950s		Marshall, 1960
Ascochyta fabae, Vicia faba	Frequent, 20 – 60% in England, 1966 – 1971	Rather low in England, 1966 – 1971		Hewett, 1973
Ascochyta pinodella, pea	Infrequent, 10% in Canada	Rather high, '4% inf. lots heavily inf.'		Wallen, 1965
Ascochyta pisi, pea	Very frequent, 85%, in Canada	High, '35% inf. lots heavily inf.'		Wallen, 1965
Aureobasidium lini, (Polyspora lini), flax	Frequent in Poland, 38%	Low in Poland average 1.4%	1:1 – 2	Pietkiewicz *et al.*, 1961
Bean common mosaic virus, bean	Frequent	High, up to 40% or up to 70%	$\frac{1:0.2}{r}$	Rohloff, 1964; Quantz, 1962
Cercospora beticola, sugarbeet	Very frequent in Central Europe, extremely infrequent in Denmark	High to very high? 1,000 – 300,000 conidia per g air-dried seed clusters		Wenzl and Krexner, 1962; Wenzl, 1959*a*
Claviceps purpurea in rye, wheat, barley	Rather frequent to frequent in most years in England, 1918 – 1957	No record		Marshall, 1960
Colletotrichum lini, flax	Rather frequent in Poland, 16%	Low in Poland, average 0.8%	1:1 – 2	Pietkiewicz *et al.*, 1961
Corynebacterium insidiosum, lucerne	Extremely infrequent	Low?		
Ditylenchus dipsaci, lucerne	Infrequent, 12% in Denmark	Low, up to 3 eelworms per 250 g seed		Nielsen, 1964
Drechslera avenae, oat	Frequent, 38% in Canada	Low in Canada, average less than 1%		Greaney and Machacek, 1946
Drechslera catenaria, Lolium perenne	Extremely frequent, above 85% in Denmark	High, up to 40% in Denmark	1:0.5	Andersen, 1959

Table 32.8 *continued*

Pathogen and host	*Degree of frequency* (per cent seed lots)	*Inoculum potential*	*SI/YR ratio*	*References*
Drechslera graminea, barley	Very infrequent in Denmark, frequent, 38%, in U.S.S.R. In Finland in 1967–1969 frequent to very frequent, 35–85% in six-rowed cv., rather frequent to infrequent in two-rowed cv., 10–30%	Low in Denmark. In Finland in 1967–1969 low in six-rowed cv., 1–5%, low in two-rowed cv., 0.3–0.6%	1:0.8	Neergaard Voitova, 1963; Halkilahti, 1971
Drechslera sorokiniana, barley, wheat, oats	Very or extremely frequent, 80 – 90%, in Canada	In Canada 1937 – 1942 in wheat high, up to 50%, with low average, 3.5%. In barley up to 95%, rather low average, 5.5%; in Sweden percentage of mean infection in barley 9 – 22% in 1964 – 1967		Greaney and Machacek 1946; Kolk, 1970
Drechslera teres, barley	Frequent, 62%, in Canada	Low, average 3%, up to high, max. 49%, in Canada		Greaney and Machacek, 1946
Fusarium nivale, wheat	Very frequent in Swedish wheat in 1960s	Rather high to very high, mean infection being about 15% in some years		Kolk, 1970
Fusarium culmorum, F. graminearum, F. nivale, cereals	Frequent or very frequent; about 50–80% of Danish lots of wheat	Rather high in England, very low in Denmark	1:0.7	Colhoun, unpublished; Neergaard
Lettuce mosaic virus, lettuce	Frequent, more than 50%; now much less	High, up to 13% (maximum about 15%)		Herold and Götte, 1957; Rohloff, 1962, Neergaard
Mycosphaerella pinodes, pea	Rather frequent, 27.5%, in Canada	High, '13% inf. lots heavily inf.'		Wallen, 1965
Peronospora tabacina, tobacco	Extremely infrequent	Extremely low?		
Phoma apiicola, celery, celeriac	In England 1935 – 1939 frequent, 44% of samples infected, 1960 – 1970, infrequent, 3% infected	In England 1935 – 1936 infection low to rather low, 1–10%, 1965 – 1967 low, 0.1 – 1%		Hewett, 1971
Phoma lingam, cabbage	Infrequent in Denmark and New Zealand, about 10%	Very low in Denmark and New Zealand		Neergaard, 1969; Lloyd, 1959*a*

Table 32.8 *continued*

Pathogen and host	*Degree of frequency* (per cent seed lots)	*Inoculum potential*	*SI/YR ratio*	*References*
Phoma medicaginis (Ascochyta imperfecta), lucerne	Frequent in Canada, 50 – 70% of samples from 5 provinces	In Canada high, up to 60%, higher in seed from Manitoba		Mead and Cormack, 1961
Plasmopara halstedii, sunflower	Infrequent? encountered in seed from many countries	Low? no information published on degrees of infection		Leppik, 1962*a*
Septoria apiicola, celery, celeriac	Very to extremely frequent, 50 – 65%, earlier up to 95% lots in England and Denmark	High to extremely high		Neergaard; Moore, 1946*a*; Jørgensen, 1934
Tilletia caries, wheat	In England rather frequent to frequent, 5 – 33%, in 1920s, but very infrequent to extremely infrequent, 0.2 – 1%, in 1950s	In England extremely low in 1950s		Marshall, 1960
Tobacco mosaic virus, tomato	Frequent, 67% of 99 Danish lots	90% seed-borne infection found in Belgium; 35% infection in Australia		Kristensen, unpublished; Van Winckel, 1965; Crowley, 1958
Tobacco ringspot virus, bud blight of soybean	Seems to be very infrequent	Up to 100% seed-borne infection		Athow and Bancroft, 1959
Ustilago avenae, Ustilago hordei, oats	Extremely frequent, almost 100% in Canada			Greaney and Machacek, 1946
Ustilago nuda, barley	(Rather) frequent	Very low to rather high in Canada in 1950s; very low in Denmark	1:0.8	Russell, 1962
Xanthomonas campestris, cabbage, cauliflower	Frequent in seed from many countries, e.g. Denmark, Holland, India	High in seed lots from India, rather high in seed lots from Holland, low in seed lots from Denmark, and several other countries (Ghana, Tanzania)		Danish Government Institute of Seed Pathology, current project
Xanthomonas phaseoli, bacterial blight of bean	Frequent, 50% in Michigan, U.S.A.	Low to extremely low		Saettler and Perry, 1972

The frequency of occurrence and the level of infection percentages may vary from one country to another, from one region within a country to another, and from one year to another. The infection percentages as well as frequency of occurrence of such pathogens as *Drechslera* in grasses, *Fusarium* in cereals, *Alternaria zinniae* in *Zinnia, Ascochyta pisi* in pea, *Alternaria porri* f. sp. *dauci* in carrot and *Phoma lingam* in cabbage, fluctuate from one year to another in Denmark, according to weather conditions. Such local and yearly variations must be taken critically into consideration before general conclusions are drawn.

Also, it must be taken into consideration that the general level of infection of a pathogen and its extent of distribution in any region may change gradually through a few decades. Introduction of resistant cultivars, the effect of certification schemes, changes in agricultural practices and other factors may outdate plant disease surveys published a few decades back. The shift in the relative prevalence of the two oat smuts, loose smut, *Ustilago avenae*, and covered smut, *U. hordei,* in the U.S.A. through the last decades provides an illustrative example. According to Holton (1970) the ratio of specimens of loose smut to covered smut from all major geographic regions of the country was the following: 1920–1930, 2 : 1; 1931–1950, 4 : 1; 1951–1965, 9 : 1. This shows that covered smut of oats is steadily declining in prevalence in the U.S.A. and is actually now little more than a curiosity.

In inoculum appraisal the relation of different cultivars of a host to seed transmission of a pathogen must be considered. Cultivars may behave quite differently and general conclusions cannot be drawn. Lettuce mosaic virus is transmitted in higher percentages in some cultivars than in others (Grogan and Bardin, 1950). The cultivars Maikönig and Attraction transmit higher proportions of the virus than most other commercial cultivars commonly grown. Smith and Hewitt (1938) who grouped 51 lettuce cultivars into four classes according to severity of induced symptoms, found the rate of seed transmission positively correlated with symptom severity. The virus is not seed-transmitted in the hypersensitive cultivar Cheshunt Early Giant (Kassanis, 1947; Couch, 1955). In Phaseolus bean the hypersensitive cultivar Top Crop does not transmit bean common mosaic virus. Barley stripe mosaic virus is seed-transmitted at lower rates in resistant barley cultivars, at higher rates in susceptible cultivars (Eslick and Afanasiev, 1955; Singh, Arny and Pound, 1960; Inouye, 1962). In two very resistant cultivars Inouye was unable to obtain seed transmission.

Also, different strains or races of a pathogen may behave differently with regard to seed transmission, either due to a different host range or due to differences in the host–pathogen interaction in seed transmission. A severe strain of lettuce mosaic virus which attacks cv. Gallega resistant to common lettuce mosaic virus, is not seed-transmitted in lettuce (Zink, Duffus and Kimble, 1973).

32.2 EXTENT OF TRANSMISSION FROM SEED TO CROP — EPIDEMIOLOGICAL RATES

Development of a seed-borne disease is dependent primarily on three principal factors: (1) amount or *rate of seed-borne inoculum,* (2) the extent of transmission of this inoculum to the seedling at any stage of its development or the amount of active inoculum under field conditions, or the *rate of seed transmission* ('seed-plant transmission'); and (3) the *rate of increase* in the subsequent development of the inoculum or disease in the field. The proportional importance of any of the two last phases in the course of infection is dependent on the other. If the rate of increase is high, as in certain downy mildews, many bacterioses, and in vector-borne viruses when the vectors are present, then even trace amounts of the initial inoculum may be of greatest importance. Any evaluation of seed transmission should be made from this angle. Finally, in seed multiplication we may also compare the amount of inoculum in seed of one generation with that in the seed of the preceding generation; this ratio may be termed (4) the *rate of re-establishment of seed-borne inoculum* (Fig. 32.1).

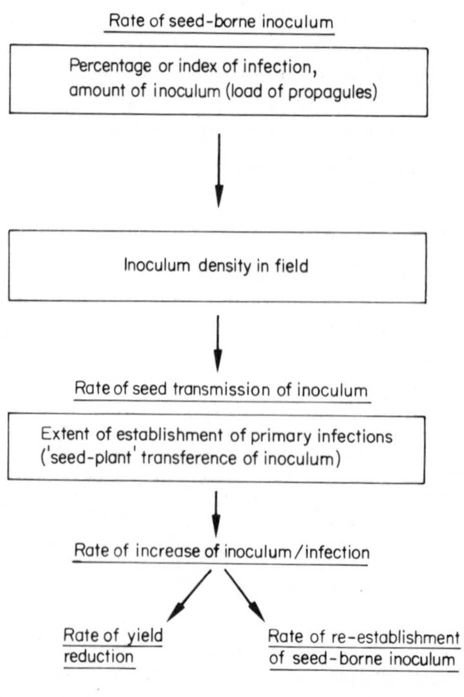

Fig. 32.1. Sequence of epidemiological rates in seed-borne diseases

(1) *Rate of seed-borne inoculum.* As discussed in Subsection 32.1.1 establishment of this rate depends greatly on the procedure by which it is determined. Quantitative determinations of seed-borne inoculum are basic, and to obtain internationally comparable and reproducible results in studies on epidemiology of seed-borne diseases, workable standard seed health testing procedures are imperative.

(2) *Rate of seed transmission.* Pathogens of many systemic diseases, for instance embryo-infecting viruses such as lettuce mosaic virus and bean common mosaic virus, are readily seed-transmitted, the seed infection/seed transmission ratio sometimes being close to 1 : 1, only actually infected seed being considered. In loose smut of barley, *Ustilago nuda,* it being a systemic disease, the transmission ratio is almost 1 : 1, again when only actually infected seed is considered.

Seed transmission of tobacco mosaic virus in tomato, on the other hand, depends on the conditions for mechanical transmission by contact of the very powerful inoculum present as contamination on the seed coat which may adhere to the seedling for some time. If the seedlings are untouched there is no transmission; if they are transplanted, inoculation usually occurs extensively and transmission is established at various rates.

Hewett (1973) found that 4–8 per cent of *Vicia faba* seeds, infected by *Ascochyta fabae*, produced diseased seedlings, while Anselme, Hewett and Champion (1970) recorded that 2–6 per cent of pea seed, infected by *Ascochyta pisi*, produced 0.5–4 per cent diseased seedlings, in addition to reduction of seedling emergence. The transmission rate as well as the rate of increase depend largely on edaphic and climatic conditions.

The seed transmission rate of halo blight, *Pseudomonas phaseolicola,* and fuscous blight, *Xanthomonas phaseoli* var. *fuscans* in Phaseolus bean is about 10 per cent, according to Taylor (1970); Wallen and Sutton (1965) also presented evidence that about 10 per cent of the infected seed produce primary foci of infection in the field. Pepper (1967) found that the rate of seed transmission of Stewart's bacterial wilt, *Erwinia stewartii* in maize, to be about 2 per cent in sterilised soil, while Schuster *et al.* (1973) found that the seed transmission rate of *Corynebacterium nebraskense* in this host was less than 1 per cent; both of these pathogens require wounding for establishment of infection. In tomato the rates of transmission from seed, infected or probably mainly contaminated by *Corynebacterium michiganense,* ranged in some cases from 1 to 4 per cent (Strider, 1969).

(3) *Rate of increase.* The rate of increase in plant diseases has been studied extensively in recent years and a new fertile branch of plant pathology, the epidemiology of plant disease, has developed (Van Der Plank, 1960, 1963, 1968, 1975; Schrödter and Ullrich, 1965; Kranz, 1968,

1974; Zadoks, 1972). Its mathematical, biological and philosophical aspects will have to be taken into account in any deep-going analysis of disease dynamics as part of assessment of seed-borne inoculum in relation to crop losses. However, this extensive and far-reaching field cannot be further dealt with here.

The sudden explosive spread of the Australian race of tobacco blue mould, *Peronospora tabacina*, in Europe and the Mediterranean area was due to escape from quarantine in England of an isolate of the pathogen introduced for research. This epidemic (see Fig. 24.3) exemplifies an extremely high rate of increase, whereas the rates of seed contamination and seed transmission of this pathogen are extremely low. That seed transmission does occur has been demonstrated by an incidence of this parasite, race unknown, having been detected in post-entry quarantine on an imported seed sample in Portugal some years before the above epidemic.

Hewett (1973) and Anselme, Hewett and Champion (1970) found for *Ascochyta fabae* in *Vicia faba* and *Ascochyta pisi* in pea, respectively, that the maximum increase in disease from the seed having levels of 2–6 per cent infected seeds as established by laboratory testing, to the harvested seed was about eight times, but other initial levels and other conditions favourable for disease spread could very well give greater increases.

Seed treatment as an attempt to remove seed-borne inoculum is usually not completely successful in this respect; only a reduction of inoculum is obtained. As it appears from the present discussion the final success of this reduction of inoculum depends on the rate of increase of the pathogen. If this is very high, a seed treatment almost fully effective, but still leaving a trace of inoculum may eventually prove to be worthless.

The rate of increase of seed- and vector-borne viruses may be considerable, examples being the viruses of bean common mosaic and lettuce mosaic (Section 32.5). Kaiser (1973) found in Iran that bean yellow mosaic virus is transmitted in broad bean, *Vicia faba*, at very low rates, up to about 1 per cent from diseased mother plants; from the primary infections in young seedlings aphids spread the virus, resulting in from 4 to 34 per cent infected plants after 15 weeks and from 93 to 100 per cent after 22 weeks.

(4) *Rate of re-establishment of seed-borne inoculum.* In seed multiplication the rate of re-establishment of seed-borne inoculum is a crucial factor in setting disease tolerances in seed certification schemes. Gray (1954) established that owing to wet weather in the flowering period of 1948 the rate of increase of loose smut in barley was especially great in 1949. A stock of cultivar Rigel contained 2 per cent in 1948, 10 per cent in 1949 and 20 per cent in 1950. Cultivar Maja, after hot water treatment, was apparently free from smut in 1948, had about one smutted head per

acre in 1949 and 0.5 per cent infection in 1950. This potential of loose smut strongly supports the policy of setting tolerance standards as low as possible.

Naturally, the rate of re-establishment of seed-borne inoculum varies considerably from year to year and from place to place, according to weather conditions. Hewett (1973) found in England, for one year, that seed lots of *Vicia faba,* carrying 1–4 per cent infection of *Ascochyta fabae,* produced commercial seed carrying up to 11 per cent. In other years there was no increase but decrease in some places, while in others there was increase from five to eight times. Another example of the rate of increase of seed-borne inoculum from generation to generation is given in Subsection 19.4.2, dealing with the stem canker of nasturtium, *Acroconidiella tropaeoli.*

32.3 RELATIVE IMPORTANCE OF OTHER MEANS OF TRANSMISSION

The concurrent effect of inoculum originating from different sources, in relation to the relative economic importance of these sources in disease development, has been discussed previously. To summarise, most smuts of cereals are predominantly seed-borne. Barley stripe fungus, *Drechslera graminea,* and *Septoria apiicola,* early blight of celery, are other pathogens predominantly seed-transmitted in North Europe. On the other hand, rare cases of seed transmission of rust fungi or downy mildews, predominantly air-borne, may have disastrous effects in giving rise to epidemics, although such incidences are exceptional. Plant refuse and overwintering plants have been shown to be more important sources of inoculum of *Cercospora beticola* in Turkey than the seed (Gassner, 1952*a*), but they seem to be considerably less so in Austria even if there they are still important (Wenzl, 1959*b*). The local climatic conditions have bearing on the relative importance of the different sources of inoculum. Soil transmission is predominant for *Sclerotinia sclerotiorum* and *Rhizoctonia solani,* but seed transmission may have serious effects and cannot be disregarded in any realistic evaluation. The severe effect of *Fusarium nivale* on field emergence of winter rye in Australia is conveyed mainly by seed, while later damage in the form of rotting, produced by the fungus under protracted snow cover, is caused by a combination of seed- and soil-borne inoculum (Neururer and Zwatz, 1965). Dwarf bunt of wheat, *Tilletia contraversa,* is seed- and soil-borne, the pathogen being introduced to new areas, new fields by seed, and inoculum being built up year after year in soil if not counteracted by crop rotation, and this soil-borne inoculum may in turn again infect crops from healthy seed. If climatic conditions result in crops having to be grown under glass, specific possibilities of carrying over of pathogens from one crop to another may arise, but there are also special possibilities of control (soil treatment, etc.).

32.4 EXTENT AND INTENSITY OF DISEASE IN THE CROP

Appraisal of losses due to seed-borne diseases must be based on well-established data on the development of the diseases in all stages of the crop from the seedling to harvest, including a fair measure of the disease intensity in terms of the rate of disease increase, r, given mathematically as percentage increase of disease or inoculum per unit of time. When in the development of an epidemic the amount of disease is plotted against time, the curve is usually sigmoid, and the slope of the curve at any time is the value of r (Dimond and Horsfall, 1960). This describes a 'regular' epidemic course. Many seed-borne diseases, however, are serious primarily in two stages of the crop, the seedling stage and the stage of maturation of the crop, leaving the crop more or less unaffected in the intervening period. In such a cycle, field inspection is useless but examination of the seed vital.

The importance of seedling diseases, including pre-emergence killing, has been extensively studied in seed treatment experiments. It has been repeatedly observed that significant increase in stands may not be followed by significant yield increase. This is usually due to an excessive seeding rate for treated seed, and most old standard seeding rates are excess if seed is treated (Chester, 1950*b*). Even the seeding rate used with untreated seed is still very often excessive. If the rate of increase of the seedling disease is low and spread does not occur to any appreciable extent beyond the seedling stage, and if the seedling disease leaves no residual weakness in surviving plants, then a fairly good stand remains and the loss is chiefly or entirely measured by the seed waste. In modern methods of precision seeding, for instance with peas for processing, and especially genetic monogerm beet seed, there is considerable loss of crop.

Some crops, particularly cereals, for instance barley, are believed to have a considerable ability to compensate for uneven spacing, with little difference observed in yields between evenly and unevenly spaced plantings (Sprague and Faris, 1931). Due to this tillering effect barley leaf stripe disease, *Drechslera graminea*, that causes a total loss in affected plants, shows merely a 3/4 per cent yield reduction for each 1 per cent of stripe. Already Lind and Ravn (1918) established in Denmark a loss of 0.7 per cent (0.6–0.8 per cent) for every 1 per cent infection of the seedling, and Suneson (1946) indicates a 25 per cent compensation for deficient stands in infected fields. Under Indian conditions 14.2 per cent of plants in the field were found affected and the loss recorded was 12.2 per cent (Mathur, Mathur and Bajpai, 1964), suggesting a plant infection/yield reduction ratio 1 : 0.86. Recent investigations by Richardson (1974) have thrown some doubt on the compensatory ability of surviving plants in cereal populations, and have, at least, substantiated that generalisations should not be made. Much of the pertaining information is based on the behaviour of cereals in the glasshouse, growth chamber or carefully nurtured plots. Under field conditions the surviving population is affected by a wide range of conditions of soil and climate as well as pathogens, including others than the one considered. This implies that although it has the opportunity to compensate by the absence of neighbours, it may not have the ability to take advantage of that opportunity.

The complexity of loss caused by one disease is well exemplified by bacterial blight of cotton, *Xanthomonas malvacearum,* as pointed out by Chester (1950*b*): 'The disease destroys seedlings, necessitating increased planting rates, seed treatment or replanting. It injures stems, depressing the vigor of growth. It partially destroys the leaves, reducing photosynthesis, with consequent reduction of fruit formation, although if defoliation from the disease is very late in the season it may be regarded as an advantage facilitating harvesting. The disease attacks the bolls, and while it does not usually destroy them, it opens the way for boll-rotting fungi to enter the bolls and weaken or destroy the lint, resulting in both lowered yields and quality of the harvested crop. Long and thorough study, and measurement of the loss fractions due to each of these components, is required before a reasonable accurate appraisal of the loss from this disease is possible.'

An illustrative approach to analysis of loss fractions has been made by Brönnimann (1968) who studied the effect of the glume blotch phase of the disease of wheat caused by *Septoria nodorum,* on yield, as measured in terms of reduction in thousand grain weight. Spring wheat cv. Hinal was inoculated at growth stage 10.5. (Feeke's scale) 'all ears out of sheath'.

At this stage the top leaf, second leaf and third leaf were inoculated independently, or two or three leaves in combination, and also ears were inoculated independently and in combination with the three leaves. The total damage as compared with untreated was 65 per cent loss, the attack on the ears and on each of the three leaves separately being responsible for percentages of losses as shown in Fig. 32.2. Brönnimann found good conformity of these results with those of Boonstra (1937) who investigated the effect on yield obtained by removal of leaves and shading of ears in wheat. In disease assessment for *S. nodorum,* growth stages showing clearest correlation with yield loss may differ between cultivars. Cooke and Fozzard (1973) found that the optimal growth stage for assessment was 10.5.2 for cv. Kolibri and 11.1 for cv. Opal.

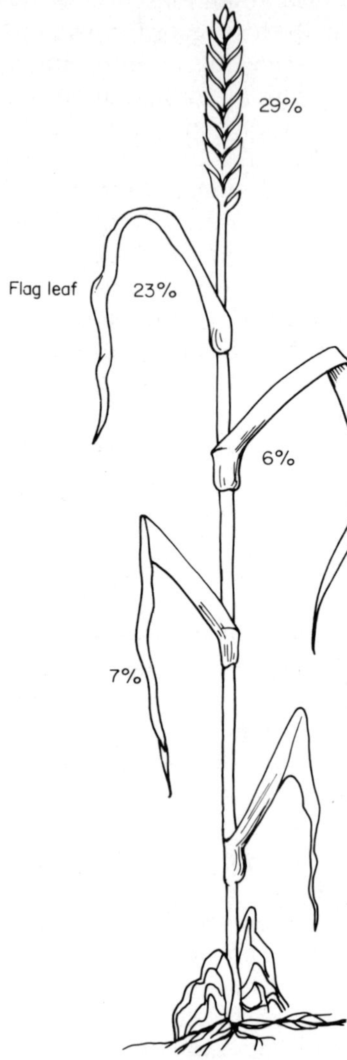

Fig. 32.2. The relative impact on yield of damage to plant parts important to photosynthesis, according to Boonstra (1937) and Brönnimann (1968). The illustration indicates shares of vital organs in a 65 per cent loss due to *Septoria nodorum,* in wheat

Differences between the pathogens of more or less the same pathogenic type in their effects at different stages of the host are found in the Ascochyta complex on peas, and clearly reflected in the extent of seed transmission and the extent of yield losses of the three related pathogens. In Canada *Ascochyta pisi* is transmitted in much higher percentages than *A. pinodes* and *Phoma medicaginis* var. *pinodella* (*Ascochyta pinodella*), but these two pathogens are more aggressive than *A. pisi* (Wallen, 1965). Of the seed infected by *A. pisi,* 25 per cent produce diseased seedlings; the remainder either do not emerge because of severe infection that kills the embryo, or emerge as seedlings with slight, suppressed infections. Secondary spread of the disease may occur from the infected seedlings. A higher proportion of seeds and seedlings infected by *A. pinodes* and *Phoma medicaginis* var. *pinodella* are killed in early stages of development. Moreover, these pathogens produce extensive lesions on the lower stem area and kill the plants before they mature to set seed. Wallen concludes that the relatively low incidence of these organisms in seed is an indication of their destructive nature.

32.5 SEED-BORNE INOCULUM IN RELATION TO YIELD REDUCTION AND CROP LOSSES

Conclusive information must be established not only in terms of disease intensity, for instance as percentage infected plants, but must be provided in terms of per cent yield reduction. Chester (1950*b*): 'When extensive data on disease intensity and yields of diseased and healthy crops are available, the relationship between disease and yield may be simply expressed as a coefficient of correlation. This not only brings out the extent to which disease is responsible for yield reductions, but also, by using partial correlations, it is possible to allocate the fractions of total yield reduction due to several injurious factors acting as a complex.' And it may be added that the relationships between seed-borne inoculum potential and disease intensity in the subsequent crop also may be expressed as a coefficient of correlation. We should endeavour to correlate infection incidences as found in laboratory seed health testing with percentage yield reduction. Some of the fragmentary information available and pertaining to this interrelation of data may illustrate this trend of thought.

Morton (1961) in yield tests in North Dakota, U.S.A., compared treated and untreated barley samples with varying percentages of loose smut, *Ustilago nuda*. A regression coefficient for the average yield loss and smut percentages showed that percentage loss increased 0.86 with each 1 per cent increase in smut. Accordingly, in Canada Semeniuk and

Ross (1942) obtained a regression coefficient of 0.85 for one locality, but 1.4 and 1.2 respectively for two other localities. In Scotland Gray (1954) compared the yield from a seed lot of barley cv. Maja, containing 0.1 per cent infection of loose smut, with the untreated stock of the same seed lot. She found that the treated seed yielded an average of 7 quarters or 56 bushels per acre, while the untreated with 15 per cent infection yielded 6–7 bushels less, or a reduction of about 12 per cent due to the 15 per cent infection. This gives a seedling infection/yield reduction ratio of 1 : 0.8. The regression coefficient for loose smut corresponds closely to the findings for *Drechslera graminea,* mentioned above, which showed a 3/4 per cent yield reduction for each 1 per cent of leaf stripe in the seedlings or mycelial infection in the pericarp (Lind and Ravn, 1918; Suneson, 1946; Mathur, Mathur and Bajpai, 1964).

Wallen (1965) found in infection experiments that *Ascochyta pisi* caused a reduction of the yield of 11 per cent, *Phoma medicaginis* var. *pinodella (Ascochyta pinodella)* 25 per cent and *Mycosphaerella pinodes (Ascochyta pinodes)* a yield reduction of 45 per cent. These reductions were not related to the infection percentages found by laboratory testing. But if we assume that 30 per cent *Ascochyta pisi* reduces the yield on average 10 per cent, then 13 per cent *Phoma medicaginis* var. *pinodella* and about 7 per cent *Mycosphaerella pinodes* may be expected to lead to the same reduction of the yield under Canadian conditions. This has bearing on setting tolerances and will be discussed later.

Phoma lingam, black leg of cabbage and other crucifers, is found in very low percentages in seed. But the seedling disease may spread rapidly, especially in seed beds. The infection may persist as weak incidences, but may attack mature plants of cabbage just before harvest; losses of 75 per cent of the heads have been encountered in the U.S.A. This is an example of a seed-borne pathogen with a low amount of seed-borne inoculum, but a high rate of increase at the beginning and towards the end of the growing season. *Septoria apiicola,* on the other hand, may have a very high seed-borne inoculum potential. This may lead merely to weak infections during the main part of the growth period, but often late in the season to heavy infections and yield reductions approximating 100 per cent in exceptionally severe cases and loss of quality in harvested crop. Pycnidia are *gritty* to eat! However, correlation between seed infection percentage and yield reduction percentage seems not to have been established for *S. apiicola* and may not be found, or may be very variable, owing to the high rate of increase of the disease.

In Danish investigations seed infection percentages found in laboratory testing of *Festuca pratensis* and *Lolium perenne* for *Drechslera catenaria* has been correlated with percentage yield reduction (Andersen, 1959). It was found that 50 per cent seed infection leads to a yield reduction of 23–29 per cent, and 33 per cent seed infection to a yield reduction of 17 per cent of total weight green matter. This suggests roughly that every 2 per cent seed infection, found in the laboratory, leads to 1 per cent reduction of the yield, or, in terms of seed infection/yield reduction, a ratio of 1 : 0.5.

Colhoun and co-workers (unpublished) found that 30 per cent seed infection of *Fusarium nivale* in the wheat cultivar Viking caused a reduction of the yield of seed by 20 per cent. This suggests a seed infection/yield reduction ratio of 1 : 0.7.

Reductions in the amount of seed-borne inoculum obtained by seed treatment may provide valuable information for establishment of the seed infection/yield reduction ratio. However, the effect of seed treatment due to factors other than those which can be attributed to specific parasites, may complicate calculations. An example may illustrate such implications.

Muskett and Colhoun (1943, 1944) obtained highly significant increases in fibre yield of flax by seed treatment against stem-break and browning (*Aureobasidium lini*). The following data are extracted from their results and, in an attempt to deduce the negative effect of the disease rather than the beneficial effect of the fungicides, the yield reduction percentages, inferred by the omission of seed treatment, have been calculated based on the yield figures obtained by the best treatment=100 per cent.

In experiment (II), 1942, a seed lot 'not heavily infected with any known parasite' was treated but there was 33 per cent lower yield in untreated seed as compared with the best treatment. It seems fairly justifiable on this basis to subtract 30 from the calculated loss percentages obtained in the two other experiments referred to in Table 32.9 (78 per cent and 55 per cent, respectively).

This leads us to the rough estimate that 35 per cent seed-borne *Aureobasidium lini* in 1941 reduced the subsequent yield by about 48 per cent, or in terms of seed infection/yield reduction with a ratio of approximately 1 : 1.5, and that in 1942 14 per cent seed-borne *Aureobasidium lini* reduced the yield by about 25 per cent or with a seed infection/yield reduction ratio of about 1 : 2.

Lettuce mosaic virus is seed-transmitted from diseased mother plants (cv. Bibb) in an average of about 8 per cent (Couch, 1955), but may occur up to a maximum of about 15 per cent. To a great extent early infected plants do not develop heads. Secondary spread by aphids is responsible for serious damage (Grogan, Welch and Bardin, 1952). A transmission percentage of 0.5 may give a total loss of crop when aphid vectors are active early in the season. Tomlinson (1962) reported that in 1959–1960, in England, lettuce crops from seed infected by lettuce mosaic less than 0.1 per cent contained 0–4.5 per cent infected plants, while crops originating from seeds with 2.2–5.3 per cent lettuce mosaic infection contained 25–96 per cent. This suggests a seed transmission/crop infection ratio of

Table 32.9. Effect of seed treatment on the amount of seed-borne *Aureobasidium lini* and the subsequent yield of flax fibre. (Data compiled from Muskett and Colhoun, 1943, 1944)

Seed lot	*Criteria*	*Best treatments*	*Untreated*	*Reduction of per cent seed-borne infection by seed treatment*	*Per cent 'yield reduction' by omission of seed treatment*
1941	Yield of fibre	1.8 – 1.9 cwt/acre	0.41 cwt/acre		78%
	Seed infection				
	Aureobasidium lini	15%	50%	35%	
1942 (I)	Yield of fibre	2.0 cwt/acre	0.9		55%
	Seed infection				
	Aureobasidium lini	2 – 8%	19%	14%	
1942	Yield of fibre	5.4 cwt/acre	3.6 cwt/acre		33%
	Seed infection				
II	Aureobasidium lini	0	0.2%	0.2%	
	Colletotrichum lini	0	1.5%	1.5%	

1 : 10–45. Under extreme conditions such a high disease increase rate, based on 0.5 per cent seed infection as mentioned above, the seed transmission/yield reduction ratio is no less than 1 : 200.

Bacterioses offer many examples of epiphytotics that develop from traces of inoculum.

In Wisconsin, U.S.A., Walker and Patel (1964*a*) demonstrated that 0.02 per cent seed infection (12 seeds per acre) could cause an epiphytotic of *Pseudomonas phaseolicola*, halo blight of bean (Fig. 32.3), the accelerating weather factors being wind and rain splashes, and Guthrie, Huber and Fenwick (1965) refer to an incidence of this disease, where one infected plant in 16,000 was enough to result in a complete loss under epiphytotic conditions. Air photographs may illustrate infection centres originating from seed-borne inoculum; Fig. 32.4 shows seed-transmitted foci of halo blight in a field of snap beans in England (Brenchley, 1968).

In Canada 0.5 per cent seed infection of *Xanthomonas phaseoli* var. *fuscans,* fuscous blight, has been enough to start an epiphytotic on bean (Wallen and Sutton, 1965). Plate V, a colour infrared air photograph, shows seed-transmitted foci of bacterial blight in field (white) beans in Ontario, Canada.

In Florida, U.S.A., a fraction of 1 per cent seed infection of *Xanthomonas vesicatoria,* bacterial spot of tomato, could lead to an epiphytotic on this host (Cox, 1966).

Tarr (1961) found in Sudan such low incidence as 0.017 per cent seed infection, or one plant in 6,000 infected with *Xanthomonas malvacearum*,

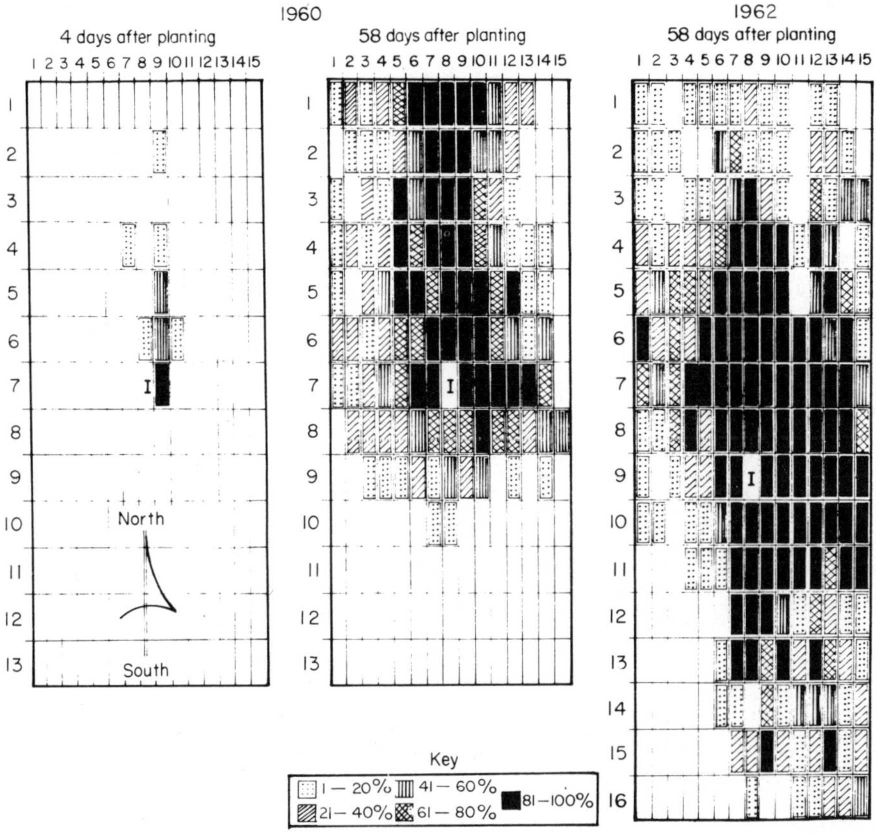

Fig. 32.3. Spread of halo blight from source of inoculum, designated I, in experiments conducted in 1960 and 1962. Results are recorded as percentages of infected plants in subplots. Direction of spread appears determined by wind during rains. (Redrawn from Walker and Patel, 1964*a*)

angular leaf spot or black arm of cotton, responsible for field loss of crop, and Faulwetter (1917) that the direction of rain splash dispersal of the black arm pathogen was determined largely by the wind during periods of rainfall. The down-wind dispersal of halo blight is also clear in Brenchley's photograph (Fig. 32.4).

32.6 DISEASE TOLERANCE IN SEED HEALTH TESTING, AND CROP LOSSES

Any insurance is based on calculation of risk in terms of probability of the number of incidences liable to occur and their individual severity. Our attempt to reduce incidences and severity of seed-borne diseases may be looked at from the same angle.

Should tolerances be fixed strictly according to the principles of insurance calculations?

Hail, a corresponding object of insurance, may occur in so many incidences out of so many possibilities of incidence; development of a disease, too. But infectious diseases are dynamic, incidences of hail are not, leaving only a stationary state of damage. But an incidence of increase of a seed-borne disease results, within variable limits according to pathogen, in increase of inoculum for the next generation of the host, while decrease of disease entails naturally a decrease of the degree of further transmission. This rate of increase and the response of the pathogen to climatic,

Fig. 32.4. Air photograph of fields taken from 1,000 ft, 30th July, 1964. Above: potato late blight, with a big and several minor foci of infection. Below: halo blight of beans, *Pseudomonas phaseolicola*. Note, in the Phaseolus bean field, foci of infection originating from seed-borne inocula. 1 mm = 3 m (i.e. scale 1:3,000). Infrared. (Courtesy of G. H. Brenchley, 1968)

edaphic and other factors are important items to be considered in calculation of disease risk in setting up disease tolerances for seed health testing.

Neergaard (1962*a*, 1962*b*) has discussed some basic principles to be adopted in setting up disease tolerances in seed health testing. These principles include consideration of quarantine requirements of the importing country, geographic destination of the seed lot, frequency of occurrence of the pathogen with the seed as well as the seeding rate and the possibilities of effective disinfection. But the principal factor to be included in such considerations is the degree of correlation between seed-borne inoculum potential and crop losses. We shall return to some of the examples discussed, in which the seed transmission/yield reduction ratio is more or less established or at least with some trends suggested.

The 'Ascochyta complex' on peas, studied by Wallen (1965), provides an example of the importance of three closely related pea pathogens in relation to seed transmission and to crop losses. The amount of seed infection may be summarised as shown in Table 32.10.

Table 32.10

Ascochyta pisi	85% seed lots	37% heavily inf.
Mycosphaerella pinodes	27.5% seed lots	13% heavily inf.
Phoma medicaginis var. *pinodella* (*Ascochyta* sp.)	10% seed lots	4.5% heavily inf.

But the three pathogens produced crop losses in terms of per cent yield reduction different from that of the degree of their severity on seed:

Ascochyta pisi	11% yield reduction
Mycosphaerella pinodes	45% yield reduction
Phoma medicaginis var. *pinodella*	25% yield reduction

Wallen did not report investigations on the percentage of infected seed correlated with yield reduction but he used equal amounts of inoculum to produce these different degrees of yield reduction. Hence the results suggest that a disease tolerance for *Ascochyta pisi* may well be acceptable at a percentage four times that for *Mycosphaerella pinodes* and twice that for *Phoma medicaginis* var. *pinodella*. Actually, Canadian tolerances for foundation seed allow 6 per cent *Ascochyta pisi* but only 2 per cent of the other two pathogens combined (Wallen, personal communication, 1970).

For *Ustilago nuda* the infection percentage found in the laboratory may easily be converted into per cent yield reduction to be expected, the ratio 1 : 0.8 being applicable (cf. Morton, 1961, see Section 32.5). In Denmark and Scotland a tolerance of 0.2 per cent is adopted, minimal in terms of crop losses to be expected, but the rate of increase from year to year is also taken into consideration.

The same view applies to *Phoma lingam* which is encountered in very low percentages in Danish seed, but which under favourable conditions may develop in disastrous proportions as already discussed. Heavy crop losses may be succeeded by soil infestation and maintenance in plant residues from one season to another under certain climatic conditions, thus adding later losses to the initial ones. The tolerance fixed for this parasite in the Danish export control of cabbage seed has therefore been zero for many years, based on 1,100 seeds per test.

Stem eelworm, *Ditylenchus dipsaci,* in lucerne, red clover and onion seed, is transmitted only in trace amounts, and establishment of infection occurs at a low rate as a slow build-up of local infestation. The disease is serious and the eelworm (lucerne race) has been included as a quarantine object for lucerne in British import regulations.

Corynebacterium insidiosum, bacterial wilt of lucerne, is another quarantine object in various countries and is also to be found in trace amounts, infestations building up rather slowly. The disease usually does not become apparent until lucerne stands are about two years old, and then progressively destroys the plants through succeeding years until the crop becomes unprofitable. This fact explains why the seed is not generally suspected by farmers as a source of infection. Once the soil has become infested, the pathogen is maintained as a soil-borne one. For such pathogens the tolerance must be zero. The problems to be tackled are those of field inspection, proper sampling, large samples being needed, and proper testing techniques.

In seed-borne downy mildews and rusts, certain bacterial and vector-borne virus diseases, and other rapidly spreading epiphytotics a trace of transmitted inoculum is the spark to bring about explosion. The ideal tolerance must be zero, and the problem again is that of proper sampling and adequate testing technique — if sampling is justifiable at all, which it is not for this category of pathogens in quarantine. In California a tolerance of only 0.003 per cent is required for lettuce mosaic virus in lettuce, i.e. at least 30,000 seeds have to be tested, the minimum necessary to ensure that the disease is kept on an acceptable low level (Greathead, 1964).

The inoculum of certain seed-borne smuts may be related to spore load per grain. The spore load critical to establishment of disease in the field depends largely on the susceptibility of the cultivar, as shown by Heald (1921) and many other works, cf. Section 17.3. In the Austrian Seed Health Certification Scheme (Anonymous, 1949; Kietreiber, 1952) the following tolerances for *Tilletia caries* and *T. foetida* were adopted:

0–50 spores/grain	= health condition satisfactory
51–1,000 spores/grain	= health condition acceptable
>1,000 spores/grain	= health condition unsatisfactory

One smut ball per 200 g wheat is equal to 800–1,000 spores per grain and this unit of inoculum was taken as the basis for the standard adopted.

The indications in Table 17.1 (Section 17.3) on percentage of smutted spikes arising from seed containing 1,000 spores of *T. caries* per grain range from about 5 in less susceptible cultivars to more than 20 in very susceptible cultivars, and show that the above tolerances were extremely lenient. In fact, later, in 1958, the Austrian tolerances were substantially reduced, seed lots with more than 20 spores/grain being regarded as unsuitable for sowing and 4–20 spores/grain as acceptable after seed treatment (Kietreiber, 1962AR; Anonymous, 1965). In 1973 (Anonymous, 1973) the tolerances were made less severe, 50 spores/grain being maximum for *Tilletia caries*/*T. foetida*, and 30 spores/grain for *T. contraversa*, the latter, however, exceptionally up to 100 spores/grain provided *special* seed treatment ('Trockenbeize 4613') is carried out.

The group of seed-borne, seedling-infecting and readily sporulating fungi of the type represented by *Fusarium, Drechslera, Alternaria* and *Stemphylium* are easily detectable in routine testing and the seed-borne inoculum potential expressed in fairly comparable and reproducible figures. These figures may be related directly to crop losses, possibly in a regular correlation. If we know that 10 per cent *Fusarium* in wheat seed means a yield reduction of 7 per cent, and that 30 per cent *Drechslera catenaria* in a seed lot of meadow fescue means about 15 per cent reduction in the output of the fresh green matter of this grass, then assessment of tolerance appears to become a matter of simple calculation. At least such should be our objective.

To adopt disease tolerances is to realise what is going on and to do something about it. One way is that of regional seed health certification schemes. The farmer cannot observe directly yield reductions such as those commonly caused by *Fusarium, Drechslera* and many other parasites with a slightly diminishing effect scattered throughout the field. The amount of such damage can be revealed only by experimental methods. It is not surprising, therefore, that a farmer will more readily accept a tolerance of 0.2 per cent for the conspicuous loose smut in barley than 5 per cent for *Fusarium nivale* in wheat or a few per cent for *Drechslera sorokiniana* or *D. teres* in barley, although actually these and other seed-borne fungi at present account for far greater damage to, for instance, Scandinavian agriculture, as clearly suggest Colhoun's investigations in England, Kolk's in Sweden and Smedegård-Petersen's in Denmark, than does loose smut in barley.